Fertility Rates: United States, 1940–2004
(Live Births per 1,000 Women Aged 15–44)

66.0

Note: Beginning with 1959, trend line is based on registered live births; trend line for 1930–1959 is based on live births adjusted for underregistration.

Source: National Center for Health Statistics.

Infant Mortality Rates: United States, 1940–2004
(Deaths per 1,000 Live Births)

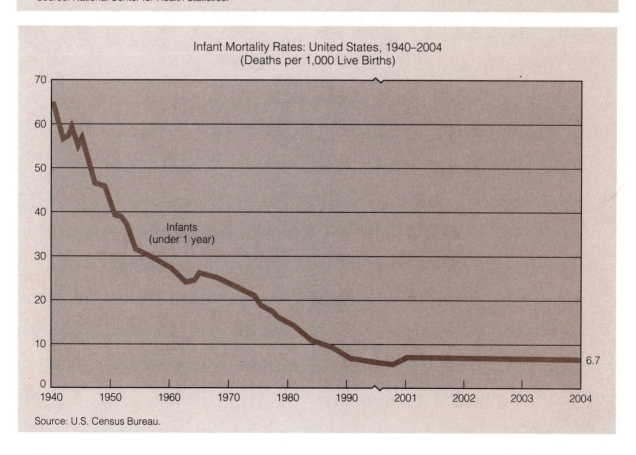

Infants
(under 1 year)

6.7

Source: U.S. Census Bureau.

10TH EDITION

HUMAN INTIMACY

Marriage, the Family, and Its Meaning

FRANK D. COX
Santa Barbara City College

THOMSON

WADSWORTH

AUSTRALIA • BRAZIL • CANADA • MEXICO • SINGAPORE • SPAIN • UNITED KINGDOM • UNITED STATES

THOMSON
WADSWORTH

Human Intimacy: Marriage, the Family, and Its Meaning, Tenth Edition
Frank D. Cox

Sociology Editor: Robert Jucha
Development Editor: Stephanie Monzon
Assistant Editor: Elise Smith
Editorial Assistant: Christina Cha
Technology Project Manager: Dee Dee Zobian
Marketing Manager: Wendy Gordon
Marketing Assistant: Gregory Hughes
Marketing Communications Manager: Linda Yip
Project Manager, Editorial Production: Cheri Palmer
Creative Director: Rob Hugel
Print Buyer: Karen Hunt

Permissions Editor: Sarah Harkrader
Production Service: G&S Book Services
Text Designer: Diane Beasley
Photo Researcher: Myrna Engler
Copy Editor: Joan Harris
Illustrator: G&S Book Services
Cover Designer: Belinda Fernandez
Cover Image: Getty Images
Cover Printer: Coral Graphic Services, Inc.
Compositor: G&S Book Services
Printer: Edwards Brothers, Incorporated/Ann Arbor

Printed in the United States of America

1 2 3 4 5 6 7   09 08 07 06 05

For more information about our products, contact us at:

Thomson Learning Academic Resource Center
1-800-423-0563

For permission to use material from this text or product, submit a request online at
http://www.thomsonrights.com.

Any additional questions about permissions can be submitted by email to **thomsonrights@thomson.com**.

Library of Congress Control Number: 2004118286

Student Edition: ISBN 0-534-62532-0

Instructor's Edition: ISBN 0-534-62533-9

Thomson Higher Education
10 Davis Drive
Belmont, CA 94002-3098
USA

Asia (including India)
Thomson Learning
5 Shenton Way
#01-01 UIC Building
Singapore 068808

Australia/New Zealand
Thomson Learning Australia
102 Dodds Street
Southbank, Victoria 3006
Australia

Canada
Thomson Nelson
1120 Birchmount Road
Toronto, Ontario M1K 5G4
Canada

UK/Europe/Middle East/Africa
Thomson Learning
High Holborn House
50–51 Bedford Row
London WC1R 4LR
United Kingdom

Latin America
Thomson Learning
Seneca, 53
Colonia Polanco
11560 Mexico
D.F. Mexico

Spain (including Portugal)
Thomson Paraninfo
Calle Magallanes, 25
28015 Madrid, Spain

Brief Contents

Contents

Chapter 3

American Ways of Love　　60

Chapter 4

Gender Convergence and Role Equity　　92

Chapter 5

Communications in Intimate Relationships   116

Chapter 8

Human Sexuality 223

Chapter 13

The Dual-Worker Family:
The Real American Revolution   397

Chapter 14

Family Crises   429

Chapter 15

The Dissolution of Marriage 464

Chapter 16

Remarriage: A Growing Way of American Life   496

Chapter 17

Actively Seeking Marital Growth and Fulfillment   523

Preface

Marriage and family seem, on the surface, to be simple age-old concepts. Yet arguments rage over how to define *family*:

- Are an unmarried single mother and her child a family?
- Should two men or two women be able to marry and create a family?
- Can a number of people in an intimate commune setting be defined as a family?
- Using an anonymous sperm donor, an egg donor, and a surrogate mother to carry the fertilized egg to birth creates a family, but to whom does the child really belong?

Everyone seems to support "family values," but there is little agreement on what, when, how, and to whom such support should be given. Most Americans believe in family values but are quick to criticize others whom they feel do not share their own particular definition of family.

As readers of past editions of *Human Intimacy* know, your author focuses on principles that lead to successful intimate relationships, regardless of what your definition of a family may be. *Human Intimacy* presents an optimistic view of the American family, concentrating on those strengths that research has found in all successful intimate relationships.

Even when the family is afloat in stormy seas and seems to be foundering, marriage and family remain the basic building blocks of a strong society. Almost all of us grow up in families, and 90 percent of us will be a marriage partner at some time in our lives. Although some people worry that the percentage of Americans marrying is dropping (and it is), if we add cohabitation figures to marriage figures we find that intimate relationships are still sought by the vast majority of Americans. Why? Because it is within the intimate love relationship we call family that individuals most often find a sense of sharing, a sense of well-being, a sense of security, a sense of fulfillment, and, perhaps above all, a meaning to one's life.

It is important to stress those characteristics of intimate relationships that help reinforce the strength and resiliency of the marital relationship and the family. Families tend to get into trouble because members are unwilling to make the effort to nourish and enrich their family relationships. By emphasizing ways by which relationships can be improved, *Human Intimacy* encourages readers to make the effort to build strong, satisfying intimate relationships.

It is the author's hope that *Human Intimacy* will contribute to the reader's ability to make intelligent, satisfying choices about intimate relationships. Individuals who can make such choices in their lives are most apt to feel fulfilled. And fulfilled people have the best chance of making their intimate relationships, their family relationships, exciting and growth producing. Fulfilled people are also more likely to contribute to society at large, thus making the general community a better place in which to live.

Why a Tenth Edition?

Continuing editions of a textbook are always exciting for an author. First, they mean that the book has been well-enough received that a new edition is warranted. Second, the author has a great deal of new input coming from those who have used the book. With such input, a new edition cannot help but be improved. Third, in a field as complex and rapidly changing as marriage and the family, the need to update is perhaps more pressing than in other fields. And last, the writer can rethink his or her basic assumptions and the writing that has gone before. It is hoped that such rethinking leads to a better textbook—more up to date, thorough, interesting, readable, and above all more exciting.

The tenth edition of *Human Intimacy* tries to realistically reflect the place of marriage and the family in today's American society. The beauty of the institution called *family* is its adaptability and flexibility. Those persons with knowledge of themselves, their family, and their culture have the opportunity in America to build a marriage and a family to their own liking. Families can change for the better; intimate relationships can become more deeply satisfying and fulfilling. It is with these goals in mind that *Human Intimacy* has been written.

Distinctive Features

Human Intimacy has several features designed to aid your reading and challenge you.

- Each chapter starts with an outline that gives you an overview of the material to follow. This is followed by a series of thought-provoking questions for you to ponder as you read the chapter.
- *Highlight* boxes supply interesting detail and add variety to the reading, much as an aside adds variety to a lecture. *What Research Tells Us* is one type of highlight that serves as a reminder of the quantity of scientific research underpinning our knowledge about intimate relationships, marriage, and the family.
- *Debate the Issues*, featured at the end of each chapter, present both sides of controversial topics. Taking a definitive stand on both sides of an issue helps make discussions both lively and thought provoking.
- *Families around the World* help readers better understand the diversity of family life.
- *What Do You Think?* are critical thinking questions that appear throughout the text to precipitate thought and discussion.
- *Making Decisions* boxes are short exercises designed to help you gain insight into such topics as your attitudes toward love, marriage, the handling of finances, and so on.
- Key terms are set in **boldface** and defined in the margin at the point of use as well as in the glossary at the back of the book.
- A short summary concludes each chapter.
- Two new appendices investigating sexually transmitted diseases and contraception are included at the end of the book.

What Is New and Unique to This Edition?

- As always, the entire book has been updated. Of the 300 new references, 80 percent are 2000 or later with many from 2003–2005. The tenth edition is the most up-to-date edition yet published. This is mainly due to the large body of information now available online, especially government statistics of all kinds.

- Tables and figures contain the most up-to-date research, data, graphs, and statistics currently available.

- The chapter order has been slightly revised in light of reviewers' suggestions. Because of its importance, the chapter on family life stages has been moved forward and placed after Chapter 10, The Challenge of Parenthood. In this way, the discussion on family life stages remains in proper order.

- The chapters on human sexuality and family planning, pregnancy, and birth have been cut down by deleting much of the biological material as well as by placing the detailed information of sexually transmitted diseases and contraception into appendices. This has helped to reduce the length of the book and to reduce the content to only the most important information.

- The discussion of HIV/AIDS has been included in Appendix A, along with a discussion of other sexually transmitted diseases. By placing it in an appendix, the instructor may bring in a discussion of HIV/AIDS at any time he or she chooses rather than having to discuss it when it appears in the chapter on sexuality.

- Because the racial and ethnic composition of the United States continues to change, more cross-cultural examples and material on ethnic and minority families has been added. A new section on American Indians is included as well as material added where appropriate.

- More emphasis is placed on weaving the positive characteristics of successful families throughout the book.

Much new material will be found throughout the book. For example:

- Chapter 1 (Human Intimacy in the Brave New World of Family Diversity) places a stronger emphasis on the similarities between peoples, groups, and their relationships while not diminishing the importance of diversity. Material on the theory and methods of scientific study has been placed in this chapter.

- Chapter 2 (Human Intimacy, Relationships, Marriage, and the Family) includes a new section on the American Indian family as well as a new Debate the Issues titled "Should Our Society Recognize Gay Marriages?"

- Chapter 3 (American Ways of Love) expands the Greek definitions of love as well as expanding theories of love to include brain studies.

- Chapter 4 (Gender Convergence and Role Equity) includes a rewritten section on the women's movement as well as a new Debate the Issues titled "Can We Create Gender-Neutral Children?"

- Chapter 5 (Communications in Intimate Relationships) has an expanded section on family life in the information age, which examines the effects video games, cell phones, home computers, and the Internet have on the family, especially children. The section on conflict and anger has been expanded.

- Chapter 6 (Dating, Single Life, and Mate Selection) expands the material on singleness, date rape, courtship violence, and cohabitation.

- Chapter 7 (Marriage, Intimacy, Expectations, and the Fully Functioning Person) includes expanded material on commitment and extramarital affairs as well as a new Highlight on Navajo marital expectations.

- Chapter 8 (Human Sexuality) has had the biological material reduced and the details about sexually transmitted diseases transferred to a new appendix. There is also a new Highlight on female genital mutilation.

- Chapter 9 (Family Planning, Pregnancy, and Birth) again reduces the biological material and adds new material on the genetic revolution as well as placing the detailed material on contraception in a new appendix.

- Chapter 10 (The Challenge of Parenthood) has the highest number of new references and has been almost completely reorganized. There are a number of new Highlights: "Diversity in Child-Rearing Values and Practices," "To Tattoo or Not to Tattoo: Out Damned Spots!" "Adolescent Hormones and Brain Functioning," and "Father Absence in African American Homes."

- Chapter 11 (Family Life Stages: Middle Age to Surviving Spouse) has been thoroughly updated with new material on grandparenting and its importance to children of broken marriages.

- Chapter 12 (The Importance of Making Sound Economic Decisions) has been greatly simplified but expands the emphasis on acquiring economic knowledge in order to have a successful marriage and family life.

- Chapter 13 (The Dual-Worker Family: The Real American Revolution) has new sections on community service and the working wife, and stay-at-home moms.

- Chapter 14 (Family Crises) has new material on defense mechanisms and ambiguous loss, as well as new sections on children and poverty and the military family in time of war.

- Chapter 15 (The Dissolution of Marriage) has all new material on the effects of divorce on children, modifying Wallerstein's popular conclusions. The chapter also takes a closer look at the many and interrelating factors that appear to relate to divorce.

- Chapter 16 (Remarriage: A Growing Way of American Life) has new sections on family law and stepfamilies, child support obligations, and custody and visitation of stepchildren.

- Chapter 17 (Actively Seeking Marital Growth and Fulfillment) seeks to end the text on a positive note, giving the reader more ideas about creating a strong and successful family life.

Supplements

Human Intimacy: Marriage, the Family, and Its Meaning, Tenth Edition, is accompanied by a wide array of supplements prepared to create the best learning environment inside as well as outside the classroom for both the instructor and the student. All the continuing supplements for *Human Intimacy: Marriage, the Family, and Its Meaning*, Tenth Edition, have been thoroughly revised and updated, and several are new to this edition. We invite you to take full advantage of the teaching and learning tools available to you.

Supplements for the Instructor

Instructor's Manual with Test Bank Written by Sylvia M. Asay of the University of Nebraska at Kearney, this supplement offers a wide range of resources to help you teach your course, including brief and detailed chapter outlines, learning objectives, key terms and concepts, and class projects. The instructor's manual also includes 75–100 multiple-choice and 25 true-false questions for each chapter, all with answers and page references. It also includes 10–15 short-answer questions and 5–10 essay questions for each chapter.

ExamView® Computerized Testing for Macintosh and Windows Create, deliver, and customize printed and online tests and study guides in minutes with this easy-to-use assessment and tutorial system. ExamView includes a Quick Test Wizard and an Online Test Wizard to guide you step by step through the process of creating tests. The test appears on screen exactly as it will print or display online. Using ExamView's complete word processing capabilities, you can enter an unlimited number of new questions or edit questions included with ExamView.

Transparency Masters for Marriage and Family 2006 Black and white tables and figures from Wadsworth's 2006 Marriage and Family texts are also available to help you prepare your lecture presentations.

CNN Today® Video Series: Marriage and Family, Volumes V–VII Illustrate the relevance of marriage and family to everyday life with this exclusive series of videos for the Marriage and Family course. Jointly created by Wadsworth and CNN, each video consists of approximately 45 minutes of footage originally broadcast on CNN and specifically selected to illustrate important sociological concepts.

Supplements for the Student

Marriage and Family: Using Microcase® ExplorIt®, Third Edition Written by Kevin Demmitt of Clayton College, this is a software-based workbook that provides an exciting way to get students to view marriage and family from the sociological perspective. With this workbook and the accompanying ExplorIt software and data sets, your students will use national and cross-national surveys to examine and actively learn marriage and family topics. This inexpensive workbook will add an exciting dimension to your marriage and family course.

Internet-Based Supplements

InfoTrac® College Edition with InfoMarks® Available as a free option with newly purchased texts, InfoTrac College Edition gives instructors and students four months of free access to an extensive online database of reliable, full-length articles (not just abstracts) from thousands of scholarly and popular publications going back as much as 22 years. Among the journals available 24/7 are *American*

Journal of Sociology, *Social Forces*, *Social Research*, and *Sociology*. InfoTrac College Edition now also comes with InfoMarks, a tool that allows you to save your search parameters, as well as save links to specific articles. (Available to North American college and university students only; journals are subject to change.)

WebTutor™ Toolbox on Blackboard and WebCT Web-based software for students and instructors that takes a course beyond the classroom to an anywhere, anytime environment. Students gain access to the rich content from our book companion websites. Available for WebCT and Blackboard only.

***Companion Website for* Human Intimacy: Marriage, the Family, and Its Meaning, Tenth Edition** The book's companion website includes chapter-specific resources for instructors and students. For instructors, the site offers a password-protected instructor's manual, Microsoft® PowerPoint® presentation slides, and more. For students, there is a multitude of text-specific study aids, including the following:

- Tutorial practice quizzes that can be scored and e-mailed to the instructor
- Web links
- InfoTrac College Edition exercises
- Flash cards
- MicroCase Online data exercises
- CNN Video exercises
- Crossword puzzles
- Virtual Explorations

And much more!

Acknowledgments

As with all such undertakings, many more people than myself have contributed to this book. The most important are the many family members with whom I have interacted all my life: parents, grandparents, aunts and uncles, siblings, cousins, and, of course, my immediate family—Pamela, my wife; Randall and Linda and their children, Alexander, Brandon, and Cameron; and Hans and Michelle and their children, Stephanie, Max, and Bella. In addition, many fine researchers and writers have contributed to my thoughts.

And then there are the important contributions of the direct reviewers of this edition, without whom *Human Intimacy* could not exist. To them I wish to extend a special thanks.

Kelley Brigman, Minnesota State University–Mankto

William Dowell, Heartland Community College

Rick Fraser, California State University–Los Angeles

Stephen M. Gavazzi, Ohio State University

Gloria Palileo, University of South Alabama

Michelle Toews, Texas State University–San Marcos

Carlos Valiente, Arizona State University

I also thank all the wonderful students who have passed through my classes. They have made me think and grow and have let me know that the American family is alive and well.

Although *Human Intimacy* has my name on it, the actual production of the book rests with: Bob Jucha, my editor; Stephanie Monzon, the assistant editor for sociology who helped guide the development of this revision; and her successor, Elise Smith. Thanks also goes out to Christina Cha, the editorial assistant who helped prepare the final manuscript for production. Once in production, the project was skillfully guided by Cheri Palmer. Finally, I extend a special thanks to Wendy Gordon, the senior marketing manager, for her labors on this edition's behalf.

I also want to thank those oft-forgotten production people who turn the final manuscript into a beautiful book and place it in the hands of the many teachers and students who use it. I am always grateful for your fine work and consider *Human Intimacy* to be *our* book, not my book.

© Richard Marks

Human Intimacy in the Brave New World of Family Diversity

Chapter

1

Questions to reflect upon as you read this chapter:

- Can you list several characteristics of a successful marriage and family?
- What are your own personal ideals about marriage and family?
- How do logic and emotion relate when making successful decisions?
- Can you suggest some things that we can do to strengthen the family?

Could *The Adventures of Ozzie and Harriet* or *Leave It to Beaver* middle-class, American family of the 1950s possibly have predicted the current changes in reproduction? In that decade, the petri dish–cultured, artificially inseminated, gene-altered, implanted egg; the surrogate mother–carried child; and, perhaps, the soon-to-be cloned child were considered to be only science fiction possibilities.

Could that ideal, middle-class American family of the 1950s possibly have envisioned and accepted the diversity of families in the twenty-first century—families comprising a stepparent or a parent that is single, single adoptive, foster, or gay/lesbian; families that are African American, American Indian, Asian American, Hispanic, or of mixed ethnicity or nationality—or any mix of the above?

The popular buzzword in today's sociological and psychological study of the American family is the term *diversity*. It is clear that the American family is, indeed, diverse. Even experts in the field of family study can no longer agree on a definition of *family*. Yet the idea of diversity (meaning differences) can be (and often is) taken to the politically correct extreme, which implies that any relationship and every behavior is acceptable, that there are no standards, and that there can be no criticism or judgment. Too often, emphasis on diversity exaggerates differences.

Successful human relationships are built on similarities and on overcoming, rather than emphasizing, differences. To lay differences aside, to minimize them, is a major step in resolving conflict. Of course, there are differences in any relationship, but discovering and emphasizing similarities leads to compromise and acceptance of differences.

In the broader sense, no society can remain vital or even survive without a reasonable base of shared values. Emphasizing similarities, rather than differences, within the general society helps to create a foundation of shared values among the populace. Shared values form the glue that holds a family and perhaps more important, a society, together.

Amazing changes are undoubtedly taking place in both reproduction technology and family functioning within the American population (Demo et al. 2000; Marks 2000; Stacey 2000). How, then, do Americans create intimate, long-term relationships and families within this new, seemingly ever-changing context? Can a child born via the new technologies into a family made up of diverse individuals learn to create long-lasting intimacy? We believe that he/she can and that is the point of this marriage and family textbook.

Building Successful Relationships

Given one wish in life, most people would wish to be loved, to have the capacity to reveal themselves entirely to another human being and to be embraced, caressed, by that acceptance. People who have successfully built an intimate relationship know its power and comfort. But they also know that taking the emotional risks that allow intimacy to happen isn't easy. Built upon the sharing of feelings, intimacy requires consummate trust. (Avery 1989, 27)

Intimate relationships—what an exciting and important field of study! By **intimate**, we mean experiencing intense intellectual, emotional, and (when appropriate) physical communion with another human being. Communicating and caring. Boyfriend/girlfriend. Husband/wife. Parent/child. Grandparent/grandchild. Family/friends. These are the relationships that give meaning to life, the relationships that give us a sense of identity, of well-being, of security, of being needed. These are the relationships that ward off loneliness and insecurity, the relationships that allow us to love and be loved. Perhaps, without intimacy, the human part of *human being* would disappear and we'd all simply *be*; we'd become automatons similar to our home computers—capable of solving problems and delivering information, but lacking in those markedly human qualities of loving, caring, and compassion. In a phrase, we'd lack the characteristics that allow human beings to become intimate. Without intimacy there is emotional isolation, and emotional isolation increases the risk of physical and emotional disorders (P. Brown 1995, 135; Ladbrook 2000; Wamboldt and Reiss 1989). Social ties of all kinds, but especially those of an intimate nature, tend to support both physical and mental health (Bramlett and Mosher 2002).

The study of intimate relationships is both essential and exciting because we live in a society where intimate relationships are important to social and emotional survival. Modern society is fast-changing. Think of the speed of technological advancement to which we all must adjust. Compare the freedom today to build an intimate love relationship with the rigidity demanded of such relationships during Victorian times. As personal freedom increases, the building of personal relationships becomes more salient. Without rigid rules of the past governing relationships, the intimate relationships an individual builds become the glue to hold marriages and families together (Jamieson 1998, 1999, 218–219). On the other hand, intimate, long-lasting relationships are more difficult to build without the rigid rules of the past. However, most Americans continue to find intimacy and satisfaction within a creative and changing family, and most still spend the bulk of their lives within marriage and family relationships. Therefore, we need to expend more energy on making marriages and families viable and fulfilling, rather than simply criticizing the institution of marriage.

Yet, disparaging marriage and advocating alternatives to current practices is easy and popular. Such criticisms tend to imply that marriage is a rigid relationship that has passed relatively unchanged into our modern culture. In reality, marriage and the family have undergone dramatic change and they continue to change as they adapt to today's world.

Changes in the family because of modernization have led some critics of marriage to long for the good old days. This suggests that there was some lost, golden age of the family such as in TV's *The Adventures of Ozzie and Harriet* of the 1950s. Studies of family history, however, have failed to uncover any such golden age. Jerome and Arlene Skolnick (1986, 17; 1991; 2000) point out that those condemning modernization may have forgotten the problems of the past. Although our current problems inside and outside the family are genuine, we should remember that many of these issues derive from the very benefits of modernization, benefits too easily taken for granted or forgotten in the fashionable denunciation of modern times. In the past, there was no problem of the aged be-

intimate
Experiencing intense intellectual, emotional, and, when appropriate, physical communion with another human being

cause most people died before they got old. Adolescence wasn't a difficult stage of the life cycle because it didn't exist; children worked and education was a privilege of the rich. Modernization certainly brings problems, yet how many of us would trade the troubles of our era for the ills of earlier times? Edward Kain (1990) suggests that the dismay at the current state of the family and the desire to return to the good old days has created the myth of family decline. The family of the past also had plenty of problems, though often different from those of the modern family.

Rather than dwelling on the problems facing families, let us examine some possible agendas for strengthening marriages and families. After all, 2004 was the tenth anniversary of the Year of the Family (NCFR 2003, 5). What better time to begin examining the strengths exhibited by successful intimate relationships, particularly within families? What would intimate relationships be like if we could make them the best possible? Even if we succeed in solving the major problems that will surely arise in any intimate relationship, can we build enduring relationships that are better than just satisfactory? Can people create intimate relationships that are secure and comfortable, yet growing and exciting at the same time? Will today's young families be able to rear children who care about themselves and the communities of which they are a part—children who will grow into adults who are capable of being intimate, caring, responsible, and loving human beings?

Having a vision of what we want for ourselves, our relationships, our families, our children, our society, and our world-to-be is of the utmost importance to human beings. The ability to visualize the ideal enables human beings to change. Without a vision of what could be, there would be little if any change. If all our behaviors were inborn, biologically determined, and preordained, then nothing could change and no vision of the ideal would be necessary to survive. To be creative in life is to see what is, visualize what could be, think of the ideal, and then work in-between to move from what is to what could be. But why discuss an ideal? Won't we all fall short of the ideal? Yes, of course we will. But ideals can be goals, and goals give us something at which to aim. They give us direction in life. They motivate us.

family of origin

The family into which we were born and grew up

We will begin our study of intimate relationships by examining the ideal qualities of strong families because it is within families that all of us learn the most (positive and/or negative) about intimate relationships. After all, our **family of origin**, the family in which we were born and grew up, is the first seat of all of our learning, and human relationships are the essence of the family. Throughout this book, we will often return to this theme: how can we build into our intimate relationships those characteristics that lead to strong individuals and successful friendships, marriages, and families? You may wonder why we began by mentioning *The Adventures of Ozzie and Harriet* and *Leave It to Beaver*. How old-fashioned! How out of touch! As already pointed out, historical families had problems and were often far from perfect. However, they did portray certain ideals that can be helpful toward the creation of successful relationships and the successful rearing of children in today's diverse society.

Data supporting the importance of the two-parent family in the rearing of children are plentiful. When compared to children raised by single parents and step-parents, children reared by their own parents are more likely to finish school, have higher grades, and attend and graduate from college. They are more likely to be employed and less likely to become single parents. In addition, they score higher on measures of competence, conduct, psychological adjustment, and long-term health (Bramlett and Mosher 2002).

Doonesbury, June 1, 1994, by G. B. Trudeau. Reprinted with permission of Universal Press Syndicate. All rights reserved.

Let us be clear that these statistics are group averages. Parenting by individuals other than the biological parents can certainly produce children who achieve as well as those reared by biological parents, though such success is less common. Remember, group statistics indicate what a group will do on the average; they do not predict what any one individual or family will do or become.

Any type of family can be successful if it understands the characteristics that make intimate relationships grow and flourish. Good decision making can build intimate relationships that are enduring and successful. Such relationships do not appear by magic. As we go through life, constant decisions must be made about trivial, daily problems; important decisions, such as about the direction our lives will take, must also be made. The better and more efficient we become at making decisions, generally the smoother our lives will flow. A flourishing, intimate relationship is built upon a foundation of ongoing, successful choices. No relationship is stagnant. Regardless of how fulfilling and happy a relationship is, decisions must be constantly made to keep it that way and to continually improve it.

Putting families first is a call that is being heard more and more throughout the United States. *Family values* is now a favored topic in government. Family support systems have become available to people in all segments of society rather than just to the poor, as was the case in the past. An increasing number of programs (both public and private) offer families the social supports that were once provided by a network of stable, extended families within the community. But despite society's growing interest in building family strengths and regardless of the number of family supports provided, the major responsibility for the creation of strong, stable, satisfying intimate relationships remains with the individual. Knowing the ideal characteristics of the successful family and making decisions that move our intimate relationships toward these ideals are the responsibilities of each person.

Most people believe that durability signals a successful marriage. This alone may or may not be true. There are unhappy, conflict-ridden marriages that last a lifetime. Couples in such marriages simply may not consider divorce or separation, whatever their reasons might be.

A successful marriage approximates the marital and family ideals held by each partner. It usually yields satisfaction and happiness for the couple. The relationship fulfills their psychological, material, social, and sexual needs. Each individual entering marriage has some ideal expectation of what the relationship will be.

Qualities of Strong and Resilient Families: An Overview

As researchers who study strong, healthy marriages and successful families point out, volumes have been written on what is wrong with the family, but little has been written about what is right in the successful family. We don't learn how to do anything by looking only at how it shouldn't be done. We learn most effectively by examining how to do something correctly and by studying a positive model. By discovering the strengths of enduring, intimate relationships, we might improve our ability to build successful marriages and create fulfilling family lives.

The basic thrust of this book is to create and develop a vision of the strong family and to weave this image throughout the book. Secondarily, we focus on how to make decisions that lead to strong, intimate relationships. Of course, we also spend time discussing family problems. But by formulating an ideal vision of what a family can be, we take the first and perhaps most important step toward resolving problems that will arise throughout our lives. This ideal will help us to understand how decisions can be made that improve our relationships.

The real question is, "*How can we make all types of intimate relationships more successful, more enduring, and more fulfilling?*"

Vera and David Mace (1983, 1985) coined the term *family wellness* to describe the strong family that functions successfully. The Maces maintain that the quality of life in our communities is, in part, determined by the quality of relationships in the families that make up the communities. Healthy families produce healthy individuals, who then help to maintain healthy community environments (Stinnett 2000; Stokols 1992). The quality of life in families is, in turn, strongly affected by the quality of relationships between the couples who founded those families.

We take the view that family wellness, in its full and true meaning, grows out of marriage wellness. A family begins when a marriage begins. We do not mean that a one-parent family cannot be a well family. It can. But since four out of five one-parent families are really in transition between marriages it is the marriage relationship that is still the foundation stone [to family success]. So the key to nearly everything else is to enable marriages to be what they are capable of being and what the people involved want them to be. (Mace and Mace 1985, 9)

Concentrating on individual and family strengths helps to counteract the prevailing social view that is absorbed with the negative—problems and failures

(Seligman 1998; Volz 2000). The idea that nothing good can come from any failure or undesirable event and that the victim is helpless in the face of adversity is self-defeating.

Science has managed to ignore the fact that undesirable events often produce extraordinary strength, growth, and creativity. Many social scientists have come to view courage, perseverance and good cheer as illusory, defensive, and inauthentic while weaknesses such as depression, greed and lust are genuine. (Seligman 1998, 11)

Such views tend to lead to the idea of victimization—the placing of blame, rightly or wrongly, on outside forces—which, in turn, leads to defeatism, inactivity, surrender, and the idea that severe trauma can't be undone.

On the other hand, strong families are optimistic and take the initiative to fight their problems, feeling they can solve them and control their lives. They are on the offensive. They do not simply react; they make things happen. Families can do a great deal to make life more enjoyable, and strong families exercise that ability.

For our discussion, we assume that basic needs such as nutrition are met, leaving family members some energy to invest in improving their lives. Obviously, in many parts of the world, discussion of family strengths and an optimistic view of life is meaningless until basic survival needs are met.

What are the relationship qualities that lead to family strength and wellness? Numerous researchers have sought answers to this question, and there has been considerable agreement among their findings (Alford-Cooper 1998; Gottman 1994; Mackey and O'Brien 1995; Robinson and Blanton 1993; Stinnett 1997, 2000; Stinnett and DeFrain 1985; Stinnett and James 2000). The research suggests the following eight major qualities shared by all strong, healthy families (each is discussed more thoroughly in later chapters):

1. *Commitment.* The major quality of strong families is a high degree of commitment. The family members are deeply committed to promoting each other's happiness and welfare. They are also very committed to the family group and invest much of their time and energy in the family. The individual family member is integrated into a relationship of mutual affection and respect. By belonging and being committed to something greater than oneself, there is less chance that individualism will turn into egocentrism. Commitment to the relationship involves wanting to stay married, feeling morally obligated to stay married, and feeling constrained to stay married (meaning that there are barriers to leaving a relationship) (Huston 2001; M. P. Johnson 1991, 1999).

2. *Appreciation.* This quality seems to permeate the strong family. The family members appreciate each other and make each other feel good about themselves. All of us like to be with people who make us feel good. Yet many families fall into interactional patterns in which they make each other feel bad. In strong families, members find good qualities in each other and can express appreciation for them. This appreciation increases a person's good behavior by rewarding it, thus making it more common, which, in turn, leads to greater appreciation from others.

3. *Good communication patterns.* Members of strong families spend time talking to each other. Sometimes families are so fragmented and busy and spend so little time together that they communicate with each other only through ru-

mor. By this we mean that families may communicate indirectly through hearsay, assumption, guesswork, and innuendo rather than directly through good communication techniques (see Chapter 5).

Strong families also listen well. By being good listeners, the family members say to each other, "You respect me enough to listen to what I have to say. I'm interested enough to listen too."

Families that communicate well also fight fairly. They get angry at each other, but they get conflict out into the open and can discuss the problem. They share their feelings about alternative ways of dealing with problems and can select solutions that are best for everybody.

empathy
The ability to understand what the other is thinking, put oneself in the other's place, and intellectually understand the other's condition without vicariously experiencing the other's emotions

Both appreciation and good communication require family members to be empathic and trustworthy. **Empathy** may be defined as the ability to understand what the other is thinking, put oneself in the other's place, and intellectually understand another's condition without vicariously experiencing their emotions (Long et al.1999, 235).

Trust is important to all successful relationships. It is essential for open communication, mutual understanding, and problem solving (F. Walsh 1998, 52). Whether it is trust between a buyer and seller, business partners, family members, or an individual and his/her government, when trust declines or is lost, the relationship is usually lost. Loss of trust leads to discomfort, skepticism, disbelief, and failure to participate. For example, many political scientists feel that Americans' low voter turnout results from distrust of the government.

4. *Desire to spend time together.* Strong families do a lot of things together. This is not a false or smothering togetherness; they genuinely enjoy being together. Another important point is that these families actively structure their lifestyles so that they can spend time together. This togetherness extends to all areas of their lives, including meals, recreation, and chores. They spend much of their time together in active interaction rather than in passive activities such as watching television. Family rituals and routines are part of the strong family's activity (Gregg et al. 1999).

Rituals and routines maintain a sense of continuity over time, linking past, present, and future through shared traditions and expectations. Routines of daily life, such as family dinner or bedtime stories, provide regular contact and order in what is increasingly a fragmented, hectic schedule for most families. (Hochschild 1997)

5. *A strong value system.* The underlying factor that adds strength to a family is a strongly held and mutually shared value system. Such a value system allows individuals to have a wider vision of life than personal success alone and enables them to reach beyond themselves. Families that share a strong value system experience *spiritual wellness*. This is a unifying force, a caring center within each person that promotes sharing, love, and compassion for others. Some will disagree, but Stinnett and DeFrain (1985) found this quality was most often expressed as a high degree of religious orientation. This finding agrees with research from the past 40 years, which shows a positive correlation among religion, marriage happiness, and successful family relationships (Lehrer 2000; Robinson and Blanton 1993, 38). In addition, the higher the importance attached to religion, the lower the likelihood of marital disruption (Bramlett and Mosher 2002). Spirituality gives us a sense of community and support. Organized religion may be advantageous to family life by (1) enhanc-

ing the family support network, (2) sponsoring family activities and recreation, (3) indoctrinating supportive family teachings and values, (4) providing family social and welfare services, and (5) encouraging families to seek divine assistance with personal and family problems (Abbott et al. 1990, 443).

These researchers also point out, however, that rigid religious doctrines that promote only traditional sex roles or negative approaches to family planning, for example, might be detrimental to family life.

Of course, organized religion has no monopoly on spirituality. Strong values can be demonstrated in many ways such as through community involvement, education, and work.

6. *Ability to deal with crises and stress in a positive manner*. Strong families have the ability to deal with crises and problems in a positive way. Such families can bounce back from adversity. They may not enjoy crises, but they can handle them constructively. Even in the darkest situations they manage to find some positive element, no matter how small, and focus on it. In a particular crisis, they may rely to a greater extent on each other and the trust they have developed in each other. Confronted by a crisis, they unite to deal with it instead of being fragmented by it. They cope with the problem and support each other.

7. *Resilience*. **Resilience** can be defined as the capacity to rebound from adversity, having become strengthened and more resourceful. It is an active process of endurance, self-righting, and growth in response to challenge and crisis. It involves more than merely surviving, getting through, or escaping a harrowing ordeal. The quality of resilience enables people to heal from painful wounds, take charge of their lives, and go on to live fully and love well (F. Walsh 1998, 4). Fortunately, research has shown that resilience can be learned by most anyone (Kersting 2003), especially children. Children seem to have an amazing capacity to withstand and recover from adversity (Masten and Berkmaier 2001). Resilience leads to a feeling of competence.

8. *Self-efficacy*. Perceived self-efficacy is a person's beliefs about his/her capacity to produce designated levels of performance that exercise influence over events that affect his/her life. More simply, ask, "What are my beliefs about how competent I am in general?" and "How competent am I, related to a given task?" Self-efficacy beliefs determine how people feel, think, motivate themselves, and behave. A strong sense of self-efficacy enhances human accomplishment and personal well-being. People with high self-assurance in their capabilities approach difficult tasks as challenges to be mastered rather than threats to be avoided. Such people approach threatening situations with assurance that they can exercise control over them. Such an efficacious outlook produces personal accomplishments, reduces stress, and lowers vulnerability to depression (Bandura 1997, 2000, 2001). "People who regard themselves as highly efficacious act, think, and feel differently from those who perceive themselves as inefficacious. They produce their own future, rather than simply foretell it" (Bandura 2004).

Regardless of the type of family, the strong presence of these eight characteristics creates a positive environment, one that is pleasant to live in because family members treat one another in beneficial ways. Members of strong families can count on each other for support and love. They feel good about themselves, both as individuals and as members of a family unit or team. They have a sense of *we*, yet this sense of belonging does not overpower their individuality. The fam-

resilience
The capacity to rebound from adversity strengthened and more resourceful

---

WHAT DO YOU THINK?

1. How many of these characteristics do you find in your family of origin?

2. Which characteristics are most important to you? Why?

3. Do you have family traditions that you'd like to carry on in your own family? What are they? Why are they important to you? How do traditions help support the characteristics of successful families?

4. How would you establish those characteristics that are important to you in your own family or future family?

ily supports and respects individuality. Perhaps strong families can best be defined as those that create homes we enter for comfort, development, and regeneration and from which we go forth renewed and charged with power for positive living (Stinnett and DeFrain 1985, 8). Within a healthy family, individuals learn how to be intimate with family members. This sets the stage for successful, intimate relationships in the future.

Families founded on the principles of equality, the inviolability of the rights and responsibilities of the individual, mutual respect, love, and tolerance are the cradle of democracy. Such families are the foundation for the well-being of individuals, societies, and nations (Sokalski 1994, 8). This important idea led the United Nations to proclaim 1994 and every 10 years thereafter as the International Year of the Family.

Can We Study Intimacy?

Can we study intimacy? We can if we study relationships that can be, and often are, intimate. We usually find intimacy within marriage and the family. Although intimacy can exist between any two people, it is within the family that most of us learn to be intimate, caring, and loving people—or not. Thus, to study the family is also to study intimacy.

The study of the family deals with many topics, as the table of contents of this book reveals. It is clear that such a study cuts across numerous disciplines: psychology, sociology, anthropology, economics, and so on (Figure 1-1). To iden-

FIGURE 1-1
Family Science

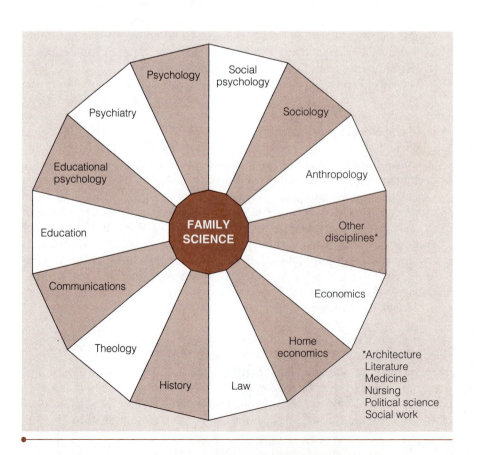

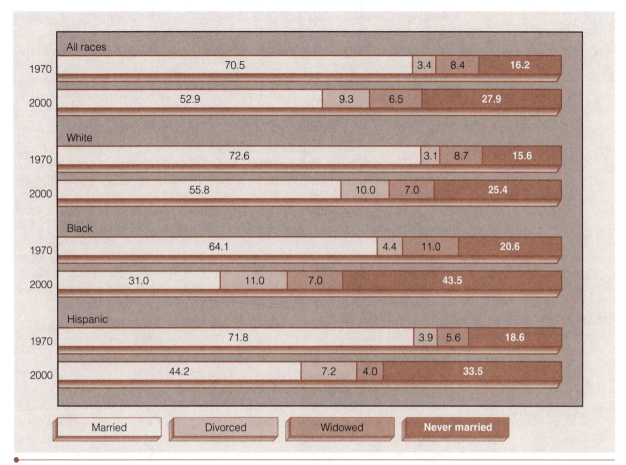

FIGURE 1-2 *Marital Status, by Race and Hispanic Origin, 1970 and 2000 (Adults in Percent)*
Source: U.S. Census Bureau, October 2003.

tify the study of marriage and family more clearly, we will use the term **family science**.

Because each of us is born into a family (the family of origin or orientation) and approximately 90 percent of us marry (Bramlett and Mosher 2002) and establish a family at some time in our lives, we usually have strong feelings about marriage, families, love, and intimacy. Figure 1-2 indicates the percentage of persons 18 years and older who are currently married, divorced, widowed, and never married.

To study family science is also to study our own feelings about the institutions of marriage and family. The statement that the birthrate in the United States in 2003 was 14.0 births per 1000 population (NCHS 2004a) may appear simple and clear, but such a statistic has little to do with an individual's personal decision about having children. Our personal experiences may or may not be represented by a scientific statistic describing the general condition of American families. That explains why, apparently, there is little agreement about how marriages and families are changing and what the changes mean.

Optimism versus Pessimism

Some Americans feel that today's families are in deep trouble because they are different from their own family of origin. Pessimists see the high divorce rate, the

family science

The study of marriage and family combining all disciplines that can shed light on marriage and family functioning large numbers of children born out of wedlock, and the devaluation of children as mothers enter the workplace as signs of family decline.

Optimists feel the wide variety of acceptable relationships now available to Americans allows people to create a family that is best for themselves. This will, in turn, improve the quality of family life. Pessimists see the high divorce rate as sounding the death knell of marriage. Optimists see it as normal behavior in a society that emphasizes personal happiness. Given the present freedom to seek happiness through divorce, individuals no longer need to endure an unhappy marriage.

Regardless of what we feel about the changes taking place in American marriages and families, such as easier divorce, all individuals interpret such data personally, based in part on their own family experiences.

But in addition to personal opinions about the family as an institution, a rich foundation of scientific information is available about this most personal and intimate of relationships. People who know the scientific facts, as well as their personal feelings about marriage and family, are in a better position to understand themselves and to build successful and satisfying, intimate relationships that will create strong, resilient families. As we have said, to study family is also to study intimacy.

Making Decisions That Lead to a Fulfilling Life

We have presented short descriptions of the major qualities and behaviors necessary to build strong and lasting relationships. We are not born with these behavioral qualities; they must be learned by the growing child through sufficient socialization by parents and the society at large. Humans learn the patterns of their culture and develop a value system as they grow into adulthood. It is the family that supplies most of this socialization.

Human beings have the longest dependency period of any mammal. Why is this long period of dependency necessary? It is needed because biology has not built many behaviors into human beings. A long socialization period is necessary for us to learn all that is needed to adapt successfully to society. A satisfying life does not automatically follow an individual's birth. Considerable learning is required for us to make decisions that lead to a satisfying life.

Because decision making is such an important part of all human existence, making successful life decisions is essential. The reader might begin thinking about such questions as the following:

- How can I learn to make the best possible decisions about life for myself as well as for those I care about?
- Will I make decisions that harm myself, my relationships, or my society?
- Will I make decisions that strengthen me, my relationships, and my society?

One person's life runs smoothly. That person seems fulfilled and happy. Another person's life is always upset. Looking carefully at each of these two people, you may discover that some people seem able to make good decisions about life, while others make poor decisions. Making good decisions is related to how a person is socialized.

A child can be socialized in both negative and positive ways. However, if a child is socialized in negative ways, antisocial behavior and negative self-feelings can result. All human beings suffer occasionally from self-doubt. If feelings about oneself are always negative, however, the potential to lead a long-lasting, fulfilling, intimate life becomes limited. A child needs loving attention to develop a positive self-image. A positive self-image leads to self-confidence and self-efficacy, which generally help the decision-making process.

If an individual grows up in a highly critical family and is seldom praised for doing well, but instead, is criticized most of the time, his/her confidence will be eroded. This, in turn, often leads to a poor self-image. Lack of confidence and a negative self-image make it very difficult for an individual to make satisfying decisions about his/her life.

How would you answer the following questions about yourself?

1. Do you feel your parents raised you in a positive manner? If so, explain. If not, explain.
2. In general, how would you rate your self-confidence?
3. How do you rate your decision-making skills?
4. Of the eight qualities found in successful families, which do you feel your family has? How about yourself?

Logic and Emotion in Decision Making

Each of us makes many decisions, perhaps hundreds each day. Some decisions are made unconsciously (without knowing or without awareness). Other decisions are made consciously (being aware and knowing). You get up in the morning, brush your teeth, and comb your hair. Probably each action is done automatically, without conscious thought. Such behaviors were taught to you by your parents when you were a small child. They are now simply automatic behaviors, controlled by decisions made long ago. However, what to wear this morning is probably a conscious decision. You ask yourself, "What am I going to do today— go to school or work, play, or visit friends?" A conscious decision will have to be made about what to wear, according to what you are going to do.

These relatively simple decisions will probably have little effect on your overall life. The failure to care for your teeth regularly will probably result in dental problems in the future; but to brush or not to brush on a given day is not an earth-shattering decision. However, each of us also makes decisions that will greatly affect our life and the lives of those around us. Who you marry, where you decide to live, whether you decide to have children or not, and what career you choose are all examples of decisions that greatly affect your life. Some people, such as world leaders, may even be called upon to make decisions that affect millions of people and perhaps the whole world.

Although you may never be required to make decisions that affect the world, you will certainly be called upon to make decisions that affect your personal world, the people in it, and your environment. Yet many societies do little in the way of teaching formal decision-making skills.

Some people learn to make decisions in a hit-or-miss manner and may not even be quite sure afterward how they came to a decision. Others may even say that they didn't make a decision at all and claim that circumstances guided their behavior. Although such a statement may seem true at the time, it probably is not. Blaming circumstances for decisions may actually be a means of avoiding responsibility, especially if there are unhappy consequences.

Decisions are also made by doing nothing at all. Such decisions can have just as much impact as those made by deliberate effort.

The process of decision making may seem very logical and rational. However, you cannot learn to make decisions in the same manner that you learn mathematics or a foreign language. It is true that you can be taught decision-making skills, but decision making requires more than simply applying logic to a set of facts. It includes your emotions as well as intellect. Skillful decision making also takes into account the emotions of others who will be affected by your decisions. A successful decision is one that strikes a balance between both the rational and emotional aspects of a given situation.

Mate selection provides a good example of the extremes of logic and emotion in decision making. In many countries, such as India, most marriages are arranged by the parents either through friends and family or a marriage broker. Parents feel young people do not have the maturity or life experiences necessary to choose the most suitable partner. Parents believe that an arranged marriage will be built upon a more rational and logical foundation than a marriage based on love alone. In an arranged marriage, if there is to be love between the couple, it will have to grow into the relationship. Such marriages may or may not be successful. In America, marital decisions are frequently based solely upon emotion. "We are in love," two young people say. The emotions of love can be so strong that they override reason.

Both methods of choosing a mate work, but only to a degree. The partners in arranged marriages may be well matched socially, but may never grow to love each other. On the other hand, marriages based on love tend to have high divorce rates. When making decisions, considering both the logical and emotional aspects of a situation is important.

Decision-Making Steps

The following are the six most important steps in the decision-making process:

1. Define needs or wants.
2. Look carefully at resources.
3. Gather information on all choices.
4. Identify, evaluate, and compare choices.
5. Make the decision, develop a plan, and get started.
6. As time passes, evaluate the decision and readjust it if necessary.

Define Needs or Wants   With almost any decision, the first step is to determine what is needed or wanted. This is not always easy. Indecision may occur due to the fear of assuming responsibility for decisions.

Consider Both Short- and Long-Term Outcomes   In making decisions, one needs to try to be fully aware of the consequences. It is especially difficult to consider long-term outcomes of a decision that gives a high degree of short-term pleasure. For example, Jamie feels he must have a stereo system; he just has to have it. He makes the purchase, agreeing to pay monthly. Within a few months, he realizes that he can't make the payments. His decision to buy the stereo system was based on short-term pleasure ("I love the sound of this system"). He did not consider the long-term consequences of paying for the system. If Jamie were married, his decision would affect his entire family.

Look Carefully at Resources   Next evaluate resources. What does one possess or know that will help get what is wanted, solve the problem, or improve the relationship? What deficiencies will work against it?

Seek Advice from Others   One important way of discovering resources is by seeking the advice of knowledgeable and respected people. Parents, relatives, friends, teachers, counselors, professionals, and experts can all help one make better decisions. Other people that one knows well can help one understand one's own abilities. They can also help one understand the resources needed to make a particular decision. Most important, others might be more objective and less emotional, especially if the question involves intimate relations. Seeking advice does not mean getting others to make one's decisions or seeking rubber-stamp approval. Rather, gaining information and learning from the experience of others helps a person make the best possible decisions.

Many decisions get made on the spur of the moment, in hit-or-miss fashion. This can happen when one doesn't take the time to judge what one's resources are or when one is under pressure by one's friends. Such decisions are often poor decisions.

Gather Information on All Choices   With all decisions, it is important to gather information first. Without correct and sufficient information, making a good decision is impossible. How a decision turns out is rarely absolutely predictable. The better the information on which one bases decisions, however, the greater are the chances of the decision turning out well.

Sort Out the Most Relevant Influences   Relevant influences are those that are important to the decision. Many factors influence the outcome of a decision, so one might want to seek all the facts and information possible. In reality, however, one needs only the most relevant information, not all the information. Information overload can contribute to indecisiveness because there is always one more thing to consider.

Use Intuition and Common Sense   Common sense and intuition are often credited for many decisions people make. **Intuition** is the immediate understanding of something without conscious reasoning or thinking about it. **Common sense** is practical intelligence, or ordinary good sense. These can be powerful decision-making tools. When examined closely, both of these reasons for decisions are usually based on much more information than is first apparent. When an intuitive decision turns out to be correct, we usually find that it has been based on relevant information, even if we didn't recognize it at the time. There is usually a perfectly logical basis for making an intuitive or commonsense decision. On the other hand, common sense and intuition are not foolproof. They may lead to illogical and poorly thought-out decisions if not based on relevant information.

intuition
The immediate understanding of something without conscious reasoning or thinking about it
common sense
Practical intelligence or ordinary good sense

Avoid Confusion between Words and Their Meanings   Most decisions are worked out in our heads by the use of words. Yet words can have different meanings. Besides the meanings found in dictionaries, words may have emotional meanings. The personal emotional meaning of a word is termed the **connotative** meaning. For example, the following sentences have primarily the same meaning, yet your emotional reaction to each sentence will differ:

connotative
The personal or emotional meaning of a word

A *slender* man of *later maturity* who was very *thrifty* bought an *inexpensive* bottle of perfume for his girlfriend.

A *skinny old man* who was very *stingy* bought a *cheap* bottle of perfume for his girlfriend.

Which sentence creates a more positive reaction? Why? Notice how the italicized words have the same dictionary meaning, but differ in their emotional content.

The connotative meaning of words can affect decisions, just as the dictionary meaning of words can. When working out decisions that affect others, be clear about what a word means to each of the persons concerned. For example, the statement "I love you" will have a different personal meaning for each individual. To the degree that two people's definitions of love vary, there may be confusion about and differing expectations of what is meant by *love* (see Chapter 3).

Identify, Evaluate, and Compare Choices   Often, people feel they have little or no freedom to make decisions or determine their own lives. They feel trapped.

"I don't have a choice of jobs because I am unskilled," one says.

Another claims, "My life doesn't go well because my mother and father fought a lot and were divorced when I was young."

Or still another complains, "I am not free to improve my life because of my race, religion, education, or finances."

Yet in almost all these circumstances, there are alternatives. People are not completely locked into prisons that allow no choice, unless they choose to be. Feelings of entrapment occur when a person is in a highly emotional state. Many of the most important life decisions are made under highly emotional circumstances, such as being in love and proposing marriage or being very angry during a dispute with your spouse and making the decision to break up the relationship. Often decisions made under highly emotional circumstances prove to be unsound.

The feeling of being trapped and without choice often indicates a lack of understanding of the possible alternatives. It takes time to identify and evaluate choices. Counselors and therapists find that the inability to identify choices is one of the major characteristics of troubled people. Yet identifying alternative choices is often crucial in making successful decisions.

Make the Decision, Develop a Plan, and Get Started   *Avoiding Snap Decisions* By basing a decision on relevant information and comparing the choices, a person can avoid making a snap decision. A *snap decision* is one that is made quickly, without considering all choices. Sometimes a quick decision must be made because time is short, but a quick decision is not necessarily a snap decision. A good decision maker will be able to make decisions more quickly than a poor decision maker. However, when time for a decision is very short, it is a good idea to see if one can buy a little extra time. Often more time is available than first believed. Of course, the opposite of making a fast decision would be to worry for so long that either no decision is made or the decision is made by others.

Once the decision is made, it is important to plan how to put it into effect and then get started.

Evaluate the Decision and Readjust It if Necessary   *Staying Flexible* It is important to remain flexible and open in making decisions. A person who bases decisions on rigid attitudes and set opinions may overlook facts and information that are highly relevant to a decision. One can never be absolutely sure that any decision will be perfect and should be open to changing even the best-made decision. Often, the outcome of a decision is not completely predictable. New information may be discovered. There may be unintentional and/or unforeseen long-term consequences. A person should remain flexible enough to change the decision, if necessary, and rather than being embarrassed or just too stubborn to change, accept the fact that all people will make poor decisions that need changing at times. A person who can remain open enough to admit mistakes is in a better position to rethink decisions.

Circumstances can change quickly, and a decision that is correct at first can quickly become incorrect. For example, when another person is involved, a perfectly good decision may turn out to be faulty because of that other person. In a basketball game, one might realize that the best shot is to make a driving layup. You start to drive toward the basket, but a much taller opposing player steps in to block the path. The decision may have been theoretically correct, but another person moved into position to make the shot fail. So you went with plan B and hit your jumper. The point is that in many situations, although one makes the best possible decision based on the facts at the time, outside forces can intervene. Part of the decision-making process means remaining flexible, evaluating your decisions, and readjusting those decisions that turn out to be incorrect.

The Gift of Choosing

Experts offer many steps and guidelines to improve decision-making skills. Yet even the longest list of guidelines cannot assure anyone of making perfect decisions. In fact, part of the challenge and adventure of life is that all of us are confronted with situations in which we must make decisions, right or wrong. We have the ability to learn and to choose much of our behavior. Choice is what makes each of us human. To give up that unique ability is to avoid decision making or to make choices by default, simply letting things take their course. Allowing others to make all our decisions or avoiding the responsibility of choice is to give away the human aspect of being a human being. We all must learn to make decisions/choices in and about our lives that will improve ourselves as individual human beings and improve our relationships, families, and the societies in which we live.

Theoretical Approaches to Family Study

Before we can apply our knowledge of family strengths and good decision-making skills, it is important to understand how our scientific knowledge of marriage and family is obtained. As we saw in Figure 1-1, family science is a very eclectic field that borrows from almost all fields of study. Thus, a textbook examining the family will also incorporate information from many theoretical points of view and from many fields of study. It is not our purpose to turn the reader into a theoretician, yet it is important to understand that research is always based on some theoretical foundation. It is equally important to understand the methods by which scientific information is obtained. Although we hear a lot about bad science, the reader will be better able to assess the information presented in later chapters by understanding the methods scientists use.

One purpose of theory is to serve as a guide for further research. A second purpose is to serve as a scientific shorthand to generalize and summarize our knowledge. A third purpose of theory is to predict. A theory need not always be factu-

WHAT DO YOU THINK?

1. How do you evaluate your own decision-making skills?

2. In what ways can you improve your decision-making skills?

3. Which courses in school do you think have helped you most to make good decisions? Why?

4. How has your family helped or hindered your ability to make good decisions?

ally correct to predict successfully. For example, one theory might be that day causes night, which is incorrect. Yet, the time at which lights must be turned on can be correctly predicted through simple observation.

Because we can view the family from so many theoretical perspectives, doing justice to all of them is impossible in a general family textbook. The various textbooks that examine only family theory cover anywhere from four to 16 general theories. David Klein and James White, in their book *Family Theories: An Introduction* (1996), examine the following six family theories:

1. Exchange theory
2. Symbolic interaction theory
3. Family development theory
4. Systems theory
5. Conflict theory
6. Ecological theory

According to these authors, the theories can be categorized by level, time, and sources of change. *Level* describes whether the theory of family behavior is mostly based on the individual (attitudes, beliefs, values, choices, and so on), the family relationships (interaction between the members of the family), family organization (the nuclear, extended, polygamous, and other family), or the institutions surrounding the family (the state of the economy, the society's marital and sexual norms, child-rearing guidelines, and so on). By *time* the authors mean a theory can be placed on a continuum between static and dynamic, based on how important the concept of time is (an episode of family violence is examined in and of itself [static] or it is viewed as a long-existing conflict that finally turns violent [dynamic]). In comparing theories by *sources of change*, the theories are examined to determine whether family change is viewed as coming from within the family or if it is being dictated by outside sources.

Another way to analyze theories is to determine whether the theory takes a realistic view or an idealistic view of the family. A realistic view claims that there are real things existing outside of human consciousness (the institutions such as government and schools, or the economy) that influence family behavior. An idealistic view suggests that it is the individual's and family's reactions to the outside forces that are crucial to understanding behavior (an individual's interpretation of an experience based on his or her attitudes, beliefs, and values is the determining factor in behavior, rather than the objective event itself).

The author can provide only a brief discussion about each theory. The author's goal is simply to alert the reader that there are numerous theories on which family science is based. It is this diversity of family theories that leads, in part, to the many debates, disagreements, and conflicting ideas about families. In a way, each theory is like a blind person feeling an elephant. The blind person who feels the tail describes the elephant differently than the blind person who feels a leg. Yet if enough blind people can feel all the parts of the elephant, their descriptions combined do, indeed, properly describe the elephant.

Exchange theory focuses on the individual level. Group phenomena, social structure, and normative culture are based on the actions of individuals. Individual choice is primary. Change results from within the family, based on these individual choices.

Exchange theory is derived from the study of economics and the workings of the marketplace. People try to maximize their rewards and minimize their costs. Perhaps a mother offers the majority of child care in a family, while the father provides the majority of economic support. In this case, both partners gain something of value from the other. Although an exchange need not be perfectly balanced, if it is too one-sided it may cause problems between the family partners. Some exchanges, however, may be quite one-sided for a while, as in the case of caring for young children. This is to be expected by the parents, whose rewards may be realized only in the future. It is usually more important that an exchange is considered fair by the parties, rather than perfectly balanced. Note that when a couple in trouble with each other is asked about it, the typical reply is expressed as an exchange "Well, he/she does support me in public (reward), but belittles me at home (cost)."

Exchange theory taken to the extreme may assume that family members are perfectly rational and thus always capable of reaching agreement. This may lead the theorist to overlook nonrational factors such as emotional attachment and commitment. Being blindly in love plays what role in a couple's bargaining? What about anger or depression?

Symbolic interaction theory tends to see the relationships between people as the ultimate determinant of behavior. Again, the source of behavior is within the family. This viewpoint examines the personal interactions in terms of meanings and symbols. A man opens a door for a woman. Is this simply symbolic behavior indicating politeness and respect, or is it a symbolic statement indicating that men are strong and women are weak?

It is within social (particularly the family) interaction that a person develops both a **self-concept** (How do I feel about myself, what are my strengths and weaknesses, what is my worth?) and an **identity** (a person's sense of who he/she is, his/her inner sameness).

An individual's reality is built upon day-to-day interactions with others and his/her world. For two people to get along well, their perceived realities must be somewhat similar. An adult raised in a family where problems were denied or put off will have interaction problems with a spouse raised in a family where problems were quickly perceived and openly discussed.

Symbolic interaction theories tend to overemphasize the individual. The goodness of a relationship is basically measured by the happiness of the partners within the family setting. Often overlooked are the effects of outside forces such as government policies, economic fluctuations, and other environmental factors that impinge upon every family.

Family development theory focuses on the family rather than the individual. It assumes that individuals, relationships, and family roles all change over time (the family life cycle). Some changes come from within the family, and some occur because of institutional changes outside the family. Time is an important marker for this theory. A good example of developmental theory is seen in Chapter 10.

Theoretically, a family is described over time. Meeting each other, marriage, having children, the empty nest when child rearing is past, middle age, retirement, grandparenting, old age, and finally, death. This perspective tends to treat all families the same, yet they are not. Cohabitation, divorce, remarriage, ever-changing pregnancy and birth technologies, unwed pregnancies, adoption, gay/lesbian families, and ethnically different families tend to have different life cycles.

Systems theory focuses on interconnectedness. The idea is that if one piece of a system (family) changes (such as unemployment of the breadwinner), then the rest of the system will change (spouse, children, friends, and others will react to

self-concept
How do I feel about myself, what are my strengths and weaknesses, what is my worth?
identity
A person's sense of who he/she is; his/her inner sameness

the unemployment and will also change). This theory emphasizes the dynamic nature of the family (action brings reaction, and together they bring change).

For an individual within a family to change, the whole family system must change. This tends to be the goal of family therapy. An 8-year-old child continues to wet the bed at night. The family therapist will want to work with the entire family, not just the child with the problem. This is important because systems theory assumes that any system works toward equilibrium. Thus, the change in one part, in this case the bed-wetting child, will also cause other parts (family members) to change their behavior.

Systems theory also looks at the rules governing a family and how those rules are made. Any system must have rules by which it functions, or the system will break down. A mechanical system is the best analogy in that if one major part (such as a transmission in an automobile) fails, the system will not work.

Conflict theory finds the normal state of the family, and of general society, to be one of conflict and change rather than harmony and status quo. For example, the ongoing debate over abortion is best viewed from this perspective. Pro-choice social activists fight for the unhampered right of each woman to decide for herself whether to have an abortion, while anti-abortion activists try to enact laws that regulate abortion (see Chapter 9). This perspective would be concerned with women's rights and civil rights movements.

Conflict is also found within the family. If we think of family as the systems theory does, then there must be times when the system is out of balance and conflict occurs in an effort to right the system. The first child is born—how will this change the roles of the new parents from what they were before the birth? The new father suggests that the mother stop working outside the home because child care is her responsibility. She objects because the family needs her income and, in addition, she enjoys her work and is due shortly for a promotion and increased salary. Obviously, there will be conflict in this situation. It can be worked out in many ways, but for the family to return to equilibrium, the solution must be accepted as equitable and fair by both mother and father.

Conflict theory also studies power relationships closely. Within the family, the partner that holds the most power tends to dominate most conflicts. If the husband earns more money than his wife can, this fact tends to dictate that he works and she stays home to care for the new child. As women have entered the workforce and become more economically equal to their husbands, the power structure within the family has changed. Wives have gained power and husbands have lost power.

Ecological theory, which is relatively new to modern sociology and psychology, places emphasis on adaptation and considers pressures from within and without the family.

The family is the microsystem in which human development takes place. However, the family comes in contact with others outside the family. Such important outside influences as employment greatly pressure the family. Over and above these family influences, there is the broader cultural system made up of norms and rules dictated by the society and enforced by authorities such as police. The society outside the family in the form of government can be supportive of the family (unemployment insurance) but can also undermine the family (drafting a parent to fight in a war).

None of these theories can completely explain the family. However, each contributes to our overall understanding of families and their interactions. In general, the study of intimate relationships, marriage and family in particular, must include: (1) the fact that marriages and families are interpersonal systems (that is, small groups), (2) the idea that spouses' psychological and physical qualities shape their individual and collective efforts to maintain a successful union, (3) that both marriage relationships and the partners themselves are dynamic (that is, they change by context and they evolve over time), and (4) that marital unions are embedded in a social context (Huston 2000, 299).

Methods of Study
The Experiment

The experimental method is the basic tool of all sciences, including family science. In its simplest form, it is made up of the following:

1. An *independent variable*—a factor manipulated by the researcher to discover what effect, if any, it has on the dependent variable.
2. The *dependent variable*—an event or behavior the researcher seeks to understand.
3. *Controls*—factors in an experimental situation that are treated in such a way as to prevent them from influencing variables.

Let us take a theoretical question and set up an experiment through which we can arrive at an answer that is more factual than just our opinion. A teacher notices that many students stay up late the night before a test. Many also mention that they drink coffee to remain alert. How does the coffee influence the student's learning efficiency? Because coffee contains the stimulant caffeine, the teacher's working hypothesis is: caffeine via coffee (independent variable) improves learning efficiency (dependent variable).

Two comparable groups of students are selected. One group becomes the *experimental group* (those subjected to the independent variable). The second group is the *control group* (a group identical to the experimental group, but which did not receive the independent variable, in this case, coffee). (See Figure 1-3.) There are many other factors in addition to caffeine that may affect learning, including fatigue, interest, intelligence, and motivation.

[handwritten margin notes: Caffeine via coffee improves learning. Not receive Caffeine]

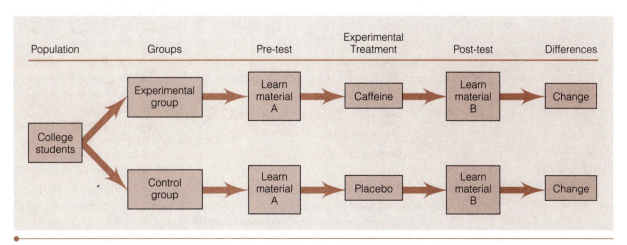

FIGURE 1-3 *Experimental Method*

By selecting comparable groups and subjecting them to identical conditions except for the application of caffeine, the researcher hopes to control all other possible influencing factors. Thus, if the experimental group's learning efficiency changes significantly more than the control group's performance between pre- and post-tests, one can assume that the change was due to the caffeine.

Methodological Problems   Unfortunately, when the experimental method is applied to the human, certain problems arise that may interfere with its success. First, humans are not always available for experimental purposes. Moral, ethical, and legal problems may arise. To avoid these problems, animals may be substituted for humans. In this case, the problem of generalizing from animals to humans arises.

A second problem is the nature of the human being. Because humans are aware, they sometimes get in the way of the experiment. For example, the person receiving the caffeine, knowing it is a stimulant, may simply learn faster because he/she thinks he/she should. In this case, the control group could be given a *placebo* (fake caffeine) so that such thoughts could be taken into account.

A third problem is simply the complexity of the human being. In our example, there are so many factors that influence one's learning efficiency that the researcher can never be certain that each possible influence has been controlled.

The Survey

The survey method serves to compliment the experimental method by making available certain areas of behavior that are closed to the experimental method. For example, the relationship between premarital sexual experiences and marital success is largely closed to experimental investigation. However, the information can be gained after the fact via the survey method. If we asked enough individuals about their early sexual experiences, we would find persons scattered along a continuum from those who had no premarital sexual experience to those who had engaged frequently in premarital intercourse. We could then select different groups of people, say those with no sexual experience prior to marriage, those with petting experience, and those with premarital intercourse experience, and by asking questions about their marital status discover if there was any relationship to marital success. Naturally, other factors that influence marital success, such as age at time of marriage, need to be controlled. Control is possible in the survey since appropriate questions can be asked to ascertain the influence of other factors.

Thus, by asking the appropriate questions after the fact (via interview or questionnaire) of a sample population, a researcher is able to gather data just as he/she does in the experiment, without having the subjects in an actual experimental situation. In the above example, the independent variable is one's premarital sexual experience, the dependent variable is the degree of marital success, and the controls are represented by questions concerning age, religion, and other factors that influence marital success.

Methodological Problems   The first problem has to do with the validity of the answers obtained in the survey. There is often little chance of checking on the truthfulness of a respondent's answers. Even if the respondent is not deliberately dishonest in his/her answers, the respondent's memory may be faulty. For example, a mother with a boy-crazy teenage daughter may respond to a question asking the length of her own engagement before marriage by saying "over 1 year." Actually, she was only engaged for 2 months. Yet she may think, "I dated my husband for over a year before we were married and that was like being engaged." She may, in fact, actually remember being engaged longer because that is what is expected of people, and besides, she didn't want her teenage daughter to know just how short her engagement was.

A second problem inherent to the survey method concerns the questions. A question may be unclear or misleading. In other cases, the questions may lead the subject to an answer desired by the researcher. This is particularly true in political polls, which are, in essence, surveys.

A third major problem has to do with the sample of subjects. This is also a problem of the experimental method, but tends to be more crucial in the survey method. In every research method, there is always the question of how representative are the data. If research subjects behave this way, will others who have not been observed also behave this way? The best a researcher can do is to sample what he/she thinks is a representative and proportional cross section of the population to which he/she wants to generalize. If a community college professor surveys his class to determine the average yearly cost of his student's college attendance, can he then generalize the results to 4-year state colleges, public universities, or private universities? Obviously, the answer is "no." When Johnny says to his mother, "Everyone can stay out until midnight, why can't I?" his mother might ask him on what sample he based his statement. His sample is probably *biased* and based only on those acquaintances whom he knows can stay out that late.

Another problem of sampling is the percentage of return from the chosen sample. Every competent researcher strives to obtain a representative sample and usually describes his sampling procedures in his research report. If a researcher is doing a mailed questionnaire to a selected representative sample, it is probably impossible to receive a 100 percent return. If a researcher can get a 70 percent return or above, he/she is doing very well. Who are the missing 30 percent? Does their absence bias the sample in any way? The smaller the percentage of return, the more cautious the researcher must be in generalizing the results.

The Clinical Method

The purpose of the clinical or applied fields of family science is to help solve practical problems of human behavior for individuals and families. The practitioner would be a family or marriage counselor, a social worker, a psychologist or psychiatrist, and others who work directly with individuals and families to solve specific problems. In the course of such work, new insights into human behavior may be gleaned. The major advantage of the clinical method, as far as science is concerned, has been its usefulness for gathering raw data and for producing ideas and theories about human behavior. Because clinicians tend to work mainly with family problems, they often have a negative outlook on marriage and family life. Their sample of clients is negatively biased, so observations from clinicians' experiences are difficult to generalize to the larger population.

Natural or Field Observation

Essentially, the researcher goes into a natural setting to observe behavior. Too often with the human being, a laboratory setting influences behavior. The success of natural observation depends on training the observer to be objective and ac-

Sherlock Holmes, Dr. Watson, and Delinquency Data

There are now available tests and scales that have enjoyed some success in predicting success or failure in college, in forecasting delinquency, even in predicting happiness in marriage. Take an example from delinquency. Among boys who have a history of truancy, a rejective family life, and other deficiencies, it may safely be said they (85 percent) will come into conflict with the police. Such an actuarial prediction reminds us of the success of safety councils in forecasting the number of traffic deaths that will occur over a holiday weekend. Insurance companies are masters at such statistical prediction.

But there is a fallacy here. To say that 85 in 100 boys having such and such a background will become delinquent is not to say that Jimmy, who has this background, has 85 in 100 chances of being delinquent. Not at all. Jimmy, as an individual, either will or will not become delinquent. There is no 85 percent chance about him. Only a complete knowledge of Jimmy will enable us to predict for sure. The fact that clinicians using an intuitive method may fail in their prediction is beside the point. If we knew Jimmy and his environment fully, we should be able to improve upon statistical forecasting, which applies only to groups of people and not to individuals.

The point at issue was once expressed by Sherlock Holmes, in the *Sign of the Four*. Holmes said to his friend Dr. Watson, "While the individual man is an insoluble puzzle, in the aggregate he becomes a mathematical certainty. You can never foretell what any one man will do, but you can say with precision what an average number will be up to. Individuals vary but percentages remain constant" (Allport 1961).

curate in his/her observations. Without training, the observer tends to see what he/she wants to see. In addition, the observers must not interfere with what they are trying to observe.

Many methods are used to keep observation unobtrusive. For example, if education students want to observe a kindergarten class, it is obvious that their presence in the classroom would cause the children to modify their behavior. The children would be looking at them, feeling self-conscious, giggling, and probably talking to each other about the visitors. In university experimental schools, the classrooms are designed so that the education students may observe the children unnoticed. This is accomplished by the use of a one-way mirror and a separate entrance into the observation area.

Another drawback to natural observation is finding what it is you want to observe at the time you want to observe it. You may be interested in observing panic behavior in a crowd. You cannot go into a movie theater and cry out "fire." If you did, you would probably be writing up your notes on prison life rather than panic behavior.

Group versus Individual Data

People are often confused by what appears to be an inconsistency between research data and their own individual experiences. For example, data indicate that the chances of a couple divorcing increase when their own parents are divorced. Yet everyone knows happily married couples whose parents were divorced. Does this mean our data are incorrect? No, not at all. For the group, the statistics are correct. But for a given individual, the statistics may or may not be correct. An individual is unique, and to ascertain how the divorce of his/her parents relates to his/her own success in marriage, we would have to study the particular individual (see "Highlight: Sherlock Holmes, Dr. Watson, and Delinquency Data").

Strengthening the Family

Families and communities are the ground-level generators and preservers of values and ethical systems. No society can remain vital or even survive without a reasonable base of shared values. . . . They are generated chiefly by the family, schools, church, and other intimate settings in which people deal with each other face to face. (Gardner as found in Etzioni 1993, 31)

If indeed the American family is in decline, what should be done to counteract or remedy this? Among the agendas for change that have been put forth, two extremes stand out as particularly prominent in the national debate.

The first is a return to the structure of the traditional nuclear family characteristic of the 1950s; the second is the development of extensive government policies (Popenoe 1991, 1996; Popenoe and Whitehead 1999).

Aside from the fact that it is probably impossible to return to a situation of an earlier time, the first alternative has major drawbacks. It would require many women to leave the workforce and, to some extent, become "de-liberated," an unlikely occurrence. Economic conditions necessitate that even more women take jobs, and cultural conditions stress ever-greater equality between the sexes.

In addition, the traditional nuclear family may not work in today's world. One must realize that the young people who led the transformation of the family during the 1960s and 1970s were raised in 1950s households. If these households were so wonderful, why didn't their children seek to copy them?

Nevertheless, the traditional nuclear family is still the choice of millions of Americans. They are comfortable with it, and for them it seems to work. It is reasonable, therefore, not to place roadblocks in the way of couples with children who wish to conduct their lives according to traditional family dictates. Women who freely desire to spend much of their lives as mothers and homemakers outside the labor force should not be penalized economically or psychologically by public policy for making that choice.

The second major proposal for change that has been stressed in national debate is the development of extensive government programs offering monetary support and social services for families, especially the nonnuclear ones. In some cases, these programs assist with functions these families cannot perform adequately; in others, the functions are taken over, transforming them from family to public responsibilities.

This is the path followed by the European welfare states, but it has been less accepted by the United States than in most other industrialized countries. The European welfare states have been far more successful than the United States in minimizing the negative economic impact of family decline, especially on children. In addition, many European nations have established policies making it much easier for women (and increasingly men) to combine work with child rearing. With these successes in mind, it seems inevitable that the United States will slowly move in this direction.

There are clear drawbacks, however, in moving too far down this road. If children are to be served best, we should seek to make the family stronger, not replace it. At the same time that welfare states are minimizing some of the consequences of decline, they also may be causing the breakup of the family unit. This phenomenon can be seen today in Sweden, where unmarried parents are highly supported by the government so that parents really don't need to marry and create a two-parent family. If the United States moves in this direction, it will be important to keep uppermost in mind that the ultimate goal is to strengthen families, not replace them (Popenoe 1991, 1996).

Americans have mixed feelings about the government formally declaring an interest in their families. Many fear such policies will lead to more government interference in their private lives. Like it or not, government influences every family. Consequently, it is important that all proposed government policies and laws be examined for their effects, if any, on the family. Those laws that harm the family in any way should be changed.

While each of the preceding alternatives has some merit, there is a third alternative premised on the fact that we cannot return to the 1950s family, nor can we depend on the welfare state for a solution. Instead, we should strike at the heart of the cultural shift that has occurred, point out its negative aspects, and seek to reinvigorate the cultural ideals of family, parents, and children within the changed circumstances of our time. We should stress that the individualistic ethos has gone too far and it has become necessary to reinvigorate the spirit of community (Etzioni 1993). Children are getting woefully shortchanged because of the *me* attitude strongly held by many Americans. In the long run, strong families represent the best path toward self-fulfillment and personal happiness. We should reintroduce to the cultural forefront the old idea of parents living together, sharing responsibility for their children and for each other.

What is needed is a new social movement with the purpose of promoting families and their values as a part of the larger community. Instead of asking what our government can do for us, it is time, as John F. Kennedy long ago suggested, to ask what we can do for our children, family, community, and country. This movement should point out the supreme importance to society of strong families, while suggesting ways the family can adapt better to the modern conditions of individualism, equality, and the labor-force participation of both women and men. Such a movement could build on the fact that the overwhelming majority of young people today still state their major life goal to be a lasting, monogamous relationship that includes procreation of children.

The family endures because it is a flexible institution with great resilience. It can be pressed, stretched, and bent, but it always seems to recover its strength, spirit, and buoyancy. Despite the many criticisms aimed at it, the family remains the basic unit of society. When functioning well, the strong, healthy family is the individual's greatest source of love and intimacy.

Summary

1. Research indicates that strong, successful families share a number of characteristics, including the following:

   Commitment to family

   Appreciation of family members

   Good communication skills

   The desire to spend time together

   A strong value system

The ability to deal with crises and stress in a positive manner

Resilience

The development of self-efficacy

2. Learning to make good decisions and choices is important to the creation of strong, successful, intimate relationships and families.

3. It is important to the study of family to understand both theoretical approaches and method of study, to separate fact from fiction.

Resources on the Internet

 ## Companion Website for *Human Intimacy: Marriage, the Family, and Its Meaning,* Tenth Edition

http://sociology.wadsworth.com/cox10e/

Gain an even better grasp on this chapter by going to the companion website to take one of the tutorial quizzes, use the flash cards to master key terms, or check out the many other study aids you'll find there. You will also find special features such as GSS data, Sociology Online, and Census 2000 information that will put data and resources at your fingertips to help you with that special project or to help you as you do some research on your own.

 ## *InfoTrac College Edition*

http://www.infotrac-college.com/wadsworth/

You can access reliable resources anytime, anywhere, with InfoTrac College Edition, the online library. This fully searchable database offers more than 20 years' worth of full-text articles (not abstracts) from almost 5,000 diverse sources, such as top academic journals, newsletters, and up-to-the-minute periodicals, including *Time, Newsweek, Science, Forbes, The New York Times,* and *USA Today.* You can conduct electronic key word searches using key terms from this chapter to supplement your reading and learning experience. To aid in your search and to gain useful tips, see the Student Guide to InfoTrac College Edition, which you can access through the companion website for this book.

© George Shelley/Corbis

Human Intimacy, Relationships, Marriage, and the Family

Chapter

2

Questions to reflect upon as you read this chapter:

- What is meant by comparing American society to a "tossed salad" rather than a "melting pot"?
- Do you think that the family is the basic unit of human organization? Why, or why not?
- Do you think the family has changed for the better or the worse since World War II?
- Can you list several unique characteristics of the American family when compared with most other families in the world?
- What are some of the primary functions of the modern American family?
- List some persons with whom you are intimate in a nonsexual way. Why are you intimate with these people and not others?
- Do you think our society should recognize nonmarital unions?

Everyone has personal beliefs about marriage and family because every person, including family science researchers, grows up in some kind of family. Although there is a great deal of scientific research in the field of family science, the goal of complete objectivity is never totally possible because everyone has this personal involvement with his/her own family, which colors their thinking. Family scholars need to consciously acknowledge how their private experience with their family affects their research and public statements about their results (Allen et al. 2000, 5). The bias of the author of this book is to supply information that helps people improve their intimate relationships, especially those within the marriage and family setting. As noted in Chapter 1, while it is important to recognize the wide diversity of American families, it is even more important to recognize those strengths and characteristics supportive of all strong families.

This chapter introduces information about today's American family. It is intended to be an overview of the contemporary family's characteristics, both strengths and weaknesses. The discussions are necessarily cursory because there is much to cover, but additional information can be found throughout the text.

In general, the most important change that has occurred in the modern American family seems to be the change from being child-centered to being adult-centered. Americans today tend to see marriage as a couples relationship, fulfilling the emotional needs of adults, rather than as an institution dedicated to raising children. Many people now expect their marriages to be a spiritualized union of souls (Whitehead and Popenoe 2003) as opposed to being child-centered.

There are two schools of thought about this change. Some people emphasize the positive results of the change, particularly for women. The new individual freedoms that adults enjoy form the core of the contemporary marriage. This emphasis should bring adults more pleasure and satisfaction from their marriages.

Conversely, some see the family as a social institution in decline. They emphasize the negative consequences of the change for children. Devaluation of children is the core of their argument (Bennett 1994; Etzioni 1993; Gallegher 1996, 2000; Whitehead and Popenoe 2003). It will be interesting for you, the reader, to evaluate your own attitude about the changes taking place in the American family. Remember that you, too, carry a core of knowledge about marriage and family, based on your own experiences within your family of origin. It is hoped that this text will help you, both to sort out your own biases and to make the best possible decisions about your own intimate relationships.

Family: The Basic Unit of Human Organization

The family is the basic unit of human organization. If defined functionally, the family is essentially universal. Its structural form and strength, however, vary greatly across cultures and time.

The term *family* is used here in a broad sense: It is defined as whatever system a society uses to support and control human sexual interaction, reproduction, and child rearing. This broad definition avoids many apparent conflicts about the meaning of changes currently taking place in family functions and structure. For example, according to some, Israeli kibbutzim are families, even though persons other than parents assume major child-rearing responsibilities in some orthodox kibbutzim. On the other hand, a definition of family will be meaningless if it is so broad that each and every relationship qualifies for family status.

History does not support the conclusion that "anything goes" in family life. If many conservatives distort the historical record in claiming the universal superiority of a particular family form or value set, many liberals distort it in the other direction. It does not follow that people can simply shrug off the weight of history to freely create new family relations and values.

Individuals who cannot or will not participate in the favored family form face powerful stigmas and handicaps. History provides no support for the notion that all families are created equal in any specific time and place. Rather history highlights the social construction of family forms and the privileges that particular kinds of families confer. (Coontz 2000, 286)

Following are several narrower but more meaningful definitions:

family
According to the Census Bureau, a group of two or more persons related by birth, marriage, or adoption and residing together

households
According to the Census Bureau, all persons who occupy a housing unit

- The U.S. Census Bureau defines **family** as a group of two or more persons related by birth, marriage, or adoption and residing together. The Census Bureau differentiates a family from a household. **Households** are defined as all persons who occupy a housing unit. They may consist of a person who lives alone or several people who share a dwelling. A cohabiting couple would be considered a household rather than a family by the Census Bureau.

- The sociologist Richard Gelles (1995, 10) defines *family* as a social group and institution that possesses structure made up of positions such as breadwinner and child rearer, and the interactions between these positions. The family typically carries out specialized functions, such as child rearing, and is characterized by biologically and socially defined kinship. It often involves sharing a residence.

- The New York Court of Appeals said the *family* can be defined by examining several factors, including exclusivity and longevity and the level of emotional and financial commitment of a relationship. "It is the totality of the relationship as evidenced by the dedication, caring and self-sacrifice of the partners."

- Mary Ann Lamanna and Agnes Riedmann, in their widely used marriage and family textbook, define *family* as any sexually expressive or parent-child or other kin relationship in which people—usually related by ancestry, marriage, or adoption—(1) form an economic unit and care for any young, (2) consider their identity to be significantly attached to the group, and (3) commit to maintaining the group over time (2003, 12–13).

- A national opinion poll asked Americans, "What constitutes a family?" A variety of answers were given. Most Americans agree that married couples with and without children are definitely families. Many also agree that a divorced parent or never-married parent living with his/her children is also a family. Some say that cohabiting couples, including homosexual couples, are also families. Thus, Americans vary in the types of intimate relationships they are willing to define as a family.

Ivan Beutler and his colleagues (1989, 806–807) suggest that any definition of the family must include ties across generations that are established by the birth process. The main difference between family and other relationships, such as those between friends or work colleagues, is the permanence of the lineage ties. Nonfamily relationships are usually entered into voluntarily and are fostered as a means to achieve specific ends. They can be broken at any time. But family ties exist as an end in themselves. A divorce may end a marriage, but it can't end generational ties between children and their parents, grandparents, and other relatives.

Part of the reason for the debate over the state of the American family is confusion between the functions of the family institution and the structure by which these functions are fulfilled. Many structures can fulfill the responsibilities of the family. In modern America, the duties, and thus the functions, of the family have been reduced. New structures, or alternative family forms, are being tried to fulfill some functions. The debate over what constitutes a family tends to be a debate over structure rather than function.

In this text, the author is concerned with intimate relationships and how to make them be of high quality and lasting. The characteristics of strong, lasting families are principles that can enhance any intimate relationship, regardless of structure. Hence, the author is less interested in structure than in the functions of the strong, successful family and how they are fulfilled.

Family Functions

The family serves both the society and the individual. Sometimes, of course, social and individual needs come into conflict. Laws mandating one child per family in urban China are an example. Although having more children may benefit a particular family, the Chinese government believes that too many children will harm the larger society. For the family to remain a viable social institution, it must meet the needs of society as well as individual family members. Thus, the family must minimize conflict between the individual and the society.

The family has handled a broad range of functions in different historical periods and societies. In some primitive societies, the family is synonymous with the society itself, bearing all the powers and responsibilities for societal survival.

The following functions are necessary for the maintenance of society:

1. Replacements for dying members of the society must be produced.
2. Goods and services must be produced and distributed.
3. Provision must be made for solving conflicts and maintaining order within the family and also within the larger society.

4. Children must be socialized to become participating members of the society.
5. Individual goals must be harmonized with the values of the society.
6. Procedures must be established for supplying intimacy and emotional gratification and for dealing with emotional crises and maintaining a sense of purpose within the family.

As the society grows more complex, social institutions are formed that assume the primary responsibility for some family functions. For example, most families are no longer production units per se. Individuals within the family often produce goods and services, but they usually do this outside the family setting, in a more formalized job setting. The family is still an economic unit, however, because it demands goods and services. It is the major consumption unit in the United States. Although the courts and police maintain external order, the family is still primarily responsible for maintaining order within its own boundaries. Formal education now trains children to become participating members of society, but the family begins, maintains, and remains primarily responsible for the socialization process.

The family plays a role in all six functions, but the contemporary American family has full responsibility for only two of the primary functions: supplying human replacements and providing emotional gratification and intimacy to its members.

1. *Providing continuing replacements for individuals so that society can continue to exist.* Despite sperm banks, surrogate mothers, and other advances in reproduction technology, most reproduction is still being carried out in the age-old, time-honored fashion. Birth follows pregnancy, which has followed sexual intercourse. The family is also the most efficient way of nurturing the human newborn, who is physically dependent during the first years of life. Without some kind of stable adult unit to provide child care during this long period of dependence, the human species would have disappeared long ago.

   Each family structure, then, has reproduction as a potential function. An extension of this function is regulation of sexual behavior. Unlike other animals, human sexual behavior may be sought and enjoyed at any time, regardless of the stage of the female reproductive cycle. Human beings are free to use sex not only for reproduction, but also for pleasure (see Chapter 8). As in so many other aspects of human life, freedom of choice is a mixed blessing. Human beings must balance their relatively continual sexual receptivity and desire with the needs of other individuals and the needs of society as a whole. They must find a system that provides physical and mental satisfaction in a socially acceptable context of time, place, and partner. In most societies, the family is the basic institution controlling sexual expression.

   No matter how sexuality is handled, human beings seem to be comfortable only when they can convince themselves their system is proper, just, and virtuous. When each culture has established a satisfactorily "correct" system of sexual control, it bolsters this system with a complex set of rules and punishments for transgressions that help the family control their members' sexuality. In most societies, but not all, sexual interaction is limited; it serves the reproduction function and binds a relationship.

2. *Providing emotional gratification and intimacy to family members and helping them deal with emotional crises, so they grow in the most fulfilling manner possible.* Expecting emotional gratification from intimate family relation-

ships seems to cause Americans more trouble than any of the other family functions. When couples report problems in their relationship, they often cite unhappiness and emotional misery as their major complaint.

The family's functions have changed over time. We can assume that its functions will continue to change. Do changing functions mean that the family as we know it will disappear? Not necessarily. The family's secret of survival over the past centuries has been its flexibility and adaptability. The family is a system in process rather than a rigid, unchanging system. As John Crosby pointed out some years ago,

No one can yet foresee what the structure of the future family will look like because no one can know with certainty what the functions and needs of the future family will be. It is likely, however, that the needs for primary affection bonds, intimacy, economic subsistence, socialization of the young, and reproduction will not yield to obsolescence. To the extent that human needs do not change drastically, the family structure will not change drastically. (1980, 40)

The American Family: Many Structures and Much Change

A free and creative society offers many structural forms by which family functions, such as child rearing, can be fulfilled.

Mom and Dad, Sissy and Junior, a dog, and maybe a cat represent the stereotypical American **nuclear family** (a married couple and their children living by themselves). Yet to talk of the American family perpetuates the myth that one family structure represents all American families. If we examine the structure of the family, we find that there are, in fact, many kinds of American families. (By **structure** we mean the parts that make up a family, and their relationships to each other.) For example, the nuclear family has a mother, a father, and their children. A single-mother-headed family has just a mother and her children. An extended family may have a mother, father, children, and other relatives.

The single-parent family has been one of the fastest-growing family structures during the past decade (Table 2-1). From 1970 to 2002, the proportion of children under 18 years of age living with one parent more than doubled, from 12 percent to 27.7 percent (J. Fields 2003).

Single-parent families tend to be temporary because the single parent usually remarries or marries. However, the single-family structure accounts for a significant number of American families at any one time. The large increase can be accounted for by the high divorce rate in America. To a lesser but growing extent, the greater social acceptance of the unmarried mother who keeps her child has also contributed to the growing number of single-parent families.

Divorce is a good example of changing family structure. The family begins as a nuclear family, becomes a single-parent family, and, in many cases, becomes a **reconstituted** or **blended family** when the single parent remarries. These kinds of changes signal the end of the initial relationship, at least as an intimate love relationship. Although the term *blended family* is popularly used to describe families in which at least one partner has been married before, it doesn't adequately describe the reconstituted family. Often, the remarried family does not

nuclear family
A married couple and their children living by themselves

structure
The parts that comprise a family and their relationships to one another

reconstituted or blended family
A family in which one or both of the partners have been married before

TABLE 2-1

What the Research Says: Percent Composition of Family Groups with Children under 18 Years, by Race and Hispanic Origin, 1970–2002

		Two-Parent	Mother-Child	Father-Child
2002	All Races	69	23	4.5
	White	75	16	4.7
	Black	39	48	4.0
	Hispanic	64	25	4.6
	Asian	69	13	2.3
1990	All Races	71.9	24.2	3.9
	White	77.4	18.8	3.8
	Black	39.4	56.2	4.3
	Hispanic	66.8	29.3	4.0
1980	All Races	78.5	19.4	2.1
	White	82.9	15.1	2.0
	Black	48.1	48.7	3.2
	Hispanic	74.1	24.0	1.9
1970	All Races	87.1	11.5	1.3
	White	89.9	8.9	1.2
	Black	64.3	33.0	2.6
	Hispanic	Data not available		

SOURCE: U.S. Census Bureau, March 1998, Table 6; J. Fields, June 2003.

blend smoothly. The remarried family is really more like a patchwork family, making many compromises as children shuttle from the newly formed family back to the biological parent for visitation, past partners work out monetary problems such as child support, and extended family members of both the past family and the newly remarried family work out their relationships.

Even in a long-lasting relationship, structure usually changes over time. A relationship begins as a couple relationship and then may broaden to include children, sons- and daughters-in-law, and finally, grandchildren. When the children leave home, the couple returns to a two-person household. Table 2-2 indicates some of the many structural forms families can take in various cultures.

It should be noted that the greatly increased life expectancy in America (it has risen from around 40 to 78 years in the past 150 years) has, in itself, caused great changes in the family. For example, when life expectancy was short, "teen marriages" were the norm rather than the exception. Marriages did not last very long, due to the short life expectancy and the high numbers of women who died in childbirth. In fact, 100 years ago, the long-term family was rare and broken families were as numerous as they are today. However, the cause of a family's breakup was death of one partner, rather than divorce. Another result of longer life expectancy is that more generations of family members are alive at one time. It is no longer unusual for a child to know his/her great-grandparents (Vandewater and Antonucci 1998, 313).

Many Americans think criticism of the family is new. Yet throughout time, criticisms of marriage and the family have been voiced. In every era, critics of the

TABLE 2-2

Different Family Structures

	Kind	Composition	Functions
Types of Marriage	Monogamy	One spouse and children	Procreative, affectional, economic consumption
	Serial monogamy	One spouse at a time but several different spouses over time; married, divorced, remarried	Same
	Common-law	One spouse; live together as husband and wife for long enough period that state recognizes couple as married without formal or legal marriage ceremony; recognized by only a few states	Same
	Polygamy	Multiple spouses	Same
	Polygyny	One husband, multiple wives	Any,[1] power vested in male
	Polyandry	One wife, multiple husbands	Any, power vested in female
	Group	Two or more men collectively married to two or more women at the same time	Any, very rare
Types of Families	Nuclear family	Husband, wife, children	Procreative, affectional, economic consumption
	Extended family	One or more nuclear families plus other family positions such as grandparents, uncles, etc.	Historically might serve all social, educational, economic, reproductive, affectional, and religious functions
	Composite family	Two or more nuclear families sharing a common spouse	Normally those of a nuclear family
	Tribal family	Many families living in close proximity as a larger clan or tribe	Usually those of the extended family
	Consensual family (cohabitation)	Man, woman, and children living together in legally unrecognized relationship	Any
	Commune	Group of people living together sharing a common purpose with assigned roles and responsibilities normally associated with the nuclear family	Can provide all functions with leadership vested in an individual, council, or some other organized form to which all families are beholden
	Single-parent family	Usually a mother and child; father and child combination less common	Same as monogamy without a legally recognized reproductive function
	Concubine	Extra female sexual partner recognized as a member of the household but without full status	Usually limited to sex and reproduction
	Reconstituted (blended)	Husband and wife, at least one of whom has been previously married, plus one or more children from previous marriage or marriages	Any
Authority Patterns	Paternalistic	Any,[1] power vested in male	Any
	Maternalistic	Any, power vested in female	Any
	Egalitarian	Any, powers divided in some fair manner between spouses	Any

[1] "Any" means that the family may assume any or all compositions or functions that have been taken or fulfilled by families in the past.

China: Families and Government Policy

N owhere is the effect of government policies on families more clearly observed than when comparing pre– and post–Communist revolution Chinese families. Pre-1949 Chinese families were usually large, extended, patriarchal families. The marital bond took a distant second place to intergenerational ties between parents and children, especially sons. Because the descent line was more important than individual family relationships, marriage took place to extend the family line, not to benefit the individuals involved (Pimentel 2000, 32).

Because China was so vast, both in land and population, there was little central control over the people. What little government control there was rested in weak local agencies, supported by occasional visits from central government bureaucrats and troops. The real control actually rested within each family.

Before the revolution, the five major functions of society were carried out by the extended Chinese family. Almost all of the economic production of China was done via the family group. In the major city of Canton, over 90 percent of the businesses were family-held in 1950. The husband-father was the chief executive officer (CEO). The wife-mother and children were the employee-workers. The family was the source of all employment. This made it nearly impossible for a family member to leave and find work outside the family. Only if there were family members in another part of China was it possible to move, because these relatives would provide work. There was no governmental support; it was up to the family, or larger clan, to care for all clan members.

The educational function was also carried out within the family because schools were largely unavailable outside of the cities. Education mostly consisted of teaching children the skills necessary to operate within the family business of farming, manu-

Traveling in style via motor scooter.

facturing, or engaging in commerce of some type. If reading was unnecessary to the family's work, the children remained illiterate.

Religious services were also carried out within the family. Ancestor worship was important. Shrines were contained in each family compound, and family members were usually the priests.

Last, but not least, the political function was also carried out by the Chinese families. In the middle of the nineteenth century, there were only about 40,000 central government officials to govern 400 million people who were spread out over 4.5 million square miles. When central government officials were called upon to settle a dispute, both of the parties were punished before a judgment would be made. This punishment was for the parties' failure to settle the conflict themselves.

With so little government throughout much of the country, exerting social control fell to the family.

Studies are important in China.

It did this by assuming responsibility for each of its members. If any member of the family transgressed social rules, the whole family suffered. This led family members to exert control over each other. Obviously, the system of family political control worked; China is the only nation that has endured throughout history.

When the Communists won control of China, it was imperative that the system of political control based on the family be changed if power was to be in the hands of the new central government. Toward this end, families were broken up, sometimes forcibly. The Communist government also sought to shift control from the extended family to married individuals. The government promoted mutual companionship as a major criteria in selecting a mate rather than fulfilling the extended family's desires.

Modern Chinese couples seem to have a relatively unromantic vision of love; they view it as being similar to close companionship. Also, despite government efforts to the contrary, parents continue to affect their children's marriages. Parental agreement in mate choice

Riding to work.

continues to be important to Chinese couples (Pimentel 2000, 45).

Under Communism, families are secondary to the state. Government policy sets strict controls on marriage, divorce, and reproduction. For example, the state enforces a quota of one child per family in urban areas. In the form of neighborhood committees, the state keeps track of pregnancies and the number of children in each family. Great pressure is placed on a pregnant woman to abort the pregnancy if she already has a child. If she decides to keep the child, she and her husband will be fined a substantial amount. An even more threatening consequence is the loss of government benefits to the family, and the child will be banned for life from working for or receiving government benefits. It is believed that the strict enforcement of the one-

child-per-family policy has led to female infanticide because sons are generally more desirable. It should be noted that multiple births, such as twins, are celebrated by all as a wonderful gift. The policy is less strict in the countryside.

Hard as it is for Americans to believe that the privacy of one's family can be intruded upon by government policy, the Communist government's successful breakup of the centuries-old extended family demonstrates in a most extreme manner that government policy can and does have an effect on family life.

SOURCE: Author's visit to China (May 1997). Conversations with C. K. Yang, author of _Family in Communist Society_ and _Chinese Communist Society_.

Shopping out of doors.

status quo have offered suggestions for new family forms. For example, in 1936, Bertrand Russell believed marriage would collapse as a social institution unless drastic changes were made:

In the meantime, if marriage and paternity are to survive as social institutions, some compromise is necessary between complete promiscuity and lifelong monogamy. Although it is difficult to decide the best compromise at a given time, certain points seem clear:

Young unmarried people should have considerable freedom as long as children are avoided so that they may better distinguish between mere physical attraction and the sort of congeniality that is necessary to make marriage a success.

Divorce should be possible without blame to either party and should not be regarded as in any way disgraceful. (1957 reprint, 171–172)

Perhaps the major contribution the turmoil of the 1960s and 1970s made to the family was to increase the available alternatives for intimacy. Families have become more differentiated. Family patterns and gender roles are now less stereotyped and rigid. It is no longer considered disgraceful not to marry, which has given rise to singlehood as a chosen lifestyle. Most Americans marry, but they don't feel that they must. There is no longer an obligation to have children, so having children can now be a more conscious decision. It is far more acceptable to leave a marriage today if either partner is unhappy. Intimate cohabitation no longer makes a person unmarriageable. Acceptance of homosexual relationships has also increased, although it is still far from universal.

Modern America is remarkably tolerant of multiple forms of intimate relationships. Other cultures have had various marital systems, but they have usually disallowed deviance from their current system. A free and creative society, however, offers many structural forms of the family to fulfill such functions as child rearing and meeting sexual needs. The reasons for America's tolerance are partly philosophic, but mainly economic. Marriage forms historically have been limited because women have been economically tied to the men who support them and their children. The women's movement, antidiscrimination laws, increased education for women, and increasingly larger numbers of women in the workforce have all helped women become more independent.

Industrialization and affluence have made it more possible for Americans to consider alternative lifestyles and marital forms. For example, affluence allows people to further their education, which exposes them to new ideas and knowledge. Affluence also brings mobility, and mobility brings contact with new people and new lifestyles. If a person discovers greater personal satisfaction in some alternative lifestyle, affluence makes it possible to seek out others who have the same interests. Affluence often enables people to postpone assuming adult responsibility (earning one's own living) and thus encourages experimentation by lessening the consequences of failure. Affluence has also given rise to the mass media, which have spread news about various lifestyles and experimental relationships and helped to create new lifestyles by portraying them as desirable, exciting, or *in*.

America's youth have a broader choice of acceptable relationships (family structures) than their grandparents did. Perhaps by choosing wisely the roles that best fit them as individuals, couples will be able to create growing intimate relationships that are more fulfilling than those of the past. On the other hand, wider choice involves greater risks, and freedom of choice does not seem to have made intimate relationships any more stable.

In the past, we knew more clearly what our roles would be. Sons often followed in their fathers' footsteps. Women became wives and mothers and ran the home. But today, the pattern has changed. Individuals now must make their own choices about their lifestyles, vocations, marriages, and families. If we do not make conscious choices, the choices are made by default. The idea that love will automatically make the decisions for us is the source of much disappointment. Wendy Shalit, in her book *A Return to Modesty*, discusses the problem of having to invent our own roles in a culture that views sexual norms and gender roles with suspicion:

If I find myself suddenly and spontaneously making up all these strange rules (as part of my proper roles), such as separate rooms, open door, no R-rated movies, always call men "Mister," it is only because I was constantly aware of the ever-present void (of rules). Even more problematic, who knows whether any particular romantic or sexual overture—from a goodnight kiss to an invitation to live together—is meant as a prelude to marriage? (Shalit 1999)

Shalit further suggests that a set of peer-generated norms are sometimes more ruthless and unforgiving than anything adults might make up. Such peer-generated rules are seldom orientated toward serving important social ends, but tend to be self-serving (Gallegher 2000, 1–4, 9–15). Gang rules for membership are an example of peer-generated norms.

The major risk in opening up choice is that errors may be made. When more choices are available, the individual must carefully consider priorities. The older, restricted system exacted a price; it often placed people in molds that would not allow maximum self-growth and social contribution. In a more open system, people run the risk of acting impulsively and making wrong choices. Thus, the price of a more open system is possible failure, so the need for a rational examination of the alternatives and improved decision-making skills is greater.

In the old system, many people were trapped in a rut; in the new one, many people seem to run constantly from one lifestyle to another, unable to choose wisely or find permanence. But the new system may also encourage experimentation that leads to better decisions. Free inquiry that leads to reasoned decisions, with opportunities to test those decisions, is the way of science. There is no reason why this method should not improve the intimate lives of people, just as it has improved their material lives. Moreover, free choice of a lifestyle can counteract the feelings of entrapment that long-married couples often express.

The characteristics of strong families described in the first chapter apply to any family structure or type of intimate relationship. Hence, particular structures are less important than the way the partners learn to define and fulfill the functions of their own relationship.

Change within Continuity and Uniqueness within Commonality

Family life involves continuity as well as change.

Each family is unique, but also has characteristics in common with all other families in a given culture.

As you try to understand the family and what is happening to it, remember that the American family covers a vast territory and is far from uniform in design.

Americans are generally free to choose how they structure their families.

Everyone is aware that family life has changed. Yet the central core of family life continues much the same as it has for many generations. Grasping these two ideas—change within continuity and uniqueness within commonality—will help you cope with the seeming riddles of the American family.

All families, whatever their form, need to develop a flexible yet stable structure for optimal functioning. Each family system maintains a preferred orderly pattern, resisting change beyond a certain acceptable range. At the same time, a family must also be able to adapt to changing developmental and environmental demands. A dynamic balance between stability and change maintains a stable family structure while also allowing for change in response to life challenges. (F. Walsh 1998, 79–80)

Some statistical examples that can be interpreted in different ways will help clarify these two principles. For example, the divorce rate has risen steadily throughout this century. In 1900, about one divorce occurred for every 12 marriages. Today, about one divorce occurs for every two marriages in a given year. The divorce rate per 1,000 population has hovered between 4.3 and 5.3 since 1975.

Some people use these statistics to support the contention that the American family is in real trouble and will soon be dead and buried. Others suggest that Americans will have better marriages and more fulfilling family lives because they no longer need to accept the dissatisfactions of empty marriages that previous generations endured.

Indeed, statistics indicate that divorced persons have high remarriage rates. Most divorced persons leave a particular mate, but not the institution of marriage. It should also be remembered that approximately the same proportion of marriages broke up in 1900 as today. The difference, as previously noted, was that the majority of marriages in 1900 ended with the death of one of the spouses, rather than with divorce. Thus, despite the high divorce rate, family permanence has continued at about the same level for the past century.

Today's families are having fewer children, which is a change from the past. But they are, indeed, having and rearing children as families have always done. This fact and the higher divorce rate help point out just how confusing and controversial the interpretation of marriage and family data can be. Furthermore, they demonstrate both principles presented in "The Family Riddle." Both the divorce rate and number of children families have are examples that show change within continuity. The divorce rate did increase dramatically in the early 1970s, which is indicative of change. Yet remarriage by people who have experienced divorce is common, which indicates the continuity of the marriage institution. All married couples share the commonality of being married. For some couples, the marriage is the first for each partner; for others, the marriage is the second for one or both partners. Thus, all married couples share the characteristic of being married, but they are unique in terms of their previous marital experience.

Some couples opt for no children; others choose to have one or more. So all couples share the opportunity to have children (except those with fertility problems, although they may adopt), but they are individual in the way they use the opportunity.

Many changes are occurring in the American family. As we discuss these changes, keep in mind the continuity. There were 2,327 million marriages in

HIGHLIGHT

The Family Riddle

Marriage, children, forever together, mom, dad, apple pie, love, Sunday softball in the park, grandmother's house for Sunday dinner—The American Family.

Cohabitation, childlessness, divorce and remarriage, stepdad, stepmom, junk food, child abuse and wife beating, spectator sports in front of the TV, Pizza Hut for Sunday dinner—The American Family.

Student: But these two descriptions can't both be of the American family.

Friend: Oh, but they are, and infinite other descriptions would also fit. American families, each and every one, are unique and representative of the individuals and their interactions within the family.

Student: But how, then, can we study the American family if each is unique?

Friend: Because they all have certain things in common. They are all alike in some ways.

Student: But they can't be. You just said they were all unique.

Friend: They are, and they are also always changing.

Student: But how can you study something that is unique and also always changing?

Friend: Because change occurs in the midst of continuity, and continuity can remain despite change.

Student: You are saying, then, that each family is unique but has things in common with all other families. And, besides, families are always changing but have continuity.

Friend: Yes.

Student: It sounds like a riddle.

Friend: It is a riddle because to understand the family, one must be able to understand two abstract principles:

1. Change can occur within continuity.
2. Uniqueness can exist within commonality.

2001, but because the population grew, the marriage rate per 1,000 population dropped from 9.8 in 1990 to 8.4 in 2001 (NCHS 2004b). So we see that marriage remains popular, even though the rate of marriage has declined.

The marriage and family statistics found throughout this book point out some of the changes in the American family. For example, we noted earlier the rapid rise of single-parent families, yet America has always had single-parent families. In fact, the proportion of such families remained at about one in 10 for the 100 years between 1870 and 1970. As was mentioned, early single-parent families resulted largely because of death rather than divorce, which is the reason today.

When the family statistics are plotted for the entire twentieth century and longer, a surprising fact emerges: Today's young people appear to be behaving in ways consistent with long-term historical trends (Skolnick 1991; Skolnick and Skolnick 1986, 6). The recent changes in family life appear deviant only when compared with what people were doing in the 1940s and 1950s. The current middle-class adults who married young, moved to the suburbs, and had numerous children were the generation that deviated from twentieth-century trends. Had the 1940s and 1950s not happened, today's young adults would appear to be behaving normally. If we focus only on the recent history of the family, we may remain blind to the broader picture of social change and continuity in family life (Kain 1990, 3–4).

Perhaps the biggest change of all is the increasing acceptance of various forms of intimate relationships. We are, after all, a pluralistic society—that is, a society made up of diverse groups. Thus, it seems natural that different family structures should become increasingly acceptable.

HIGHLIGHT

Tricky Statistics: Cohabitation and Rejection of Marriage

The news media usually interpret cohabitation statistics to mean that, increasingly, young couples are living together in a sexually and emotionally intimate relationship without being married. However, the Census Bureau clearly states in all of its publications covering cohabitation data that the census takers do not ask questions about the nature of the couple's personal relationship.

In fact, the Census Bureau goes further and states that many cohabiting households undoubtedly contain couples with no emotional or sexual involvement. For example, a young male college student who rents a room in the home of an elderly widow causes that household to show up under cohabitation data. Whenever times are hard economically, many people double up in their living arrangements. Thus, a man and a woman who live together only for economic reasons would also show up in the cohabitation data.

In 2000, the Census Bureau reported 11 million households with two unrelated adults (U.S. Census Bureau 2000). Eighty-nine percent (9.7 million) of these housemates are opposite sex couples, while 11 percent (1.2 million) are same-sex couples. Yet we really do not know what percentage of the housemates are involved in sexually intimate relationships.

Despite the bureau's clear statement that it says nothing about the couple's personal relationship, the media tend to present cohabitation data as a rejection of marriage by American young people. Yet just the percentage of cohabiting couples who see their relationship as a permanent rejection of marriage is impossible to determine. Studies of cohabitation indicate that most intimately cohabiting couples either break up or marry with time (see Chapter 4).

The number of opposite sex couples cohabiting at any one time represents only about 7 percent of the total number of couples living together. The remaining 93 percent are married couples. However, the number of people having had a cohabitation experience before marriage is close to 50 percent (Alternatives to Marriage Project 2004). Again, we do not know the nature of their cohabiting experience. About half of all couples who marry are registered at the same address at the time of the wedding (Gilmore et al. 1999), which would seem to imply that their cohabitation prior to marriage was an intimate experience.

Family: A Buffer against Mental and Physical Illnesses

The family becomes increasingly important to its members as social stability decreases and people feel more isolated and alienated. Indeed, the healthy family can act as a buffer against mental and physical illnesses.

Early in the chapter, we noted that one of the remaining family functions was providing emotional gratification and intimacy to family members. Ideally, each family member helps other family members in solving their individual problems. Spouses can offer support to each other in times of stress and turbulence. In general, as the pace of life quickens, some people cannot keep up and may become increasingly isolated from their larger society. In this case, the family becomes more important as a refuge and source of emotional gratification for its members. The family has been called the shock absorber of society—the place where bruised and battered individuals can return after doing battle with the world, the one stable point in an ever-changing environment. If you do not belong to a family, where do you turn for warmth and affection? Who cares for you when you are sick? What other group tolerates your failures the way a devoted wife, husband, mother, or father does? The family can serve as portable roots, anchoring its members against the storm of change. Furthermore, the family can provide the security and acceptance that lead to the inner strength required to behave individually, rather than always in conformity with peers. The family can be a source of security and a protective shield against environmental pressures.

Of course, to be all of this, the family must be healthy and strong. As we all have learned from the recent media emphasis on family violence, all problems faced outside the family can also penetrate into the family. An unhealthy family may not supply nurturing and love. It can be a place from which individual members may wish to escape, rather than a place to which they return for warmth and security. Nevertheless, even an unhealthy family may still provide stability and safety for its members.

As families have become smaller, intimate relations within the family have become more intense, more emotional, and more fragile. For example, if a child has no other significant adults to interact intimately with besides his or her parents, then this emotional interaction becomes crucial to the child's development. If this interaction is positive and healthy, the child develops in a healthy manner; if it is not, then the child is apt to develop in an unhealthy manner. The family, in a sense, is a hothouse of intimacy and emotionality because of the close interaction and intensity of relationships. It has the potential to do either great good or great harm to its members. Because of the potential for harm, it becomes even more important to understand how to create a strong and healthy family that can help individual members grow and expand in healthy directions.

We will all spend a good part, if not all, of our lives in a family unit. Within this setting, most of us will achieve our closest intimacy with other persons—the shared human intimacy that promotes feelings of security and self-esteem. Such feelings lead to improved communication, and good communication tends to be therapeutic. According to Carl Rogers, "The emotionally maladjusted person is in difficulty first, because communication within himself or herself has broken down, and second, because, as a result of this, communication with others has become damaged" (1951, 1; 1972). To the degree that our family can help us become good communicators, it can help us toward better life adjustment.

Robert Coombs (1991) reviewed more than 130 empirical studies on a number of well-being indices and found that married men and women are generally happier and less stressed than the unmarried. When compared to single, divorced, separated, or widowed people, married persons appear to be physically and mentally the healthiest (Anson 1989; Bramlett and Mosher 2002, 3; Horwitz and Howell-White 1996; Joung et al. 1997; Ladbrook 2000; Murphy et al. 1997; Verbrugge 1983, 1986; Verbrugge and Madans 1985; Waite 2000; Waite and Gallagher 2000). Research findings also suggest that improving marital quality over time is associated with decreasing physical illness of family members (Wickrama et al. 1997).

In Charlotte Schoenborn's summary of marital status and health in the United States, she reported the following:

Regardless of population subgroup or health indicator, married adults were generally found to be healthier than adults in other marital categories. Marital status differences in health were found in each of the three age groups studies (18–44 years, 45–64 years, and 65 years and over), but were most striking among adults aged 18–44 years. The one negative health indicator for which married adults had a higher prevalence was overweight. (Schoenborn 2004)

Interpretation of such data is complex and controversial. Research suggests that the benefits of marriage may be partially due to a selection effect and partially due to true benefits gained from being married as opposed to being unmarried (Bramlett and Mosher 2002, 3; Waite 2000). The differences between various marital groups, however, are substantial enough to suggest that they are real rather than simply chance. The fact that married persons generally appear most healthy, and divorced and separated persons least healthy, suggests that families have a strong influence on health. The successful family operates to improve its members' health, while the unsuccessful family may do the opposite. If this is true, it becomes a matter of health to work toward improved family functioning and increased levels of intimacy.

The Need for Intimacy

Seeking physical, intellectual, or emotional closeness with others seems to be a basic need of most people (Fromm [1956] 1970; Maslow 1971; Morris 1971; Murstein 1974). To feel close to another, to love and feel loved, to experience comradeship, and to care and be cared about are all feelings that most of us wish and need to experience. Such feelings can be found in many human relationships. It is within the family, though, that such feelings ideally are most easily found and shared. Families that do not supply intimacy are usually families in trouble; often, these families disintegrate because members are frustrated in seeking meaningful, intimate relationships. Some successful families are not close, but studies of strong families generally find that such families can still supply intimate relationships to family members, and through this intimacy, contribute to their health.

intimacy
Experiencing the essence of one's self in intense intellectual, physical, and/or emotional communion with another human being

The term **intimacy** generally covers all the feelings mentioned previously; it is also a common euphemism for sexual intercourse. Intimacy is something we all can recognize, yet it is an illusive concept and difficult to define. For our purposes, we will use Carolynne Kieffer's definition: "Intimacy is the experiencing of the essence of one's self in intense intellectual, physical and/or emotional communion with another human being" (1977, 267). Intimacy combines six processes: communication of personal feelings, acceptance of personal limitations, respect for personal feelings, affirmation of each other, sharing of hurts and fears of being hurt, and forgiveness of errors (Sperry 1999, 34). Note that intimacy does not preclude the negative. The sharing of hurts and the forgiveness of errors is an integral part of the successful intimate relationship.

Note also that these six processes must include the characteristic of empathy that was briefly discussed in Chapter 1. An individual must be able to recognize what his/her partner is feeling and communicating. Researchers have had some success in training romantically involved couples to become more empathetic (Long et al. 1999); to the degree that the training is successful, the couple's relationship may become increasingly intimate in the broad sense.

The primary components of intimacy are choice, mutuality, reciprocity, trust, and delight (Calderone 1982). Two people like each other and make overtures toward establishing a closer relationship. They make a choice. Their act, of course, must be mutual for an intimate relationship to develop. As confidence in the other grows, each reveals more thoughts and feelings. Reciprocity means that each partner gives to the relationship and to the other, sharing confidences and feelings back and forth. This sharing nurtures acceptance and trust, which, in turn, increase the sharing, which eventually leads to the delight in each other that true intimates always share.

B. J. Biddle (1976) suggests that intimacy must be considered on each of three dimensions: breadth, openness, and depth. Breadth describes the range of activities shared by two people. Do they spend a great deal of time together? Do they share occupational activities, home activities, leisure time, and so on? Openness

HIGHLIGHT

Why Do We Avoid Intimacy?

To seek and find intimacy with another is highly rewarding. Yet people often avoid intimacy for many reasons. To open ourselves to another invites intimacy, but also risks hurt. What if we open ourselves to others and trust them to reciprocate, but they do not? Each of us has probably had such an experience. Who hasn't liked someone and been rejected by that person? We may now be able to laugh at some of our early failures with intimacy (label them puppy love, and so forth), yet each time we fail at intimacy, we become more guarded and apprehensive.

Fear of rejection is one of the strongest barriers to intimacy. Each time we are hurt by another, it becomes more difficult to be open, trusting, and caring in a new relationship. To be the first to share our innermost feelings—to say, "I like you" or "I love you"—leaves us open to rejection. The first steps toward an intimate relationship are especially hard for the insecure person who lacks self-confidence. To build an intimate relationship, we must first be intimate, accepting, and comfortable with ourselves. To the degree that we are not these things with ourselves, we will probably be fearful to enter an intimate relationship.

Intimacy demands active involvement with another. Often, passive spectator roles seem more comfortable—let the other person supply the intimacy. Our society teaches us to be spectators via television. Society often conditions us to play roles, to always please others, to deny our feelings. For example, macho males don't cry or show caring emotions.

Anger can be another barrier to intimacy if it is not dealt with openly. When we suppress, deny, and disguise anger, we do not rid ourselves of it. Rather, the anger lingers as growing hostility. Of course, we all become angry with our most intimate loved ones on occasion. Anger does not destroy intimacy. Suppressed anger, though, leads to hostility and will, over time, tend to destroy intimacy. Remember that intimacy implies openness between intimates. Suppressed anger is unexpressed and thus keeps us closed rather than open. Suppressing anger also implies lack of trust in the partner. Without trust, there cannot be intimacy.

Fear of rejection, nonacceptance of ourselves, spectator roles, and unexpressed anger are four of the strongest barriers to intimacy.

implies that a pair share meaningful self-disclosures with each other. They feel secure enough and close enough to share intellectually, physically, and emotionally. They trust each other enough to be honest most of the time, and this encourages further trust in each other. Depth means that partners share really true, central, and meaningful aspects of themselves. Self-disclosure leads to deeper levels of interaction. In the ultimate sense, both individuals can transcend their own egos and fuse in a spiritual way with their partner (Dimitroff 1999). Such an experience is difficult to attain, yet many believe that it is in the deepest, intimate experiences that love and potential for individual growth are found.

The discussion thus far makes it sound as though intimacy is something a person has or doesn't have. But this is not true. Intimacy is something that must be built between persons. Not only that, but in the most intimate of relationships, the quality of the intimacy will vary occasionally and over time. For example, in young couples the physical part of intimacy (sexuality) tends to predominate. In older couples, intimate communication, respect, and general caring for the partner tend to predominate. No couple can act in an intimate manner all the time, because intimacy does not occur in a vacuum. The distractions of work, children, the general social milieu, how one feels physically and/or psychologically, and the activities of the moment all act to increase or decrease intimate interaction (Carlson and Sperry 1999).

Kieffer (1977) adds to Biddle's three dimensions the age-old idea of intellectual, physical, and emotional realms of action. A totally intimate relationship would have breadth, openness, and depth in each activity realm. Table 2-3 analyzes a highly intimate relationship. Kieffer cautions that such a description is simplistic and does not include the numerous psychological processes that char-

TABLE 2-3

Intensity Matrix for the Analysis of an Intimate Relationship

	Intellectual	Physical	Emotional
Breadth (range of shared activities)	Telling of the meaningful events in one's day Participating in a political rally Years of interaction resulting in the sharing of meanings (phrases, gestures, etc.) understood only by partners Decision making regarding management of household	Dancing Caressing Swimming Doing laundry Tennis Shopping Gardening Sexual intercourse Other sensual/sexual activities	Phone calls providing emotional support when separated Experiencing grief in a family tragedy Witnessing with pride a child's graduation from college Resolving conflict in occasional arguments
Openness (disclosure of self)	Disclosing one's values and goals Discussing controversial aspects of politics, ethics, etc. Using familiar language Not feeling a need to lie to the partner Sharing of secrets with the partner and using discretion regarding the secrets of the partner	Feeling free to wear old clothes Grooming in presence of the other Bathroom behavior in presence of the other Nudity Few limitations placed on exploration of one's body by the partner Sharing of physical space (area, possessions, etc.) with few signs of territoriality	Describing one's dreams and daydreams Feeling free to call for "time out" or for togetherness Maintaining openness (disclosure regarding one's emotional involvement with other intimates) Telling of daily joys and frustrations Emotional honesty in resolving conflict Expressing anger, resentment, and other positive and negative emotions
Depth (sharing of core aspects of self)	"Knowing" of the partner Having faith in the partner's reliability and love Occasional experiencing of the essence of one's self in transcendental union Working collectively to change certain core characteristics of the self and of the partner	Physical relaxation, sense of contentment and well-being in the presence of the other	Committing oneself without guarantee, in the hope that one's love will be returned Caring as much about the partner as about oneself Nonjealous supportiveness toward the other intimate relationships of the partner

SOURCE: From *Marriage and Alternatives: Exploring Intimate Relationships* by Roger W. Libby and Robert N. Whitehurst. Copyright © 1977 by Scott, Foresman and Company. Reprinted by permission.

acterize the interaction of the partners or that brought them to their level of involvement. In addition, she reminds us that intimacy is a process, not a state of being. Also the closeness of a truly intimate relationship can raise problems and create stress that might not occur in a less intimate relationship (Antonucci et al. 1998). For example, my mate becomes seriously ill and because we are so intimately close, I become very emotionally distraught. With a more distant acquaintance, I might feel sorry that he/she is ill, but I would not become nearly so upset.

In the past, intimacy was partially maintained throughout an individual's life by the geographical and physical proximity of the family. As economic patterns in this country changed and increasing geographical mobility separated people from their families, social emphasis shifted from family closeness to individual fulfillment. Many people have found intimacy more difficult to achieve because of this shift (Etzioni 1993). Froma Walsh (1998) suggests the term *intimacy at a distance* to describe parents and adult children who live at some distance apart, but still value close contact with each other.

Life today, however, offers many different opportunities for fulfilling intimacy needs. Marriage is no longer seen as the only avenue to intimacy. If we examine our lives, we will probably find we have a *patchwork intimacy* (Kieffer 1977). By this, Kieffer means that most people are involved in many intimate (not necessarily sexual) relationships of varying intensity.

If intimacy is as rewarding as we have suggested, and if American society is allowing each person to seek it in ways other than marriage, we need to develop an ethic for intimates. For example, how do we keep the quest for individual intimacy and fulfillment within boundaries acceptable to our spouse? How do we keep the quest from lapsing into the selfish pursuit of always doing your own thing? Questions such as these must be considered by all couples seeking intimacy.

> **WHAT DO YOU THINK?**
> 1. Is intimacy a goal for you?
> 2. If so, is it difficult for you to be intimate?
> 3. In what action realm (intellectual, physical, emotional) do you share intimacy most easily?
> 4. Which realm is most difficult? Why?
> 5. In what ways is your relationship with your parents intimate?
> 6. In what ways can you not be intimate with your parents?

The Family as Interpreter of Society

The attitudes and reactions of family members toward environmental influences are more important to the socialization of family members than are the environmental influences themselves.

A family with small children is watching the evening news describe a protest at an abortion clinic. The parents either support or criticize the actions of the protesters. The parental reaction to this social event is more influential on the children than is the event itself.

We have seen that the family supplies the society's population by reproducing children. But the family also physically and psychologically nurtures the offspring into adulthood. As noted, the family is the main avenue for **socialization** of young children to their culture. Formal education takes over part of the job of socializing the young when they reach school age, but the family retains the greatest overall influence on young children. The family's continuing formal and informal socialization of its children may, indeed, supply the most deeply lasting lessons. However, with increasing numbers of mothers joining the workforce when their children are young, formal preschool child-care programs and television are assuming ever more influence over American children.

socialization
The physical and psychological nurturing of children into adulthood

Without socialization by adults, children cannot internalize society's rules, mores, taboos, and so forth. Society cannot exist unless its members internalize at least a minimum socialization level. In fact, people who fail to be socialized to their society are known as **sociopaths** or **psychopaths**. Such people are characterized by inadequate conscience development, irresponsible and impulsive behavior, inability to maintain good interpersonal relationships, rejection of authority, and inability to profit from experience.

sociopaths
Persons who fail to be socialized to their society
psychopaths
Older term for persons who fail to be socialized to their society
modeling
Learning by observing other people's behavior

Social learning theory has long pointed out the importance of modeling in learning, especially for young children (Bandura 1969, 1997). **Modeling** (observational learning) is learning by observing other people's behavior. Parents and other family members are the most significant models for young children. How they react to their society is often more important than what they formally teach

WHAT DO YOU THINK?

1. Which people in your life do you find it easiest to be intimate with?

2. Why is it easier with each of these persons than with others?

3. Is it easier for you to be intimate with men or women? If there is a difference, what do you think that difference is?

4. Is your family of orientation generally close and intimate?

5. Do your friends generally consider you to be a person with whom they can be intimate? Why?

their children about it. If, for instance, a father teaches his children the importance of obeying rules and then asks them to watch for police officers when he exceeds the speed limit, he is teaching them that not getting caught breaking the rules is more important than obeying them.

Although society often blames families for the problems of their individual members, it must be remembered that social problems—economic depression, inflation, unemployment, civil unrest, and so on—also greatly influence families. Each family will have an opinion about such problems, based on their own value orientation. These opinions may lead the family to take certain actions when the social controversies touch the family or its members. For example, if a child is to be bused 20 miles into a new neighborhood and school to promote racial equality, there will certainly be consequences within the family. The parents will lose some control over the child. The distance to school may mean that the parents cannot participate in school activities such as PTA (Parent-Teacher Association) and class parties. They will not know the families of their child's playmates. The child will be exposed to different social mores and expectations that the parents may or may not find acceptable.

More important than environmental change will be the family's reaction to it. Will the family accept and support the change? Will the family decide that the overall social benefits of having their child sent across town for schooling outweigh the inconveniences? Will it picket, riot, and protest? Will it transfer children from public to private school? Each family's reaction teaches its children values about minority groups, racial prejudice or tolerance, law, and authority.

An additional function of the family, then, is to help family members interpret social influences. This function also brings the possibility that an individual family may teach an interpretation that is unacceptable to others in the society as a whole. However, if family values share many of the general society's values, the chances are increased that the family will help its children grow into responsible community members. Thus, by strengthening families, the community strengthens itself.

Unique Characteristics of the American Family

The American family, especially the middle-class family, has certain characteristics that make it unique. Among these, the following stand out:

1. *Relative freedom in mate and vocational selection.* Historically, most cultures believed that mate and vocational selection were too important to be left to inexperienced young people. Decisions that would influence an individual's entire life were often made long before the child reached puberty. To this day, in many countries, children enter the labor market early. Whether tending the family's goats in Morocco or tying the thousands of knots in an Iranian carpet, children contribute to their family's economic well-being and learn their lifetime vocation early.

Modern Western society, especially the United States, has rejected child labor. Childhood (a historically recent concept) has increasingly become a protected period of freedom from adult responsibility. Today, young people are encouraged to seek out their own vocations. Families may make suggestions about possible vocations, but the final decision is usually left to the individual.

This freedom is even more evident in mate selection. As we shall see in Chapter 6, unstructured dating as a method of mate selection is a relatively new American contribution to the mate selection process. When Americans are asked why they marry, invariably they list love as their most important reason. Marrying for romantic love is another American contribution to the mate selection process. Although romantic love has always been recognized historically, it has almost never served as a basis for marriage. Marriages were contracted by parents for their children. The contracts were made for economic, political, and prestige reasons, not for love. If love were to appear in the contracted relationship, generally it would have to grow as time passed. The lyrics to the song "Do You Love Me?" from *Fiddler on the Roof* exemplify a contracted marriage.

Tevye to His Wife Golde

Do you love me?

Do I what?

Do you love me?

Do I love you? With our daughters getting married and this trouble in town, you're upset, you're worn out. Go inside, go lie down. Maybe it's indigestion.

Golde, I'm asking you a question: Do you love me?

You're a fool!

I know, but do you love me?

Do I love you? Well, for twenty-five years I've washed your clothes, cooked your meals, cleaned your house, given you children, milked the cow. After twenty-five years why talk about love right now?

Golde, the first time I met you was on our wedding day. I was scared, I was shy, I was nervous.

So was I.

But my father and mother said we'd learn to love each other, and now I am asking, Golde, do you love me?

I'm your wife.

I know, but do you love me?

Do I love him? For twenty-five years I've lived with him, fought with him, starved with him. Twenty-five years my bed is his; if that's not love, what is?

Then you love me.

I suppose I do.

And I suppose I love you too.

It doesn't change a thing, but even so, after twenty-five years it's nice to know.

2. *Relative freedom within the family, fostered by a high standard of living compared to other countries, physical mobility, lack of broader familial responsibilities, and the pluralistic nature of American society.* Freedom of vocational choice and mate selection stem from the general freedom that exists for the young within the American family and society. We have already examined the role that societal affluence plays in allowing broader choice of family structure.

As we saw, affluence tends to increase the amount of education people receive, and the proliferation of mass media widens the experience of American youth.

HIGHLIGHT

What the Research Says

In 1890, the size of the average American household living together was 5.6 persons. By 1998, the size had dropped to 3.18 persons. Average family sizes differ for various subgroups—3.92 for Hispanic families; 3.02 for White families; and 3.42 for Black families (U.S. Census Bureau 1998c).

Married couples account for 53 percent of households, while people living alone account for 25.5 percent of households. The majority of those living alone are women, often elderly widows (over 65 years of age) (J. Fields and Casper 2001).

Such experiences increase a youth's freedom, both within the family and in the life choices that all youth must make. Of course, such freedom increases the need for decision making, because along with freedom comes responsibility for one's actions. Without responsibility, freedom becomes anarchy.

Freedom also means freedom to make mistakes. Thus, the freedom that American youth enjoys tends to increase anxiety and insecurity. This increased anxiety and insecurity may partially explain the interest of some young Americans in cults or gangs, most of which ask their members to live by strict rules.

3. *An extremely private character.* The private character of the American family is another result of the general affluence of our society. Throughout most of the world, housing is in short supply. Many families are fortunate if they have more than one room. Living quarters often house not only the nuclear family, but also more distant relatives.

The average size of American families living together has decreased over the years. This decrease is accounted for mainly by the loss of the additional relatives and roomers that used to live with families.

Shrinking household sizes and growing numbers of persons living alone are possible only in a society affluent enough to supply abundant housing. An American family living in a 1,400-square-foot, three-bedroom, bath-and-a-half house uses enough space to house approximately 20 persons in a country such as Afghanistan. This ability of the American nuclear family to live separately from relatives and neighbors has allowed it to become private. Such privacy is a luxury of economic affluence. Our society's privacy brings both advantages and disadvantages. Living with fewer people increases individual freedom but also increases the duties and responsibilities of the adults. Living alone certainly increases privacy, but it also increases the chances of loneliness.

Family: The Consuming Unit of the American Economy

The family is the basic economic unit in modern America.

The members of early American farm families were also production workers. The Industrial Revolution removed much of the economic production from the American family home into factories and offices. The money earned by work outside the home now supports the family. Thus, the family's economic well-being is controlled more by the whim of the anonymous marketplace than by the work of individual family members. Rather than each family producing what it needs to survive, the family goes to the marketplace and buys what is needed. The family is the consuming unit of society's agricultural and industrial output.

Of course, the family still provides services to its members, such as meal preparation and home maintenance, which require productive work from family members. However, such production is largely unpaid and unrecognized by the larger society.

Family consuming—consumer spending—is important to the health of the economy. When consumer spending drops, as it did during the recession of 2001–2003, the economy suffers. Thus, the family also acts as an economic foundation block of American society.

American Families: A Great Diversity of Types

America is made up of a wide diversity of peoples, thus creating a diversity of family types.

The United States is made up of a great variety of peoples. The 2000 census found over 31 million out of a population of approximately 282 million Americans were foreign born. All races and nationalities and regions of the world are represented in the American population (Figures 2-1, 2-2, 2-3, and Table 2-4; Martine and Midgley 2003; Pollard and O'Hare 1999).

Although the American dream of a social melting pot is often espoused, the principle of **homogamy** is dominant in most relationships. This principle states that people are attracted to others who share similar objective characteristics such as race, religion, ethnic group, education, and social class. In other words, like tends to marry like. However, as America's population diversity has grown, intermarriage between different groups is on the rise. For example, interracial marriage has grown at least 500 percent since 1970. The largest proportion of intermarriage occurs across various Asian American communities, where rates can be as high as 50 percent in some cities. African American/Caucasian intermarriages represent the highest number of intermarriages because the African American population is large.

Over 6.5 million people (2.4 percent) reported themselves as "two or more races" (U.S. Census Bureau 2001d). This refers to people who chose more than

homogamy
The principle that people are attracted to others who share similar objective characteristics such as race, religion, ethnic group, education, and social class

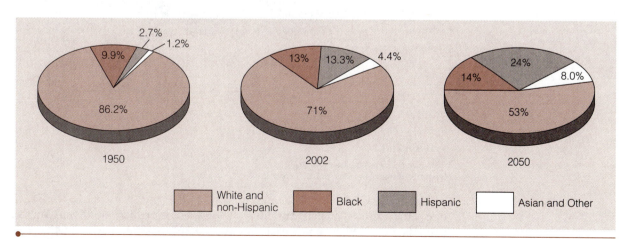

FIGURE 2-1 *U.S. Racial and Ethnic Composition, 1950–2050*

NOTE: Numbers may not sum to 100 percent due to rounding.
SOURCE: U.S. Census Bureau, 2004.

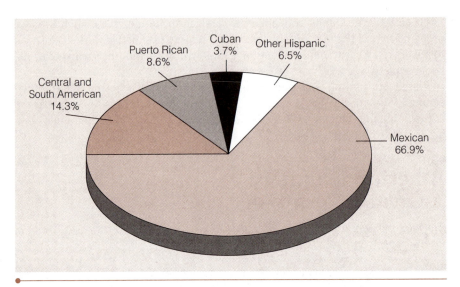

FIGURE 2-2 *Hispanics by Origin, 2002 (in Percent)*

SOURCE: U.S. Census Bureau, Annual Demographic Supplement to the March 2002 Current Population Survey; Ramirez and De La Cruz 2003.

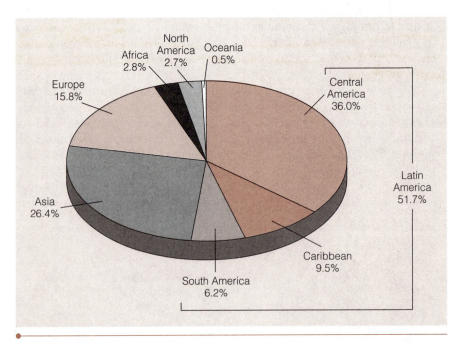

FIGURE 2-3 *Percent Distribution of the Foreign-Born Population by World Region of Birth, 2000*

NOTE: Adds to 99.9 percent due to rounding.
SOURCE: U.S. Census Bureau, Census 2000, December 2003.

one of the six race categories in the 2000 census: White, Black or African American, American Indian and Alaska Native, Asian, Native Hawaiian and Other Pacific Islander, and some other race.

Thus, rather than a melting pot, the United States resembles a tossed salad, with numerous familial groups: the Chinese American family, the African Amer-

TABLE 2-4

Some Demographic Variables for Four Family Groups

Variable	Asian American	African American	Latino	Caucasian
Percent population*	4.3% (11.9 million)	13% (36 million)	13.3% (36 million)	70% (211 million)
Married, spouse present[5]	60%	30%	44%	59%
Fertility rates per 1,000 women, 15−44 years[1, 6]	64.8	67.4	94.4	57
Out-of-wedlock births[1]	14.9%	68.2%	43.5%	28.5%
Children under 18 living with[5]				
2 parents	70%	39%	64%	75%
Mother only	13%	48%	25%	16%
Father only	3%	5%	5%	4%
Divorced persons[7]				
Men	3%	10%	6%	9%
Women	5%	13%	9%	11%
High school graduates[2]	87%	79%	57%	85%
Median family income with children[3, 9]	$50,000	$30,339	$33,075	$60,861
Median income married couples[3, 9]	High	$55,734	$40,541	$71,102
Persons below poverty level[3]	11%	22%	21%	11%
Children below poverty level[4]	Low	33%	31%	10%
College graduation[2]	47%	17%	11%	27%
Youth death rates: 15−24 per 1,000 population of this age group[9]	59	187	125	105

*The population of the United States was approximately 293 million on May 8, 2004.

[1] Martin et al. 2003.

[2] Department of Commerce 2003.

[3] Lichter and Crowley 2002.

[4] Lu and Koball 2003.

[5] J. Fields 2003.

[6] Whitehead and Popenoe 2003.

[7] U.S. Census Bureau 2003f.

[8] U.S. Department of Labor 2004.

[9] U.S. Department of Health and Human Services 2003.

ican family, the Hispanic family, and so forth. The families in each group share most characteristics with all American families, but also have specific characteristics that are unique to that group (see Table 2-4).

Figure 2.3 shows changes in the U.S. foreign-born population in more detail.

Although this book tends to focus on aspects of family life common to all families, regardless of race or nationality, it is important to note that *each* family is not *all* families. Table 2-4 compares a few demographic characteristics for four different family groups: Asian Americans, African Americans, Hispanics, and Caucasians. This overview is followed by a brief discussion of each family group (other than the Caucasian) as well as American Indians. These discussions are not meant to be exhaustive. They are aimed at alerting the reader to the fact that the United States is made up of diverse family groups representing many subcultures. Further detailed discussion of these various subgroup families will be found throughout the remainder of the book. Note also that the table gives statistics about an entire group of people and overlooks differences within the groups. For example, the table shows that African American income is lower than in some other groups, but does not reveal that the African American middle class has been growing faster than the middle classes in the other groups. Remember, the table only presents averages, which may or may not describe individual families.

African American Families

In Census 2000, 36.4 million people, or 12.3 percent, reported being Black. Another 1.8 million or 0.6 percent reported being Black and also one of the other races (U.S. Census Bureau 2001b). African Americans made up the largest minority in the United States until they were surpassed by the Hispanic population in 2003. The Black population increased slightly faster (15.6 percent between 1990 and 2000) than did the total population (13.2 percent) during this time. Just as there is no one family type in the United States, there is no single type of African American family. The proportion of married women and men in the Black population is considerably lower than in the White population. Only 39.8 percent of Black women are married but even fewer Black men are married—31.1 percent (McKinnon 2003).

Much past literature has focused on African American families near or beneath the official poverty level. Yet, the more important economic story of the last 40 years for African Americans has been the emergence of the middle class, whose income gains have been real and substantial. Note also that the median income for Black married couples ($55,734) is far higher than the median Black family income ($30,339) (Lichter and Crowley 2002; U.S. Department of Health and Human Services 2003). This is largely true because both married partners contribute economically, whereas many Black families are headed by a single adult. Approximately 33 percent of all Black families earn more than $50,000 a year, compared to 57 percent of White families. However, 52 percent of Black married families had incomes over $50,000, compared to 64 percent of White counterparts (McKinnon 2003, 5–6).

The improved economic condition of America's African American population is partly attributable to higher educational attainment, with 79 percent of African Americans now graduating from high school and 17 percent graduating from college (U.S. Census Bureau 2003f).

Despite economic gains, unemployment and poverty affect many more Black families than White families. Black families headed by a female with no spouse present are likely to have incomes less than $25,000 (58 percent) (McKinnon 2003, 6). The Black unemployment rate for March 2004, was 10.2 percent, whereas for Whites it was 5.1 percent (U.S. Department of Labor 2004). It is important to realize that the terms African American family or Black family encompass many kinds of families, both economically and racially.

In other than economic areas, the Black family has also become stronger. The proportion of unwed births among Black women has dropped from 70.4 percent in 1994 to 68.2 in 2001. The percentage of Black children living in two-married-parent families increased from 34 to 39 percent between 1996 and 2002 (Whitehead and Popenoe 2003, 3).

Hispanic Families

The 2000 census found 35.3 million (or 12.5 percent) of the population to be Hispanic. This subpopulation has been growing rapidly, especially in the border states such as California, Texas, and Florida. The Hispanic population increased by more than 50 percent since 1990. It surpassed the Black population in 2003 and is expected to reach one quarter of the U.S. population by the year 2050. Hispanics mainly comprise persons from Mexico (58.5 percent; see Figure 2-2). They probably constitute a higher percentage of population than the census figures indicate because an unknown number, especially those from Mexico, enter the country illegally and do not participate in the census.

It should be noted that there is conflict over what Hispanic families should be called. The Census Bureau uses both Latinos and Hispanics. Some subgroups prefer names that identify their country of origin, such as Mexican American or Cuban American.

The fertility rate for the Latino population is greater than that found in the general population. The **fertility rate** is defined as the number of women who report having a child in a 12-month period per 1,000 women aged 15 to 44 years. This rate was 94/1,000 among Latino women, 57/1,000 among White women, 68/1,000 among Black women, and 65/1,000 among Asian American women in 2003. Also, Hispanic women tend to have children earlier in their lives than women in the general population. The median age of women at first birth for the U.S. population as a whole is 25.1 years; it is 22.4 years for Mexican-American (Martin et al. 2003).

fertility rate
The number of women who report having a child in a 12-month period per 1,000 women aged 15 to 44 years of age

The more recent Hispanic immigrants, mainly those from Central America and Mexico, are often entering the United States illegally. Illegal immigrants generally have less schooling and less English language ability; therefore, they are less able to find economic success. Unemployment among Hispanics is 7.4 percent compared to 5.1 percent among Whites (U.S. Department of Labor 2004). Those of Hispanic origin who have been in the United States for a long period of time, often for generations, are economically much more successful and have a long history of contribution to their communities.

Lack of schooling and English language ability deter the recently arrived Hispanic population from improving their economic position. Note that the median income for a Hispanic married couple is considerably lower ($40,542) than for the other three family groups. In 2003, 57 percent of Hispanics graduated from high school and 11 percent graduated from college (U.S. Census Bureau 2003f). Here again, Table 2-4 masks the differences found between the many different Latino families. For example, 47 percent of Mexican Americans graduate from high school, while 60 percent of Puerto Ricans graduate from high school (Del Pinal and Singer 1997, 32; Pollard and O'Hare 1999).

The Hispanic population is considerably younger than the U.S. population as a whole. While 25.7 percent of the U.S. population was under 18 years of age in 2000, 35 percent of Hispanics were less than 18 years of age. The median age for Hispanics was 25.9 years, while the median age for the entire U.S. population was 35.3 years (U.S. Census Bureau 2001c).

In general, the extended family plays a larger role in Hispanic everyday life than in other groups. For example, about 8 percent of Hispanics aged 30 and over lived with their grandchildren, although not all were caregivers (U.S. Census Bureau 2003c). In addition, 26.5 percent of family households in which a Hispanic was the head of the household consisted of five or more people. In contrast, only 10.8 percent of White family households were this large (Ramirez and De La Cruz 2003, 4).

Asian American Families

In 1965, all discriminatory immigration quotas against Asians were lifted. Such quotas originated in 1882, when Congress passed the infamous Chinese Exclusion Act. This act remained in force until 1943, when China and the United States became allies during World War II. Since 1965, a large number of Asians,

© Tony Freeman/PhotoEdit

Do you think the children in this family will carry on traditions of both their parents?

mainly those dislocated by the wars in Vietnam and Cambodia, have entered the United States. Census 2000 reported 11.9 million, or 4.2 percent of the population, to be Asian.

The major Asian American groups, in order of size, are the Chinese, Filipinos, Asian Indian, Korean, and Vietnamese (Barnes and Bennett 2002, 9). The most recent Asian immigrants have largely kept to themselves, setting up their own subcommunities and maintaining many of their family traditions. As with any group of immigrants, however, Asians who came early to this country (mainly Chinese, Japanese, and Koreans) have slowly become acculturated. Thus, their families have become increasingly similar to American families.

Despite being discriminated against in earlier years, Asian Americans are now often perceived as a model minority because, as a whole, they are better educated (44 percent of Asian women and 51 percent of Asian men have a bachelor's degree or higher). They tend to hold higher occupational positions and they earn more than the general U.S. population; 40 percent of Asian families earn $75,000 or more, compared to 35 percent of White families (Reeves and Bennett 2003, 5). Asian families generally tend to exhibit the characteristics of strong, successful families that we have described.

Again, Table 2-4 does not discriminate between different Asian American subgroups. For example, the birthrates for the various subgroups vary considerably. The fertility rates are very low among Chinese and Japanese women, but are quite high among Vietnamese and other Southeast Asians.

The more recent Asian immigrants, such as the Cambodians, Hmong, and Laotians, are having a more difficult time than the earlier Asian immigrants, due to language and educational shortcomings. There is some evidence that these populations are becoming part of the American underclass, which comprises people who lack the social and technical skills to find and keep jobs, and thus experience more unemployment and poverty (S. Lee 1998, 34–35). However, second- and third-generation Asian Americans will have better education and English language skills, which should translate into better job opportunities.

The American Indian and Alaska Native Population

Census 2000 indicated there were 4.1 million, or 1.5 percent, American Indian and Alaska natives in the U.S. population. Although this is a small population compared to the other three groups discussed above, this population proportionately increased faster than the total population between 1990 and 2000. The overall birthrate was lower than either the Hispanic or Black birthrates, but higher than Asian and White rates. Depending on how the American Indian population is measured, it increased from 26 to 110 percent, compared to the overall population increase of 13 percent. Seventy-nine percent of the respondents indicated a tribal affiliation. The five largest American Indian tribal groupings were, Navajo (298,000), Cherokee (281,000), Latin American Indian (181,000), Choctaw (159,000), and Sioux (153,000). The largest Alaska Native tribal group is Eskimo (55,000) (Ogunwole 2002, 3, 8).

Although many Americans believe most American Indians live on reservations, the majority actually live in urban areas, where they tend to be invisible. This is because they are dispersed, rather than residentially clustered as with many immigrant groups. Their urban households are quite fluid. For example, one household group in the San Francisco Bay area lived in a two-bedroom apartment rented by a woman, her aunt, and their children. In the course of a month, at least 38 people—a shifting set of relatives, male friends, and their children—used the apartment. Some nights, as many as 18 people stayed over, but never the same set of people for more than a few nights.

Many friends who are not biologically related come to think of themselves as brothers/sisters or uncles/aunts if they come from the same tribe. Married couples with children represent about 33 percent of American Indian family households—less than Hispanic, Asian, or White households (Pollard and O'Hare 1999). Children are often traded between relatives and friends; thus, this percentage is questionable. The fluidity of the urban American Indian living arrangements suggests a census undercount of the population. "Urban Indian communities are networks of relationships rather than geographic location or neighborhoods. The fluid and flexible nature of the urban Indian community contributes to its resiliency and persistence, as well as to its invisibility from an outside perspective" (Lobo 2002).

Most American Indian families live in urban areas.

The rural American Indian population living mainly on reservations seems to fare less well than American Indians living in urban areas. Unemployment rates of 70 percent or more are found on some reservations, where health care is insufficient. Suicide rates among young American Indian men are the highest of any demographic group in America. Substance abuse and alcoholism are widespread. Despite such problems, many American Indian communities are vigorously working to combat their problems (Bensen 2003).

Because American Indian lands are self-governing and thus not accountable to local governing agencies, American Indians have been able to establish a number of income-producing activities, the most important of which has been gambling casinos. In general, the proliferating casinos across the country have brought increasingly large proceeds into the various tribes, as well as employment. In some cases, poverty has been completely eradicated. As with any such endeavor, the way in which the casinos are managed, and who manages them (tribal members or outside paid management), influence how successful the casinos are in bettering the lives of tribal members.

Summary

1. The family is the basic unit of human organization. If defined functionally, the family is essentially universal.
2. A free and creative society offers many structural forms by which family functions, such as child rearing, can be fulfilled.
3. Family life involves continuity and change.
4. Each family unit is unique, but also has characteristics in common with all other families in a given culture.

5. The family becomes increasingly important to its members as social stability decreases and people feel more isolated. Indeed, the healthy family can act as a buffer against mental and physical illness.

6. The attitudes and reactions of family members toward environmental influences are more important to the socialization of family members than are the environmental influences themselves.

7. The American family, especially the middle class, has certain characteristics that make it unique. Among these, the following stand out:

   Relative freedom in mate and vocational selection.

   Relative freedom within the family fostered by a high standard of living compared to other countries, physical mobility, lack of broader familial responsibilities, and the pluralistic nature of American society.

   An extremely private character.

8. The family is the basic economic consuming unit in modern America.

9. America is made up of a great diversity of peoples. Americans come from different ethnic backgrounds and races. Therefore, when we speak of the American family, we are actually speaking of many kinds of families.

Resources on the Internet

Companion Website for *Human Intimacy: Marriage, the Family, and Its Meaning,* Tenth Edition

http://sociology.wadsworth.com/cox10e/

Gain an even better grasp on this chapter by going to the companion website to take one of the tutorial quizzes, use the flash cards to master key terms, or check out the many other study aids you'll find there. You will also find special features such as GSS data, Sociology Online, and Census 2000 information that will put data and resources at your fingertips to help you with that special project or to help you as you do some research on your own.

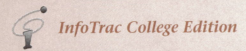

InfoTrac College Edition

http://www.infotrac-college.com/wadsworth/

You can access reliable resources anytime, anywhere, with InfoTrac College Edition, the online library. This fully searchable database offers more than 20 years' worth of full-text articles (not abstracts) from almost 5,000 diverse sources, such as top academic journals, newsletters, and up-to-the-minute periodicals, including *Time, Newsweek, Science, Forbes, The New York Times,* and *USA Today*. You can conduct electronic key word searches using key terms from this chapter to supplement your reading and learning experience. To aid in your search and to gain useful tips, see the Student Guide to InfoTrac College Edition, which you can access through the companion website for this book.

Should Our Society Recognize Gay Marriages?

At this juncture in time (2005), only one state, Massachusetts, allows gay couples to marry, although this right may be rescinded in 2006 when put to the voters. The federal Defense of Marriage Act of 1996 supports the right of states not to recognize marriages from other states. Accordingly, 38 states now have legislative bans on marriage rights for gays and on recognizing other states' gay marriages. Four states have constitutional amendments barring marriage to gays, and other states are considering such an amendment (Sullivan 2004).

YES

What's the problem? My partner and I have been together for 14 years. We own a home, hold responsible positions in our community, volunteer our time, and parent our 10-year-old daughter. We love and are committed to each other. How are these things different from how happily married heterosexual couples act? Since we act the same way, why are we denied some of the privileges and legal protections enjoyed by our heterosexual married neighbors? One of the criticisms often leveled against gay persons, gay men in particular, is that they are highly promiscuous. Might allowing gay marriage reduce this type behavior? Some religious groups and churches object to homosexual behavior as a matter of religious belief. Those churches would not be required to perform marriage ceremonies for gay couples; there is always the alternative of civil marriage available to such couples. Gay marriages would have no effect on heterosexual marriage, so what is the big deal?

NO

Marriage is a sacred relationship between a man and a woman. Marriage is partially predicated on procreation and the ensuing legal protection of the children derived from the marital relationship. Gay couples are free to live together, but since they can't procreate together, what would be the purpose for their marriage? It is true that they can adopt a child, and lesbians can be artificially impregnated so as to become a parent. Yet, the male or female homosexual couple cannot present to this child a well-rounded socialization from having both a male and female parent. A gay couple might refute this argument by pointing to heterosexual single parents who successfully socialize their children. This overlooks the fact that most single parents have someone of the opposite sex in their lives much of the time. Also, most single parents remarry in time.

Perhaps more important, if the society allows gay marriage because some gay persons feel victimized without it, what will society be able to say to other groups who feel victimized because the behavior they want is banned by society? Giving in to gay demands for marriage may well open the door to polygamy, polyandry, sex with children, and so forth. Because gay persons can live together if they so desire, what is the big deal about gaining the right to marry?

What Do You Think?

1. Should society allow same-sex persons to marry? Why?
2. What are the advantages of living together without legal sanction? The disadvantages?
3. What are the advantages to legal marriage? The disadvantages?

© Grace/zefa/Masterfile

American Ways of Love

Chapter

3

Questions to reflect upon as you read this chapter:

- Can you list several ways Americans define love?
- What is the difference between loving and liking?
- What is the difference between love and friendship?
- Describe some of the stages one goes through in learning to love.
- What is meant by companionate love?
- Define jealousy and describe how a person may feel when jealous.
- Can you list several actions to reduce jealous feelings?

W*hen God would invent a thing apart from eating or drink or game or sport, and yet a world-restful while in which our minds can melt and smile. He made of Adam's rib an Eve creating thus the game of love.* (PIET HEIN)

For many Americans, love and marriage are like hand and glove, apple pie and ice cream, bacon and eggs—they belong together. Of course, we all know of marriages without love, and romantic literature is full of examples of love without marriage. But the traditional ideal, the ultimate in human relationships for most Americans, is the steady, time-honored, and sought-after combination of love and marriage.

When Americans are asked, "Why do you want to marry?" they often reply, "Because I love. . . ." So we Americans marry for love. But, doesn't everyone? What other reasons could there be? No, not everyone marries for love. In fact, love as a basis for marriage may be a unique American contribution to the world. As Ralph Linton pointed out many years ago, "All societies recognize that there are occasional violent emotional attachments between persons of the opposite sex, but our present American culture is the only one which has attempted to capitalize on these and make them the basis for marriage" (Linton 1936, 175).

In many societies, love has historically been an amusing pastime, a distraction, or, in some cases, a god-sent affliction or even an addiction. For example, courtly love, or love as an amusing distraction, began as a diversion among the European feudal aristocracy. It exalted both chastity and adultery. Courtly love glorified love from afar and made a fetish of suffering over love affairs. It made love into a great game, where men proved their manliness on the jousting field in the name of love and a woman's honor. Adultery was an integral part of courtly love. The intrigue and excitement of adultery added to the sport and made the love even sweeter. Marriage was not considered the proper place for courtly love. Married love was too mundane and unexciting.

The story of the knight Ulrich von Lichtenstein highlights courtly love, which seldom found consummation in marriage. At an early age, Ulrich pledged his love and admiration to an unnamed lady. He accepted every challenge in an effort to prove himself worthy of serving his love. He was filled with melancholy and painful longings for his lady, a condition that he claimed gave him joy. The heartless lady, however, rejected his admiration, even after his 10 years of silent devotion and his many feats of valor. Undaunted, perhaps even inspired by her rebuffs, he undertook a stupendous journey in 1227 from Venice north to Bohemia, during which he claimed to have broken the incredible total of 307 lances fighting his way to Vienna and his lady love. It comes as something of a shock when, by his own statement, he stopped off for 3 days to visit his wife and children. For the fact is that this lovesick Galahad, this kissless wonder, this dauntless knight-errant had long had a wife to lie with when he had the urge, and a family to live with when he felt lonely. He even speaks of his affection for his wife, but, of course, not his love; to love her would have been improper and unthinkable to the ethic of courtly love (Hunt 1959, 132–139).

On the other hand, ancient Japan regarded love as a grave offense if it was not properly sanctioned, for it interfered with proper marriage arrangements. Etsu Sugimoto describes this in *A Daughter of the Samurai*:

When she was employed in our house, she was very young, and because she was the sister of father's faithful Jiya, she was allowed much freedom. A youthful servant, also of our house, fell in love with her. For young people to become lovers without the sanction of the proper formalities was a grave offense in any class, but in a samurai household it was a black disgrace to the house. The penalty was exile through the water gate—a gate of brush built over a stream and never used except by one of the Eta, or outcast class. The departure was public and the culprits were ever after shunned by everyone. The penalty was unspeakably cruel, but in the old days severe measures were used as a preventative of law-breaking. (1935, 115–116)

Surprising as it may be, such attitudes are still common (Hatfield 1993), although a study of 167 diverse cultures found that 87 percent recognized romantic love, but not necessarily as a basis for marriage (Janowiak and Fischer 1992). Marriage in many cultures has been and still is based on considerations other than love. In India, the Hindu place responsibility for finding a suitable mate on the parents or older relatives. The potential mate is judged by his/her economic status, caste, family, and physical appearance. These criteria are not by any means simple snobbery; they reflect the couple's prospects for rapport, financial stability, and social acceptance—all valid concerns in marriage. Even in the United States, such considerations are often hidden in the ephemeral concept of love. Although an American may find it difficult to believe, most young adults in India feel strongly that it is best for them if their marital partner is selected by their parents or a marriage broker. Students interviewed at a women's teachers college in northern India felt that older adults would make a much better selection of a mate for them than they could, at their young age. The young women also felt that the American system of mate selection was degrading because it forced women to compete with their friends for potential mates. In addition, many of the married Indians espoused the same feelings, having had their mate selected by their parents, perhaps with the help of a marriage broker (F. Cox 2004). For young Americans, parental interference in the mate selection process would undoubtedly cause a rift in the family.

The American Myth: Romantic Love Should Always Lead to Marriage

John Crosby (1985, 295) has suggested that the American idea of romantic love leading to marriage is a myth. The myth implies that love alone is the one indispensable ingredient that should determine whom and when a person marries. A corollary to the myth states that love will overcome all, never mind the obstacles.

Crosby goes on to say that the reader may protest, "Oh, but nobody really goes to that extreme! People don't really marry just because they fall in love, and if they do, they know there are other important factors in the selection of a mate." Yes, it appears that people are more thoughtful than the myth suggests, yet when it comes to behavior, our emotions tend to cancel out and override our reason. Who in the midst of falling in love is always reasonable? Indeed, the myth suggests that if you are always reasonable about being in love, you probably aren't really "in love."

Because Americans tend to believe the romantic myth that love must lead to marriage, numerous marriages have little other than love going for them—so little that, within a short time, the union dissolves because the couple has no basis on which to build a lasting relationship. Indian couples, when chosen as mates by their parents, have many characteristics in common. They may not have had much, if any, interaction before their marriage, so if there is to be love between them, it must be built on the basis of their similarities. For Crosby, the belief that love is the only basis for marriage is folly.

Lest you feel that such strong criticism means that love should play no role in marriage, it must be pointed out that love does, indeed, play a role in all intimate, lasting relationships. Even in the arranged marriages of India, many couples who marry without feeling intense love often build a loving relationship over time.

Hindu children are taught that marital love is the essence of life. So men and women often enter married life enthusiastically, expecting a romance to blossom. As Hindus explain it, "first we marry, and then we fall in love." (Mace and Mace 1959, as found in F. Cox 2004; H. E. Fisher 1992, 83)

Much of the confusion about the myth of love as a basis for marriage has to do with the many definitions of love. Love does not take just one form; it assumes many forms. I love my wife. I love my child. I love Mozart. I love a T-bone steak. Love is obviously many things. It is romantic love as a basis for marriage that Crosby criticizes.

Defining Love

Trying to define love has kept poets, philosophers, and sages busy since the beginning of history (see "Highlight: Love Is . . ."). What is this phenomenon that causes two people to react to one another so strongly? Is it physical? Chemical? Spiritual? A mixture of the three? Why does it only occur with some people and not others? Is it the same as infatuation? The questions are endless, yet we speak of love and seek it. "Couples can identify its absence or presence but pinpointing its mercurial mechanics can be like asking for a description of air—elusive" (O'Hanlon and O'Hanlon 1999, 249).

The ancient Greeks divided love into a number of elements: **ludus** (game-playing love), **storge** (friendship love), and **mania** (possessive/dependent love). Ludus and mania types of love tend to produce negative consequences. Storge is based on repeated contacts, leading to a high level of familiarity; it most resembles the attachment between parents and children.

For the Greeks, the three most important types of love leading to more successful intimate relationships are eros (carnal or physical love), agape (spiritual love), and philos (brotherly or friendly love). **Eros** is the physical, sexual side of

ludus
Game-playing love

storge
Friendship love

mania
Possessive/dependent love

eros
Physical, sexual side of love

HIGHLIGHT

Love Is . . .

Love is such a tissue of paradoxes, and exists in such an endless variety of forms and shades that you may say almost anything about it that you please, and it is likely to be correct.
Henry Fink
Romantic Love and Personal Beauty

Love is patient and kind; love is not jealous or boastful; it is not arrogant or rude. Love does not insist on its own way; it is not irritable and resentful; it does not rejoice at wrong, but rejoices in the right. Love bears all things, believes all things, hopes all things, endures all things.
I Corinthians 13:4–7

How do I love thee? Let me count the ways.
I love thee to the depth and breadth and height
My soul can reach, when feeling out of sight
For the ends of Being and ideal Grace.
I love thee to the level of everyday's
Most quiet need, by sun and candlelight.
I love thee freely, as men strive for Right;
I love thee purely, as they turn from Praise.
I love thee with the passion put to use
In my old griefs, and with my childhood's faith.
I love thee with a love I seemed to lose
With my lost saints—I love thee with the breath,
Smiles, tears, of all my life!—and if God choose,
I shall but love thee better after death.
Elizabeth Barrett Browning

Love is the triumph of imagination over intelligence.
H. L. Mencken

Let me not to the marriage of true minds
Admit impediments. Love is not love
Which alters when it alteration finds,
Or bends with the remover to remove.
O, no! It is an ever-fixed mark,
That looks on tempest and is never shaken;
It is the star to every wandering bark,
Whose worth's unknown, although his height be taken.
Love's not Time's fool, though rosy lips and cheeks
Within his bending sickle's compass come;
Love alters not with his brief hours and weeks,
But bears it out even to the edge of doom.
If this be error and upon me proved,
I never writ, nor no man ever loved.
William Shakespeare
"Sonnet 116"

Love birds burn in the sky,
The flame of Passion carries them high.
Beaks touching as one,
Wings beginning to melt, as they ride the crest of the flames.
With nothing to hold them in their flight
They fall
Into the inferno
Of their own passion.
Cynthia Moorman

Love is a game of exaggerating the difference between one person and everybody else.
George Bernard Shaw

---

love. It is needing, desiring, and wanting the other person physically. Although Eros was a Greek god, the Romans called him Cupid; as we know, Cupid shoots the arrow of love into our hearts. Eros is that aspect of love that makes our knees shake, upsets our routine, and causes us to be obsessed with thoughts of our lover.

agape
Altruistic, giving, nondemanding side of love

Agape is the altruistic, giving, nondemanding side of love. It is an active concern for the life and growth of those whom we love. Agape is an unconditional affirmation of another person. It is the desire to care for, help, and give to the loved one. It is unselfish love.

Theologian Paul Tillich sees the highest form of love as a merging of eros and agape:

No love is real without a unity of eros and agape. Agape without eros is obedience to moral law, without warmth, without longing, without reunion. Eros without agape is a chaotic desire, denying the validity of the claim of the other one to be acknowledged as an independent self, able to love and to be loved. (1957, 114–115)

Philos is the love found in deep and enduring friendships. It is also the kind of love described in the injunction "Love thy neighbor as thyself." It can be deep friendship for specific people, or it can be a love that generalizes to all people. Philos is often nonexclusive, whereas eros and agape are often exclusive.

philos
Love found in deep, enduring friendships

The philos element of love is most important to a society's humanity. A caring person creates loving relationships, and enough loving relationships make a loving society. A society that has a high level of philos among its members fosters the other elements of love. Lack of this kind of love creates a society of alienated, isolated individuals. When people are alienated and isolated from each other, the chance of dehumanized conflict between them escalates. U.S. crime statistics make it clear that the command to "Love thy neighbor" (philos) is often ignored in our society. Philos turns strangers into friends, and it is more difficult to perpetrate a crime against a friend than against a stranger.

It is the family that primarily teaches the moral values that foster philos love in a community. No society can function well unless most of its members behave most of the time, voluntarily heeding their moral commitments and social responsibilities. Philos love is, therefore, most important in creating a successful society (Etzioni 1993, 30).

Theories of Love

Although the myth of romantic love has discouraged the development of adequate psychological theories of love relationships (Sharpe 2000, 31), there are probably as many theories of love as there are persons in love. However, it should be noted that such theories do not always enjoy much empirical support. Nevertheless, even if only superficially, examining a few of them is worthwhile. Learning how other thoughtful people have theorized about love will help us understand our own feelings about love. The better we know our own attitudes and definitions of love, the better we will become in making long-lasting, intimate relationships.

In his classic book *The Art of Loving*, Erich Fromm defines love as an active power that breaks through the walls that separate people from each other. In love, we find the paradox of two beings becoming one, yet remaining two (Fromm [1956] 1970). Like the Greeks, Fromm discusses several kinds of love, including brotherly and maternal love. *Brotherly love* is characterized by friendship and companionship with affection. *Maternal love* is characterized by an unselfish interest in your partner; you place yourself second to your partner's needs. To Fromm, mature love includes attachment plus sexual response. More important, it includes the four basic elements necessary to any intimate relationship: care, responsibility, respect, and knowledge (all attributes of strong successful relationships). People who share all elements of mature love are pair-bonded. The relationship is reciprocal. Fromm goes on to suggest that a person's need to love and be loved in this full sense arises from feelings of separateness and aloneness. Love helps us escape these feelings and gain a feeling of completeness, of being united.

Expanding on Fromm's ideas, Lawrence Casler (1969) considers that love develops, in part, because of our human needs for acceptance and confirmation. These needs are heightened in a society as competitive and individualized as ours. Thus, we are relieved when we meet someone whose choices coincide with our own, who doesn't try to undermine us in some way. We tend to attach ourselves to such a person because he or she offers us validation, and such validation is an important basis for love.

Casler points out that love feelings in dating partners may be more a by-product of American dating than the result of some innate attraction. For example, a per-

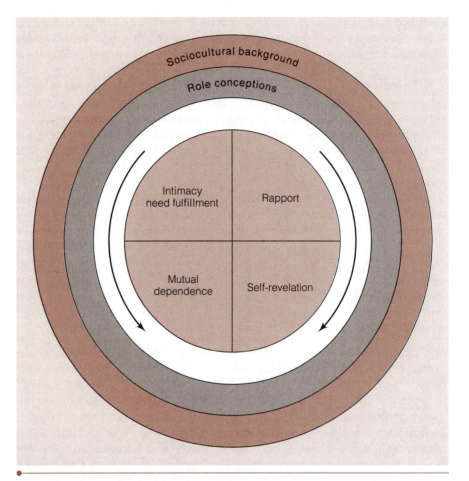

F I G U R E 3-1   *The Wheel as a Model of Love*

Source: From *Family Systems in America*, 4th ed., by Ira L. Reiss and Gary R. Lee. Copyright © 1988 by Holt, Rinehart and Winston. Reproduced by permission of the publisher.

son pretends to like his/her date more than he/she really does for any number of reasons, including simple politeness. The date, also seeking validation, responds favorably. This favorable response makes the first person feel good, and he/she feels fondness for the individual who has made him/her feel good. As one person's positive feelings increase, the other's are likely to also. Thus, love is based on progressive, positive emotional interaction between two persons. Obviously, it is easier to love someone who loves you than someone who is indifferent.

Ira Reiss (1988, 100–106) suggests the wheel (Figure 3-1) as a model of love and intimacy:

Stage 1: In the rapport stage, the partners are struck by the feelings that they have known each other before, that they are comfortable with each other, and that they both want to deepen the relationship. Social and cultural background, education, style of upbringing, and other broad background factors operate to enhance or inhibit feelings of rapport.

Stage 2: During the self-revelation stage, the partners share more intimate thoughts and feelings. This sharing deepens the relationship because such sharing is only done with special people. Self-disclosure is associated with increased commitment, mutual trust, and feelings of love. Women are more likely than men to reveal their thoughts and feelings. Self-revelation is higher among couples with relatively equalitarian attitudes toward gender roles. For example, gay and lesbian long-term couples tend to be more egalitarian than heterosexual couples and appear to have more similar self-revelation levels (Nardi and Sherrod 1994; Peplau and Spalding 2000).

Stage 3: As the sharing becomes more intimate, a feeling of mutual dependence develops. Falling in love provides a sense of very rapid expansion of the boundaries of self. The partner is included within one's self and thus is needed to feel complete (Hendrick and Hendrick 2000). There comes a feeling of loss when the partner is absent.

Stage 4: The partners experience more intimacy need fulfillment as they deepen their relationship. Reiss suggests that perhaps the initial rapport itself was caused by the hope of having these deeper needs met. The common characteristic of deeper needs is that they concern intimacy.

They are needs which, as they are fulfilled, express the closeness and privacy of the relationship. For example, the needs for emotional support and sympathy are expressions of the underlying need for intimacy. We have other needs, but they are not related to intimacy and are thus less relevant for explaining love relationships. (Reiss 1988, 102)

These four processes are in a sense really one process for when one feels rapport, he/she reveals him/herself and becomes dependent, thereby fulfilling his/her personality needs. The circularity is most clearly seen in that the needs being fulfilled were the original reason for feeling rapport. (Reiss 1960, 143)

In Figure 3-1, note that the outermost ring is labeled sociocultural background. An individual's general background factors (education, religion, and so on) account for the second ring, role conceptions. What do I perceive love to be? What is my role in expressing love? What role do I expect my partner to play in an intimate relationship? What will my role be as a lover, husband or wife, father or mother? These two rings influence how quickly the wheel of love is formed, if it is formed at all.

Helen Fisher (2004) studied love in two ways: She first used a 54-item questionnaire in both the United States and Japan. Secondly, she studied the reaction of the brain in various love situations via functional magnetic resonance imaging (fMRI).

With the questionnaire, Fisher found only small differences in answers between men and women, old and young, homosexuals and heterosexuals, and those of various races and religions. There were some differences between the Americans and Japanese, but the 12 questions that showed fairly large differences appear to have rather obvious cultural explanations. She concluded that certain characteristics of being in love are fairly universal.

She posited the following general characteristics and behaviors when a person is in love:

Special meaning—The beloved becomes unique, novel, and all-important.

Focused attention—The love-possessed person focuses attention on the beloved, often to the detriment of everything else.

Aggrandizing the beloved—The beloved's characteristics are magnified and exaggerated.

Intrusive thinking—The love-possessed person can't get thoughts of the beloved out of his/her head.

Emotional fire—There is a torrent of intense emotions.

Intense energy—There is a surge of energy when in the presence of the beloved.

Mood swings—The moods of the love-possessed person are closely tied to the behavior of the beloved.

Yearning for emotional union—The craving to merge with the beloved pervades the love-possessed person's thoughts and desires.

Looking for clues—Dissecting the beloved's behavior for clues to their feelings about the love-possessed person.

Changing priorities—Lovers rearrange their lives to accommodate each other.

Emotional dependence—Lovers become very dependent on the relationship.

Empathy—Lovers feel great empathy toward each other.

Sexual interest—Lovers are sexually interested in each other and desire sexual exclusivity.

Although a detailed investigation of brain activity is beyond the scope of this text, it is clear that certain parts of the brain of those in love show increased activity. First, parts of the caudate nucleus, a large C-shaped region that sits deep near the center of the brain, become particularly active as a lover gazes at the photo of his/her sweetheart. Second, there is activity in the ventral tegmental area, a central part of the reward circuitry of the brains. This is a major area of dopamine-making cells. Elevated levels of dopamine produce focused attention, strong motivation, and goal-directed behaviors, which partially characterize those persons in love. (For a detailed discussion of this research, see Bartels and Zeki 2000; H. E. Fisher 2004, Chapter 3.)

Robert Sternberg (1986, 1988, 1998) suggests three elements within love: intimacy, passion, and commitment. He visualizes these in the form of a triangle (Figure 3-2). He then oversimplifies the various types of love on the basis of these three elements.

- *Nonlove:* Absence of all three elements
- *Liking:* Intimacy without passion or commitment
- *Infatuation:* Passion without intimacy or commitment
- *Romantic love:* Intimacy and passion without commitment
- *Companionate love:* Intimacy and commitment without passion
- *Fatuous (foolish) love:* Passion and commitment without intimacy
- *Empty love:* Commitment without passion and intimacy
- *Consummate love:* Combination of intimacy, passion, and commitment

In reality, a person will experience some mix of these eight possible kinds of feelings toward their partner. For example, the next section discusses romantic love, which Sternberg views as a combination of intimacy and passion, but without much commitment—at least at first. The individual experiencing romantic love has probably gone from liking to infatuation to romantic love. Perhaps given more time in the relationship, romantic love will become companionate love.

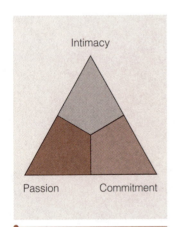

FIGURE 3-2
The Triangle of Love
Source: Sternberg 1986.

There are many other theories, but these give you an overview of some of the thinking about love and how it is formed. Note that all of these theories find that love has numerous parts and/or develops through a series of stages over time.

Romantic Love

Many Americans' thoughts about attraction and intimacy are profoundly influenced by the idea of romantic love. This concept of love encompasses such ideas as love at first sight, the one and only love, lifelong commitment, "I can't live without him/her," and the perfect mate. In addition, once the couple is committed to each other, romantic love says the following:

We should be everything to each other.

We should always be faithful to each other.

We should do everything together.

We should have the same goals, values, and beliefs.

We should never feel disappointed in each other.

We should never disagree and should strive to be of one mind.

We should have a romantic, passionate relationship (Sharpe 2000, 144).

In essence, the concept of romantic love supplies a set of idealized images by which we can judge the object of our love and also the quality of the relationship. Unfortunately, such romanticized images usually bear little relationship to the real world. Often, we project our beliefs onto another person, exaggerating the characteristics that match the qualities we are looking for and masking those that do not. We transform the other person into an unreal hero or heroine to fit our personal concept of a romantic marital partner. *Thus, we often fall in love with our own romantic ideas rather than with a real human being. We fall in love with love.*

Romantic love is only a set of attitudes about love (see "Making Decisions: Are You a Romantic or a Realist?"). The confusion between romance and love causes great trouble in forming long-lasting, intimate relationships. For example, the traditional romantic ideals dictate a strong, confident, protective role for a man and a charming, loving, dependent role for a woman. A woman holding this stereotype will tend to overlook and deny dependent needs of her mate. She will tend to repress independent qualities she discovers in herself. The man who has traditional romantic ideals will ignore both his own dependence needs and the woman's need for independence.

Those who fall in love with love in this way will suffer disappointment when their partner's real person begins to emerge. Rather than meeting this emerging person with joy and enthusiasm, partners who hold romanticized ideals may reject reality in favor of their stereotypical images. John Robert Clark provides the following description of this process:

In learning how to love a plain human being today, as during the romantic movement, what we usually want unconsciously is a fancy human being with no flaws. When the mental picture we have of someone we love is colored by wishes of childhood, we may love the picture rather than the real person behind it. Naturally, we are disappointed in the person we love if he/she does not conform to our picture. Since this kind of disappointment has no doubt happened to us before, one might suppose we would tear up the picture and start all over. On the con-

MAKING DECISIONS

Are You a Romantic or a Realist?

This Love Attitude Scale is intended to assess the degree to which you are a romantic or a realist in terms of love and does not relate to being a happy or mature person.

Instructions: Circle the response you believe is most appropriate:

1 = strongly agree (SA)
2 = mildly agree (MA)
3 = undecided (U)
4 = mildly disagree (MD)
5 = strongly disagree (SD)

	SA	MA	U	MD	SD
1. Love doesn't make sense. It just is.	1	2	3	4	5
2. When you fall head-over-heels-in-love, it's sure to be the real thing.	1	2	3	4	5
3. To be in love with someone you would like to marry but can't is a tragedy.	1	2	3	4	5
4. When love hits, you know it.	1	2	3	4	5
5. Common interests are really unimportant; as long as each of you is truly in love, you will adjust.	1	2	3	4	5
6. It doesn't matter if you marry after you have known your partner for only a short time as long as you know you are in love.	1	2	3	4	5
7. If you are going to love a person, you will "know" after a short time.	1	2	3	4	5
8. As long as two people love each other, the educational differences they have really do not matter.	1	2	3	4	5
9. You can love someone even though you do not like any of that person's friends.	1	2	3	4	5
10. When you are in love, you are usually in a daze.	1	2	3	4	5
11. Love at first sight is often the deepest and most enduring type of love.	1	2	3	4	5
12. When you are in love, it really does not matter what your partner does because you will love him/her anyway.	1	2	3	4	5
13. As long as you really love a person, you will be able to solve the problems you have with that person.	1	2	3	4	5

trary we keep the picture and tear up the person. Small wonder that divorce courts are full of couples who never gave themselves a chance to know the real person behind the pictures in their lives. (1961, 18)

Dating and broad experience with the opposite sex can help correct much of this romantic idealism.

Sometimes, people cling to their romantic expectations and try to change their partner into their romantic ideal. Changing one's partner is difficult, however. The person being asked to change may resent the demand and/or may not wish to change.

Because romantic love constantly seeks passion (emotional arousal), such love becomes increasingly difficult to maintain within long-term relationships such as marriage. Fixing dinner every night, writing the checks to pay the bills, changing diapers, going to work each morning, and doing the household chores are not romantic activities. Yet in the long run, these are the activities that make a relationship succeed. Couples must be able to find love within the sharing of

14. Usually there are only one or two people in the world whom you could really love and be happy with.	1	2	3	4	5
15. Regardless of other factors, if you truly love another person, that is enough to marry that person.	1	2	3	4	5
16. It is necessary to be in love with the one you marry, to be happy.	1	2	3	4	5
17. Love is more of a feeling than a relationship.	1	2	3	4	5
18. People should not get married unless they are in love.	1	2	3	4	5
19. Most people truly love only once during their lives.	1	2	3	4	5
20. Somewhere, there is an ideal mate for most people.	1	2	3	4	5
21. In most cases, you will "know it" when you meet the right one.	1	2	3	4	5
22. Jealousy usually varies directly with love; that is, the more you are in love, the greater your tendency to become jealous.	1	2	3	4	5
23. When you are in love, you do things because of what you feel, rather than what you think.	1	2	3	4	5
24. Love is best described as an exciting, rather than a calm, thing.	1	2	3	4	5
25. Most divorces probably result from falling out of love, rather than failing to adjust.	1	2	3	4	5
26. When you are in love, your judgment is usually not too clear.	1	2	3	4	5
27. Love often comes but once in a lifetime.	1	2	3	4	5
28. Love is often a violent and uncontrollable emotion.	1	2	3	4	5
29. Differences in social class and religion are of small importance as compared with love in selecting a marriage partner.	1	2	3	4	5
30. No matter what anyone says, love cannot be understood.	1	2	3	4	5

SOURCE: D. Knox, *The Love Attitudes Inventory.* Copyright © 1983 Family Life Publications, Saluda, NC.

Note: 30 = lowest possible score—the most romantic response; 150 = highest possible score—the most realistic response.

such daily tasks as well as within the excitement of romance. But when an individual confuses romance and love or believes the two to be synonymous, true love can never emerge. The moment the fires of romantic love begin to die down, those trapped by their belief in romantic love turn away from their partner in a never-ending search for the ideal mate with whom they can share a lifetime of romantic love. The fact that no such ideal mate exists means that they will be doomed to continue the search forever.

Generally, romantic love's rose-colored glasses tend to distort the real world, especially the mate, thereby creating a barrier to happiness. This is not to deny that romantic love can add to an intimate relationship. Romance will bring excitement, emotional highs, and color to the relationship. Indeed the spouse needs to be viewed, at times, as the most wonderful person in the world if a couple is to experience those intense sensations associated with passionate love. Idealizing, in this respect, is like a love potion. It should only be administered in small doses. Too much idealizing can stifle a relationship and strangle passion. Without any idealizing, there is no spark to ignite passion (Sharpe 2000, 159–160). As emotional, intellectual, social, and physical intimacy develop, romance becomes one of several aspects of the relationship, not the only one.

Infatuation

Romantic love and infatuation are often confused. Some say that they are actually the same thing. Lovers use the term *romantic love* to describe an ongoing love relationship. The couple's feelings, physiological reactions, and interactions when they are romantically in love are the same as those that occur in infatuation. The difference may be only semantic. The term *infatuation* is used to negate past feelings of love that have now changed. Love is supposed to last forever, so falling out of love means that the feeling for the other person was not really love; it must have been something less—namely, infatuation. "You were infatuated with him (not in love) before you met me." According to this line of thinking, perhaps infatuation can only be distinguished from romantic love in retrospect.

Others suggest that infatuation may be the first step toward love. The feelings of physical attraction, the physiological arousal, the intense preoccupation with your partner—all characteristics of infatuation and romantic love—are also the precursors to real love. Infatuation can grow into a more mature love that incorporates basic trust, care, support, and long-term commitment to the partner. Some suggest that love is grown-up and infatuation, childish. Yet playfulness and childishness are acceptable and important parts of any intimate relationship. It is apparent that we probably cannot agree completely on the difference between love and infatuation.

How can we avoid the pitfalls of romantic love and infatuation? How can we be sure that we love someone just as they are and that we're not being dazzled by our own romantic image of them? Perhaps we can't. Love is learned, and part of learning to move toward a mature, realistic love may be simply trial and error.

Furthermore, the most important prerequisite for true love may be knowing and accepting ourselves, complete with faults and virtues. If we cannot deal with our own imperfections, how can we be tolerant of someone else's? As Erick Fromm puts it,

Love of others and love of ourselves are not alternatives. On the contrary, an attitude of love toward themselves will be found in all those who are capable of loving others. Love, in principle, is indivisible as far as the connection between "objects" and one's own self is concerned. Genuine love is an expression of productiveness and implies care, respect, responsibility and knowledge. ([1956] 1970, 59)

Loving and Liking

In the play *Who's Afraid of Virginia Woolf?* Edward Albee depicts a couple who are in love but who also dislike, and at times hate, each other. Can one really be in love with someone whom one dislikes? The answer is that one can be passionately in love with someone one doesn't like, but the love probably will not evolve into a mature, enduring love. Instead, it will diminish as the dislike grows and ultimately leads to the breakdown of the relationship. In Albee's play, the dislike leads to hate and horrendous fighting.

It is probably more important to like someone than it is to love them, if one is to live together over an extended period of time. Roommate situations clearly demonstrate this. Living closely together on a day-to-day basis is difficult. Sharing cooking, eating, cleaning, and the many mundane chores that make daily living successful is next to impossible if one dislikes one's roommate. On the other hand, one doesn't have to love one's roommate to live successfully together.

Distinguishing between liking and loving in concrete terms is difficult. It is possible that the emotional aspects of these two states differ. On the other hand, perhaps the emotional aspects are the same and all that differs is your interpretation of the situation (Hendrick and Hendrick 1992, 88) (also see "What Research Tells Us: Love and Friendship").

Positive reinforcement and positive associations are important to the maintenance of a liking relationship (Guerrero and Anderson 2000, 177–178). It is obvious that someone you feel good being with is easier to like than someone with whom you feel rotten. It is also important that the liked person be associated with positive experiences. A person needs to associate his/her mate with the pleasurable areas of the relationship, not the negative areas. For example, doing fun things that both partners enjoy, such as going to movies, sports, out to dinner, and so on, are always important to maintaining a positive relationship.

Older couples still "in love" will show companionate or enduring love.

The Double Cross

Strange but true, those traits and characteristics that attract us to our mate in the first place sometimes become the very traits and characteristics that cause us the most concern and trouble later in our relationship. George Roleder (1986) defines this circumstance as love's *double cross*.

Here is one witness's story:

When we were going together, I loved my boyfriend's spontaneity. He was never boring because I never knew what to expect. There was constant excitement. He might arrive on Friday night for our movie date and tell me to get ready for a weekend ski trip to the local ski resort. I'd have to rush around, gather my equipment, and break plans with others that I might have made for the weekend, and we'd rush off to the mountains within the hour.

We've been married for three years now and his constant inability to plan ahead drives me crazy. I just wish that for once, we could sit down and make plans for the next week and stick to them. When there is a plan, there is something to look forward to and that is half the fun. I used to think that his spontaneity was fun but now I realize that he is unable to make a decision and stick to it. We both work and with the arrival of our daughter, I cannot just pick up and go on the spur of the moment. He doesn't seem to understand that having a family means that you need to make plans and follow through with them if you are to be mature and responsible. I don't think he will ever grow up.

Notice how her husband's trait of spontaneity has changed from being fun and exciting to being immature and irresponsible. How is this possible? It is clear that dating a person and living constantly with the same person are very different experiences. The personality traits and characteristics that make an evening's date enjoyable may, in fact, be very difficult traits to live with. A spontaneous, energetic person can certainly be a lot of fun in a short-term relationship. On the other

WHAT RESEARCH TELLS US

Love and Friendship

We have discussed many kinds of love and love in many contexts in this chapter. Keith Davis (1985) published the results of some fascinating research he did in an effort to describe how love and friendship, two essential ingredients to a fulfilling and happy life, differ.

Davis suggests that friendship includes the following essential characteristics:

- **Enjoyment.** Friends enjoy one another's company most of the time even though there may be temporary anger or disappointment.

- **Acceptance.** Friends accept one another as they are, without trying to change each other.

- **Trust.** Friends share basic trust, and each assumes that the other will act in light of the friend's best interest.

- **Respect.** Friends respect each other in the sense of assuming that each exercises good judgment in making life choices.

- **Mutual assistance.** Friends are inclined to assist and support one an-

other. They can count on each other in a time of need.

- **Confiding.** Friends share experiences and feelings with each other.

- **Understanding.** Friends have a sense of what is important to the other and tend to understand each other's actions.

- **Spontaneity.** Friends feel free to be themselves in the relationship.

Davis thinks that romantic relationships would share all the characteristics of friendship but would also have additional characteristics, over and beyond friendship. He identifies two broad categories unique to love relationships. The first he terms the "passion cluster" set of characteristics.

Passion Cluster

- **Fascination.** Lovers tend to be preoccupied with each other, obsessed with each other, and desirous of spending all their time together.

- **Exclusiveness.** Lovers are so intensely occupied with each other that

their relationship precludes having a similar relationship with another. Romantic love is given priority over all other relationships in one's life.

- **Sexual desire.** Lovers desire physical intimacy with their partners.

Davis terms the second cluster of characteristics related to romantic relations the "caring cluster."

Caring Cluster

- **Giving the utmost.** Lovers care enough to give the utmost, even to the point of self-sacrifice, when their partner is in need.

- **Being a champion/advocate.** Lovers actively champion each other's interests and make positive attempts to ensure that the partner succeeds.

The accompanying figure shows the way in which friendship and love are theoretically related. In general, the reports of Davis's subjects support this love-friendship model. However, there are some interesting results that differed from the model.

hand, a quiet, less energetic person may be very nice to come home to after a hard day's work.

You will find that there are many such double crosses in love. You are drawn to a very physically attractive person. Yet that very attractiveness brings her/him a great deal of attention from others, which makes you insecure and perhaps jealous. You find that the high intelligence and quick mind of your mate really attracted you when you first met. Now you find that you can't keep up mentally and often suffer from feelings of stupidity when you are with him/her. It was great fun to visit your boyfriend's apartment when he cooked one of his special gourmet meals. Now, however, with two small, hungry children, he is only a nuisance in the kitchen and you have to remind him constantly that small children prefer plain food.

Such examples should alert one to the fact that falling in love and staying in love may be two rather different processes. The first is based mainly on immediate attraction, while the second will require negotiation, compromise, and a far better understanding of both oneself and one's partner. Falling in love is grand; staying in love is hard work.

The caring cluster did not show as strong a difference between lovers and friends as had been expected. In fact, the "being an advocate/champion scale" did not show a difference at all. Friends championed each other as much as lovers did.

Another unanticipated finding is that best friendships were seen as more stable than spouse/lover relationships. Perhaps this is related to the finding that the level of acceptance was significantly lower among spouses and lovers than among friends. In other words, lovers tend to be more critical and less tolerant of each other than they are of friends.

Davis concludes that typical love relationships will differ from even very good friendships in having higher levels of fascination, exclusiveness, and sexual desire (the passion cluster); a greater depth of caring about the other person; a greater potential for enjoyment; and other positive emotions. Love relationships will also have a greater potential for distress, ambivalence, conflict, and mutual criticism, however.

One clear implication of these differences is that love relationships tend to have a greater impact on both the satisfaction and frustration of the person's basic needs.

What Do You Think?

1. Why might lovers be more critical of one another than good friends?
2. How important is it that one's lover also be a friend?
3. Can men and women be friends and lovers at the same time?
4. Are there other aspects of friendship that you feel are important but were not listed by the researcher?
5. For a long-lasting relationship, do you feel that the friendship factors or the added love factors are more important?
6. Why do love relationships have a greater potential for distress and conflict?

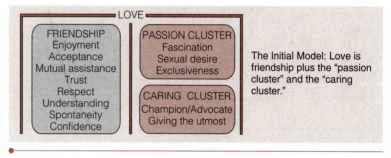

The Initial Model: Love is friendship plus the "passion cluster" and the "caring cluster."

Reprinted with permission from *Psychology Today.* Copyright © 1985 Sussex Publishers, Inc.

Love Is What You Make It

The more one investigates the idea of love, the harder it becomes to pin down. Everyone is quick to describe it, most have experienced it, and all know the mythology of the romantic ideal, even though many disclaim their belief in it. Although there does seem to be some agreement on at least a few aspects of love, much of what love is appears to be unique to each person. We each define love for ourselves. This may lead to problems for a couple if the partners define love somewhat differently. In a lasting relationship, the partners must come to a mutual understanding of what each means by love. It is important to understand your own concept of love so that you can recognize and accept differences between what you and your partner may mean by love. So, if you both can agree on what love is and on what loving acts are, and can act on this agreement most of the time, their chances are greater of maintaining love in their relationship.

Most people agree that there is a strong physical attraction between lovers, at least during the early stages of the relationship. This attraction is often accompanied by a variety of physiological reactions, such as more rapid breathing, in-

© 1977 Cathy Guisewite. Reprinted by permission of Universal Press Syndicate. All rights reserved.

creased pulse rate, and muscular tension. In other words, the person in love experiences general emotional arousal when thinking of the loved one or when in his/her presence.

Of course, such a reaction could be just sexual attraction and infatuation, rather than love. But if the physical attraction is accompanied by a strong and growing emotional attachment, and if there is a marked tendency to idealize and be preoccupied with the person, then the reactions are more indicative of love.

Generally, lovers experience a feeling of openness. Both feel they can confide in the other. Both believe the other likes them as they really are, so they can be more open, honest, and communicative—that is, more intimate—than in non-love relationships. One way of viewing love may be as intimate self-disclosure.

Such open sharing of your true feelings can be risky. A person in love is easily hurt, as all lovers will attest. However, the greater the risk, often the greater the return. Loving, intimate interaction increases the potential for both joy and sorrow in the relationship. One researcher terms this dual potential the *intimacy dilemma* (Prager 1999, 109). Thus, to love is always an adventure because danger is involved.

Indifference is the opposite of love. When hurt occurs, a lover may react to the pain with hostility, anger, and, at times, hate. Someone who doesn't care, who isn't hurt, and who isn't in love will be indifferent. The indifferent person, the person who cares least in a relationship, exercises more control over the relationship.

This is called the *principle of least interest*. The most loving person in the relationship is more vulnerable and, therefore, often goes to great lengths to placate and please his/her mate.

Another way to think of love is to ask whether the love experience is leading to personal growth. As discussed, most people in love experience an expansion of self. Being loved by another leads to feelings of confidence and security. Many people are encouraged to venture into new, and perhaps unknown, areas of themselves. "I feel more emotions than I've ever felt before." "I used to feel awkward meeting new people, but when I'm with Mary, I have no trouble at all." "Bill makes me feel like I can do anything I want."

Most Americans agree on the characteristics of love discussed thus far: physical attraction, emotional attachment, self-disclosure and openness, and feelings of personal growth. Yet individuals express these characteristics in a variety of ways, and such variance can lead to communication breakdowns. To say and mean "I love you" is one thing. We all may recognize the word, but it has been used so indiscriminately throughout time that it is difficult to identify love with certainty or clearly tell what it represents.

It is not uncommon for one partner to feel loving toward the other at the same time that the other feels unloved. It is as if they are on different emotional wavelengths. "But you never tell me that you love me," one may say. "I shouldn't have to tell you. I do loving things for you," the other replies. Stereotypically, the husband may believe that bringing home a paycheck, fixing broken appliances, and avoiding arguments are the "loving things" that he does. In his wife's eyes, these are merely things any good husband does routinely. She defines evidence of love as words of endearment, intimate sharing, touching, and tenderness—the kinds of behavior that may make her husband uncomfortable. He knows he loves his wife, but she is not getting the message. In other words, love is more than emotion. Love is also an intellectual concept. It is what one thinks it is. Its definition may differ between the partners, as previously noted.

Intimate love connections can be expressed in many areas of a person's life: friendships, sexual relations, companionship with another, business partnerships, and so on. If a disconnection about the expression of love is not resolved and continues, a couple will have relational problems that might lead to a future breakup (O'Hanlon and O'Hanlon 1999, 249–250).

Jamey and Phil were a couple who appeared to have all of the attributes of a loving pair: a good sex life, enjoyment of each other's company, and shared interests. Yet the supposedly unromantic area of finances tripped them up. They did not agree on how to budget and spend their earnings. In the end, financial disagreement cost them their relationship.

The concept of love has become feminized in America (Cancian 1990). The woman's concept of love (emotional expressiveness, intimate sharing, discussion of the relationship, saying "I love you") is the defining concept of love for most Americans. American men tend to show their love in more practical ways, such as by being a good family provider. When accused of lack of love, American men often resort to listing all of the things they do for the family. Since love is identified by American women as the proper emotional display, how American men display love is often discounted, and they may be accused of not being loving. "Let's talk about us" probably creates more misunderstanding between women and men than does any other single phrase (Wood 2000a, 308) (also see "Highlight: Gender Differences in Love").

HIGHLIGHT

Gender Differences in Love

Although popular thought tends to believe that women are more romantic than men, Knox and Schacht (1995, 2000) found that men were slightly more romantic than women in answering questions on their love attitudes survey. Men tend to fall in love more quickly than women, but women tend to form a more intense and lasting bond with the object of their love (A. Walsh 1991). Men, however, have been found to more easily separate sex from affection and love (Blumstein and Schwartz 1983). Women more often place sexual activity within a relational context than do men, although men indicate that their most erotic sexual experiences occur in a relational context (Barbach 1984). Bernard Murstein (1986) reminds us of the common saying: "Women give sex to get love, men give love to get sex." Young men tend to say that sex is more important than love, while women tend to say the opposite (Peplau and Gordon 1985; Vohs and Baumeister 2004).

The following differences seem to be supported by research (Murstein 1986):

- Men are less willing to marry without being in love than are women.

- Once a woman commits herself to her partner, she tends to become more expressive than he is.

- Men tend to enjoy love now and reflect on it later, whereas women tend to reflect more now, hoping to enjoy love later.

- Women tend to take more time to love and commit themselves to a relationship than do men.

- Women generally seek emotional relationships, whereas men tend to initially seek physical relationships.

- Once serious courtship is under way, the woman's love may be a better predictor of the course of the relationship. Even if she is more involved with him, she may break up the relationship if she thinks it is going nowhere. The man is more likely to enjoy the fruits of involvement and not be so concerned about where the relationship is going.

In general, since the invention of the pill for birth control and the sexual revolution, women's sexual behaviors have moved in the direction of men's. As we shall see in our discussion of cohabitation in Chapter 6, men more often tend to enter into such a relationship for the sexual pleasures, whereas women tend to enter into a cohabitation relationship for the broader love and intimacy aspects of it.

What Do You Think?

1. If your parents' marriage is intact, do you see a difference in the way your father and mother individually express their love for each other?
2. Have you ever felt loving toward an intimate other only to be told you weren't loving? Explain.
3. In what other ways do you feel men and women differ in their demonstrations of love?

Because we usually assume that our meaning of love is the same as our partner's, we may often feel unloved when we are, in fact, simply failing to recognize our partner's expression of love. The emotional script in this case is:

I, like every man and woman, want to be loved. But I have my own idea, grounded in my personality and attitudes and experience, of what loving and being loved means. Moreover, locked in the prison of my own ways of thinking and feeling, I assume that my definition of love is the only correct one. As a result, I want and expect to be loved in the same way that I love others, with the same responses that I interpret as the evidence of lovingness.

But I am not loved in that way. Instead (and quite logically, if one could be logical about love), I am loved the way my partner thinks and feels about love, the way he or she understands and expresses it. In my own distress, I do not recognize that my partner is experiencing the same incongruity in reverse. Puzzled, hurt, unable to communicate our confusion to each other, we both unreasonably feel unloved. (Lasswell and Lobsenz 1980, 15)

In a truly loving relationship, each partner's concept of love grows to include the other's concept. As our personal definitions of love move closer together, our chances of feeling loved increase.

Love in Strong Families: Appreciation and Respect

As we saw in the Highlight about love and friendship, it is astounding to realize that people often treat those they claim to love worse than they might treat a stranger. Most of us would excuse ourselves if we bumped into a stranger in a crowded department store. Yet, when we bump into our brother at home, we may say, "Why don't you look where you are going? You are so clumsy!" In strong families where there is a mature loving relationship, the expression of appreciation permeates the relationship. Mature love will always include appreciation of the loved ones, including the spouse, children, parents, grandparents, or simply good friends. Stinnett (1986) defines appreciation with the help of a story he calls "Dirt and Diamonds":

Diamond miners spend their working lives sifting through thousands of tons of dirt looking for a few tiny diamonds. Too often, we do just the opposite in our intimate relationships. We sift through the diamonds searching for dirt. Strong families are diamond experts.

Many young people report that they can be good and obey all the family rules for weeks on end, yet when their conduct finally lapses, that seems to be all their parents remember. As one teenager described:

When I started to date, my parents set 11 p.m. as the time I had to be home from a date. For the first six months I dated, I was always home on time. Finally, I got home one Friday evening at midnight and they blew a fuse. It was all I heard about for the next three months. It was as if my six months of getting home on time counted for nothing. One slip had cost me the entire previous six months. It made me wonder why I had even bothered to get home on time during the prior months.

Many spouses and families take each other for granted until something happens to upset their routine. The routines of daily living often dull appreciation for each other. For example, parents often forget to let their children know how much their good behavior is appreciated. Too often, children fail to let their parents know how nice it is to come home to a good meal, a clean house, a regular allowance. Each forgets to find the diamonds. It is amazing how far a little appreciation goes in a family where there is little of it. Parents in strong families emphasize positive behaviors and attitudes exhibited by their children. They avoid overemphasizing minor negative transgressions. When discussing appreciation, the author sometimes suggests to his students that a simple phone call home, to say that you had a good day in school and that you really appreciate your parents' support, can work wonders.

The ability to appreciate others starts with appreciating oneself. When we don't feel good about ourselves, it is difficult to feel good about and love another. And, of course, we learn to feel good about ourselves by having others appreciate and love us. Thus, strong families seem to start a circular process: "I appreciate and respect you; you learn to appreciate and respect yourself, which leads you to appreciate and respect me." Strong families start this pattern early in their children's lives. Unfortunately, some families start what we call a vicious-circle pattern of behavior instead: "I don't appreciate you; you don't learn enough self-respect to appreciate others." A **vicious circle** is a pattern of behavior in which a negative behavior provokes a negative reaction, which, in turn, prompts more negative behavior.

vicious circle

A pattern of behavior in which a negative behavior provokes a negative reaction, which, in turn, prompts more negative behavior

Respect is another important quality of strong families. We can't appreciate what we don't respect. Curran (1983, 90–111) suggests that respect is shown in many ways in strong families. First, the family respects individual differences. Parents, for example, don't expect all their children to be just the same or to be carbon copies of themselves. They respect each other's privacy. The family accords respect to all groups, even those with whom they may not agree. McCubbin and colleagues (1988) describe a support network around strong families that emphasizes the positive aspects of relationships. Family members also respect those outside the family, the property of others, and society's institutions. For example, parents who ridicule the school system as inadequate can hardly expect their children to go happily to school. Constructive criticism of society's institutions and also the family and its members can and should be made, as long as appreciation is also shown.

Appreciation includes respect for privacy. Family teamwork does not necessarily mean that everything about every member of the family is shared. A parent may have a particular problem that is shared only with the spouse. A child may do something wrong that does not affect others in the family, so the parent may keep it just between the child and him- or herself. Even in large families, each member needs some place and times for privacy. Fathers and mothers employed outside the home rarely have any time to themselves at home. When they are home, someone else is always there. In strong families, parents may take turns allowing each other some private time around the house by taking the rest of the family elsewhere.

We have spoken mainly about giving appreciation, but members of strong families realize that the ability to receive appreciation gracefully is also important. When we offer a sincere compliment to another person only to have it rejected, we feel stupid. For example, you may tell your brother, "My, you look handsome today." If he answers, "Well, I feel ugly today," what do you say? Inability to give or accept compliments stifles the flow of appreciation.

Learning to Love

We learn the meaning of love and how to demonstrate it from those around us—our parents, siblings, and peers—and from the general culture in which we are raised. For example, such a simple thing as family size may influence our definition of love. An only child becomes accustomed to a great deal of adult attention. Later, as an adult, such a child may feel unloved if his or her spouse (who is the third of four children and received relatively little adult attention) does not attend to him/her all the time.

Our experiences mold our attitudes and behaviors. Thus, the way we express love and what we define as love are the results of our past experiences.

Actions and Attitudes

The attitudes a person brings to love, to dating, and later to marriage have developed over many years. Because no two people experience the same upbringing, it is not surprising to find great attitudinal differences between people, even when they are in love. Many of the difficulties we experience in our interpersonal relationships stem from conflicting attitudes and unrealistic expectations, rather than from specific behavior. For example, a few socks on the floor are not as up-

PEANUTS reprinted by permission of United Feature Syndicate, Inc.

setting to a new wife as is her husband's general attitude toward neatness. Does he expect her to wait on him? Frequently, it is necessary to change underlying attitudes if behavior is to change, but this is difficult to do. We take our attitudes for granted, often without being aware of them. For example, a young man roundly criticized the double standard. But a day later, when discussing spouses' freedom to be apart occasionally, he stated that he certainly deserved time out with the boys, but women probably shouldn't go out with their girlfriends after marriage. He suggested that this might be misunderstood as an attempt to meet other men.

Attitudes generally include three aspects: affective, cognitive, and behavioral. The affective aspect is an individual's emotional response resulting from an attitude, such as "I like blondes." The cognitive aspect consists of a person's beliefs and/or factual knowledge supporting the particular attitude, such as "blondes have more fun." The behavioral aspect involves the person's overt behavior resulting from the attitude, "I date blondes."

Unfortunately, these aspects are not necessarily consistent. For example, attitudes may not be founded on fact. A person may not act on his/her attitudes. Or a person may voice attitudes that he/she really doesn't believe. The young man above is an example of someone who voices one attitude (against the double standard), but who favors another (for the double standard). Where do such attitudes come from? We are not born with them. We learn our most basic attitudes as we grow up. A society passes its values on to new members, beginning at birth. This process is termed **socialization**. This is why we must trace our attitudes toward sex, love, and marriage from early childhood to understand adult behavior. The capacity to form and maintain intimate relationships in adulthood is predicated on the fulfillment of needs and the attainment of competencies during earlier developmental stages (Prager 2000, 237).

socialization

Passing society's values on to new members beginning at birth

Developmental Stages in Learning to Love

Children pass through various stages of development as they grow to adulthood. Such stages are actually arbitrary classifications set up by theorists, but they are useful in efforts to understand development. Sigmund Freud delineated four psychosexual stages leading to adult sexual and love expression: self-love, parental identification, group or same sex, and heterosexual adult stage (Rieff 1994).

Self-Love Stage: Infancy and Early Childhood
The self-love stage occurs during infancy and early childhood. During their early years, young children are so busy learning about their environment that almost all of their energy, including sexual energy, is focused on themselves and exploring the environment. During these early years, it is important for the child to receive stimulation, including physical fondling. As Ashley Montagu writes, "By being stroked, and caressed, and carried, and cuddled, by being loved, the child learns to stroke, and caress, and cuddle and love others" (1972, 194). Montagu concludes that involvement, concern, tenderness, and awareness of others' needs are communicated to the infant through physical contact in the early months of life. It is here that the infant/child develops attachments to significant others in his/her environment. It is this early attachment that may be the basis for later adult love feelings (Feeney and Ward 1997; Karen 1998; Shaver et al. 1988). The infant/child begins to learn the meaning of love and to develop attitudes about intimacy, although the infant cannot intellectually understand these concepts. The child who is deprived of early physical contact may later be unable to make attachments based on love and caring because he or she has not experienced loving relationships. "The affirmation of one's own life, happiness, growth, and freedom is rooted in one's capacity to love, i.e., in care, respect, responsibility, and knowledge. If an individual is able to love productively, he loves himself too; if he can only love others, he cannot love at all" (Fromm [1956] 1970, 59–60). In order to love, we must be loved. Thus, even as small children, we learn about love from the way in which we are loved.

Parental Identification Stage: Early and Middle Childhood
The parental identification stage occurs during early and middle childhood. During this stage, children learn the masculine or feminine role that goes with their biological gender. In many respects, children act in a neuter way until they are about 5 or 6 years old, even though they start learning their gender roles much earlier. However, children are usually 5 or 6 years old before they make a commitment to the gender role, by identifying with the same-sexed parent.

The gender role that we learn is an important determinant of our definition of love and an even more important influence on how we display love. If a boy learns that it is not masculine to show tenderness, for example, tenderness may not be part of his style of love.

Group Stage: Late Childhood and Preadolescence

The group or same-sex stage occurs during late childhood and preadolescence. This stage coincides fairly well with the usual elementary school years in our society. It is called the group stage because each gender tends to avoid the other, preferring to spend time with groups of same-sex friends.

© Frank Cox

Learning to love starts early.

The main tasks of this stage are consolidation of the socially appropriate gender role and adjustment to cooperative endeavor and formal learning. The gang, or peer group, helps the child learn cooperative behavior and the give and take of social organization. This give and take helps children learn how to approach each other and how to show love and caring in many acceptable ways, according to the individuals with whom they interact.

The onset of puberty usually signals the end of the group stage. The age at which puberty begins varies so widely in our culture that each child will probably enter it at a slightly different time. Thus, the primary importance of the group diminishes only gradually, as the members one by one begin to turn their attention toward the opposite sex. The girls' groups dissolve first because, on the average, they reach puberty 2 years ahead of boys. Girls have been ahead biologically all along, but the distance is greatest at the onset of puberty.

Heterosexual Adult Stage

Most young persons entering the heterosexual adult stage quickly turn their attention toward the opposite sex and soon begin dating. Each preceding stage of development becomes the foundation for the next stage. And we must assimilate the lessons of all stages to become mature, loving individuals. People who have passed successfully through the first three stages of development reach the heterosexual stage with a positive attitude toward themselves, a fairly clear idea of their roles as men or women, a heightened interest in the opposite sex, and some knowledge of what love means and how to display it. They have learned several different kinds of love in their relationships with their parents and their peers. Their definitions of, and attitudes about, love are fairly well set, as is their own personal style of loving. Thus, long before a young adult falls in love, he or she has developed what researchers call a love map; this is the map that guides one to fall in love with one person rather than another (H. E. Fisher 1992, 45; Money 1986).

Children develop these maps during the early developmental stages in response to their environmental experiences, usually with significant others. For example, as a child gets used to the turmoil or tranquility in his/her home, the way mother scolds or father jokes, whether his/her parents show affection to each other, the child absorbs certain of these behaviors. And gradually, the memories accumulated during growing up take on a pattern of likes and dislikes, shaping an unconscious map of what love is and how it is displayed, and suggesting the characteristics of the person to be loved.

It should be noted that gay and lesbian persons do not necessarily follow these precise stages. For example, some may not leave the group (same-sex) stage. There is dispute about the origins of gay and lesbian behavior. Some claim it is biological and inborn. Others suggest that it is a learned behavior. The nature-nurture, either-or argument probably masks how individual gay and lesbian behavior develops. Be that as it may, theoretical stages of development are just that—theoretical—and do not necessarily apply to every individual, although they may well describe general development.

Love over Time: From Passionate to Companionate Love

Many of the characteristics of love described thus far can be labeled passionate or romantic love, and they are most apparent early in a love relationship. Being in love at age 20 with your new mate will probably be quite different from the love experienced with your mate after 20 years of marriage. However, it is important for a newly married couple to spend enough time alone together to establish a strong couple bond and identity. Numerous behaviors can prevent the formation of an adequate couple identity, including: marrying because of pregnancy or just before a long separation (as may occur in military marriages); beginning a relationship after a divorce or death too quickly; and starting schooling or a demanding career that requires most of the couple's time, attention, and energy (Sharpe 2000, 43–44).

companionate love

A strong bond that includes tender attachment, enjoyment of the other's company, and friendship

In an enduring relationship, romantic love tends to become companionate love. **Companionate love** is a strong bond that includes tender attachment, enjoyment of the other's company, and friendship. It is not necessarily characterized by wild passion and constant excitement, although these feelings will be experienced from time to time. The main difference between romantic and companionate love is that the former thrives on deprivation, frustration, a high arousal level, and absence. The latter thrives on contact and requires time to develop and mature. With time, the emotional excitement of passionate love tends to fade into a lower-key emotional state of friendly affection and deep attachment.

As a partnership progresses, the couple will grow together and find their emotional attachment becoming stronger. Some of the passion may give way to a more enduring, caring, and comfortable relationship. If and when children arrive, a couple's love broadens to include them. There will then be a greater proportion of agape, or selfless love.

Where might love go from here? It is hoped that it will increasingly become a mixture of romantic, selfless, and companionate love. Rollo May (1970) calls this mixture "authentic love." Erich Fromm ([1956] 1970) uses the phrase "mature love" to describe healthy adult love. Mature love preserves the integrity and individuality of both persons. It is an action, not just a passive emotion. Giving takes precedence over taking. Yet this giving is not felt as deprivation, but rather, as a positive experience.

One must be able to accept and also give love if a relationship is to remain loving. Loving is a reciprocal nurturing relationship. By accepting your partner's love, you affirm that person as an accepted and valued companion.

The expression of love has social, physical, intellectual, and emotional aspects. At different periods in an individual's life, one or another of these aspects may be dominant, and thus, the way in which love is shown will vary from time to time. Such changes do not mean an end of love, but may only signal a changing interplay of all of its factors.

Thus, the route love might take in a relationship that lasts a lifetime is basically the pattern of dependency, mutuality, passion, caring, and respect, and then perhaps dependency again in old age. Of course, there are other courses that love can take when couples find their love has diminished or died.

American culture presents some obstacles to an enduring love relationship. Our culture, especially through mass media, exalts passionate or romantic love, which is usually linked closely to sexual expression. For those who equate true love only with passionate love, intimate relationships are doomed to end in disappointment. It is impossible to maintain a constant state of highly passionate love with its concomitant elation and pain, anxiety and relief, altruism and jealousy, and constant sexual preoccupation. Certainly, no one newly in love wants to hear that the flame will burn lower in time. Who, newly in love, preoccupied from morning until night with thoughts of their love, can think, much less believe, that the feelings they are experiencing so strongly will ever fade? On the other hand, who could tolerate being in this highly charged emotional state forever? The fact that love changes over time makes it no less important, no less intimate, no less meaningful when some of the passion is replaced by a warm, deeply abiding and caring, quieter companionate love.

Equating sex with love may cause other important personal and interpersonal potentials to be neglected and may diminish the overall quality of love in a relationship. Luther Baker (1983, 299) points out "that sex is not the *piece de resistance* of the good life, and our present concentration upon it often prevents people from developing other aspects of personal functioning which will produce a good life." The best sex, he says, flows:

Spontaneously out of a life and a relationship filled with love, joy, struggle, growth and intimacy. Good sex is a by-product and, like happiness, is most likely to occur when one is not worried about having it. We may find that when we come to know one another better in non-sexual ways, the expression of sexual intimacy will take on a new and more fulfilling dimension.

Following Baker's suggestion to de-emphasize the sex-is-love philosophy would make it easier for two people passionately in love to accept the changes that slowly transform their love feelings into a subtler, more relaxed, companionate love style.

Another obstacle to an enduring love relationship is our culture's emphasis on the individual, which leads to a preoccupation with oneself, self-improvement, self-actualization, and so forth. As Amitai Etzioni (1983, 1993) suggests, in this age of the individual ego, marriage (love) is often less an emotional bonding than a breakable alliance between two self-seeking individuals. Love must involve a strong feeling of community: the caring and loving of mate, children, and other intimates within a person's environment. Passionate love tends to preclude all but oneself and the object of one's love. Indeed, if all Americans remained passionately in love, there would never be a sense of community. Passionate love ul-

HIGHLIGHT

Love and the Loss of One's Self-Identity

As time passes, couples often report escalating conflict between sharing love together and the feeling of loss of identity. Robert Bellah and his colleagues, in *Habits of the Heart* (1985), discuss conflict between individualism and commitment in American life. One woman reported that, during the first years of her marriage, she acted out the role of good wife, trying continually to please her husband. "The only way I knew to be was how my mother was a wife. You love your husband and this was the belief I had, you do all things for him. This is the way you show him that you love him—to be a good wife." Trying so hard to be a good wife, she failed to put her "self" into the relationship. She put aside her willingness to express her own opinions and act on her own judgment. As time passed, she felt the loss of herself more strongly. "All I thought about was what he wanted. The very things I was doing to get his approval were causing him to view me less favorably."

Losing a sense of who one is and what one wants can make an individual less attractive and less interesting. To be a person worth loving, one must assert his/her individuality. Having an independent self is a necessary precondition to joining fully in a relationship.

Love creates a dilemma for Americans. In some ways, love is the quintessential expression of individuality and freedom. At the same time, it offers intimacy, mutuality, and sharing. In the ideal relationship, these two aspects of love are perfectly joined—love is both absolutely free and completely shared. Such moments of harmony among free individuals are rare, however. The sharing and commitment in a love relationship can seem, for some, to swallow up individuality. Paradoxically, since love is supposed to be a spontaneous choice by free individuals, someone who has "lost" him/herself cannot really love. Losing a sense of one's self may also lead to being exploited, or even abandoned, by the person one loves.

---

timately is selfish love. As it is transformed into companionate love, the selfish element is reduced, and there is once again time for other aspects of the loving relationship besides the passionate and sexual. As the preoccupation with the loved one subsides, the individual can again attend to the broader community.

It should be remembered that companionate love does not preclude passionate love (Hendrick and Hendrick 2000, 204–205). Ideally, there will always be times of high passion in a loving relationship, no matter how long the relationship has lasted. These times are to be savored and enjoyed. Because they will become less frequent with the passage of time in even a successful, loving relationship, one must not value the loving less.

The myth of couple self-sufficiency also works to derail love relationships (Sharpe 2000, 33–34). This myth ties into the emphasis on individual (the American idea of rugged individualism) satisfaction. It suggests that a couple can accomplish everything necessary to the success of their relationship, without any outside help. Because Americans can move about the country with ease, many young couples start their life together by moving away from their roots, that is, from family, friends, and community ties. Thus, there are no mothers, fathers, aunts and uncles, and friends or church members to come over and help when the new baby arrives, or when there is sickness or other emergency needs. The couple alone is expected to successfully take on all of life's challenges. There are times, however, that such challenges may overwhelm even the strongest couple unless there is additional help from others who care about the couple.

From our discussions, it is clear that it takes little effort to fall passionately in love, even if we don't exactly understand what such love is or just why we fall into it. It is equally clear that thought and work are required to remain in love. Love neglected will not survive in most cases. Falling passionately in love is not the end. It must be a beginning that, when nurtured and cared for, will endure, albeit in a somewhat changed form. To love lastingly and well, to like and care for another over a period of time, is a unique gift that humans can give each other. The idea that enduring love will simply occur because one is passionately in love is a false belief that may destroy a relationship. If enduring love is to evolve from passionate love, the enduring love relationship must be worked at and tended so that it will not wilt and die with the passage of time.

Love's Oft-Found Companion: Jealousy

Jealousy, the green-eyed monster, is an unwelcome acquaintance of nearly everyone who has ever been in love. Because it is a personal acquaintance, it is described in many different ways, almost as many as the descriptions of love. **Jealousy** may be defined as the state of being resentfully suspicious of a loved one's behavior toward a suspected rival. Jealousy is an aversive emotional reaction evoked by a relationship involving an individual's current or former partner and a third person. It is a multidimensional concept (Buunk 1991, 1997; Buunk and Dijkstra 2000; Buunk et al. 1996). Jealousy refers to the belief that a relationship is in danger of being lost.

Jealousy is sometimes used to mean envy. Envy, however, represents a discontent with oneself and/or a desire for the possessions or attributes of another. Jealousy is characterized by a sense of feeling lonely, betrayed, afraid, uncertain, and suspicious. Envy elicits more shame, longing, guilt, denial, and a sense of inferiority (Parrott and Smith 1987).

Jealousy relates as much, if not more, to an individual's own feelings of confidence and security than it does to the actions of a loved one (Buunk and Dijkstra 2000). As Margaret Mead (1968) has said, "Jealousy is not a barometer by which the depth of love can be read. It merely records the degree of the lover's insecurity. It is a negative miserable state of feeling having its origin in the sense of insecurity and inferiority." Jealousy usually involves feelings of lost pride and threatened self-esteem—a person's feelings that his/her property rights have been violated. Whether the threat or violation is real, potential, or entirely imaginary, the jealous feelings aroused are real and strong.

Jealousy is also related to what an individual's culture teaches about love and possession inasmuch as the culture prescribes the cues that trigger jealousy.

On her return trip from the local watering well, a married woman is asked for a cup of water by a male resident of the village. Her husband, resting on the porch of their dwelling, observes his wife giving the man a cup of water. Subsequently they approach the husband and the three of them enjoy a lively and friendly conversation into the late evening hours. Eventually the husband puts out the lamp, and the guest has sexual intercourse with the wife. The next morning the husband leaves the house early in order to catch fish for breakfast. Upon his return he finds his wife having sex again with the guest. The husband becomes violently enraged and mortally stabs the guest. (Hupka 1977, 8)

This story seems unintelligible to an American. How can the husband condone his wife having sexual intercourse with another man on one occasion and be enraged by it a short time later? The answer lies in the cultural ways of the Ammassalik Eskimo. In this culture, the husband would be seen as inconsiderate if he did not share his wife sexually with his overnight guest. "Putting out the lamp" is a culturally sanctioned game that acts as an invitation for the guest to have sex with the host's wife. Yet it is not unusual for an Ammassalik Eskimo hus-

jealousy
The state of being resentfully suspicious of a loved one's behavior toward a suspected rival

band to become so enraged as to try to kill a man who has sex with his wife outside the prescribed game rules.

Such behavior illustrates that jealousy is, to some extent, rooted in social structure. Hupka (1981) and colleagues (1985) identified several characteristics that differentiate cultures with jealousy from those without. Cultures low in jealousy discourage individual property rights and view sexual gratification and companionship as easily available. Such cultures place little value on personal descendants or the need to know whether the children in the family are an individual's own progeny. Marriage is not required for economic survival, companionship, or recognition of the individual as a competent member of society (Salovey and Rodin 1989, 241). Such cultures are obviously the opposite of American culture. Also, western countries with more liberal sexual attitudes, such as the Netherlands, report lower levels of jealousy (Buunk et al. 1996).

In American society, we are also taught the rules of love. In the past, love meant exclusivity, monogamy, and lifelong devotion and faithfulness. Lovers were possessions of each other. Historically, adultery was grounds for divorce in every state. Jealousy is still usually nearby when a young American falls in love.

And for most who experience it, what an unpleasant condition jealousy is! Its characteristics include suspicious feelings that seem to manufacture their own evidence to support the jealousy at every turn; compulsive preoccupation with the perceived infidelity; anger, sorrow, self-pity, and depression all wrapped into one continual emotional upheaval; eating and sleeping problems; and an assortment of physical ills evolving out of the continual emotional turmoil.

Researchers shed some light on the manifestation of jealousy among Americans. The following characteristics have emerged from various studies and are partially summarized by Buunk and Dijkstra 2000, 324–329:

- Jealousy goes with feelings of insecurity and an unflattering self-image (Buunk 1991; Cano and O'Leary 1997).
- People who feel jealous because of a mate's real or imagined infidelity are often themselves faithless (Salovey 1985).
- People who report the greatest overall dissatisfaction with their lives also feel jealous most often.
- Happy or not, jealous people feel strongly bound to their mates. Desire for exclusivity is the strongest predictor of jealousy (Salovey and Rodin 1989, 233).
- Younger people report jealousy more often than older people.
- Concealing jealousy from others is difficult.
- Jealousy seems to cause women greater suffering than men.
- Women tend more often to try to repair the damaged relationship, whereas men try to repair their damaged egos (self-esteem).
- Marriage therapists report that jealousy is a problem in one-third of all couples coming in for counseling (Pines 1992).
- Men are more apt to give up a relationship in which jealousy is triggered by the woman's infidelity than are women in the reverse situation.
- Women are more apt to induce jealousy in their partner to test the relationship, bolster their own self-esteem, gain increased attention, get revenge, and/or to punish the mate for some perceived offense (G. White 1980).

Jealousy is more closely tied to an individual's own perceptions than it is to the reality of a situation:

Dieter asks his steady girlfriend out for Friday night. She says she's sorry, but she has to go out of town to visit her grandparents. (She is actually going to do this.) Dieter becomes jealous and angrily accuses her of having a date with someone else. No matter how she reassures him that she is indeed visiting her grandparents, he remains unconvinced. When she returns from her visit, he is even more angry and upset because in his view she has refused to tell him the "truth" as he perceives it.

We can see that Dieter's behavior is being dictated by his own subjective view of the world, not reality. Yet even though both we and his girlfriend know that his view is incorrect, the difficulty is just as real as if he were correct. Because he is angry, they may fight and not speak to each other for a time. If Dieter finds out from others that she was, indeed, visiting her grandparents, he will be apologetic and sorry that he has acted this way. Remember, we act on our perceptions, and they are not necessarily the same as objective reality.

One last point: To the degree that a person's culture teaches the cues that trigger jealousy, changes in the culture may also change the characteristics and incidence of jealousy. As American sexual mores have loosened, as increased divorce has eroded long-term monogamy, as cohabitation and premarital sexual intimacy have become more acceptable and marriage postponed, the need for jealousy has also changed. Although the evidence is unclear at this time, greater sexual freedom may be leading to decreased sexual jealousy for some people. For example, cohabiting couples appear to engage in sexual relations outside their relationship more often than married couples.

Is there anything that one can do to manage and control irrational jealous feelings? This question is difficult to answer, especially for a person in the midst of a jealous emotional reaction. Naturally, any steps we can take toward becoming confident, secure individuals will help us cope with our own jealousy. A person can try to learn what is provoking the jealousy. What exactly are his/her feelings, and why is he/she feeling that way? An individual can try to keep jealous feelings in perspective. Remember that jealousy is also a sign that a person's relationship is meaningful and important. A person can also negotiate with his/her partner to change certain behaviors that seem to trigger jealousy. Negotiation assumes that an individual is also working to reduce unwarranted jealousy. Choosing partners who are reassuring and loving will also help reduce an individual's irrational jealousies. Unfortunately, following such advice is difficult because jealousy is so often irrational and unreasonable and, for the jealous partner at the moment of jealousy, all too often uncontrollable. It remains one of the puzzling components of love relationships.

Summary

1. American youths are given relative freedom to choose a mate. American mate selection is usually based on the nebulous concept of romantic love.

2. Love is a difficult concept to define. One helpful approach is the view of the ancient Greeks, who classified love into several types such as eros (sexual love), agape (spiritual love), and philos (brotherly or friendly love).

3. Our attitudes about, and personal definitions of, love guide mate selection and lead us to form a set of idealized expectations about the kind of mate we desire.

4. Attitudes about love and marriage develop through a number of stages as an individual grows from infancy to adulthood.

5. Such stages are only theoretical and will vary with cultures and individuals.

6. It is important that we love and also like our mate if a relationship is to endure.

7. Passionate love will gradually turn into longer lasting companionate love as time passes. Those persons who believe that only passionate (romantic) love is real love are usually doomed to failure because no relationship can sustain passionate love forever.

8. Jealousy is an oft-found yet unwanted companion of love. It is closely related to an individual's own security and self-confidence.

Resources on the Internet

Companion Website for *Human Intimacy: Marriage, the Family, and Its Meaning,* Tenth Edition

http://sociology.wadsworth.com/cox10e/

Gain an even better grasp on this chapter by going to the companion website to take one of the tutorial quizzes, use the flash cards to master key terms, or check out the many other study aids you'll find there. You will also find special features such as GSS data, Sociology Online, and Census 2000 information that will put data and resources at your fingertips to help you with that special project or to help you as you do some research on your own.

InfoTrac College Edition

http://www.infotrac-college.com/wadsworth/

You can access reliable resources anytime, anywhere, with InfoTrac College Edition, the online library. This fully searchable database offers more than 20 years' worth of full-text articles (not abstracts) from almost 5,000 diverse sources, such as top academic journals, newsletters, and up-to-the-minute periodicals, including *Time*, *Newsweek*, *Science*, *Forbes*, *The New York Times*, and *USA Today*. You can conduct electronic key word searches using key terms from this chapter to supplement your reading and learning experience. To aid in your search and to gain useful tips, see the Student Guide to InfoTrac College Edition, which you can access through the companion website for this book.

Are Love and Marriage Good for You?

This seems like a strange question to ask of Americans. Of course, love and marriage are good for everyone. However, as noted early in the chapter, there are cultures where love is not important when marriage is contemplated. And as we saw with courtly love, there have been times when love was reserved for relationships outside of marriage.

Love and Marriage Are Good for Everyone

Anyone who has ever been infatuated or in love knows the happy, ecstatic feelings that accompany being in love. Granted, one cannot maintain these initial love feelings forever. But such feelings appear to translate themselves into both better physical and mental health for couples who follow love feelings into marriage.

Many studies report on the health benefits of marriage (Joung et al. 1997; Murphy et al. 1997). Bryce Christensen (1988, 1994) goes so far as to suggest that marriage is so good for people's health that the government could greatly reduce health care costs by encouraging marriage and the establishment of families.

Mental health problems are also lower for married persons. This may partly stem from the fact that two people are intimately acquainted. Each can serve as a check and balance to the other's behavior. A person living alone may not have immediate feedback on the success or lack of success of his or her various behaviors. For example, a person suffering the early stages of a manic-depressive illness may be unaware of the symptoms for a much longer time when living alone than when he or she has an intimate partner who recognizes and calls attention to the partner's behavioral changes. Early recognition of any type of illness, physical or mental, is important to successful treatment.

Love and Marriage Are Not Good for Everyone

The writer Liz Hodgkinson (1994) comments: "The 'commitment' we expect from a marriage partner is similar to a commitment to jail: it prevents our being free individuals." The requirement in marriage that a person is expected to devote his/her life to another individual, husband or wife, is stifling to the individual. Life expectancy has greatly increased so that, instead of spending perhaps 20–30 years with your spouse, married couples are expected to spend 50–60 years together.

It is often noted that long-married couples become more alike. The reality is often that one spouse, usually the woman, loses his or her individuality to the stronger partner. Learned helplessness may occur on the part of the partner who loses individuality. The depression that an individual sometimes feels when married is probably really anger and resentment turned inward because of the loss of the individual's self-identity.

Violence statistics indicate that marriages can be dangerous to an individual's health and safety. Spousal violence is common in marriage, with wives and husbands abusing each other about equally. Because men are stronger, severe violence and even death are more often suffered by wives than husbands. However, just about as many wives kill their husbands as vice versa. If marriage is such a happy, healthy institution, how is one to account for the seemingly high levels of violence therein?

What Do You Think?

1. How do you account for the fact that married people are generally both physically and mentally healthier than unmarried people?

2. Can a person marry and still maintain his/her individuality?

3. Can a person stay in love with his/her spouse over a lifetime of marriage?

4. Why do you think there is so much violence between spouses?

© Jess Alford/Photodisc Green/Getty Images

Gender Convergence and Role Equity

Chapter
4

Questions to reflect upon as you read this chapter:

- What is meant by role equity?
- What three factors determine a person's gender identity?
- What is people liberation?
- How do you think people liberation will affect families?
- What effect have strengthened sexual harassment laws had on the workplace?
- Do you feel society oppresses women? Does it oppress men? If so, how?
- How is parenting affected by both parents working outside the home?

S omewhere out in this audience may even be someone who will one day follow in my footsteps and preside over the White House as the President's spouse. I wish him well! BARBARA PIERCE BUSH

Male and female, man and woman, boy and girl, masculine and feminine—this duality is easily recognized in the human species. Almost everyone's first conscious thought upon meeting another human being is "he or she." Radical feminists chide us for automatically making this differentiation. Others say, "Vive la difference!" Still others suggest that the duality is false and that there are multiple genders (Roscoe 1994, 2002).

Most men and women recognize this duality, but where did it come from? Are men and women really different biologically? Psychologically? Intellectually? Socially? If so, are such differences inborn (based on a different biological foundation) or learned? Are men the stronger sex because on the average they are physically stronger? Are women the stronger sex because they are more resistant to illness and generally live longer than men? Can a woman really be like a man? Can a man really be like a woman? Why have men traditionally been breadwinners and women homemakers? Is the relationship between men and women one of equality? One of dominance and submission? One of independence and dependence? Complementary? What would happen if a society did away with gender-role differentiation (culturally assigned behaviors based on a person's sex)—if it could—and we all became unisex?

Controversy about the answers to such questions swirls around us. Regardless of how we answer the questions, the gender meanings assigned to man and woman are changing. The differences between the sexes are lessening, and the similarities are increasing.

Male = Masculine and Female = Feminine: Not Necessarily So

Simply stated, whether an individual is male or female sexually is biologically determined. The behaviors or roles that go along with being male or female (gender roles), however, are largely learned from an individual's society. For example,

American women tend to be more expressive and socially aware than American men, often discussing relationships and their feelings (Eagly 1995; Feingold 1994; Wood 2000a, 2001). American men tend to be more assertive and competitive than American women, preferring to discuss politics, sports, business, and other relatively impersonal topics (Wood 2000a, 308; 2001). We call this socially learned behavior in our society *masculine* for the male and *feminine* for the female. A person's **gender identity** is how that person views him-/herself: "Am I a man or woman, masculine or feminine?"

An individual's sex is determined by the chromosomal and hormonal influences that lead to the anatomical differences between the sexes. A person's gender includes his/her biological sex and all of the attitudes and behaviors (masculine and feminine) that an individual learns from his/her culture. Thus, sexual identity includes both physiologically prescribed sex behaviors and culturally prescribed gender behaviors. Nature and nurture, as in most behaviors, are both important (Azar 1997).

Norms and Roles

Before we discuss gender development, it is important to define the terms norm and role. **Norms** are accepted and expected patterns of behavior and beliefs established either formally or informally by a group. Usually, the group rewards those who adhere to the norms and punishes those who do not. In the United States, sanctions against cross-gender behavior are greater for boys and men than for girls and women. For example, an American woman can dress in pants and male clothing without censure. An American man dressing in a skirt, blouse, and other female clothing might well be arrested. These sanctions start at an early age and operate through the influence of parents, peers, and others, including social institutions such as schools and churches as well as the media.

Roles involve activities demanded by the norms. A husband working to support his family is fulfilling the traditional role as husband and thereby fulfilling the social norm of a husband as economic supporter of the family. Because gender norms have been changing in America, it might be that a man is a stay-at-home husband; thus, his role may now be caring for the children and household while his wife is the family's major breadwinner.

Because there are many norms in a society, a person plays many roles. A traditional married woman may fulfill the roles of sexual partner, cook, mother, homemaker, financial manager, psychologist, and so on. If she works outside the home, she also fulfills career roles. A traditional married man may fulfill the roles of breadwinner, sexual partner, father, general repairman, and so on. The point is that all people play a number of roles at any given time in their lives. Conflict between roles is frequent because so many roles are necessary to live successfully in a complex society.

Each of us brings to a marriage, or to any intimate relationship, a great number of expectations (Chapter 7) about what our roles and those of our partner should be. Many disappointments in marriage stem from frustration of the role expectations we hold either for ourselves or for our spouses. A traditional husband may assume that the role of wife is restricted to caring for him, the house, and the children. His wife may believe that the role of wife can also include a career and that the role of husband can include household duties and care of children. Such conflicting role expectations will undoubtedly cause difficulty for this couple.

Accepting norms and roles tends to smooth out family functioning. Problems occur when roles and norms are not accepted or when they are unclear. In many

gender identity
How one views oneself as either a man or a woman

norms
Accepted and expected patterns of behavior and beliefs established either formally or informally by a group

roles
Activities that norms require

societies, marriage involves very definite goals, such as increasing the family's land holdings, adding new workers (children) to the family, or even bringing new wealth into the family in the form of the bride's dowry. But our society does not set definite goals other than the vague "living happily ever after" one sees in the movies and romantic fiction.

In societies where roles and norms are stable, people enter marriage with clear ideas of each partner's rights and obligations. In American society, almost all norms and roles are being questioned—none more than those associated with masculinity, femininity, and sexuality. Uncertain gender roles make it more difficult to establish a successful intimate partnership. With changing roles, it is difficult to predict what an individual's gender role will be in a new relationship (Bateson 2000). Yet some flexibility of gender roles also holds out the hope of individual couples building more satisfying partnerships than traditional American gender roles may allow (see "Role Equity," page 98).

How Sex and Gender Identity Develop

Three factors determine gender identity:

1. Sex is genetically determined at conception.
2. Hormones secreted by glands directed by the genetic configuration produce physical differences.
3. Society defines, prescribes, and reinforces the gender-role aspect of sexual identity (masculinity and femininity).

Biological Contributions

Every normal person has two sex chromosomes, one inherited from each parent, that determine biological sex. Women have two X chromosomes (XX); men have an X and a Y chromosome (XY). If a man's X chromosome combines with the woman's X, the child will be female. If a man's Y chromosome combines with the woman's X, the child will be a male. Thus, the male's Y chromosome biologically determines the sex of the child.

All embryos have the potential to become either sex; that is, the already existing tissues of the embryo can become either male or female. For a male to be produced, the primitive, undifferentiated gonad must develop into testes rather than ovaries. The male hormone (chemical substance) testosterone spurs development of testes, while another hormone (Müllerian-inhibiting substance) simultaneously causes the regression of the embryonic tissues that would otherwise become the female reproductive system. In the absence of male hormones, the female organs develop. The hormones have already started working by the time the embryo is 6 millimeters long (about 2 weeks). By the end of the eighth to twelfth week, the embryo's sex can be determined by observation of the external genitalia.

The male appears to develop only with the addition of the male hormones. These are stimulated by the Y chromosome. Since only a female will develop without the addition of male hormones, some researchers conclude that the human embryo is innately female (Kimura 1985).

During puberty, hormonal activity increases sharply. In girls, estrogen (one of the female hormones) affects such female characteristics as breast size and the filling out of the hips. Estrogen and progesterone (another female hormone) also begin the complicated process that leads to the first menstruation, followed about a year later (or sometimes immediately) by ovulation. In boys, testosterone induces secretory activity of the seminal vesicles and the prostate and the regular production of sperm in the testes. Testosterone also affects such male characteristics as a large body size, more powerful muscles, less body fat, deeper voice, and the ability of the blood to carry more oxygen (LaCayo 2000). **Castration** (removal of the testes) generally leads to obesity, softer tissues, and a more placid temperament because of reduced testosterone production. Note that both sexes produce male and female hormones. The direction of development is determined by the relative balance of these hormones.

Because every human starts with the potential of becoming either male or female, as a fully differentiated adult each still carries the biological rudiments of the opposite sex. For example, the male has undeveloped nipples on the chest, and the female has a penis-like clitoris. In a few rare individuals (1 to 2 percent) even though the sex chromosomes set the sexual direction, the hormones fail to carry out the process. Such persons have characteristics of both sexes (though neither is fully developed) and are called **intersexual** (the earlier term was hermaphrodite). Such people are rare and should not be confused with transsexuals or transvestites. A **transsexual** is a person who feels psychologically that he/she is actually of the opposite gender, and who may have undergone a sex-change operation. A **transvestite** is a person who gains sexual pleasure from dressing like the opposite sex.

The early adaptability of the tissues that grow into mature sexual organs is quite amazing. Ovaries and vaginas transplanted into castrated male rats within the first 24 hours after birth will grow and function exactly like normal female organs. This plasticity quickly vanishes as the hormones cause further differentiation. Transplants of female organs into male rats more than 3 days old are unsuccessful.

Environmental Contributions

Once a baby is born, society begins to teach the infant its proper gender role and reinforce its sexual identity. In the United States, we name our children according to gender; we give girls pink blankets and boys blue blankets; and at Christmas and birthdays, boys receive masculine toys such as mechanical games, and girls receive feminine toys such as dolls.

In keeping with cultural prescriptions, we exhibit different attitudes toward the two sexes, and we expect different behaviors of each. For example, we generally encourage boys to engage in rough-and-tumble activities and discourage girls from engaging in similar activities. This may, in part, explain men's greater interest in contact sports. Parents provide guidance to help the child assimilate the proper role.

Parents say to boys:

"Be strong."

"Be competitive."

"Be tough."

"Be assertive."

"Be sportive."

"Be a man."

castration
Removal of the testes

intersexual
A person who has biological characteristics of both sexes

transsexual
A person who feels psychologically that he or she is actually of the opposite gender

transvestite
A person who gains sexual pleasure from dressing like the opposite sex

Notice that a girl who has the above behaviors will probably be termed a tomboy. This is much less true today as young women are being encouraged to engage in sports such as soccer and basketball, which are certainly contact sports.

Parents still say to little girls more often than not:

"Be nice."

"Stay clean."

"Be gentle."

"Think of others."

"Be kind."

"Be thoughtful."

"Be a lady."

Notice that a boy who has these behaviors may be termed effeminate or a sissy.

Nothing inherent within the child will give rise to socially sex-appropriate behaviors. Each child must learn—from parents, relatives, teachers, friends, and the media—the appropriate masculine and feminine behaviors for their own culture.

It is easy to see the way in which learning affects gender roles by observing masculine and feminine behaviors in different societies. Dramatic shifts in masculine and feminine behavior over time have been described among a certain group of biological men in the Middle Eastern country of Oman (see "Highlight: The Xanith of Oman").

Gender Differences

Theories about the development of gender roles range from biological determinism, as found among sociobiologists and evolutionary psychologists (Buss 1994, 1995, 1996, 1999; Wright 1994), to straight social learning and identification the-

HIGHLIGHT

The Xanith of Oman

The Xanith are biologically male. They sell themselves in passive, homosexual relationships, but they also are in great demand and earn a good wage as skilled domestic servants. Their dress is distinctive; it resembles the long tunic worn by men, but is made of pastel-colored cloth. Although they retain male names, the Xanith violate all the rules that govern female seclusion. They may speak intimately with women in the street without bringing women's reputations into question. They sit with the segregated women at a wedding, and they may see the bride's unveiled face. They may neither sit nor eat with men in public. Their manners, perfumed bodies, and high-pitched voices make them appear effeminate, even though their bodies have not been changed medically.

Because the feminine gender of the Xanith is socially selected, several possibilities are open to them. Should they wish to become men again, they need only marry and prove them-

selves able to perform heterosexual intercourse with their brides. Some Xanith never choose this path and remain women until they grow old, at which time their anatomical sex places them in the category of "old men." Some actually become women, then men, then women again until old age places them back into the "old man" role.

Because Omani women are off limits to all men but their husbands, the Xanith offer a sexual outlet for the single men in the society. Because Xanith are socially women, to have intercourse with them (as long as the man purchasing their services is the one who penetrates) in no way casts a man in an unfavorable light. His manhood remains unquestioned. The ease with which these transformations can be made clearly points out the strength of the society in determining gender roles.

Source: Adapted from Archer and Lloyd 1985.

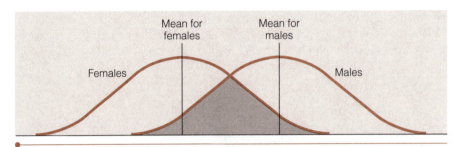

FIGURE 4-1 *Normal Curves of Most Gender Characteristics Show Overlap Whether it is height, weight, reading, math scores, or how far one can throw a baseball, the normal curves for men and women overlap, suggesting that similarities between men and women are more important than differences.*

WHAT DO YOU THINK?

1. Do you think gender role is more learned or more formed by biology? Why?

2. In light of your answer to question 1, do you agree or disagree with those homosexuals who claim that their gender is biologically determined and thus nurture has no role in their behavior?

3. Some suggest that an important method of reducing violence is to raise boys in a manner that limits their robust behaviors, such as chasing and wrestling with each other. Do you think such efforts will successfully reduce male violence? Why or why not?

ories (Tavris and White 2000). The former suggest that gender roles are essentially inborn, or formed by many generations of men and women interacting with the environment in differing ways depending on their biological differentiation. The latter suggest that gender roles are essentially learned. This old argument, about whether environment and learning or genetics and biology determine gender-role behavior, should not be stated in this either/or form. It is the interaction of these two great molders of behavior that determines gender-role behavior.

The "Stereotypical Sex-Role Differences Compared with Research Findings" lists some of the differences between the sexes. The differences based on cultural roles appear to be lessening in America as men and women assume a greater variety of roles (T. Adler 1989; Moses 1991; Rutter and Schwartz 2000; Tavris 1992). Studies done on biological differences support some differences between the sexes (Eagly 1995). In every case, however, the differences between persons of the same sex on a given characteristic are greater than the average differences between the sexes (see Figure 4-1). As one researcher puts it:

The major problem with biological (brain) theories of sex differences is that they deflect attention from the far more substantial evidence for sex similarity. The finding that men and women are more alike in their abilities and brains than different almost never makes the news. Researchers and the public commit the error of focusing on the small differences—usually in the magnitude of a few percentage points—rather than on the fact that the majority of women and men overlap. (Tavris 1992, 54)

Reduction in the use of gender-role stereotypes—or any behavioral stereotypes—is a worthy goal. Furthermore, to argue endlessly over the relative influence of biology versus environment is a waste of energy. Men and women are similar in more ways than they are different. However, "neither women nor men can be considered 'superior' or 'inferior' to the other, any more than a bird's wings can be considered superior or inferior to a fish's fin or a kangaroo's legs" (Buss 1995, 167). Because we know that culture does influence gender roles to a great extent, modifying those roles taught by the culture is certainly possible.

Role Equity

When men and women are freer to choose gender roles for themselves, especially within intimate relationships like marriage, their chances for success and fulfillment increase. If intimate partners can share the decisions and responsibilities

WHAT RESEARCH TELLS US

Stereotypical Sex-Role Differences Compared with Research Findings

Stereotype		Findings
Perceptual Differences		
Men have:	better daylight vision	Mild but in direction of stereotype
	less sensitivity to extreme heat	"
	more sensitivity to extreme cold	"
	faster reaction times	"
	better depth perception	"
	better spatial skills	"
	more ability to rotate three-dimensional objects in their heads	Moderate
Women have:	better night vision	Mild
	more sensitivity to touch in all parts of the body	"
	better hearing, especially in higher ranges	"
	less tolerance of loud sound	"
	better manual dexterity and fine coordination	"
	more ability to read people's emotions in photographs	"

Aggression and Assertiveness

Males are more aggressive and assertive.	Strong consistent differences in physical aggression
Females are less aggressive and assertive.	Milder differences in assertiveness

Dependency

Females are more submissive and dependent.	Weak differences that are more consistent for adults than
Males are more assertive and independent.	children

Emotionality

Females are more emotional and excitable.	Moderate differences on some measures; overall findings
Males are more controlled and less expressive.	inconclusive

Health

Females suffer more depression and phobic reactions.	Moderate differences
Males have more heart disease and are more at risk for disease in general.	Strong differences

Verbal Skills

Females excel in all verbal areas including reading. Males are less verbal and have more problems learning to read.	Moderate differences, especially for children

Math Skills

Males are better in mathematical skills. Females are less interested and do less well in mathematics.	Moderate differences on problem-solving tests, especially after adolescence

SOURCE: This table has been constructed using several sources: Maccoby and Jacklin 1974; Frieze et al. 1978; Goleman 1978; McGuiness and Pribram 1979; Durden-Smith and DeSimone 1982; Archer and Lloyd 1985; Caplan et al. 1985; Kimura 1985; Gallagher 1988; Adler 1989; Hines 1989; Eagly and Wood 1991; Gorman 1992; Kimura 1992; Tavris 1992; Oliver and Hyde 1993; Feingold 1994; Buss 1995; Eagly 1995; Gazzaniga 1995; Hyde and Plant 1995; Shaywitz 1995; Witelsen et al. 1995.

of their relationship in ways that feel right for them, their satisfaction will be greater than if each is forced into stereotyped behaviors that may or may not fit. Free choice should help reduce the feelings of oppression and dissatisfaction that often appear over time in intimate relationships. Reducing gender-role rigidity and stereotyping does not eliminate gender roles nor make the sexes the same, but it does promote freedom to choose a more personally and socially fulfilling lifestyle. Increasing freedom to choose is a proper goal of a free society.

role equity

The roles one fulfills are based on individual strengths and weaknesses, rather than on a set of preordained stereotypical differences between the sexes

Role equity means the roles a person fulfills are based on individual strengths and weaknesses, rather than on a set of preordained stereotypical differences between the sexes. (Granted, there are biologically influenced differences between the sexes.) Equity implies the fair distribution of opportunities and restrictions without regard to gender. For example, one spouse may have more interest or better skills in arithmetic and bookkeeping, so he or she manages all the family finances. If such an arrangement is freely chosen and believed to be fair by both partners, it is then an equitable arrangement.

Equity between the sexes in family life embraces variation—having many relationship models, not just one. True equity between the sexes implies freedom to establish roles within the relationship that accent the unique personality of each partner. Each is free to fulfill his/her own capabilities to the greatest possible degree. In the past, the term **androgyny** described the blending of traits associated with the sexes by society. Popularly, it was taken by some to mean making gender roles identical for men and women. In reality, identical gender roles really make no sense. They overlook real differences between individuals and limit relationships, just as rigid traditional gender roles did in the past.

androgyny

The blending of traits associated with the sexes by society

It is the concept of role equity that really focuses on gender-role transcendence. Role equity in a relationship might mean both the man and the woman fulfilling traditional gender roles, mixing traditional gender roles, or even reversing traditional gender roles. In addition, new roles may be created based on individual strengths. Role equity emphasizes both partners fulfilling as much as possible their own individual interests and skills, yet maintaining an intimate relationship. It is an agreement freely made between partners about the arrangement or role(s) that works best for each of them.

Some people who suggest new gender roles are really only suggesting that we move from rigid, traditional roles to different, but equally rigid, gender roles. For example, some feminists downplay motherhood as demeaning to women. Yet the majority of women become mothers at some point in their lives, thus making the role of mother important to their self-image. Whether a mother assumes the traditional role of motherhood, works full time, or gives much of the mothering role to her husband is up to her and the relationship she has with her partner. To demean motherhood is to substitute another rigid gender role, nonmother, for the traditional role.

Gender roles can never be exactly equal, because individuals differ. In real life, enduring marital and family bonds are asymmetrical and often out of balance in a variety of ways. The ability to negotiate and live with asymmetries, imbalances, and at times even inequities is necessary in any lasting relationship (Sprey 2000, 25). Balancing roles and negotiation of roles are the foundation blocks of lasting, intimate relationships. Role equity allows the couple to arrive at gender roles that are freely negotiated to bring out the best in each person as often as possible, given their life circumstances.

It is true that when roles are tightly prescribed by society and few, if any, deviate from them, people tend to feel safe and secure. Some people feel threatened by too many alternatives. But American society encompasses a great deal of diversity. People may find that their neighbors have different lifestyles. Children may point to imperfections in their parents' marital life and question whether people need to get married. Couples may find the mass media criticizing a relationship they have never seriously questioned. They may read articles praising alternative living arrangements that they were taught are immoral. Such experiences lead to confusion, insecurity, anxiety, resentment, and sometimes an even firmer commitment to the status quo. Indeed, the extreme attacks on the traditional American marriage have undercut a thoughtful and constructive approach to change within the family roles. Most American men and women want to improve their marriage, do a better job in the work world, better cope with the stress of fast-paced modern life, and/or ease parental burdens. To offer only criticism of traditional gender roles is of little help in answering such questions.

To think of intimate relationships in terms of role equity opens each relationship and each individual within a relationship to the highest personal fulfillment. Whether the roles chosen are contrary to past traditions or incorporate those traditions is immaterial, as long as the roles are freely chosen, felt to be fair and acceptable by each partner, and increase relationship satisfaction.

Traditional Gender Roles

The depth of most people's belief in gender-role stereotypes is often overlooked. Many people simply take the various traditional gender-role behaviors for granted. A husband may have a need to express warmth and closeness to male colleagues, yet the masculine stereotype forbids him to do so. A wife may be a natural leader, yet she may suppress her leadership behavior because the stereotype says it is not feminine.

Traditional roles historically emphasize the woman's childbearing functions and the man's greater physical strength and need to defend his family. A male's status today is still partially determined by his physical prowess, especially in sports during the school years. The traditional role stresses masculine dominance in the society and also within the family. The traditional role also allows men more sexual freedom, while severely limiting women's sexuality (double standard).

As we discuss traditional roles for men and women in American society, it must be understood that these roles also vary. Different ethnic groups, age groups, social classes, and regions have variations of the traditional roles. For example, African American men may define their family roles differently because of racism's past impact on their lives. A single, African American mother may define her role as being far more independent than would a White, middle-class, married mother. Hispanic women are not as supportive of abortion rights as White women because of their Catholicism, but they are highly supportive of equal employment and day care because they generally have more children than White women and live at an economic level that requires work outside the home.

The traditional feminine role is essentially the complement of the masculine role. Man is active, so woman is passive and submissive. Wives help their husbands run the home and family, and work outside the home only if necessary. Much of their personal identity is derived from their husbands. Women are the source of stability, strength, and most of the love and affection within the home.

WHAT DO YOU THINK?

1. Why do you think the sexual double standard lasted for so many years?

2. Do you think the sexual double standard still exists? Why or why not?

3. Do you think women and men should have identical sexual behavior standards? Why or why not?

The traditional role for women is found throughout the historical American society and in almost all other societies. For example, in the Koran (A.D. 630), Muhammad wrote:

Men have authority over women because Allah has made one superior to the other, and because they spend their wealth to maintain them. Good women are obedient. They guard their unseen parts because Allah has guarded them. As for those from whom you fear disobedience, admonish them and send them to beds apart and beat them. Then if they obey you, take no further action against them.

The main reason that the traditional role for women has been found in almost all cultures and throughout history is that women bear the children. Even today in modern America, this fact tends to color thinking about women. For example, many businesses still regard a woman as a temporary worker, who works until she decides to bear children. Even though most American mothers are now in the work world, the fact that they still bear most of the responsibility for their children means that they will be absent from work more often than their partners (J. Collins and Thornberry 1989). Such facts, and other influences, combine to hold down women's pay compared with men's. Women tend to follow their husbands if they change jobs because the husbands can earn more. This fact also fosters the perception that women are less reliable long-term employees.

In sociological terms, the traditional masculine character traits are those labeled *instrumental*. Such traits enable one to accomplish tasks and goals. Aggressiveness, self-confidence, adventurousness, and dominance are examples. Feminine character traits are those labeled *expressive*. They include gentleness, expressiveness, lovingness, and supportiveness. Notice how the traits often complement one another. For example, the man exhibits aggressiveness and the woman provides the opposite, gentleness. Evolutionary psychologists suggest that these differences stem from the fact that women bear children. Therefore, women must exhibit such traits as gentleness and lovingness to be successful mothers. They must also be able to make a lasting relationship with a man (preferably the father) so that they and their children are protected and supported. Men, on the other hand, have historically been called upon to support and defend the mother and their children. Obviously, self-confidence and aggressiveness help men succeed in this role (Buss 1995).

We find relative agreement between American men and women when they assess the advantages and disadvantages of their gender roles. Essentially, the perceived advantages of one sex are the disadvantages of the other. Masculine disadvantages consist of obligations and responsibilities, with a few prohibitions. The disadvantages of the female role arise mainly from prohibitions, with a few obligations. Thus, women complain about what they can't do, men about what they must do. Such a description is no longer as valid as it once was. Men and women generally list similar ideal characteristics for both sexes, which suggests that stereotypical gender roles are weakening. Also, as more women choose nontraditional occupational roles, masculine and feminine stereotypical characteristics continue to be reduced.

Gender roles are so deeply embedded within society that they are probably impossible to escape completely. The conflicts over changing masculine and feminine roles and the real world of our culture cause many people to feel confusion, ambiguity, and frustration. For example, women are told to expand themselves by seeking careers outside the home, but are made to feel guilty if they do not devote themselves to their families. Many women who opted for careers in the 1970s and 1980s are now having children at a later age, so they abandon their

WHAT DO YOU THINK?

1. What do you think, in general, are women's best characteristics? What are their worst?

2. What do you think, in general, are men's best characteristics? What are their worst?

3. How differently do you think men and women might answer the above questions?

careers, at least temporarily, to fulfill the motherhood role. This can be seen in the large number of women having their first babies after age 40. The birthrate for women aged 40–44 years was 47 percent higher in 2001 than it was in 1990. For women aged 45–49 years, the birthrate has more than doubled between 1990 and 2001 (Hamilton et al. 2003, 5–6). Women who have concentrated on their careers but find the biological clock ticking down and so decide to have a child are often caught in a catch-22, or no-win, situation. They must have a child while they still can biologically, but that usually means putting their career on hold, at least for a short time.

Fathering teaches children gender roles as does mothering.

Changing Male and Female Roles

If Americans do succeed in becoming freer from past gender-role stereotypes, much credit must be given to the women of the United States. The women's movement has focused attention on gender inequalities and has energized efforts to reduce these inequalities. It has brought about profound changes in the relationships between men and women and within the American family.

As women change their roles, the masculine role also changes. Women's liberation, to the extent that it has succeeded, has also meant men's liberation. For example, as more women have entered the workforce, more families have become two-paycheck families. No longer is the husband solely responsible for fulfilling the role of breadwinner for the family. Shared economic responsibility means more freedom for men and, thus, more time for men to spend with the family. Economic participation by women means that they can become more economically independent.

It is possible that a male revolt against the economic burdens of being the sole source of family support actually predated the feminist movement. Postponement of marriage, increasing divorce, fewer children, a decreasing percentage of men in the workforce, and failure to pay child support after divorce are all cited as evidence of this revolt by men against the breadwinner role. However, it was the women's movement that focused public attention on stereotypical gender roles and the harm that such stereotypes can do to both men and women.

Not all the changes fostered by the women's movement and its allies have been readily accepted. The failure of the Equal Rights Amendment (ERA) to win ratification, the strength of the pro-life attack on abortion, and the antifeminist movement all bear witness to the conflicts that are stirred up when traditional gender roles are challenged.

Women and the Law

The women's movement has fostered many new laws granting women increased rights. American laws are extremely complicated and have often worked to the detriment of both sexes. (It is important to remember that most laws governing marriage and gender roles are state laws and, hence, do not apply throughout the country.)

In the past, laws have considered women to be weaker and less responsible than men and, therefore, in need of protection. For example, in one state, a married

WHAT DO YOU THINK?

1. If you are a female, do you consider yourself a feminist? Why or why not?

2. If you are a male, do you support the feminist movement? Why or why not?

3. What do you think are the greatest gains for women made by the feminist movement?

4. What effect has the feminist movement had on men? Any gains or losses?

5. What effect has the feminist movement had on intimate relationships between men and women?

The Dani of Irian Jaya (Indonesian New Guinea)

I t wasn't until the 1930s that the Dani people, living in the Highlands of Irian Jaya, were discovered by the outside world. Often described as a Stone Age culture, the people today live much as they did thousands of years ago. Their expert gardening skills are now helped by steel rather than stone tools, but their daily behavior, families, and villages carry on almost as they always have.

Although Dani now sometimes wear Western cloths brought by missionaries, in the more remote villages the children are naked, and single women wear only grass skirts. After marriage, their skirts are made of strings of seeds worn just below the abdomen. The men are naked except for a gourd covering their penis, and perhaps some decorations such as shell beads or feathers.

The Dani people represent the majority of cultures around the world, with clearly defined gender roles that match closely with what Americans think of as traditional roles for men and women. Essentially, the man is in control of his family and of his village group. The woman tends the home, the children, and the gardening. Approximately 85 percent of their diet is made up of many kinds of potatoes, especially sweet potatoes. The men usually do the heavy work of clearing the land so that the women may plant and harvest crops. The men cut the wood; build the family home, fences, and other needed village buildings; and make the tools. They also defend their families, villages, and others in the extended group.

Because the women spend much of their time working in the fields, the daytime care of small children is usually undertaken by young girls (as young as 6 years of age). The young children have little or no adult supervision during the day. Thus,

This woman is carrying a number of hand-woven nets (Noken) in which are sweet potatoes and leaf vegetables she has harvested, and, perhaps, she may also be carrying an infant on her back.

the young girls assume the major responsibility for child rearing at an early age, and thereby learn their roles as mother and family caretaker. The Dani boys are busy playing with bows and arrows, learning to make and throw spears, and physically competing with each other for power and control, thereby learning their adult roles.

The sexes are rigidly divided; the men sleep in the men's grass house (Honnay), while the women live with young children and pigs in the women's long houses (Wewumah). The pigs are an im-

A village meeting.

portant part of village life. Before money was introduced by missionaries, the pigs and shells were important sources of wealth. The pigs were so important that care of them by the women was a central function of village life. If necessary, a woman might even nurse a weak piglet hoping that her milk would bring it back to health. A man must still buy a wife by offering her family the proper number of pigs and other gifts, according to the woman's status and desirability.

Dani mothers nurse their children for up to 5 years. During this time a man may not sleep with his wife. This practice leads to polygamy, but only well-off men such as chiefs may have several wives. Usually, sexual intercourse is done somewhat privately in the garden.

Leisure time is generally spent with members of an individual's own sex. The Dani men are very concerned with ceremonial dress for sing-sings (ceremonial gatherings for weddings, funerals, and pig exchanges), and even the strongest warriors can sit for hours making feather headdresses and shell necklaces.

Gender roles for the Dani are unquestioned, and village life revolves around men and women doing what has been expected of them since the earliest times.

A clothing factory.

A Dani warrior.

Town meeting.

woman had to use her husband's home as her legal address, and until a recent change in the law, she could not buy or sell stocks or property unless her husband consented and, thus, accepted responsibility for her actions.

Since the 1970s, many laws have been enacted guaranteeing both women and men more protection against unfair discrimination and unequal treatment. Laws concerning living together, divorce, child custody, and support have been modified. New laws covering rape and sexual harassment in the workplace have been passed. Other laws, such as equalizing benefits from Social Security, have been passed that narrow the differences in treatment of men and women.

Changing a law to make the sexes more equal is a worthy goal, yet new inequities may result. For example, divorce laws that require equal division of property may leave older women who have little work experience worse off than their former husbands. Indeed, the growing number of female-headed, single-parent families falling beneath the poverty level has resulted in what many researchers are now calling the *feminization of poverty*. The divorced woman may receive half of the property and still be unable to support herself. Rehabilitative alimony and the Displaced Homemakers' Relief Acts are legal responses to these new problems created by changing laws. The idea is that the former husband should provide financial help to his ex-wife so that she can retrain herself to become self-supporting.

Many changes are making work and economic participation fairer to women. Some state statutes now make it an unfair employment practice to discriminate on the basis of pregnancy, childbirth, or medically related conditions. Such laws also require employers who provide disability insurance for their employees to include disability for normal pregnancy as a benefit.

In the criminal courts, women have successfully strengthened rape laws. Punishments for rapists have become more stringent. More important, many states have eliminated the humiliating defense tactic of cross-examining the victim about her previous sexual conduct. In 1977, Oregon passed a statute that made it a criminal offense for a husband to rape his wife, and some other states have followed suit. This is a complete departure from earlier marital law, where there was no such thing as rape within marriage.

Sexual harassment in the workplace is another area in which the law is helping women (and some men). Women have won numerous cases against employers who used the woman's need for work to obtain sexual favors. A few men have also won such cases.

Sometimes the enthusiasm that gets a new law passed leads to the extreme. For example, sexual harassment has been broadened far beyond the traditional

HIGHLIGHT

"Firsts" for Women in the Past Three Decades

- In 1972, Sally Priesand was ordained the first female rabbi. Since then, many others have been ordained.

- In 1977, Dr. Olga Jonasson was named the first female head of a major surgical department at Chicago's Cook County Hospital.

- Lauded for her eloquence during the 1974 Watergate hearings, Texas Congresswoman Barbara Jordan appeared at the Democratic National Convention 2 years later and became the first woman ever to deliver the keynote address.

- Janet Guthrie is the first woman to have driven in the Indianapolis 500. Guthrie took part in the 1977, 1978, and 1979 Memorial Day classics and had her best race in 1978, when she finished in ninth place.

- In 1981, Sandra Day O'Connor became the first female justice on the U.S. Supreme Court. Said O'Connor, "Women have a great deal of stamina and strength. It is possible to plan both a family and a career and to enjoy success at both." Ruth Bader Ginsburg joined O'Connor on the Supreme Court in 1993.

- In 1984, Congresswoman Geraldine Ferraro became the first female vice presidential candidate of a major party when she was picked as running mate to Walter Mondale, the Democratic candidate.

- In 1985, Sally K. Ride became the first American female astronaut in space. "I was not an active participant in women's liberation," Ride once said, "but my career at Stanford

[where she earned a Ph.D. in physics] and my selection as an astronaut would not have happened without the women's movement."

- In 1986, Nellie Speerstra became NATO's first female fighter pilot.

- During the Persian Gulf War in 1991, women served in many military capacities. Major Marie Rossi appeared on television during the war, indicating that national defense had become sex-blind; she later became the best known of the women killed in combat.

- In 1993, the number of women in state legislatures was 1,516—up from 424 in 1973.

- In 1994, the U.S. Navy placed women on combat ships (*Time* 1995) and the U.S. Air Force allowed women to become combat pilots.

- In 1995, the government administration comprised approximately 40 percent women, including such offices as heads of Health and Human Services and the Environmental Protection Agency.

- In 1999, women enrolled in college outnumbered men (approximately 56 percent of college enrollees were women). In addition, they outnumbered men in such traditionally male fields as law and medicine (U.S. Department of Education 1998, 1999, 196).

- In 2005, the first Black woman became secretary of state.

interpretation, which was the seeking of sexual favors in return for promotion or other job advantages. A recent case charged that the presence of a *Playboy* magazine in a fire station was sexual harassment because there was a woman assigned to the station. The court found against this because the magazine was read privately by others at the station. There is debate about just how far supposedly creating a hostile workplace can go before it will clash with the First Amendment right of free speech. Is telling a sexist joke at work the creation of a hostile workplace?

Gender-Role Stereotypes

Intimate relationships built only on stereotypical gender roles may limit freedom and impair the growth of both individuals and their relationships. Although many married couples function well by fulfilling traditional gender roles (that is, husband-provider, wife-homemaker), such roles may limit a couple who adheres to them too rigidly. A woman who limits herself to a child-centered, home-centered, husband-centered life may come to feel isolated and restricted. She may have to repress other aspects of herself to conform to this role. Her husband may find himself married to someone who depends on him for the fulfillment of all her needs, for all decisions, and so on. What began as an ego trip (the helpless, idolizing wife and the strong, responsible husband) can become a heavy burden for the husband. Few honest men today would deny that such overwhelming responsibility is unpleasant and restricts their own lives.

A husband may also be limited by conforming to stereotypical gender roles. In our competitive society, where success is measured by individual productivity and achievement, a husband must often manage two marriages: one to his career and the other to his wife and family. When conflict arises between the two, he may have to place his work first if he is to be economically successful. This can be difficult for his wife to accept, especially if she is family-oriented. She may feel cheated and rebuffed by her husband because so much of his energy is expended outside the family. She wants him to be successful, but by encouraging his efforts outside the family, she also loses some of his interest and presence in the family. Many men locked into the traditional provider role early in their lives complain later that their emphasis on work deprived them of family life; they missed their children growing up and lost out on the benefits of close emotional family ties.

As long as the economic system is partial to men, escaping the provider role will be difficult for them. The unfortunate antagonism between male economic success and family life is difficult to resolve. As increasing numbers of women become career-oriented, family and work roles also come into conflict for them. Indeed, a wife may have to put even more energy into her career than her husband puts into his to overcome society's remaining resistance to women seeking high-level careers.

Most of our discussion thus far has focused on the middle-class family. In many ways, the traditional working-class family has even more rigidly stereotyped marital roles. The wife is expected to be in the home most of the time unless she is working. Her work is often not considered a career, but rather, simple

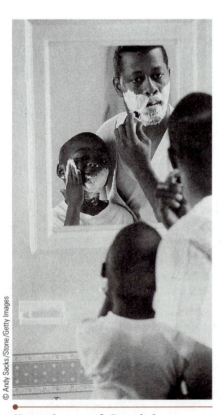

© Andy Sacks/Stone/Getty Images

How to be a man? Copy dad.

earning of survival money. Her husband may find much of his social life in relations with his male friends. Thus, sharing activities and joint family participation appears less important than in middle-class families. Segregated roles may leave the husband and the wife with little to say to each other. There may also be an advantage—each partner may be given a sense of competence and independence.

The working-class husband may be threatened when his wife works because she will contribute a much higher proportion of the family income than her middle-class counterpart. By maintaining rigid traditional roles in marriage, the husband is better able to maintain control and a sense of pride and importance. Working wives in these families often express job satisfaction, while their husbands do not. The wife's working broadens her world and opens choices to her. Often, her husband's own lack of skills entraps him in a low-pay, low-prestige job without a future. He may feel he is in a dead-end situation and may blame his family responsibilities for his unfavorable circumstances.

On the surface, a marriage based on stereotypical gender roles may give a couple security and reduce conflict because both partners know their place and role. Yet, the rigidity of the roles can make the marriage more fragile and less resistant to new strains and pressures.

Growing Up in Two Cultures   The U.S. population is made up of many immigrant groups that often have various gender ideas about what it is to be a man or woman. For example, Fatima's parents are native-born Saudi Arabians who migrated to the United Sates when Fatima was 10 years old. They migrated to the United States to improve business opportunities, but chose to maintain Saudi social and cultural traditions. Both parents came from middle-class Saudi families. Their marriage had been arranged by their traditional Saudi parents.

Because Fatima's parents found American children undisciplined and often unruly, they decided to homeschool Fatima. However, as Fatima grew into adolescence, it became increasingly difficult to keep her isolated from the popular American culture. She was allowed to visit with some neighborhood children whose parents had become friends with her family. She listened to American pop music and saw American life depicted on MTV. The older she became, the more desirous she became to have at least some of the freedoms that young people are accorded in many American families.

Fatima finally came into direct conflict with her parents as her interest in boys emerged. She heard about dating from friends and wished that she, too, could join her friends on mixed-sex outings. Her parents made it clear that they would certainly allow no interaction with men because that would ruin her reputation and, hence, her ability to be wed to the properly acceptable mate.

It is clear that for many immigrant children growing up in America, there will be conflict between their parents' ideas about proper gender behavior and ideas of the general American society. The first-generation immigrant child growing up in the United States will usually be trapped in this conflict.

For a relationship to foster growth, each mate must be committed to the ideas of seeking equity in the relationship and to communicating openly any feelings of inequity. This philosophy requires that the couple be willing to experiment and change if first solutions fail.

This kind of exploring can lead to greater satisfaction than limiting oneself to prescribed roles that may or may not fit with one's interests and skills. With true equity, each couple has to sort out responsibilities in agreed-upon ways that maintain love and respect between them. "Making Decisions: The Couple's Inventory"

WHAT DO YOU THINK?

1. How would you handle family values that differ from the values in a friend's home?

2. What if you were Fatima's parents? What would be your conflicts, living in a new culture with different family values?

3. Do you have a friend who faced these kinds of value conflicts? Describe.

(in Chapter 7) can help you explore your own and your partner's gender-role attitudes.

Even with equity, a couple may be encumbered by the stereotypical gender roles held by society. For example, Consuela and Hector may agree that Consuela will handle the investments, but may find that the stockbroker always asks to speak with Hector. Nevertheless, each couple can work to realize more freedom within marriage. A society that has gone beyond narrow ideas of femininity and masculinity to the ideal of the fully functioning person offers the widest possible range of choices for its members. The real issue is the establishment of a society in which men and women have equal opportunity to fulfill their hopes and dreams, unhampered by oppressive stereotypes.

The Movement toward Gender Equality

When the ERA (Equal Rights Amendment) died in 1982, many people felt that the women's liberation movement had accomplished its goals and the amendment was unneeded. The amendment read:

Equality of rights under the law shall not be denied or abridged by any state on account of sex. The Congress shall have the power to enforce, by appropriate legislation, the provisions of this article.

There have, indeed, been many gains for women during the past 30 years. Professions long closed to women, such as law and medicine, are now open. More women than men earn college degrees. Women—married and unmarried, with children and without—are entering the workforce in vast numbers. Pay inequality between the sexes is decreasing. Maternity and parental leaves no longer cost women their jobs. Flexible work schedules help parents better cope with family and parental responsibilities. Good child-care facilities are increasingly recognized as important to the working family. Women's school sports are gaining equality with men's sports. Relationships between men and women are becoming more equitable.

Many of these changes started with Betty Friedan, whose book *The Feminine Mystique* (1963) helped ignite the modern women's movement. Although this is not the place to present a complete history of the women's rights movements, suffice it to say that the early movements (1850–1925) were concerned with abolition, temperance, and voting rights. In the early years of the modern movement, commencing in the 1960s, such groups as radical and lesbian feminists were strident and adversarial, often taking an "against all males" stance. These militant feminists described in vitriolic terms a gender war in which men worked to keep women cowering and submissive, in which lesbian rights seemed to take precedence over caring for children, in which a court decision allowing women to go topless on New York subways was hailed as a great victory for women, and in which there was more pro-choice support than support for better child care.

The belittling of motherhood by radical feminists has also turned away many American women (Crittenden 1999; Graglia 1998; Shalit 1999). Sylvia Ann Hewlett (1987, 1991) describes how many feminists attacked her and other women who wanted to gain equality in the work world yet share the joys of family and motherhood at the same time. She suggests that a movement that looks away from the central fact of most women's lives—motherhood—will not win widespread support. Today, probably because of this early militancy, some women

HIGHLIGHT

Men as the Oppressed Sex

The feminist movement's central proposition has been that women have historically been oppressed by a patriarchal society, that is, men. "Man the enemy" has been a subtle message given to American women by the more radical feminist faction. Warren Farrell (1993, 1994), Aaron Kipnis (1991), and Christina Sommers (2000), in their books about men, ask: if all men are chauvinist pigs and oppressors of women, as depicted by radical feminists, how can the following be true?

- Women are often spoken of as one of the minorities, yet they outnumber men by almost 10 million.

- Women live, on the average, 6 years longer than men.

- Men suffer from stress-related problems such as stomach ulcers in far greater numbers than women, although stress-related problems in women are increasing as they more frequently enter traditional men's occupations.

- Males have higher age-adjusted death rates (approximately 20 percent) than females (NCHS 2004a).

- There are 6.8 work-related deaths for males per 100,000 population compared to only 0.7 deaths for females (NCHS 2004a).

- During past wars, men have been drafted into military service, whereas women have not.

- Men are only slightly less likely to die from prostate cancer than women from breast cancer, but breast cancer research receives approximately six times more funding than prostate cancer research.

- Men are twice as likely as women to be victims of violent crime, even counting rape. Men are also three times more likely to be murdered.

- Historically, approximately as many wives kill their husbands as vice versa.

- Men convicted of murder are 20 times more likely to receive the death penalty than women convicted of murder.

avoid being labeled a feminist, although they say that they support many of the feminist goals such as better pay, better child care, and better jobs (Kaminer 1993; Wood 2001).

In 1981, Friedan wrote *The Second Stage*, a critique of the movement that pointed out the need to move in new, more positive directions. She suggested that it was time for the women's movement to abandon strident militancy and work to consolidate women's gains. It was time to move away from the negative messages and help both men and women move toward more caring, respectful, and supportive relationships. Julia Wood suggests that there is now a third wave of the women's movement that does have as one of its defining features an explicit interest in affirming and improving connections between women and men (Wood 2001, 85).

We have suggested the concept of equity between sexes as a way to improve relationships between the sexes. To try and say that men and women are equal in all ways overlooks the advantageous, and disadvantageous, differences between the sexes. But to say that one sex is better than the other because of those differences is stereotyping at its worst. Masculinity and femininity are not polar opposites; indeed, most of us are both masculine and feminine, depending on the situation. This flexibility has proven to be a good thing. People who are rigidly masculine or feminine across all situations appear to be less healthy, mentally and physically, than people who can adopt the best qualities of both sexes. Under some conditions, the qualities we label feminine are good for both sexes; under other conditions, the qualities we label masculine are good for both sexes. In the family, for example, the positive qualities of traditional femininity—compassion, nurturance, warmth, and so on—are associated with marital satisfaction in both sexes. Naturally, everyone wants a spouse who is affectionate and caring, whether he/she is a man or a woman. At work, the positive qualities of masculinity—assertiveness, competence, and self-confidence—are associated with job satisfaction and self-esteem in both sexes. Naturally, everyone wants a co-worker who is self-confident and capable (Tavris 1992, 293).

Personal liberty and equality cannot be won by one sex at the expense of the other. They can be won for each sex only by both sexes working in partnership. The women's movement must be a women's and men's movement if a full measure of liberty and equality of sexes is to be realized (Carbone 1999; Emery 1999; Friedan 1981; Hewlett 1987, 1991; Roiphe 1993; Sommers 1994; Tavris 1992; Wolf 1993).

It should be noted that the discussion of the women's movement often treats it as a single entity. In reality, like all social movements, the women's movement is not monolithic. It is clear that many philosophies are encompassed within the modern women's movement, ranging from conciliatory philosophies to extreme philosophies as espoused by lesbian feminists who declare that only women who don't orient their lives around men can be truly free. Being a feminist need not be in conflict with motherhood and being feminine (Wood 2001, 5).

The women's movement has helped to bring increased equity between females and males. It has helped to destroy many of the stereotypes that held women back from realizing their full potentials. It has also helped men by encouraging the breakdown of the extreme masculine or macho image that disallows men to share the emotional side of themselves. It has encouraged the participation of men in parenting. For example, as late as the 1960s, many hospitals banned fathers from participating in the birth of their children. Today, most hospitals encourage father participation in the birth process. Studies have found that participating fathers bond earlier and more closely with their newborns.

Thus, each change in the feminine role also brings about changes in the masculine role. Overall, the changes wrought by the women's movement have been positive for both men and women. The gains toward equality for women are well discussed—better education, better jobs, better pay, more freedom of choice, more family support, and the like—but with these gains have come some losses.

The losses appear to revolve around sexuality, childbearing, and child rearing. Hand in hand with the women's movement came the sexual revolution. "The pill" brought reliable contraception to women and allowed couples to separate sexual intercourse from procreation. The gains for women were significant. A woman could reliably plan pregnancy, and thereby participate in the working world on different terms. She found new sexual pleasure as the fear of unwanted pregnancy was reduced. She became sexually more equal as the double standard broke down. Above all, she gained the freedom to say yes to her own sexuality.

Women's sexual freedom came so fast that within a few years, young women were as embarrassed to admit virginity as they had formerly been to admit to premarital sexual relations. Men no longer needed to promise love and commitment to get sex. The female who refused sex wasn't liberated, free, or "with it." To say no to sex was to say no to the women's movement, to modern society. Hence, many young women, having won the right to say yes to sex, found they had lost the right to say no. And the sex they were saying yes to was often without love and commitment.

The combination of the sexual revolution and the women's movement had the practical result of sexually liberating men more than women. Today's young women do not complain about the double standard or about feeling sexually. Instead, they complain about not finding men who care and who will make commitments, respect them, and share responsibility for birth control, pregnancy, and their children.

And what of unwanted premarital pregnancy? Men have long been morally obligated to provide for children they help conceive. The shotgun wedding traditionally forced the man to provide if he tried to dodge his obligation. Today the term shotgun wedding sounds prehistoric.

If a woman gets pregnant, the man, who 30 years ago might have married her, may feel today that he is gallant if he splits the cost of an abortion. He is legally obligated to support the child, yet if he will not acknowledge the child as his, a woman can do little aside from initiating a paternity suit, which is costly, painful, and not always successful.

Even the expectation of child support (no guarantee itself, today) is far from the legal norm that holds both parents responsible for supporting their children. Moreover, many divorced women who are awarded child support from the fathers receive money only spasmodically or not at all.

Thus, a combination of the pill, free-choice abortion, the sexual revolution, the women's movement, and nonenforced child-support laws have freed (liberated?) American men from responsibility for their sexual behavior. Of course, failure to assume responsibility may leave a man isolated and alienated from what could be a loving family.

Single-parent, female-headed families fall disproportionately below the poverty line and cost society a disproportionate share of public monies via the various welfare programs that help them survive. Poverty is being feminized as the responsibility for children has shifted increasingly to women and away from men. Female-headed families represent over half of all poor families. Examined in these terms, the movement toward sexual equality for women appears to have freed the American man, not the American woman. When we look at the American woman only in terms of sexuality, childbearing, and child rearing, a good case can be made that she has lost status and freedom, even though she has gained them in many other areas of her life.

It seems that it is now time to take stock. We need to reevaluate past gender roles so that we can keep those parts that are positive (rather than labeling all traditional roles as bad and throwing them out). We also need to evaluate honestly some of the changes already made so that we can keep the beneficial ones and rectify those that have proved to be damaging. It is now time for *people liberation*. Only when men and women work together in mutual respect and with love, care, and commitment to each other will the American family again be strong. This does not mean that it will be the American family of old or some rigidly idealized new family form of the future. It should be any family form in which men and women can realize their individual abilities and work together to rear children who become responsible adults, willing to make commitments and assume the responsibilities that make a society great.

Summary

1. *Equity between the sexes, not sameness, is the goal to seek*. Yet, achieving such a goal is not easy. First, an individual is born male or female. Second, the individual learns from society the roles (masculinity and femininity) that go with his/her gender.

2. *Two important stumbling blocks to people's liberation that are caused by stereotypical gender roles are the economic deprivation of women and laws that discriminate between the sexes.* Until women earn the same wages as men for the same work, it will be difficult for couples to change the traditional roles of "man the provider" and "woman the homemaker." In addition, many kinds of discrimination are built into our legal system.

3. *The elimination of gender-role stereotypes from society means that couples are free to establish the most satisfying relationships they can.* It means that they can select the gender roles that best fit them. It also means freedom of choice within the marriage. The best marital roles are those that fit the individual best.

4. *As positive as the movement toward gender equality has been, it has also caused some losses, especially for women.* For women, these losses center on sexuality, pregnancy, and child rearing. Both sexes have experienced the loss of commitment, security, and stability within intimate relationships. It is hoped that such losses are transitional.

Resources on the Internet

 Companion Website for *Human Intimacy: Marriage, the Family, and Its Meaning,* Tenth Edition

http://sociology.wadsworth.com/cox10e/

Gain an even better grasp on this chapter by going to the companion website to take one of the tutorial quizzes, use the flash cards to master key terms, or check out the many other study aids you'll find there. You will also find special features such as GSS data, Sociology Online, and Census 2000 information that will put data and resources at your fingertips to help you with that special project or to help you as you do some research on your own.

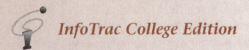 ***InfoTrac College Edition***

http://www.infotrac-college.com/wadsworth/

You can access reliable resources anytime, anywhere, with InfoTrac College Edition, the online library. This fully searchable database offers more than 20 years' worth of full-text articles (not abstracts) from almost 5,000 diverse sources, such as top academic journals, newsletters, and up-to-the-minute periodicals, including *Time, Newsweek, Science, Forbes, The New York Times,* and *USA Today.* You can conduct electronic key word searches using key terms from this chapter to supplement your reading and learning experience. To aid in your search and to gain useful tips, see the Student Guide to InfoTrac College Edition, which you can access through the companion website for this book.

Can We Create Gender-Neutral Children?

YES

The Xanith of Oman (page 97) indicates that various gender behaviors can be changed. Will Roscoe's study (2002) of berdaches (men who adopt woman's dress and do woman's work, and the opposite in far fewer cases) found that, in nearly 150 North American Indian tribes, gender is not necessarily tied to an individual's biological sex. One or two percent of children are born without clear sexual identity, sometimes having partially developed organs of both sexes. Obviously, without a clear sexual identity, gender identity will probably be unclear. During the 1960s, the idea of blending gender traits associated with each sex to create an androgynous person was briefly popular. In fact, some couples attempted to raise their children in a gender-free manner (Bem 1998) by trying to eradicate all references to gender differences between the sexes. Theoretically, it might be possible to raise a totally androgynous person if all cultural references to gender difference could be controlled. If this were done, men and women might understand each other better, get along better, and be culturally equal. Conflict between the sexes might then disappear.

NO

It is clear that, except for those few people born with undeveloped organs of both sexes, an individual's biological sex is the foundation of gender differences. It is equally true that gender behavior is learned and thus may vary within differing cultures. But to find a culture that does not recognize biological sex differences and thus creates an androgynous population, is impossible, both historically and today. Socialization without regard to gender is impossible because males and females respond differently to the same socialization. J. Richard Udrey (1994) lists the following clinical syndromes of hormonal anomaly to show how the hormones that direct sexual development also, in part, direct gender behavior. Under normal circumstances, biology determines an individual's sex and together with an individual's culture, gender identity is created.

1. Human females exposed fetally to abnormally high levels of androgens show distinctly masculinized behavior beginning in childhood and extending through adulthood.

2. Girls who, because of a genetic anomaly, lack all sex hormones (even female sex hormones) grow up unusually feminine.

3. Females exposed as fetuses to physician-administered androgenic hormones for the mother's therapy show masculinized behavior in childhood, even though they show no masculinization of anatomy.

4. Women with high adult androgen levels show masculine-skewed behavior, as compared to women with low androgen levels.

What Do You Think?

1. Do you think it is possible to create a gender-neutral population?

2. If such a population could be created, would you want to be a member of such a society? Why or why not?

3. Do you see any advantages (disadvantages) to such a gender-free society?

Communications in Intimate Relationships

Chapter

5

Questions to reflect upon as you read this chapter:

- What are the foundation blocks of successful communication?
- Describe the five skills necessary to successful communication.
- What is meant by nonverbal communication?
- Can you list some basic differences in the way men and women communicate?

I see communication as a huge umbrella that covers and affects all that goes on between human beings. Once a human being arrives on earth, communication is the largest single factor determining what kinds of relationships he/she makes with others and what happens to him/her in the world. VIRGINIA SATIR

"You need to learn how to communicate better and how to fight fairly." This advice seems to be given to couples in conflict more often than any other. The suggestion is absolutely correct, but the question remains, "How does one learn to communicate better and fight fairly?"

communication

The sending and receiving of messages, intentional and unintentional, verbal and nonverbal

Communication is the sending and receiving of messages, intentional and unintentional, verbal and nonverbal. Communication is basic to every human relationship. Without it, there could be no relationship. The more intimate the relationship, the more important high-quality communication becomes. Compared with unhappy couples, successful couples spend more time communicating about their relationship and personal feelings. They use humor and seem to laugh more often at themselves and their problems (Gottman and Notarius 2000; Gottman et al. 1998, 17; Noller and Fitzpatrick 1991). They compliment and express appreciation for each other more often. Good communication skills are always evident in strong families (Mackey and O'Brien 1995; Wallerstein and Blakeslee 1995). Communicating and listening appear to be the primary traits of healthy families. This finding is echoed throughout the research on strong, healthy, successful families (Alford-Cooper 1998; Gottman and Notarius 2000; Gottman et al. 1998; Stinnett 1997; F. Walsh 1998).

In the past, communication skills were less important because families were largely developed around gender roles, as was discussed in Chapter 4. The definition of a good wife and a successful husband was clear. A good wife, a successful husband, and a strong family resulted when all family members, including children, fulfilled their clearly specified roles. Fulfilling clearly defined roles spoke for itself; thus, communication was less important.

Today, gender roles are no longer clear-cut. The family's major function has become relational rather than role-fulfilling. The meeting of emotional needs in families has become more important than the meeting of physical needs. Because emotional needs are so individual, they are much harder to define and understand than physical needs. A proper role for a family member cannot be easily prescribed when that role is judged by the nebulous standard of "meeting emotional needs."

Stating that meeting emotional needs is more important than meeting physical needs in the modern American family assumes that the family is affluent enough to meet their physical needs. For families in real poverty, the meeting of day-to-day physical needs is still most important. It is much more difficult to communicate well and live happily with your partner and your children if your stomach is continually empty.

An intimate relationship that supplies emotional gratification to both partners and helps them deal with crises and grow in a fulfilling manner can only be achieved by supportive interaction between the partners (Cohan and Kleinbaum 2002; Lorenz et al. 2002). Successful problem solving is probably the strongest predictor of marital satisfaction and stability.

Successful communication has become increasingly difficult in recent decades. Challenges posed by high-pressure, dual-earner family life; changing gender roles; post-divorce coparenting; remarriage; job changes; and the fast pace of technological change have made good communication more complex and difficult to achieve (F. Walsh 1998, 106). Skills in talking, listening, negotiating, and problem solving are more necessary than ever for building strong, intimate relationships.

In a successful relationship, each partner's viewpoint is appreciated. The ability to compromise is essential because seldom, if ever, can one person always have his/her way in a relationship. In fact, the appealing idea that a good marriage is conflict-free and avoids the expression of anger is false. In reality, there is no such thing as a conflict-free relationship or one in which anger is never expressed. As desirable as good communication, negotiation, and compromise are to a successful relationship, they may not always help an individual deal with the real sacrifices entailed in a relationship or with the serious disappointments that a person will sometimes suffer, whichever way a decision is made.

Living closely with another person while bringing up children and making way for the needs, wishes, and even whims of other family members inevitably creates frustrations. And who among us is able to give up independence and selfishness without a struggle? Conflict goes with the territory of marriage. (Wallerstein and Blakeslee 1995, 143–145)

Good Communication: A Basic Strength of Successful Families

Successful communication is the cornerstone of any relationship. Good communication skills contribute significantly to how well relationship functions are fulfilled, and how much happiness people derive from their relationships (Burleson et al. 2000, 248; Meeks et al. 1998). Such communication must be open, realistic, tactful, caring, and valued. Developing and maintaining this kind of communication is not always easy unless all family members are committed to the belief that good communication is important to life satisfaction. This may sound simple, yet couples in marital trouble almost always list failure to communicate as one of their major problems.

Good communication is both especially important and especially difficult in marriage because an intimate relationship arouses intense emotions. High emotional levels, especially negative emotions, tend to interfere with clear communication. If you have ever had trouble communicating when you are calm and collected, imagine the potential problems when you are excited and emotionally aroused. Yet only through clear communication can each partner hope to under-

stand the needs and intentions of the other. When conflicts arise, the partners can resolve them only if each is able to communicate fairly about the problem, define it clearly, and be open to alternative solutions. Good communication also helps minimize hostilities. For example, unexpressed dissatisfaction tends to create hostility, but fairly expressed dissatisfaction allows the other partner to understand the problem and act to reduce the first partner's unhappiness and deflate the hostility.

Although everyone agrees that good communication is vital to all relationships, it is difficult to define just what it is. Family members often differ in their perceptions and priorities about communication. For example, Nathan's parents constantly push him to be open and share his thoughts and behaviors with the whole family. They stress that open communication is important to the fully functioning, happy family. Nathan views their request as prying and intrusive of his privacy. Do you think Nathan is correct? How do you relate to your family's requests for openness and sharing from you?

Conflict management is essential to all intimate relationships. When two people marry, they have already experienced many years of socialization and have acquired multiple—and perhaps, quite different—attitudes. The surprising thing is that they can get along at all on an intimate day-to-day basis! What each person has learned about handling interpersonal relationships is, to a large extent, particular to that individual. One partner may have learned early on that you do not talk about problems until everyone is calm and collected. As a consequence, sometimes a problem is not discussed for several days. The other partner has learned that you always speak your mind immediately. He/she has been taught that waiting only makes things worse. Differing beliefs about when and how to communicate will probably cause difficulties in the relationship.

Communication is also affected by the general society. American men are taught to disclose less about their feelings than are American women. The traditional American masculine role is to be strong and silent. Expressiveness, sensitivity, and tenderness are considered feminine traits. But these traits are obviously important to good communication.

It is important to emphasize the influence of the general society and its institutions on intimate relationships. Most people encountering marital problems tend to believe that their difficulties are strictly personal. They believe that the problems are unique to themselves, their spouse, their families, or the immediate circumstances. Placing the source of the problems solely upon the partners and their marriage reinforces the myth that marriage to the *right* person will solve all problems and result in a constantly happy family life. Certainly, many problems are unique to a given family, but it is equally true that many family problems arise because of social pressures. Unemployment, for example, will bring stress and strain to a family.

The myth of the right partner sometimes leads a troubled couple to attempt to end their problems by changing partners, when it is, in fact, their relational skills that need changing. A person whose lack of communication skills finally leads to the demise of his/her marriage will soon find that any new relationship is also in trouble.

Part of good communication is a couple's ability to relate to each other in a manner relatively free from cultural influences. In other words, partners should attribute meaning to each other's communications based on their personal knowledge of each other, rather than on traditionally accepted interpretations. Jane calls her husband, Jim, pigheaded, but instead of becoming angry at a traditionally accepted put-down, he laughs and hugs her. Calling him pigheaded is Jane's special way of complimenting him for being properly assertive. Does your family have negative names for any family members that are used in positive ways?

As the example of Jane and Jim also makes clear, for good communication to occur, the partners must know and agree on the unique meanings of each other's communications. To react with a hug rather than with anger, Jim needed to know that his wife's use of a traditionally negative label had a positive meaning. This example also indicates that successful communication takes time to develop. It requires a learning process between two people and develops gradually. The ultimate outcome of successful communication is the achievement of interpersonal understanding.

Such understanding between people in close relationships also implies a certain richness in their communication. *Richness* means that they have many ways of communicating with each other, even many ways of conveying the same thing. Happily married couples have more communication styles than unhappily married couples.

One of the real changes in the American family has been the widespread entrance of wives and mothers into the workforce. The dual-earner family has now become the norm rather than the exception. When both partners work outside the home, two important factors underlying strong and successful families— time spent together and conversation—tend to be weakened. Strong families do spend a lot of time conversing, both about trivial aspects of their lives and about deeper, more important issues. Disclosure of negative thoughts, problems, and hostilities and also positive thoughts and emotions by family members is crucial to family success.

Disclosure remains strongly linked to satisfaction in both men and women even when marriages last for 15 years or more, suggesting that disclosure and satisfaction remain mutually transformative from the beginning to the end of a relationship. (Baxter 1987; Finkenauer et al. 2004, 207)

Communication of support, affirmation, appreciation, caring, respect, and interest in other family members is the lifeblood of the successful family system. McCubbin and associates (1988, 123–130) found many factors that related to communication skills:

- Trusting and confiding in one another
- Trying new ways to deal with problems
- Working together to solve problems
- Expressing caring and affection for each other daily
- At least one family member talking to his/her parents regularly
- Sharing feelings and concerns with close friends
- Parents spending time with teenagers for private talks
- Checking in and out with each other

Almost all couples who come in for counseling, start divorce proceedings, or indicate serious problems in their relationship begin by describing their inability to communicate well. "He never listens to me." "All she does is nag." "I never know what he is feeling." "She expects me to read her mind." "He never talks to me."

A husband who doesn't listen when his wife talks about her day, or a parent who doesn't listen when a child talks about his/her problems at school, seems to be expressing disinterest, which signals a lack of respect and appreciation. The wife, husband, or child who doesn't share his/her thoughts and feelings also seems to be expressing a lack of trust in other family members. Parents who do not attend to the day-to-day activities of their children shouldn't be surprised when their children choose not to discuss major problems, such as drugs or sex, with them. Parents frequently say, "Why didn't you come to me with this before it became a problem?" Yet when parents turn a deaf ear to a child's day-to-day trivia, such as who said what to whom, or how the tennis class or the history exam went, the child may easily conclude that his/her parents probably won't listen to problems either.

Poor communication within a family is manifested in a number of ways. First, the family members suffer from a constant feeling of frustration, of not being understood, of not getting their message across. This frustration sometimes leads to preoccupation with escape—the need to go out, to be constantly on the phone, to have the television set on all the time, or to keep headphones in place during family time. Poor communication leads to the following oft-quoted parent-child interaction:

"Where are you going?"

"Out."

"What are you going to do?"

"Nothing."

Poor communication is often evidenced in sharp words, quarrels, and misunderstandings. The poorly communicating family tends to bicker and engage in conflict frequently. Sometimes this behavior becomes so unpleasant that family members simply cease to talk, and silence becomes the norm. Some people confuse silence with lack of communication, but silence certainly sends messages and can be a devastating form of communication. Think back to the last time someone gave you the *silent treatment*. How did you react?

What Causes Communication Failure?

Communication failure means communication isn't accomplishing what it is intended to accomplish. It doesn't convey what a person wants to convey. This is not always an accident. The failure may be intentional, though not necessarily overtly or consciously so. People sometimes choose not to communicate.

Often, when a marriage is disintegrating, the partners say, "We just can't communicate" or "We just don't talk to each other anymore." They identify a deficiency of talk with a lack of communication and believe the failure to talk is the cause of their problems. Actually, the nontalking is not a lack of communication, but rather, a form of communication that sends a surplus of negative or aversive messages that hasten the disintegration of the relationship. Remember, silence also communicates.

Failure to communicate well with others (interpersonal communication) sometimes begins with a breakdown of intrapersonal (internal) communication. Anger, emotional maladjustment, stress and strain, and faulty perceptions can all lead to blind spots and overly strong defenses, which interfere with good interpersonal communication. When individuals' self-images become unrealistic, they tend to filter all communications to fit the faulty self-picture. Under such circumstances, communication that accomplishes shared meaning is not likely.

Aversive Communication

One basic principle of human behavior is that people seek pleasure and avoid pain. But, actually, people first avoid pain and then seek pleasure because, in the hierarchy of human needs, pain avoidance is basic to survival. This point is crucial to understanding communication failure because motivation to communicate will be most urgent when the intention is to avoid pain. Unfortunately, the quickest means of changing the behavior of the partner who is giving us pain is to threaten him/her—that is, to combat pain with pain (aversive communication). But this threat causes pain to the spouse and is counterproductive because the threatened partner will most likely respond at the same or higher level, creating a vicious circle. This kind of communication is a power struggle in which the winner is the person who generates the most aversion. The loser feels resentful and may engage in typical loser behavior such as deceit, procrastination, deliberate inadequacy, sarcasm, and sullenness. Any verbal input by the loser thereafter is likely to be an emotional discharge of the resulting resentment and hostility. Generally, the loser stops talking because communication has become so painful. Or the loser may deliberately attempt to mislead the winner. The loser may begin to seek fulfillment outside the relationship, at which point noninteraction will become an intentional goal.

An example of aversive communication is one that starts out dealing with one conflict, but soon turns into a confusing kaleidoscope of other marginally related disagreements that are connected only by the pain each partner causes the other:

"Where did you put my socks?"

"You mean the socks I have to crawl under the bed to find every time I do the laundry?"

"Yes, every time you do the laundry—the second Tuesday of every month."

"With the machine that's always broken, that nobody fixes."

"Because I'm so busy working to pay for tennis lessons and ladies' luncheons."

"And losses at Tuesday night poker bashes."

"I'm sick of this argument."

"Well, you started it."

"How?"

"Well, I can't remember, but you did."

It is important to remember that negative thoughts usually need to be communicated, but there are positive ways to communicate them. Positive responses to negative thoughts are ways of saying to the partner, "Yes, I'm listening." Other explicit listening indications are "I see," "Tell me about it," "This seems important to you," and "Okay, let's work on it together." Substituting more positive responses for negative responses shows care for the partner and the relationship.

Inappropriate reassurance may be received negatively when the other person has shared a problem. Your attempt to make him/her feel better by downplaying the problem may backfire. For example, saying, "Don't worry about it—everyone has that problem at some time in their lives" may effectively stifle further discussion. On the other hand, reassurance and praise after the partner has sufficiently explored all the feelings that are associated with the problem are appropriate and will have positive outcomes.

Remember that communication serves many purposes and takes many forms. Communication among intimates and family members will include both content and relationship dimensions. "Go to your room," spoken by a parent to a child is both content (requesting an action) and relational (I am your parent and have authority over you).

Communication is also both expressive and instrumental. "I just love you so much. Let's go out for a romantic dinner tonight." In this case, the spouse is expressing feelings (affection toward his/her spouse) while suggesting task-oriented behavior (going out to dinner).

Communication Can Be Used for Good and Bad Purposes

This chapter is based on the popular assumption that improved communication is always healthy and helpful to personal relationships. Mass media, popular authors, and marriage and family therapists all recommend improving communication to help troubled families.

It is important to realize that the relationship between communication skills and marital satisfaction is far from simple. For example, negative communication behaviors frequently observed in distressed spouses may result more from ill will than poor skill (Burleson and Denton 1997, 897). We must also remind ourselves that communication is a double-edged sword. It can be used to express honest feelings; convey facts and observations; share helpful information; enhance another person, ourselves, or a relationship; and for other positive purposes. Conversely, it can be used to manipulate feelings; convey falsehoods; hide information; or degrade another person, oneself, or a relationship.

We see communication being used to mislead. We hear lies being told. We observe language being misused to persuade people to buy something they don't need or to destroy something or someone. A good con artist is usually a good communicator who can lie with sincerity and twist the truth with conviction. We see communication used as propaganda to hide or distort realities. We are bombarded with meaningless slogans designed only to arouse us emotionally. Just exactly what do phrases like "America: Love it or leave it!" or "Power to the people!" mean?

Learning better communication skills does not necessarily mean that the improved skills will be used to further a relationship; they could just as easily be used to destroy it.

Hence, our discussion of communication is set within the context of creating improved personal and family relationships. Good communication skills are one of the major strengths of strong, successful families and other important relationships.

The Foundation Blocks of Successful Communication

Three general conditions must be met to achieve the kind of communication that builds strong, positive relationships:

1. Commitment. The partners must be committed to make their relationship healthy and strong.
2. Growth orientation. The partners must accept the fact that their relationship will always be dynamic and changing, rather than static.
3. Noncoercive atmosphere. The partners must feel free to be themselves, to be open, and to be honest.

These three conditions represent the foundation blocks of successful communication.

Commitment

Simply stated, commitment means making a pledge or binding oneself to a partner and includes the idea of working to build and maintain the relationship. As we know, commitment to others is one of the characteristics of strong, successful families. Couples often fail to seek marriage counseling until so much pain and suffering have occurred that the commitment of one or both marital partners has been destroyed. At that point, there is little hope of resolution. Often, one partner is still committed enough to want counseling, while the other has covertly given up hope and resists counseling for fear of being drawn back into a painful trap. Sometimes an uncommitted partner will seek therapy only in the hope that a dependent spouse will become strong enough to survive without the marriage. This situation adds complexity, especially if the dependent partner, sensing abandonment, avoids getting stronger in the hope that the uncommitted partner will stay in the marriage.

Many uncommitted partners allow the other spouse to plan activities, anticipate problems, make necessary adjustments, and so on. Sometimes this kind of relationship works. More often than not, however, the partner with the greater commitment becomes resentful of the uncommitted partner's lack of involvement. Paradoxically, the one who cares the least controls the relationship. This is termed the **principle of least interest**. Although commitment is a precondition to effective communication, each partner's commitment to the relationship can be increased through use of the five communication skills, to be discussed later.

principle of least interest
The one who cares the least controls the relationship

Growth Orientation

Individuals change over time. The needs of a 40-year-old person usually differ from the needs of a 20-year-old. Marriage must also change if individual needs are to be met within the ongoing, dynamic marriage framework. Yet change is often threatening. It upsets comfortable routines and may be resisted. Such resistance will be futile because relationships do not stand still. Thus, it is wise to accept and plan for change. An individual oriented to growth incorporates the inevitability of change into his or her lifestyle. This implies that an individual would be wise not only to accept change, but also to create intentional and orderly change in a chosen direction. For example, Gina enjoyed attending large social gatherings, but her husband Art dreaded them. Both were oriented to growth, however, and Art wanted to become more comfortable in social situations. They decided to work together on helping Art change his dread of social interaction to enjoyment. At Art's suggestion, Gina began to invite one or two couples for evenings at their home, where Art felt more comfortable. He soon began to enjoy these small get-togethers, which Gina then gradually expanded. Although Art is still not perfectly at ease in large groups away from home, he is much more comfortable than before and accompanies Gina most of the time.

Had Gina and Art not been growth-oriented, their differing social needs might have caused them to grow apart or become resentful of each other. Gina might have chastised Art for his social inabilities and gone out without him, leaving him home to brood and resent her. On the other hand, Art might have been dictatorial and forbidden Gina to go to large social gatherings, which would probably have made her feel trapped and rebellious. Because neither was afraid of change and they agreed that a change in Art's attitude would be beneficial, they set out deliberately to encourage this process. Notice that both were committed to the change. If only one partner desires a change, particularly a change in the other partner, and the other disagrees, conflict can result.

Noncoercive Atmosphere

The goals of marriage will usually be lost if either partner is subjugated by the other. Free and open communication cannot exist in a one-sided, totalitarian relationship. When any two people share a common goal, the issues of responsibility and authority arise. The situation in marriage is not unlike that in government. A marriage can be described as *laissez-faire* if both partners have total freedom of choice and action, *democratic* if responsibility is shared and authority is delegated by equitable agreement, or *autocratic* if both have responsibilities but authority is assigned to a single leader.

Most Americans say they want to establish a democratic marriage. In reality, this pattern is the most difficult to maintain, although it is ultimately the most satisfying. Most marriages end up being some mixture of all three types.

When power is invested in a single authority, a coercive relationship is usually the result. At least one partner, and often both, loses freedom. The subjugated person feels unfree and coerced by the partner with the power. The person with authority may also feel unfree because he/she may be held responsible for all the couple's decisions. Free and open communication usually cannot coexist when one or both partners are coerced.

Mike and Daphne have been married 6 months. She pleads with him to give up his weekly evening handball game because she is afraid to be alone. He feels unfairly restricted and calls her a child. She, in turn, calls him an immature jock who can't give up playing with the boys. One thing leads to another, and Mike walks out of the room and slams the door. Daphne tearfully runs to the bedroom and locks herself in, leaving him to nurse his guilt. Although he apologizes profusely, she won't come out. So Mike stays home; his physical coercion (walking out and slamming the door) has been overpowered by her emotional coercion. He is resentful and she is resentful. The intimacy that they experienced in their courtship is being dissolved by the acid of resentment created by their coercive acts. At this moment, it is difficult to say just who has the power in this relationship. However, it is clear that successful communication will not survive long in such an environment. Sharing responsibilities, giving up control voluntarily, and feeling relatively free in a relationship greatly facilitate positive communication.

Instead of the foregoing sequence, Mike and Daphne might have worked out a compromise that would have satisfied them both. For example, knowing how much Mike enjoys his weekly game, Daphne might have supported him in this pleasure in the hope that he would then respond to her fear of being alone. He might reduce the number of times he plays handball; she might plan to visit her parents or a friend that evening. If each saw the other as noncoercive and supportive, compromise solutions would then be far easier to work out.

Developing a Smooth Flow of Communication

The three conditions for successful communication—commitment, growth orientation, and noncoercive atmosphere—all contribute to creating a sense of order in which messages flow back and forth smoothly between sender and receiver. The first step in a communication flow is to encode the message and send it to the receiver via some verbal, nonverbal, or written communication channel. The receiver then decodes the message and, to ensure that it has been correctly received, must feed back what has been decoded to the sender. The sender verifies the message; or, through correction, negotiation, and problem solving, he/she resends the message, and the process is repeated. This circular pattern may have to be repeated several times before the sender's intention and the receiver's interpretation coincide (see Figure 5-1).

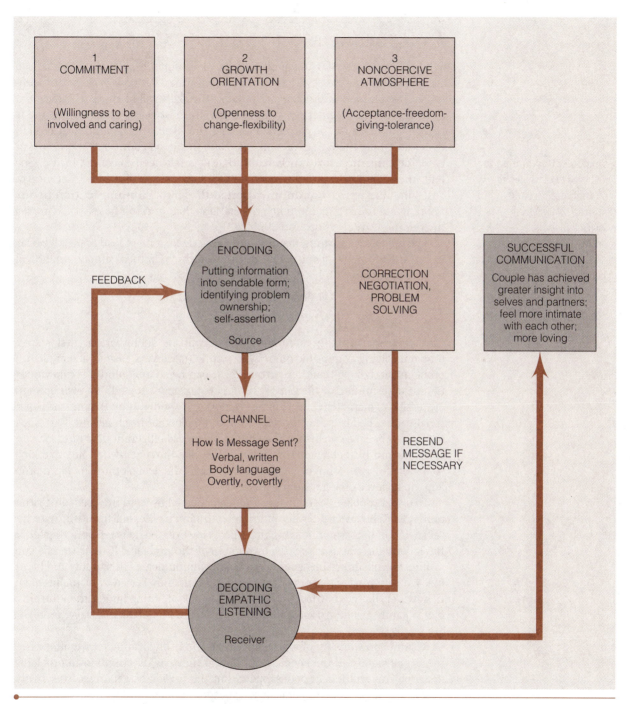

FIGURE 5-1 *Communications Flowchart, Showing the Three Conditions Necessary for Successful Communication*

Communication Skills

When all is going well with couples, they usually don't think about their communication skills. Yet the best time to build communication skills is when disruptive forces are minimal. If problems arise later, the skills will be there. Conflicts that appear irreconcilable destroy relationships. In most cases, conflicts become irreconcilable because the partners have failed to develop the communication skills that would help them resolve conflicts. Note that good communication must include **communication motivation**, that is, a felt need to communicate. Second, the individual must have the skill to accomplish his/her communicative goals, in other words, **communication skill**. Third, **communication behavior** refers to the verbal and nonverbal actions that a person takes to accomplish his/her motivations (Burleson 2000, 249).

The following activities, if successfully practiced, tend to lead to positive, clarifying, and problem-solving communications, which, in turn, support an individual's relationship with his/her partner.

communication motivation
A felt need to communicate
communication skill
The ability to accomplish one's communication goals
communication behavior
Verbal and nonverbal actions that a person takes to accomplish his/her goals

Identifying Problem Ownership

Clarifying responsibility, or problem ownership, is an important first step in problem-solving communication. This step is not always easy to accomplish. A problem can belong to either partner, or it can be shared jointly. The key question in determining ownership is, who feels tangibly affected? To own or share ownership of a problem, a person must know and openly admit that he/she is personally disturbed by it. For example, Janice thinks that her husband, Ray, is losing friends because he drinks heavily and becomes belligerent. But Ray isn't concerned about his behavior. He thinks people like him even when he's drinking, so he refuses ownership of the problem. Because he doesn't think he has a problem, he believes change is unnecessary.

To own a problem, one must be tangibly affected by it. In the preceding situation, if Ray's behavior does not interfere with Janice's life, her friendships, or her sense of self, it does not tangibly affect her. However, if she is tangibly affected by Ray's behavior, she then shares ownership of the problem. If their friends stop calling, her needs are being affected. If she communicates this to Ray and he refuses to admit to it, as paradoxical as it may seem, only Janice owns the problem. If, however, Ray acknowledges that his behavior is affecting Janice, then the problem is jointly owned. In most intimate relationships such as marriage, problems are jointly owned.

Assuming ownership of a problem is extremely important, yet we often shun the responsibility. Some people think that if they pay no attention to problems, the problems will lessen or disappear, but the reverse is generally true. In the long run, unattended problems usually become worse. Denial effectively cuts off communication and prevents change.

Modern American society too often encourages individuals to cop out of problems by supplying scapegoats on whom responsibility can be erroneously placed and/or encouraging individuals to claim victimization:

> "I'm unaffectionate because my parents rejected me as a child."

> "My father beat my mother when he drank; now I can't relate to men because of my deep hostility toward him."

Calvin and Hobbes

by Bill Watterson

Calvin & Hobbes © Watterson. Reprinted by permission of Universal Press Syndicate.

"American society is racist; I am African American (or Jewish or Hispanic or . . .), and therefore all my problems are caused by society."

The list could go on, but these examples and the cartoon give an idea of what must be avoided if one is to solve personal problems. For instance, it may be true that you have problems because your parents rejected you, but they are still your problems. Rejecting parents are not suddenly going to become loving parents just to solve your problems. Your parents obviously have problems of their own, but you can't make them solve their problems; you can only work to solve your own. The first step toward solution is assuming ownership and responsibility for your problems.

We must add an important caution to this discussion of problem ownership. Forcing problem ownership on one's partner through blaming is highly destructive. Some people use problem ownership—an aspect of good communication—in a negative manner; that is, they lay blame on the other person, saying that he/she is responsible for the trouble. "Our whole problem is your apathy" or "If only you took some pride in your appearance." By saying, "It's your problem, not mine" one escapes responsibility. We can make the same kind of comment about the next skill we will discuss, self-assertion. In essence, one partner may attack the other under the guise of being self-assertive, thus perverting a positive communication skill into harmful communication.

If the problem is mine, then I will use the skill of self-assertion. If my partner owns the problem, then I will try to use the skill of empathic listening. If we both own the problem, we will alternate these two communication skills.

Self-Assertion

Self-assertion is the process of recognizing and expressing one's feelings, opinions, and attitudes while remaining aware of the feelings and needs of others.

Some people are nonassertive. They fail to make their feelings and thoughts known to others. This makes full, successful communication almost impossible. Louisa thinks Walker is becoming less affectionate toward her. He doesn't hug her and touch her as much as he used to. Louisa always liked the close physical

self-assertion
The process of recognizing and expressing one's feelings, opinions, and attitudes while remaining aware of the feelings and needs of others

contact and misses it. But she finds it hard to talk about her physical desires with anyone, much less a man, so she says nothing. Her anxiety and discomfort grow. She believes that if Walker really loved her, he would recognize how miserable she feels and give her more physical contact. She becomes increasingly hostile toward him until one day he asks, "What's the matter?" She replies, "You ought to know; it's your fault." If a person doesn't talk about what is bothering him/her, how is the partner to know if it is his/her behavior that might need changing? Walker is completely in the dark. Louisa's nonassertive behavior has precluded successful communication and thus has foreclosed solving the problem.

Some of the personal reasons for nonassertive behavior are fear, feelings of inferiority, lack of confidence, shyness, and embarrassment. Society and its traditional role expectations may also influence nonassertive behavior. Louisa may have accepted the earlier, traditional, American, feminine-role expectation of passivity and nonassertiveness, of expecting the man to solve her problems. Whatever the reasons, if needs go unexpressed, they will not be magically recognized and fulfilled by others.

In contrast to nonassertive individuals, aggressive individuals completely bypass tact and recognition of others' needs in expressing their feelings. They demand attention and support and insist on having whatever they want at the moment, often overriding the rights and feelings of others in the process. In fact, aggressive individuals often seem unaware that other people have rights and feelings. Consequently, they may hurt other people without even being aware that they are doing so.

Not all aggression is destructive, however. Used in a constructive manner, aggression can offer emotional release, and/or let a partner know how intensely the other partner is feeling. Rather than anger, it is contempt, belligerence, and defensiveness that are the destructive patterns during conflict resolution (Gottman et al. 1998).

Self-assertive people feel free to express themselves, but are aware of the feelings and needs of others. They will comment on their own observations and feelings, rather than criticize another person's actions. Contrast the statement, "This kind of behavior is hard for me to handle and makes me angry, even though I don't want to be" with "You make me so mad!" The second statement is almost useless to successful communication. It judges and places blame on the other person. Such a statement usually provokes a defensive comment on the part of the listener. This, in turn, elicits yet another aggressive comment from the first person, and so on, until the situation spirals out of control.

It may seem surprising that people building a relationship often need special help with self-assertion. Many people believe that what is needed is less self-assertion. In their view, self-seeking assertions and selfishness are the basis of marital difficulty. Remember, though, that our definition of self-assertion includes awareness of others' feelings and needs. We certainly need to be less destructively aggressive in our relations, but we must not confuse that with being nonassertive.

It is important to be aware that recognition and expression of needs do not necessarily lead to their fulfillment. For example, I may recognize and express my desire to smoke, which is self-assertive. If other people in the room find cigarette smoke unpleasant and tell me so, they are also being assertive. This conflict can be successfully resolved. I may go outside to smoke, thus satisfying my needs without interfering with the needs of other people; or, I may recognize that smoking would be more unpleasant for them than pleasant to me and decide to

MAKING DECISIONS

You-, I-, and We-Statements

I-statements locate the feelings or concerns inside the person who is making the statement, rather than placing the problem on the partner. Gordon (1970) suggests that I-statements are less apt to provoke resistance and are less threatening than statements about the other person, and therefore, promote openness. Good I-messages may include "feeling," "when," and "because" parts. For example, "I feel upset when the living room is in a mess because guests may then think I'm a sloppy housekeeper."

Wesley Burr (1990) suggests adding we-statements to the communication process. We-statements place the problem in the group or relationship, rather than on one of the individuals in the relationship.

An example of the three different ways of defining problems in families will help illustrate the differences between you-, I-, and we-statements. If members of a family are having difficulty with the amount of affection in their relationships, they might say:

You-statement: "You're not giving me enough affection."

I-statement: "I'm not getting enough affection."

We-statement: "We don't show each other enough affection."

In intimate relationships, you-statements are probably least effective in problem-solving situations. They tend to create distance between people by placing blame on the other. Placing blame usually arouses defensiveness and resistance on the part of the person blamed. You-statements also tend to create and maintain an adversarial "I versus you or them" relationship. As a result, you-statements often start or expand conflicts and controversies, rather than moving the family toward peaceful, loving, harmonious solutions to problems.

Most of the time, I-statements are more effective and helpful than you-statements. I-statements locate the problem within the person who is suggesting there is a problem. They subtly communicate a warmer and more accepting concern for the individuals whom the "I" person views as the focus of the problem. Thus, I-statements tend to reduce defensiveness and resistance. I-statements may, however, create a problem for the other person because they make that person aware that he/she is central to something that creates a problem for the "I" person.

We-statements center the problem in both persons or in the group. They emphasize mutuality and connectedness. This emphasis tends to reduce emotional and relationship distance, rather than exaggerate it. We-statements place the responsibility for doing something about the problem on both people or on the group, but do not imply that any one person has more responsibility than another. Such statements also reduce the power imbalance in the relationship. Rather than implying that "I have power over you," they suggest that "We are in this together." They imply "I'm bringing up the problem, but it is something that we need to deal with, rather than a situation where I already know what the problems and solutions are."

We-statements are more useful when someone wants to enhance the togetherness aspect of the relationship or family and wants to build a sense of cooperation. Like all communication principles, however, we-statements can also be used in negative ways. For example, a parent who says something like "We don't believe that in our family" may stifle differences of opinion. By combining simple I-statements and we-statements, such a problem can be avoided. For example, by saying "I don't think we believe that in our family," the parent will introduce a subjective quality that allows others to disagree but still keeps the focus primarily on family belief.

How you say something is often more important to good communication than what you say. I- and we-statements are usually more likely to produce problem solutions than are you-statements. When things are going well, it is probably unnecessary to differentiate between you-, I-, and we-statements.

---

forgo smoking in their presence. In either case, I have been assertive—I have recognized and expressed my need, though I may not have fulfilled my desire. *Self-assertion does not mean getting your way all the time.*

The definition of self-assertion includes recognizing needs and inner feelings. This recognition is not always easy.

For example, Alicia has to conduct a PTA meeting tomorrow morning, and she is both anxious and resentful about it. She has put off planning the meeting until there is almost no time left. Brad comes home and says, "How about going to the movies tonight?" Alicia blows up and says irritably, "I still have dinner to cook and dishes to do. As the old saying goes, 'Man works from sun to sun, but woman's work is never done.'" Brad responds, "If women have it so tough, why do men have all the ulcers and heart attacks?" And so the argument spirals on without chance of resolution, at least partially because Alicia has not recognized the real source of her irritation—namely, her anxiety about the PTA meeting. But why doesn't she? Perhaps she doesn't want to admit to herself that she is afraid of conducting the meeting. She avoids thinking about the meeting (denial) and thereby avoids the fear that arises when she does. Or maybe she knows she is afraid but doesn't want Brad to see this weakness, so she covers it by starting an irrelevant argument.

Self-assertion requires self-knowledge; that is, successful communication depends, in large part, on knowing oneself. Any relationship will have known and unknown dimensions. One way of diagramming and discovering these dimensions is called the Johari window (Figure 5-2). You do not know everything about yourself or about your partner. And you do not necessarily share everything that you do know about yourself with your partner. As a relationship grows and becomes increasingly intimate, however, you usually learn more about both yourself and your partner.

When you begin a new relationship with another person, one in which you both are committed to helping the other grow, you might both draw Johari windows. The easiest square to fill in is common knowledge. At the beginning of the relationship, common knowledge might include only such things as physical appearance, food preferences, political persuasion, and family composition. As the relationship continues, you will be able to fill in some of the other's blind spots, perhaps helping him/her become aware of such things as insensitivity, selfishness, nervous laughter, bad breath, snoring, and so forth. Your partner can do the same for you. As you continue to develop trust in each other, you can begin to fill in your secrets, such as feelings of inferiority or being afraid to be alone at night. By now you may be discovering aspects of each other's unknown selves; these may be hidden potentials, talents, or weaknesses that interaction between the two of you has uncovered. As you get to know each other better, the blind spots and secret areas will become smaller; more information is moved into the common knowledge window for both of you.

Of course, the Johari window is not just for new relationships. It can also illuminate behavior and knowledge in ongoing relationships. In fact, the process of exploring new dimensions of the self and the other person is exciting and never ending. Unfortunately, many couples share little common knowledge because

FIGURE 5-2
Johari Window

Things about myself that I . . .

		do know	do not know
Things about myself that the other . . .	does know	common knowledge	my blind spots (such as an irritating mannerism I'm unaware of)
	does not know	my secrets (things I've never shared about myself)	my unknown self (things neither you nor I know about myself)

each partner has little self-knowledge. People who are afraid to learn about themselves usually block communication that might lead to self-knowledge. As we saw earlier, aversive reactions are excellent ways to stifle communication. Jim thinks his wife Tamara looks unusually good one evening and says, "I really feel proud when men look at you admiringly." Tamara replies angrily, "When you say that, I feel like a showpiece in the marketplace." "Why don't you get off that feminist trip?" he retorts. Tamara's aversive reply to Jim's statement will probably make him more reluctant to express his feelings in the future. In essence, her aversive response punished him for expressing his feelings. But why did she respond aversively? She may simply have been in a bad mood. Or perhaps she has negative feelings about her appearance and thus needs to deny Jim's statement because it is inconsistent with her self-image. Her attractive appearance might be a blind spot in her Johari window. However, another interpretation of Jim's statement makes Tamara's reaction more understandable. She may be objecting to his implied sense of ownership—to his describing her as if she were a shiny, new sports car envied by all his friends.

So, we can define the communication skill of self-assertion as learning to express oneself without making the other person defensive. Neither Jim nor Tamara is particularly good at this aspect of self-assertion.

Empathic Listening

Of all the skills we have been discussing, none is more important—and often more difficult—than that of being a good listener. Most people are ready to offer their own thoughts and opinions, but are not so ready to listen to those of others. Often, a person is so preoccupied with his/her own thoughts or with preparing his/her replies that he/she does not listen to what the other person is saying. Being a good listener is an art and is much appreciated by most people. It is surely one of the most important ingredients of intimate relationships.

The research on strong, healthy families finds that members of these families are listened to by other family members. They feel understood and accepted because family members not only listen well, but also recognize and respond to nonverbal messages. Good listening implies a real empathy for the speaker. Empathy means not only understanding what the speaker is saying, but also responding to and feeling the speaker's nonverbal communications and emotions.

Real listening keeps the focus on the person who is talking. The empathic listener actively tries to reduce any personal filters that distort the speaker's messages. Most listeners usually add to, subtract from, or in other ways change the speaker's message. Perhaps someone says, "I wrapped my car around a tree coming home from a party last night." You might hear these words and think, "Well, you probably were drinking (inference), and no one should drive when drinking (value judgment)." Although you have no evidence that the speaker was drinking, you have immediately assigned your own meaning to the statement. If the speaker recognizes your nonverbal signals, he or she may react with anger or cease talking.

In contrast, empathic listening is nonjudgmental and accepting. To the degree that we are secure in our self-image, we can listen to others without filtering their message. When we really listen to and understand another, we open ourselves to new self-knowledge and change. This can be frightening to persons who feel insecure and unsure of themselves. George is unsure of Liz's love for him. At dinner she comments, "I really like tall men with beards." George rubs his clean-shaven chin and begins to feel insecure. He doesn't hear much of the ensuing conversation because he is busy trying to decide whether Liz would like him to grow a beard, or if she has a crush on another man. He has completely filtered out the part about tall men. He is tall and was, therefore, not really threatened by that part of her comment. George has put his own inferences on Liz's comment and becomes increasingly upset as he ponders not what she actually said, but what he thinks she meant.

It is interesting to note that it is sometimes easier to listen empathically to a relative stranger than to someone with whom you have a close relationship. Married people sometimes say that some other person understands them better than their spouse. This may well be true because emotions often get in the way of listening. For example, when George's co-worker also says she likes tall men with beards, George doesn't even think about her comments but simply replies, "That's nice." Also, when a couple has spent years together, each partner often assumes that "of course, we understand one another," so there is no reason to listen carefully.

Empathic listening tells the speaker that the listener hears and cares. In essence, the speaker feels nonthreatened, noncoerced, and free to speak—some of our preconditions for successful communication.

Empathic listening has several components. Obviously, the listener must feel capable of paying close attention to the speaker. If we are consumed by our own thoughts and problems, we cannot listen empathically to another. But when we know that we cannot listen well at a particular moment, we should point that out (self-assertion) and, perhaps, arrange to have the discussion at another time. For instance, family members sometimes bring up problems at inopportune times—perhaps at the end of a busy day, when others are likely to be tired and hungry. They then become upset when no one is willing to listen. Most people need time, a quiet environment, and a peaceful mind to be good, empathic listeners.

We also need to listen fully. Instead, we often use *selective attention* to filter out what we do not want to hear. This means we hear only what we want to hear. Teenagers are often particularly good at not hearing something a parent has told them. Selective retention also plays a role. People may attend to something at the time of communication, but later will remember only what they want to remember.

We may also need training and practice to become good listeners. As children, we receive plenty of training to improve our verbal skills, but we seldom receive training for our listening skills. Just as our verbal skills can be improved through training and practice, so can our listening skills.

Nonverbal Communication If we are to be good listeners, part of our attention needs to be directed to the nonverbal communications of the speaker, often referred to as **body language**.

The influence of nonverbal behavior probably is due to its relatively involuntary character. People's facial expressions, voice tones, postures, and gestures can reveal unspoken emotions and intentions and can override efforts at impression management. It is much more difficult to lie nonverbally than to lie with words. (Prager 2000, 232)

People in a good marriage read body language as well as words (Wallerstein and Blakeslee 1995, 243). Emotions are reflected throughout the body. A hand gesture, a frown, and eye contact or avoidance of eye contact are all meaningful communications. Although bodily communication generally is culturally based,

body language
Nonverbal communication expressed via the body

certain gestures are more idiosyncratic (personal and unique) than verbal communications. Nonverbal communications are also often more descriptive of a person's feelings. Pamela habitually uses her hands in an erasing motion to wipe away unpleasant thoughts. Rick pulls at his ear and rubs the back of his neck when being criticized. The nonverbal message usually represents a more accurate picture of how a person is feeling emotionally than do his/her verbal communications. Pearson (1989, 74) cites many research findings in concluding that, when nonverbal and verbal communications conflict, most adults believe the message conveyed nonverbally. I am late for my date. My girlfriend says that's all right, she understands, and she is not angry. Yet when I reach out to hug her, she pulls back and rejects my touch. Her face is set and unsmiling. Do I believe her words or her actions? People tend to process the emotional messages first because agreement on content cannot be reached unless each other's feelings about the content are known. It takes time to learn what an individual's personal, nonverbal communication patterns mean, but the empathic listener will notice and interpret such patterns.

Distance between people is also a nonverbal communication indicator. The more intimate our relations with another person, the closer we can physically be with that person and feel comfortable. I put my arm around my wife or girlfriend, but not around the store clerk, who is a stranger. Each of us has a sense of our personal space, but this space varies according to both the culture and the intimacy of the relationship. North Americans tend to prefer a greater distance between people in their casual personal relations than do South Americans. An interesting personal experiment is simply to halve the normal distance between you and people with whom you interact and watch their reaction. They'll almost always back away.

Besides facial and bodily expressions and spatial considerations, nonverbal vocalizations such as sighs and even how one dresses give communication clues to others.

In reviewing the literature on nonverbal communication, L'Abate and Bagarozzi (1993) found that nonverbal communication is subject to many interpretations, such as when crying is seen as sadness and vulnerability, or as being used to manipulate a partner who is very uncomfortable around anyone who cries. Unhappy partners tend to read nonverbal communications more negatively than do happy couples.

Feedback   Another key component of empathic listening is feedback. Empathic listeners periodically check their perceptions with the speaker by paraphrasing the speaker's words (Markman et al. 1994, 64–66). This rephrasing reassures the speaker that she/he is being listened to and interpreted accurately. It also provides the speaker with an opportunity to correct the listener's perceptions, if necessary, and to hear his/her ideas from the listener's perspective. Some examples of feedback are the following:

Speaker: "You're always working."

Listener: "You're right, I do work a lot. Is it interrupting something we need to do?"

or

Speaker: "Why don't we get out of this town?"

Listener: "You feel like leaving. I guess you don't like it here."

Note that empathic listening involves an effort on the listener's part to pick up feelings as well as content. Feeding back feelings is, however, more difficult than feeding back content.

The desire to give advice is one of the major deterrents to empathic listening and clear feedback. Advice—especially unsolicited advice—always has a negative side. It says to your partner, "You don't know what to do and I do," which could be interpreted as "Boy, are you stupid." There is no better way to turn off communication than to stir up someone's defenses. Instead of giving advice, offer alternatives. This approach increases the speaker's options but leaves the responsibility of making the decision with the speaker. Simultaneously, it indicates respect for the speaker's intelligence (Strong 1983, 137).

Despite the emphasis on the importance of good listening skills, conflict makes the use of such skills difficult. Even couples in stable, happy marriages do not always listen actively (Gottman and Notarius 2000; Gottman et al. 1998, 17). In fact, this study suggested that active listening occurred infrequently in the resolution of marital conflict (Gottman et al. 1998, 17–18).

Much communication is of a different nature than just problem solving. Communication is for play and fun, to establish contact, or to impart information. Using problem-solving skills in inappropriate situations can actually create problems. For example, paraphrasing is unnecessary if someone says "Please pass the butter." It would be inappropriate to reply (feedback), "Oh, I understand that you'd like to have some butter on your bread." Such a remark will probably invoke a muttered comment like, "There you go using that dumb psychobabble again." Knowing how to listen and give feedback are invaluable skills, but they can be misused. Knowing when to use these skills is just as important as knowing how to use them.

Negotiating

If a problem is jointly owned, the situation calls for negotiation. In negotiation, the partners alternate between self-assertion and empathic listening. Usually, the more distressed partner starts with self-assertion. But because the problem is jointly owned, the listener's feelings are also involved, and it is imperative to switch roles relatively often to ensure that each person's communications are understood by the other. Set a time period—say, 5–10 minutes—for each partner to speak. Remember that in these situations, listening will take extra effort to avoid the temptation to think about your side of the problem while your partner is speaking, which interferes with empathic listening.

When roles are exchanged, the partner who was listening should first restate (feedback) the assertive partner's position, so that any necessary corrections can be made before going on to his/her own assertions. Knowing that you must restate the speaker's position to his/her satisfaction before you can present your own position works wonders to improve listening ability.

This simple procedure of reversing roles and restating the other's position before presenting your own is also amazingly effective in defusing potential emotional outbursts. Frustration usually builds up because partners do not listen well to each other and therefore often feel misunderstood. But if the negotiation process is used, even when you and your partner strongly disagree, at least you both will be assured that your partner has heard and understood your position.

No matter how well a couple can negotiate, some problems and conflicts will be unresolvable. In this case, the couple may finally agree to disagree. Such a solution demands tolerance from each partner and acceptance of the idea that

one's partner may never be just the way you want him/her to be (Markman et al. 1994, 96). Compromise is often a part of intimate relationships.

Problem Solving

When a partner has established ownership of the problem, begun to clarify the problem, and discharged some of the emotion surrounding the problem, he/she is ready to solve the problem. Of course, by now, many problems will have been greatly diminished and may even have disappeared. If one or both partners still think there is a problem, however, it is now time to apply the seven steps to scientific problem solving (see the discussion on decision making in Chapter 1).

1. Recognize and define the problem.
2. Set up conditions supportive of problem solving.
3. Brainstorm for possible alternatives (establishing hypotheses).
4. Select the best solution.
5. Implement the solution.
6. Evaluate the solution.
7. Modify the solution, if necessary.

The first two steps have already been accomplished if you have used the skills discussed. Step 3, **brainstorming**, broadens the range of possible solutions. The purpose of brainstorming is to produce as many ideas as possible within a given

brainstorming

Producing as many ideas as possible within a given time period in an effort to solve a problem time period. In other words, you select a time period, such as half an hour, and during that time you and your partner suggest or write down possible solutions to the problem without pausing to consider whether they are ridiculous or realistic. Negative judgments stifle creative thinking, so it is important to suspend any evaluating until both of you have run out of ideas or reached the end of the time period. Just jot down ideas as they occur; don't judge them.

Once you have run out of ideas or time, you can begin to select the best solution. Be sure to consider all ideas. Then use the skills of self-assertion and empathic listening to evaluate the likely ideas. Setting an amount of time for defending and judging each idea can be helpful.

Once you agree on the best idea, you must implement it. If the problem has been serious, it is a good idea to schedule future discussions about how well the solution is working (Owens 1994; Warren 1995). If the solution works well, you will not need to use step 7. If the two of you continue to experience difficulties, you may have to modify the solution, using the information that comes from your evaluation sessions. Or you may have to go back to the possible alternatives generated during the brainstorming session and select another solution to test.

Men and Women: Do They Speak the Same Language?

Although this section concentrates on communication differences between men and women there are far more similarities than differences (Canary and Dindia 1998). Communication differences between the sexes are probably exaggerated because the sexual world is so easily reduced to a dichotomy, which is viewed as reality.

From the earliest ages through adulthood, boys and girls create different (language) worlds, which men and women go on living in. It is no surprise that women and men who are trying to do things right in relationships with each other so often find themselves criticized. We try to talk to each other honestly, but it seems, at times, that we are speaking different languages—or at least different genderlects (Tannen 1990, 279).

Deborah Tannen and other sociolinguists suggest that some of the misunderstanding between men and women arises because of differences in communication styles. There are gender differences in accepted behavior and communication styles, just as there are cultural differences. Boys and girls grow up in different psychological worlds, but men and women usually think they are in the same world and tend to judge each other's communications by their own standards. Hence, many a discussion between women and men has ended with the man's comment, "Can't you get to the point?" because feminine speech is more detailed. Since men tend to move directly to the point, with little emphasis on feelings and without much detail, a woman often asks of a man, "Why don't you tell me how you were feeling and what else was going on?" (Wood 1988, 2000b, 2001).

In general, Tannen suggests that men grow up in a competitive world. To men, life is a challenge, a confrontation, a struggle to preserve independence and avoid failure, a contest in which they strive to be one up on colleagues. In their conversations, men attempt to establish power and status. In the world of status, independence is the key.

Women, on the other hand, approach the world seeking connection and intimacy, close friendships, and equality with their friends. In their conversations, they seek to give confirmation and support and to reach consensus. Interactions

between two women tend to be higher in depth of intimacy than do interactions involving two men (Prager 2000, 240). For women, intimacy is the key in the world of connection. These gender differences do not tend to hold up in gay and lesbian relationships (Nardi and Sherrod 1994).

In a relationship, if the woman emphasizes connections and intimacy and the man emphasizes independence and status, conflicts and misunderstandings are bound to arise. For example, many women feel it is natural to consult their partners, while men automatically make more decisions without consultation. Women, in general, expect decisions to be discussed first and to be made by consensus. They appreciate discussion itself as evidence of involvement and caring. But many men feel oppressed by lengthy discussions about what they see as minor decisions. They feel unfree if they can't act without a lot of talking first. Women may try to invite a freewheeling conversation by asking, "What do you think?" Men may take the question literally and think they are being asked to make a decision, when in reality, their partner only wants discussion.

Wives often accuse their husbands of being noncommunicative: "He never talks." In a Blondie comic strip, Blondie complains, "Every morning all you ever see is the newspaper! I'll bet you don't even know I'm here!" Dagwood reassures her, "Of course, I know you're here. You're my wonderful wife and I love you very much." With this, he unseeingly pats the paw of the family dog, which Blondie has put in her place before leaving the room. In fact, withdrawn, noncommunicative husbands are the primary complaint of many wives who seek marital counseling (Markman et al. 1994, 35).

Yet, research does not support the idea that men are less communicative than women. It depends on the situation. Men actually talk far more than women in meetings and in mixed-group discussions. They are quicker to offer advice and direction, even when it is not desired by the other person. In a public situation, they talk about themselves more often than women do. In fact, many women complain that when they go out with a new male acquaintance, all he does is talk about himself. He never seems interested in what she has to say. To women, much of this talk seems to be bragging and one-upmanship. The man talks about his achievements, his interests, sports, and politics. He often seems to be telling her how to do something, giving directions, teaching, or preaching. Of course, if Tannen is correct, this style fits in with men's emphasis on independence and the achievement of status.

If the woman is more interested in intimacy and connection, then her conversation will not emphasize her status or how she is better than others. Instead,

HIGHLIGHT

How Men and Women Can Communicate Better in Intimate Relationships

Lillian Glass (1992), after listing some 105 ways men and women tend to communicate differently, suggests some communication techniques that both sexes can use to improve their intimate relationships.

What Men Can Do to Have a Better, Intimate Relationship with a Woman

1. Provide more facial expression and more nonverbal feedback, such as smiling and head nodding.
2. Use more eye contact.
3. Talk less about themselves, interrupt less, and become better listeners.
4. Be more gentle in their touch and caress, cuddle and hug more.
5. Show more emotion and become more comfortable with giving and receiving praise.

What Women Can Do to Have a Better, Intimate Relationship with a Man

1. Don't expect a man to be a mind reader. Tell him what you want in a loving, yet direct, manner.
2. Don't bring up things from the past. This is confusing to men who prefer to keep to the issues at hand.
3. Don't nag a man into opening up. Instead, let him know that you are there for him, should he care to talk to you about sensitive issues.
4. When asking a man for help with a problem, expect him to tackle the problem directly, rather than discussing your feelings about the problem.
5. Understand that men's conversations tend to dwell less on feelings and emotions and more on activities such as sports, cars, and jobs.

This information is partially adapted from Deborah Tannen, *You Just Don't Understand: Men and Women in Conversation* (New York: Ballantine, 1990).

it will emphasize fitting in, being equal with the others, and sharing and giving support to those with whom she is talking. For women, talk is the glue that holds relationships together. Women tend to be more tactful than men and are less direct in their conversations, so as not to offend. Men may mistake this indirectness for weakness and lack of power.

A woman complains to a woman friend about something going on in her life. Her friend may then share something negative in her own life, in an effort to show that she understands the problem and is supportive of her friend's feelings. Men, however, on hearing the complaint usually offer answers and solutions. They take the complaint as a challenge to their ability to think of a solution.

Many men feel that women often complain without taking action to solve the problem that is bothering them. Women think that men are often insensitive to their (women's) problems and are therefore unsupportive. As one man commented, "Women seem to wallow in their problems, wanting to talk about them forever. Most men I know want to get them out of the way, solved, and be done with them."

For many men who work in competitive positions, the comfort of home means freedom from having to prove themselves and impress others through verbal display. At last they are in a situation where talk is not required, and they are free to remain silent. But for the woman, home is the place where she and her partner are free to talk. Especially for the traditional homemaker, the return of her husband from work means they can talk, interact, and be intimate. In this situation, the woman is likely to take her husband's silence as a rejection, while he takes her need to talk as an invasion of his privacy.

Although communication styles of men and women differ in many respects, it is a mistake to think that one style is better than the other. What is important is

SALLY FORTH/ by Greg Howard

Reprinted with special permission of King Features Syndicate.

to learn how to interpret each other's messages and explain your own message in a way that your partner can accept. There is not one right way to listen, talk, or have a conversation or a relationship. Although a woman may focus more often on intimacy and rapport and her partner more on status, this difference need not lead to misunderstanding—if the couple accepts that such differences do not imply that one partner's communication style is correct and the other's is wrong.

Communication and Family Conflict

Every intimate relationship will have periods of conflict. There is no such thing as a conflict-free relationship, marriage, or family. Families, however, view and handle conflict in different ways. Some families thrive on conflict, others use conflict constructively, and still others view it as a necessary evil. In some families, it is avoided; in others, it is a common occurrence. Conflict may lead to destruction of one family unit; it may be the manner by which problems are solved in another, and thus be viewed as essential for the continuation of a satisfying family life (Pearson 1989, 287, 305).

Research suggests that conflicts are equally present in happy and unhappy marriages (Markman et al. 1994; Wallerstein and Blakeslee 1995). The difference is that, in successful marriages, the partners have learned how to handle their conflicts and even use them to improve their relationship. If conflict is suppressed, it can cause stagnation, failure to adapt to changed circumstances, and an accumulation of hostility that may erode the couple's relationship.

Tolerance for conflict allows for overt disagreement and acknowledgement of differences. The best predictor of marital success is not the absence of conflict, but its management, that is, how differences are handled and resolved. Conflict avoidance is dysfunctional over time, heightening the risk for later marital dissatisfaction and divorce. (F. Walsh 1998, 122)

Conflict occurs when one or more family members believe that what they want is incompatible with what one or more other family members want. Realistic conflicts result from frustration over specific needs, whereas nonrealistic conflicts are characterized by the need for tension release by at least one of the partners. By exploring the process of conflict and how it can be used constructively, we can better manage it. Successful management of conflict solves problems and helps a good relationship evolve into an even better one. Recognizing nonrealistic conflict and successfully coping with it can reduce tensions and change the nonrealistic conflict into realistic conflict. Realistic conflict is usually easier to handle.

Conflict becomes damaging to a relationship when it is covert or hidden. If an individual cannot confront and work with real conflict, it is nearly impossible to resolve a problem. Hidden conflict generally relies on one of the following communication strategies (Galvin and Brommel 1982, 1986):

- *Denial.* One partner simply says, "No, I'm not upset; there is no problem," when, in fact, he or she is upset and there is a problem. Often, the person's body language contradicts the words.

- *Disqualification.* A person expresses anger and then discounts it. "You make me so angry! Sorry, I am not feeling well today." Of course, this discounting could be true. It only becomes disqualification when the person intends to cover an emotion and deny that a real conflict exists.

- *Displacement.* The person places emotional reactions somewhere other than the real conflict source. For example, John is really angry at his wife, but he yells at the children. Thus, the source of the conflict is kept hidden.

- *Disengagement.* Family members avoid conflict simply by avoiding one another. This avoidance keeps the conflict from surfacing. Unfortunately, the conflict remains below the surface, creates anger that can't be vented, and increases the tension in the relationship. It appears that asking for change in a spouse is often accompanied by disengagement of the spouse so asked (Klinetob and Smith 1996).

- *Pseudomutuality.* Family members appear to be perfectly happy and delighted with each other. In this style of anger, no hint of discord is ever allowed to spoil the image of perfection. Only when one member develops ulcers or a nervous disorder or acts in a bizarre manner does the family reveal a crack in its armor of perfection. Anger remains so far below the surface that the family members lose all ability to deal with it directly. Pretense remains the only possibility.

Overt conflict can also be destructive to a relationship, even though the chances of dealing with it are greater than with covert conflict. This chapter cannot cover the many destructive patterns of overt conflict. In general, however, attacking one's partner verbally or physically is devastating to the relationship and will make it extremely difficult to ever establish a fully healthy relationship. For example, whenever Kimberly becomes frustrated with her husband, Mark, she heaps verbal abuse on him: She says he has no drive, he is too dumb to get ahead in his job, and he is a lousy lover. The name-calling greatly affects Mark's self-esteem, and he slowly loses self-confidence. He doubts himself, which, in turn, causes him to act in ways that are self-defeating. What began as name-calling and negative labeling has gradually become his reality. The old saying "Sticks and stones may break my bones, but names will never hurt me" is not true. Verbal abuse such as "You idiot," "Liar," or "I hate you" attacks a person's self-respect and integrity. Verbal abuse can cause real psychological damage if it is continued for an extended period. Most people can forgive an occasional verbal attack during an outburst of anger, but if the attacks come often or become the norm for handling conflict, the partner and the relationship will be damaged. John Gottman (R. Edwards 1995, 6) suggests that "one 'zinger' counteracts 20 positive acts of kindness."

Five negative types of couple interaction are predictive of relationship failure. These five are (Gottman et al. 1998):

1. *Contempt*—indicating one's partner is inferior or undesirable
2. *Criticism*—making disapproving judgments about one's partner
3. *Defensiveness*—not listening, but defending oneself against attack
4. *Stonewalling*—refusing to listen to a partner's complaints
5. *Belligerence*—being provocative and challenging a partner's power and authority

Physical violence causes even more relationship problems and tends to lead to more violence. Physical violence generally occurs in families that lack communication skills. One member may be increasingly frustrated in the relationship and be unable to communicate with the partner or the children about the conflict. The frustration finally becomes so great that the family member loses control and strikes out physically. Anything that lowers an individual's inhibitions or frustration tolerance, such as alcohol or drugs, increases the possibility of physical violence.

Successful conflict management is certainly a characteristic of strong, healthy families. For some conflicted families, conflict resolution training can help members to better resolve ongoing conflicts (Blakeway and Kmitta 1998). On the other hand, answering yes to most of the following questions is a warning signal that a relationship is probably in trouble.

1. Do routine discussions erupt into destructive arguments?
2. Do you or your partner often withdraw or refuse to talk about important issues?
3. Do you or your partner often disregard what the other says, or do you often use put-downs?
4. Does it seem as if the things you say to your partner are often heard much more negatively than you intended?
5. Do you feel that there has to be a winner and a loser when you disagree?
6. Does your relationship often take a backseat to other interests?
7. Do you often think about what it would be like to be with someone else?
8. Does the thought of being with your partner a few years from now disturb you (Markman et al. 1994, 5–6)?

Anger

"For better or worse, till anger do us part." Certainly, uncontrolled anger will, in the long run, destroy an intimate relationship. Yet, anger also has a positive side at times. Mild anger episodes, as compared to extreme anger outbreaks, appear to strengthen relationships in some cases. Research finds that a large percentage of people (40–50 percent) report positive outcomes after an angry interaction (DeAngelis 2003b; Kassinove and Tafrate 2000; Tafrate et al. 2002). Anger tends to be negative because it is associated with aggression and violence in many people's minds. Yet, "anger seems to be followed by aggression only about 10 percent of the time and lots of aggression occurs without any anger (Kassinove and Tafrate 2002).

Constructive anger can alert a person's partner to a problem within the relationship. Anger expression is hard to overlook and usually focuses a person's partner on a problem that may have been overlooked by the partner for some time. When people frame anger in terms of solving a mutual problem, rather than as a chance to vent feeling, the question becomes, what can we do to solve the problem (Tavris 1989)?

Repressing anger over a long period of time tends to devitalize a relationship. It often leads to emotional detachment and indifference. It also disallows working toward a solution to the problem that is causing the repressed anger because the conflict is never brought to the fore, thus denying discussion and possible solution.

Anger also serves as a motivator, energizing people to tackle a problem. For example, anger served an empowering function following the events of September 11, 2001. It tended to provide people a sense of certainty and prepare them for action (Lerner et al. 2003).

Most normal people experience mild anger a few times a week. Some people experience more intense anger so often that it negatively affects their relationships and may interfere with their work and health. In this case, they may need help in anger management.

Women and men tend to differ in their expression of anger (D. I. Cox et al. 2003). Men tend to be more physically aggressive and more impulsive in dealing with their anger. They also have a revenge motive more often and are more coercive of other people. Women tend to be angry longer, more resentful, and less likely to express their anger. Women tend to use indirect aggression by "writing off" people, intending never to speak to them again (DiGuiseppe in Dittmann 2003).

Over What Topics Do Couples Conflict?

Intimate couples obviously conflict over topics that are unique to their relationship. However, various studies report some topics that appear to be fairly common points of conflict. Handling money, dividing household tasks, relative relationships, jobs, social activities, alcohol use, moodiness, handling anger, and children all appear to be possible areas of conflict for many couples. It is important to recognize that a couple's areas of conflict change with time. For example, very few couples report sexual conflicts during the first year or so of marriage. By the fifth year of marriage, couples rank sex as the third most important area of conflict, after time and attention and household tasks. Conflict over the division of household labor appears especially strong in women (Kluwer et al. 1996).

Paul Amato and Stacy Rogers (1997) found that conflicts over sexual infidelity, jealousy, drinking, spending money, moodiness, not communicating, and anger tend to increase the odds of divorce. Extramarital sex was a particularly powerful predictor of divorce. This is consistent with the research finding that, in at least one-third of divorce cases, one or both spouses had been involved with another person (South and Lloyd 1995) (see Chapter 7).

Lawrence Kurdek (1994) studied areas of conflict for gay (male), lesbian, and heterosexual couples. The areas of conflict fell into six clusters: power, social issues, personal flaws, distrust, intimacy, and personal distance. The gay and lesbian couples argued more about distrust, especially about previous lovers, than did heterosexual couples. The heterosexual couples argued more about social issues than either the gay or lesbian couples. There were no significant conflict differences between gay, lesbian, and heterosexual couples in the other four clusters. He did find with heterosexual couples that the husband's marital satisfaction was more frequently affected by how their wives resolved conflict (Kurdek 1994) than was the opposite.

Some research suggests that wives, more than their husbands, show increased levels of stress hormones during conflict. Also, wives tend to recall marital conflicts more vividly and reminisce more often about important relationship problems than do their husbands. These memories also induce stress-related physiological changes in the women (Gottman et al. 1998; Kiecolt-Glasser et al. 1997).

John Gottman and his colleagues (1998) have found that men's refusal to share power with their partners is a major source of conflict. "Our data suggest that only newly married men who accept influence from their wives are winding up in happy and stable marriages (Gottman et al. 1998, 19)." Richard Mackey and Bernard O'Brien (1995), in their study of long-term marriages, concluded that successfully married couples reported increasingly satisfying and effective handling of conflict and increased understanding and sensitivity toward their partners. In other words, these couples were acquiring better communication skills as their relationships continued. This suggests, again, the need to strive to improve one's marital relationship or any intimate relationship. Intimate relationships do not just take care of themselves. However, note that working at marriage is not the same as making work of marriage.

Family Life in the Information Age

The fast-changing world of communication technology impacts everyone (*Time* 2000b). Almost all American families have telephone service (94 percent), a television (98 percent), and a VCR/DVD (82 percent) (see Chapter 10). More recently, computers, cellular phones, pagers, and universal access to the World Wide Web via the Internet have speeded information sharing even more. Sixty-two percent of married-couple households use online services. For the first time, online access surpassed newspaper subscriptions in 2000 (Woodard and Gridina 2000). In the year 2000, over half of homes with children had online access. This technological sophistication raises many questions for families, especially those families with children (see Chapter 10).

The Internet and its offspring, The World Wide Web, can now supply information to consumers that was previously available only from experts, libraries, schools and universities, and books. Charles Smith (1999) suggests that "The Web is a public information gateway, an individual's information liberation (31)." For example, doctors who are accustomed to controlling information now have patients who arrive in their offices with printouts of the latest information about their illness, possible treatments, and the prescribed drugs that might be used to fight the illness. The car buyer arrives at a dealership with a printout of the manufacturer's costs for the vehicle he/she wants to buy.

The web has brought about a true democratization of knowledge, in that anyone can now give and get information on the web. This exposes the web user to the best and the worst of information. The open nature of the web means that the most valid and ethical information may appear next to deceitful and unethical rumor mongering.

Electrical communication is captivating. Learners who might otherwise be intimidated by more dominating colleagues are more likely to express their opinions and feelings on an electronic forum than they might face-to-face. Race, appearance, age, and sex mean little in this setting. What you say and how you say it are what counts in cyberspace. (C. Smith 1999, 32)

There are obviously problems that can arise with Internet use. A husband or wife may now have a virtual reality sexual affair via an Internet chat room. An intimate other may now seek sexual gratification via pornography via his/her computer. Children can stumble across graphic pornography. Such behaviors are likely to create conflict within a marriage or any meaningful relationship.

Despite some potential negatives of the communication revolution, the new technologies have also brought wonderful benefits. For example, there are many exciting and motivating educational programs for children; all topics from arithmetic and reading to reasoning and evaluation can be found. A child needing to do a class report on any subject can seek out the necessary information on the web. An adult needing to purchase an automobile can make the best possible deal by first getting all the pertinent information online before approaching an auto dealer. A child with a cell phone and/or pager can keep in contact with home, no matter where he/she is or what he/she is doing. A person shut in because of illness or infirmity has immediate access to all kinds of information and assistance. Such a person can shop for food, books, records, and so on, and can also send flowers and stay in communication with friends and relatives via e-mail.

Because not everyone has access to the web and/or a cell phone or pager, there is increasing political discussion about the *digital divide* between the information rich and the information poor. The rural poor, rural and central city minorities, young households, and female-headed households are the most likely not to have access to information technology (Hughes et al. 1999; Stanton and Chang 2000, 10–12). Through efforts such as supplying all schools with online access, private and governmental agencies are attempting to narrow this divide.

The exact effect that all of these technological advances will have on the family remains to be seen. Certainly, one possibility is to use the communication technology to draw family members closer together. When a parent works with a child on a computer learning project, it is exciting and rewarding for both parent and child. E-mail allows contact with family members at any distance. Websites containing information about family relations, parenting, self-help groups, and other pertinent topics can help the family and its individual members to function better.

Despite the many advantages created by the communications revolution, care must be taken in the use of this technology to avoid negative consequences. For example, Kraut et al. (1999) found that the use of the Internet by the 100 families he and his colleagues studied led to decreased face-to-face communication among family members and a slight increase in feelings of loneliness. Parents concerned about their children viewing inappropriate material on TV or on the web have begun to use blocking devices to filter what their children view. Despite the need for watchfulness over the new technologies, it is clear the overall benefits from proper use far exceed the possible negatives.

Summary

1. Good communication is one of the major characteristics of strong, successful families. Couples having marital trouble almost always report communication failure as a major problem.

2. Although communication problems are often the result of personal problems partners may have, society can also facilitate or hinder good communication.

3. When most people talk about failure in communication, what they mean is that communication has become so aversive that it causes discomfort. In other words, the problem is too much negative and hurtful communication, rather than no communication.

4. Three basic conditions must be met before good communication can be ensured. There must be a commitment to healthy communication. The partners must be oriented to growth and to improving the relationship. Neither partner must try to coerce the other through communications.

5. When these basic conditions are met, problem-solving skills can be called into play. Basically, five skills are involved in successful problem-solving communication. The partners must identify the ownership of the problem; each partner must be willing to speak up and state his/her position and feelings (self-assertion); each must be a good listener (empathic listening); each must be willing to negotiate; and each must be willing to use problem-solving methods, if needed.

6. Men and women tend to communicate differently from each other.

7. Using problem-solving skills and avoiding both physical and verbal abuse will enhance any relationship and keep it alive and growing.

8. The technological communications revolution affects all individuals and families. It is hoped that the positive outcomes of these changes will far outweigh any negative influence for most people.

Resources on the Internet

Companion Website for *Human Intimacy: Marriage, the Family, and Its Meaning,* Tenth Edition

http://sociology.wadsworth.com/cox10e/

Gain an even better grasp on this chapter by going to the companion website to take one of the tutorial quizzes, use the flash cards to master key terms, or check out the many other study aids you'll find there. You will also find special features such as GSS data, Sociology Online, and Census 2000 information that will put data and resources at your fingertips to help you with that special project or to help you as you do some research on your own.

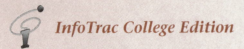

InfoTrac College Edition

http://www.infotrac-college.com/wadsworth/

You can access reliable resources anytime, anywhere, with InfoTrac College Edition, the online library. This fully searchable database offers more than 20 years' worth of full-text articles (not abstracts) from almost 5,000 diverse sources, such as top academic journals, newsletters, and up-to-the-minute periodicals, including *Time*, *Newsweek*, *Science*, *Forbes*, *The New York Times*, and *USA Today*. You can conduct electronic key word searches using key terms from this chapter to supplement your reading and learning experience. To aid in your search and to gain useful tips, see the Student Guide to InfoTrac College Edition, which you can access through the companion website for this book.

Is Honesty Always the Best Policy?

YES

The more open and honest the self-disclosure, the better the communication and thus the better the relationship. Withholding important information and keeping secrets from your partner builds a relationship on a dishonest basis. If the hidden information is discovered by a person's partner, trust between the couple is broken. Trust is essential for open communication, mutual understanding, and problem solving (F. Walsh 1998, 52). Once trust is broken, it is very hard to rebuild.

Often the partner withholding information becomes suspicious of the other partner. "If I am hiding something from my partner, then he/she is probably also hiding something from me." This is the old defense mechanism of projection at work. Complete trust requires complete honesty.

Generally, the more openness a couple can tolerate and agree upon between themselves, the better the relationship's chance of success. With the advent of sexually transmitted diseases such as HIV and herpes, honesty about infection with potential partners becomes not only important, but also legally mandatory (F. Cox 2000). Numerous states now have laws making transmission of a communicable disease, such as HIV, a crime. In addition, numerous civil suits have been won against HIV- or herpes-infected persons who did not share this information with their uninfected sexual partners.

NO

Some researchers suggest that both too much and too little self-disclosure can reduce satisfaction with a relationship (Galvin and Brommel 1982). In addition, Schuman and his colleagues (1986) suggest that simply increasing disclosure does not necessarily increase marital satisfaction. Instead, selective disclosure seems to increase marital satisfaction (Sillers et al. 1987). Others suggest that being completely honest in a relationship is not possible (B. Duncan and Rock 1993).

Many nonadaptive responses for both husbands and wives come as a result of open sharing of feelings about violated expectations. Contrary to popular belief about the benefits of such sharing, this open communication does not always lead to improved relations. For example, disclosure of an extramarital affair is almost always destructive to a marriage, yet not disclosing the affair can also have negative consequences (Van Matre 1992). Disclosure that is negative in affect, is hostile or nagging, or expresses excessive concern about oneself is also counterproductive. In these cases, less may be better.

Schuman and his colleagues (1986) suggest that the quality of the self-disclosure is an important variable that has been overlooked. They found that when there was high regard for the partner and this regard was apparent in the self-disclosure, the more disclosure the better. When there is low regard for the partner, greater openness seems to create less marital satisfaction.

The real question is how to disclose, rather than whether to disclose. What is said (the content), how positive or negative it is, and the levels on which it is said (superficial to deeply meaningful, intellectual to emotional) all have to be taken into account.

Certainly, both partners must share a willingness to converse. But equally important, they must learn how to communicate successfully and remain aware of each other's weak and sensitive points (Bandura 1997, 13).

Thus, how one discloses thoughts and feelings, especially negative ones, strongly influences how the communication will be accepted. Those who believe it is important to keep everything out in the open and to always express themselves to the fullest must also concern themselves with how this expression is conveyed and with the feelings of their partner. Otherwise, they may find such openness backfiring and actually weakening communication, rather than enhancing it.

There is a fine line between encouraging openness in one's partner and invading his/her privacy. Respect of each other's privacy is an important element of a successful relationship. Each couple has to work out the appropriate balance between openness and privacy in the relationship. Unilaterally crossing over this line will usually be offensive to one's partner. Rather than becoming more open, the offended partner may withdraw (becoming cool and distant) or attack (becoming sarcastic and resistant to ideas) (Strong 1983, 240). In these cases, openness reduces, rather than enhances, the chances for successful problem solving.

What Do You Think?

1. Do you think that a relationship can work when there is complete honesty in every aspect between the partners?

2. Have you ever been dishonest with a girlfriend or a boyfriend? Why?

3. Can you keep trust without basic honesty between partners?

4. How do you feel when you discover you have been lied to by a friend or partner?

5. How does tact relate to honesty?

© Mark Leibowitz / Masterfile

Dating, Single Life, and Mate Selection

Chapter
6

Questions to reflect upon as you read this chapter:

- Describe briefly several reasons for dating.

- List several factors that seem to influence mate selection.

- Which of these factors is most important to you?

- What questions might you want to answer before deciding to engage in premarital intercourse?

- List several possible problems associated with premarital sexual relations.

- List a number of questions you would want to answer before deciding to cohabit.

- Why does cohabitation seem more unfair to the woman than the man?

- What are the functions of engagement?

Perhaps the characteristic of mate selection in America that most distinguishes it from the rest of the world is that it is run by the participants themselves. There are no arranged marriages, no marriage brokers, no chaperones, no child-hood betrothals, no family-alliance marriages, and not even many marriages of convenience. In America, this important decision is made by the participants themselves.

Every society has a system, formal or informal, by which mates are selected and new families are started. In the United States, mate selection is carried out by relatively unrestricted interaction (dating) among unmarried persons. The selection process is fairly informal. However, once a couple decides that each is his/her choice for a future mate, the system becomes more structured, and engagement and marriage usually follow.

The topic of mate selection has become increasingly complex as American lifestyles have changed. Increasing numbers of Americans are delaying marriage, cohabiting, and divorcing; some never marry at all. In addition, as the life span increases, growing numbers of widows and widowers are also rejoining the ranks of the single. Because most people marry at some time in their lives and divorced persons tend to remarry, there is a large group of older people seeking mates. Certainly, mate selection at age 18 is different from mate selection at age 40, which is still different from mate selection at age 70. The selection of a mate requires some measure of dating by the couple involved. Hence, dating plays a meaningful role at some time in most people's lives.

Premarital American Dating

Dating is usually the precursor to mate selection. Modern dating in America developed largely after World War I, when it was encouraged by both the emancipation of women and the new mobility afforded by the automobile. The population moved from the countryside into the cities, where youth found many more opportunities to meet.

Modern youth are so accustomed to having access to each other that it is difficult for them to imagine how hard it was for a young man to meet a young woman at the turn of the century. Although America never had an arranged marriage system or a system of chaperones, meeting the opposite sex was still more difficult than today. After finishing elementary school, boys and girls went to separate schools. Often, an introduction of a young man to a young woman's parents had to be arranged before they could interact, and this was not always easy. Even if the parents approved of the young man, the couple could spend little leisure time together because most of their time was usually occupied by activities with other family members and work.

Modern youth now meet at parties, make dates via the telephone, and go off alone to spend time together on the town. American dating is participant-run for most young people. However, youth from immigrant families and also more traditional religious orders may not have this dating freedom. But in contrast to the closed choice of arranged-marriage societies, the American system is nearly open choice for the participants. With open-choice mate selection, the ensuing marriages are usually subservient to individual needs and desires. Individual attraction is the main guide to early dating. There is probably no completely open-choice dating and mate selection in the world, but America appears to come closer to it than any other country.

Open choice of a mate does not mean that the parents of young adults have no influence at all. Parents often try to influence their children's dating choices which, in turn, influences mate selection, through monitoring their children's dating activities. Studies find that parental monitoring, as contrasted with coercive control, is an important basic foundation for young people's later choices outside of parental purview (Longmore et al. 2001). Children, in turn, try to predispose their parents favorably toward their dating partners. In general, parents try harder to influence a daughter's choices. The more serious the relationship, the more parents and children try to influence one another. It appears that parents, and even roommates (college students), are more accurate in predicting the stability of a dating relationship than are the dating persons (MacDonald and Ross 1997). However, this fact seldom carries much weight with the dating person.

Why Do We Date?

We date for many of the following reasons:

1. *Dating fills time between puberty and marriage.* It is often simple recreation, an end in itself.
2. *Dating is a way to gain social status based on whom and how often one dates.* The American dating system makes it possible for certain persons to be rated highly desirable, and in this way, to raise their status within the peer group. Thus, dating can change a person's status in an open-choice system. The classic example of a change in status is the young man who dates and then marries the boss's daughter.
3. *Dating is an opportunity for the sexes to interact and learn about each other.* Because Americans live in small nuclear families, young people may have had little opportunity to learn about the opposite sex, especially if they have no opposite-sex siblings near their own age. Dating is also an avenue to self-knowledge. Interacting with others gives a person a chance to learn about his/her own personality and the personalities of others. Dating allows experience with a succession of relationships and the opportunity to learn something about gender, marital, and familial roles.

4. *Dating meets ego needs.* A person needs to be understood and considered important. Being asked for a date and being accepted yields importance, and a successful date usually involves understanding each other to some extent.
5. *Dating leads to mate selection for most individuals.* Most young people do not begin each date by asking, "Would I marry this person?" Yet sometime in the course of their dating experience, they will ask this question.

In early adolescence, dating is mostly considered fun. The older a person gets, the more serious dating becomes, and the more concerned that person becomes with mate selection. Thus, dating patterns can be placed on a continuum, leading from casual dating to marriage (Figure 6-1).

M. Zusman and D. Knox (1998) found that daters listed poor communication and lack of commitment as their top two problems. Remember that commitment and good communication are two of the major building blocks of successful, intimate relationships. Building commitment is really important if a relationship is to become permanent and lasting. Although the Zusman and Knox study involved premarital dating, mainly among college students, divorced and widowed persons also report more difficulty with building commitment than any other single facet of a relationship. Whether single, divorced, or widowed, women tend to complain about not finding commitment more than men do.

The emphasis on sexual intimacy in America today seems to lead to the neglect of other important aspects of a relationship that promote commitment (see "Changing Sexual Mores"). Social compatibility, development of shared interests, and increased knowledge of each other and of oneself are all important, especially if the relationship is to become permanent. If early sexual involvement overrides these concerns, it may actually work against the furthering of a relationship. Some studies suggest that having a number of premarital sexual partners elevates the risk of later marital disruption (Binstock and Thornton 2003).

Premarital Dating Patterns

Young people's interactions have become increasingly more individualized in the past few years. This great variation makes it difficult to describe a common American dating pattern. With the fluidity of the adolescent period and the fact that dating is usually participant-run, each new generation redefines the dating codes and norms. Modern dating has changed from the more formal dating of yesteryear in these three important ways:

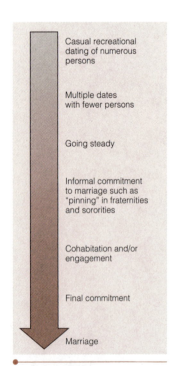

Casual recreational dating of numerous persons

Multiple dates with fewer persons

Going steady

Informal commitment to marriage such as "pinning" in fraternities and sororities

Cohabitation and/or engagement

Final commitment

Marriage

FIGURE 6-1
Dating Continuum

1. There is greater opportunity for informal, opposite-sex interaction.
2. Subsequent dating is less formal.
3. There no longer seems to be a set progression of stages (as in Figure 6-1) from first meeting to marriage.

Formal dating—where a boy approaches a girl beforehand and arranges a meeting time, place, and activity for them—has declined, especially in the larger urban areas. Formal *going steady*—where a girl and boy exchange class rings and letter jackets—has also declined in these areas.

More traditional, formalized dating is still found in small towns and rural parts of America. The man asks for the date, plans the date, and usually pays for the activities. The woman waits to be asked out, waits to be picked up, and usually follows the man's date plan, at least early on in the relationship (Laner and Ventrone 2000). Young people cruise Main Street on Friday and Saturday nights. Dates are prearranged and afterward, everyone meets at a drive-in restaurant or another popular spot. In small-town America, many still go steady for a good portion of their high school years.

In urban areas, relationships are less formal, more relaxed, and casual. And once a couple is past a period of almost total preoccupation with each other, there seems to be more group activity—going out with other couples or with friends.

Dating in high school differs from dating in college. Generally, high school dating involves a person's friends more than college dating does. High school friends tend to inform each other about who is interested in whom. An individual's reputation is more on the line according to who is being dated; "date a nerd, become a nerd" is the byword. Cliques play an important role in high school. There is more gossiping among high school students about who is dating whom. Dating casually or dating many partners in high school is harder, because the gossip tends to put a couple into a steady dating relationship, whether their dating is serious or not. There is a lot more difficulty and embarrassment in asking for dates.

High school students do more group dating for several reasons. On a group date, an individual is not so "on the spot." The whole group helps start conversations. Group dating reduces the chances of friends assuming that the couple is going steady. Activity costs are usually shared. The chances of unwanted sexual contact is also reduced in the group date situation.

College Dating: Hanging Out, Hooking Up, Joined at the Hip*

Dating among college students tends to be even more relaxed, with fewer rules. In fact, the term *dating* might not really be applicable for much of the male-female interaction among unmarried college students. Only 50 percent of senior college women reported having been asked out on six or more formal dates by men since entering college.

The past policy through which colleges and universities assumed some of the responsibility for college students (*in loco parentis*) via separate dorms for men and women, dorm mothers and fathers, and strict rules and curfews has largely disappeared. Many colleges and universities now have mixed-sex dorms, with males and females sometimes on the same floor and occasionally sharing bathrooms.

In fact, for many college men and women, dating is synonymous with "hooking up." **Hooking up** is defined as a sex-without-commitment interaction. It is

hooking up
Sex without commitment

---

*Much of the discussion in this section is based on the research by Glenn and Marquardt (2001).

a deliberately ambiguous phrase and can mean a couple kissed, performed oral sex, or had sexual intercourse without long-term commitment. Dating partners tend to know each other less well when they first get together than did dating partners in high school. There is usually a greater diversity of people in college and thus the possibility of greater dating diversity.

In part, because of the close proximity of men and women in unsupervised, coed dorms and mixed-sex, off-campus housing, a college couple who is dating is sometimes in a fast-moving, highly committed relationship that includes sleeping at each other's room most nights, studying together, sharing meals, and so on, but rarely going out on dates. The researchers termed these couples **joined at the hip**. Again, because of the close proximity of living conditions between college men and women, there is a great deal of hanging out, where time is loosely spent together without explicit interest in one another. This is often done in a group setting.

As a result of these changes in dating among college students, the culture of courtship—a set of social norms and expectations that once helped young people find the pathway to marriage—has largely disappeared. This is even truer among young adults who do not go on to higher education.

As a relationship continues and becomes more serious, the couple gradually limit their dating to each other. They enjoy the security of knowing they have a more permanent relationship. They feel comfortable together, and find it's a relief not to have to face the insecurity of finding and meeting a new potential partner.

Going together requires a higher degree of commitment than casual dating. This ability to commit oneself becomes the foundation of later marriage. Going together helps the couple understand what kind of commitments are necessary to marriage.

But going together also creates problems. It tends to add pressure to the sexual conflicts experienced by the young couple. As they see each other more frequently, it becomes increasingly difficult to avoid sexual intimacy (see "Changing Sexual Mores"). Young Americans also tend to be possessive. He/she may regard any attention or compliments paid to his/her girlfriend/boyfriend as threatening, so may try to restrict her/his social interaction. Resentment of this may lead to arguments.

On the whole, starting early and remaining in a steady relationship throughout adolescence, or being joined at the hip throughout college years, is probably disadvantageous to later adult relations. The young person who has always gone steady is unlikely to have had enough experience with a broad cross section of the opposite sex to have developed his/her interpersonal abilities to the fullest. Going steady early usually leads to earlier marriage, which tends to be less stable than later marriage. If dating is to work as a method of mate selection, it is important to date a variety of people to ensure a good mate choice.

On the other hand, hooking up or hanging out with many people is also dysfunctional. The young person never gets any practice in the give-and-take of long-term relationships.

Ideally, then, one must date enough people to understand the many individual differences that exist. At the same time, experiencing some longer-term relationships helps a person gain knowledge of the commitments and compromises necessary to maintain a relationship over time.

Dating varies with social class, ethnicity, and cultural background. Most dating is quite homogenous; that is, an individual tends to date people of similar race/ethnicity, religion, education, social class, and age.

joined at the hip
Couples who spend much of their time together but rarely go out on formal dates

Regardless of dating style, everyone faces the problems of finding dating partners, coping with bad dates, and avoiding exploitation. The fact that American dating and mate selection today are so informal and without rules, compared with mate selection processes in the past, means that each young person has to make his/her own decisions about what is best. Of course, many social and cultural pressures will influence these decisions, but these pressures vary and are not always clear-cut. With the increasing postponement of marriage and high divorce rates, it is clear that a large proportion of many people's lives will be spent as a single. And for most singles, regardless of age, dating is significant.

The important thing to remember is that any dating pattern, if it is to contribute to the mate selection process, must give each person sufficient experience with the opposite sex to make good decisions about intimate relationships. Knowledge about the opposite sex comes largely through interaction with the opposite sex. Knowledge about intimate relationships generally comes from being in such relationships. Often, dating couples do not recognize that the problems they are having when dating affect their chances for successful marriage. For example, high levels of premarital conflict are predictive of high levels of marital conflict. Working out problems in a satisfactory way during dating is good practice for the necessary give-and-take that helps make a lasting, intimate relationship work. This is why premarital counseling, and couple training in relationship enhancement and communication skills, help couples prevent relationship problems from arising; it is hoped that training will carry over into the marital relationship.

Dating and Extended Singleness

Thus far, we have examined the dating patterns of young Americans, while assuming that dating is a form of courtship that eventually leads to marriage. Not all dating Americans want to marry, however, and not all dating Americans are young. Because the average age of first marriage has increased, the period of singleness before marriage is longer today than it was in the recent past. The much higher divorce rate has also increased the number of older single persons, even though many divorced persons may have their children living with them. These two factors have combined to cause a dramatic increase in the number of people living alone. Despite the increase in singleness, it is still a transitory state for most Americans and usually ends with marriage or remarriage.

Yet single life is becoming more popular. The increasing emphasis on self-fulfillment as a major life goal and society's greater tolerance of differing lifestyles both reduce the pressure to marry. Personal growth and change have become popular goals, and they conflict directly with the traditional goals of long-term commitment to marriage. The women's movement has also contributed to singleness by telling women, "There are other roles you can fulfill that do not necessarily include being a wife and mother."

Greater educational opportunity also acts to postpone marriage, because going to school usually postpones economic independence. Greater acceptance of sexual relations outside marriage has also removed one strong reason for marrying. Despite these social changes, the idea of marriage remains strong in America. In a recent national survey, 83 percent of college women agreed that "Being married is a very important goal for me" (Glenn and Marquardt 2001).

There are a number of social and legal reasons why certain persons may not want to marry. For example, a woman may not want to change her name. Although

WHAT RESEARCH TELLS US

Women and Singleness

Time ran as its cover story "Who Needs a Husband?" (2000d). The authors found that there are 43 million adult females currently single (40 percent, up from about 30 percent in 1960). Looking at women aged 25 to 55 years, 83 percent were married in 1963; by 1997, that figure had dropped to 65 percent.

This is somewhat misleading, because some 10 percent are cohabiting with their lovers and another small per-centage are being more open about gay relationships.

The report found that nearly 60 per-cent of single women owned their own home, buying them faster than single men; that single women fuel the home-renovation market; and that unmarried women are making up half of the adven-ture travelers and two out of five busi-ness travelers.

While the birthrate has fallen among teenagers, it has climbed 15 percent among unmarried women in their 30s who apparently feel they have the where-withal to rear a child on their own.

Although the article is strongly posi-tive about remaining single, it does cau-tion that women may set themselves up for disappointment, putting off marriage until their 30s or later, only to find them-selves unskilled in the art of compatibil-ity and surrounded by male peers who are looking at women in their 20s.

she can marry and keep her maiden name, transacting daily business is often awkward and difficult if she does so. Two people may want to keep their eco-nomic assets separate. In certain years, taxes have been higher for married couples than for unmarried people (the *marriage penalty tax*). Older people may lose Social Security benefits if they marry. If one stays single, another person does not have to cosign for him/her to enter into various business transactions; another's driving record does not affect one's insurance rates; and another's credit history does not affect one's credit.

What effect will remaining single longer have on marriage? Obviously, both partners will be older. Older women face greater risks during pregnancy and childbirth. Couples will also be older when they have chil-dren, so the age difference between themselves and their children will be greater. Older couples will probably have fewer children, which will affect schools, the baby food and clothing industries, and so on. Older persons, used to a long period of singleness, may become more set in their ways. They may, then, find the compromises of the marriage relationship more difficult.

On the other hand, data suggest that the older a couple is at marriage (within limits), the greater the chances of marital success. Later mar-riages should also mean better economic circumstances for the couple. The maturity of older couples may also make them more willing to work toward a positive experience in marriage.

But what of the single life? What are the advantages? Freedom is probably the major one. A single person need only worry about him-/ herself. Obligations are undertaken voluntarily, rather than being dic-tated by tradition, roles, or law. One has time to do what one wants, when one wants. Expenses are lower than for a family. A single person can change jobs, or even cities, more easily. Independence can be maintained, and the conflicts about activities and lifestyle that arise in marriage don't occur.

Possible loneliness, failure to relate intimately, and a sense of mean-inglessness are potential disadvantages of singleness.

© PhotoNetwork/PictureQuest

Let's get in the swing of things.

WHAT RESEARCH TELLS US

Singles in America

1. Approximately 25.5 percent of adults aged 18 and over lived alone in 2000, compared to 17.1 percent in 1970. The majority were women (14.8 percent) (J. Fields and Casper 2001, 3–4).

2. There is a considerable difference in the age composition between men and women who live alone. Men tend to be much younger (median age, approximately 43 years) than women (median age, 66 years).

3. The trend to postpone marriage is increasing the number of young singles and thus the length of the dating period before marriage.

4. From 1970 to 2000, the divorce ratio (number of divorced persons per 1,000 persons who are married and living with their spouse) tripled. Divorced women represent about 13 percent of the adult population.

5. Another group that tends to be forgotten in discussions of dating are the widowed singles, who make up 12.5 percent of the adult population (widowed women, 10 percent; widowed men, 2.5 percent) (J. Fields and Casper 2001, 10; Kreider and Simmons 2003).

6. Never-married women account for 25 percent of adult women. Never-married men account for 31 percent of unmarried adult men. Obviously, most of these persons are young persons (Kreider and Simmons 2003).

7. Combining never-married single persons with the unremarried divorced and widowed singles, one can see there is a large pool of persons who might engage in dating behavior.

Changing Sexual Mores

In the past, mate selection in America was viewed as movement down two paths: the path of commitment, and the path of physical intimacy. At first, commitments are superficial. People think "Let's get together for an evening." At the end of the path is deep-seated commitment. They say, "Let's spend our lives together." The physical intimacy road runs from casual hand-holding to a full and continuing sexual relationship.

In recent years, the road to commitment seems secondary to the road to physical intimacy. In fact, today's youth, especially women, often complain that those in whom they are romantically interested will not make a commitment. Early physical intimacy in a relationship appears to make commitment growth more difficult. Around 1184–1186, in his book *The Art of Courtly Love*, Andrew Capellanus wrote that "the easy attainment of love (physical intimacy) makes it of little value; difficulty of attainment makes it prized" (as quoted in Murstein 1986, 104). Might this long-familiar knowledge explain some of the complaints of sexually active youth about the difficulty of finding commitment?

Puberty signals the beginning of adult sexuality. Children become biologically able to reproduce, and the male-female relationship takes on an overtly sexual nature. Adolescent years in most Western cultures are a time of sexual stress because, even though biology has prepared the individual for sexual intercourse and reproduction, Western society has traditionally denied and tried to restrain these biological impulses. For most people, sexual stress lasts until marriage. It is primarily in marriage that American culture allows its members to engage freely in sexual interaction. Assuming males enter puberty at about age 14 years and marry at an average age of 27 years, they must wait 13 years after biology prepares them for adult sexuality before society condones sexual intercourse (in marriage). There is a comparable 13-year period for females, assuming that puberty begins at approximately age 12 years and they marry at an average age of

25 years. Thus, as we have seen, much interaction among American young people revolves around sexual aspects of their relationships.

American society is not only more permissive than in the past; it actually encourages early contact between the sexes. Makeup, adult fashions, and adult activities are advertised as ways of increasing popularity. The teenage market is huge, and business creates and caters to the tastes of adolescents. Much advertising is based on sex appeal, thus heightening the tensions of this period.

Popular music is an especially strong influence in the lives of teenagers; major themes tend to be drugs, sex, and violence. Considering Americans' concern about such violence as rape and murder, it is ironic that much of the music American teenagers listen to romanticizes violence and the degradation of people, especially women. Musical groups such as Snoop Doggy Dogg, 2 Live Crew, and Tupac Shakur romanticize not only sex, but also bondage, sexual assault, and murder.

The stress of emerging sexuality is compounded by the extended opportunities a young American couple have to be alone together, as previously noted. The automobile not only revolutionized transportation and contributed to the highly mobile American way of life; it also facilitated early sexual experimentation. A boy and girl can be alone at almost anytime in almost anyplace. The feeling of anonymity and distance from social control is increased because no one is present who might comment or report on their behavior.

What we find in America is a society that supposedly prohibits premarital sexual relations, yet actually encourages them through the mass media and the support of early boy-girl relations. Young people are often left to their own resources to determine what their sexual behavior will be. In the end, they will make the decision about the extent of their sexual relationships, based on their attitudes, peer influences, and the pressures of the moment. The days of hard-and-fast rules about sexual behavior are gone.

It is now up to the individual couple to decide how far they will go. In the past, it was the female partner who controlled a couple's sexual behavior. However, as was indicated in the discussion of hooking up among college students, changing sexual mores have lessened her control of her sexual interactions by indicating that sexual behavior is okay.

The escalation toward physical intimacy as the couple dates longer may cause young people to become too centered on sex. Sexual intercourse does not end the preoccupation with sex, but rather, often exaggerates it.

The historical double standard that tacitly allowed sexual relations outside of marriage (before marriage, not within marriage) for males, but forbade them for females, has gradually disappeared. Slightly more than half of females have now engaged in sexual intercourse by their eighteenth birthday, compared to nearly two-thirds of males (see Table 6-1). More than twice as many females 14–16 years of age are sexually active today, as compared to just 15 years ago (Moore et al. 2000). Boys who have had sex in the past are unanimous in wanting to continue to have it, while about one quarter of the girls who have had sex do not intend to have sex again, at least in the next year (Gillmore et al. 2002, 895). This suggests that an early sexual experience for a young girl is not as satisfying as it is for a boy. Also, girls who want to have sex are more successful in implementing this decision, suggesting that finding a willing partner is less of a problem for them than for boys (Gillmore et al. 2002, 895).

TABLE 6-1

Gender and Ethnic Differences in the Timing of First Sexual Intercourse

Characteristic	% Distribution	% Who Have Had Sex	Median Age at First Sex
Total	100.0	26.6	16.9
Gender			
Male	53.4	29.7	16.6
Female	46.6	23.3	17.2
Ethnicity			
White	25.9	27.0	16.6
Black	11.1	45.6	15.8
Hispanic	48.7	24.4	17.0
Asian American	10.7	14.2	18.1
Other	3.6	34.2	17.4
Family Structure			
Both biological parents	57.9	19.4	17.6
Single parent	26.0	35.9	16.4
Stepfamily	12.6	36.2	16.2
Other	3.5	44.4	16.6

NOTE: Data obtained from Los Angeles County.

SOURCE: Upchurch et al. 1998, Table 1.

The majority of teenagers who have engaged in sexual intercourse report that "it just happened," usually in the home of one of the partners. Other reasons given are: "They were curious and wanted to experiment," "They wanted to be more popular or impress their friends," "They were in love," and "They were under pressure from those they were dating." For females, the first sexual partner is, on the average, several years older. For males, the first partner averages less than 1 year older. First intercourse is usually, but not always, consensual. The younger the first experience, the greater the likelihood that first intercourse was nonvoluntary (Moore et al. 1989, 2000). Sixteen percent of girls whose first intercourse was before age 16 years report that it was not voluntary (coerced, not necessarily rape), compared to just 3 percent of women whose first intercourse was at age 20 years or older (CDC 1997).

It is clear that American sexual mores have become increasingly more liberal. However, the increased incidence of sexually transmitted diseases (such as acquired immune deficiency syndrome [AIDS] and herpes); growing social concern about unwed pregnancy and the single, never-wed, female-headed family; the high number of children from single-parent families growing up in poverty; and increasing female dissatisfaction with sexual exploitation by men may lead to an increased societal concern about sexual permissiveness.

The ultimate decision about sexual behavior will be made by young couples themselves. It is best to supply young people with as much good information as possible, so that their decision will be based on a firm foundation of knowledge, rather than on ignorance. In a nutshell: Knowledge breeds responsible sexual behavior; ignorance just breeds.

Deciding for Yourself

An unmarried couple contemplating sexual intercourse should consider (1) personal principles, (2) psychological principles, (3) social principles, and (4) religious principles.

Every person has a set of *personal principles* that guide his/her life. The following are some personal questions you should ask yourself if you are contemplating nonmarital intercourse:

1. Is my behavior going to harm the other person or myself, either physically or psychologically? Will I still like myself? What problems might arise? Am I protecting my partner and myself against sexually transmitted diseases and pregnancy?
2. Will my behavior help me become a good future spouse or parent? Do I believe sex belongs only in marriage?
3. Is my sexual behavior acceptable to my principles and upbringing? If not, what conflicts might arise? How will I deal with these?

Of course, questions such as these have no generally correct answers. Each individual will have different answers, but young adults should ask themselves such questions before they engage in sexual experimentation.

Psychological principles may be the hardest to uncover. Because the socialization process begins at birth and continues throughout a person's lifetime, it is difficult to remain completely unaffected by society and family. Many of our attitudes are so deeply ingrained that we are unaware of them. When our behavior conflicts with these attitudes, we usually feel stress and guilt. The following are some of the psychological questions that you must grapple with:

1. Can I handle the guilt feelings that may arise when I engage in nonmarital sex?
2. How will nonmarital sex influence my attitudes and the quality of sex I experience after marriage?
3. What will I do if I (or my partner) get(s) pregnant? Can I handle an abortion? Would I be able to raise a child? Could I give the baby up for adoption? Am I ready for marriage?

Other questions arise when one considers general *social principles*. Our society has long supported certain rules about sexual behavior. If enough people break these rules, pressure is placed on the society to change the rules. Thus, each person who decides to act against the established code adds his/her weight to the pressure for change. Before you make such a decision, you should ask yourself the following:

1. What kind of behavior do I wish to prevail in our society? Is nonmarital sex immoral? Will nonmarital sex contribute to a breakdown of morals? Is this desirable?
2. What kind of sexual behavior do I believe would make the best kind of society? Would I want my friends to follow in my footsteps? Would I want my little brother or sister to emulate me?
3. Am I willing to support the social rules? What will happen if I don't?

Questions concerning *religious principles* also need answering. Most of us have had religious training, and we have learned attitudes toward sexual behavior from that training. In an informal study of several hundred community college students' attitudes toward nonmarital sex (F. Cox 1992), 90 percent of those who were against it gave religion as their primary reason. The following are some questions a religious person needs to ask:

1. What does my religion say about sexual conduct? Do I agree?
2. Am I willing or able to follow the principles of my religion?
3. Do I believe there is a conflict between the sexual attitudes of my church and society? Are the attitudes of my church and my friends at odds?

Freedom of Choice and Sexual Health

Freedom of sexual choice for the young, unmarried individual can be much more threatening than for the married person. The sexual mores in many parts of American society have changed from supporting postponement of sexual intercourse until marriage to pressuring young people to engage in nonmarital sex (see Chapter 8).

Splendor in the grass.

Yet, part of a healthy model of sexual behavior is the freedom to choose whether to participate in sexual relations. Being forced into sexual relations will not lead to a healthy experience. As noted, studies find that the younger the person involved in nonmarital sex, the greater chance that coercion was involved (Moore et al. 2000, 3).

Healthy sexual expression is a primary part of human intimacy. As we try to make good decisions about our sexual intimacy, it is helpful to think about what we mean by healthy sex. This topic is much debated. Some maintain that no sexual involvement before marriage is healthy. Others argue that complete sexual freedom is healthy. In-between are a variety of less extreme viewpoints.

Possible Problems Associated with Premarital Sexual Relations

As American mores have relaxed, the distinction between nonmarital and marital sexual activities has blurred. However, a number of problems are clearly more closely related to nonmarital sex than to marital sex.

Sexually transmitted diseases (STDs) (see Appendix A) is more prevalent among unmarried participants in sexual activities. Their chances of having more than one sexual partner are greater than married people's. They also are more likely to have short-duration sexual encounters in which communication is less open. Such encounters may lead to later discovery of an STD.

Unwanted pregnancies are a problem. Despite the increased availability of birth control, nonmarital pregnancies have increased (see the section "Children Having Children" in Chapter 9). It should be noted that the actual birthrate among teenage mothers was higher in the 1950s than it is now, but it reflected the lower age of marriage at that time. In fact, teenage births declined from 37.5 per 1,000 women in 1990 to 23.2 per 1,000 women in this age group in 2002 (Sutton et al. 2004, Figure 3). In 1970, 7 of 10 teen births were within marriage. Today, the opposite is true; 8 of 10 teen births are outside of marriage (NCHS 2000, 4).

Early commitment and isolation are frequent partners of premarital sexual involvement. Sex is such a powerful force in young people's lives that it can override other aspects of a relationship. Sexual involvement often reduces social, intellectual, and other areas of involvement. Sexual activities can also promote exclusivity in a relationship, thus narrowing a young person's interpersonal experiences. Such relationships may lead to sexual commitment alone rather than to a total relationship. An early commitment based on only one aspect of a relationship, especially sex, is usually an unstable basis for any long-term relationship.

The *quality of sex may be impaired* by nonmarital sexual experience. Because nonmarital sexual intercourse is becoming more prevalent, an increasing number of American youth may be initiated into sexual intercourse under adverse circumstances (Moore et al. 2000, 3). Having sexual relations under circumstances that arouse fear, conflict, or hostility can cause sexual problems later in life. Masters and Johnson (1966) point out that these three mental states most often cripple the sex life of both men and women.

Nonmarital sex for the very young in our culture is often of relatively poor quality for two major reasons: a negative environment and the sexual ignorance of the young couple. The environment of early sexual experiences is seldom conducive to relaxed, uninhibited sexual activities. For example, some sexual en-

---

HIGHLIGHT

Dawn and Nonmarital Pregnancy

Dawn is a 20-year-old college junior, majoring in business administration. She has gone with Louis for 2 years. He is a senior in premed and has been accepted to medical school. She recently discovered that she was pregnant. In the past, both she and Louis agreed to postpone marriage until they had finished school and Louis had established a medical practice.

Louis is angry with her for becoming pregnant. "How stupid of you to forget to take your pill," he mutters. "Are you sure that you didn't do it on purpose? You could have told me, and I'd have used a condom. But I think it is really up to the girl to protect herself."

Dawn doesn't quite agree. "You're having sex with me, too," she responds with feeling. "I don't see why birth control must always be my responsibility. You know as well as I do that I didn't want to get pregnant. It wasn't my fault that I forgot. After all, you didn't have to make love to me. You were the one who was hard up and pushing, not me. But now that I am pregnant, I'm going to keep the baby. A lot of my friends are doing it. Having a baby before you are married isn't half as bad as it used to be. I don't really care what you or our parents think; it's the modern thing to do."

Louis objects strongly: "Well, I'm not going to marry you under these conditions. We agreed to wait until I finished medical school, and you blew it. The only thing you can do is get an abortion. They're easy to get now. There is nothing to them physically, and I'm willing to pay for it. Keeping the baby will

just foul up our lives and tie us down. We'll have plenty of time to have children when we finish school."

"No abortion for me," replies Dawn. "I don't think they're right. Besides, I don't expect you to marry me. I wouldn't want you to feel forced into anything. I've only a year of school left, and I'm sure my folks would watch out for the baby until I'm working and independent. Then, with child care, I'll be able to take care of the child just fine, without you."

"Well, if you're that stupid, I'm glad I found out now," retorts Louis. "I won't be a part of such a dumb plan. Unless you do the smart thing and get an abortion, I'm through with you, pregnant or not."

What Do You Think?

1. How will Dawn and Louis resolve their conflict? What would you advise?
2. What would you do if you were Dawn? What if you were Louis?
3. What kind of attitudes show through Dawn's statements? What is Louis's attitude?
4. How would your parents react to such a situation?
5. What alternatives are open to the couple?
6. How do you feel about illegitimacy? How do you think society feels today?
7. Whose responsibility is birth control? Why?

MAKING DECISIONS

Is My Sexual Behavior Healthy?

The following questions may help you discover foundations in your basic attitudes that can promote healthy sex.

1. *Does my sexual expression enhance my self-esteem?* Behavior that adds to me, increases my self-respect and my positive feelings about myself, and helps me to like myself better is behavior that is apt to be healthy. Behavior that creates negative self-feelings and causes loss of self-esteem is better avoided. Low self-esteem creates many problems, especially in intimate relationships.

2. *Is my sexual expression voluntary (freely chosen)?* Obviously, rape is neither voluntary sexual expression nor health-enhancing. Other situations, however, are not always as clear-cut. Is my behavior voluntary if I have sex because I fear I'll lose my boyfriend/girlfriend if I don't? Perhaps it is, yet the element of fear raises a doubt. Does the fear make me think I must do it? Does the fear rob me of voluntary choice? Does the fear cause me to overlook the broader question: if I will lose him/her only because I will not have sex at this time, is he/she really someone with whom I want to have an intimate relationship? If my decision is really mine, independent of peer and social pressure, then the chances increase that my chosen behavior will be healthy.

3. *Is my sexual expression enjoyable and gratifying?* This may sound like a strange question to ask. Isn't all sex fun and enjoyable? We often find that it is not. Many people report disappointment with their early sexual encounters. For most, however, positive answers to the first two questions will help answer this question positively. At times, sex for its own sake is gratifying; but, in general, enjoyable and gratifying sexual expression tends to occur most often within the context of intimate relationships.

4. *Will my sexual expression lead to an unwanted pregnancy?* Sexual activity leading to wanted children is healthy. Sexual relations using birth control methods and thereby avoiding unwanted children can also be healthy. Some people disagree with the latter statement for religious reasons. For them, healthy sex might mean abstinence if children are not desired at a given time. Taking steps to prevent unwanted pregnancy is an important element in healthy sexual expression.

5. *Will my sexual expression pass a sexually transmitted disease to my partner or myself?* Obviously, healthy sex in the medical sense does not promote sexually transmitted diseases (STDs). Thus, knowledge of STDs and taking precautions against them must be part of healthy sexual expression (see Appendix A).

---

counters take place in a car. Having to hide each time another car passes doesn't help the couple relax and feel secure. Both relaxation and security are important psychological prerequisites for satisfying sex. The general environment is especially important to a woman's ability to find satisfaction.

Many women report that their first sexual intercourse was not very enjoyable. This is not to deny that first intercourse can be pleasurable and exciting, but some young women clearly are disappointed. Such disappointment may breed later problems in their attitude toward sex. Because much modern literature, movies, and television depict sexual satisfaction for the female as a wild and violent climax, the woman may interpret her lack of enjoyment as a personal shortcoming. She may begin to think that something is wrong with her sexually, a belief that will, in turn, increase her anxiety and make her less capable of finding satisfaction. Thus, an early negative sexual experience may start a vicious circle of behavior.

In most cases, nothing is wrong with the young woman. If her early sexual encounters take place in a secure environment with a man who appeals to her psychologically and physically and offers her intellectual rapport, warmth, and a feeling of self-respect and if her expectations are realistic, she will probably achieve a great deal of satisfaction. Although we have mainly discussed the problems of the disappointed woman, if she shows her disappointment, the man will often feel threatened. He may react with feelings of inferiority of his own because he has apparently been unable to satisfy her. His ability to satisfy is one of his chief masculine ego boosters, and he is highly vulnerable to insecurity in this area, especially if he is very young. Yet, he needs to hear and understand what his partner is experiencing.

Most young men reach orgasm very quickly, compared with young women. Often, the male has ejaculated before the female has even become aroused. Premature ejaculation is the major problem of the young male, while failure to achieve orgasm is the major problem of the young female. The two problems are complementary and, therefore, contribute greatly to dissatisfaction with early sexual experiences.

The problems described in this section, although prevalent, do not always appear.

Date Rape and Courtship Violence

The term *date rape* has suddenly become popular in the mass media. A better term that encompasses rape but also speaks to broader problems is *acquaintance violence*. Although the figures vary, numerous studies report 15 to 40 percent of persons have been involved in receiving or inflicting violence while dating. Unfortunately, violence is not always defined in the same way. In some studies it means actual physical violence, whereas in other studies it may mean verbal abuse. In contrast to common beliefs, several studies report the prevalence of dating violence to be about equal for males and females (Larimer et al. 1999; Symons et al. 1995), although women are more likely to be victims of physical force. Acts of aggression by women tend to be overlooked, compared with aggressive acts by men, because men usually are physically bigger and stronger than women and their aggressive acts are more apt to produce injury.

Generally, women of all ages, races, and marital status report that most incidents of violence are committed by an acquaintance, friend, or intimate other, rather than by a stranger (Berkman 1995, A14). This also holds true for homosexual men who are raped (Hodge and Cantor 1998). Many couples dismiss minor aggression, either as a sign of love or as brought on by outside circumstances, such as stress. Women tend to react more passively than men to courtship violence. Many don't report the incident and ignore it while maintaining the relationship.

Although rape on a first date involves a more pathological, antisocial personality on the part of the rapist, rape later in a relationship may, to some degree, involve gender differences in socialization toward sex. As we shall see in the sexuality chapter (Chapter 8), men tend to develop sexual intensity earlier than women do. Also, the sexual double standard (although not as strong as in the past) has encouraged men to be sexually active at an earlier age than American women. Traditionally, males have controlled dating by initiating a date, and the use of force may be, partially, the exertion of that control.

Today's more liberal sexual attitudes also play a role in the incidence of violence by leading both men and women to believe that sexual relations between couples, regardless of marriage, are now the norm. After reviewing numerous studies, Shotland (1989) concludes that "the available evidence strongly supports the proposition that males view females in more sexual terms than females view males." Men and women see the relationships among sexual interest, love, and plain friendship differently. Men often appear to confuse friendly with sexually interested behavior. This perceptual mismatch is important because it is structured so that if a miscommunication around sexual intent occurs within a couple's relationship, the man will likely perceive sexual intent when the woman feels that she communicated none. Combine this scenario with alcohol or drugs, and the stage is set for an episode of date rape.

Researchers have found that males who engage in date rape have different characteristics from those who do not. They place a higher value on sexuality and feel greater sexual deprivation. They tend to be more traditional in their sexual values, perhaps believing more in the sexual double standard, viewing women as property, and so on.

Misunderstandings about sex are likely to arise between dating couples because of the following:

1. Many men and women do not discuss their sexual intentions openly and frankly.
2. Differences exist in perceptions of sexual intent.
3. The use of token resistance by some women may create a belief in some men that protest encountered is not really meant by the woman.
4. There are differing expectations concerning the stage of the relationship when sexual intercourse is appropriate.
5. Miscommunication is more common when alcohol or drugs are being used.

Until the couple honestly discuss their sexual wants and desires, or until a sufficient history is in place by which future behavior is predictable from past behavior, miscommunication and misperception are likely to occur.

Other researchers point out that dating partners who experience aggression, either in the dating situation or in other situations such as at home, are more apt to be aggressive themselves. Sexual aggressiveness by a person in one's peer group tends to make all group members more aggressive. To some degree, people learn to be aggressors or victims by observing parents and peers. Witnessing a parent hitting the other parent or being struck by an adult are positively associated with the perpetration of dating violence in both males and females. It is interesting to note that being hit by a mother is positively associated with future dating violence in girls, but not boys (Foshee et al. 1999).

Most persons tend to view all violence between dating couples as damaging to a further relationship, yet relationships often continue, even after episodes of date rape. Researchers have also consistently found that physical aggression is more common in cohabiting couples than in married couples (Magdol et al. 1998; Stets 1991, 669).

With the greater concern about date rape, there are increasing reports of false accusations. The FBI reports that about 8.5 percent of reported rapes (not date rapes) are unfounded (Dershowitz 1994a, 275). An unhappy, vindictive ex-lover can inflict great harm on a partner by accusing him of rape or date rape. Individuals can protect themselves to some degree from such episodes by clearly explaining their expectations and limits to potential dating partners.

There is a rising incidence in the use of date rape drugs GHB (gamma hydroxybutyrate) and Rohypnol (flunitrazepam). These are akin to the old time "mickey" slipped into a person's drink. They sedate the victim and cause short-term memory loss. GHB is now controlled since the Date-Rape Drug Prohibition Act of 2000 was signed into law on February 18, 2000.

Statutory rape (adults having consensual sex with a minor) must not be forgotten. Although for many years such laws have been ignored, the high teenage

statutory rape
Adult having consensual sex with a minor

pregnancy statistics, with pregnancies often caused by considerably older men, have inspired new steps toward enforcing such laws. In 1995, California started prosecutions. California state law forbids anyone to have consensual sex with a minor who is 3 years younger than them. Successful conviction can bring a prison term of up to 10 years. Although most prosecutions are of men, there are some prosecutions of women.

Cohabitation: Unmarried-Couple Households

cohabitation
A couple living together without being married

A courting couple's desire to spend more time together may lead them to cohabit, that is, live together without marriage. The increasing incidence of heterosexual **cohabitation** is related to the increase in the single lifestyle due to the older average age at marriage. Between 1970 and 2000, the median age of first marriage for women increased 4.3 years, to 25.1 years; for men, the increase was 3.6 years, to 26.8 years (J. Fields and Casper 2001). The high divorce rate also contributes to increased numbers of individuals living singly. Approximately 26 percent of households consist of people living alone (T. Simmons and O'Neill 2001). The number of unmarried-partner, opposite-sex households has grown from 523,000 in 1970 to 4.9 million in 2000, which accounts for 5.2 percent of all households (T. Simmons and O'Neill 2001). (An unmarried partner is now defined by the Census Bureau as a person unrelated to the householder, but who shares living quarters and has a close personal relationship with the householder.)

Looking at cohabitation figures in a cumulative sense, nearly 60 percent of unions formed in the early 1990s began with cohabitation (Bumpass and Lu 2000; Teachman 2003). It is estimated that about a quarter of unmarried women ages 25–39 are currently living with a partner (Bumpass and Lu 2000; Popenoe and Whitehead 2002; Whitehead and Popenoe 2003).

Although much is made of the rapid increase in cohabitation, it overlooks the fact that prior to the nineteenth century, especially in the unsettled western frontier, young couples lived together without being married. Basically, this was because there was no place to obtain a marriage license and no public official available to marry the couple. Most of those couples expected to marry and usually held themselves out as married. Most states legally recognized them as married after a set number of years (seven in most states). This legal change of status from cohabiting to married is called **common law marriage**. All of the laws pertaining to marriage applied to the couple, including the need to legally divorce if they wished to break up their common law marriage. As legal marriage became more available, most states dropped common law marriage. Today, only 13 states recognize common law marriage (Willetts 2003, 940–941). (Check your own state's marriage laws to see if common law marriages are recognized.)

common law marriage
In some states, a couple living together for more than a certain number of years can be treated as legally married

So many couples are now openly living together in intimate relationships that some theorists consider living together to be an ongoing part of the mate selection process for a growing minority of couples. If cohabiting couples are thought of as families, then the postponement of marriage we are witnessing doesn't really mean a longer period of premarital singleness. Young cohabiting couples actually make up for the lowered marriage rate (Seltzer 2000).

There are a number of reasons for the increase in nonmarital heterosexual cohabitation, as follows:

- Society's increased tolerance toward nonmarital sexual relations makes intimate cohabitation more acceptable.

- Higher education, especially for women, and the increasing entry of women into the workforce have lessened women's dependence on marriage as a way of economic survival.

- Increasing urbanization leads to increased anonymity and fewer restrictions on individuals.

- The high rate of divorce may make young people more wary of rushing into a marital relationship.

The Nature of Cohabiting Relationships

People choose to live together for many of the following reasons, some of which may not be true for every cohabiting couple:

1. Many couples consider these experiences to be no more than *short-lived sexual flings* and only gradually drift into a cohabitation relationship.
2. Some couples live together for *practical reasons*. They are essentially no more than opposite-sex roommates. In this case, the couple live together without necessarily having a deep or intimate relationship. Having a member of the opposite sex as a roommate affords certain advantages. A woman who lives in a high-crime area may feel safer living with a man. The couple can learn to share skills; for example, she might teach him to cook, while he teaches her automobile maintenance. Generally, the couple simply lives together as two same-sex roommates would; they date others, keep their love lives out of their living quarters, and react to each other as friends. Each gains something from living together in partnership rather than separately. The reason most often given for such an arrangement is to save money. This is especially true of elderly and/or retired cohabitants. In light of the new Census Bureau definition, this is not true cohabitation and will not be counted as such, as it has been in the past.
3. Some couples see cohabiting as a true *trial marriage*. As one young woman explained, "We are thinking of marriage in the future, and we want to find out if we really are what each other wants. If everything works out, we will get married." Some of these couples go so far as to set a specific time period. About 50 percent of first-time cohabitations lead to marriage within 5 years (Bumpass and Lu 2000; Ciabattari 2004, 118).
4. Other couples view cohabiting as a more *permanent alternative to marriage*. For example, African Americans tend more often than Caucasians to cohabit as a long-term alternative to marriage (S. L. Brown 2000).

   Some permanent cohabitants have philosophically rejected the marital institution as being unfavorable to healthy and growing relationships. They are especially critical of the legal constraints imposed on partners' rights in marriage. Many couples live together to avoid what they perceive to be constraining, love-draining formalities associated with legal marriage. They say, "If I stay with my mate out of my own free desire rather than because I legally must remain, our relationship will be more honest and caring. The stability of our relationship is its very instability."
5. Some couples, especially those in which one or both have been divorced, live together out of *fear of making the same mistake again*. These couples also have the vast majority of children in such relationships.

Is the Woman Exploited in Cohabitation?

Obviously, the answer to this question will depend on the individual relationship. However, some interesting statistics bear on this question in a general way. A number of studies indicate that males seem less committed than their female partners in living-together relationships (Nock 1995, 59). A larger proportion of cohabiting women than noncohabiting women want to get married, yet only a minority of men report marrying the woman with whom they cohabited.

Many cohabitation relationships are short-lived and tend to revolve around the sexual part of the relationship. In fact, men tend to list sex as their major reason for cohabiting. If the man tends to be less committed to the relationship than the woman, it is a reasonable assumption that he is often the one who ends the relationship, if not directly then indirectly by refusing equal commitment.

Spanier (1986) also points out that cohabiting men are much less likely to be employed than married men, but cohabiting women are much more likely to be employed than married women. This might suggest that men are using women economically to help themselves get through school or to pursue their own interests.

Many cohabiting couples share expenses equally, even though the woman's earnings are usually less than the man's. A man earning $40,000 a year who shares expenses equally with his live-in girlfriend earning $20,000 a year is at a distinct monetary advantage.

Although cohabitation appears to be avant-garde, suggesting that those who are involved are liberated, many of their attitudes are quite conventional (Ciabattari 2004; DeMaris and MacDonald 1993). For example, the division of labor in the household is often traditional; that is, the woman does the cooking, cleaning, and other household work. Women increase their housework when they cohabit with men, whereas the opposite pattern holds true for men (Ciabattari 2004, 119; Gupka 1999).

The proportion of cohabiting women who experience some kind of abuse is larger than the proportion of married, divorced, or separated women who feel they have been mistreated.

Judith Krantz (1984), novelist and commentator about modern American women, suggests that from the woman's point of view, cohabitation is a losing proposition. She lists the following misconceptions commonly held by some women:

1. He is not ready to make a total commitment, but once we are living together I know it is only a matter of time until we marry.
2. How can you really get to know a person unless you live with them?
3. We're in love, and I trust him completely. Why bring marriage into it? It's just a piece of paper.
4. I'm simply not ready to get married yet, but I don't like being alone and need someone to give my love to.

The young woman who early on enters a cohabitation relationship is robbed of her early years of independence, living alone, and learning about her own individuality. The best time to learn about oneself is before taking on the responsibilities of marriage. This is a short but precious time and should not be given away lightly. To lose this time to an early marriage is bad enough, but to lose it to a meaningless cohabitation experience is even worse.

Perhaps a couple considering cohabitation should carefully examine the nature of the intended relationship before entering it and answer the "Questions for Couples Contemplating Living Together" on page 175.

The Relationship between Cohabitation and Marriage

Many young people argue persuasively that living together provides a good test for future marriage. Living together, they say, is like a trial marriage, without the legal framework required to end the relationship if it doesn't work. Cohabitants embrace a variety of other arguments in support of cohabitation, as follows:

- It provides an opportunity to try to establish a meaningful relationship.
- It can be a source of financial, social, and emotional security.
- It provides a steady sexual partner and companionship, thus providing the central pleasures of marriage without so much commitment and responsibility.
- It provides a chance for personal growth—a chance to increase self-understanding while relating to another person on an intimate basis.
- It enables cohabitants to develop a more realistic notion of their partners and, generally, less-romanticized ideas about the relationship.
- It gives cohabitants a chance to get beyond typical courtship game playing.
- Long periods of intimate contact provide an opportunity for self-disclosure and concomitant modification and/or realization of personal goals.

Although these arguments seem logical and reasonable, research to date on the quality of marriage after cohabitation experiences has found little, if any, relationship between cohabitation and the degree of satisfaction, conflict, emotional closeness, or egalitarianism in later marriage. Although most cohabiting couples state their intent to marry at some time in the future, over 40 percent never make it to the altar. Of those who do, more than half divorce within 10 years, compared with 30 percent of married couples who did not live together (Wetzstein 1998). Marital instability is greatest with serial cohabitants (Cohan and Kleinbaum 2002; DeMaris and MacDonald 1993, 399). Perhaps as with divorced persons, serial cohabitants have experienced breakup and find subsequent breakup (divorce) less threatening and easier to do.

Researchers (Cohan and Kleinbaum 2002; DeMaris 2001; Dush et al. 2003; S. Fields 1999; Labi 1999; Popenoe and Whitehead 2002; Qu 1999; Wetzstein 1998) have found the following negative effects of cohabitation:

- Cohabiting partners are more unfaithful and fight more often than married couples.
- Cohabiting couples have higher rates of domestic violence than married couples.
- Cohabiting couples are more often estranged from family support than are married couples.
- Marriage after cohabitation tends to be more unstable.

One might predict that as cohabitation has become increasingly common, many of the early-found negative consequences for later marriage might be moderated. However, this has not proved true.

Even though many cohabiting couples view living together as a way to assess compatibility prior to marriage the evidence strongly suggests the opposite is true.

Among married individuals, premarital cohabitation is related to lower marital satisfaction, less time spent together in shared activities, higher levels of marital disagreement, less supportive behavior, less positive problem solving, more reports of marital problems, and a greater likelihood of marital dissolution. These characteristics hold true for remarriages as well as first marriages. (Dush et al. 2003)

These effects are generally stronger for couples who have cohabited for longer periods before marriage and for those who have cohabited more than once. Couples who intend to marry before entering a cohabitation relationship with their would-be marital partner appear to have fewer of these problems after their marriage.

Perhaps a contributing factor to the higher divorce rate for couples who cohabited prior to marriage is the finding of increased sexual contact outside of marriage for such couples. Approximately 20 percent of women who cohabited before marriage have had a secondary sexual partner while married, compared with only 4 percent of married women who had not cohabited (Forste and Tanfer 1996).

Another factor that appears to influence cohabitation stability is having a child/children with the cohabiting partner (Wu 1995, 1996). This seems to increase stability, whereas children of one cohabiting partner (stepparent cohabiting families) may reduce stability of the cohabiting relationship (S. L. Brown 2004). However, transitions in and out of cohabiting relationships are hard on children, and when added to marital breakups, increase current family instability by about 30 percent for Caucasian children and 100 percent for African American children (Raley and Wildsmith 2004).

Susan Brown (2004) found that even children living in cohabiting families with two biological parents experience worse life outcomes on average than those residing with two biological married parents. However, among children 6–11 years of age, economic and parental resources attenuate these differences. Among adolescents 12–17 years of age, parental cohabitation is negatively associated with well-being, regardless of the level of resources.

One apparently inexplicable phenomenon found among cohabitants who do marry is the number of couples who divorce shortly after marrying, even though they may have a long cohabitation history. Although little data are currently available on this behavior, the following two hypotheses might explain it:

1. Cohabitation teaches the couple to withhold total commitment. One of the reasons for cohabitation is to postpone or avoid the total commitment of marriage. Thus, when a cohabiting couple do marry, they may find themselves unable to live with the commitment imposed by marriage.

   The ephemeral nature of cohabitations observed in everyday life may undermine the notion that intimate relationships are lasting and permanent. The experience of cohabitation may foster less conventional attitudes regarding marriage and family life. Nonfamily living may lead to individualist, self-centered values and attitudes among American adults (Qu 1999).

2. The partners may still retain the stereotypical, romantic ideals that most Americans have about marriage as a rite of passage. For earlier generations, marriage signified the rite of passage to adulthood. Sexual relations became accepted as proper. Economic independence from parents, setting up an in-

WHAT RESEARCH TELLS US

Children and Cohabitation

About 35 percent of cohabiting couples have children under the age of 15 living with them.

Approximately 5% (3.5 million) of U.S. children currently live in cohabiting families. This represents a 50 percent increase since 1990 (S. L. Brown 2002, 2004; Bumpass and Lu 2000; Casper et al. 1999).

Approximately 40 percent of U.S. children will spend some time in a cohabiting family.

About 40 percent of all births to single mothers are actually to unmarried, cohabiting parents (S. L. Brown 2004; Bumpass and Lu 2000).

About 20 percent of single-mother families include a cohabiting male partner in the household (S. L. Brown 2004; London 1998).

The transition from cohabitation to marriage does not seem to be motivated by the desire to have children. Once nonpregnant cohabitors marry, the timing of the marital first birth is similar to that of married women who never cohabited (W. Manning 1995).

Cohabiting White women in their 20s who become pregnant are more likely than single White women to legitimate their first birth via marriage.

However, among teenage White women and Black women of all ages, pregnant cohabiting women are no more likely to marry before the child is born than are single women (Dunifon and Kowaleski-Jones 2002; W. Manning 1994). In general, Black women and relatively disadvantaged White women are just as likely to have children while cohabiting as when married (Dunifon and Kowaleski-Jones 2002; Loomis and Landale 1994).

---

dependent household, making decisions for oneself based on discussion with the new partner (not parents), and learning to become a partner in a team were all exciting steps one took toward adulthood within a first marriage.

Cohabitants marry after some years of cohabitation and then wait for the magic to happen. But since they have already done everything that married couples do, the anticipated excitement of a new status fails to materialize. When the excitement doesn't materialize, they express disappointment: Something is missing. As one disappointed, newly married couple who had lived together for the previous 3 years said, "Many of our friends wondered what the big to-do about a wedding was, even though we were excited. We then went to a romantic spot for our honeymoon, but nothing really seemed any different. In fact, it was all kind of boring."

People who divorce are more apt to cohabit after the divorce than never-married people. One study found that 47 percent of divorced women and 56 percent of divorced men under 35 years of age cohabit after their divorces (Wu and Balakrishman 1994, 729).

Perhaps cohabitants may unknowingly be creating a self-fulfilling prophecy. They enter a relationship partly to see if it will last, and, by keeping the possibility of discontinuing the relationship in mind, they actually help cause the breakup. It is possible that some relationships might turn out differently if the partners entered them with an unshakable determination to succeed, but the cohabiting experience may teach them to withhold this kind of total commitment. Living together can be a way for young people to avoid responsibility. It is easier to "play house" than to be married. Such an attitude might be carried into a subsequent marriage and contribute to marital dissatisfaction.

Divorced couples who cohabit and then go on to marriage fare better than single cohabiters who marry. Remarried couples who lived together before marriage report significantly higher degrees of happiness, closeness, concern for the

WHAT DO YOU THINK?

1. Do many of your friends cohabit? What kinds of problems do they experience in the cohabitation relationship?

2. Does the man or the woman usually seem most committed to the relationship?

3. In cohabitation relationships that have broken up, has the man or the woman initiated the breakup?

4. Who benefits more from cohabitation, the man or the woman? Why? What are the benefits?

partner's welfare, and positive communication, and they perceive more environmental support than remarried couples who had not cohabited. In general, divorced couples are older and more experienced than never-married couples. They have also gone through a marriage, not just a cohabitation experience.

The decision to cohabit should not be taken lightly. The ramifications of such a decision should be considered carefully. Legal ramifications, which we will discuss in a coming section, must also be considered—now more than ever.

Breaking Up

The breakup rate is high among cohabiting couples. After all, part of the reason a couple decides to live together rather than marry is the ease with which the relationships can be broken if one or another partner becomes unhappy. By the end of 4 years, about 75 percent of all cohabiting couples break up.

Society, including the cohabiting couples themselves, has largely ignored one of the philosophical foundations of cohabitation: Breaking up is easy if the relationship fails. Yet, for partners who are highly committed to a relationship, a breakup can be as emotionally devastating as a divorce. Moreover, society has support systems for divorced people, but no such support systems exist for persons leaving a cohabitation relationship.

Because ease of leaving a cohabiting relationship is one of the reasons for being in such a relationship, friends of the couple who break up often feel little sympathy. "Why are you making such a big deal out of breaking up?" they wonder. Since the breakup rate among cohabiting couples is high, the number of people who must cope with the trauma of breakup by themselves, or perhaps with only a friend's shoulder to cry on, will increase as the number of cohabiting couples increases.

Living Together and the Law

Many readers may be surprised to learn that cohabitation is illegal in some states. Indeed, sexual intercourse between unmarried persons (fornication) is still a crime in a few states, although such laws are on the way out and seldom enforced if still on the books. These are termed *sodomy laws*. In states where such laws are in effect, there are numerous ramifications for the cohabiting couple. For example, living together can be grounds for losing one's job. Because membership in professional associations and licensing may be conditional on demonstration of moral fitness, such privileges could be denied to someone cohabiting outside of marriage. In some states, a divorced person who is receiving alimony may lose it due to cohabiting. A divorced woman and her children who have been given the right by the court to remain in the family home until the children are 18 years old (at which time the home is to be sold and the proceeds divided) may lose that right if the woman cohabits.

Seeking to rent or buy housing is another problem for the unmarried couple, as is obtaining utility services. Opening joint bank accounts or shared credit cards is also problematic.

Laws making cohabitation a crime are so seldom enforced, however, that they are not nearly as meaningful to the cohabiting couple as several recent court decisions concerning property distribution and the obligation of support. The term *palimony* was coined to describe settlements made to a nonmarried, live-in partner. The term has no legal significance, but is descriptive of what a court can do.

MAKING DECISIONS

Questions for Couples Contemplating Living Together

Note that other than question 10, these are also good questions for couples contemplating marriage.

1. How did you make the decision to live together?

Good signs: We gave considerable thought to the decision, including the advantages and disadvantages of living together.

Concern signs: We kind of fell into it. We didn't really talk about the advantages and disadvantages.

2. What do you think you will get out of the relationship?

Good signs: We are concerned about learning more about ourselves and each other through intimate daily living. We want to test our commitment to the relationship.

Concern signs: It's nice to be free of our parents. It is convenient in many ways, including having easy sexual access. It makes me feel like an adult.

3. What roles will you and your partner fulfill in the relationship?

Good signs: We've thought a lot about our roles—how we'll divide the housework, contribute food money, and so on, and agree on the roles each of us will fulfill.

Concern signs: We'll just naturally fall into the proper roles. I love my partner and will do what is necessary to make our life together pleasant. We do disagree rather often about our expectations of each other.

4. What were your earlier dating experiences, and what have you learned from them?

Good signs: We have both dated and feel we have had a variety of rich experiences with others. We have positive perceptions of ourselves and of the opposite sex and are aware of what we have learned from previous relationships.

Concern signs: I've never dated much. I feel more secure just being with one person. I have some negative perceptions of myself or of the opposite sex. I don't feel that I learned much from dating others.

5. What are your partner's physical and emotional needs? To what degree do you think you can fulfill them?

Good signs: I think I have a clear understanding of my partner's needs, and I want to try to meet these needs.

Concern signs: I really don't know my partner's needs, but I'll find out about them when we move in together.

6. What are your physical and emotional needs? To what degree do you think your partner will satisfy them?

Good signs: Most of my needs are currently being met and are likely to continue to be met in a cohabiting relationship.

Concern signs: Although my needs are not being met in the present relationship, I'm sure they will be met when we live together.

7. To what degree can you and your partner honestly share feelings?

Good signs: We can be open and honest with each other without difficulty.

Concern signs: We do not believe expressing feelings is important. We have difficulty expressing feelings to each other.

8. What are your partner's strengths and weaknesses? To what extent would you like to change your partner?

Good signs: I am usually able to accept my partner's feelings. I can accept my partner's strengths and weaknesses.

Concern signs: I have difficulty understanding my partner. I can accept the strengths but not the weaknesses, but I'll probably be able to change him/her when we live together.

9. How do you each handle problems when they arise?

Good signs: We express feelings openly and can understand and accept our partner's point of view. We are able to solve problems mutually.

Concern signs: We have difficulty expressing feelings openly or accepting our partner's point of view. It is better to avoid problems. We often fail to solve them mutually.

10. How will your family and friends react to your living together?

Good signs: We are aware of the potential repercussions of family and friends when they learn of our cohabiting relationship. We have considered how we will handle them. Our family and friends are supportive of the cohabiting relationship.

Concern signs: We don't know how our family and friends will react to our living together. They probably won't support our cohabiting relationship, so we won't tell anyone about it.

SOURCE: Adapted from Ridley et al. 1978, 135–136 by permission of the author.

If a case goes to court, the court may see the union as *meretricious* (payment for sex), so any implied contracts would be illegal. The bottom line is that a given court will have the final word about a cohabiting couple's obligations to each other if they break up.

Despite the fact that courts are slowly awarding unwed cohabitants some legal rights, cohabitation still offers few legal safeguards for either partner. A sanctioned marriage has legal rules and forms that provide otherwise unprepared married couples with an efficient system for dealing with the unexpected; thus, the legal system helps a married couple minimize hardships that result from unforeseen events. For example, if a spouse is killed, the state has rules that allocate that person's property in the absence of a will and guarantee a share of the property to the surviving spouse. By contrast, if one member of a cohabiting relationship dies without a will, the surviving partner has no legal rights at all. Legally, cohabiting couples are at a distinct disadvantage, compared with married couples. Couples who regard the marriage license as simply another piece of meaningless paper do not understand the legal ramifications of the marriage contract.

Domestic partnership agreements for unmarried couples are now accepted in many localities. Such agreements bring some lawful protection to cohabiting partners (see Chapter 7). It should be recognized that many cohabiting partners cohabit to avoid the legal entanglements of marriage. Such couples do not want to have legal obligations tied to their relationship and resist the legal movement to give such couples increased protection under the law.

Finding the One and Only: Mate Selection

Love is the major reason Americans give for marrying, and dating is the major method in America of finding a marital partner. Yet, is love really the magic wand that directs our mate selection? Certainly, we fall in love or think we do at the time. Why we are attracted to a particular individual, however, is a complicated process that is not completely understood. Critics who argue against using love as a basis for marriage believe that Americans are seduced by the romantic ideal into ignoring practical considerations that help ensure a successful marriage. Such factors as compatible social and economic levels, education, age, religion, ethnicity, prior marital experience, race, and so forth may be ignored. Yet this is not really true. Our social system does take these factors into consideration. One doesn't fall in love with just anybody.

First of all, there is a field of *desirables*, or people to whom one is attracted. Within this field is a smaller group of *availables*, those who are free to return one's interest. You can meet with them, they are not in love with someone else, they are unmarried, and so forth.

Availability is closely related to how we live. Our communities are organized into neighborhoods according to social class, and in America this generally means by economic level. So the people who live nearby will be socially and economically like us. In fact, we very often marry someone living quite close geographically. This marital choice variable is called **propinquity**. Actually, the phenomenon is broader than just residential propinquity. Institutional propinquity exists as well; people meet in the workplace, in social organizations, and at school and church.

propinquity
Dating and marrying someone living quite close geographically

We also tend to choose individuals similar to ourselves; this tendency is termed **homogamy**. Thus, middle-class Caucasians tend to marry middle-class Caucasians; lower-class Caucasians tend to marry lower-class Caucasians; Catholics tend to marry Catholics; African Americans tend to marry African Americans; and so on. Similarly, we tend to marry within our own group, which is called **endogamy**.

There are exceptions to the rule of endogamy, but generally, society aims us toward loving and marrying someone similar to ourselves. If an individual doesn't marry the girl or boy next door, he/she is likely to marry someone he/she meets at school. The American system of neighborhood schools and selective college attendance makes education one of the strongest endogamic factors directing mate selection. Or, an individual may find a future mate through his/her family's social circle, a job, or his/her church. All the groups an individual joins limit their membership to people who are somewhat similar in socioeconomic status and in their more obvious reasons for being part of the group. Strange are the ways of love, but in reality, a banker's college-educated daughter seldom falls in love with the uneducated son of a factory worker. There are many theories of marital choice, but none seems totally satisfactory or predicts with much success who will make a good marital partner. There are theories that examine background factors, individual traits and behaviors, and interactional processes (see Larson and Hickman 2004; Larson and Holman 1994 for a thorough review of these factors).

homogamy
The tendency of people to marry persons similar to themselves
endogamy
The tendency of people to marry within their own group

Background Factors

Generally, the couple's background factors are the least predictive of marital quality and stability. However, numerous studies find that the higher the marital quality in the parents' marriage, the higher the quality in their adult children's marriages. Generally, people who come from successful families are better able to make successful marriages. This may be, in part, because the married children maintain relationships with their families, who are supportive of their children's marriages. Married couples whose own parents are divorced and perhaps remarried live in a much more complex and difficult environment. "To whose home do we go for Christmas? How do I handle Mother's or Dad's constantly putting down my other parent?" Thus, the unhappy or broken family often continues to put stress on their children, even after the children leave home and marry.

Other background factors, such as age at marriage, education, economic class, race, and the like, also have some influence on marital success. For example, many studies indicate a strong correlation between early age of marriage and marital instability.

Research is mixed on the effects of education on subsequent marital quality. For example, wives' education appears to be unrelated to marital stability, whereas well-educated husbands have more stable marriages than less-educated husbands (Whyte 1990).

Other background factors—such as occupation, friends, and religion—have some marital predictive value. Large numbers of good friends who support a couple's marriage are helpful. Religious beliefs and participation is one of the background factors found in strong, successful-families research. Financial success through gainful employment is predictive of stable marriage. Occupational stress is a negative influence on success of marriage. The *marital gradient* in mate selection is the tendency for husbands to be more advanced than their wives with regard to age, education, and occupational success.

Individual Traits

Jeffrey Larson and Thomas Holman (1994, 231) describe the following four conclusions about the predictive effects of personality on marital success:

1. Some fairly stable personality traits and mental health factors influence marital success. Generally, good emotional health, lack of depression, sociability, and positive attitudes toward marriage and family are predictive of marital success. Impulsivity has been shown to be negatively related to marital stability.
2. There are few apparent sex differences in personality effects on marital success.
3. Personality factors generally account for more variance in marital functioning than do background factors.
4. Personality traits and couple interpersonal processes are both important and interrelated predictors of marital success.

Numerous cross-cultural studies find that women seek men with good economic resources who are willing to commit to the relationship. Men tend to prefer young, healthy mates who can bear children. Evolutionary psychologists suggest that these differences stem directly from the need for species survival (Avery 1996).

Interactional Processes

Similar attitudes, values, and beliefs are related to marital success. Similar gender-role expectations are also predictive of marital success.

Generally, the longer a couple know one another, the greater the chance of successful marriage. One to 2 years' acquaintance before marriage appears to be the optimum.

Theoretically, two principles guide mate selection: endogamy and exogamy. As we have seen, endogamy is marriage within a certain group. Race, religion, ethnicity (nationality), and social class are the four major types of groups. The mores and taboos against crossing groups vary in strength. For example, only about 5 percent of Americans cross the intergroup line when *marrying* (*Population Today* 1999) but 20 to 25 percent of them cross religious lines (see Figure 6-2).

miscegenation
Marriage or interbreeding between members of different races

Until recently, most states had prohibitions that enforced racial endogamy. These prohibitions were generally termed **miscegenation** laws, because they prohibited interracial marriages. Most of these laws were originally aimed at preventing Caucasian–Native American marriages, but they were also used to prevent Caucasian–African American and Caucasian-Asian marriages. As recently as the end of World War II, 30 states had such laws. Gradually, some states declared them void; in 1967, the U.S. Supreme Court finally declared all such laws unconstitutional. By chance, the defendant's surname was Loving, so the name of the case, *Loving v. Virginia*, fits in well with the American romantic ideal.

exogamy
The tendency of people to marry outside their group

Exogamy, on the other hand, is a requirement that people marry outside their group. In our culture, requirements to marry outside your group are limited to incest and same-sex prohibitions (although this appears to be changing); that is, you may not marry a near relative or someone of your own sex. All states forbid marriage between parents and children, siblings (brothers and sisters),

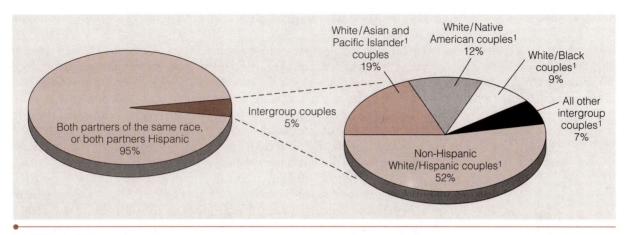

FIGURE 6-2 *Intergroup Married Couples, 1998*

The number of married couples who are of different racial/ethnic groups (White, Black, Hispanic, Asian/Pacific Islander, and Native American) has nearly doubled from 1.5 million in 1980 (3 percent of the total) to 2.9 million in 1998 (5 percent). Slightly more than half (52 percent) of intergroup couples in 1998 were made up of non-Hispanic Whites and Hispanics, followed by those made up of non-Hispanic Whites and Asians/Pacific Islanders. White/Black couples, by contrast, constituted less than one-tenth of intergroup couples. These figures reflect the degree of "social distance" between the various groups in American society.

NOTE: The "White," "Black," "Asian and Pacific Islander," and "Native American" categories do not include Hispanics.

[1] Percentages represent the percentages of all intergroup couples.

SOURCE: *Population Today* 1999.

grandparents and grandchildren, and children and their uncles and aunts. Most states forbid marriage to first cousins and half-siblings, although some states do not. About half the states forbid marriage between stepparents and stepchildren, and about the same number prohibit marriage between a man and his father's former wife or his son's former wife.

The other rule of exogamy, which until quite recently was taken for granted, is the requirement that we marry someone of the opposite sex. This helps ensure reproduction and continuance of the species. In the past few years, attempts by homosexuals to obtain more rights have led to some "marriage" ceremonies for homosexuals that do not have legal standing. In 2004, the Massachusetts Supreme Court ruled that the state law banning homosexual marriages is unconstitutional, thus allowing such marriages to be legal in the state for the first time. Other states are also moving toward legalizing homosexual marriage. The opposition to such marriages is strong, however, and this Massachusetts ruling will, no doubt, reach the U.S. Supreme Court. The ruling has also provoked a controversial movement toward a constitutional amendment banning homosexual marriages.

From First Impressions to Engagement

Let us look more specifically at the mate selection process for the individual. First impressions are usually belittled as superficial and unimportant. Yet without them, the process cannot begin. A favorable first impression must be made, or no further interaction will take place.

Physical attractiveness, although subjective, tends to create the first appeal. This is true for people of all ages, from children to the elderly, but it appears to be a stronger factor for males (Buss et al. 2001; Knox et al. 1997; Subramanian 1997).

It is interesting, too, that there is a halo effect operating in regard to physical attractiveness. The **halo effect** is the tendency for first impressions to influence succeeding evaluations. Physically attractive persons are imbued with other positive qualities that may not actually be present. Attractive persons are seen to be more competent as husbands and wives and to have happier marriages. They are seen as more responsible for good deeds and less responsible for bad ones. Their evaluations of others appear to have more impact. Other people seem more socially responsive to them and more willing to work hard to please them.

halo effect

The tendency for first impressions to influence succeeding evaluations

Thus, to be physically attractive is usually an advantage at the first-impression stage of a relationship. This is not always true, however. Some persons tend to avoid highly attractive individuals, to enhance their own chances of acceptance and to reduce the chances that their mate will be sought after by others.

The importance of attractiveness is, no doubt, the reason that so many Americans express dissatisfaction with their body images (Newport 1999). It is no wonder that young women and men express such dissatisfaction when one examines the media images of humans. "Thin is in" defines the ideal woman, and a muscular, hard torso defines the ideal man. At least for women, the media images are so persuasive that eating disorders are common among young women.

Meeting via the Internet has, however, lessened the importance of attractiveness as an early factor in meeting another person. In a chat room, for example, a relationship and rapport can be established without the influence of attractiveness because the parties cannot see each other, although new technologies may change this. Meeting via the Internet is also helpful to the shy individual who has trouble approaching others in a public situation (see Chapter 16).

Your impressions of *cognitive compatibility* (how the other thinks, what his/her interests are, and so on) are also important first impressions. In general, similarity seems to go with attraction because (1) another's similarity is directly reinforcing, (2) another's similar responses support the perceiver's sense of self-esteem and comsfort, and (3) such responses indicate the other's future compatibility. When values differ, the dissimilarity often signifies not just difference, but wrongness. Knowing that another believes as we do tends to reinforce the correctness of our own viewpoint, which makes us feel good.

We usually try to further a relationship by various attraction-seeking strategies. To attract a potential partner's attention, we attempt to buoy the other person's esteem by conveying that our regard for him/her is high. We probably try to do things for the person. We agree with

Caring gestures are important in any relationship.

© David Hanover/Stone/Getty Images

him/her and attempt to make ourselves look good. In our efforts to further a relationship, we sometimes overgeneralize and exaggerate, which may lead to later misunderstandings and disappointments if the relationship progresses.

Attraction tends to lead to self-disclosure. Intermediate degrees of disclosure may, in turn, lead to greater attraction, but this depends on the desirability of the information disclosed. Self-disclosure tends to lead to reciprocity. This phase occurs early in a relationship, when a lot of exciting, mutual sharing takes place. "Is that the way you feel about it? That's great, because I feel the same way." As more is disclosed, excitement increases, because the potential for a meaningful relationship seems to be increasing rapidly. However, the speed with which one discloses personal aspects is also important. Disclosure that seems overly quick may arouse suspicion, rather than trust.

As we have seen, predicting relationship success is complex and difficult. Let us look at two relationships: one built on similar needs, and one built on complementary needs. In a relationship of similar needs, the partners begin with similar interests, energy levels, religion, socioeconomic background, age, and so on. They share many characteristics and hence seem well suited for each other. Ideally, their relationship will lead to mutual satisfaction, but this is not always the case. If, for example, the partners are competitive, the results can be disastrous.

Bruce and Gail are alike in most respects, including being computer programmers, but Gail has advanced more rapidly. Bruce is competitive, feels like a loser, and begins to resent his wife. He subtly puts her down and gradually becomes more critical of women in general. The more success Gail has, the poorer their relationship becomes because Bruce cannot accept, let alone find joy in, her success. To him, it only points up his weaknesses.

By contrast, in a relationship of complementary needs, each partner supplies something that the other lacks. For example, an extrovert may help an introvert become more social; an organized partner may help bring structure to the life of a disorganized partner; or a relaxed partner may create an environment that helps ease the stress of a tense partner. Such complementarity can be very beneficial, but it cannot be counted on to happen because such differences often polarize the partners, rather than drawing them together. Consider Carol and Greg:

Carol is so organized that Greg counts on her to pay the bills, make social arrangements, find the nail clippers, and so on. The more Greg depends on her for these things, the more organized Carol feels she needs to be, and she begins to try to organize Greg. For his part, Greg resents this pressure and becomes passively resistant (procrastinates and becomes forgetful and careless). Polarization has occurred.

Unfortunately, there are no entirely satisfactory techniques for selecting marriage partners for lasting relationships. As we saw, even trial marriages through cohabitation have proved relatively ineffective.

Engagement

Marriage is the culmination of courtship. Traditionally, the final courtship stage has been engagement. Until recently, this has been a fairly formal stage. But with the increased age at first marriage, higher number of remarriages, and greater cohabitation among unmarried couples, engagement has become much less formal for some people.

Engagement means that simple dating for the couple is over and they have decided to marry at a future date. Qualities that were important in a date have gradually changed into qualities that are desirable in a spouse. The importance of physical attractiveness will diminish in favor of such traits as responsibility, trustworthiness, and caring.

In a typical engagement, the couple make a public announcement of their intention to marry and begin active marriage preparation. Once the engagement is announced, the couple usually make concrete plans for when the wedding will be, what type of wedding they will have, who will be invited, where they will live after marriage, what level of lifestyle they can afford, and so forth.

The families usually arrange to meet if they have not already done so. During the engagement, the couple often spend more time with each other's families and begin to be treated as kin. Marriage is the union of two families as well as two individuals. Above all, the two people begin to experience themselves as a social unit. Families and friends, and the public in general, react to them as a pair rather than as separate individuals.

Types of Engagements

The *short, romantic engagement* lasts from 2–6 months. Time is typically taken up with marriage plans, parties, and intense physical contact. Normally, such a short engagement period fails to give the couple much insight into each other's personalities. Indeed, so much time is taken up with marriage preparation that the couple may not have enough time for mutual exploration of their relationship.

The *long, separated engagement*, such as when one partner is away at college or must serve away from home in the military service, also presents problems. There are two distinct philosophies of separation: "Absence makes the heart grow fonder" expresses one, and "Out of sight, out of mind" sums up the other. In reality, the latter often prevails. Prolonged separation tends to defeat the purposes of the engagement and raises the question of exclusivity in the relationship. Does one date others during the separation? Dating others may cause feelings of insecurity and jealousy, whereas separation without dating is lonely and may cause hostility and dissatisfaction. In general, the separated engagement is usually unsatisfactory to both members of the couple because they can't spend time together, which, as we have seen, is a major component in strengthening a relationship.

Another possibility is the *long but inconclusive engagement*. Here, the couple puts off marriage because of economic considerations, deference to parental demands, or just plain indecisiveness.

About one in four engaged couples break up temporarily, while some couples break their engagement permanently. The causes of breakups appear to be simple loss of interest, recognition of an incompatible relationship, or the desire to reform the prospective mate. Sometimes the parents of one or both engaged persons work to break up the relationship and forestall a marriage. The major areas of disagreement tend to be matters of conventionality, families, and friends. Broken engagements can be considered successful in one sense: At least during this formal commitment period, the couple had the time to look more closely at each other and to realize that marriage would not succeed.

Functions of Engagement

What should an engagement do? How can engagement help the couple achieve a better marriage? Basically, the couple should come to agree on fundamental life arrangements. Where will they live? How will they live? Do they want children? When? How many? Will they both work?

The couple need to examine long-range goals in depth. Do they want similar things from life? Are their methods of obtaining these things compatible? Do their likes and dislikes blend? What role will spirituality play in their lives? How will each relate to the other's family and friends? Work associates? They may not be able to answer such questions completely, but they should at least agree on some tentative answers. The most important premarital agreement may be an agreement about how answers to such questions will be worked out in the future. A couple with a workable, problem-solving approach to life is in a good position to find marital success (see Chapter 5).

An important part of the engagement is the premarital medical examination, which serves several useful functions. For instance, one of the partners may have a general health problem that will require special care by the other partner. Marrying a diabetic, for example, means that the partner will have to understand and participate in careful control of diet and periodic administration of insulin. Both persons need to be checked and cured of any possible STD. The Rh factor in each partner's blood needs to be determined, since it will be of major importance in future pregnancy. Information on mutually acceptable methods of birth control can be obtained at this time, if the couple has not already chosen such methods. And each partner will have the opportunity to talk over questions about the coming marriage with the physician.

Premarital counseling helps illuminate a little of the blindness that comes with being in love. It can be helpful to discuss ideas and plans with an objective third person, such as a minister, marriage counselor, or mutual friend. Various instruments are available that a couple can use to assess their relationship. For example, Prepare (the Premarital Personal and Relationship Evaluation), by Olson et al. (1986), and Starting Your Marriage, by Roleder (1986), help guide couples contemplating marriage to a meaningful examination of their relationship. The Catholic Church has long made premarital counseling a prerequisite for marriage in the church.

A truly successful engagement period leads either to a successful marriage or to a broken engagement. An unsuccessful engagement, in all likelihood, will lead to marital failure.

Summary

1. *American dating is a unique form of courtship and mate selection.* American dating has become increasingly less formal, especially in the last 20 years, when it can perhaps better be termed getting together or hanging out, or hooking up.

   American dating is controlled by the youth themselves and involves relative freedom for the young man and woman to be alone together. This intensifies the pressure for the pair to move toward premarital intercourse as a means of handling their sexual drives.

2. *Marriage is being postponed, thus increasing the length of time during which a person dates before marriage.*

3. *The onset of puberty signals the beginning of adult sexuality.* In America, the age of sexual maturity does not coincide with social acceptance of overt sexual behavior, especially sexual intercourse. There is a period of several years during which conflict rages between the dictates of biology and of society.

4. *There appears to be increasing acceptance of premarital intercourse in American society.* However, because there are still social mores against premarital intercourse, those engaging in it often face conflict within themselves.

5. *Cohabitation has increased in the past 20 years to such an extent that some theorists now consider it to be a stage in the courtship process.* The law is becoming increasingly interested in the cohabitation relationship, but the relationship remains legally precarious.

6. *Mate selection is an involved process that is not yet fully understood.* However, it is fairly clear that similarities of socioeconomic backgrounds and energy levels help increase the chances of marital success.

7. *Mate selection usually leads to engagement, a formal expression of marital intentions.* The engagement period is used to make: wedding plans, philosophical decisions about the relationship, practical decisions about where and how to live, and so forth. A successful engagement leads to either marriage or the couple's breakup.

Resources on the Internet

Companion Website for *Human Intimacy: Marriage, the Family, and Its Meaning,* Tenth Edition

http://sociology.wadsworth.com/cox10e/

Gain an even better grasp on this chapter by going to the companion website to take one of the tutorial quizzes, use the flash cards to master key terms, or check out the many other study aids you'll find there. You will also find special features such as GSS data, Sociology Online, and Census 2000 information that will put data and resources at your fingertips to help you with that special project or to help you as you do some research on your own.

InfoTrac College Edition

http://www.infotrac-college.com/wadsworth/

You can access reliable resources anytime, anywhere, with InfoTrac College Edition, the online library. This fully searchable database offers more than 20 years' worth of full-text articles (not abstracts) from almost 5,000 diverse sources, such as top academic journals, newsletters, and up-to-the-minute periodicals, including *Time, Newsweek, Science, Forbes, The New York Times,* and *USA Today.* You can conduct electronic key word searches using key terms from this chapter to supplement your reading and learning experience. To aid in your search and to gain useful tips, see the Student Guide to InfoTrac College Edition, which you can access through the companion website for this book.

Does Sex Education Prevent Pregnancy or Encourage Promiscuity?

In many cultures, sex education takes place in the relaxed atmosphere of the family. This was once the case in America. A century ago in America, many children were raised on the farm, where close proximity to animals mating and giving birth was a natural part of life. Living quarters throughout most of the world and in early America were small and cramped, and children often shared a sleeping room with their parents. Hence, parents often could not conduct their sexual life in complete privacy. Only with the advent of the Victorian suppression of sexuality, and sufficient affluence to provide children with sleeping quarters separate from their parents, did sex education begin to lose some of its informality. As sex education became more formalized, it became less efficient, and some parents simply avoided giving their children formal sex instruction. Some people began to advocate that the schools should take charge of more sex education.

Sex Education Reduces Youthful Unplanned Pregnancies

Sex education is more necessary today than ever before. Sex education must reflect a changing society. As Table 6-1 makes clear, American young people are engaging in sexual behavior earlier than ever before, whether adults like it or not. It is equally clear from the statistics on teenage pregnancies (see the section "Children Having Children" in Chapter 9) that many of our sexually active youth are becoming pregnant. American schoolchildren are sexually active, and the need for formal sex education in schools is obvious!

If society is realistic, it will recognize these facts and deal with them. To pretend that young Americans are not sexually active, and to deny them education that will help them use their sexuality responsibly, is to bury our heads in the sand. The onset of AIDS and the rapid spread of other STDs add urgency to the need for formal sex education provided by schools.

Formal sex education works. There has been a decline in the proportion of teenagers who are sexually experienced, which supports the effectiveness of sex education. The teen birthrate is also declining, with Black teens showing the sharpest drop (NCHS 2003b).

Sweden has mandatory sex education at all grade levels and provides free contraceptives to adolescents. Explicit contraceptive usage is taught. Teachers are trained to answer children's questions about sex from preschool on. Although Swedish teenagers have their first sexual relations at an earlier age than American teenagers, both pregnancy and abortion rates are far lower, thus proving formal sex education works. Sweden's sex education experience clearly informs us that American education must do a better job educating our youth about sexual behavior.

Sex Education Promotes Sexual Activity among American Youth

Having students learn to place condoms on bananas is hardly going to control adolescent sexuality. Providing sex education and making contraceptives available to school-age children simply encourage sexual activity. If condoms may be obtained from the school health clinic, children receive the implicit message that sex is okay.

Most young children are not sexually aware until adolescence, although they may have simple questions such as, "Where do babies come from?" Giving preadolescent children thorough, explicit sexual instruction is unnecessary and only exaggerates their interest in sex. This obviously leads many children to increased sexual exploration. It is certainly appropriate to answer their questions at a simple level; it is unnecessary to provide more detail than is required to satisfy their curiosity.

Sex education properly belongs in the home, where the family can present sexual information in a moral context. School sex education tends to be neutral, teaching reproductive plumbing. But sexual plumbing presented outside of a value framework does not help children make sexual decisions and establish a set of ethics about their own sexual behavior. It certainly doesn't make adolescents have less interest in sex!

The emotional debates over pregnancy and abortion overlook the fact that there is choice, regardless of your position on these issues. The choice that needs to be discussed is the choice about when, where, and under what circumstances one will be sexually active. Finding oneself pregnant or causing a pregnancy means that a reasonable choice has not been made.

Without a sense of values, sexual choices cannot be made. It is easy to teach sexual facts and anatomy. What is difficult is the teaching of sexual mores and values.

Unless sex education teaches values and how to make sexual decisions, sex education is only promoting unthinking sexual activity. The teaching of sexual plumbing without placing it in a value context does not lead to either reduced sexual activity or to reduced premarital pregnancies. Despite years of sex education in many schools, contraceptive usage has not risen much among high school students; teenage pregnancy, although down a bit, is still a major problem.

What Do You Think?

1. Did you have sex education in school? At what age? How explicit was it?

2. Do you feel that such education encouraged you to be more sexually active?

3. Do you feel that you were able to make better sexual decisions for yourself because of the sex education you received from your schools?

4. When would you ideally start sex education for your children?

5. Do you think such education should be given by schools, or do you feel it should only be given at home?

6. Who is responsible if parents are not educated well enough or don't wish to provide adequate sex education to their children?

© Frank Cox

Marriage, Intimacy, Expectations, and the Fully Functioning Person

Chapter

7

Questions to reflect upon as you read this chapter:

- What do you consider to be the two best reasons for marrying?
- What do you consider to be the hardest transition from single to married life?
- Why are your expectations about your marriage and future spouse important to the success or failure of your marriage?
- What is meant by the self-fulfilling prophecy?
- Briefly describe the characteristics of a mentally healthy individual.
- Why is the marriage certificate more than "just a piece of paper"?

Marriage Matters

Marriage is the broadest and most intimate of all human interactions. At its heart, marriage is an interpersonal relationship between a man and a woman. Regardless of how marriage is defined and what exact form it may take, it is within marriage that most adults try to fulfill their psychological, material, and sexual needs. To the degree that they are successful, the marriage is successful.

Marriage and the family ideally act as a haven from which individual members can draw support and security when facing the challenges of our rapidly changing, technological society. Ideally, a good marriage acts as a buffer against mental health problems such as alienation, loneliness, unhappiness, and depression (see Chapter 2).

Among the primary functions of the contemporary American family are providing emotional gratification to members, helping family members deal with emotional crises, and helping them grow in the most fulfilling manner possible. In a word, a successful marriage can be therapeutic, a curative to the problems of its members. However, an unsuccessful marriage can and will create problems for its members.

As we saw in Chapter 6, finding "the one and only" usually is a romantic myth. In fact, a number of persons can become satisfactory and long-lasting partners for an individual. The building and maintenance of a relationship is more crucial to a relationship's success than the selection of the imaginary perfect mate. The marriage ceremony is really a beginning, not a culmination.

In perusing the vast amount of marital and family research accomplished over the past 30 to 40 years it is apparent that "Marriage is an important social good associated with an impressively broad array of positive outcomes for children and adults alike" (*Why Marriage Matters*, 2004).

About Men:

Married men have longer life expectancies than single men.

Married men earn between 10 and 40 percent more than single men with similar education.

Marriage increases the likelihood fathers will have good relationships with their children.

About Women:

Married mothers have lower rates of depression than single or cohabiting mothers.

Marriage significantly reduces poverty rates for both mothers and their children.

Married women appear to have a lower risk of domestic violence than cohabiting or dating women. Even after controlling for race, age, and education, women who cohabit are three times more likely to report violent arguments than married women.

About Children:

A successful marriage increases the likelihood that children will graduate from college and achieve high-status jobs.

Children who live with their own two married parents enjoy better physical health, on average, than children in other family forms. The health advantages of married homes remain even after taking into account socioeconomic status.

Parental divorce approximately doubles the odds that adult children will end up divorced.

About Society:

Adults who cohabit are more similar to singles than to married couple in terms of physical health and disability, emotional well-being and mental health, and assets and earnings. Their children more closely resemble the children of single people than of married people.

Marriage appears to reduce the risk that children and adults will be either perpetrators or victims of crime. Boys raised in single-parent homes are about twice as likely (and boys raised in stepfamilies three times as likely) to have committed a crime leading to incarceration by the time they reach their early 30s (*Why Marriage Matters*, 2004).

Most of the research supporting these and other favorable characteristics of married persons tries to tease out *selection effects*, or the pre-existing differences between individuals who decide to marry or divorce, and so on. For example, does divorce cause poverty, or are poor people more likely to divorce? Even though selection effects may never be completely ruled out, the research makes it clear that marriage does create favorable effects for many couples and their children.

The Transition from Single to Married Life

Married life is, indeed, different from single life. Marriage brings duties and obligations. Suddenly, an individual is no longer responsible just for him-/herself, but shares responsibility for two people and perhaps more, if children arrive.

Furthermore, an individual's very identity changes with marriage. No longer is the person simply an individual; he/she is now a wife or husband, and may become a mother or father. The individual becomes interdependent with the others (in-laws, and so on) in the family and loses much of the independence of singlehood.

The transition from dating to establishing a home and family is a large step for both partners. The couple may have cooked some meals together during court-

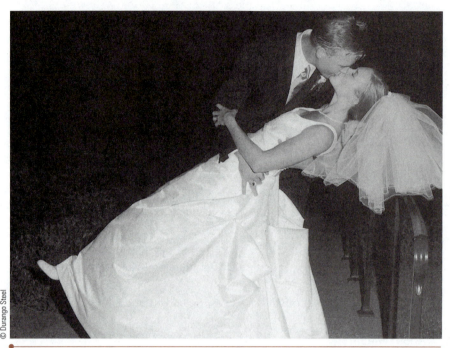

Ain't love grand?

ship, but preparing 1,095 meals a year for two on an often-tight budget is a far greater challenge. Planning a month's finances is certainly more difficult than raising money for a weekend of skiing. In fact, living within some kind of family budget is a difficult transition for many newly marrieds. Before marriage, a person is used to spending his/her income as he/she wishes. Now another person must be consulted, and the partners' incomes must be shared.

Leisure time activities, which are often spontaneous and unplanned when an individual is single, now must be planned with another person. Compromises must be worked out. Both Mark and Tanya were active singles until they met, fell in love, and married in their mid-20s. Mark had enjoyed trips to the desert to ride his motorcycle with other dirt bike enthusiasts. Tanya had spent a great deal of time with her office colleagues, bowling on the office team. In marriage, Mark and Tanya find themselves somewhat bored and uncomfortable with the other's activities. Unable to compromise their activities, they gradually spend less time together. Although they are still married, they do almost nothing together. Neither is willing to compromise and share each other's activities, or find a new activity they can share. Their friends wonder if they ever see each other.

New relationships must be developed with both sets of parents-in-law after marriage. The primary relationship is now with a spouse, rather than with parents, and couples must learn to relate with in-laws. Failure to build a satisfactory in-law relationship makes married life more difficult in that the relationship with one's parents is important to each spouse.

The sexual relationship may also involve a transition. Premarital sexuality may have been restrained, covert, and only partially satisfying. Ideally, the marital sexual relationship will become fully expressive and satisfying to both partners. Judith Wallerstein (Wallerstein and Blakeslee 1995, 28) lists "establishing a rich and pleasurable sexual relationship and protecting it from the incursions of the workplace and family obligations" as one of the essential tasks of the happily married couple.

There are many other transitions that must be made if a marriage is to be successful. All are transitions from the self-centeredness of childhood to the other-centeredness of adulthood. To consider the likes and dislikes of a partner, to compromise desires, to become a team that pulls together rather than in opposing directions, to become a pair that has more going for it than the sum total of the two individuals—these are a few of the things necessary for a successful marriage. If one or both partners cannot make these transitions, the chances for a successful marriage decline.

The transition from single to married status does not end the need for change. As time passes, the marriage will change and will require of both partners further transitions. The coming of children, for example, places a whole new set of demands on the married couple.

As important as the formation of a united team and the development of joint interests are, individual interests and identities must also be maintained. To build togetherness by creating intimacy, while carving out each partner's autonomy, is central throughout marriage. It is especially important at the outset of a relationship. *Building togetherness and maintaining autonomy is one of several important balancing acts that partners must manage if a marriage is to be successful.*

In general, the couple contemplating marriage seldom realize the extent of the changes necessary to make that marriage successful. They often believe that love will take care of all of the transitions from single to married life. "We get along now; of course we'll get along after we marry." But that is not necessarily so. If there is a defining characteristic of marriage, as opposed to singleness, it is the continuous need of successful give-and-take between the married partners.

Marriage: A Myriad of Interactions

Each day, we interact with numerous people. On the job, we talk with colleagues, receive instructions from superiors, and give orders to those who work for us. At stores, we talk with salespeople. On the way home, we interact with other drivers or with people on the subway. All these interactions are relatively simple. For example, when buying things at a store, I simply want to make my purchases. I don't need to know how the salespeople are feeling, how their children are doing in school, or how they feel about their jobs. I need only relate to them as a customer. The interaction begins and ends at a superficial level.

Not so with marriage. Perhaps more than any other institution, the family is an arena of intimate and complex interaction. Literally hundreds of interactions take place within a family each day, and they vary in infinite ways. We can think of these interactions as ranging from purely intellectual to strongly emotional. Note in the following conversations how the interactions move from superficial and distant to caring, committed, and emotionally involved:

"Good morning. How are you feeling?"

"Morning. Fine. How are you?"

"Good morning. How are you feeling? Did you sleep well?"

"So, so. But I hope you slept well."

"Good morning. I'm really happy to see you looking so well this morning! I'm glad that headache went away."

"Morning. I feel better. Thanks so much for reminding me to take those aspirins; they really did the trick! Don't know what I'd do without you!"

"Good morning! You know, it's always wonderful to wake up in the morning with you!"

"It makes me feel so good when you say that! I'm so happy with you!"

The first interaction is superficial—a general morning greeting. The next, though still superficial, demonstrates more concern and awareness of the other person. The concern deepens in the next interaction: The first speaker remembers the other's headache of the previous day and is happy that the partner is out of pain. The partner expresses gratitude for the concern and for the aid. And the last interaction demonstrates a deep emotional level of sharing between the partners.

Let's also look at some of the following role interactions that go on in a marriage:

man	↔	woman
husband	↔	wife
lover	↔	lover
friend	↔	friend
provider	↔	provider
provider	↔	provided for
provided for	↔	provider
father*	↔	child*
child*	↔	mother*
child*	↔	child*
taker	↔	giver
giver	↔	taker
teacher	↔	learner
learner	↔	teacher
learner	↔	learner
worker	↔	employer
employer	↔	worker
worker	↔	worker
colleague	↔	colleague
leaning tower	↔	tower of strength
tower of strength	↔	leaning tower

(*These are not interactions between a real child and real parent, but rather, interactions in which one partner behaves like an authority, or parent, and the other partner reacts as a dependent child. In the child ↔ child interaction, the partners both act like children with each other.)

Obviously, the list of possibilities is endless! And think of the complications that can arise when children enter the picture. Remember also that cross-interactions can occur, such as between lover and friend or lover and teacher. For example, the friend can interact with the woman, the lover, and so on, down the list. Thus, when I wake up in the morning and say to my wife, "How do you feel?" it means far more than saying the same thing to a passing acquaintance. It could mean "I am concerned about you" as a friend; "I'd like to have sex with you" as a lover; "Will you be able to work today?" as an employer; and so forth. To further complicate matters, the spouses may not agree on what the meaning is. The wife may take the husband's question to mean he'd like to make love before getting up, whereas he really is inquiring about how she is feeling because she had been sick the day before. Complicated, isn't it?

To manage successfully the hundreds of interactions that occur daily in marriage would be a miracle, indeed. But to manage them better each day is a worthy and attainable goal. If people can be successful in marriage, chances are they will also be successful in most other interpersonal relations because other relationships will almost certainly be simpler than their marital relationship.

Fulfilling Needs in Marriage

Marriage can supply love and affection, emotional support and loyalty, stability and security, companionship, friendship, sexual fulfillment, and material well-being. This is a big order to expect of any relationship and explains, in part, why there is a high rate of marital disruption in American society. Americans ask a great deal more of marriage than they have in the past and a great deal more than do many other people in the world. Such high expectations must certainly lead to disappointments.

Society recognizes that fulfilling sexual, material, and psychological needs is a valid responsibility of the marriage institution. In fact, so important is the meeting of these needs for individuals that all states recognize failure to do so as a legitimate reason for divorce.

Sexual Needs   Marriage is the only legitimate outlet for sexual energies recognized by American society. Indeed, sexual intercourse is a state-mandated part of marriage. If sexual needs are not fulfilled in a marriage, the marriage can be dissolved.

Material Needs   "Room and board" is a part of every marriage. Breadwinning and homemaking are essential to survival. Material needs also affect how successfully psychological and sexual needs are met. Marital disruption is considerably higher among families in economic trouble than among families satisfactorily meeting material needs.

Psychological Needs   As you may remember from Chapter 2, one of the basic principles of this book is that the strong family becomes increasingly important to its members as social stability decreases and people feel more isolated and alienated.

Mobility, increased anonymity, ever-larger and more bureaucratic institutions, and lack of social relatedness all contribute to feelings of loneliness and helplessness. Because of these feelings, our psychological need for intimacy has increased greatly, and most people hope to find this intimacy in marriage.

Today, men and women seek not only love, but also emotional survival within the context of their intimate relationships. A loving spouse and/or family can supply personal validity and relevance—a confirmation of one's existence.

WHAT DO YOU THINK?

1. What kind of marital role expectations do you think you have learned?

2. Are they similar to the roles your parents fulfilled in their marriage?

3. What do you think are the two most important roles your spouse should fulfill?

4. What would you do if your spouse disagreed with you and didn't wish to fulfill one of these two roles you consider important?

Marriage in Japan

Marriage in Japan has been surrounded by centuries-old formalities and traditional beliefs. Marriage is particularly important in Japan, and pressure is strong to settle down and start a family by one's mid-20s. Despite Japan's modernization, arranged marriages are still important, although self-selection of a mate is becoming more popular. Indeed, some businesses are even offering employees lessons in how to date and find a mate.

The marriage Japanese young people enter has goals far different from those of their American counterparts. Americans stress the importance of building an intimate, emotionally fulfilling marriage. Marriage in Japan is based on the wife agreeing to provide a home and raise the children almost completely on her own. The husband's role is to work hard and provide the family with financial security.

Modern Japan demands so much of the man in the workplace that his first obligations must be to his company, his co-workers, and his clients. After putting in a full day's work, he is often called upon to socialize with his colleagues. He returns home late, often having eaten dinner, and has little, if any, energy left for family interaction.

Japanese marriage ceremony.

© Kenneth Hamm/Photo Japan

Although this lack of marital emotional intimacy is strange to Americans, it appears to be what many Japanese couples desire. The commitment is not about relating or developing a deeper, broader, and more mature relationship though intimacy and personal interaction. There is no time and energy for such things. Intimacy is not part of the deal (Bonhaker 1990, 12).

Interestingly, Japanese couples appear to have two marital relationships: the public and the private relationship. In public, the wife is properly deferential to her husband. His position is superior, and she caters to him at all times. However, within the home, she usually holds the power. Historically, the oldest woman in the household, often the husband's mother, controlled the household and had the privilege of eating with the men, while the younger wife and mother ate with the children. To some extent, Japanese wives, even today, gradually adopt the role of mother toward their husbands. Most husbands turn over their paychecks to their wives who, in turn, give their husbands a weekly allowance.

Americans might think that such nonintimate marriages will not last. Yet, there is strong pressure in Japan not to divorce. Japan's population is approximately one-half America's population. Yet, Japan has only about 15 percent as many divorces as the United States. A number of reasons ap-

Dinner with television.

Walking the babies.

pear to account for the low divorce rate. It is considered disgraceful for a Japanese woman to divorce. Wage discrimination against females makes it difficult for a single woman to earn a living. Unmarried women older than 25 to 30 years of age are ridiculed and considered to be failures.

Although Japan is a highly industrialized, affluent, modern country, marital relationships are considerably different from those here in the United States. The Japanese marital relationship would be unsatisfactory to most Americans, but it appears to work for Japanese couples. If one measures marital success by a low rate of marital breakup, Japan's marriages might be said to work better than American marriages.

SOURCES: W. Bonhaker, *The Hollow Doll* (New York: Ballantine Books, 1990); C. Nakane, *Japanese Society* (Tokyo: Charles E. Tuttle, 1989).

Family stroll.

Need Relationships   It is obvious that psychological needs are closely related to both sexual and material needs. As was pointed out, marital disruption tends to be highest among those least well off materially.

The relationship between sexual and psychological needs is more complicated than the relationship between material and psychological needs. A satisfying sexual relationship certainly fulfills many of one's psychological needs. However, a satisfying sex life also often reflects a relationship's general success at meeting needs, not just its sexual success. Lauer and Lauer (1985), in their study of successful marriages, found that agreement about and satisfaction with one's sex life was far down the list of reasons the respondents gave for a good marriage. Fewer than 10 percent of the respondents felt that a good sexual relationship kept their marriage together. On the other hand, unhappy couples or those from broken marriages often list sexual relations as one of their major problems. When other psychological needs are not being met, the sexual relationship is often the first place in which trouble appears. A couple's sexual relationship often acts as a sensitive indicator of the health of their relationship. Couples having serious marital problems seldom find that their sex life is unaffected.

Defining Marital Success

Because the vast majority of Americans marry at some time in their lives, it is important to understand how to build a successful marriage. Chapter 1 introduced the characteristics found in strong, successful families, and these are discussed throughout the book.

Marital success remains difficult to define, partly because it is often confused with marital adjustment, permanence, and happiness. These three concepts, however, are not necessarily the same. A neurotic relationship of mutual misery (as powerfully depicted by George and Martha in the play *Who's Afraid of Virginia Woolf?*) may be just as binding as a healthy relationship of mutual support, esteem, and love. Thus, a couple may be said to have achieved marital adjustment, although the adjustment has not led to happiness or marital success in the broadest sense (Heaton and Albrecht 1991). On the other hand, if marital success is defined as permanence only, it might be said that the miserable couple in *Who's Afraid of Virgina Woolf?* has marital success, as long as their misery does not lead to dissolution of the marriage. And if success is defined as happiness only, no marriage could be considered successful because no family is happy all the time.

Over the years, many factors—such as socioeconomic level, years of schooling, presence of children, length of premarital acquaintance, role congruence, communication patterns, religious affiliation, unemployment rate, family of origin status, race, and ethnicity—have been studied and related to marital success. Yet, the results of such research remain vague and inconclusive (Bramlett and Mosher 2002; Fincham and Kemp-Fincham 1997).

For our purposes, *marital success will be defined broadly to include adjustment, happiness, and permanence.* Thus, a successful marriage is one in which the partners adjust to the relationship, are in relative agreement on most issues of importance, are comfortable in the roles that they assume, and can work together to solve most of the problems that confront them over time. Each partner expresses satisfaction and happiness with the relationship, and the marriage lasts.

Marriages in which partners positively interact with each other most of the time lead to feelings of happiness in the spouses. In fact, marital interaction and partner happiness reciprocally affect each other. A happy marriage produces liking, which generates more positive interaction, which tends to lead to more marital satisfaction (Gottman 1994; D. Johnson et al. 1992; Wallerstein and Blakeslee 1995; Zuo 1992). A couple's perception of the amount of warmth (or hostility) in their partner also greatly influences marital stability. It is interesting that the wife's perceptions are a better predictor of marital stability than are her husband's (Matthews et al. 1996).

As we saw, marriage requires the partners to play many roles. Some research suggests that couples who maintain a balance across their many roles score higher on various measures of well-being (Marks and MacDermid 1996).

Understanding how your definition of love influences your evaluation of your partner's love is also important. It is not surprising to find that compatible love styles and definitions increase marital stability (Hendrick and Hendrick 1997, 2000; Sokolski 1994).

Unfortunately, and despite all the research, it remains impossible to give a definitive list of factors that influence a couple's ability to attain marital success. Research does pinpoint age at marriage and the presence of children as two factors closely related to marital satisfaction and success. In general, the younger the age at marriage, the greater the risk of marital dissatisfaction and failure. Also, a great deal of research finds that marital satisfaction is lower when children are present than in pre- or postparental stages of marriage (see Chapter 10).

Numerous studies on marital instability also shed light on marital stability. For example, such things as extramarital affairs (Buunk and Dijkstra 2000); financial stress; family illness, accident, and death (Harvey and Hansen 2000); and physical or sexual aggression (Christopher and Lloyd 2000) all affect family stability.

Couples probably vary so much (just as individuals do) that an important factor for one couple may be unimportant to another. This is especially true in a society such as ours, where individuals not only have great freedom in mate selection, but also great freedom to build the kind of relationship that they desire. As always, freedom is accompanied by responsibility. Thus, the freedom to create the kind of marital relationship a person wants means that he/she is largely responsible for the success or failure of that relationship. Of course, intimate relationships are also affected by outside factors, such as the state of the economy; acceptance by mainstream society; and presence or absence of social support from family, friends, and neighbors.

Strong Relationships and Families

Most Americans build their intimate relationships on a foundation of love. And persons who are deeply and newly in love have an almost insatiable desire to spend time with the object of their affection. In fact, it can be agonizing for the partners to be apart for any length of time. Thus, the desire to spend time together is found in most intimate couples' relationships, at least in the early part of their relationships, when the couples are dating and courting. The desire to spend time together is also one of the major characteristics of the strong, resilient family.

Stinnett and DeFrain (Stinnett 1997; Stinnett and DeFrain 2000) do an exercise with their subject families, entitled "The Journey of Happy Memories."

They direct their subjects to close their eyes and spend 5 minutes wandering through childhood memories; then they ask them to describe the happiest memory. The researchers found that the memories recalled almost always involved time spent in family activities. Your author tried this exercise with his students. Some of their childhood memories included the following:

My favorite memory is climbing the big rocks that were in the campground where my grandfather would take me and my cousins camping for two weeks each summer.

The family would go to the sand dunes by the beach, and Dad would always be disappearing, making us play hide and seek with him. We'd be walking along and suddenly he wouldn't be there and we'd run off to find him. We'd try to keep an eye on him, but he was very good at disappearing.

At Christmas time, Dad and Mom would always get out the old 8mm family movies and although we'd all say, "Oh, not the movies again," we always loved to see Mom and Dad as little kids and especially ourselves growing up.

It is clear that family time together is of great value, at least in retrospect. Yet, as time passes and the partners' relationship becomes more enduring, as a couple's romantic love turns into companionate love, spending family time together seems to become increasingly difficult because of demands made by children, work, and the general maintenance of the family.

The major revolution in family structure in recent years has been the advent of the dual-earner family, the family in which both partners are employed outside the home (see Chapter 13). It is obvious that in such a home, family time together will be at a premium.

As family time has become scarcer, many have argued that it is not the quantity, but the quality, of time together that counts. There is certainly truth in the argument. I may be with my partner physically, but does that do much good if I

 HIGHLIGHT

A Hectic Schedule

Denise and Murray have two children, aged 5 and 7. Denise remained home with the children until the second child started preschool, at age 2, and has worked since that time. She arises about 6:30 a.m. to get everyone organized. Their first child leaves for school at 8:00 a.m., and Denise drops the second child off at preschool in time to get to her work at 9:00 a.m. Murray leaves about 7:45 a.m. to commute to his 9:00 a.m. job. The 1 hour and 15 minutes that they are all together in the morning passes quickly, and with little communication, because everyone is preoccupied with getting ready for his/her day. A neighbor watches the oldest child in the afternoon after school is out, and Denise picks up the second child about 5:15 p.m. Murray arrives home about 6:00 p.m. After a long day, the family members are tired, yet they must prepare the evening meal and do homework and other family chores. The youngest child is in bed by 8:00 p.m., and the oldest child follows at around

9:00 p.m. Thus, the whole family is together for a maximum of 2 to 3 hours at the end of the day, and there is clearly little quality time to communicate in the midst of household chores. Only after 9:00 p.m. do Murray and Denise have time for themselves. It is little wonder that the family feels that time is their most sought-after commodity.

What Do You Think?

1. If you live in a two-parent home and both parents work, how do they divide up household chores?
2. If you live in a single-parent home and your parent works, how much work do you put in around the house?
3. If you are married or cohabiting, how pressed for time do you and your partner feel? How do you attempt to handle the time pressure? Does time pressure cause problems in your relationship?

am not there mentally? However, the argument often seems to ignore the fact that quality and quantity of time together are interrelated. There must be enough time together for quality to surface. It seems, too, that the argument is sometimes used to soothe the guilty conscience of a spouse or parent who is spending very little time with the family (see Chapter 10).

When we examine the daily pattern of many dual-earner families, the difficulties become clear. Whether there can be quality family time with such a daily routine is open to serious question. Without quality family time, it is doubtful a marriage can survive.

The description of Denise and Murray's routine fails to take into account the many other kinds of activities that may drain a family's time together. Most families spend time on such things as youth sports, PTA meetings, children's music lessons, church socials, and money-raising charity events. It is true that time can be spent together while engaging in such activities, but for most families, only one spouse is involved in a given activity.

A family needs time to play together. Periodically, the family must get away from work and responsibility and simply play and enjoy life. The couple needs to do this alone—away from the children—and also with the children. The family must be careful not to work at playing. Some families plan and organize their playtime to such a degree that it turns into work, rather than play. Happily married couples do not hide behind "Someday we will . . . ," "When we get time, we'll . . . ," or "When we finally have the money, then we'll. . . ." They deliberately make the time to play, to be together, to work on their relationship. They make time to grow and improve themselves and their marriage.

Strong families seem to have the ability to work, play, and vacation together without smothering each other. During the 1950s, there was a great deal of criticism of the idea of *togetherness* in marriage. Too much togetherness between partners meant that individuality was lost. Today, the opposite problem is more prevalent. There seems to be too little togetherness and too much emphasis on individuality to maintain family togetherness and strength.

Spending time together allows a family to develop as a team—a group unity—and to create a sense of family history. This feeling of belonging helps family members find an identity. The need for adolescents to turn to their peer group for support and identity is reduced if they have a strong sense of family.

Besides spending time together in different family activities, almost all strong families indicate that at least one meal a day is reserved as a time of family togetherness (McCubbin et al. 1988, 123–130). That meal together is usually the evening meal, and everyone in the family living at home is expected to attend on a regular basis.

Perhaps the importance of spending enough quality time together is best summed up by this recent student comment:

My brother and two sisters are my best friends and my Mom and Dad are not only my Mom and Dad but my pals as well. I always look forward to doing things with the family because we have so much fun together. We have fun, but I also know that if I need help they will all pitch in and give me the help I need.

It is clear that the desire to spend time together, felt by all couples newly in love, must somehow be maintained as the relationship becomes permanent. If it can be maintained and expanded into marriage and into the growing family if and when children arrive, then a key element to maintaining family growth and stability is in place.

Marital Expectations

There are no longer any hard-and-fast rules about what a marriage should be, about the roles that each partner will play, or even about the primacy of the relationship. The expectations that one brings to the marital relationship play an important role in the success or failure of the relationship. To a large degree, we are disappointed or satisfied in life, depending on how well what is happening matches what we expect—what we think should happen. Expectations play a crucial role in determining our level of satisfaction in an intimate relationship such as marriage. Thus, an examination of the role that personal expectations play in marital success is perhaps more important than an examination of a long list of factors that may or may not affect a particular couple's marital success.

In a very real sense, human beings create their own world. The wonderful complexity of the human brain allows us to plan, organize, and concern ourselves not only with what is, but also with what we think should be. We predict our future and have expectations about ourselves, our world, our marriages, our spouses, and our children. In a way, expectations are also our hopes about the future, and hope is an important element to our well-being. We often hear about the person who has given up hope, and we know that this can be dangerous to both physical and psychological well-being. The average American certainly enters marriage (at least the first marriage) full of high hopes for success and happiness. Who can be more hopeful than the couple in love, newly married, and off on their honeymoon?

In our earlier discussion of love (Chapter 3), we noted that love often acts like the proverbial rose-colored glasses, in that we don't see the people we love as they really are, but as we wish or expect them to be. A spouse's failure to meet these expectations can lead to disappointment in marriage.

In essence, the world is as we perceive it. Our perceptions are based, in part, on the input of our senses and, in part, on how we personally accept, reject, interpret, change, or color that input. The study of how people experience their world is called **phenomenology**. It is important to realize that most people react to their perceptions of the world rather than to what the world really is. A simple example may clarify this point. If we insert a straight, metal rod halfway into a pool of clear water, the rod appears bent or broken because of the refraction of the light waves by the water. How would we react if we knew nothing about light refraction and had never seen a partially submerged object before? We would see the rod as bent and would act on that perception. But we who have measured and examined the rod know that it is straight. Moreover, most of us have learned that light waves will be refracted by the water and appear bent, so we assume that the rod is straight, even though our eyes tell us it is bent. In other words, we know that our perceptions do not always reflect the objective world. We have learned that appearances can be deceptive.

How does this relate to marriage? In our interactions with other people, we often forget that our perceptions may not reflect reality or that our spouse may have different perceptions. Brandon's perception is that he has greatly reduced the time he spends with his friends and that he spends much more time with Tracy. Tracy feels that Brandon spends too much time with his friends and too little time with her. Who is right? Certainly, their perceptions of the situation differ, but is one right and the other wrong? This is probably one of those situations in which both partners are right, at least in their own minds.

phenomenology
The study of how people experience their world

HIGHLIGHT

Carol: The Perpetual Seeker

Carol's father died when she was 8 years old. Her mother never remarried because she felt no other man could live up to her deceased husband. Through the years, she told Carol how wonderful and perfect her father had been, both as a man and as a husband.

As a teenager, Carol fell in love often and quickly, but the romances also ended quickly—usually, she said, because the boy disappointed her in some way. The only romance that lasted was with a young man whom she met while on a summer vacation in Florida. Even though they had known each other for only a month, they continued their romance via mail and the Internet the following year. They met again the next summer. Carol was then 19. Three months after they parted at summer's end and after many letters declaring their love and loneliness, Carol's friend asked her to come to Florida and marry him. Carol assured her mother that he was the right man—strong, responsible, and loving—just as her father had been.

They married and everything seemed to go well. By the end of the first year, however, Carol was writing her mother letters and complaining that her husband was changing for the worse—that he wasn't nearly the man Carol had thought him

to be. She questioned whether he would ever be a good father, and she noted that he didn't have the drive for success that her father had at his age.

Two years after marrying, Carol divorced her husband. She had met an exciting, wonderful man who had given a talk to her women's sensitivity group. He was all the things her husband wasn't and, incidentally, all the things her father had been. She married him not long after her divorce became final.

Carol is now 30 years old and married to her third husband, about whom she complains a great deal.

What Do You Think?

1. Why does Carol seem to find so little satisfaction with her husbands?
2. Why did Carol's romance with her first husband persist so long before marriage, whereas all of her earlier relationships dissolved quickly?
3. What are the qualities of your dream spouse?
4. How does your relationship with your parents influence your marriage ideals?
5. How can Carol find satisfaction in marriage?

The Honeymoon Is Over: Too High Expectations

The honeymoon is over. This stage is very important in most marriages. It usually means that the partners are reexamining their unrealistic, overly high expectations about marriage and their mate, created by romantic love. In a successful relationship, it means that subjective perceptions are becoming more objective, more realistic. It also means that the partners are, at last, coming to know each other as real human beings, rather than as projections of their expectations.

One morning, after Peter and Holly have been married for about a year, Holly awakens and realizes that Peter is no longer the same man she married. She accuses him of changing for the worse: "You used to think of great things to do in the evenings, and you enjoyed going out all the time. Now you just seem to want to stay home." Peter, of course, insists that he has not changed, that he is the same person he has always been, and that he has always said he looked forward to quiet evenings alone with her.

In your past relationships, did you ever feel that your partner suddenly changed? Unfortunately, some people throw away the real person in favor of their own idealizations. In essence, they are in love with their own dreams and ideals, not with the person they married. Originally, their mate became the object of their love because he/she met enough of their expectations that they were able to project their total set of expectations onto the mate. But when living together, day in and day out, it is only a matter of time until the partners are forced to compare their ideals with the flesh-and-blood spouses. The two seldom coincide exactly.

Romantic ideals and expectations lead us to expect so much from our mates and from marriage that disappointment is almost inevitable. How we cope with this disappointment helps determine the direction our marriage will take when the honeymoon is over. If we refuse to reexamine our ideals and expectations and blame our mates for the discrepancy, trouble lies ahead. If, on the other hand, we realize that the disappointment is caused by our own unrealistic expectations, we can look forward to getting to know our mates as real human beings. Further reflection also makes it clear that recognizing our mates as real human beings with frailties and problems, rather than as perfect, godlike creatures, greatly eases the strain on our marriages and ourselves. Realizing the humanness of our partners also allows us to relax and be human. But consider what can happen if we are unwilling to give up our expectations of a perfect mate. Is Carol's idealization of her father destroying her relationships?

Carol is clearly in love with her idealized image of her father. Since she never was or will be able to know her father, she has no way to correct her idealized image of him. A real, live husband has no chance to live up to her expectations, because they are based on someone who can't make mistakes. The chances of Carol ever being happy with a spouse are low unless she can stop comparing her spouse with her romanticized images of her father.

Note that when our expectations are unconscious, uncommunicated, or unrealistic, we can feel betrayed, even when we haven't been. We feel betrayed because our spouse has not recognized our needs as we feel he/she should have. Yet, betrayal implies that a promise has been made and then broken. If we have not clearly communicated our expectations, how can our spouse break a promise that was never made? A wife expects her husband to take her out to dinner this coming Friday. Her husband sees them having a romantic dinner at home. Neither shares his/her expectation with the other. What do you think will happen on Friday night?

From the perspective of normal development, couples' patterns of idealizing and de-idealizing have their own life and gradually move the couple toward greater acceptance of each other. We are disappointed when our partner does not meet our idealization, yet acceptance of him/her as a real person helps to move us toward a more mature love (Sharpe 2000, 1280).

Romantic Love or Marriage?

We can see why disappointment with marriage is almost inevitable if we consider that our prevailing cultural view of marriage, as expressed in the mass media, is one of everlasting romance. One of the greatest disappointments that newly married couples face is the fading of romantic love. Romantic love depends on an incomplete sexual and emotional consummation of the relationship. Physical longing is a tension between desire and fulfillment. When sexual desire is fulfilled, romantic love changes to a feeling of affection that is more durable, though less intense and frenzied, than romantic love. Because we have been flooded with romantic hyperbole by the media, we are often unprepared for this natural change in the emotional quality of the relationship.

Renee fell in love with Hernandez when she was 16 and he was 17. Neither of them had ever been in love before. After 2 years of marriage, Renee finds that she is not nearly as excited by Hernandez as she was earlier. The couple is struggling to make a living, and she feels that the bloom is fading from the rose. One day, as Renee listens to a lecture by her art history teacher, she feels a lump in her throat, a thumping in her heart, and a weakness in her knees. This was the way she felt about Hernandez 2 years ago, and she suddenly realizes that she doesn't have these feelings for him anymore. She can't tell Hernandez, and she is afraid of her feelings for the teacher, so she drops the class and tells Hernandez it's time for them to have a baby. Do you think having a baby will rekindle her original romantic feeling toward Hernandez?

It's clear that Renee is confusing feelings of sexual attraction with love. Since American folklore says you can't love more than one person at a time, she must either suppress these feelings of tension and longing, as she does, or decide that she is no longer in love with Hernandez and, perhaps, consider a divorce. Defining love narrowly, as only the great turn-on, means that all marriages will eventually fail because that intense feeling usually fades with time. In successful marriages, however, it will continue to appear, though less often.

Differing Expectations

Even if their expectations about marriage are realistic, spouses may have different expectations about marital roles, especially the roles that each expects the other to play. Traditional gender roles are no longer clear-cut (see Chapter 4). For example, the husband is no longer the sole breadwinner and the wife the homemaker. However, many of our expectations about role behavior come from our experience with our parents' marriage; often, we are unaware of these expectations. Consider the possibilities for conflict in the following marriage.

In Randy's traditional Midwestern family, his father played the dominant role. His mother was given an allowance with which to run the house. His father made all the major money decisions. In Chen's sophisticated New York family, both her mother and her father worked hard at their individual careers. Because they both worked, they decided that each would control his/her own money. Both contributed to a joint checking account used to run the household, but paid for individual desires from personal funds. There was seldom any discussion about money decisions because each was free to spend his/her own money.

Randy and Chen marry after Chen has taught school for a year, and Randy gets a good position with a New York bank. Randy believes that, because he has a good job, Chen no longer needs to work; he can support them both, which he believes is the proper male marriage role. He sees no reason for Chen to have her own bank account. After all, if she wants something, all she has to do is ask him for it. Do you think these differing expectation will cause this couple problems? If so, how would you resolve the differences?

When roles are not clear-cut and accepted by everyone, the chances of conflicting role expectations increase greatly. A man and a woman from a culture in which roles are clearly specified find it easy to define the meaning of a good spouse. If the spouse is fulfilling the assigned roles well, he/she is automatically a good spouse. Where role specification is not socially assigned but is left up to the individual couple, confusion may occur. Each spouse may want to fulfill a specific role and compete with the other to do so; or each may automatically think the other is fulfilling a certain role and come to realize later that neither is.

Eighty Percent I Love You; Twenty Percent I Dislike You

Human beings are complex, and for any two people to meet each other's needs completely is probably impossible. If a couple could mutually satisfy even 80 percent of each other's needs, it would be a minor miracle.

The expectation of total need fulfillment within marriage ruins many marital relationships. As time passes, the spouse with the unmet needs will long to have them satisfied and will accuse the partner of failure and indifference. Conflict will grow because the accused spouse feels unfairly accused and defensive. Life will increasingly revolve around the unfulfilled 20 percent, rather than the fulfilled 80 percent. This is especially true if the partners are possessive and block each other from any outside-need gratification. Unless such a pattern of interaction is broken, a spouse may suddenly fall out of love and leave the mate for someone else. These sudden departures are catastrophic for all parties. The ensuing relationship often fails because of the same dynamics. For example, the dissatisfied spouse finds a person who meets some of the unfulfilled needs and, because these needs have become so exaggerated, concludes that, at last, he/she has met the right person. In all the excitement, the person often overlooks the fact that the new love does not fulfill other needs that have long been met by the discarded spouse. In a few years, the conflicts will reappear over different unmet needs, and the process of disenchantment will recur. Note that Carl and Allison focus on the 20 percent unmet needs rather than on the 80 percent of needs that are met.

Such dynamics are prevalent in "love" marriages. This is true because love blinds, reducing the chances of realistic appraisal and alternative seeking. To the person who believes that love is the only basis for marriage, the need to seek realistic alternatives is a signal that love has gone. But there is another way, as Lederer and Jackson point out in their classic book *The Mirage of Marriage*:

The happy, workable, productive marriage does not require love or even the practice of the Golden Rule. To maintain continuously a union based on love is not

---

HIGHLIGHT

Carl and Allison: The Perfect Couple

Carl and Allison are very much in love. Their friends are amazed at how compatible and well suited they are to each other. Each expects total fulfillment within the marriage. Allison enjoys staying up late and insists that Carl, who likes to go to bed early to be fresh for work, stay up with her because she hates to be alone. At first Carl obliges, but he gradually returns to his habitual bedtime. Of course, he can't sleep well because he feels guilty leaving Allison alone, which, she reminds him, demonstrates his lack of love and uncaring attitude. She tries going to bed earlier, but also can't sleep well because she isn't sleepy; so she simply lies there resenting Carl. She begins to tell him how he has failed her, and he responds by listing all the good things he does in the marriage. Allison acknowledges these good things but dismisses them as the wrong things. He doesn't do the important things that show real love, such as trying to stay up later.

Naturally, sex and resentment are incompatible bed partners, and Allison and Carl's sex life slowly disintegrates. Each begins to dislike the other for not fulfilling their needs, for making them lose their identity in part, and for making their sex life so unsatisfactory.

Then Carl, working overtime one Saturday, spontaneously has intercourse with a co-worker and is soon in love. Sex is good, reaffirming his manhood, and the woman loves to go to bed early. He divorces Allison and marries his new "right" woman. Unfortunately, they also divorce 3 years later because Carl can't stand her indifference to housekeeping and cooking. He reminds her periodically during their marriage that Allison kept a neat house and prepared excellent meals.

What Do You Think?

1. What suggestion can you make to keep Carl and Allison from focusing on the 20 percent of their relationship that each feels is missing, instead of focusing on the 80 percent that is right about the relationship?

2. Can you reasonably expect another person to fulfill all of your needs?

3. What percent of your needs would a partner have to fulfill to make you feel that he/she is the right and only partner for you?

feasible for most people. Nor is it possible to live in a permanent state of romance.
Normal people should not be frustrated or disappointed if they are not in a con-
stant state of love. If they experience the joy of love for ten percent of the time they
are married, attempt to treat each other with as much courtesy as they do dis-
tinguished strangers, and attempt to make the marriage a workable affair—one
where there are some practical advantages and satisfactions for each—the
chances are the marriage will endure longer and with more strength than so-
called love matches. (1968, 59)

The Expectation of Commitment:
A Characteristic of Strong and Successful Families

The expectation that a relationship will grow, be strong, and last—that is, a strong
commitment to the relationship and to your partner—is the single most important factor influencing the success of a relationship.

Commitment in the strong family is multifaceted. First of all, there is commitment to your partner. The expectation that the relationship will endure despite problems leads to a commitment to work with your partner in finding ways
to cope with those problems. If you are committed to a relationship, problems become something to be surmounted, not something that will destroy your relationship. Overcoming the problems will strengthen the relationship.

Commitment to the family unit itself is also important. Support is offered to
other family members because the family recognizes that if one member is in
trouble and hurting, all members will be in trouble and hurting. One way each
family member experiences commitment is by trusting that the family can be de-
pended upon for support, love, and affirmation. Commitment is also experienced
as willingness to support the family if trouble arises. The well-being of all family
members is a major goal of the family. The squeaky violin concert presented by
the third-grader just learning and the professional playing of the college student
both receive enthusiastic support.

It is not just parents who are committed to offer support; all members of the
family offer support to each other. The strong family has a high degree of cohesion and togetherness, though not so much that individuality is lost. Thus, another

cathy® by Cathy Guisewite

facet of commitment is that family members are committed to each other as individuals, and also to the family as a whole. Overworked parents may experience this commitment as the children pitching in to help them clean up the house and yard. Perhaps a brother or sister will give up plans to help a sibling in need. All members of the strong family accept the general idea that the family is a team of individuals working together to achieve individual and team goals. Cooperating as a team is often the most efficient way to achieve individual and also group goals.

Another facet of commitment is that it is long-lasting. It is this quality that creates family stability. Family members can count on support today, tomorrow, and next year. "I love you today" is a wonderful sentiment, but it is not worth much if the people to whom it is said have no idea whether they will be loved tomorrow. If they cannot be reasonably sure tomorrow, next week, or next year, they can only feel insecure and fearful of what tomorrow may bring. The relationship can have no stability—only fear of instability. Commitment that isn't long-lasting really isn't commitment at all because it robs people of their enthusiasm for the future, their ability to plan for the future, and most important, their ability to commit to the relationship. A person who is robbed of his/her ability to commit will be robbed of the possibility of an enduring, intimate relationship.

Another facet of commitment for the strong family is that the commitment to family overrides all other commitments, even the commitment to work. Stinnett and DeFrain relate the following story, told by a member of one of the 3,000 strong families they studied:

I was off on my usual weekly travel. Business took me away from home three or four days a week. I'd left a teenager disappointed because I would miss her dance recital. My wife felt swamped. She'd described herself as a "de facto" single parent. I had a growing sense of alienation from my family; sometimes I missed chunks of their lives. Indignantly, I thought, "Yeah, but they don't mind the money I make. I have work to do. It's important!" Then the flash of insight came. What frontier was I crossing? I wasn't curing cancer or bringing world peace. My company markets drink mixer. Drink mixer! Granted we sell it all over Ohio and are moving into other markets, but how many gallons of mixer for my family? I didn't quit. I enjoy sales and it's a good job. I make good money. I did learn to say "no" to some company demands. And I plan my travel to leave more time at home. Sometimes now I take my wife or a child along. In a few years I'll retire and within a few months I'll be forgotten in the mixer business. I'll still be a husband and a father. Those will go on until I die. (1985, 27–28)

This does not mean to imply that other things can't, at times, take precedence over the family. The key phrase is *at times*. As important and necessary as work is to family well-being, a parent married to his/her job will have a difficult time being married to his/her family. Everyone understands that, at times, something else will be more important than a given relationship. But if these times become all the time, others in the relationship will begin to feel secondary and question the depth of the commitment. Knowing that your mate or parent always puts other things ahead of you can only erode your self-esteem and confidence.

Strong commitment to one's family should not be confused with loss of individuality and belonging to others. The healthy family is committed to helping family members maintain their individuality. Strong families support individuation of their members. Individuation means that individual family members are encouraged to have independent thoughts, feelings, and judgments. It includes a firm sense of autonomy, personal responsibility, identity, and boundaries of the self.

Commitment creates responsibility. At times, family members help each other shoulder these responsibilities. At other times, they will allow a family member to live with the consequences of irresponsibility. Without responsibility, commitment becomes meaningless. "If I can't depend on myself, if others can't depend on me, then protestations of commitment aren't helpful."

In summary, commitment has many sides. Basically, the members of a strong family experience commitment to the family as trust, support, affirmation, acceptance, belonging, love, and enduring concern about their personal well-being and the well-being of the family.

The Expectation of Primariness: Extramarital Relations

In large part because American marriage is rooted in Judeo-Christian principles, one's sexual and emotional outlet is limited to one's spouse. The ideal of monogamy is an important part of American marriage. It is interesting to note that in Murdock's (1950) study of 148 societies, 81 percent of them maintained taboos against adultery, so America is not alone in its expectation of primariness (monogamy) in marriage.

Over 80 percent of all Americans believe in faithfulness in marriage (Michael et al. 1994). Putting numbers on adulterous behavior is difficult because such behavior tends to be covert. Adultery figures vary greatly. The famous Kinsey studies (1948, 1953) reported that 26 percent of American women and about 34 percent of American men were unfaithful. During the sexual revolution in the 1960s

And they lived happily ever after.

© Richard Marks

and 1970s, many sociologists felt these percentages went up considerably, with perhaps as high as 50 percent of married men and women being unfaithful.

Yet, the recent studies of American sexual behavior have reported lower percentages of unfaithfulness. Robert Michael and associates (1994, 104–106) found that 80 percent of women and 65 to 85 percent of men of every age reported they had no sexual partners other than their spouse while married. Wallerstein (Wallerstein and Blakeslee 1995, 262) reported that 16 percent of the women and 20 percent of the men in their study of long-term, happy marriages had brief affairs. However, sex with other than one's partner in dating and cohabiting relationships was found to occur more often (Buunk and Dijkstra 2000).

The expectation of primariness in sexual relations in marriage is supported by approximately 90 percent of the general public, who say it is always or almost always wrong for a married person to have sex outside of marriage (Thornton and Young-DeMarco 2001; Treas and Giesen 2000). This figure is actually higher than it was during the 1980s, which counters the prevailing trend toward freer sexual expression.

As noted in Chapter 6, the changing sexual mores have definitely increased premarital sexual relations, and premarital intercourse tends to correlate with extramarital affairs. Some research data indicate that younger wives (under age 30) in marriages contracted since 1970, and especially those that were involved in a cohabitation arrangement before marriage, now engage in extramarital relations slightly more often than their husbands (Forste and Tanfer 1996), which is a change from past data. This may relate to the greater numbers of women entering the workforce. It appears that wives of all ages who work outside the home have a higher incidence of extramarital relations than that of wives whose work is within the home.

One of the reasons that married men (considering all ages) tended to have more extramarital affairs in the past than married women did is the overabundance of available, single women. For some single women, the unavailability of single men makes an affair with a married man the only alternative to celibacy. In the case of some career women who prefer to remain unmarried, companionship and a sexual relationship may be all that they are seeking, and these needs can be fulfilled by a married man. A number of factors correlate with greater acceptance of extramarital sexual relations, as follows:

- Being male
- Being young
- Being nonreligious
- Being highly educated
- Believing in the equality of the sexes
- Being politically liberal
- Being unmarried
- Being premaritally sexually permissive
- Being in a cohabitation situation

The long-time extramarital affair must be distinguished from the short-term, one-night-stand, extramarital experience. The long-time affair usually includes a strong emotional attachment in addition to sexual involvement. Such relationships are more likely to lead to the breakdown of a marriage. Affairs involving the wife are more likely to lead to divorce because women tend to be more emotionally involved in sexual relationships than men. This leads to a detachment from their spouse that is hard to overcome (Glass 1995, 6).

Generally speaking, men tend to take their extramarital affairs more lightly than women do, and they center more on the sexual aspect than on the emotional aspect. They tend to associate their affairs with an increase in marital satisfaction, due to a decrease in boredom and tension. Older women tend to have affairs that last longer, become more emotionally involved, and associate their affairs with decreasing marital satisfaction. The women often report that the companionship and emotional support derived from the affair are of greater importance than the sex.

The reasons for extramarital sexual affairs are many and varied. Some engage in an affair out of simple curiosity and the desire for some variation in sexual experience. Others yearn for the romance that has been lost in their marriages or search for the emotional satisfaction they feel is missing in their lives. Some simply fall into an affair out of friendship for someone. An adulterous affair can also be a rebellion or retaliation against the spouse.

Most threatening to a marriage is an affair that stems from falling in love with the new partner. Marriage counselors indicate that it isn't so much the notion of the spouse having sex outside the marriage that causes a rift between couples, as the idea that the spouse is emotionally involved with someone else, along with the sex (Masters et al. 1988, 405). Overall, women have been found to experience more emotional distress about infidelity then men do. However, women are more apt to be upset about emotional infidelity, while men are more likely to be upset about their partner's sexual infidelity (Carroll 2005; Nanninc and Meyers 2000).

We tend to equate extramarital relationships with sexual relationships. Yet, many kinds of extramarital relationships with the opposite sex do not include sex. How do people feel about such relationships? David Weis and Judith Felton (1987) showed a written set of activities with the opposite sex to a sample of college women and asked which of the activities would be acceptable for a spouse take part in. They found that their subjects' attitudes varied greatly. For example, 82.3 percent felt that it would be all right for their spouse to go to a movie or the theater with a friend of the opposite sex. Only 30.5 percent felt it would be all right for their spouse to dance to the stereo with another partner when they weren't present. As soon as the activities took on a directly sexual nature, such as necking or petting, very few accepted the behavior (3.7 percent of those sampled).

The researchers concluded that the diversity of attitudes would create great potential for marital conflict over the issue of exclusivity. Some persons are more afraid of nonsexual extramarital ties that have an emotional basis than they are of extramarital sex without emotional interaction. Others are afraid that nonsexual extramarital relationships will lead to sexual involvement.

Extramarital affairs are often difficult for a spouse to combat. Early in the affair, the cheating couple is very aroused and can usually manage only a very limited time together. The cheating partners are on their best behavior and make special efforts to be attractive and appealing. Expectations and excitement are high because of the clandestine nature of the affair. This kind of excitement is hard to bring back into a long-standing marriage.

The positive aspects of an affair are often overshadowed by practical and emotional problems that accompany the maintenance of a secret affair, including the need to have a private place to meet; the need to be careful in telephoning, writing, e-mailing, and seeing the extramarital partner; and the probability that partners will experience guilt and anxiety, jeopardize their primary relationships, and risk contracting a sexually transmitted disease (Buunk and Dijkstra 2000, 323).

Although the expectation of primariness is sometimes broken, it remains an important expectation. Breaking it for whatever reason causes most spouses to become very upset and often ends the marriage. We have seen that mutual trust is one of the cornerstones of strong and enduring relationships. The discovered clandestine affair damages this trust. Even the undiscovered affair damages trust. The spouse engaging in the affair tends to project his/her behavior onto the partner; that is, "I lie to my partner about my affair; therefore, my partner probably lies to me." Once basic trust in your partner is damaged, it is very difficult to regain. Feelings of inadequacy and self-doubt often arise in the spouse who discovers the other's extramarital affair.

Such reactions may not be the case when the extramarital relationship is consensual—that is, when both spouses agree to it. During the 1970s, the media reported a great deal of consensual adultery. Estimates ranged from 15 to 26 percent of married couples having a relationship that allowed sex outside the marriage, under limited circumstances. Blumstein and Schwartz (1983) found that 28 percent of cohabiting couples and 65 percent of gay male couples also reported such agreements. (See Sponaugle 1989; Thompson 1983 for reviews of the research on extramarital relations during the 1980s.) Such consensual adultery has greatly diminished during the past decade.

The Self-Fulfilling Prophecy

Expectations that an individual holds about another person tend to influence that person in the direction of the expectations. Thus, holding slightly high expectations about another person is usually productive, as long as the expectations are close enough to reality that the other person can fulfill them. Remember, though, that expecting a different behavior from a person implies that one doesn't approve of the person as he/she is. Having expectations of another's behavior that are clearly out of that person's reach tells that person that he/she is doomed to failure because the expectations can't be met. This often happens to children. Sometimes children feel so frustrated by their inability to meet their parents' expectations that they deliberately do the opposite of what their parents desire, to free themselves from impossible expectations. "All right, since you are never satisfied with my schoolwork, no matter how hard I try, I'll stop trying," they may think.

Such dynamics may also be found in marriage. If a mate constantly expects something of the other that cannot be fulfilled, the other will feel incompetent, unloved, and unwanted. On the other hand, positive and realistic expectations about the spouse (or anyone else) are more likely to be fulfilled, because they make the other person feel wanted and valuable and the person then acts on this positive feeling.

Clearly, the closer to objective reality our expectations and perceptions are, the more efficient our behavior will generally be. If we expect impossible behavior from our mates, we doom both our mates and ourselves to perpetual failure and frustration. If, on the other hand, we accept ourselves and our mates as we are, we have the makings for an open, communicative, and growing relationship. This does not mean that people cannot change for the better. It means that expected changes must be realistic and accomplishable.

The Self-Actualized Person in the Fully Functioning Family

To have a healthy, intimate marriage, one needs to be a psychologically healthy person. A person must be able to respect and genuinely like him-/herself, admit error and failure and start again, accept constructive criticism, and be self-supportive rather than self-destructive. Of course, if these steps were easy, we would all be living happily ever after. Although much is known about helping people live more satisfying lives, a great deal remains to be learned. Individuals are complex and vary greatly; no single answer will suffice for everyone. We need many paths by which people can travel to self-actualization. How a person attains psychological well-being so that he/she will be a successful marriage partner and later, a nurturing parent, is an important question. And what are the characteristics of good mental health?

Characteristics of Mental Health

The National Association for Mental Health has described mentally healthy people as generally (1) feeling comfortable about themselves, (2) feeling good about other people, and (3) being able to meet the demands of life.

Feeling Comfortable about Oneself

Feeling Comfortable about Oneself   Mature people are not overwhelmed by their own fears, anger, love, jealousy, guilt, or worries. They take life's disappointments in stride. They have a tolerant, easy-going attitude toward themselves and others, and they can laugh at themselves. They neither underestimate nor overestimate their abilities. They can accept their own shortcomings. They respect themselves and feel capable of dealing with most situations that come their way. They get satisfaction from simple, everyday pleasures.

Notice that this description recognizes that people's lives have negative aspects: fear, anger, guilt, worries, and disappointments. Mentally healthy people can cope with such negative aspects. They can accept failure without becoming angry or considering themselves failures because of temporary setbacks. Moreover, they can laugh at themselves, which is something maladjusted people can seldom do. A person comfortable about him-/herself has taken the first step toward a successful, intimate relationship.

Feeling Good about Other People

Feeling Good about Other People   Mature people are able to give love and consider the interests of others. They have personal relationships that are satisfying and lasting. They expect to like and trust others, and they take it for granted that others will like and trust them. They respect the many differences they find in people. They do not push people around or allow themselves to be pushed around. They can feel part of a group. They feel a sense of responsibility to their neighbors and country.

As you have probably noticed, this description is largely based on common sense. Certainly, we would expect people who are considerate of other people's interests to have lasting relationships. As we noted earlier in our discussion of self-fulfilling expectations, if we approach people in an open, friendly manner, expecting to like them, they will be warmed by our friendliness and will probably feel friendly toward us. If, on the other hand, we approach people as if we expect them to cheat us, the chances are that they will be suspicious of us and keep their distance.

Family fun via motor home.

© Frank Cox

Another aspect of this description is that it recognizes that we are gregarious or, as the song says, "people who need people." Mature people recognize this need and are also aware of the responsibilities that people have toward one another.

Feeling Competent to Meet the Demands of Life Mature people do something about problems as they arise. They accept responsibilities. They plan ahead and do not fear the future. They welcome new experiences and new ideas and can adjust to changed circumstances. They use their natural capacities and set realistic goals for themselves. They can think for themselves and make their own decisions. They put their best effort into what they do, and get satisfaction out of doing it. This means that such a person will also put their best effort into building and maintaining her/his intimate relationships. Note also that successful intimate relationships help a person become self-actualized, personally fulfilled, and better functioning.

Self-Actualization

Abraham Maslow (1968) spent a lifetime studying people, especially those he called self-actualized people. These were people he believed had reached the highest levels of growth, people who seemed to be realizing their full potential, people at the top of the mental health ladder, people who can create successful intimate relationships. The following are some of the characteristics they share:

1. *A more adequate perception of reality and a more comfortable relationship with reality than average people have.* Self-actualized people prefer to cope with even unpleasant reality, rather than retreat to pleasant fantasies.
2. *A high degree of acceptance of themselves, of others, and of the realities of human nature.* Self-actualized people are not ashamed of being what they are, and they are not shocked or dismayed to find foibles and shortcomings in themselves or in others.
3. *A high degree of spontaneity.* Self-actualized people can act freely, without undue personal restrictions and unnecessary inhibitions.
4. *A focus on problem-centeredness.* Self-actualized people seem to focus on problems outside themselves. They are not overly self-conscious; they are not problems to themselves. Hence, they devote their attention to a task, duty, or mission that seems peculiarly cut out for them.
5. *A need for privacy.* Self-actualized people feel comfortable alone with their thoughts and feelings. Solitude does not frighten them.
6. *A high degree of autonomy.* Self-actualized people are, for the most part, independent people who are capable of making their own decisions. They motivate themselves.

7. *A continued freshness of appreciation.* Self-actualized people show the capacity to appreciate life with the freshness and delight of a child. They can see the unique in many, apparently commonplace, experiences.

Sidney Jourard sums up the qualities of a mentally healthy individual as follows:

Healthy personality is manifested by the individual who has been able to gratify his basic needs through acceptable behavior such that his own personality is no longer a problem to him. He/she can take him/herself more or less for granted and devote his/her energies and thoughts to socially meaningful interests and problems beyond security, or lovability, or status. (1963, 7)

As noted in Chapter 1, Albert Bandura (1997, 2000) describes successful people as those with a strong sense of self-efficacy. They have high self-assurance in their capabilities to approach life's tasks. They feel that challenges are to be mastered, rather than threats to be avoided.

Living in the Now

Marriages are constantly troubled because one or both spouses cannot live in the present. Are the following remarks familiar? "I'm upset because this Christmas reminds me of how terrible you were last Christmas." "This is a nice dinner, but it doesn't compare with the one I want to fix next week." "This never happened in my last marriage."

All phases of time—past, present, and future—are essential for fully functioning people. The ability to retain what has been learned in the past and use it to cope with the present is an important attribute of maturity. The ability to project into the future, and thereby modify the present, is another important and unique characteristic of humans. The healthy person uses the past and future to live a fuller, more creative life in the present.

Although retention and projection of time can help us cope with the present, they can also hamper present behavior. In the preceding comments, the present Christmas is being ruined because of a past Christmas. One spouse may now be perfectly pleasant, yet the other is unhappy because he/she is dwelling on the past, rather than enjoying the present. People who do not learn from the past are doomed to repeat mistakes, yet people must develop the capacity to learn from the past without becoming entrapped by it.

Much the same can be said of the future. Planning for the future is an important function. Yet, people may also hamper their behavior by projecting consequences into the future that keep them from acting in the present. For example, "I know that the economy next year will slump, so I won't invest in anything now." The problem is that no one can be certain of the future, so if this uncertainty keeps a person from doing anything to prepare for the future, if it freezes his/her actions, the future remains something to fear.

As we have seen, some people live frustrated lives because their expectations of the future are unrealistic. Remember Carol and her three husbands (page 201)? Carol projects idealized expectations onto her husbands. Rather than letting them be the persons they are, she expects them to act in a certain manner (as she imagines her father did). She is so preoccupied with expecting her husband to behave as she thinks he should that she derives no satisfaction from his actual behavior.

The Goals of Intimacy

Clearly, building a satisfying, intimate relationship is a difficult and complex task, yet such relationships are much sought after. True intimacy with others is one of the highest values of human existence; there may be nothing more important for the well-being and optimal functioning of human beings than intimate relationships (Prager and Roberts 2004, 43). Many factors will influence the success of such relationships. Is it possible to prepare oneself to meet the problems so often found in intimate relationships and learn the skills of open communication and problem solving? If a person doesn't let hostilities and inability to communicate stand in his/her way, he/she has a better chance of maintaining and fulfilling intimate relationships. Marriage can be treated as a complex vehicle to personal happiness; as with any vehicle, preventive maintenance and regular care will minimize faulty operation.

Marriage counselors see many couples whose marriages are so damaged that little, if anything, can be done to help them. But most American couples begin their marriages in love. They do not deliberately set out to destroy their love, each other, or their marriage. Yet, it is often hard to believe that the couples in the marriage counselor's office or the divorce court ever felt love and affection toward each other. Too often they are bitter, resentful, and spiteful. Their wonderful "love" marriage has become a despised trap, a hated responsibility, an intolerable life situation. They seem to have forgotten the positive characteristics that led them to fall in love in the first place.

Why? Marriages often become unhappy because we are not taught the art of getting along intimately with others, or the skills necessary to create a growing and meaningful existence in the face of the pressures and problems of a complex world. Far more could be done by many couples to handle marital difficulties intelligently and successfully; failure to teach these skills, both before and after marriage, is a shortcoming of contemporary treatments of the subject of marriage. How one handles conflict is the single most important predictor of whether a marriage will survive (R. Edwards 1995, 6; Notarius and Markman 1993).

To get along intimately with another person and create a fully functioning family, we need to be clear about ourselves, our expectations, and the basic goals of intimacy. In their most general terms, the goals of intimacy are identical to some of the functions of the family that we discussed in Chapter 1. In particular, they involve (1) helping family members deal with crises and problems, (2) helping members grow in the most fulfilling manner possible, and (3) providing emotional gratification to members. The following examples show how a marriage can fulfill these functions:

Dealing with crises. When Pete has a crisis in the office, he tells Gail about it. She listens, offers support, asks questions, and only occasionally, offers advice. When he is through telling her about it, Pete often understands the situation better and feels more ready, either to accept or to change it.

Growing in a fulfilling manner. Gail wants to be less shy. Pete encourages her to be more assertive with him and to role-play assertiveness with others. Gradually, she finds she can overcome her shyness—in large part, because Pete encourages and supports her.

Providing emotional gratification. Pete loves to tinker with things, and he feels important and emotionally gratified if he can fix something in the house for Gail. Gail loves to knit and Pete loves sweaters, so Gail feels worthwhile and appreciated when she knits him a sweater. Gail knows she can sound off when she gets angry, because Pete understands and doesn't put her down. Pete feels emotionally supported by Gail because she backs him when he wants to try something new. Both partners are having their own emotional needs met, and also meeting many of the needs of the other.

Certainly, it does not seem to be asking too much of a marriage for partners to help each other deal with crises that arise, to encourage each other to grow in a personally fulfilling manner, and to supply emotional gratification to each other. Yet, marriage often fails this assignment or, to put it more accurately, marriage partners often fail to create a marriage in which these positive elements thrive.

You and the State: Legal Aspects of Marriage

Every society has some kind of ceremony in which permanent relationships between the sexes are recognized and given status. The society (or the state) sets minimum standards for marriage in the interest of order and stability. In Western societies, the state is interested in (1) supporting a monogamous marriage, (2) assuring the legitimacy of issue (that children are born of legally recognized relationships), (3) protecting property and inheritance rights, and (4) preventing marriages considered unacceptable (such as those between close relatives).

In the United States, marriage laws are established by individual states. Although the requirements differ, all states recognize marriages contracted in all other states except gay marriages. California law is typical of many state laws relating to marriage, which it defines as follows:

Marriage is a personal relation arising out of a civil contract, to which the consent of the parties capable of making that contract is necessary. Consent alone will not constitute marriage; it must be followed by the issuance of a license and solemnized as authorized. (West 1989)

Marriage in the United States is a contract, with the rights and obligations of the parties set by the state. Like all contracts, the marriage contract must be entered into by mutual consent, the parties must be competent and eligible to enter into the contract, and a prescribed form must be followed. All states set minimum age requirements, and most require a medical examination and a waiting period between the examination and the issuance of the license. Unlike most contracts, which are between two parties, the marriage contract involves three parties: the man, the woman, and the state. The state prescribes certain duties, privileges, and restrictions. For example, the contract can only be dissolved by state action, not by the mutual consent of the man and woman.

States set a number of other standards for marriage. The state may consider a marriage invalid under certain circumstances: for example, if consent to marry is obtained by fraud or under duress or if either party is already married, suffers from mental incapacity, or is physically unable to perform sexually.

No specific marriage ceremony is required, but the parties must declare, in the presence of the person solemnizing the marriage, that they take each other as husband and wife, and the marriage must be witnessed, usually by two persons. Although some states are quite specific, there is a general trend away from uniform marriage vows. Traditional vows reflect the permanence society expects of marriage: ". . . to have and to hold from this day forward, for better or worse, for richer, for poorer, in sickness and in health, to love and to cherish, till death do us part."

Up to this point, we have discussed the legal aspects of marriage established by the state. However, religious faiths regard marriage as a sacrament. Most marriages in the United States take place in a church. The state vests the clergy with the legal right to perform the marriage ceremony. God is called on to witness and bless the marriage: "Those whom God hath joined together let no man put asunder." Some religions feel so strongly that marriage is a divine institution that should not be tampered with by humans that they do not recognize civil divorce (the Roman Catholic Church is one example).

The marriage ceremony commits the couple to a new status. It sets minimum limits of marital satisfaction. Typical of these directives are the following:

- A husband is required to support his wife and family. It is his duty to support his wife, even though she has an estate of her own.

- A husband has no duty to support his wife's children from a previous marriage.

- Married persons' conjugal rights include enjoyment of association, sympathy, confidence, domestic happiness, comforts of dwelling together in the same habitation, eating meals at the same table, and profiting by joint property rights, as well as intimacies of domestic relations.

- One of the implied conditions of the marriage contract is that the wife shall give her husband companionship and the society of home life without compensation.

The laws and statutes of the state of California concerning marriage and divorce cover approximately 1,000 pages of the civil code. Marriage, then, is much more than just a piece of paper. It commits the couple to a new set of obligations and responsibilities. In essence, the couple marries the state, and it is the state to which they must answer if the prescribed responsibilities are not met.

Some companies, cities, counties, and states have set up domestic partnership ordinances. A **domestic partnership** recognizes as valid some unmarried heterosexual and homosexual couples' relationships. These ordinances typically define partners as two financially interdependent adults who live together and share an intimate bond, but are not related in the traditional sense of blood or law (W. Duncan 2001; Willetts 2003, 939).

Couples seeking licensure are required to complete an affidavit stating that they are not already biologically or legally related to each other or legally married to someone else, that they agree to be mutually responsible for each other's welfare, and that they will notify the proper authorities if there is a change in the relationship such as dissolution or legal marriage (Willetts 2003, 939).

Such policies are controversial. Many people feel that such recognition undermines marriage (Blankenhorn 1999; Tettamanzi 1998). Others who prefer cohabitation over marriage feel that any kind of legal recognition of cohabitation threatens their freedom and involves them in obligations they prefer to avoid by cohabiting rather than marrying. Many businesses fear greatly increased costs if they adopt a domestic partnership policy (Embree and Griego 2000).

Domestic partnership laws, although they apply to both heterosexual and homosexual couples, are seen by some as attempts by the gay rights movement to

domestic partnership
The law recognizes as valid some unmarried and homosexual couples' relationships

gain the right of marriage for gays. As noted earlier, in 2004 Massachusetts became the first state to grant the right of marriage to homosexual couples. However, other states have passed legislation barring such unions. As of 2005, there has been one congressional attempt to pass a constitutional amendment barring such unions. Domestic partnership policies raise a number of difficult questions (Wisensale and Hechart 1993, 202–203):

- How should domestic partnerships be defined, and how can they be proved to exist?
- How extensive should benefit coverage be?
- What are the additional costs of such policies?
- How should the tax structure respond to domestic partnerships?
- Should there be a national domestic partnership law?
- Since marriage is already legally recognized as signifying an intimate partnership, why bother with a domestic partnership policy?
- Would such policies invade the privacy of an intimately cohabiting couple who have deliberately chosen to avoid the legal entanglements of marriage?
- Some radical feminists see marriage as a sexist institution that denies women their rights and fear that gay/lesbian domestic partner laws may take away individual rights of the partners.

Writing Your Own Marriage Contract and Prenuptial Agreement

As we have seen, marriage is a formal contract between the couple and the state and has a set form in each state. The standard state contract does not meet every couple's needs, however, and an increasing number of couples are writing their own **marriage** or **prenuptial contracts**. Such a personal contract cannot take the place of the state marriage contract, nor can it legally override any of the state contractual obligations. As a supplement to the state contract, it can afford a couple the freedom and privacy to order their personal relationship as they wish. It can also enable them to escape, to some degree, the sex-based legacy of legal marriage and to move toward an egalitarian relationship. The couple can formulate an agreement that conforms to contemporary social reality. Personal contracts can also be written by couples who wish to have a relationship, but not one of marriage (such as a cohabiting couple or a couple barred from marriage, such as a homosexual couple).

 Besides its legal advantages, a personal contract facilitates open and honest communication and helps prospective partners clarify their expectations. Once the contract agreement has been reached, it serves as a guide for future behavior. Contracts also increase predictability and security by helping couples identify and resolve potential conflicts in advance.

 Personal contracts must be constructed carefully if they are to be legal. For example, when one partner brings to a marriage a great deal of wealth that has been accumulated before the marriage, the couple may want to sign a contract that keeps this property separate from the property that accumulates during the marriage. This is particularly difficult to do in community property states. Such a contract needs to be drawn by an attorney who understands the state laws gov-

marriage or prenuptial contract

Working out the details of a couple's relationship before they wed erning community property. Any topic may be handled in a personal contract, but such contracts generally cover the following:

- Aims and expectations of the couple
- Duration of the relationship
- Work and career roles
- Income and expense handling and control
- Property owned before and acquired after the contract
- Disposition of prior debts
- Living arrangements
- Responsibility for household tasks
- Surname
- Sexual relations
- Relations with family, friends, and others
- Decisions regarding children (number, rearing, and so on)
- Religion
- Inheritance and wills
- Resolving disagreements
- Changing and amending the contract
- Dissolution of the relationship

It may seem unromantic to sit down before marriage with one's intended partner and work out all of these details of the future relationship. Some people feel it demonstrates a lack of trust in a future mate. In some cases, work on a prenuptial agreement may cause real problems that actually break up the relationship before marriage takes place.

But it is a worthwhile exercise, even if one doesn't plan to have a written legal contract. People's expectations about each other and their relationship are important determinants of behavior. If one partner's expectations differ greatly from the other's and are left unexpressed and unexamined, the chances are great that the couple will experience conflict and disappointment. By going through the steps of working out a personal contract, the couple can bring their attitudes and expectations into the open and make appropriate compromises and changes before major problems arise. Answering the questions in "The Couple's Inventory" (page 219) is a good start toward understanding each others' attitudes and expectations.

HIGHLIGHT

Navajo Marital Expectations

In many traditional cultures, spouses know before their wedding day exactly what their rights are. In essence, they have a prenuptial contract. For example, a Navajo child is born into his/her mother's clan, so everyone knows who "owns" the child (child custody), should the couple separate. Land and goods are not negotiable either. Navajo women own their own property; men own theirs. As a result, despite the traumas of a divorce, there are no child custody or property disputes when a marriage fails.

MAKING DECISIONS

The Couple's Inventory

Personal goal: To look at how role behavior influences decision making, autonomy, and intimacy in your relationship with your partner.

Directions: Both partners fill out separate inventories and then compare statements.

1. I am important to our couple because _____.

2. What I contribute to your success is _____.

3. I feel central to our relationship when _____.

4. I feel peripheral to our relationship when _____.

5. The ways I show concern for you are _____.

6. The ways I encourage your growth are _____.

7. The ways I deal with conflict are _____.

8. The ways I have fun with you are _____.

9. I get angry when you _____.

10. I am elated when you _____.

11. The way I get space for myself in our relationship is _____.

12. The ways I am intimate with you are _____.

13. The ways I am jealous of you are _____.

14. I have difficulty being assertive when you _____.

15. You have difficulty being assertive when I _____.

16. The strengths of our relationship are _____.

17. The weaknesses of our relationship are _____.

18. Our relationship would be more effective if you _____.

19. I feel most masculine in our relationship when I _____.

20. I feel most feminine in our relationship when I _____.

21. I trust you to do/be _____.

22. I do not trust you to do/be _____.

23. I deal with stress by _____.

24. You deal with stress by _____.

25. The division of labor in household tasks is decided by _____.

26. Our finances are controlled by _____.

27. The amount of time we spend with our relatives is determined by _____.

28. Our vacation plans are made by _____.

29. Our social life is planned by _____.

30. Taking stock of our relationship is done by _____.

31. I am lonely when _____.

32. I need you to _____.

SOURCE: Sargent 1977, 87.

Summary

1. Marriage is the generally accepted mode of life for most Americans. *Legally, marriage is a three-way relationship that involves the man, the woman, and the state.*

2. *Marriage is constant interaction between family members and fulfillment of many roles within the family relationship.*

3. *Expectations that are too high and/or too low can lead to relationship problems.* Both the expectation of permanence (commitment to the relationship) and the expectation of primariness (monogamy) are important building blocks of a strong, enduring relationship.

4. *Self-actualizing people essentially are people who feel comfortable about themselves and others. They can meet most of the demands of life in a realistic fashion.* Self-actualizing people tend to use their past experiences and ideas about their future to enhance the present, rather than to escape from it.

5. *The basic goals of intimacy in a marriage are providing emotional gratification to each partner, helping each deal with crises, and helping each grow in a fulfilling manner.*

6. *The marriage ceremony commits a couple to a new status, with certain privileges, obligations, and restrictions.* Besides state-mandated marital obligations, some couples are writing their own marriage contracts which state goals, obligations, and responsibilities that they wish to be a part of their marriage.

Resources on the Internet

Companion Website for *Human Intimacy: Marriage, the Family, and Its Meaning,* Tenth Edition

http://sociology.wadsworth.com/cox10e/

Gain an even better grasp on this chapter by going to the companion website to take one of the tutorial quizzes, use the flash cards to master key terms, or check out the many other study aids you'll find there. You will also find special features such as GSS data, Sociology Online, and Census 2000 information that will put data and resources at your fingertips to help you with that special project or to help you as you do some research on your own.

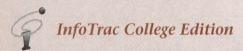

http://www.infotrac-college.com/wadsworth/

You can access reliable resources anytime, anywhere, with InfoTrac College Edition, the online library. This fully searchable database offers more than 20 years' worth of full-text articles (not abstracts) from almost 5,000 diverse sources, such as top academic journals, newsletters, and up-to-the-minute periodicals, including *Time*, *Newsweek*, *Science*, *Forbes*, *The New York Times*, and *USA Today*. You can conduct electronic key word searches using key terms from this chapter to supplement your reading and learning experience. To aid in your search and to gain useful tips, see the Student Guide to InfoTrac College Edition, which you can access through the companion website for this book.

Is Marriage a Dying Institution?

Cohabitation, domestic partnership laws, gay marriage, out-of-wedlock birth, single-parent families—because any consensual intimate relationship is now available to Americans, why marry at all?

Marriage Is Dying

Marriage is old-fashioned. It traps people into a legal relationship that stunts the partner's growth by stereotyping gender roles. In a country based on "liberty and freedom for all," citizens should be able to form any kind of intimate relationships upon which the partners agree.

Since the relationship is freely chosen and consensual, there would be less conflict, no messy divorces, no need of lawyers and courts dictating the terms of a settlement, and no punishing of one partner through the requirement of alimony. Any conflict that arises could be settled through mediation, leading to fair and agreed-upon division of property and care of any children resulting from the relationship.

Marriage Is Changing, but Change May Not Mean Decline

Bemoaning the state of marriage has been fashionable throughout history. The question to ask of all the doom-and-gloom sayers, concerning the American marriage, is, "How do we explain that, despite all of the problems, the vast majority of Americans still marry, have children, and at least half of them stay married the rest of their lives?"

It is pie-in-the-sky thinking to believe that just any intimate relationship can serve to keep two people together and to rear children successfully so that society as a whole remains healthy.

If any intimate relationship is acceptable, how does the society control relationships that are harmful to one or both partners and/or their children? Does acceptance of any and all relationships mean that a person may have sex with children, either their own or others? What about a person who is human immuno-deficiency virus (HIV)-positive having sex with another unknowing individual?

Indeed, marriage is changing just as it always has throughout history. But the negative changes seem not to affect the vast majority of Americans, who are passionate about being married and are trying to be the best parents that they can be to their children. Most families are working to make life better than the way it was in the families in which they grew up. The changing American marriage has helped the wife/mother become more independent and equal to her husband, thus increasing her status and self-respect—both qualities that help a person have successful intimate relationships. The idea that a person can be smothered in a marriage has lost validity with the increased emphasis today on meeting individual needs and also family needs.

Overall, Americans can now choose from a variety of marriages, rather than being trapped in a single kind of marriage. Free choice in a free society should allow the future building of more successful marriages and families. For example, reduction in the size of the family may not reflect a dislike of children, but rather, a growing ability to regulate and space children to provide a better quality of family life.

What Do You Think?

1. Do you think that the American marriage is on its way out, or is it only changing?

2. If the American family does die, what do you see taking its place?

3. If the American family collapses, how will children be raised?

4. What suggestions can you make that would strengthen the American family?

5. What kind of family would you like to build for yourself?

Human Sexuality

Chapter

8

© Thinkstock/Getty Images

Questions to reflect upon as you read this chapter:

- In what ways does human sexual behavior differ from animal sexual behavior?

- What are the major differences between male and female sexuality?

- What are some suggestions for keeping the sexual excitement alive in long-term, intimate relationships?

- In what ways can drugs interact with a person's sexuality?

- What steps can a person take to avoid contracting a sexually transmitted disease?

Two friends were talking about sex one day. Do you have any answers for their questions?

FRIEND: What is sex?

STUDENT: Everyone knows what sex is! Sex is for having babies—you know, reproduction.

FRIEND: I know, but if sex is only for reproduction, why don't humans mate like other animals, once a year or so? Why don't human females go into heat to attract males?

STUDENT: Well, human females are more sexually receptive at certain times during their monthly cycle, aren't they?

FRIEND: The evidence on that is mixed, but even if it were true, why are humans interested in sex all the time?

STUDENT: Perhaps sex is for human pleasure.

FRIEND: But if sex is for fun, why are there so many restrictions on sexual behavior? Why does society try so often to regulate sexual expression? Why are there so many sexually transmitted diseases (STDs)? STDs certainly aren't fun. Why does religion try to focus sexual behavior toward some higher purpose?

STUDENT: Well, then, perhaps sex is for love, and love will limit the number of sexual partners.

FRIEND: But, what is love exactly? Does sex always mean love? If masturbation is sex, does it mean I love myself if I do it? Can I love more than one person at the same time?

STUDENT: Love is emotional closeness that allows you to communicate at an intimate level. Love also makes you feel good about yourself—it enhances your ego. So, if sex is love, it does all these things too.

FRIEND: Certainly, sex can be for all of the things you mention. But isn't sex sometimes just for biological release? This doesn't sound much like love or ego enhancement, does it?

STUDENT: No, but sex can and should be an expression of love.

FRIEND: Ah, yes, but what it sometimes is and what it should be are often two different things.

STUDENT: What do you mean?

FRIEND: Well, is sex an expression of love when it is used to possess another person, such as when a woman is considered to be a man's property? Or when it is used to gain status, such as when a king marries the daughter of another king to increase his holdings and thereby, his prestige? Or when it is accompanied by violence, as in rape? Or when it is a business, as in prostitution? Or when it is used indirectly, as in advertising, where promotions based on sex are used to sell many different products?

STUDENT: Now I'm really confused. Just what is sex?

From this short discussion, it is obvious that sex is many things and, at times, is something of a riddle. If sex were only for reproduction, or only an expression of love, or only for fun, there would be little controversy about it and no need to control it. However, sex isn't for one purpose but rather, for many. It is this fact that causes people to be concerned and, at times, confused about the place of sex in their lives. Our sexuality is a strong drive that causes great pleasure and also great pain and trouble at times. Much is written about our sexuality, from how-to books to scientific treatises that attempt to explain all aspects of our sexuality. Newsstands teem with magazines offering articles on the subject, such as the special issue of *Time* (2004) entitled "Love, Sex and Health."

We talk about sexuality rather than just sex because it involves the entire person, the whole life course, not just sexual acts. (Herdt 2004)

Of all the splendidly ridiculous, transcendently fulfilling things humans do, it's sex—with its countless permutations of practices and partner that most confound understanding. Why in the world are we so consumed by it? The impulse to procreate may lie at the heart of sex, but like the impulse to nourish ourselves, it is merely the starting point for an astonishingly varied banquet. Bursting from our sexual center is a whole spangle of other things—art, song, romance, obsession, rapture, sorrow, companionship, love, even violence and criminality—all playing an enormous role in everything from our physical health to our emotional health to our politics, our communities, our very life spans. (Kluger 2004)

There is more about sex, especially pornography, on the Internet than about any other single subject. *Cybersex* is the term coined for those who find some sexual arousal and/or outlet via the computer, using chat rooms or viewing pornography.

Marriage is society's sanctioned arrangement for sexual relations. A happy, satisfying sex life is a characteristic of the healthy family. In fact, a higher proportion of married than cohabiting or single men and women report being extremely emotionally satisfied with sex (Lauman et al. 1994; Waite and Joyner 2000). Sex is one of the foundations of most human, intimate relationships. Sex is the basis of the family-procreation—and the survival of the species. Sex is communication and closeness. It can be pleasuring in its most exciting and satisfying form. Certainly, it is proper to study marriage by viewing humans as the sexual creatures they are. It is important to understand the biological foundations if we are to fully understand sexuality and the male-female bond. But it is our thoughts, attitudes, and values about sex that are the most important part of human sexuality. Our involvement in sex can vary from absolutely superficial—where two people are just triggering reflexes in each other's bodies—to the point of profound meaning. It is our mental and emotional processes, how we feel about what we are doing sexually, that are the largest determinants of overall sexual satisfaction, rather than the tactile maneuvers described in so many how-to books. Sex can serve to bond a couple even closer together in an intimate relationship. Indeed, the term *intimate* usually implies a sexual relationship and also a close emotional relationship.

MAKING DECISIONS

Sex Knowledge Inventory

Sex is a subject that most people think they know a lot about. Let's see if we do. Mark the following statements true or false. The answers are given at the end of the survey.

1. Women generally reach the peak of their sex drive later than men.
2. It is possible to ejaculate without having a total erection.
3. Sperm from one testicle produce males and from the other, females.
4. A person is likely to contract a sexually transmitted disease when using a toilet seat recently used by an infected person.
5. If a person has gonorrhea once and is cured, he or she is immune and will never get it again.
6. Certain foods increase the sex drive.
7. Premature ejaculation is an unusual problem for young men.
8. The penis inserted into the vagina (sexual intercourse) is the only normal method of sex.
9. It is potentially harmful for a woman to take part in sports during menstruation.
10. A woman who has had her uterus removed can still have an orgasm.
11. During sexual intercourse, a woman may suffer from vaginal spasms that can trap the male's penis and prevent him from withdrawing it.
12. The cause of impotence is almost always psychological.
13. For a certain time period after orgasm, the woman cannot respond to further sexual stimulation.
14. For a certain time period after orgasm, the man usually cannot respond to further sexual stimulation.
15. Taking birth control pills will delay a woman's menopause.
16. The size of the penis is fixed by hereditary factors and little can be done by way of exercise, drugs, and so on to increase its size.
17. If a woman doesn't have a hymen, this is proof that she is not a virgin.
18. As soon as a female starts to menstruate, she can become pregnant.
19. About 80 percent of women infected with gonorrhea show no symptoms.
20. The penis of the male and the clitoris of the female are analogous organs.
21. A woman can't get pregnant the first time she has intercourse.
22. A good lover can bring a woman to orgasm, even when she doesn't want to have an orgasm.
23. Herpes is easily curable with antibiotics.
24. Thus far, over one-half of persons with AIDS have died.
25. One can become infected with herpes simply by kissing an infected person.

24.T 25.T
1.T 2.T 3.F 4.F 5.F 6.F 7.F 8.F 9.F 10.T 11.F 12.T 13.F 14.T 15.F 16.T 17.F 18.F 19.T 20.T 21.F 22.F 23.F

Because of the sexual revolution, all Americans are supposed to know everything about sexuality. See if you do by taking the "Sex Knowledge Inventory" before you continue reading the chapter.

Human Sexuality in the United States

No society has ever been found where sexual behavior was unregulated. True, regulations vary greatly: one spouse, multiple spouses, free selection of sexual partners, rigidly controlled selection, bisexuality, homosexuality, and so forth. Actually, if one takes a cross-cultural view of sexuality, the specific regulations include almost any arrangement imaginable. Within a given culture, however, the regulations are usually strictly enforced through taboos, mores, laws, and religious edicts. Transgression may bring swift and sometimes severe punishment, as in some Middle Eastern cultures where an adulterous woman may be stoned to death.

Why do human beings surround sex with regulations? Certainly, sex is controlled among lower animals, but the controls are usually identical throughout the species, dictated by built-in biological mechanisms. Human beings regulate sex precisely because their biology has granted them sexual freedom of choice, and sexual behavior can occur at any time. Among animals, with a few exceptions, sexual behavior occurs only periodically, depending on the estrus cycle of the female in mammals below primates. For mammals, sexual behavior is usually for reproduction. Thus, sexual responsiveness is tied directly to the period of maximum fertility in the female.

In lower animals, sexual behavior is controlled by lower brain centers and spinal reflexes activated by hormonal changes. In general, the larger the brain cortex, the higher the species and the more control the animal has over its own responses. So we come to human beings with their large cortex and what do we find? Earth's sexiest animal, an animal with few built-in restraints and hence many variations in sexual behavior. Without built-in biological guidelines, human sexuality is dependent on learning, and because different societies and groups teach different things about sexuality, there are many variations in sexual attitudes and behavior. Sexual compatibility, in part, depends on finding another person who shares your attitudes about sex. Because sexual expression in humans is less tied to reproduction than it is in lower animals, sexual expression serves other purposes for humans, beyond reproduction.

Human sexuality differs significantly from that of other animals in several other important ways besides the greater freedom from instinctive direction. It appears that human females are the only females capable of intense orgasmic response. The sexual behavior of the human male, however, still resembles the sexual behavior of male primates; it depends largely on outside perceptual stimuli and is under partial control of the female in that she usually triggers it.

Another important difference between humans and other animals is that, unlike most other animals, human females are not necessarily more sexually responsive during ovulation. There seems to be no particular time during the menstrual cycle when all women experience heightened sexual desire. A few women seem to become more sexually aroused at midcycle, when they are most fertile. For other women, sexual desire peaks just after the menstrual flow begins; possibly, this pattern is related to a reduced fear of pregnancy. Some women peak just before the menstrual flow. It has been suggested that these differences, coupled with the development of the orgasm in females, may mean that humans are the only species to derive pleasure out of sexual behavior without becoming involved in its reproductive aspects.

The major difference between humans and animals is that much of human sexuality depends on attitudes, values, and what the individual thinks, rather than on biology. Compared with other species, human sexuality is:

- Pervasive, involving humans psychologically as well as physiologically.
- Under conscious control, rather than instinctual, biological control.
- Affected by learning and social factors and thus more variable within the species.
- Largely directed by an individual's beliefs, values, and attitudes.
- Less directly attached to reproduction.
- Able to serve other purposes in addition to reproduction, such as pair bonding and communication.
- More of a source of pleasure.

HIGHLIGHT

The Sambians of Papua

The Sambians of Papua, New Guinea, have sexual customs so completely different from American sexual customs as to be almost unbelievable to Americans. Women are feared and shunned, and husbands must be cleansed through bleeding after every act of intercourse. Young boys aged seven to 10 years are brought to live in the "men's house," separate from women, until they are ready for marriage. During this time, they are strengthened and made ready for battle by being inseminated by older males in ritualized, homosexual acts. The sperm are considered the essence of the "male force," standing for bravery, honor, and strength. The Sambians believe there is a limited pool of sperm in the world and that to pro-

duce their own, they first have to be inseminated by other males. Once they marry, sexual relations become heterosexual.

In the Sambian culture, homosexual acts between men and boys are the norm, whereas such acts are rare in the United States. The Alan Guttmacher Institute (2003) found that approximately 2.3 percent of men in their 20s and 30s report they have had sex with another man, while only 1.1 percent report being exclusively homosexual. Robert Michael and his colleagues (1994) found about the same percentages. Such statistics are probably not definitive because the samples tended to be small.

It is these differences that have led human beings to create such a variety of sexual standards and practices. Because human sexuality is not totally under biological control, a free society will also experience changes in sexual attitudes and behaviors. Such changes have been occurring rapidly during the past several decades in the United States. Behaviors and expressions of human sexuality that were once considered out of the range of normalcy in American society are seen by some as more acceptable variations in human sexual behavior (Szuchman and Muscarella 2000, xii). Homosexual relationships, once hidden (relegated to the closet), are now accepted by many in American society. Recent studies have found that gay-couple households are found throughout the country (*New York Times* 2001). By the time you are reading this textbook, it may well be that legal marriage between same-sex partners will be possible. Popular media now often discuss people whose identity is gay, lesbian, bisexual, transgender, or questionable. Some sexuality books and also marriage and family texts use the acronym GLBTQ in referring to such persons (Carroll 2005, 314).

Changing Sexual Mores

For better or worse, sexual attitudes and behavior in American society have changed rapidly over the past 40 years. Generally, sexual expression has become freer, more diverse, and more open to public view. The infamous *double standard*, which promoted sexual expression for men while limiting it for women, has broken down for many Americans.

Better understanding and acceptance by women of their sexuality may be one of the revolutionary changes affecting the family and all intimate relationships. As noted in Chapter 6, the sexual revolution has really been a revolution for American women, in that their sexual behaviors have become more like men's.

Women are now able to channel sexuality into their lives in their own way and at their own pace. If each person is free to express him-/herself sexually with a partner and respects that same freedom for the partner, the chances of sexual exploitation of one partner by the other are reduced. Without exploitation and manipulation, the chances for sexual fulfillment and enjoyment are greatly increased. Freedom of sexual expression includes the freedom to say no to sexual interaction.

In a sense, greater sexual diversity and freedom both create and solve problems. Freedom means responsibility. An individual must assume personal responsibility for his/her actions if he/she is free to choose those actions. In the past, when sexual expression was surrounded by mores, taboos, and traditions, responsibility was partially removed from the individual. One could always blame the rules for lack of satisfaction, failures, and unhappiness. But in the past few years, America has rapidly removed the rules from sexual expression. Now more than ever, sexual decisions are up to each individual, and this can be frightening. This means that personal responsibility for one's sexual behavior is the norm, not some outside rule of conduct. Yet, the removal of the traditional rules makes it increasingly difficult for American youth to know how to handle their sexuality.

As noted in Chapter 6, the majority (63 percent; Boonstra 2003) of Americans become sexually active in their teenage years. Slightly more than half of females and two-thirds of males have now engaged in sexual intercourse by their eighteenth birthday (Moore et al. 2000). Estimates are slightly higher for minority adolescents and adolescents of lower economic status (Bremern et al. 2002; Miller 1999, 85–86). About 12 percent of males and 3 percent of females have engaged in sexual intercourse by age 12 (Meschke et al. 2000). Despite the high level of sexual activity among teenagers, pregnancy rates have fallen about 30 percent from their high point in 1991 (NCHS 2003b). However, this is probably due to the high teen abortion rate rather, than a reduction in rates of sexual intercourse. Although teenage abortion rates dropped during the 1990s, American adolescents have the highest abortion rate among developed nations (Alan Guttmacher Institute 2003; CDC 2000a).

Modifying Sexual Behavior

As we have seen and will continue to see throughout this book, American sexual attitudes and practices have been considerably liberalized. Starting in the 1980s, however, a number of factors began to modify some of the changes brought about by freer sexual expression.

The most important modification factor has been the epidemic return of STDs that has accompanied freer sexual life (see Appendix A). With the emergence of herpes simplex virus, type II, and a resurgence of all the historical diseases, Americans began to curb their sexual experiences and to question just how far sexual liberation could go. The appearance of the deadly, acquired immune deficiency syndrome (AIDS) has had an even greater impact on American sexual behavior.

A second factor may be that recreational sex alone becomes dull and boring. Boy meets girl; boy and girl have sex; they part. They never discover the excitement of the chase, the enjoyments of sex within a wider relational context, the bonding that sex can create (see the section on marital sex, pages 240–243), the expression of care and intimacy that sex can be, and many of the other roles that sex can play in an intimate relationship. One young woman declared, "I have sex with a new date as soon as I can to get the hassle out of the way." Such a person has not yet discovered many of the pleasures and relational enhancements sex can provide.

A third factor may be that increasingly more young women report that they are tired of dating just to have sex. They complain that their partners are interested only in the bedroom aspect of a date. If they say no, the date doesn't call back. If they say yes, he leaves after the sex. Some young women are asking of young men, "Aren't you interested in anything else besides my body?"

As noted in Chapter 4, many of the women who were early supporters of the women's movement and had more liberal sexual attitudes are starting to criticize some of the practical outcomes of freer sex. They feel that they have gained the right to say yes to their sexuality, but lost the right to say no. They believe men are the big winners, in that men can now have sex whenever they want (because the liberated woman supposedly won't say no), without the necessity of commitment to a more meaningful relationship or even the responsibility for birth control or ensuing pregnancies. Numerous young women who have found themselves pregnant and have sought counseling indicate that the father-to-be doesn't know of their pregnancy. When asked why, they say that they don't want to bother him with it. This attitude is leading many older women to think that they may have been misled by the sexual revolution—that they were tricked into playing the male's game of easy sex.

Freedom to have uncommitted sex, freedom to become pregnant, and freedom to be a single parent were not exactly what women had in mind when they voiced support for freer sexual mores and women's rights. In a way, it has been the men who have been freed of the responsibilities that, in the past, went with an intimate sexual relationship. And it is the newly liberated women who now must bear the responsibilities. Stanley Graham (1992), in his presidential address at the American Psychological Association's 99th Annual Convention, said:

The emancipation of women was paralleled by the abandoning of responsibility by men. What I am saying is that the social and economic revolution as it pertained to women gave them a lot more work, a lot more responsibility, and sexual freedom to some degree, but it gave men much more sexual freedom at least until the spread of herpes and AIDS.

Women who have been unable to establish a long-lasting, intimate relationship with a man, despite their liberated ways, are beginning to express bitterness. "I feel used and manipulated by the men I meet," they say. Or, "I really am getting tired of wondering whose bed I'm in and whether he cares who I am." "I'd like someone to at least offer to share in birth control responsibilities."

Many women suggest that affection, tenderness, and cuddling are more important to their happiness than sex. For some people, then, sexual liberation may have become as much of a trap as the old Victorian constraints on sexuality.

Saying no to sex when one wants to say yes, because one thinks one shouldn't, is sometimes heroic. But saying yes when you want to say no, because you think you should, is merely weakness. If anything in human life should be voluntary and spontaneous, it is erotic behavior.

Another factor modifying America's liberalized sexual behaviors may be that, despite all sex education efforts, the practical outcome of more sex is more pregnancies and subsequently, more abortions and unwanted children. Young, unwed women who are forced to become single parents usually find themselves in economic difficulty, because many fathers do not or cannot assume responsibility for their children. Too often, unwanted children are neglected or abused, and ultimately become unsocialized and troubled adults who cost society, rather than contributing to it. In most cases, society as a whole must pick up the monetary and social costs for the young, unmarried woman and her child.

HIGHLIGHT

"What Role Should Sex Play in My Life?" Beth-Ann Asks

It's been a perfect evening. The moon is full and casting a soft glow on the water. The air is thick with your perfume and sea air. You and your date have just finished a wonderfully romantic dinner and you now sit on the beach. A soft kiss goodbye when he drops you home will finish the evening—WRONG! Back to reality, girl! This young man just bought you an expensive dinner and roses instead of putting a new stereo in his car. Now you owe him, or at least that's what he'll tell you when he's pleading for you to come back with him to his apartment.

So what do you do? I mean this guy's really cute, and you really want him to call again. So do you have sex with him, even if it's only your first date? It's this constant pressure that guys put on girls that really bothers me the most. If a guy takes you out on a date, why do you need to perform sexual favors to make the night worthwhile to him?

Men seem to feel that sex is the natural end to a date. Of course, there is the occasional sensitive man who likes you for your mind, but he's becoming a real rarity. Now when a guy calls and wants to take you out, you have to doubt his reasons. This isn't right. When two people go out, it should be to enjoy each other's company and have a good time, and this can be accomplished without getting under the sheets.

Sex is not a form of payment for a night out. Sex is not a way of saying "thank you." Nowhere is it written that you have to have sexual intercourse with a guy because he spent money on you. It's your body and ultimately, your word.

And as if that isn't enough, if you finally do give in and have sex with him, the next day you're known as a cheap, easy "lay," and no guy wants you because you're "too easy." So ultimately, you lose either way. Either you are a prude or a cheap slut.

What Do You Think?

1. Do you feel that a woman owes a man sex if he spends money taking her out for a nice evening? Explain.
2. At what point in a relationship do you think sexual intercourse would be appropriate?
3. Do you agree or disagree with the idea that most women feel men are only interested in their bodies, rather than their companionship? Explain.

In a way, the very emphasis on sex and liberation has made sex education more difficult. Everyone is now supposed to know all there is to know about sex. If you don't, you're not cool, with it, or liberated. Better not to ask; better instead to feign knowledge, rather than to show ignorance by asking. As a result, sex is still a subject that is rife with misunderstanding and misinformation.

Such troubling factors, as we have discussed, will not cause a return to earlier sexual mores, but they may moderate sexual behavior.

In the past, sexual intimacy was usually accompanied by commitment to a broader relationship. As sex has become more an end in itself, commitment, caring, and the broader aspects of a meaningful intimate relationship have often been lost. When these components of sexuality are lost, it isn't long until sex for sex alone becomes boring. More Americans seem to be searching for these lost relational elements.

Differences between Male and Female Sexuality

Figure 8-1 diagrams some general differences in sexual drive between men and women across their life span. The source of these differences is the subject of considerable debate. Some biologists suggest that they stem from inborn differences in biological makeup between males and females. Most sociologists and psychologists believe that the differences stem from the socialization processes that teach men and women their gender roles and the place of sexual behavior in their lives. The truth probably lies somewhere between these two views, with

HIGHLIGHT

A Precoital Contract

The following letter in a student newspaper was obviously written by a coed with her tongue in her cheek (a little bit).

Editor:

Recently, a mere acquaintance presumed I'd be thrilled to have casual sex with him. I told him to buzz off. Next time, I am going to be prepared. I don't know about you, but as a woman, I am forced to be exceedingly responsible about my reproductive capacities. What is casual sex for some is a headache for me. How can sex be casual when I run the risk of getting pregnant or contracting any of a variety of ghastly diseases?

Being a die-hard romantic, I've got to temper my idealism with a good dose of realism. So I've devised the following PRE-COITAL CONTRACT, to stifle any attempt by my hormones to sabotage my good intentions:

1. All prospective lovers must submit a signed medical report that proves they are free from sexually transmitted diseases.

2. All prospective lovers must submit proof of attending a sex education and contraception class and must be fully prepared to participate in preventing pregnancy.

3. All prospective lovers must post a bail of $300 in case of an accidental pregnancy because no method of birth control, no matter how diligently used, is 100 percent effective.

Presenting this PRECOITAL CONTRACT is going to be a problem. Do I slip it under the door and demand signature when a prospective lover arrives to pick me up for our first date? Do I pick him up and drive to a distant place and demand his signature and bond before I'll give him a lift back to town? Do I wait for a lull in the conversation somewhere between the peas and the prune Danish to spring the contract on him? What would happen if I procrastinated until after our first embrace—would my prospective lover be so overcome by desire that he would sign anything—would I want him to be?

Sex in the '90s demands a PRECOITAL CONTRACT—I just haven't worked out the logistics—YET.

both nature (biology) and nurture (culture) combining in some manner to create the differences between the sexes.

Because puberty begins, on the average, about 2 years earlier in females than in males, young girls develop an interest in sex earlier than boys (area A in Figure 8-1). This interest is usually described as *boy craziness*. During this period, most boys remain essentially uninterested in girls. When puberty does arrive in young men, their sexual interests and desires soar above those of girls of similar age. From age 15 through the 20s, males are at the height of their sexual drive. Most females of similar age perceive their male counterparts as being preoccupied with sex during this period (area B in Figure 8-1).

Masturbation is common among young men at this time in their lives. Among Americans aged 18 to 59, about 60 percent of men and 40 percent of women are found to have masturbated in the past year (Michael et al. 1994, 158). About 25 percent of men and 10 percent of women report they masturbate at least once a week. Most men have masturbated in their early youth, whereas relatively few women have. Women's masturbation rates climb slowly over the life span, whereas the men's rates are highest early in their lives (Schwartz 1999).

During his teens and 20s, the male is simply a much more sexual creature than the female. The female's sexual drive increases gradually, reaching its peak when the woman is between 30 and 40 years of age. Because of the females' multiorgasmic capability, their sexual drive may become even stronger than the males in later years (area C in Figure 8-1).

Figure 8-1, taken as a whole, indicates that males and females are somewhat incompatible across their sexual lives. If we consider only sexual drive, older men and younger women and older women and younger men make the most compatible partners sexually. Margaret Mead, the famous anthropologist, suggested that a sexually compatible culture would be one in which older men married very

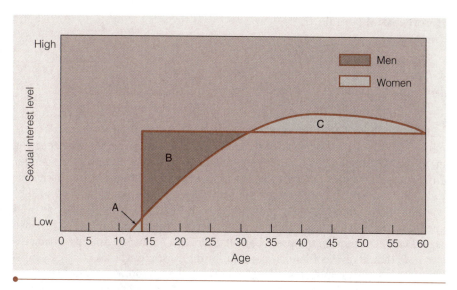

FIGURE 8-1 *Intensity of Sex Drive across the Life Span*

young women, who, on the death of their first husbands, would be left economically secure. These women, in turn, could marry young, sexually compatible men whom they could help to a good economic start. Although such an idea might solve the problem of differing sex-drive intensity between men and women at different ages, it is not likely to find wide acceptance in our culture. Thus, men and women will have to work out compromises as their sex drives vary over time and clash periodically.

Males generally are aroused more easily and directly than women by visual stimuli or mental imagery, such as pictures of nude women and pornography. Women's reactions to various sexually provocative materials are more complex, depending to a larger extent on the type of material. Some people suggest that an important distinction exists between pornographic material and erotica. They see pornography as based on the exploitation of women, featuring sex without emotional involvement and, all too often, sexual violence. Erotica is sexually explicit material that has emotional and romantic overtones of sensuality and caring. Women are more apt than men to react negatively to sexually explicit pornography. Women react more positively to erotica. Here again, however, women and men tend to be coming closer together in their reaction to sexually explicit materials (*Time* 2004).

Sexual fantasies play an important part in both men's and women's sex lives. Such fantasies can be important to sexual arousal. Gathering information on people's sexual fantasies is difficult, because they are regarded as private and intimate. Fantasies are influenced by personal experiences and also by societal mores regarding sex. Although the details of people's fantasies vary greatly, certain general themes have been found in both sexes (Kealing 1990; Rokachi 1990). Masters and his associates (1988, 350) report the following as the five leading fantasy contents for heterosexual men and women:

Heterosexual Male

1. Replacement of established partner
2. Forced sexual encounter
3. Observation of sexual activity
4. Homosexual encounters
5. Group-sex experiences

Heterosexual Female

1. Replacement of established partner
2. Forced sexual encounter
3. Observation of sexual activity
4. Idyllic encounters with unknown men
5. Lesbian encounters

Probably the most outstanding feature of these lists is their similarity. In fact, more recent research suggests that men's and women's fantasies are becoming more similar (Block 1999). Their general similarity masks more specific gender differences, however. For example, women tend to surround sexual fantasies with more romantic images than men do. Masters and his associates (1988) point out that fantasy content will change with time, personal experience, and one's culture. For example, women are reporting more graphic and aggressive sexual fantasies than in the past (Carroll 2005, 287). Masters and his associates also note that analysis of fantasy content for diagnostic purposes is usually nonproductive. For example, it is sometimes said that a person who fantasizes about same-sex experiences may be a latent homosexual. Yet, both heterosexual men and women report such fantasies, and both homosexual men and women report fantasizing about heterosexual relations. We do not label homosexuals who fantasize about heterosexual relations as latent heterosexuals.

Another important difference between men's and women's sexuality is that women have to learn how to reach orgasm, but men do not. This is probably where the idea originated that a woman must be awakened to her sexuality.

The capacity of females for orgasm differs more widely between individual women than it does between individual men. Some women never achieve orgasm, and some only when they are 30 to 40 years of age. At the other extreme, some women have frequent, multiple orgasms. Neither of these extremes is true for males (Michael et al. 1994).

Another difference is that females tend to have a cyclical increase in sexual desire related to the menstrual cycle, although the pattern varies among individual women. There is no counterpart of this cyclically heightened desire in the male.

Physiology of the Sexual Response

Masters and Johnson (1966) pioneered sexual research using human subjects which has led to an understanding of the physiology of the human sexual response. Although controversial, this research opened a new field of study, gave us a new understanding of the human sexual response, and paved the way for programs to help people with sexual difficulties.

Masters and Johnson divide the sexual response of both men and women into four phases: excitement, plateau, orgasm, and resolution. The responses in all the stages for both females and males are usually independent of the type of stimulation that produces them, whether it is by manual manipulation or penile insertion.

Female Sexual Response

Sexual response begins with the *excitement phase*, which may last anywhere from a few minutes to several hours. Vasocongestion (increased blood supply) causes the breasts to swell, the skin to become flushed, the nipples to possibly become

HIGHLIGHT

Female Genital Mutilation

Female genital mutilation, the practice of cutting off part or all of a girl's genitalia (also known as the euphemistic "female circumcision") is a widespread practice in many parts of the world.

There has been a worldwide outcry against this practice. The United States criminalized female genital mutilation (FGM) in 1997. The law requires federal authorities to inform new immigrants from countries where the custom is widespread, that they face up to 5 years in prison for performing the procedure or arranging for it for their daughters. In addition, over two dozen countries have issued official statement or laws against the practice. On March 5, 2004, Amnesty International launched a global campaign to eradicate FGM.

There are different types of mutilation ranging from mild—such as removing only the hood of the clitoris—to severe removal of the clitoris, the labia minor, and the surface of the labia majora, which are then stitched together to form a cover over the vagina, leaving a small hole to allow urine and menstrual blood to escape.

FGM is practiced for any number of the following reasons:

1. Cultural identity and tradition are most often cited as reasons for the custom. Among many groups, it is simply thought of as the normal thing to do. A girl becomes an adult though initiation ceremonies that include FGM.

2. Creating gender identity to differentiate females from males. The clitoris and labia are viewed by some as the male parts of the girl's body; thus, their removal enhances her femininity, often synonymous with docility and obedience.

3. Control of a woman's sexuality by reducing her desire for sex is another reason for FGM. It is difficult, if not impossible, in some cultures for a woman to marry if she has not undergone mutilation. To be a desirable marriage partner, she must be a virgin. FGM is an attempt to guarantee her virginal status.

4. Beliefs about hygiene, aesthetics, and health also play roles in FGM. Popular terms for mutilation include purification and cleansing. In some FGM-practicing societies, unmutilated women are regarded as unclean, and are not allowed to handle food or water.

5. Religion, in certain limited instances, is used to support FGM.

FGM occurs at varying ages, from infancy through adolescence, depending on customs. There are numerous health and psychological risks to the various procedures. Painful intercourse (dyspareunia) is an obvious consequence of the more severe types of FGM.

SOURCE: K. Chalkiey, "Female Genital Mutilation," *Population Today* October 1997. Washington, DC: Population Research Bureau.

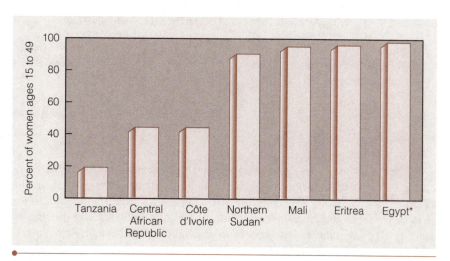

FIGURE 8-2 *Prevalence of Female Genital Mutilation, Selected Countries, 1990 to 1996*

*Ever-married women only.

SOURCE: *Demographic and Health Surveys, 1990–1996.* Calverton, MD: Macro International.

MAKING DECISIONS

What Are Your Biggest Problems and Complaints about Sex?

See if you can identify which of the following comments were made by a male and which by a female (the answers are given below). These comments were made anonymously by students in my marriage and family classes, in response to the following question: "What are your problems or complaints in the area of sex?" It is interesting to note that women write much more and describe many more problems than the men. The fact that women are more expressive in answering this question may indicate that they are more troubled by their sexuality. On the other hand, they may simply be more open about it. Or maybe they just write more and better.

1. Many people just want to jump right into bed and "do it." The preliminaries are totally skipped. All my partners seem to want is a piece of the action. Is all romance dead?

2. My problems in the area of sex are the moral questions: is it right to have sex before marriage? I will, probably, but not until I am in a lasting relationship, and some love is there. I find myself sexually attracted to members of the opposite sex before I know them emotionally. This is difficult, because I'm not sure if there is also any emotion in it.

3. Many of the times when I have had opportunities to have sex, anxiety seems to set in and I back out. Too much pressure on being "good."

4. I really hate it when members of the opposite sex expect us to start and do everything. They want equality, but want us to start all of the seduction games.

5. After dating steadily for a while, I start to feel this pressure to go further than just necking and petting. I have broken off a lot of relationships because of this pressure.

I am amazed at how many members of the opposite sex expect you to do it, like it isn't any big deal.

6. My biggest complaint about sex is that, to me, it's such a definite step toward total commitment to a person. When I do have intercourse, I want it to be with the person I marry. I want this part of me to belong to only one person. Sometimes this bothers me, because it's such a hard thing for my dates to accept.

7. Basically, the problem of the still-existing double standard. Men are allowed to enjoy casual sex without acquiring a negative label, whereas a woman enjoying sex for the sake of sex can risk her "good reputation" unless she is very discreet in her choice of sexual partners.

8. My biggest complaint is pressure from friends. "Hey, we're doing it, why aren't you?" I'm not out for a quickie.

9. I can't get enough, can't last all night, am afraid of VD and possible pregnancy.

10. Not enough responsibility taken for birth control by my partner. Not enough communication between partners. There should be more sex education for younger men and women.

11. The hang-ups people have about sex. If both partners want it, then what's wrong? I never get enough.

12. After having sex, my partners want a commitment that I don't feel is necessary to give just because you sleep with someone.

13. I never get enough holding, cuddling, and talking before or after sexual intercourse.

14. My partner never seems interested in sex, never starts anything sexual, and seems more interested in talking about our relationship.

1.F 2.F 3.M 4.M 5.F 6.F 7.F 8.F 9.M 10.F 11.M 12.M 13.F 14.F

F indicates that a female and M indicates that a male made the comment.

---

clitoris
A small organ situated at the upper end of the female genitals that becomes erect with sexual arousal; homologous with the penis

erect, and the **clitoris** to become engorged (tumescent). General muscle contractions may occur in the thighs, back, abdomen, and throughout the body. The vagina walls begin to sweat a lubricating fluid that facilitates the entrance of the penis. The inner portion of the vagina balloons, and the uterus may undergo irregular contractions. The labia minora (inner vaginal lips) increase in size. Blood pressure, heart rate, and breathing rate increase. Women who have had children may have more rapid vasocongestion (Carroll 2005, 280).

The *plateau phase* lasts from only a few seconds to about 3 minutes. Tumescence and the sex flush reach their peak. Muscle tension is high, and the woman becomes totally absorbed, both physically and emotionally, with the impending climax. The clitoris withdraws beneath its hood and can only be stimulated indirectly. (The idea that direct stimulation of the clitoris is necessary for female orgasm is untrue. Indirect stimulation is effective, and the heightened clitoral sen-

TABLE 8-1

Frequency of Orgasm (%)

	Always	Usually	Sometimes	Rarely	Never
Men	75	20	3	1	1
Women	29	42	21	4	4

SOURCE: Michael et al. 1994, 128.

sitivity during this phase may, in fact, make direct stimulation uncomfortable.) Muscle rigidity reaches a peak as shown by the facial grimace, rigid neck, arched back, and tense thighs and buttocks. The labia minora experience a dramatic change in color. Blood accumulates in the arteries and veins around the vagina, uterus, and other pelvic organs. This pelvic congestion is relieved by the orgasmic phase.

The third phase, **orgasm**, is the most intense. During orgasm, most of the built-up neuromuscular tension is discharged in 3 to 10 seconds. Orgasm is so all-absorbing that most sensory awareness of the external environment is lost. The whole body responds, although the sensation of orgasm is centered in the pelvis (see Table 8-1).

orgasm
The climax of excitement in sexual activity

The most dramatic of the widespread muscle responses is caused by the muscles that surround the lower third of the vagina. These muscles contract against the engorged veins that surround that part of the vagina and force the blood out of them, thus releasing the vasocongestion. These contractions also cause the lower third of the vagina and the nearby upper labia minora to contract a number of times.

In the last phase, *resolution*, the body returns to its prestimulated condition, usually within 10 to 15 minutes. If orgasm does not take place, the resolution phase may last 12 hours or more. Women have the capacity of repeating the four-

HIGHLIGHT

Describing Orgasm

Previous research has shown that expert judges were unable to distinguish reliably between written reports of male and female orgasms. The sex of the person describing orgasm in the following examples from our files may surprise you.

1. Like a mild explosion, it left me warm and relaxed after searing heat that started in my genitals and raced to my toes and head.
2. Suddenly, after the tension built and built, I was soaring in the sky, going up, up, and up, feeling the cool air rushing past. My insides were tingling and my skin was cool. My heart was racing in a good way, and breathing was a job.

3. Throbbing is the best word to say what it was like. The throbbing starts as a faint vibration, then builds up in a wave after wave, where time seems to stand still.
4. It's either like an avalanche of pleasure, tumbling through me, or like a refreshing snack—momentarily satisfying, but then I'm ready for more.
5. My orgasms feel like pulsating bursts of energy, starting in my pelvic area and then engulfing my whole body. Sometimes I feel like I'm in free fall, and sometimes I feel like my body's an entire orchestra playing a grand crescendo.

SOURCE: Masters et al. 1988, 91.

1.M 2.F 3.M 4.F 5.F

phase cycle immediately after resolution, and can experience multiple orgasms if stimulation is continued. For some women, the clitoris is extremely sensitive for several minutes after orgasm; thus, direct stimulation may be unpleasant during this time.

Male Sexual Response

The male undergoes the same changes as the female during the four stages of the sexual response cycle, but the four phases are not as well defined as in the female. During the excitement phase, the penis (and also the breasts and nipples) become engorged with blood until the penis erects, usually within 10 to 15 minutes. The sperm also begin their journey to the penis.

The major difference occurs during the orgasmic phase. The male reaches orgasm by the ejaculation of the semen and sperm through the penis. Once the male ejaculates, penile detumescence (loss of erection) usually follows quickly in the resolution phase, though complete detumescence takes longer. Unlike the female, who can experience multiple orgasms, the male usually experiences a *refractory* (recovery) period, during which he cannot become sexually aroused. This period may last only a few minutes or up to several hours, depending on such factors as age, health, and desire. Some men are able to keep an erection or partial erection after they ejaculate and continue to enjoy sex for a while. Research has found that some men are able to teach themselves how to have multiple orgasms (Carroll 2005, 284; Chia and Abrams 1997). Because, in our society, the male has usually been taught that it takes time to become sexually aroused again after ejaculation, the refractory period may be psychological besides physiological. Although orgasm and ejaculation occur simultaneously in the male, it is possible for the male to learn to withhold ejaculation, even though orgasm occurs. Certain Eastern yogis and others claim this ability, although exactly how they acquire it is unclear.

The male who is sexually aroused for a length of time without ejaculating may experience aching in his testicles and a general tension. These sensations can last for an hour or two. Such tension can be relieved by masturbation if a sexual partner is unavailable.

Variations in Sexual Response

Although the early work of Masters and Johnson suggested that all persons follow the four-phase pattern of sexual response, others suggest that there may be more individual variation than first thought. This is especially true for women.

Although much research has been done on the female orgasm, we still do not have a definitive answer about its nature. What is clear is that females have a greater diversity of sensation and reaction to sexual stimulation and are more capable of multiple orgasms than men. The range of feeling with orgasm is much greater in females than males. Therefore, when a woman reads a popular description of what she should be feeling and how she should be reacting to sexual stimulation, she must be aware that her responses may not fit the description. If they do not, she should not necessarily label her own feelings as inadequate. Feeling and intensity of orgasm are matters of perception, and satisfaction is influenced by many psychological and physiological factors. Again one's sexual attitudes and values are usually even more important to sexual satisfaction than the physiological factors alone (assuming that one's sexual physiology is working properly). This is especially true for the female.

Hormonal balances play a large role in sexual interest and arousal in both male and female. A drop in a man's testosterone levels will lead to gradual loss of interest in sex. Many believe that the same will be true of a woman if her estrogen level drops. Nevertheless, although estrogens influence a woman's attractiveness by keeping her skin soft, her hair full and shiny, and her breasts firm and help sex to be pleasurable by stimulating vaginal lubrication, they seem to have little to do with desire. Interestingly, testosterone seems to influence women's level of desire, just as it does in men. Long after estrogen levels drop with menopause, women remain interested in sex. But if a woman's testosterone level drops, she loses interest just as a man does. Conversely, women treated for medical disorders with synthetic testosterone typically experience a surge in sexual desire that may come as quite a surprise (Angier 1994).

Some Myths Unmasked

The research by Masters and Johnson put to rest a number of myths about human sexuality. First, it established beyond question that women can have multiple orgasms. It also established that, within normal ranges, the size of the penis and vagina have little to do with the experience of orgasm. For example, most of the stimulation occurs in the front third of the vagina, the labia minora, and the clitoris. A larger penile circumference may increase a woman's sexual sensation by placing pressure on the vaginal ring muscles, thus causing pleasurable tugging of the labia minora. Further, a longer penis may heighten sensations by thrusting against the cervix. On the other hand, too large or too long a penis may be uncomfortable for the woman and detract from her sensual enjoyment.

Masters and Johnson also revealed that it is not essential to stimulate the clitoris directly for orgasm to occur, though such stimulation produces the quickest orgasm for most women. And they found that the myth that the female responds more slowly is not necessarily true. When she regulates the rhythm and intensity of her own sexual stimuli, the female reaches orgasm in about the same time as the male. The female's much-discussed slowness in arousal is probably due to cultural attitudes rather than to some physiological difference. It is also clear that the woman is less genitally oriented than the man. As noted, her sexuality is more diverse and more dependent on her feelings and attitudes than is the male's.

Partners often believe that to achieve complete sexual satisfaction, they should experience orgasm simultaneously. As pleasant as this may be, there is no reason why partners must reach orgasm at the same time. In fact, doing so may hinder spontaneity and may prevent each person from being fully aware of the other's pleasure. The couple may find it equally satisfying to reach orgasm at different times in the lovemaking sequence.

Besides these specific findings, Masters and Johnson's work changed our ideas about other general aspects of sexuality. For example, many sexologists now consider masturbation to be an important and necessary part of sexual expression, rather than a taboo behavior. Some women who have never experienced orgasm can learn how by first learning the techniques of masturbation. Once they have learned how to reach orgasm, they can transfer what they have learned to sex with their partners. The rationale behind this is that so many emotions surround sex (such as shame and guilt about the failure to achieve orgasm, disappointment, and perhaps insecurity on the part of the partner) that it becomes almost impossible for some women to enjoy sex or change their behavior. The direct stimulation of masturbation encourages orgasm, and being alone may ease the emotional tension associated with sex.

WHAT DO YOU THINK?

1. Should a person's sexual conduct be judged as long as no one is harmed? Explain.

2. Can sexual behavior be labeled normal or abnormal? Explain.

3. What standards of conduct would you use to apply such labels?

4. How many sexual partners do you personally feel a person might have before you would label him or her "promiscuous"?

5. Why and how did you choose this number?

Americans' attitudes toward sexual expression have generally become more open as they have learned more about their sexuality. Rather than ignoring or hiding sexuality, understanding it is important so that one can use it to enhance one's life and bring greater joy and pleasure to intimate relationships. Unfortunately, America's greater acceptance of sex has also led to problems such as more sexually transmitted disease and increased numbers of unwanted pregnancies.

Does Sexual Addiction Exist?

Some researchers suggest that sexual addiction is a problem for some Americans, whereas others find the idea of sexual addiction to be just another myth. Sexually addicted men and women report being unable to control or stop sexual behaviors that lead to harmful consequences. When sex is separated from love and care, it becomes a block to intimacy. Addictive sex does not open up feelings, but is carried out in an attempt to hide them. "It is skin touching skin in search of a 'high.' After the high, participants feel lonely, empty, and often disgusted" (Kasl 1989, 10).

Often, the sexually addicted person feels unloved and uncared for and mistakes sexual interaction for love and care. "This person wants to have sex with me, therefore I am attractive, I am loved, I have a relationship, I am no longer lonely," one might mistakenly think. Such thoughts and the sexual interaction lead to an instant high and momentarily hide a person's sense of loneliness. However, the relief is only temporary, and feelings of shame and low self-esteem return when the person realizes that the sex meant none of these things (Earle 1990).

On the other hand, some researchers suggest that the notion of sexual addiction represents a pseudoscientific idea of cultural values, rather than a valid diagnosis. There is nothing intrinsically pathological in the conduct labeled as sexual addiction. These researchers believe that sexual addiction is a myth for two reasons. The first stems from the cultural relativity of sexual conduct (Levine and Troiden 1989). The very behavior described as sexual addiction in our culture is considered perfectly normal in other societies. For example, among the people of the Polynesian island of Mangaia, casual sex with numerous partners is seen as perfectly normal. American sexual mores term such behavior *promiscuous*. Using masturbation to help young boys and girls sleep is normal among certain peoples in India. In America, such sexual stimulation of children would be considered sexual addiction and a crime.

Second, sex is the only type of addiction in which the so-called addict does not have to give up the abusing behavior as part of treatment. As long as sex is confined to the culturally appropriate context (that is, marriage or a committed relationship), there is no addiction. Also, the so-called characteristics of sexual addiction—preoccupation, ritualization, sexual compulsivity, and despair—can just as well describe the intense passion of falling in love.

Marital Sex: Can I Keep the Excitement Alive?

The sexuality question most often asked by married couples is, "How can we maintain the excitement and interest that we've had in our sexual interactions through years of marriage?" Robert Michael and his colleagues (1994) conclude the following in their study of American's sexual behavior: "[America] is a nation of people who are for the most part content, or at least not highly dissatisfied, with the sexual lot they have drawn." In another sexual behavior survey, Mark Clements (1994) found that 67 percent of married persons expressed satisfaction with their sex lives, as compared with only 45 percent of single persons.

Certainly, many factors in married life do act to reduce sexual interaction and excitement. Monetary concerns, job demands, household chores, and children all conspire to rob a couple's sexual life of spontaneity and time. Both the quantity and quality of the average couple's sex life are diminished by the daily chores of maintaining a family. Many studies find that the quantity of sexual relations declines as time passes. However, the reduction in quantity seems less important to most couples than the reduction in quality. Quality sex takes time and concentration, both of which are lacking in a busy family, especially today, when both partners are likely to be employed.

Couples with children find it especially hard to be alone together, and even when they are alone, concern about the children is very disruptive. "We don't make love until the kids are in bed and asleep. Then it's 'Be quiet,' or 'Don't make so much noise,'" say some married couples. Postponing sexual relations until the children are asleep often means that both parents are tired and sleepy themselves. Who wants to be romantic, build a fire, listen to music by candlelight, and spend 3 hours making love and attending to each other when it's 11 p.m. and you both must rise at 6:30 a.m., get the children off to school, and be at work on time?

Husbands tend to complain more than their wives about their children interfering with their sex lives. Wives tend to dismiss the impact of child rearing as a temporary inconvenience or attribute their sexual problems to other factors, such as fatigue. "Jennifer has taken a lot of time away from us. It seems like on the weekend, when we would normally like to sleep in, or just have lazy sex, Jennifer wakes up and needs to be fed. But I'm sure that will pass as soon as she gets a little older. We're just going through a phase," says one young mother. Perhaps because women are generally more involved in mothering than men are involved in fathering young children, they tend to dismiss the interruption in their sex lives caused by children more easily than do new fathers.

Rather than continuing to discuss the various difficulties (jobs, housework, and so on) that can plague married couples' sex lives, let's examine a more ideal role that sex can come to play in a long-term, healthy, and growing relationship. After all, happily married, monogamous couples can be said to enjoy greater sexual freedom because they are free from the guilt associated with violating one's sexual standards; worry about out-of-wedlock pregnancy, AIDS, and other sexually transmitted diseases; fear of comparison with other partners; and fear of losing one's mate to another lover.

One of the dangers of the more liberal sexual practices may be that when sex becomes too casual, too taken for granted, or too available, the bonding qualities described by this husband (see "Highlight: One Husband's Sexual Life") may be lost.

Therapists report that increasingly, they are treating couples for **inhibited sexual desire**. This may be defined as pervasive disinterest in sex, with genital contact occurring less than twice a month, which is distressful to one partner. The problem with such a definition is that it implies some normal standard or level of sexual interaction. The only real standard is that both partners are happy and satisfied with their sexual interaction.

inhibited sexual desire
A pervasive disinterest in sex

Many sexually active young adults report increasing dissatisfaction with their sexual relationships. Few seem to have deeply meaningful sexual relationships.

HIGHLIGHT

One Husband's Sexual Life

I find that after 15 years of marriage, the quality of our sex life has increased immensely, even as the quantity has decreased. Sex between us seems to serve as a bonding agent. Sexual relations are embedded in our total relationship now, rather than existing in and of themselves as they did at first. Perhaps I can best explain this situation by telling you what turns me on sexually now, as compared with when my wife and I first met. You may be surprised.

Like most young men, I was at first attracted to her physical attractiveness, her smile, her body (wow!), her walk, and other physical qualities, and these attributes still attract me. Today, however, I can take a shower with her and not necessarily become sexually aroused. I couldn't even imagine such a thing at the start of our relationship. A tiny glimpse of her nude, and I was turned on. If seeing her nude does not necessarily turn me on today, what does?

We take a ski trip with the children. I watch them happily skiing down the hill with grace and skill. Suddenly, I feel sexually aroused toward my wife. I'd like nothing better than to throw her down in the snow and make mad, passionate love to her. Why? Because she is the mother of these wonderful children. She bore them and cared for them and helped (along with me, I hope) to make them what they are. I admire and respect her and love her and want to tell her this. What better way to communicate this than to physically get close to her, feel her, share our love together?

We have a fight, and she makes the first efforts to resolve the conflict. I love her for still making the effort after 15 years—for wanting to make the effort, for caring to make the effort. How lucky I am. How sexually attracted I am.

She is sound asleep. She looks peaceful and contented. She looks like an angel. I think back over the ups and downs of our relationship. I think of the things she really doesn't like about me, but tolerates and accepts. I think about her encouragement when I tried something new or difficult. And suddenly, I am sexually attracted to her and want to hold and cuddle her and tell her "thank you."

Down inside, I've been lusting for an expensive new car, which I know we can't afford and which I really won't buy. The new model that I have been reading about for 3 years finally is in the showrooms, and she suggests we go look at it. It is beautiful, but out of financial reach. She says, "You've worked hard, honey. Buy it, you deserve it. I'll help make ends meet." Whether I really buy it or not is immaterial. I want to take it for a drive to the hills and make love to her in the back seat because of her thought.

Of course, we have "quickie" sex. Of course, we have sex for sex. Of course, we have sex when there is no time, when we can't concentrate. But then there are times when our souls meet, when sex becomes the ultimate communication, when it stands for all the things she means to me, when it transcends all our differences, all our problems, when it becomes the ultimate expression of our love. Once you experience sex in this way, avant-garde discussions of number of orgasms, lovemaking techniques, sex as an end in itself, multiple partners, and the like pale in comparison.

Of course, I'm titillated at times by such thoughts. Of course, I'm attracted to other women at times. Of course, the sexual excitement is sometimes missing, but I wouldn't trade in our sex life. After all, it took years to build it into a meeting of our souls. Would I really trade that in on a one-night stand because I was horny? No way!

Perhaps, when a person knows a relationship is only transitory or fears that it may be, he/she guards against the impending loss by reserving total commitment, by holding back from too much closeness. If such reservations are a constant part of an individual's premarital sexual life, they may carry over into his/her married sexual life. In this case, premarital sexual activity may work against fulfilling, broadly meaningful, marital sex.

Married couples must work to maintain a fulfilling sex life. First of all, it is imperative to find time alone together, to concentrate on each other without distraction. When children are present, babysitting money is a couple's best expenditure in helping maintain sexual happiness. Having a place that is comfortable, romantic, private, and fairly soundproof will also help couples to relax. Constant worry that the children will hear is inhibiting to sexual activity.

One technique that some couples have found helpful is to start dating each other again, with the partners alternating responsibility for the dates. To add interest, some of the dates may be surprises. One partner asks the other

HIGHLIGHT

Sex and Physical Disability

The physically disabled (more than 11 million people) are sometimes thought to be uninterested in, or unable to engage in, sexual behavior. Generally, however, the disabled are every bit as interested in sexual behavior as anyone else. Although their disability may prevent them from engaging in certain kinds of sexual behavior, this does not mean that they are less interested in sex than they might otherwise be. The love and intimacy that can be expressed in various forms of sexual interaction are as important to the disabled person as to anyone else. In fact, sexual expression may be even more important to the disabled, because it verifies their acceptability to others.

"People in wheelchairs have to stop thinking that their sex life is over," said one spinal cord–injured patient. "While your genitals may or may not be dead, your emotions are very much alive. And you can still express your emotions without your genitals—your eyes, hands, fingers, lips, and tongue still work" (Knox 1994, 394).

Various diseases can also interfere with sexual expression. For example, a person with rheumatoid arthritis—whose afflictions include swollen and painful joints and muscular atrophy—may experience pain when interacting sexually with a partner, depending on the positions used. The couple may have to select sexual positions carefully and choose times when the arthritis is quiescent. If the arthritis responds to heat, the individual may need to plan sex after a warm bath.

Sexual enrichment aids are also available, often from mail-order catalogs, which can help disabled persons who have sexual difficulty.

What is clear is that sexuality is important to all people and important to the building of intimate relationships. If someone is injured or has a disease that interferes with normal sexual expression, the person and his/her partner must be creative and learn different ways of expressing sexuality.

---

out for a coming night, but doesn't say what the date will be, indicating, perhaps, only the appropriate dress. Going back to places you visited when you met and fell in love, listening to music that rekindles memories, and doing things together that you both enjoy and make you laugh and relax all help to reawaken sexual interest.

Occasionally, there are physical reasons for loss of sexual desire. For example, some medicines for high blood pressure lower desire. If such a reason is suspected, the couple should check with their physician. In general, however, the couple that makes working on their overall relationship a part of their everyday lives will stand the best chance of maintaining a satisfying sexual life.

Sex and the Aging Process

For some inexplicable reason, the myth has grown that for older persons, sex is a thing of the past. Yet, the natural function of sex endures as we age, just as other natural functions do, albeit in changing forms. We don't expect to run as fast at age 70 or to have the physical strength we had at age 20. Yet, we accept these changes and don't give up jogging or exercise. Likewise, changes in our sexual functioning don't mean that we shouldn't still enjoy it. *Time* recently described the increase in elders' sexual activity due, in part, to the introduction of erection-enhancing drugs (*Time* 2004). Men and women who have been sexually active early in life tend to remain so, even in their 80s and 90s, although the frequency of intercourse is limited by their physical health and by social circumstances, including lack of an available partner (Bretschneider and McCoy 1988; Byer and Shainberg 1994; Mulligan and Moss 1991; Schiavi et al. 1990). Robert Michael and his colleagues (1994, 116) found that 66 percent of men and 49 percent of women aged 50 to 59 years had sex a few times or more a month. N. K. Edwards

*One is never too old for a
loving relationship.*

(2000) found that over half of all Americans 60 years and older report that their sex lives are as good as, or better than, when they were younger. The majority of elderly persons maintain some sexual activity—even those in nursing homes (Walker and Ephross 1999).

Masters and Johnson (1981) have found three criteria for continuing sexual activity regardless of age. First, one must have good general health. Second, an interesting and interested sexual partner is necessary. Third, past 50 years of age, the sexual organs must be used. "Use it or lose it" is their advice to aging people, especially males.

Knowledge of such changes is important for both the man and his partner (see Table 8-2). If a man does not anticipate and understand the changes associated with aging, he may develop fears about his sexual performance that may rob him of his sexual desire. If his partner does not understand these changes, she may question her own sexuality when confronted by them. For example, she may interpret his slower erective response as loss of interest in her. If he doesn't ejaculate regularly, she may be concerned that he doesn't desire her. The introduction of Viagra in 1997 to help men maintain an erection brought publicity and new life to the idea of continued sexual interaction throughout one's life. Other drugs are also being developed. Two new erectile drugs were approved by the FDA in 2004 and are now marketed in the United States, Cialis and Levitra. Positive relations with a significant other are important to aging men's continued sexual activity and satisfaction (Schiavi 1999).

The aging process also changes the woman's sexual facility. The older woman produces less lubricating fluid and at a slower rate. Because the vaginal walls lose some of their elasticity, sudden penile penetration or prolonged coital thrusting can create small fissures in the lining of the vagina. As with the aging male, more time should be allocated for precoital stimulation. If neither partner evidences a sense of urgency in sexual interaction, erections and lubrication usually develop satisfactorily. Even with these physiological changes, it is important to remember that the psychologically appreciated levels of sensual pleasure derived from sex continue unabated.

Menopause

menopause (climacteric)

The cessation of ovulation, menstruation, and fertility in the woman; usually occurs between ages 46 and 51

The cessation of the menstrual cycle in women is termed **menopause**, or the **climacteric**. The median age is 51 years, but menopause can occur as early as 45 years or as late as 58 years (North American Menopause Society 2003). A number of factors may influence an earlier start of menopause: being without a spouse or significant other, smoking, and a history of heart disease (Brett and Cooper 2003; Carroll 2005; M. K. Whiteman et al. 2003).

Although sperm count reduces as men age, leveling off around 60 years of age, men do not entirely lose their ability to reproduce, as women do.

Some women find that they become even more interested sexually after menopause, because the fear of pregnancy is gone. Those women who have their ova-

TABLE 8-2

Physical Changes in Older Men and Women

In Men

1. Delayed and less firm erection
2. More direct stimulation needed for erection
3. Extended refractory period (12 to 24 hours before arousal can occur)
4. Reduced elevation of the testicles
5. Reduced vasocongestive response in the testicles and scrotum
6. Fewer expulsive contractions during orgasm
7. Less forceful expulsion of seminal fluid and a reduced volume of ejaculate
8. Rapid loss of erection after ejaculation
9. Ability to maintain an erection for a longer period
10. Less ejaculatory urgency
11. Decrease in size and firmness of the testes, changes in testicle elevation, less sex flush, and decreased swelling and erection of the nipples

In Women

1. Reduced or increased sexual interest
2. Possible painful intercourse due to menopausal changes
3. Decreased volume of vaginal lubrication
4. Decreased expansive ability of the vagina
5. Possible pain during orgasm due to less flexibility
6. Thinning of the vaginal walls
7. Shortening of vaginal width and length
8. Decreased sex flush, reduced increase in breast volume, and longer postorgasmic nipple erection

Source: Carroll 2005, 308.

ries removed usually report a drop in libido. Treatment with testosterone seems to revive sexual interest in these women. However, in the same study, those women receiving a placebo also reported a revived sexual interest. This clearly shows the importance of both one's psychology and one's biology as far as sexual behavior is concerned (Lemonich 2004).

Menstruation does not stop suddenly, but usually phases out over a period of time, usually about 2 years. As long as a woman has any menstrual periods, no matter how irregular, the possibility of ovulation and therefore conception remains.

Because of the changing hormonal balances during menopause, a woman may experience some unpleasant symptoms such as hot flashes, excessive fatigue, dizziness, muscular aches and pains, and emotional upset. These symptoms may last considerably longer than 2 years. Newer research seems to indicate that the most prevalent problems with menopause are not these classic symptoms. It is the lack of close interaction, affection, cuddling, and sexual contact with a partner that is most missed (Carroll 2005; von Sydow 2000). One needs to remember that menopause also has an effect on a woman's partner.

Estrogen replacement therapy (ERT) may be prescribed to help reduce negative symptoms of menopause. Because of this use, estrogen gained a reputation during the 1960s for slowing the general aging process, although this is un-

estrogen replacement therapy (ERT)
Supplying estrogen to menopausal women

true. ERT lost favor in the 1990s after a number of studies related its use to an increased risk of cancer of the uterine lining (endometrium). By adding progestin to the therapy, however, the uterine cancer risks are apparently reduced. Estrogen has been the number one prescription drug in America, used by about 40 percent of menopausal women (Anstett 2002).

Earlier studies indicated that ERT might improve a woman's ratio of so-called good cholesterol (HDL) to bad cholesterol (LDL) and to maintain the pliability of the blood vessels, thereby reducing the risk of blockage. However, more recent research disputes these results and indicates the opposite. Unfortunately, the landmark Women's Health Initiative (WHI) study of 16,608 women was canceled before being completed, thus leaving the debate over ERT's affect on heart disease unclear and confusing. The study also left unclear the effect of ERT on potential breast cancer (D. Smith 2002, 52). This research has served to elevate the controversy over the use of ERT, so that many women are confused and re-thinking its use (Quindlen 2002, 64).

Estrogen does alleviate some unpleasant menopausal symptoms and also appears to:

osteoporosis
Progressive deterioration of bone strength

- Be an effective means of preventing **osteoporosis** (the thinning of bones that make older women vulnerable to fractures).
- Reduce the incidence of colon cancer.
- Help preserve skin elasticity (*Time* 1995b, 48).
- There is some tentative evidence that estrogen may shield against cognitive decline during aging and may even fend off Alzheimer's disease (Carpenter 2001, 52–53).

On the negative side, ERT means that a woman may take the drug for the rest of her life, which can be decades. The evidence is mixed, but there is a suspected link between ERT and breast cancer (Colditz et al. 1995; F. Stewart 1998, 83), although the Women's Health Initiative (WHI) study indicated a definite link. There may also be a link between the long-term use of estrogen and the risk of ovarian cancer. There is a continuation of menstrual bleeding, if the ovaries have not been removed. There may also be a continuation of premenstrual-type symptoms such as tender breasts, irritability, and fluid retention.

Because ERT appears to have many positive effects but also negative effects, a menopausal woman is faced with a difficult choice. As our research data become more complete, it is hoped that the best choice will become more apparent.

Too many older men and women are losing their sexual involvement far earlier than necessary, because little effort has been made to educate them to the physiological facts of aging sexual function. As we saw in the last section, older people are, and should continue to be, sexually responsive human beings.

Sex and Drugs

aphrodisiac
A chemical or other substance used to induce erotic arousal or to relieve impotence or infertility

People have long sought the ideal **aphrodisiac**, a substance that would arouse sexual desire. Although many substances have been tried in the past, it is only recently that modern pharmacology has created chemicals for men with erectile dysfunction (ED), such as Viagra (Handy 1998). In this case, it is not desire that is increased; rather, Viagra can intercede to correct an erectile problem. Thus,

Viagra is not a true aphrodisiac because it will not work in the absence of desire. Yet, it is obvious that a man who can be assured of becoming erect will have a greater sexual desire than a man who is fearful of erectile failure. To achieve an erection, arteries in the penis must expand to allow sufficient blood to flow into the spongy, erectile tissue so that the veins that drain blood from the penis are squeezed shut. Thus, blood is trapped within the penis, which leads to erection. Viagra works by prolonging the effects of the natural chemical that controls arterial expansion. There are possible negative side effects to Viagra. About one in 10 men will develop a headache and, in a few cases, blackouts can occur if Viagra triggers a sudden drop in blood pressure. Also, nasal stuffiness may occur (Gorman 1998).

Alcohol is the most widely used sexual stimulant in America, but in reality, it is a depressant and inhibits the sexual response in males if ingested in large amounts. Long-term alcohol consumption increases the production of a liver enzyme that destroys testosterone, thereby reducing sexual desire (Klassen and Wilsnack 1986; Lang 1985). Women who drink large quantities of alcohol seem to have more difficulty achieving orgasm and experience less orgasmic intensity. However, women believe (self-reports) that they experience increased sexual arousal and heightened pleasure. Alcohol's reputation as an aphrodisiac apparently stems from its psychological effects: it loosens inhibitions, thereby indirectly stimulating sexual behavior. Perhaps Shakespeare's phrase about "much drink" best sums up the effect of alcohol on sex: "It provokes the desire, but it takes away the performance."

Marijuana has mixed effects on sexuality. There is no evidence that it heightens physical reactions, but it does cause some sense distortion that probably increases sexual sensitivity, especially sensitivity to touch. Most likely, it enhances the user's perception of sexual enjoyment (Byer and Shainberg 1994, 256). As a true aphrodisiac, however, it is a failure in that it appears to have neither a positive nor a negative effect on sexual desire. Some evidence indicates that those who use marijuana heavily for prolonged periods have a higher incidence of impotence than nonusers, probably because testosterone levels drop.

LSD, by distorting time, may seem to prolong the sexual experience. It is not, however, linked with enhanced sexual response. A bad "trip" (frightening hallucinations, and so on) can have disastrous effects on an individual's sexuality.

Amphetamines (speed) act as stimulants, in that the male can maintain a prolonged erection, but their long-term use destroys general health and sex drive and ultimately, leads to impotence.

Cocaine has a drying effect on vaginal secretions, so prolonged intercourse may be uncomfortable for women unless extra lubrication is supplied. Direct injection of cocaine seems to prolong a man's erection (Rubinstein 1982). Kolodny (1985) found, as did Weinstein and Gottheil (1986), that 17 percent of a sample of cocaine users had episodes of erectile failure when they used the drug, and 4 percent experienced *priapism* (painful, persistent erections) at least once during or immediately after use of cocaine. The dangerous practice of "freebasing" cocaine consistently leads to sexual disinterest and situational impotence (Yates and Wolman 1991). Kolodny (1985) studied 60 male crack-cocaine users and found that more than two-thirds were sexually dysfunctional, while 23 of 30 female crack users reported decreased sexual interest and responsiveness. It is interesting to note that because cocaine is expensive, offering it to a person of the opposite sex is almost always secondarily an invitation to have sexual interaction. Cocaine opens the doors of sexual access for many males and provides a convenient excuse for many females who otherwise might pass on having instant sex with a partner they hardly know.

The Roman philosopher Seneca knew the best aphrodisiac: "I show you a philtre [potion], without medicaments, without herbs, without witch's incantations. It is this: If you want to be loved, love."

anaphrodisiac

A drug or medicine that reduces sexual desire

Anaphrodisiacs are drugs that decrease sexual desire and activity. Perhaps the best known in popular folklore is saltpeter (potassium nitrate). Saltpeter acts as a diuretic, and frequent urination may deter sexual activity. However, saltpeter has no known direct physiological effect on sexual behavior.

Four groups of drugs impair sexual functioning. *Sedatives* such as barbiturates and narcotics, can suppress sexual interest and response. *Antiandrogens* are drugs that counter the effect of androgen (a male hormone) on the brain and thus, diminish sexual responsiveness. *Anticholinergic* and *antiandrogenic drugs* work to diminish sexual response by blocking the blood vessels and nerves connected to the genitals. These drugs are used to treat diseases of the eye, high blood pressure, and circulatory problems. Two drugs commonly used to treat hypertension, reserpine and methyldopa, cause loss of sexual interest and erectile incompetence. *Psychotropic drugs,* such as tranquilizers and muscle relaxants, may cause ejaculatory and erectile difficulties. Some of the psychiatric drugs reported to cause erectile dysfunction include Tofranil, Vivactil, Pertofrane, and Nardil. Those reported to impair or delay ejaculation include Librium, Haldol, and Elavil (Carroll 2005, 449). It is important for an individual to check with his/her doctor when using drugs for medical purposes, to understand their effects on sexual behavior.

Regular use of opiates such as heroin, morphine, and methadone often produces a significant drop in sexual interest and activity in both sexes. Nicotine has the same result in men (Byer and Shainberg 1994, 254).

As this discussion shows, there are a number of drugs with known anaphrodisiac qualities. Drugs with aphrodisiac qualities are less understood and are often surrounded by unsubstantiated folktales.

Drug use is also playing an increasingly important role in transmission of STDs (F. Cox 2000; Hatcher et al. 1998).

Sexually Transmitted Diseases

Unfortunately, no discussion of human sexuality is complete without mention of the most social of human diseases, that oft-found bedmate, STD (see Appendix A). STDs result in billions of dollars in preventable health-care spending.

In the past, the incidence of such diseases in the United States was drastically reduced by the use of antibiotics, such as penicillin. However, STDs remain a critical health problem, especially for individuals under age 25, who account for a majority of cases. Approximately 4 million teens get an STD every year. Experts estimate that as many as one in three sexually active young people will have an STD by age 24 (*Journal of the American Medical Association* 1999). For example, in 2002, young females aged 15 to 24 years had the highest rates of gonorrhea, the most commonly reported communicable disease in the United States. However, the rate has decreased for females in this age group by about 12 percent since 1998 (Centers for Disease Control, February 2, 2004; Centers for Disease Control, 2003). The rate for men is less than half the rate for women in this age group. In addition, the numbers of different STDs have increased to include 20 organisms and syndromes. At least three factors have played a role in the resurgence of STDs:

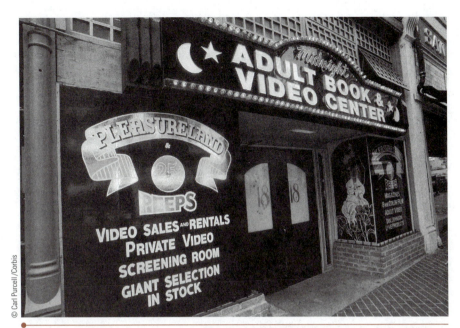

© Carl Purcell /Corbis

Public sexual portrayals often forget to remind viewers of the possibility of sexually transmitted diseases.

1. Because the pill became a major method of birth control in the 1960s and 1970s, use of the condom, with its built-in protective barrier against STD infection, decreased. However, commencing in the mid-1980s, with the appearance of human immunodeficiency virus (HIV)/AIDS, condom usage as a part of safer-sex education and practice has risen.

2. Antibiotics themselves have lulled people into apathy. "Who cares about STDs—they're easy to cure" appears to be a common attitude. This belief is partly true, but cure depends on prompt treatment; furthermore, new forms of antibiotic-resistant strains of the infecting organisms are appearing.

3. The increased sexual activity among the young, especially the increased number of sexual partners, has contributed to the spread of STDs.

It is important to recognize that women suffer more severe long-term consequences from STD infections, including infertility, ectopic pregnancy (the fertilized egg implanting in a Fallopian tube), cervical cancer, and chronic pelvic pain. Because semen is introduced into the woman's body during intercourse, she is more likely to acquire an STD infection than is her male partner (Centers for Disease Control 2003; Hatcher et al. 1998, 78).

If an STD is suspected, it is important to have a medical examination as soon as possible. In most cases, the disease does not go away, even though some of its symptoms may change or even disappear. If treatment is begun early, it is effective; if it is delayed, the disease may recur or become more dangerous. Anyone seeking treatment will be treated with confidentiality. If an STD is diagnosed, all persons who have had recent sexual contact with the carrier should be notified and examined. Most unmarried, young Americans underestimate the incidence of STDs and their own risk of infection. Just under half feel they are not at risk at all, a third see a slight risk, and the remainder see themselves at only moderate risk. Interestingly, teenage boys are more likely than teenage girls to believe they are at risk (*Journal of the American Medical Association* 1999). With increasing public awareness, it is hoped that the STD incidence can reduced.

Summary

1. *Sexuality pervades the lives of humans.* This is mainly because human sex is, to a great extent, free from instinctual control. Although sexuality is a biological necessity, much of the way in which sex is manifested is learned, unlike sexual behavior that is instinctual in lower animals. For this reason, human sexual behavior exhibits far more variations than the sexual behavior of other animals.

2. All human societies try to control sexual expression, but the controls vary from one society to another. Because of the variability of sexual expression and the often-conflicting teachings about sexuality, confusion about sexuality exists within both societies and individuals.

3. In all societies, human sexuality serves purposes other than procreation, such as communication, strengthening the male-female bond, increasing intimacy, pleasuring, having fun, and generally reducing tension.

4. In the United States, sexual behavior has become freer since the late 1960s. However, the appearance of herpes and AIDS has brought about a reevaluation of Americans' sexual behavior.

5. Our understanding of sexual physiology has increased greatly as science and medicine have expanded our knowledge about the functioning of the body. Males and females are physiologically similar, having developed their sexual organs from common structures, but they also differ in some respects. One difference involves timing: Young men tend to have higher sexual intensity than young women, reaching the peak of sexual intensity in their late teens and early 20s. Women tend to peak sexually during their 30s and 40s. Another important difference is the cyclical preparation for pregnancy that the female goes through each month, but the male does not. Males also tend to be more genitally oriented than females.

6. Both males and females share the same basic physical responses during sexual activity. These are called the excitement, plateau, orgasmic, and resolution stages. The major difference between the sexes takes place during the orgasmic stage—the ejaculation by the male. The male also usually goes through a refractory period after the resolution stage before he can have another erection.

7. Satisfactory sex at older ages is much more prevalent than many people believe. Good health and partners desirous of maintaining sexual interaction throughout their lives are necessary. Individuals who remain sexually active will continue to have sex into their later years.

8. Unfortunately, STDs too often accompany sexual activity. All STDs seem to be on the increase, and many new STDs, such as herpes and AIDS, have appeared in recent years.

Resources on the Internet

Companion Website for *Human Intimacy: Marriage, the Family, and Its Meaning,* Tenth Edition

http://sociology.wadsworth.com/cox10e/

Gain an even better grasp on this chapter by going to the companion website to take one of the tutorial quizzes, use the flash cards to master key terms, or check out the many other study aids you'll find there. You will also find special features such as GSS data, Sociology Online, and Census 2000 information that will put data and resources at your fingertips to help you with that special project or to help you as you do some research on your own.

InfoTrac College Edition

http://www.infotrac-college.com/wadsworth/

You can access reliable resources anytime, anywhere, with InfoTrac College Edition, the online library. This fully searchable database offers more than 20 years' worth of full-text articles (not abstracts) from almost 5,000 diverse sources, such as top academic journals, newsletters, and up-to-the-minute periodicals, including *Time, Newsweek, Science, Forbes, The New York Times*, and *USA Today*. You can conduct electronic key word searches using key terms from this chapter to supplement your reading and learning experience. To aid in your search and to gain useful tips, see the Student Guide to InfoTrac College Edition, which you can access through the companion website for this book.

Pornography

Sexually explicit material has always existed, but 50 years ago in the United States, such material was difficult to find. Today, it is easily available to everyone. With the advent of the VCR/DVD, the computer, and the Internet, sexually explicit materials have found effective new methods of reaching the public. It is suggested that the porn industry grosses about $12 billion a year. Sex via the Internet is now the biggest grossing area of Internet offerings. It is estimated that there were 260 million pages of pornography online as of July 2003. This represents an increase of 1,800 percent since 1998 (Paul 2004) and has reenergized the debate over sexually explicit materials.

Pornography Is Harmful to Women and Children and Is Destructive to Society

Many concerned people express outrage over the constant pornographic depiction of women as sexual objects, to be demeaned and brutalized. They especially abhor the violence in pornographic material. Many find that pornographic films are characterized by aggression, objectification of people, humiliation, and sexual exploitation. There appears to be a brutal disregard for the value of women as human beings.

Men exposed to such materials tend to become more tolerant of the idea of violence and, in some cases, actually become more aggressive toward women. Many rape counselors report that rapists often expect women to like uninvited sex, as is sometimes depicted in pornographic materials. Many rapists confess to being stimulated by pornographic materials. Again, some researchers suggest that people become increasingly desensitized by pornographic material and that this insensitivity carries over to real life. In addition, increasing desensitization leads to the need for increasingly more explicit and graphic pornographic materials to turn on sexually. Thus, Internet pornography (cyberporn) for some, especially men, can become addictive (cybersex addiction). *Cybersex* is defined as "typewritten communication for the sole purpose of achieving arousal and orgasm, generally between two people on line" (Smith Bailey 2003, 20).

Another potential problem with easily available pornographic material is that children may get their sexual information from it, rather than from valid sources that can supply correct knowledge. For example, young

men viewing several men in a "gang-bang" situation may come to believe that is what women want, and then act out those beliefs. The Internet Online Summit held in 1997 in Washington, D.C. revealed that 70 percent of children viewing pornography on the Internet are doing so in public schools and libraries. In a 2001 poll by the Kaiser Family Foundation, 70 percent of 15- to 17-year-olds said they had accidentally come across pornography online (Paul 2004). It seems naive to believe that the large amount of obscene materials available in the United States can be kept out of the hands of children.

The high numbers of rapes and the violence toward women seem to be evidence of the harm that can be wrought by the easy availability of pornographic material, in which sex is associated with violence toward women. The Meese Report (1986) supported this conclusion and recommended numerous steps to fight the problem, such as the banning of obscene television programming and dial-a-porn telephone services. In June 1989, however, the U.S. Supreme Court found that a state law banning dial-a-porn calls violated the First Amendment.

In 1992, the Canadian Supreme Court decided that pornography can cause violence against women. It concluded that when freedom of expression clashes with the rights of women and children to be protected against violent sex crimes, the rights of women and children come first. In addition to upholding the obscenity code, the Court redefined obscenity "to mean that which subordinates or degrades women." It is interesting to note that there is little about women and their use and reaction to pornography in the various studies. The studies seem to be all about the effects that pornography may have on men as they relate to women (Paul 2004).

Pornography Can Be a Source of Pleasure and a Stimulant to an Unsatisfactory Sex Life

Pornography is defined as any form of communication intended to cause sexual excitement. If this broad definition is taken literally, banning pornography would also mean banning sex. How can people engage in sexual relations without somehow trying to cause sexual arousal? Historically, in the United States as well as in other societies, the graphic depiction of sexual inter-action was an accepted art form and was, indeed, used to create sexual excitement. From ancient Greek vases to high-relief carvings on Indian temples, portrayals of the sex act have been perfectly acceptable. The ancient Indian love manual *Kama Sutra*, dating from about A.D. 400, did not have to be hidden by its readers. To-day, Japanese films are notorious for their depictions of violence, bondage, and rape, often of young girls—yet Japan has one of the lowest rates of rape of any indus-trialized nation. Even in the United States, the gov-ernment's blue ribbon Commission on Obscenity and Pornography reported that it was unable to find evi-dence of direct harm caused by pornography. Instead, the roots of rape tend to be in aggression, violence, power, and control, rather than in the sex itself. The Meese Report (1986), which declared pornography to be harmful to the family and society, has been taken to task by many researchers for stacking the evidence and drawing unwarranted conclusions (Nobile and Nadler 1986).

In some ways, sexually explicit material can be of help to both society and to individuals. For example, poor though it may be, such material can serve an ed-ucational purpose by teaching persons who have had no sex education something about the sexes and sex (McCaw 1994).

Masters and his colleagues (1988, 374) list a number of ways in which sexually explicit material can be use-ful. Erotica can trigger the imagination and thus help people deal with forbidden or frightening areas in a con-trolled way. It gives people the opportunity to imagina-tively rehearse acts that they hope to try or are curious about. It can provide pleasurable entertainment sepa-rate and apart from its sexual turn-on effect. But its most important use may be to stimulate a more active sex life between partners who have experienced some loss of interest. Exposure to sexually explicit material also leads to more open communication about sex. "What do you think about that?" "I'd like to try that." "That's not for me." Many sex therapists recommend explicit sexual material to clients who experience loss of interest in their sexual relationship, as a means of stimulating their sexual interest in each other.

Although in the past, men have been thought to respond more frequently and strongly to erotica, the availability of sexually explicit videos and cyberporn have increased women's participation in the use of erotic materials. About half of the erotic videos are now checked out by women.

What is the difference between pornography and erotica as portrayed on this Indian temple?

© Frank Cox

Perhaps one might conclude that evil is in the eye of the beholder. What one person considers an artwork and/or a reasonable adjunct to his/her sex life, another person may find unacceptable. The U.S. Supreme Court seems to agree. In 1997, the Supreme Court struck down the Communications Decency Act of 1996, which attempted to ban sexually explicit speech on the Internet (Weiner and Richman 1998). Thus far, the Supreme Court has refused to uphold attempts to censor pornographic materials on the Internet. How-ever, many groups are protesting the Internet availabil-ity of hard-core pornography to children. In the future, there may be controls developed that block availability of such material to children.

What Do You Think?

1. Do you think reading or looking at pornography is generally harmful to adults or not? Explain.

2. How much and how explicit do you feel sex scenes in nonpornographic movies or television programs should be?

3. Do you think magazines that show nudity should be banned? What about hard-core sex magazines? Explain.

4. Should X-rated movies and videos be banned? What about Internet sexual materials? Explain.

5. Should they be available to adults only? Why, or why not?

6. If you do not believe such materials should be banned, what role do you see them playing in the American culture? In your life?

© Pamela Cox

Family Planning, Pregnancy, and Birth

Chapter

9

Questions to reflect upon as you read this chapter:

- Are you ready for children? If so or if not, explain.

- What are the most important questions you would want answered before deciding to have a child?

- Why do you think there are more unwanted pregnancies when contraceptives are, in fact, easier than ever to obtain?

- If you were sexually active, would you use a contraceptive? If so, which one and why? (See Appendix B.)

- What is your position on abortion, and what are your reasons for taking that stance?

- What relation do sexually transmitted diseases have to infertility?

- What are your feelings about the technological advances in the science of conception and birth?

Family planning means just that—intelligently planning one's family. It means controlling one's sexuality to avoid unwanted pregnancies. It means creating children when they are wanted and when the family in which the children will grow up is the healthiest possible. Family planning means, in part, responsible sex. Family planning means becoming pregnant and giving birth only when one wants, is ready to have children and can properly care for them. Thus, family planning involves both the avoidance of pregnancy and the creation of pregnancy. It means having children by informed choice.

Children by Choice

Reduced infant mortality, through greater control of disease and pestilence and improved medical techniques, has removed the necessity for most American families to have large numbers of children. Because a much larger percentage of children now reach adulthood, a family may safely limit the number of children they produce to the number they actually desire. Hence, the historical pressure on the American family to reproduce continually has disappeared. The fertility rate in the United States has been at or below zero population growth for the past decade. If this trend continues and immigration is controlled, the population will reach a high in the year 2015 and start downward around 2020. See Table 9-1 for fertility rates in other countries.

One major change in America's fertility rate (number of births per 1,000 women aged 15 to 44 years) is the postponement of parenthood. Whereas the highest fertility rate for many years was in women aged 20 to 24 years, the highest rate today is in women aged 25 to 29 years. In fact, the birthrate has fallen for women below 30 years of age, but has actually increased for women over 30 (Hamilton et al. 2003, Figure 3). This relates to the later age of marriage and the increased number of career women.

A couple's conscious or unconscious reasons for having children strongly affect the way the child is treated and reared. Because species survival, political gain, and economic factors are no longer relevant to the decision to have children

TABLE 9-1	
Total Fertility Rate in Selected Countries	
Nation	**Total Fertility Rate**[1]
Germany	1.3
China	1.7
United States	2.0
Brazil	2.2
Mexico	2.8
India	3.1
Saudi Arabia	5.7
Nigeria	5.8

[1] The average number of children a woman will have throughout her childbearing years (15–49).

SOURCE: Population Reference Bureau 2003.

in modern Western societies, people's personal reasons for wanting children have become much more important. Unfortunately, with too many young people, pregnancy is not based on thoughtful reasoning about having children and becoming parents. Rather, the pregnancy is an accident, and may be based on reasons that make it difficult to become a good parent.

It is helpful to list some of the wrong reasons to have a child, as follows:

1. To save or strengthen a relationship. In fact, research suggests that children add to relational problems, rather than helping to solve them.
2. To please parents or friends. Parents and friends are not the ones who will be obligated for the next 20 or more years to care for the child.
3. Because everyone has children. Becoming a parent is a personal decision, not one to be made by group consensus.
4. To escape from the outside working world. Today, unlike the past, most mothers work outside the home.
5. To relieve loneliness. A newborn will certainly take up one's time, but loneliness may be intensified because there is little free time for adult interaction.
6. To do a better job than one's parents did. Resentment of any kind, including resentment toward one's parents, is the poorest of reasons to have a child. The resentment will almost always spill over onto the child.
7. To prove to the world that one is a real man or woman. Having to prove oneself by producing a child usually means doing so at the expense of the child.

Of course, there are many personal reasons for having children. To know and understand our reasons for having children is a small step in the right direction. It is also important to remember that the resultant child is an individual in his/her own right. This new individual has no responsibility to fulfill parents' needs. It is the parents' job to fulfill the child's needs.

Are We Ready for Children?

The ideal family is one that allows children to grow up to become healthy adults. Proper family planning leads to the first question a couple should answer when thinking about having a family: do we really want children? The answer requires a resolution of many questions, such as the following:

- Are we physically and psychologically healthy enough to give a child the care, attention, and love he/she needs?
- Can we afford to provide our child with the food, clothing, and education needed for at least the next 18 years and perhaps beyond?
- How much time do we want for just each other and for establishing a home?
- How much more education do we want or need for the jobs and income we want?
- What if we don't get the girl (boy) we want?
- What if the child is disabled in some way?

Most family planning experts advise young couples to wait awhile before having their first child. Waiting gives them time to make important adjustments to each other, to enjoy each other's individual attention, and to build some economic stability before adding the responsibility of a child. Yet, many couples don't or can't wait. For them, it is already too late for good family planning. For example, Karen, 16 years old, and James, 19 years old, are spending this Saturday night watching television in the hospital with their 1-day-old infant, Susan Marie. "When I get out we'll celebrate," says the new mother. "We're too young to go to a bar, so I guess we'll hang out at the mall." James has offered to marry Karen, but she has thus far refused his offer. Do you think Karen and James should keep their baby, or perhaps give her up for adoption? Note also that James is 3 years older than Karen and she is underage. Should he be prosecuted for statutory rape? How will they care for Susan Marie if they decide to keep her?

The decision to have children is one of the most important family decisions that can be made, yet it is often made haphazardly or by default. Karen and James hadn't even made a decision about their relationship, much less about becoming parents. "It just happened." Having children is too important to let it "just happen." Having a child generally means assuming long-term responsibilities, usually for 18 to 20 years or more. In fact, once a parent, always a parent. We can divorce a spouse, but we cannot divorce our children. There is almost no time when children do not make heavy demands on their parents, sometimes even long after they have become adults. True, having children brings many rewards. There is joy in watching a child grow, become competent, and assume a responsible role in society. There is joy in learning to know another person intimately. Doing things together as a family is just plain fun. Responsible family planning makes it easier to experience these joys and to create a healthy family.

Children Having Children

With many teenagers now sexually active, some high schools have begun dispensing birth control devices and establishing day-care facilities on the premises. Karen's solution to child care is more common: Her mother will look after the baby until the high school junior graduates. "I think Karen will think twice about having sex now," says her mother who, like many parents, believes that offering birth control measures in the schools only encourages students to be sexually active. "A single girl Karen's age shouldn't need contraception. I don't want her on birth control—she's only a teenager." (See "Debate the Issues," Chapter 6.)

A paradox has arisen in America during the past 20 years. Contraceptives are now sold openly in drugstores. Prescription contraceptives, like the pill or diaphragm, are easily obtained from organizations like Planned Parenthood. Yet, despite the availability of contraceptive devices, many unplanned pregnancies occur. Society has worked hard to get the family planning message across. What has gone wrong is a matter of growing debate.

The percentage of children conceived out of wedlock has risen substantially between 1960 and the present, but the percentage of women marrying before the birth has dropped significantly. Childbearing by unmarried women peaked in 1994 and has changed very little since. Rates actually declined for teenagers, dropping to 84.5 pregnancies per 1,000 for women aged 15 to 19 years—the lowest rate reported since 1976. Teenage pregnancy rates declined about one-third each for Caucasian and African American teenagers, and by 15 percent for Hispanic teenagers (Ventura et al. 2004). This lower rate is due, in part, to the fact that about 45 percent of teen pregnancies are ended through abortion. Women ages 20 to 24 years have the highest nonmarital birthrate (Hamilton et al. 2003, 7).

Society's greater acceptance of out-of-wedlock births and of the trend to postpone marriage contributes to the greater numbers of such births than in the past.

In 1950, for example, an out-of-wedlock pregnancy led to a quick marriage in most cases. Many women may have become pregnant before they were married, but these pregnancies did not necessarily become a statistic because a quick marriage allowed the women to have the child within marriage.

Teenage pregnancy has been around as long as there have been teenagers, but the dimensions and social costs of the problem are greatly increased when the new mother remains unmarried. Teen unwed pregnancy imposes lasting hardships on two and sometimes three generations: parent and child and perhaps grandparents, if caring for the child either physically or economically falls to them.

- Teen mothers are many times more likely than older mothers to live below the poverty level.
- Only half of those who give birth before age 18 complete high school (compared to 96 percent of those who postpone childbearing).
- On the average, unwed teen mothers earn half as much money and are far more likely to be on welfare.

As infants, the offspring of teen mothers have high rates of illness and mortality. Later in life, they often experience educational, behavioral, and emotional problems (J. A. Levine et al. 2001, 355). Some are the victims of child abuse at the hands of parents too immature to understand why their baby is crying or has developed a will of its own. Recently, the media have reported numerous cases of infanticide, wherein the young mother abandons her newborn, who is left to die. Finally, daughters of young, unwed mothers, especially Caucasian mothers (Musick 2002) are prone to drop out of school and become teenage mothers themselves.

With disadvantage creating disadvantage, it is no wonder that teen pregnancy is widely perceived as the very hub of the U.S. poverty cycle. Much of the so-called feminization of poverty starts with teenagers having babies. Teenage pregnancy ranks near the very top of issues facing Americans, and has been made a central political issue by Congress. The prevalence of teenage pregnancy in America was brought more forcibly to light when the Alan Guttmacher Institute released the results of a 37-country study (Guttmacher 1984). Its findings: The United States leads nearly all other developed nations in its incidence of pregnancy among girls aged 15–19. Looking in detail at Sweden, the Netherlands, France, Canada, and Britain, the researchers found that American adolescents were no more sexually active than adolescents in these countries, but became pregnant in much greater numbers. The efforts to reduce teenage birth since the Guttmacher study have worked, as noted, but teenage sexual activity has not been reduced.

To understand the nature of the problem, one must look beyond the statistics and examine the dramatic changes in attitudes and social mores that have swept through American culture during the past 30 years. The teenage birthrate was actually higher in 1957 than it is today, but that was an era of early marriage, when nearly one-quarter of 18- and 19-year-old females were married. Thus, the overwhelming majority of teen births in the 1950s occurred in a marital context, and mainly to girls over age 17.

All of this has changed. Today, if a girl does not choose to abort her pregnancy, chances are she will keep and rear her baby without the traditional blessings of marriage. Because teen marriages are two or three times more likely to end in divorce, parents may figure, "Why compound the problem?"

Unfortunately, unwed motherhood may even seem glamorous to impressionable teens. The mass media showcase celebrities such as Goldie Hawn for having their lover's children and expressing their happiness. Yet, an out-of-wedlock pregnancy for one of these women hardly leads to the dire financial poverty in which most teenage, unwed mothers will find themselves.

Social workers are almost unanimous in citing the influence of the popular media—television, rock music, videos, and movies—in propelling the trend toward precocious sexuality. That a sexual relationship may lead to a pregnancy or sexually transmitted disease never seems to be mentioned in most popular songs, videos, and other media that push sex.

Instead, young people are barraged by the message that to be sophisticated, they must be sexually hip. And yet, for all of their early experimentation with sex and their immersion in rock, rap, and the erotic fantasies on MTV, one thing about American teenagers has not changed: they are, in many ways, just as ignorant of the scientific facts of reproduction as they were in the days when Doris Day, not Madonna or Britney Spears, was their idol. For example, studies show that young teenagers generally wait about 12 months after first becoming sexually active before they seek contraception. Unable to grasp the situation when they become pregnant, they often wait too long to consider an abortion. The gravity of the situation completely eludes them. "I was going to have an abortion, but I spent the money on clothes," said one.

For many young girls, another less tangible factor is in the sequence of events leading to parenthood: a sense of fatalism, passivity, and, in some cases, even pleasure at the prospect of motherhood. "Part of me wanted to get pregnant. I liked the boy a lot, and he used to say he wanted a baby." For young girls trapped in poverty, life offers few opportunities apart from getting pregnant. Pregnancy brings recognition. Many young women with a child born out of wedlock report that being a mother makes them someone: It gives them an identity, if not prestige. "Before I was pregnant, I was nothing. Now I'm somebody, I'm a mother."

At the time of the Guttmacher study, African American and Hispanic teenagers had the highest fertility rate of any teenage population in the entire world. As noted, however, between 1990 and 2000, birthrates for African American and Hispanic teenagers dropped considerably. The National Urban League has declared teenage pregnancy its primary concern. Says past Urban League president, John Jacob, "We cannot talk about strengthening the black community and family without facing up to the fact that teenage pregnancy is a major factor in high unemployment, the numbers of high school dropouts, and the numbers of blacks below the poverty line."

Needy girls of any color or race who imagine that having a baby will fill the void in their lives are usually in for a rude shock. Hopes of escaping a dreary existence, of finding direction and purpose, generally sink in a sea of responsibility. With no one to watch the child, school becomes impossible, if not irrelevant. And despite the harsh lessons of experience, many remain careless or indifferent about birth control. About 15 percent of pregnant teens become pregnant again within 1 year, and 30 percent do so within 2 years.

The problems faced by the children of such parents begin before they are even born. Only one in five girls under age 15 receives prenatal care at all during the vital first 3 months of pregnancy. Teenage mothers are more likely to smoke during pregnancy (NCHS 2003a, Table 11). The combination of smoking, inadequate medical care, and poor diet contributes to a number of problems. Teenagers are much more likely to have anemia and also complications related to premature births, than older mothers are.

These factors add up to twice the normal risk of delivering a low-birth-weight baby (one that weighs under 5.5 pounds), a category that puts an infant in danger of serious mental, physical, and developmental problems that may require costly care (Mathews et al. 2003, 5).

The teenager is a person with newly active hormones that make sexual activity especially hard to resist. This fact, combined with the overwhelming amount of sexual stimulation provided by the mass media and more permissive attitudes, means that teenagers will continue to get pregnant unless parents and other social institutions act to help them handle their newly emerging sexuality (East 1999).

Notice in the preceding discussion that little has been said about the fathers of these children. In the recent past, society seems to have taken little interest in who fathers a teenage girl's child, or in holding the father responsible. As noted earlier, statutory rape laws (making intercourse with a girl under age 18 illegal) are beginning to be used again to curb intercourse with underage girls. Contrary to popular opinion, most of the fathers are not teenagers themselves, but rather, men in their 20s.

Family Planning Decisions

Modern birth control techniques, though far from perfect, have made better family planning possible. Women are having fewer children and are postponing both marriage and childbirth. Most women are now having their children between the ages of 25 and 34.

The idea of birth control has long been a part of human life. The oldest written records mentioning birth control date back to the reign of Amenemhet III in Egypt, around 1850 B.C. Women were advised to put a pastelike substance in the vagina to block male sperm from reaching the egg. Pliny's *Natural History*, written in the first century A.D., lists many methods of birth control, including potions to drink, magical objects, primitive suppositories and tampons, and physical actions such as jumping to expel the semen. These early birth control methods were generally unsuccessful. In the United States, Margaret Sanger and others waged a long, hard battle for birth control that lasted well into this century. The Puritan morality in the United States opposed birth control. Although contraceptives were recommended for the highest ethical reasons—that is, to prevent poverty, disease, misery, and marital discord—their promoters were accused of immorality and were often brought to trial and fined. In 1873, Congress passed the Comstock Law, which prohibited the distribution of contraceptive information through the mail. Numerous states also passed repressive laws. For example, physicians were forbidden to prescribe contraceptives in Connecticut until 1965, when the state law was overthrown by the U.S. Supreme Court.

Of course, family planning is possible without the aid of mechanical contraceptive devices. Postponing marriage acts as an effective birth deterrent, providing illegitimacy is controlled. Withdrawal and abstinence also reduce fertility rates. France, for example, has long had a relatively stable population of about 50 million people. Inasmuch as contraception is prohibited by the Roman Catholic Church, the church most French people belong to, this population stability has been achieved mainly by withdrawal (coitus interruptus), even though it is not completely reliable. Other methods of sexual outlet—such as masturbation, oral sex, and homosexuality—also serve a contraceptive purpose.

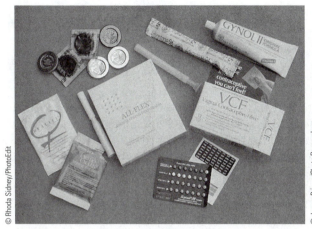

Numerous contraceptive choices are available, although none is perfect.

Condoms are now more popular because they help protect against STDs like AIDS as well as acting as a contraceptive.

Although various chemical and mechanical means of birth control are widely used in the United States, approval of such methods is by no means universal. The Roman Catholic Church has official doctrines banning "artificial" methods of birth control. Some minority-group members discourage birth control in an effort to increase their proportion of the population. Others who are concerned about the world population explosion advocate birth control (Townsend 2003). Nevertheless, the vast majority of Americans practice birth control at times, using many of the specific methods we will describe.

Deciding on a Contraceptive

Although family planning in itself is generally healthful, some methods of implementing it may have side effects that can be detrimental to the user's health. Appendix B summarizes the various contraceptive devices, indicating their effectiveness, advantages, and disadvantages (including possible side effects).

Planning a family requires couples to decide how many children they want and how far apart the children should be. The couple must also answer questions about birth control, including "Who will be responsible for using a birth control method? What method will be used? How will the method chosen affect our sex lives? What will be the cost?"

An *ideal* **contraceptive**—which does not yet exist, though research continues—would:

- Be harmless.
- Be reliable.
- Be free of objectionable side effects.
- Be inexpensive.
- Be simple.
- Be reversible in effect.
- Be removed from the sexual act.
- Protect against venereal disease and AIDS.

contraceptive
Any agent used to prevent conception

Although the contraceptives discussed in Appendix B do not fulfill all of these goals, they do meet many of them.

Abortion

In the United States, the 1960s witnessed mounting interest in induced abortion as a method of birth control. Although each state had restrictive legislation against abortion, many illegal abortions were performed. Expectant mothers sometimes died or were harmed because of nonsterile or otherwise inadequate procedures.

Other countries successfully use abortion to control population and to help women avoid unwanted pregnancies. Japan, for example, was plagued by overpopulation for years until 1948, when it enacted the Eugenic Protection Act. In essence, that act allowed any woman to obtain a legal abortion. In conjunction with a massive educational campaign to make the populace aware of the need to reduce family size, this measure reduced the birthrate from 34.3 to 17 per 1,000 population by 1956. In 2003, the Japanese birthrate was 1.3 children per fertile woman (Population Reference Bureau 2003). The availability of abortion varies considerably from country to country.

After a number of states liberalized their abortion laws, the U.S. Supreme Court (*Roe v. Wade* and *Doe v. Bolton*, January 22, 1973) made abortion on request a possibility for the entire country. Essentially, the court ruled that the fetus is not a person as defined by the U.S. Constitution and therefore does not possess constitutional rights. "We do not resolve the difficult question of when life begins. When those trained in the respective disciplines of medicine, philosophy, and theology are unable to arrive at any consensus, the judiciary . . . is not in a position to speculate as to the answer," the Court held.

More specifically, the Court said that in the first trimester (12 weeks) of pregnancy, no state may interfere in any way with a woman's decision to have an abortion, as long as it is performed by a physician. In the second trimester (13 to 25 weeks), a state may lay down medical guidelines to protect the woman's health. Most states that allow abortion by choice in the second trimester permit it only through the twentieth week. After that time, medical evidence must clearly show that the mother's health is endangered, or that the baby will be irreparably defective. Only in the last trimester may states ban abortion, and even then an abortion may be performed if continued pregnancy endangers the life or health of the mother.

The proponents of abortion on request believed they had won a final victory, yet this conclusion proved premature. Abortion is one of the most emotional issues that the nation faces. On one side are the crusaders known as "right-to-lifers" or "pro-life" people, who argue on religious and moral grounds that abortion is murder

In your opinion, who's right?

© Mark Richards/PhotoEdit

WHAT RESEARCH TELLS US

Who Has Abortions?

White	46%	Protestant	37%
Black	29%	Catholic	31%
Hispanic	20%	No religion	24%

SOURCE: Brimelow 1999, 110.

and therefore should be outlawed. On other side are the "pro-choice" crusaders, who contend that every woman has the right to control her own body, which includes the right to have an abortion if she so chooses.

One reason the conflict continues is that the American public is fairly evenly divided on the question. Most Americans agree that abortion is acceptable under some circumstances such as rape, danger to the mother's health, and if a fetus is malformed, but most Americans also object to using abortion simply as a birth control method.

Death associated with legal, induced abortions occurs rarely (one or fewer deaths per 100,000 abortions). Complications from anesthesia, infection, and hemorrhage are the major contributing factors. It is estimated that one in five American women of reproductive age has had an abortion, and that almost one-half of American women will have had at least one abortion by the age of 45 (Cates and Ellertson 1998, 679–700). However, the abortion rate has declined, from a high in 1980 of 35.9, to 25.6 abortions per 1,000 live births in 1999. In 1999, Caucasian women had 17.7 abortions, African American women had 52.9 abortions, and Hispanic women had 26.1 abortions, per 1,000 live births. Estimates of the total number of abortions vary. The Centers for Disease Control suggest that there were 862,000 legal abortions in 1999. The Alan Guttmacher Institute estimates that there were 1,315,000 that same year (NCHS 2003a, Table 16).

The pro-life movement's ultimate goal is to add an amendment to the constitution that would reverse the Supreme Court's decision. The Human Life Amendment has been introduced several times in Congress, to do what the Court said it could not do—namely, decide when life begins. The amendment reads: "For the purpose of enforcing the obligation of the States under the Fourteenth Amendment not to deprive persons of life without due process of law, human life shall be deemed to exist from conception." The amendment would allow states to pass laws defining abortion as murder. To date, the amendment has not been passed.

Many states have passed restrictions controlling second-trimester abortions as allowed by the Court. However, the restrictions vary from state to state and may include the following:

- Women seeking abortions must be told about fetal development and alternatives to ending pregnancies.

- Women must wait 24 hours after receiving that information before proceeding with an abortion.

- Physicians are required to keep detailed records, subject to public disclosure, on each abortion performed.

- Unmarried girls under age 18 and not supporting themselves are required to get consent of one of their parents or the permission of a state judge who has ruled that the girl seeking the abortion is mature enough to make the decision on her own (N. Adler et al. 2003).

Note that not every state has such restrictions.

The Supreme Court decision in *Roe v. Wade*, suggesting that the fetus has no rights under the law, is now being questioned by various recent court decisions. In South Carolina, a pregnant cocaine addict is convicted of child neglect. In Oklahoma, a drunk driver is found guilty of manslaughter after causing the death of a fetus in a vehicle collision. In South Dakota, a couple is allowed to sue a frozen-food company after salmonella poisoning caused the death of their 7-week-old fetus (*Trial* 1997, 13). The California Supreme Court (1997) ruled that a 3-year-old can sue her mother's former employer for injuries received while she was in utero, because the workplace had a toxic amount of carbon monoxide breathed by the mother, with subsequent damage to the child she was carrying. In April 2004, a federal bill was signed into law indicating that an unborn fetus killed in a murder of the mother has the same rights as the mother, and the perpetrator can be charged with fetal murder. Thus, the fetus retroactively is assigned the attributes of an American person and can be assigned a lawyer. If, as the *Roe v. Wade* decision concluded, the fetus is not a person with legal rights, how can the fetus be given legal rights in these cases, and even be assigned a lawyer, yet have no right to life?

The partial-birth abortion controversy also raises interesting questions. If it is all right to help the child die immediately at birth (partial-birth abortion), how can young women who let their newborn die immediately at birth be prosecuted for murder?

Another aspect of the debate concerns the right of the father to participate in the abortion decision. Current laws exclude him from the decision, on the ground that it is the mother-to-be's body and, therefore, under her control. In the past, the Supreme Court has struck down state provisions that required married women to tell their husbands about their plans for abortion. In the second trimester of pregnancy, though, a hospital may require the husband's consent in some states. This remains a controversial aspect of the abortion law. Should the father of an unborn child be totally barred from participating in the abortion decision? What happens if the mother-to-be and the father cannot agree on an abortion? In general, it has been found that when the father participates and supports the mother-to-be in the abortion decision, the woman shows more positive responses after the abortion (Carroll 2005, 422).

Improving medical technology also influences the abortion debate. In 1973, the earliest that a fetus could survive outside the womb was at about 28 weeks. Now that threshold has dropped to 22 weeks, and many physicians predict that it will soon be 20 weeks or less. Only about 1.5 percent of all abortions are done later than 21 weeks of gestation (NCHS 2003a, Table 16). What does this mean for a woman contemplating abortion later than 20 weeks? What happens when a saline abortion is performed late in pregnancy, but the fetus, although damaged, can survive with the new technology? Many states hold physicians to the same standard of care for fetuses born live from induced abortions as for any other premature infant. This occasionally results in the paradox of a physician performing an abortion, and then fighting to keep the infant alive.

The divisive abortion controversy has raged since *Roe vs. Wade* was passed.

Only time will tell where the abortion debate will end, if it ever ends at all. One thing seems sure: the question of abortion will remain controversial for some time.

The debate over induced abortion tends to cloud the physical act of abortion itself. First of all, spontaneous abortion occurs by the end of the tenth week in 10 percent or more of all diagnosed first pregnancies. If undiagnosed first pregnancies are taken into account, the figure is probably closer to 25 percent. A high percentage of spontaneously aborted embryos are abnormal. There is evidence that emotional shock also plays a role in spontaneous abortions.

Induced abortion is a fairly simple procedure, though it is unpleasant. One method of legal abortion is **dilation and curettage (D&C)**. In this method, the cervix is dilated by the insertion of increasingly larger metal dilators, until its opening is about as big around as a fountain pen. At this point, a curette (a surgical instrument) is used to scrape out the contents of the uterus. An ovum forceps (a long, grasping, surgical instrument) may also be used. The woman is instructed not to have sexual intercourse for several weeks, and she may take from 10 days to 2 weeks to recover fully from the procedure.

Vacuum aspiration is now the preferred method of abortion (about 96 percent of all induced abortions) because it takes less time, involves less loss of blood, and has a shorter recovery period than a D&C. In this method, the cervix is dilated by a speculum (an expanding instrument), and a vacuum-suction tube is inserted into the uterus. A curette and electric pump are attached to the tubing, and suction is applied to the uterine cavity. The uterus is emptied in about 20 to 30 seconds. The physician sometimes also scrapes the uterine lining with a metal curette. The procedure takes 5 to 10 minutes. The woman may return home after resting a few hours in a recovery area. She is usually instructed not to have sexual intercourse and not to use tampons for a week or two after the abortion.

The FDA gave approval to U.S. marketing of the drug RU-486 (mifepristone and misoprostol) in 2000, which is produced by a French pharmaceutical company. The drugs can abort pregnancies of up to 7 weeks' duration. This chemical abortion procedure is currently used in most of Western Europe, China, and Israel; it might make the preceding methods obsolete. The synthetic prostaglandin RU-486 is placed in the vagina in suppository form or given in pill form (*Newsweek* 1995; Roan 2000; Zitner and Savage 2000).

The popular impression of RU-486 as a one-step, hassle-free procedure is wrong. The first pill overrides the pregnancy hormones and breaks down the lining of the uterus. A second pill, taken 48 hours later, induces contractions and expulsion of the embryo, usually in about 4 hours. This stage is painful and accompanied by cramping and bleeding. The potential side effects may also include nausea and vomiting.

Methotrexate—a drug used to treat breast cancer, psoriasis, and rheumatoid arthritis—when given in combination with a prostaglandin, can also cause a drug-induced miscarriage (Bygdeman and Danielson 2002).

After the fourteenth to sixteenth week, a different method called the **saline abortion** may be used; the fetus is too large to be removed by suction, and performing a D&C now may cause complications. The uterus is stimulated to push the fetus out; in other words, a miscarriage is induced. When a fetus dies, the uterus naturally begins to contract and expel the dead fetus. In order to kill the fetus, the mother is given a local anesthetic, and a long needle is inserted through her abdominal wall into the uterine cavity. The amniotic sac is punctured, and some of its fluid is removed. An equal amount of 20 percent salt solu-

dilation and curettage (D&C)
An abortion-inducing procedure that involves dilating the cervix and scraping out the contents with a curette

vacuum aspiration
An abortion-inducing procedure in which the contents of the uterus are removed by suction

saline abortion
An abortion-inducing procedure in which a salt solution is injected into the amniotic sac to kill the fetus, which is then expelled via uterine contractions

WHAT DO YOU THINK?

1. Do you believe that a fetus is a living person, with all of his/her individual rights under the U.S. Constitution? Explain.

2. Under what circumstances would you personally have an abortion?

3. Do you think that underage women should have the right to an abortion without parental knowledge? Why or why not?

4. Should we think of abortion as simply another method of birth control?

5. What role should the father have in an abortion decision?

6. What are acceptable alternatives for a pregnant woman who does not want the child?

7. What are the advantages and disadvantages of each alternative you suggest?

tion is then injected into the sac. The injection must be done slowly and carefully, to avoid introducing the salt solution into the woman's circulatory system. She must be awake so that she can report any pain or other symptoms. Once the fetus is dead, the uterus begins to contract in about 6 to 48 hours. Eventually, the amniotic sac breaks and the fetus is expelled. In up to 50 percent of cases, the placenta does not come out automatically, and a gentle pull on the umbilical cord is necessary to remove it. In about 10 percent of cases, a D&C must be performed to remove any remaining pieces of the placenta. The saline procedure should be carried out in a hospital, and the woman should remain there until the abortion process is complete. The recovery period is longer than for other forms of abortion, and complications are more frequent.

The decision to have an abortion may not be easy and should not be made lightly. Whenever possible, it is wise for a woman considering an abortion to discuss the decision with the prospective father, her parents, physicians, counselors, or knowledgeable and concerned friends. But even though abortion should be a considered decision, the decision to have an abortion should be reached as quickly as possible: The later in the pregnancy the abortion is done, the greater are the risks of complications.

There is also debate over how women react psychologically after an abortion. Obviously, there will be many kinds of reactions. Positive emotions may include relief and happiness. Socially based reactions may include shame, guilt, and fear of disapproval. Other emotions may include regret, anxiety, depression, doubt, and anger—depending on how the woman felt about the pregnancy (Carroll 2005, 420). It is suggested that some women will cycle through all of these reactions (Thorp et al. 2003).

Abortion is not recommended as a means of birth control. The sexually active, responsible man and woman who are well informed about sex and contraceptive methods will not be faced with having to make an abortion decision. (See pages 280–281 for the various fetal developmental stages.)

Infertility

Although many people think only of contraception when they hear the phrase *family planning*, problems of infertility are also important aspects of family planning. **Infertile** means that a man is not producing viable sperm, or a woman is either not producing viable eggs or has some other condition that makes it impossible to maintain a pregnancy. The term *infertility* is usually applied when conception does not occur after 1 year of regular sexual intercourse without the use of any type of contraception. About 17 to 20 percent of married couples have problems conceiving, and approximately 7.1 percent of married couples are infertile (Berkow et al. 2000; G. Stewart 1998a, 654). About 33 percent of all married couples conceive the first month they try, and about 60 percent conceive within the first 3 months (Hatcher et al. 1998).

The World Health Organization describes the following three types of infertility:

1. *Primary infertility*. The couple has never conceived, despite having unprotected intercourse for at least 12 months.
2. *Secondary infertility*. The couple has previously conceived, but is subsequently unable to conceive within 12 months, despite having unprotected intercourse.

infertile

Unable to produce viable sperm if a man or become pregnant if a woman

3. *Pregnancy wastage*. The woman is able to conceive, but unable to produce a live birth.

The treatment of infertility involves three phases: education, detection, and therapy. Young couples should persist in trying to conceive for at least 1 year. If they are not successful, they should seek help from a physician. Often, when couples learn more about how conception occurs and the possible reasons for failure to conceive, they will feel less tense and anxious and thus increase the chances of conception.

Prerequisites of Fertility

For a couple to produce a child unassisted by the new fertility technology, both partners must be **fecund**, meaning they must have the capacity to reproduce. For the male, fecundity includes the following:

fecund
Having the capacity to reproduce

1. He must produce healthy, live sperm in sufficient numbers. Ordinarily, a single, normal testicle is all that is required, although both usually assume an equal role in producing sperm. To function properly, a testicle must be in the scrotal sac.
2. Seminal fluid (the whitish, sticky material ejaculated at orgasm) must be secreted in the proper amount and composition to transport the sperm.
3. An unobstructed seminal passage must exist from the testicle to the end of the penis.
4. The man must be able to achieve and sustain an erection and to ejaculate within the vagina.

For the female, fecundity involves the following:

1. At least one ovary must function normally enough to produce a mature egg.
2. A normal-sized uterus must be properly prepared to receive the developing fetus by chemicals (hormones) fed into the bloodstream by the ovary.
3. An unobstructed genital tract must exist from the vagina, up through the Fallopian tubes, to the ovary to enable passage of egg and sperm.
4. The uterine environment must adequately nourish and protect the unborn child until he/she is able to live in the outside world.

It is not uncommon for couples seeking fertility help to conceive before treatment begins and for adoptive parents to conceive shortly after they decide on adoption. Clearly, emotional and psychological factors are tremendously important to the process of conception.

Causes of Infertility

Males account for about 40 percent of infertility problems, as do females. Problems with both members of a marriage account for another 20 percent of infertility problems (G. Stewart 1998a). With men, for example, **impotence** (the inability to gain or maintain an erection) precludes sexual intercourse and thus conception. Often, impotence is psychological in nature, though it can be caused by alcohol, general fatigue, or a debilitating disease. Low sperm count is another possible reason for infertility. An ejaculation that contains fewer than 100 to 150 million sperm limits the possibility of conception. Alcohol, tobacco, and/or marijuana use have all been implicated in lower sperm motility and count. Infectious

impotence
Inability to gain or maintain an erection

diseases, such as mumps, can damage sperm production. Sterility can also occur if the testes have not descended into the scrotum, because the higher temperature of the body inhibits the production of healthy sperm. A prolonged and untreated sexually transmitted disease can cause permanent sterility in both men and women.

If a couple consults a physician, as they should, about their apparent infertility, it is easier to test the man first because fertility tests for the male are much simpler than those for the female. Basically, the tests involve collecting a sample of ejaculate, determining the number and activity level of the sperm, and checking for abnormal sperm.

Unfortunately, some men associate fertility with manhood. To them, an examination for possible fertility problems is an attack on their manhood, and they may be unwilling to cooperate. Both partners, however, must share in the search for a solution to infertility.

A number of factors have combined to increase infertility problems. Many more women are postponing childbearing into their 30s due to work commitments. The effects of age alone on fertility are moderate and do not begin until the late 30s. Lowered fertility in the late 30s may be connected to the fact that a woman is born with her entire supply of eggs and produces no new ones. As she ages, the remaining eggs also age, and the number of viable eggs is reduced. In addition, as time passes, there is increased risk of exposure to sexually transmitted diseases and other infections that can impair fertility.

A woman must ovulate if she is to conceive. Almost all mature women menstruate, but in about 15 percent of a normal woman's cycles, an egg is not released. In a few women, ovulation seldom occurs, which makes them almost infertile.

A woman may have a problem conceiving if the tract from the vagina through the uterus and Fallopian tubes to the ovary is blocked. This is the most common

MAKING DECISIONS

How Old Is Too Old?

Joni Mitchell gave birth to a healthy baby in 1992. Nothing startling, you say. But she was 53 at the time and had gone through menopause several years earlier. Her successful pregnancy represents one of the recent achievements of infertility science (*People* 1994, 36–41).

By treating older clients with hormones to prepare the uterine lining for implantation, using an egg from a younger woman and sperm via artificial insemination, pregnancies can result in successful births almost regardless of the woman's age. In fact, the world record as of 1999 is a 63-year-old mother giving birth (Gosden 1999, 211). In 2005, a 68-year-old woman gave birth, but the details are uncertain at the time of this writing.

The procedure is controversial, because the health risks of pregnancy increase with age, as do the chances of fetal abnormality. There are also ethical and social concerns. Mrs. Mitchell will be 70 years old when her child graduates from high school. A communications gap may develop between much older parents and their adolescent children. On the other hand, increased life expectancy means that Joni can expect as many years with her child as her grandmother had with her last child. Also, no one seems too concerned when men father children while in their 50s, 60s, or older.

Older parents are more likely to be emotionally and financially stable. If couples want a child badly enough to go through the procedure and pay the $10,000 to $20,000 costs, the babies will probably be greatly appreciated and loved.

What Do You Think?

1. For a woman, how old do you think is "too old" to bear a child? What about for a man?
2. What problems, if any, do you foresee for parents who are much older when they have a child?
3. Do you see any difference between much older parents or grandparents who rear their grandchild, as compared to younger parents?

reason for infertility in women. If the egg and sperm cannot meet, conception cannot occur. It is possible to determine if the Fallopian tubes are open by filling them with an opaque fluid and X-raying them.

Both vaginal infections and ovarian abnormalities can cause infertility. Sexually transmitted diseases (STDs) are the leading cause of preventable infertility, and account for the dramatic jump in infertility problems among young women that has occurred in recent years (G. Stewart 1998a, 659–660).

Physicians place some of the blame for this increase on more liberalized sexual attitudes. Increased sexual activity and more sexual partners have contributed to increases in genital infections (see Appendix A), especially pelvic inflammatory disease (PID). Such infections scar the delicate tissue of the Fallopian tubes, ovaries, and uterus. As noted, about half of these cases of PID result from chlamydia, whereas gonorrhea accounts for another 25 percent of pelvic inflammatory infections.

Another possible problem can arise from the chemical environment of the woman's reproductive organs. Too acid an environment quickly kills sperm. The chemical environment may also make implantation of the fertilized egg into the uterine wall difficult or impossible. In the latter case, the woman may conceive and then spontaneously abort (miscarry) the embryo.

Methods of Treatment: Designing Babies

Roger Gosden begins his 1999 book, *Designing Babies: The Brave New World of Reproductive Technology*, with:

We have seen fertilization by sperm injection, freeze-banking of embryos, postmenopausal motherhood, and genetic tests for embryos, not to mention transgenic animals and, of course, Dolly the sheep.

People choose eggs and sperm from "prize" donors advertising on the internet, and surrogate wombs are offered for rent . . . genetic counseling and screening before conception are more widely used, acceptance of routine screening and genetic diagnosis to check the condition of the fetus and to have the option to terminate the pregnancy are advisable.

Perhaps one day they will be able to determine the color of their children's eyes . . . and what will the attitude be if it becomes possible to clone and engineer the best possible child?

We cannot possibly discuss in detail all of the new, assisted reproductive technologies (ART) that seem to be appearing almost daily. Suffice it to say that couples having reproductive problems have an increasing number of methods by which to overcome their problems—to conceive and deliver a child. Tables 9-2 and 9-3 and Figure 9-1 give you some idea of a few of the technologies presently available.

Artificial Insemination   Artificial insemination (AI) is one of the oldest techniques promoting conception. Sperm banks have been established where sperm are frozen and stored for later use. Sperm that have been frozen for up to 3 years have been used for successful human fertilization.

However, even with the simple process of AI, controversies arise, especially if the sperm come from someone other than the husband. Questions of legitimacy and parental responsibility have arisen. For example, after a divorce, can a sterile

TABLE 9-2

Reproductive Alternatives

Method	Advantages	Risks
Fertility pump	Physiologic, mimics what occurs naturally. No known side effects. Works when Clomid (fertility pill) doesn't. Much less costly than in vitro fertilization. No surgery. Normal incidence of multiple births	Inflammation of vein
In vitro fertilization	Best if not only way for a couple to have their own baby, if woman's tubes irreparably blocked. Potential for correcting genetic defects in embryo	Surgery and anesthetic. Multiple births. Potential genetic manipulation for malevolent as well as benevolent ends
Embryo transfer	No surgery or anesthetic. Avoids genetic disorders carried by mother. Pregnancy possible for women who have no ovaries or ovaries that don't function	Not being able to retrieve fertilized egg from donor
Cryopreservation (freezing embryos for later implantation)	A second chance without surgery for a woman whose first attempt at in vitro fertilization fails. Banks of fertilized ova available for transfer to infertile women who are willing to bear another woman's child. Potential for correction of genetic defects	Slight increase in risk of abnormalities. Potential for genetic manipulation for malevolent as well as benevolent ends
Microsurgery	Very effective in reversing female sterilization. Very effective in removing scar tissue around Fallopian tubes. Effective on tubes blocked at end near uterus	Surgery and anesthetic
Surrogate mothering	Woman whose infertility is not correctable can rear a child fathered by her husband and carried by another woman	Wife must adopt baby. Legal questions of legitimacy, inheritance, adultery, financial responsibility, and rights of the biological mother
Artificial insemination	Best sperm can be used. Sperm is deposited directly into uterus	Cramping. Slight risk of infection. When sperm is not husband's, legal questions of legitimacy, inheritance, financial responsibility, adultery, and rights of biological father
Clomiphene citrate (fertility pill)	Very effective in women with normal or high estrogen level	Multiple births. Effective only 50 percent of time
Pergonal (fertility injections)	Very effective for low estrogen levels	Multiple births

husband deny his financial responsibility for a child conceived with sperm from another man? A court in Italy let such a father disown his child conceived by the AI of another man's sperm (Reuters 1994). Conversely, can such a husband be denied visitation rights because he is not the biological father of the child? Biologically, however, AI is a perfectly acceptable manner of overcoming a man's infertility.

Egg Freezing   More recently, experimental egg freezing has been tried. To date, only about 100 births have been reported from frozen eggs. There are a number of reasons that a woman might want to have some of her eggs frozen. Perhaps she has cancer and the treatments may render her infertile. She may not want

TABLE 9-3

The Costs of Infertility Tests, Treatments, and Adoption[1]

	Procedure	Fees
Diagnostic procedures	Initial visit, interview, physical exam (female)	$200, plus any lab work ordered
	Initial visit, interview, physical exam (male)	$100–150, plus any lab work ordered
	Semen analysis	$75–125 per test
	Sperm antibody test	$75–150 per test
	Hysterosalpingogram	$400–500 for radiologist and hospital charges
	Various hormonal blood tests (male and female)	$75–100 per test
	Testicular biopsy	$800–2,000 depending on whether performed in physician's office or hospital
	Endometrial biopsy	$250 for physician's and laboratory charges
Corrective surgical procedures	Donor insemination (DI)	$225 per insemination, plus sperm
	Laparoscopy (diagnostic) or laparoscopic surgery (reparative)	$8,000–8,500 for surgeon, anesthesiologist, and outpatient hospital charges
	Major surgery for removal of tubal blockages, adhesions	$5,000 or more for surgeon, assistant surgeon, anesthesiologist charges, plus $7,000 or more for hospital charges, depending on length of stay, time in operating room, medications, and so on
	Vasectomy reversal	$2,500–6,000
	Varicocele surgery	$1,500–4,000
Fertility/pregnancy procedures	"Fertility drug" treatment with Pergonal	$1,500–2,500 per cycle for drug and monitoring
	Donor insemination (DI)	$225 per insemination, plus sperm
	IVF (in vitro fertilization)	$8,000–10,000 per attempt, depending on procedures and individual program
	GIFT (gamete intrafallopian tube transfer)	$8,000–10,000 per attempt
	ZIFT (zygote intrafallopian transfer)	$8,000–10,000 per attempt
	ICSI (intracellular sperm injection)	$1,000–2,000 per attempt
	Frozen embryo transfer	$1,000 per attempt
Adoptions	Independent adoption	$8,000–30,000, depending on birth mother's and newborn's needs for living, counseling, and medical costs; plus $3,000–4,000 for attorney's fees and other legal costs
	Agency adoption	Varies from several hundred dollars for some types of public agency adoptions to thousands of dollars for some private agency and international adoptions
	Surrogate arrangements	From about $12,000 for an arrangement based on the independent adoption model to $50,000 or more when fees for surrogate and agency services are added

[1] Only 12 states mandate insurance coverage for infertility treatments (Arkansas, California, Connecticut, Hawaii, Illinois, Maryland, Massachusetts, Montana, New York, Ohio, Rhode Island, and Texas). In these states, only certain treatments and procedures are covered.

SOURCE: Carroll 2005, 361.

FIGURE 9-1 *New Ways of Creating Babies*

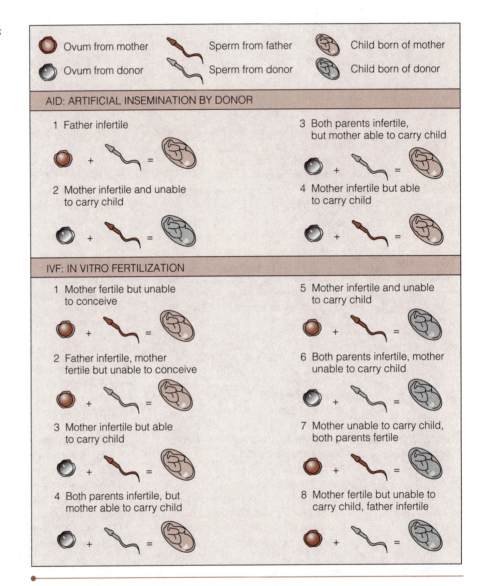

children until she is much older and may worry about problems old eggs could cause. Thus, she might save some of her eggs when she is young and plan to have them implanted at a later age. The approximate dollar cost for egg extraction and freezing is currently $10,000. It is still unknown what, if any, problems that exposing the sensitive egg cells might cause to babies born through this procedure (J. Graham 2004).

Fertility Drugs   Fertility drugs such as clomiphene citrate (Clomid) and menotropins (Pergonal) have long been used to stimulate ovulation. The drugs often overstimulate ovulation, resulting in multiple births. The twin birthrate has risen 38 percent since 1980, to 31 births per 1,000 births in 2002 (NCHS 2003b). The number of triplet and other higher-order multiple births has shown a remarkable rise over the past 2 decades, up from 1,337 births in 1980, to 6,737 in 1997 (NCHS 1999c). Two factors account for this. First, women are postponing pregnancy and thus having their children at later ages than in the past. Women over age 30, for some unknown reason, tend to have more multiple births. About one-third of the increase in multiple births reflects older maternal age. The remaining multiple births appear related to increased use of ovulation-enhancing drugs. However, there was a slight drop in 2002 in multiple births (NCHS 2003b).

Interest in (and concern about) the relation between fertility drugs and multiple births was greatly heightened with the birth of septuplets (seven) in 1997. This was the first set of septuplets to live past the first few days or weeks.

The use of fertility drugs has been wonderfully beneficial to women who have ovulation problems and are seeking to have children. Before the advent of such drugs, many women who had ovulation problems would have remained childless. Thus, the septuplets graphically demonstrate the promise of modern fertility treatments. Conversely, they also show the perils of such treatment.

Mortality rates for multiple births are higher than for single births. For example, infant mortality rates for triplets are 12 times higher than for single births. Gestation is, on the average, 7 weeks shorter for triplets and 9 weeks shorter for quadruplets than single births. Premature birth is related to increased chance of problems with the newborn.

Because multiple-birth infants are at greater risk, when the number of fetuses is high, "selective reduction" is usually recommended. By aborting some of the fetuses, the others stand a better chance of being born healthy. In addition to possible problems with multiple-birth infants, there is the increased stress and strain on the parents. Even if multiple birth infants remain healthy and grow trouble-free into adulthood, troubling questions remain about the high number of multiple births caused by fertility drugs (Gosden 1999).

Artificial Fertilization and Embryo Transfer (In Vitro Fertilization)

For the woman who cannot become pregnant, it is now possible for a human egg to be fertilized outside her body and then implanted within her uterus. This pro-

> **WHAT DO YOU THINK?**
>
> 1. Is it proper for humans to bear whole litters of children at once? Explain.
>
> 2. Is it ethical to bring multiple-birth children into the world, when the chances are high that some or all will suffer from health problems and even be disabled? Explain.
>
> 3. Is it ethical to abort some fetuses to save others? Explain.
>
> 4. Would you personally opt to have a multiple birth of four or more children? Why or why not?
>
> 5. In general, what are your feelings about the increasing intrusion of technology into the conception and birth processes?

LOOK, LADY — YOU'RE THE ONE WHO ASKED FOR A FAMOUS MOVIE STAR WITH DARK HAIR, STRONG NOSE AND DEEP SET EYES...

Cartoon by Mike Peters. Reprinted by special permission of King Features Syndicate.

cess is called in vitro fertilization (IVF). Louise Brown, the first such *test-tube baby*, was born in July 1978, in England. The first baby in the United States conceived by the IVF technique was born in December 1981. There have now been over 300,000 such births worldwide.

The first step in IVF is the retrieval of a mature female egg. The physician inserts a laparoscope through a small incision, to see the follicle inside the ovary where the egg is being produced. Then a long, hollow needle is inserted through a second incision; it gently suctions up the egg and surrounding fluid. The ovum is carefully washed and placed in a petri dish containing nutrients, and the dish is placed in an incubator for 4 to 8 hours. The second step is to gather sperm from the husband and add them to the dish containing the egg. If all goes well, the egg is fertilized and starts to divide. When the embryo is about eight cells in size, it is placed in the woman's uterus.

It takes 2 to 3 weeks to know whether the pregnancy will be viable, and the chance of success is only about 20 percent. The major problem is getting the fertilized egg to implant into the uterine wall. With newer techniques, such as inserting a mix of sperm and eggs directly into the Fallopian tube (gamete intrafallopian transfer [GIFT]), success rates have doubled. In this procedure, the egg is only out of the woman's body for a few minutes, and sperm are ensured of reaching it.

Even when pregnancy ensues, about one-third of the women will miscarry within the first 3 months. The current cost of this procedure is approximately $10,000 to $15,000. The procedure often has to be repeated several times before it works, thus increasing the costs dramatically.

surrogate mother
A woman who becomes pregnant and gives birth to a child for another woman incapable of giving birth

Surrogate Mothers   When it is impossible to correct a woman's infertility, another woman may be hired to bear a child for the couple. Such a woman is termed a **surrogate mother** and is usually paid between $10,000 and $20,000. Normally, she is fertilized via AI, using the husband's sperm. Once the baby is born, it is legally adopted by the infertile couple. A number of organizations throughout the United States bring infertile couples together with potential surrogate mothers. The organizations screen and match the parties and handle the complex legal problems of adoption and payment. Because it is illegal in every state to buy or sell a child, payment made to compensate the surrogate mother is not for the child, but for such things as taking the risk of pregnancy and childbirth and for the loss of work because of pregnancy. The contract in each situation is unique, and may include such specifications as the following:

- The surrogate agrees to terminate her maternal rights and allow adoption by the couple.
- The surrogate agrees to abort the fetus if an abnormality is discovered during the pregnancy.
- The surrogate agrees not to drink, smoke, or use drugs during the pregnancy.
- The couple will take out an insurance policy on the surrogate mother and on the biological father (husband), with the child named as beneficiary.
- The couple agree to compensate the surrogate at a reduced rate if she miscarries.
- The couple will accept all babies should multiple births occur.
- The couple will accept the child should it be born abnormal.

HIGHLIGHT

The Ultimate Breakthrough

Egg farming through frozen egg banking, pre-implantation genetic diagnosis (PGD), artificial uteruses so babies can be made outside the body, and DNA manipulation at the time of conception all pale beside the possibility of a man becoming pregnant and bearing a child. Given the facility with which embryos can implant, doctors could probably establish a pregnancy in a male belly by injecting an embryo through the skin with a syringe. It is possible that an embryo would implant, triggering the hormonal changes that women go through during pregnancy. Men are as sensitive as women to the hormones estrogen and progesterone, which are produced in large quantities during pregnancy; changes in men's bodies would occur similar to in pregnant women's bodies. For example, hormone action would encourage fat to accumulate on the male thighs, buttocks, breasts, and so on. Obviously, birth would have to take place surgically. Farfetched? Yes. Totally impossible? Maybe not (Gosden 1999, 195–196).

The potential legal problems are complex. Even though a contract is signed, what actually happens when a surrogate mother decides to keep the child? Or what happens when the parents-to-be decide they don't want the child if, for example, the surrogate mother gives birth to a mentally defective child? If a surrogate mother contracts to bear a child, does she have the right to smoke or drink in defiance of the couple's wishes? Does she still have a right to an abortion? Does a child born to a surrogate mother have a right to know its biological mother?

The use of a surrogate mother is a method of last resort. Informally, however, the procedure is as old as history. In days past it was not unusual for a woman to have a child and give it to another couple, infertile or not. This often happened with an illegitimate birth that was hidden from public view. For example, a daughter might have an illegitimate child, who was then passed off as her new brother or sister.

In a modern twist, the mother of a daughter born without a uterus became a mother and grandmother at the same time. In this case, the daughter's fertilized eggs were placed in her own mother's uterus. Her mother carried the twins to term and then delivered her daughter's biological twins via cesarean section. Thus, she became the grandmother to her daughter's children at the same time she became the mother via the birth (*Santa Barbara News Press* 1991). This is an example of the type 7 family in Figure 9-1.

The Genetic Revolution: Genomics The gradual decoding of the human genome (the study of the functions and interactions of all the genes in the chromosomes) is leading to a revolution in the treatment of disease. In the past, treatment was based largely on overt symptoms of the living person. As genetic understanding increases, prevention can take place before birth by manipulating an individual's chromosomes. Thus, the ultimate future designer baby may be designed through genetic manipulation at the time of conception. The idea of predicting an individual's potential biological life, as well as actually designing it (which seemed farfetched only a few years ago), may become a reality in the future (for better or worse) (F. S. Collins 1999; Jeffords and Daschle 2001; Patenaude et al. 2002).

One seldom-discussed problem common to all these new techniques is the potential reaction of the person who finds that he/she was born through one of these methods. Just as adopted children may become curious about the identity

HIGHLIGHT

Who, in Fact, Are the Parents of Jaycee?

Jaycee Louise Buzzanca is a parentless 2-year-old, the product of the new pregnancy technology. Her genetic parents are a sperm donor and an egg donor, anonymous and unrelated. Next came the gestational mother, a surrogate who was paid to carry the child to term. The parents-to-be, the Buzzancas, signed a contract saying that upon birth, they would get the baby. However, while Jaycee was still in the womb, Mr. Buzzanca, the intended father, changed his mind and left Mrs. Buzzanca, the intended mother. When Jaycee was born, Mrs. Buzzanca brought her home and became the rearing mother. Mr. Buzzanca claimed

he was not the father and owed nothing for support of the child; that was upheld by the court. He is not the biological father or an adoptive father, and is not married to the surrogate mother (*California Lawyer* 1997; Goodman 1994b).

What Do You Think?

1. Who are the parents of Jaycee?
2. What would you tell Jaycee when she is older, about who her parents are?

---

of their biological parents, so too may a child born by using the various methods described.

Suppose a child is conceived from sperm and egg of anonymous donors, carried by a surrogate mother, and raised by a family to whom the child is not even biologically related. Should the child be told about such things? What will be the child's reaction if he/she finds out? Who, in fact, are the parents of the child?

Pregnancy

That we start life inside another's body is one of the most extraordinary facts of our biology, although it is so familiar to us that we seldom stop to question the body's wisdom (Gosden 1999, 181).

How does a woman know if she is pregnant? Because the union of the egg and the sperm does not produce any overt sensations, the question "Am I pregnant?" isn't easy to answer in the early stages of pregnancy. What are some common, early symptoms of pregnancy?

- *A missed menstrual period*. Although pregnancy is the most common reason for menstruation to stop suddenly in a healthy woman, it is certainly not the only reason. For example, a woman may miss a period because of stress, illness, or emotional upset. In addition, about 20 percent of pregnant women have a slight flow or spotting, usually during implantation of the fertilized egg into the uterine wall.

- *Nausea in the morning (morning sickness)*. Early in the first trimester, a pregnant woman often experiences nausea in the morning, although vomiting usually doesn't occur. The nausea usually disappears by the twelfth week of pregnancy.

- *Changes in shape and coloration of breasts*. The breasts usually become fuller, the areolae (the pigmented areas around the nipples) begin to darken, and veins become more prominent. Sometimes the breasts tingle, throb, or hurt because of the swelling.

- *Increased need to urinate*. The growing uterus pressing against the bladder and the hormonal changes that are taking place may cause an increase in

HIGHLIGHT

Am I Really Pregnant?

The following is a conversation that you as a woman might have with yourself. It captures some of the feelings you may have as the early signs of pregnancy begin to appear.

"My period is late."

"Well, there's nothing unusual in that," you tell yourself.

"I just need a good night's sleep, or maybe I'm catching a cold."

"I'm exhausted," you say. "That's why it's late."

After a week has gone by, it begins to look as if your period is not just late—it is altogether absent. Even that is not so unusual. You have heard of many women who have skipped periods completely during times of stress or illness.

You begin to search your memory for things you have heard that bring on delayed menstruation—like hot baths, running up five flights of stairs, jumping off porches, or taking laxatives.

But mixed in with the hearsay and old wives' tales, you cannot force out of your mind one hard fact: You had unprotected intercourse last month, so the odds are more than even that you are pregnant.

This is somehow unthinkable if you have not planned to be pregnant, and especially if you have never been pregnant before. It is your body—known, familiar. You realize in an abstract way that it is equipped for pregnancy, but the idea that it should suddenly begin to function in this strange and unfamiliar way, without your willing or intending it, seems utterly unreasonable. How can it happen to you?

---

the need to urinate. Somewhere around the twelfth week of pregnancy, the uterus will be higher in the abdomen and will no longer press against the bladder, so urination will return to normal.

- *Feelings of fatigue and sleepiness.* Because of the hormonal changes that are taking place, some women find that they are always tired during the first few months of pregnancy, and need to sleep more often and for longer periods.

- *Increased vaginal secretions.* These may be either clear and nonirritating or white, slightly yellow, foamy, or itchy. Such secretions are normal.

- *Increased retention of body fluids.* Increased body fluids are essential to pregnant women and growing babies; thus, some swelling of the face, hands, and feet is normal during pregnancy.

Some women do not experience any of these symptoms; some experience one or two; only a few experience them severely.

Pregnancy may also have some cosmetic effects. Skin blemishes often abate, leaving the complexion healthy and glowing. In the latter stages of pregnancy, pink stretch marks may appear on the abdomen, although most of them will disappear after birth.

Pregnancy Tests

Usually, a physician can tell if a woman is pregnant by a simple pelvic examination. It is possible to feel the uterine enlargement and softening of the cervix by manual examination after 6 to 8 weeks of pregnancy. However, because most women want to know if they are pregnant as soon as possible, chemical tests are used to discover pregnancy earlier.

Physicians usually perform a test of *agglutination*, the clumping together of human chorionic gonadotropin (HCG), to assess whether pregnancy has occurred. This process takes only a few minutes and can be used 1 to 14 days after

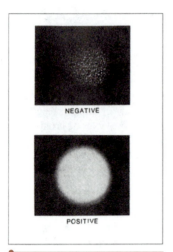

FIGURE 9-2 *Negative and Positive Pregnancy Test Reactions*
In a negative test, agglutination (clumping) will be visible within 2 minutes. In a positive test, no agglutination will occur at 2 minutes.

false pregnancy (pseudocyesis)

Signs of pregnancy occur without the woman actually being pregnant

the time menstruation should have started. The test involves taking a morning urine specimen, placing a drop of it on a slide, and adding the proper chemicals. In a negative reaction, agglutination will be visible in 2 minutes. If the woman is pregnant, no agglutination will occur at 2 minutes (Figure 9-2). Several tests examine the woman's blood serum for HCG. A radioimmunoassay (RIA) test can detect pregnancy within eight days of conception.

The tests are considered 95 to 98 percent accurate, but can give inaccurate results if they are performed too early (before enough hormone shows up in the urine) or if there are errors in handling, storing, or labeling the urine. Sometimes, it may take several tests to determine if a woman is pregnant because she may produce very low levels of hormone, and thus have too low a concentration in her urine to give a positive result, even though she is pregnant. It is rare for a pregnancy test to give a positive result when the woman is not pregnant.

False pregnancy (pseudocyesis), in which the early physical signs are present though the woman is not really pregnant, is also possible. Inexpensive pregnancy tests clarify the situation and usually end the symptoms (unless there is a physical problem causing them).

Because the menstrual cycle is easily affected by one's emotions, uneasiness about engaging in sexual activity may disrupt a woman's monthly cycle enough to delay her period or even cause her to skip a period altogether. Unfortunately, emotional reaction and physical reaction interact to increase her problems in this case. For example, if a woman has sex and worries that she might be pregnant, the worry may actually postpone her period, causing her increased worry, which further upsets her menstrual timing, and so on. A pregnancy test is one way to resolve the worry over possible pregnancy.

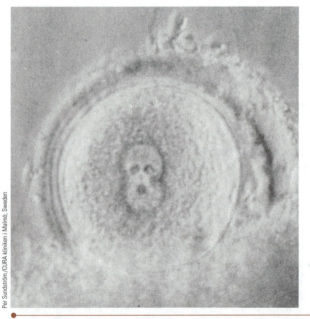
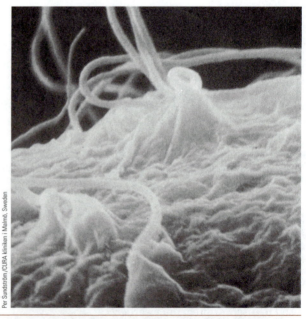

Left: Fertilized human egg showing two nuclei containing chromosomes from the sperm (father) and egg (mother) respectively. The round egg can be seen emerging from a group of cumulus-corona cells, which at ovulation completely surround the egg. On the egg's surface are thousands of spermatozoa, which look like small needles. Right: Spermatozoa in early stages of penetration on the moonlike landscape of the egg's shell.

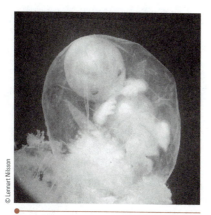

Human embryo in amniotic sac,
approximately 8 weeks old.

At 4½ months (just over
7 inches). When the thumb
comes close to the mouth,
the head may turn, and lips
and tongue begin their suck-
ing motions—a reflex for
survival.

The average duration of pregnancy is 266 days, or 38 weeks, from the time of conception. For the first 2 months, the developing baby is called an **embryo**; after that, it is called a **fetus**. The change in name denotes that all the parts are now present. The sequence of its development is shown in Table 9-4.

Environmental Causes of Congenital Problems

Although the developing fetus is in a well-protected environment, negative influences from outside may still affect it. Sometimes, these outside elements cause birth defects, which are called **congenital defects**. (These should not be confused with **genetic defects**, which are inherited through the genes.)

The developing fetus gets its nourishment from the mother's blood through the **umbilical cord** and **placenta**. There is no direct intermingling of the blood, though some substances the mother takes in can be transmitted to the fetus. Considering the extremely small size of the fetus during its early months, it is easy to see how a small amount of a substance can do a lot of harm.

Drugs If the mother uses narcotics, such as heroin (or even methadone) or cocaine, the child will be born addicted and will suffer withdrawal symptoms if he/she is not given the drug and then gradually withdrawn from it.

Furthermore, most common prescription drugs affect the fetus. Some antihistamines may produce malformations. General anesthetics at high concentrations may also produce malformations. Cortisone crosses the placenta and may cause alterations in the fetus. Antithyroid medications may cause goiter in infants, and tetracycline may deform babies' bones and stain teeth. It is best to avoid all drugs during pregnancy. If this is impossible, check with a physician to make sure that there will be no harmful effects on the fetus.

Infectious Diseases Certain infectious diseases contracted by the mother, especially during the first 3 months of pregnancy, may harm the developing fetus. The best known of these is German measles (rubella), which can cause blindness, deafness, or heart defects in the child. Some women who contract German measles elect to have an abortion, rather than risk having a deformed child.

STDs also affect the fetus. Herpes (see Appendix A), for example, can cause a spontaneous abortion, inflammation of the brain, or other brain damage. STDs

embryo
The developing organism from the second to the eighth week of pregnancy, characterized by the differentiation of the organs and tissues into their human form

fetus
The developing organism from the eighth week after conception until birth

congenital defect
A condition existing at birth or before, as distinguished from a genetic defect

genetic defect
An abnormality in the development of the fetus that is inherited through the genes, as distinguished from a congenital defect

umbilical cord
A flexible cordlike structure connecting the fetus to the placenta and through which the fetus is fed and waste products are discharged

placenta
The organ that connects the fetus to the uterus by means of the umbilical cord

TABLE 9-4

Prenatal Development

Time Elapsed	Embryonic or Fetal Characteristics	Illustrations
28 days 4 weeks 1 month	¼–½ inch long. Head is one-third of embryo. Brain has lobes, and rudimentary nervous system appears as hollow tube. Heart begins to beat. Blood vessels form and blood flows through them. Simple kidneys, liver, and digestive tract appear. Rudiments of eyes, ears, and nose appear. Small tail.	
56 days 8 weeks 2 months	2 inches long. $\frac{1}{30}$ ounce in weight. Human face with eyes, ears, nose, lips, tongue. Arms have pawlike hands. Almost all internal organs begin to develop. Brain coordinates functioning of other organs. Heart beats steadily and blood circulates. Complete cartilage skeleton, beginning to be replaced by bone. Tail beginning to be absorbed. Now called a fetus. Sex organs begin to differentiate.	
84 days 12 weeks 3 months	3 inches long. 1 ounce in weight. Begins to be active. Number of nerve-muscle connections almost triples. Sucking reflex begins to appear. Can swallow and may even breathe. Eyelids fused shut (will stay shut until the sixth month), but eyes are sensitive to light. Internal organs begin to function.	
112 days 16 weeks 4 months	6–7 inches long. 4 ounces in weight. Body now growing faster than head. Skin on hands and feet forms individual patterns.	

Eyebrows and head hair begin to show.

Fine, downy hair (lanugo) covers body.

Movements can now be felt.

140 days
20 weeks
5 months

10−12 inches long.

8−16 ounces in weight.

Skeleton hardens.

Nails form on fingers and toes.

Skin covered with cheesy wax.

Heartbeat now loud enough to be heard with stethoscope.

Muscles are stronger.

Definite strong kicking and turning.

Can be startled by noises.

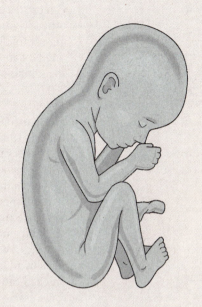

168 days
24 weeks
6 months

12−14 inches long.

1½ pounds in weight.

Can open and close eyelids.

Grows eyelashes.

Much more active, exercising muscles.

May suck thumb.

May be able to breathe if born prematurely.

196 days
28 weeks
7 months

15 inches long.

2½ pounds in weight.

Begins to develop fatty tissue.

Internal organs (especially respiratory and digestive) still developing.

Has fair chance of survival if born now.

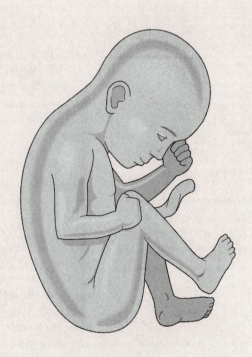

224 days
32 weeks
8 months

16½ inches long.

4 pounds in weight.

Fatty layer complete.

266 days
38 weeks
9 months

Birth.

19−20 inches long.

6−8 pounds in weight (average).

Of full-term babies born alive in the United States, 95% will survive.

can be contracted by the newborn baby. If the mother has syphilis, the baby will have the symptoms of the second and last syphilitic stages (see Appendix A). Gonorrhea can affect the newborn's eyesight; as a precaution, silver nitrate or another prophylactic agent is applied to the eyes of all newborns. This safeguard has almost totally eradicated the problem. HIV/AIDS can be passed on to the unborn child by an infected mother (see Appendix A). HIV/AIDS can also be passed on to an infant through the infected mother's milk.

Smoking   Smoking adversely affects pregnancy. Fortunately, the number of mothers reporting they had smoked during pregnancy has declined from 19.5 percent in 1989 to 12 percent in 2002 (NCHS 2004c, Tables 12, 61, 62). Smoking increases the risk of spontaneous abortion, of premature birth, and of low-birth-weight babies carried to term. Low birth weight is one of the major predictors of infant death. In a health department (NCHS 2004c, 2004d) comparison of pregnant mothers who smoked, it was found that Caucasian mothers were more likely to smoke than African American mothers; and those African American mothers who did smoke, smoked less than the Caucasian mothers. Hawaiian and Native American women had the highest rates of smoking while pregnant. Hispanic (especially mothers from Mexico, Cuba, and Central America) and Asian mothers smoked the least during pregnancy.

Alcohol   Fetal alcohol syndrome (FAS), first identified in 1973, affects babies born to chronically alcoholic mothers. FAS includes retarded growth, subnormal intelligence, and slow motor development. Affected infants also show alcohol-withdrawal symptoms such as tremors, irritability, and seizures. Even when the children are of normal intelligence, they often have a disproportionate amount of academic failure. Although the concern over FAS relates to alcoholic mothers, drinking of any kind during pregnancy may cause problems.

Radiation   Radiation penetrates the mother's body, so it will also reach and affect the fetus. The worst abnormalities occur if the mother is X-rayed during the first 3 months of pregnancy, when the embryo's major organs are developing. Even one pelvic X-ray can cause gross fetal defects during this period. Thus, if pregnancy is suspected, a woman should avoid all X-rays, even dental X-rays, especially early in the pregnancy.

Rh factor
An element found in the blood of most people that can adversely affect fetal development if the parents differ on the element (Rh negative versus Rh positive)

Rh Blood Disease   The **Rh factor** (named for the rhesus monkey, in whose blood it was first isolated) is a chemical that lies on the surface of the red blood cells in most people. People with the chemical are considered Rh positive; those without are Rh negative. Only about 15 percent of Caucasian Americans and 7 percent of African Americans are Rh negative. If a child inherits Rh positive blood from the father, but the mother is Rh negative, the fetus's Rh positive factor is perceived as a foreign substance by the mother's body. Like a disease, it causes the mother's body to produce antibodies in her blood. If these enter the fetus through a capillary rupture in the placental membrane, they destroy red blood cells, which can lead to anemia, jaundice, and eventual death unless corrective steps are taken. Only small amounts, if any, of the child's antibody-stimulating Rh factor reach the mother through the placenta during pregnancy, so the first child is usually safe. However, during delivery, the afterbirth (placenta and remaining umbilical cord) loosens and bleeds, releasing the Rh positive substance into the mother, which causes her system to produce antibodies. Once these are produced, she is much more easily stimulated to produce them during future pregnancies involving Rh positive children. Each succeeding child will be more affected than the previous one. With complete replacement of the child's blood at birth, many can be saved. Another commonly used treatment is injecting into the mother an Rh immunoglobin that blocks the mother's immune system, thereby preventing production of the antibodies that attack the red blood cells.

Controlling Birth Defects

Once a woman learns that she is pregnant, she should arrange for regular visits to a physician. Regular prenatal care will help avoid birth defects and dispel any fears she may have.

Diet   The expectant mother should eat a well-balanced diet, with plenty of fluids. Because the mother's diet has a direct effect on the fetus, she should consult her doctor if there is any doubt about the adequacy of the diet. Protein and vitamin deficiencies can cause physical weakness, stunted growth, rickets, scurvy, and even mental retardation in the fetus. Poor diet can also cause spontaneous abortions and stillbirths. The pregnant woman may find that she cannot eat large meals early in pregnancy if she feels nauseous. In late pregnancy, the uterus takes up so much room that she may again be unable to eat large amounts. In both cases, she should eat small amounts more often and avoid going for long periods without food.

An inadequate diet leaves the mother more prone to illness and complications during pregnancy, both of which may cause premature birth or low birth weight. As we have seen, premature and low-birth-weight babies are more prone to illness and possible death than are normal-term babies. Women whose finances are inadequate for them to eat well during pregnancy and nursing can get help from the Women, Infants, and Children program (WIC), a supplemental food program sponsored by the government. This program provides milk, fruit, cereal, juice, cheese, and eggs. (Contact your local public health office for information about this program.)

Amniocentesis   Fetal monitoring has taken the secrecy away from the womb. There are now several monitoring procedures, one of which is *amniocentesis*. This procedure has been developed for detecting genetic defects, such as Down syndrome (formerly called mongolism), amino acid disorders, hemophilia, and muscular dystrophy. It involves taking a sample of the amniotic fluid and studying sloughed-off fetal cells found in it. Amniocentesis should be done between the fourteenth and sixteenth weeks of pregnancy. The test can be performed in a physician's office, though the laboratory work will take another 14 to 18 days to complete. The test also reveals the fetus's sex.

It is good to have the test if a woman has already had a child with a hereditary biochemical disease, if she (or her husband) is a carrier of hemophilia or muscular dystrophy or other hereditary genetic problems, and if she is over age 40 (the risk of having a child with a genetic abnormality increases with age). If the test indicates the presence of a birth defect, the woman, her husband, and the physician can discuss their options, including possible abortion.

Amniocentesis is not risk-free. The technique can induce miscarriage, but this occurs in less than 1 percent of cases. Other problems, such as infection and injury to the fetus, also occur in about 1 percent of cases.

ultrasound
Sound waves are directed at the fetus that yield a visual picture of the fetus; used to detect potential problems in fetal development

Ultrasound in Obstetrics   **Ultrasound**, generally considered safe and non-invasive, has become a major means of obtaining data about the placenta, fetus, and fetal organs during pregnancy. It can replace the X-ray as a method of viewing the developing child in the uterus, thus avoiding radiation exposure for both mother and child. Ultrasound is also simpler than X-rays because the picture is immediately available.

Ultrasound works on the principle that different tissues give off different-speed echoes of high-frequency sound waves directed at them. Moving a transducer (sound emitter) across the mother's abdomen creates an echogram outline of the fetus and its various organs. Using a real-time transducer that gives off several simultaneous signals from slightly differing sources will produce a picture showing movement of the different organs, such as the heart. The *sonograph* (picture produced by ultrasound) allows the physician to learn about the position, size, and state of development of the fetus at any time after about the first 10 weeks of pregnancy. For example, the procedure can tell a physician if the fetus will be born in the normal headfirst position or in some problem position.

fetoscopy
Examining the fetus through a small viewing tube inserted into the mother's uterus

Fetoscopy   **Fetoscopy**, a delicate procedure usually performed some 15 to 20 weeks into pregnancy, allows direct examination of the fetus. First, an ultrasound scan locates the fetus, the umbilical cord, and the placenta. The physician then makes a small incision in the abdomen and inserts a pencil lead–thin tube into the amniotic sac. The tube contains an endoscope with fiber-optic bundles that transmit light. This light-containing tube enables the physician to see tiny areas of the fetus. By inserting biopsy forceps into the tube, the physician can take a

Sonograph of a fetus.

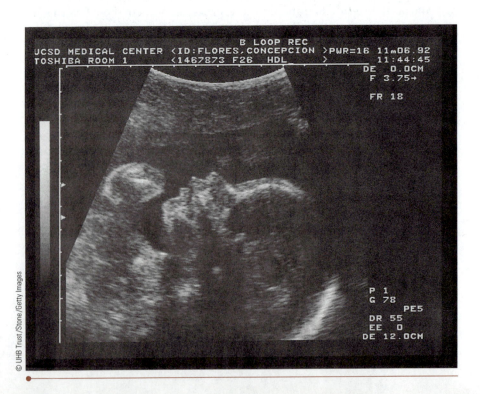

© UHB Trust/Stone/Getty Images

skin sample from the fetus. A blood sample can also be drawn by inserting a needle through the tube and puncturing one of the fetal blood vessels lying on the surface of the placenta. The technique induces miscarriage in about 5 percent of cases, whereas the rate for amniocentesis is less than 1 percent. Since ultrasound has come into use, the need for this procedure has lessened.

Birth

By the time 9 months have passed, the mother-to-be is usually anxious to have her child. She has probably gained 25 to 35 pounds. This extra weight is distributed approximately as follows:

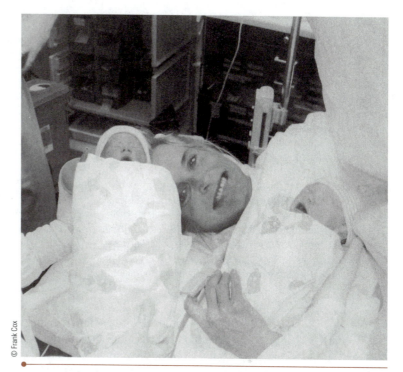

Surprise! Two for the price of one.

- Amniotic fluid: 2 pounds
- Baby: 7 to 8 pounds
- Breast enlargement: 2 pounds
- Placenta: 1 pound
- Retained fluids and fat: 6 pounds
- Uterine enlargement: 2 pounds

There are two objections to gaining too much weight during pregnancy: excessive weight can strain the circulatory system and heart, and many women find it difficult to lose the extra weight after the baby is born. Fetal weight above 9 pounds complicates labor and delivery and increases the risk of postpartum hemorrhage. On the other hand, dieting during pregnancy to remain within an arbitrary weight-gain limit is also risky because, as we pointed out earlier, good nutrition is important during pregnancy.

Maternal weight gain has its most visible impact on the infant's birth weight. Babies born to mothers who gain 31 pounds or more are at considerably reduced risk of low birth weight, compared with mothers gaining less than 21 pounds (NCHS 2000e).

Although the average length of time a child is carried is 266 days, the normal range varies from 240 days to 300 days. It is, therefore, difficult for the physician to be exact when estimating the time of delivery. In fact, there is only about a 50 percent chance that a child will be born within a week of the date the physician determines. The expected birth date may come and go without any sign of imminent birth. This can be wearisome for the expectant mother, but it is perfectly normal.

Cesarean Birth

From as early as 1882, under certain circumstances, such as when a baby is too large to pass through the mother's pelvis or when labor is very long and hard, the baby has been removed via a cesarean birth. In this operation, an incision is made

HIGHLIGHT

Infant Mortality Rates

The Centers for Disease Control indicated that the 2001 *infant mortality rate* (IMR) (infants dying in the first year of life) in the United States reached a record low of 6.8 per 1,000 live births. The three leading causes of infant death were congenital malformations, low birth weight, and sudden infant death syndrome.

While all groups showed improvement overall, major disparities by race and ethnicity still exist in America: In 2001, rates ranged from 3.2 per 1,000 live births for Chinese mothers to 13.3 for African American mothers. Between 1995 and 2001, the overall IMR declined 10.5 percent; rates were down 9 percent for infants of African American mothers and down 14 percent for infants of Hispanic mothers. IMRs were higher for infants whose mothers had no prenatal care, were teenagers, had less education, were unmarried, or smoked during pregnancy.

IMRs are higher for infants of women born inside the United States, compared with women outside the United States.

- IMRs are higher for male infants, multiple births, and infants born preterm or having low birth weight.

- IMRs also vary greatly by state. Rates are generally higher for southern states and lowest for western or northwestern states.

Source: Mathews et al. 2003.

through the abdominal and uterine walls, and the baby is removed. The recovery period is longer than for a normal birth. Contrary to widespread belief, it is possible for a woman to have several babies in this manner or to have one by cesarean birth and the next normally. The name of the operation derives from the popular, but erroneous, legend that Julius Caesar was delivered surgically.

The U.S. cesarean rate dropped 8 percent between 1991 and 1996 (from 22.6 to 20.7 per 100 births), but then increased again by 4 percent between 1999 and 2000 (22.9). It was once thought that after having a cesarean birth, natural vaginal births should be avoided, but this has proved false. The rate for vaginal births after a cesarean birth was 20.7 per 100 births in 2000. Negative outcome for these births is no higher than for natural births to women who have never experienced a cesarean birth (NCHS 2001).

The frequency of cesarean births has led to controversy because the operation is more traumatic to the mother's body than natural birth. At the same time, it is far less traumatic to the child because there is no prolonged pressure on the child, as there is in the normal birth process.

Hospitals contend that better monitoring of the child just before birth and throughout the procedure leads to early recognition of possible problems, and that many of these can be headed off by a cesarean birth. They admit that monitoring has led to a higher proportion of cesarean births, but point out that it has also reduced infant mortality. However, it appears that some of the increase in cesarean births stems from the desire of parents to have a perfect baby and the ensuing fear on the part of the delivering physician that legal action will be taken against him/her if the baby is less than perfect.

Birth Pain

The uterine contractions, which are simple muscle contractions, usually don't cause pain, though prolonged or overly strong contractions can cause cramping. The majority of pain arises from the pressure of the baby's head (the largest and hardest part of the baby at the time of birth) against the cervix, the opening into the birth canal. In the early stages of labor, the contractions of the uterus push the child's head against the still-contracted cervix, and this point becomes the major source of pain. By trying to relax at the onset of a contraction, by breathing more shallowly to raise the diaphragm, and by lying on one side with knees somewhat drawn up, a woman can reduce labor pain to a minimum.

Once the child's head passes through the cervix, the woman experiences little or no pain as the baby passes on through the vagina. Hormonal action has softened the vagina to such an extent that it can stretch up to seven times its normal size. An additional difficulty may occur at birth, as the child passes out of the mother into the outside world. There is often a slight tearing of the perineum, or skin between the vaginal and anal openings, because the skin may have to stretch beyond its limits to allow the infant to exit. In most cases, the physician will make a small incision called an **episiotomy** so that the skin does not tear. The incision is sewn after the delivery. Sometimes, massage can be used to stretch the skin enough so tearing will not occur, especially if the infant is small.

episiotomy
A small incision made between the vaginal and anal openings to facilitate birth

Natural Childbirth

In recent years, many women have sought an alternative to the automatic use of anesthesia and the rather mechanical way many American hospitals have handled childbirth in the past. Many years ago, Grantly Dick-Read coined the phrase **natural childbirth** and suggested in his book, *Childbirth without Fear*, that understanding of birth procedures by the mother could break the pattern of fear, tension, and pain too often associated with childbirth (Dick-Read 1972). Natural childbirth means knowledgeable childbirth, not simply childbirth without anesthesia or at home.

natural childbirth
Birth wherein the parents have learned about the birth process and participate in exercises such as breathing techniques to minimize pain and, therefore, the use of drugs

General anesthesia for childbirth has become much less popular, even in hospitals, because it slows labor and depresses the child's activity, making the birth more difficult even though less painful. All systemic drugs used for pain relief during labor cross the placental barrier by simple diffusion. Although such drugs affect the fetus, so do pain and stress experienced by the mother. Withholding medication from a tense, anxious, laboring woman may not accomplish the intended goal of reducing fetal problems (Olds et al. 1998, 692). Generally, however, using systemic drugs for women in labor should be minimized. An epidural block is sometimes used in both first and second stages of birth. Anesthetic injected in a lower area of the spinal column causes loss of feeling in the pelvic region.

Paracervical anesthesia involves injection of Novocain® or a similar painkilling substance into the area around the cervix. This quickly deadens the area, blocking out the pain. The anesthetic action is localized and has little, if any, effect on the baby; the mother is completely conscious and able to participate in the birth.

Hypnosis is also being used more frequently during labor and delivery, to help relax the mother and reduce her sensations of pain. It is particularly useful for women who cannot tolerate the drugs used in anesthesia. It cannot be used with everyone, and requires a knowledgeable physician. Those who use it report relaxed and relatively uncomplicated deliveries.

Today, a growing number of physicians are letting the woman decide whether she wants an anesthetic and, if so, what type. It is important that the expectant mother is informed of the benefits and disadvantages of the available forms of anesthesia, so that she can make her choice intelligently.

As mentioned earlier, hospitals, Red Cross facilities, county medical units, and evening adult schools often provide childbirth-preparation classes. The classes provide information on the birth process, what to expect, and how to facilitate the natural processes. The woman is also taught physical exercises that will help prepare her body for the coming birth. She learns techniques of breathing, to help the natural processes along and to reduce the amount of pain she would otherwise experience.

We have already noted that one of the basic principles underlying natural childbirth is that knowledge reduces fear, and reduced fear means less tension and pain. The other basic principle of natural childbirth is to make the mother and father active participants in the birth of their child, rather than passive spectators. For instance, controlled breathing (with the father helping to pace the breathing) supplies the right amount of oxygen to the working muscles, giving them the energy they need to function efficiently. Voluntarily relaxing the other muscles helps focus all energy on the laboring muscles. Another aspect of breathing exercises is to focus attention on responding to the contractions, which keeps attention from focusing on the pain.

Since the couple has decided to have the child together, it is also important that they learn together about the processes involved and that the father is not simply a spectator during labor and delivery. The couple will want to know if the hospital they plan to use will allow the father into the labor and delivery room, so that he can give psychological support and comfort to the mother. During the early stages of labor, he will be able to remind her of what to do as the contractions increase. He can keep track of the time intervals between contractions, monitor her breathing, remind her to relax, massage her (if she finds that helpful), and keep her informed of her progress. In other words, he can help by sharing the experience. Several birthing methods (including Bradley and Lamaze) urge the father to learn about the birth process and actively participate. This also facilitates father-child bonding.

Once the baby has passed through the cervix into the birth canal (vagina) and the physician is present, events happen so fast and the woman is so involved with the imminent birth that the father's presence and help may be less important than during labor.

Don't forget about breast feeding.

© Frank Cox

Rooming-In

Increasingly, more hospitals are allowing *rooming-in*, which means that the mother is allowed to keep her baby with her, rather than having the child remain in a hospital nursery. Rooming-in is especially helpful to the mother who breastfeeds. Both mother and newborn benefit from the physical closeness, and the child will cry less because he or she will get attention and be fed when hungry.

Alternative Birth Centers

Some hospitals have created *alternative birth centers*, homelike settings for childbirth. Relatives and friends are allowed to visit during much of the childbirth process. Barring complications, childbirth takes place in the same room that the mother is in during her entire stay at the center. Couples interested in such birth centers usually must apply in advance. The mother-to-be must be examined, to be as certain as possible that she will have a normal delivery. In addition, the birth center usually requires that the couple attend childbirth classes. Birth centers are a compromise between normal hospital birth and home birth. They are part of the continuing trend by hospitals to make childbirth less mechanical and help couples participate as fully as possible. There are also numerous out-of-hospital birth centers now operating.

Home Births

Some women prefer *home birth* to having their children in a hospital. Less than 1 percent of all births take place in a private home. At home, the woman is in familiar surroundings, can choose her own attendants, and can follow whatever procedures soothe and encourage her (such as to have music playing). At home, birth is also a family affair. Social support, especially that of the father-partner, is especially important during pregnancy and birth. The majority of those women who have had a home birth say that they would also prefer to have future births at home.

Is home birth safe? In Europe, where home birth is common, statistics indicate that home birth is just as safe as hospital birth (England and Horowitz 1998). American physicians do not agree, and cite many instances of tragedy that could have been avoided in a hospital. The risk can be reduced by careful prenatal screening of the mother and by providing backup emergency care. If prenatal screening indicates conditions that might involve a complicated delivery, the woman should have her baby in a hospital that has the facilities to deal with possible problems.

The *midwife* is an honored professional, used by 80 percent of the world's population to attend childbirth (Olds et al. 1998, 8–9). Midwives are used in home delivery in many modern countries such as Sweden and the Netherlands; they were used in the United States until the turn of the century. Yet, the organized medical profession and many others believe that delivery in the hospital by physicians is much safer. In addition, they believe that it would be difficult to enforce standards of competence if home delivery by midwives became widespread. England, on the other hand, has set high standards and has trained midwives since 1902. About 75 percent of all births are midwife-assisted. In England, the obstetrician is the leader of the birth team, and the midwife does the practical work. In Holland, home birth is considered so safe that the national health insurance, which pays for most births, will not pay for a hospital birth unless medically indicated.

In the United States, certified nurse-midwives (CNMs) are registered nurses who practice legally in the hospital setting. Some states do not allow them to perform home births.

As interest in home births increases and as hospital costs soar (it now costs over $5,000 to have a baby with standard hospitalization and delivery), the idea of midwifery is returning in the United States. Some states now have licensed training programs for midwives.

Postpartum Emotional Changes

The first few weeks and months of motherhood are known as the postpartum period. About 60 percent of all women who bear children report a mild degree of emotional depression (baby blues) following the birth, and another 10 percent report severe depression.

Most women experience this mild, postpartum depression between the second and fourth days after delivery. This may be due, in part, to the changing chemical (hormonal) balance within the woman's body as it readjusts to the nonpregnant state. This readjustment takes about 6 weeks. For most women, this depression passes quickly. The general emotional reactions to parenthood, especially with the first child, continue for a much longer period. Postpartum counseling before birth can help couples understand and cope with any normal feeling of depression that might follow birth.

Some of us are high, some are mellow, some of us are lethargic and depressed, or we are irritable and cry easily. Mood swings are common after birth. We are confused and a little scared, because our moods do not resemble the way we are accustomed to feel, let alone the way we are expected to feel. If this is our first baby we may feel lonely and isolated from adult society. The special attention and consideration many of us receive as expectant mothers shift to the baby and we too are expected to put the baby's needs before our own. (Boston Women's Health Book Collective 1992)

Although definitive research is unavailable, it appears that postpartum hormonal changes may also play a positive role in how the mother interacts with her new baby. Certain increased hormonal levels seem to influence how attracted and attentive to her baby the mother is (Azar 2002b; Carter et al. 2001; Corter and Fleming 2002; Maestripieri 2001).

The father is often a forgotten person immediately after birth. Not only does the mother occasionally suffer mild depression after birth, but the father may also suffer feelings of neglect, jealousy, and simply "being left out" after childbirth. The stronger their partnership, the more a couple can share in the whole process of conception, pregnancy, and childbirth—which means greater satisfaction for both the father and mother.

Summary

1. Family planning is an important part of marriage. In general, the couple who plan their family realistically and control reproduction increase their chances of having a happy marriage and a fulfilling sex life.

2. Birth control has a long history, but only with technological advances such as the pill, the condom, and others have such methods been effective. The perfect contraceptive, however, has not yet been invented.

3. Sterilization as a form of birth control has become increasingly popular.

4. Although abortion laws now make abortion upon demand a reality, there remains a great deal of controversy over just how available abortion should be.

5. Assisted reproductive technologies have become increasingly sophisticated. New ways to produce children and, perhaps, design them to parents' specifications are continuing to advance.

6. The egg and the sperm cells are among the smallest cells in the body, yet they contain all of the genetic material necessary to create the adult human being.

7. Pregnancy takes several weeks before it becomes recognizable, though some pregnancy tests can determine pregnancy within 7 to 10 days of conception.

8. The average pregnancy is 266 days, or 38 weeks.

9. Increasing emphasis is being placed on participation of both the mother and father in the birth process, to the extent possible.

Resources on the Internet

 ## Companion Website for *Human Intimacy: Marriage, the Family, and Its Meaning*, Tenth Edition

http://sociology.wadsworth.com/cox10e/

Gain an even better grasp on this chapter by going to the companion website to take one of the tutorial quizzes, use the flash cards to master key terms, or check out the many other study aids you'll find there. You will also find special features such as GSS data, Sociology Online, and Census 2000 information that will put data and resources at your fingertips to help you with that special project or to help you as you do some research on your own.

 ### InfoTrac College Edition

http://www.infotrac-college.com/wadsworth/

You can access reliable resources anytime, anywhere, with InfoTrac College Edition, the online library. This fully searchable database offers more than 20 years' worth of full-text articles (not abstracts) from almost 5,000 diverse sources, such as top academic journals, newsletters, and up-to-the-minute periodicals, including *Time*, *Newsweek*, *Science*, *Forbes*, *The New York Times*, and *USA Today*. You can conduct electronic key word searches using key terms from this chapter to supplement your reading and learning experience. To aid in your search and to gain useful tips, see the Student Guide to InfoTrac College Edition, which you can access through the companion website for this book.

To Clone or Not to Clone

Early in March 1997, a landmark paper was published in the journal *Nature* by Dr. Ian Wilmut, a Scottish scientist at the Roslin Institute near Edinburgh, Scotland. The mad scientist was suddenly real and alive, except for the fact that the Frankenstein monster was not a monster at all, but Dolly, the lamb created from a single adult cell. The delicate process is as follows:

1. A cell is taken from an adult sheep and placed in a culture with so few nutrients that the cell ceases dividing.

2. At the same time, an unfertilized egg is taken from a female lamb. The egg's nucleus (with its DNA) is sucked out, leaving an empty egg cell, which still contains all the machinery necessary to produce an embryo.

3. The two cells are placed next to each other, and an electric pulse causes them to fuse together. A second pulse copies the burst of energy at natural fertilization, starting cell division.

4. A few days later, the resulting embryo is implanted into the uterus of a female sheep.

5. After a normal gestation period, the pregnant sheep gives birth to a lamb (Dolly) that is genetically identical (an exact carbon copy) to the original cell donor.

Dolly's birth was followed by the birth of calves, Charlie and George, at the University of Massachusetts. Not only were Charlie and George cloned, but they were also genetically modified (Saltus 1998). Even more recently, the Japanese produced the clone of a cloned bull (Gato 2000). The primary purpose of this experiment was to produce tasty beef, consistently. The researchers wanted high-quality bulls to be available for propagation (Gato 2000; Gosden 1999). Cloning reduces the amount of time needed for reproduction. Cloned beef has been on sale in Japan since 1997.

In 2001, scientists cloned human embryos by replacing egg nuclei with mature nuclei from adult cells; the cloned cells divided briefly before they died. In February 2004, Korean scientists cloned 30 human embryos and grew them to the blastocyst stage. They then harvested them for stem cells and created single stem-cell colonies (*Science* 2004).

And who knows what is to follow? Suddenly, it becomes possible to reproduce a carbon copy of Michael Jordan, the basketball player; Madonna, the singer-actress; the president of the United States; or Einstein. Or a serial killer? Or YOU? Though the description of what the scientists did to produce Dolly sounds simple, it is not. Out of 277 tries, the researchers produced only 29 embryos that survived longer than 6 days; only one, Dolly, was actually born. While living, Dolly reproduced through normal pregnancy.

Why Not Clone if We Can?

These experiments have given rise to any number of scenarios such as the following:

1. Your 4-year-old daughter must have kidney transplants sometime in the next few years, to survive. Because a compatible donor is unlikely to be found, you decide to use one of her cells to clone a genetic copy. Several years later, the clone's kidneys are harvested and given to your daughter, who grows into adulthood and lives a long, healthy life.

2. Geniuses such as an Einstein are rare. Before he dies, the society decides, with his agreement, to use one of his cells and clone him. Because Einstein has donated so much to human knowledge, the society decides to clone several hundred of him, to speed future scientific breakthroughs.

3. I have a big ego and very much want future generations to remember me. In the past, I might have built a monument in my honor or named a library or school after me. Far better, however, is simply to create a new me. Then, of course, clone another me when the first clone ages, and so on, into perpetuity. Thus, I will last forever.

Supporters of cloning have included high-profile celebrities such as Christopher Reeve and Michael J. Fox. Fox suffers from physical problems, as did Reeve before his death. Both have suggested that animal cloning would be beneficial to humans, even if human cloning is deemed unacceptable (Gosden 1999; T. Murray 1998; Saunders 2000). For example, it is expected that cows cloned and genetically modified will secrete pharmaceutical drugs in their milk—at lower cost, in many instances, than conventional manufacture. There could be herds of identical cattle (or various poultry) with improved meat and milk characteristics, or with genetic traits making them resistant to various diseases such as "mad cow" disease (Saltus 1998).

To Create a Human Being Is Unethical and Dangerous

Arguing strongly on ethical, moral, and religious grounds, those against cloning of human beings (Kass 1998; Kilner 1998; Marty 1998; O'Conner 1998) have made an impact; in January 1998, 19 European nations signed a treaty stating that cloning people was a violation of human dignity and a misuse of science (Schuman et al. 1998). In August 2001, the United States permitted limited federal funding of stem-cell research, using only stem-cell lines that had already been derived from human embryos. Most scientists working on cloning feel that the existing lines are too few for meaningful further research.

In a *Time*/CNN poll, 90 percent of the sample found cloning a human to be a bad idea. Ninety-three percent would not clone themselves. Sixty-nine percent found the cloning of a human to be against God's will. Even 67 percent found that the cloning of animals was also a bad idea (*Time* 2001, 55).

Some negative scenarios are, of course, possible, such as the following:

1. As it becomes clear that Germany will lose World War II, fanatic Nazis decide to clone a number of Hitlers so that in future years, the Nazi movement can arise again and be led by their god, Adolf Hitler. By cloning several Hitlers, his followers can better ensure that at least one will survive the war, to lead again.

2. A despot ruler of a country decides to clone himself, so that he can maintain control over the country after he dies.

3. The kidneys of a clone are transplanted into a child who suffers severe kidney problems. Does this mean the clone must die if both of its kidneys are removed for transplantation?

Despite the loud debate that erupted about the possible cloning of a human being, and no matter who or how many people claim it should or should not be done, knowing humans, it will one day be accomplished— even if it is done merely to see if it can be done.

What Do You Think?

1. Can you support any of the above scenarios? Which ones? Why?

2. Which scenarios can you not support? Why?

3. If my clone is genetically identical to me, will that clone be the same person as I am when it grows up? Explain.

4. What about the role of the environment in shaping behavior? Identical twins may be genetically the same, but they certainly have some differences when they are adults. Unlike with most identical twins, clones will probably grow up in environments much different from those of the donors, so just how similar do you think an adult clone's personality will be to the original person's?

5. Do you think we will ever clone a human being, or is this just a pipe dream that society will never allow? Explain.

6. Would you clone yourself, or someone you love and don't want to lose? Why, or why not?

7. What do you think about cloning a beloved pet?

The Challenge of Parenthood

Chapter

10

Questions to reflect upon as you read this chapter:

- In what ways do children change the marital relationship?
- Do you think American families are neglecting their children?
- How has television changed the way in which children are raised?
- Do you think television is a positive or negative influence on children? Video games?
- If you could not have children, would you consider adoption?
- Do you think a single-parent family can successfully raise a child?
- What problems might the single-parent family face that a two-parent family would not?

A mother listening to her 6-foot son explain the finer points of football, or a father escorting his lovely 26-year-old daughter down the marital aisle find it difficult to remember the beginnings: mother and father falling in love; their own marriage; their lovemaking; a missed menstrual cycle that led her to think, "Maybe I'm pregnant"; the thrill of feeling a tiny, unseen kick; the scary feelings when labor started; the dramatic rush to the hospital; holding the red, wrinkled, 7-pound newborn son or daughter and counting the fingers and toes to make sure they were all there; the long discussion over the name; the wet spot on dad's suit after hugging the baby goodbye before going to work; the first tooth; the first sickness; the first step; the first bicycle; the first day of school; the first date; high school graduation; and now marriage—and perhaps soon, the new cycle when a little one hugs them and says, "Hi Granddad, hi Grandma."

Parenting is an indispensable civic activity, not simply a set of private joys and responsibilities. Being a good parent is not a private activity, but rather, a public and societal activity. Good parents ensure that a child becomes a well-adjusted person who succeeds in life and becomes a productive and responsible citizen, contributing both to those in his/her community and to the society as a whole (Hewlett and West 2002, xx, xxi). Children reflect the strength and values of the society that produces them; their well-being exemplifies the country's well-being; their future is the country's future (Bianchi 1990, 36). It is clear that a family's effect on its children is crucial to the rearing of successful (in all its meanings) adults. Such things as family size and structure, birth order, parental relationship, the closeness of extended family members such as grandparents, family economic circumstances, and family educational level all influence the family's children for better and worse. This book is based on the idea of building strong families, and nowhere is a strong family more important than in child rearing. Because children are the future of our society, every child must function at a level commensurate with his/her potential, and every child must contribute his/her best to the society as a productive, fully functioning adult. Parenthood is not about children; it is about creating mature, responsible, caring adults. Childhood is a passing period of life; adulthood lasts a lifetime.

Overall, those who find childbearing most rewarding are those who are married and have positive feelings about their pregnancies (Groat et al. 1997). Unwanted pregnancies can certainly lead to wanted children, but unwanted pregnancies too often lead to unwanted children; this, in turn, starts the parents off with negative feelings toward parenting. Thus, family planning (see Chapter 9) plays an important role in creating happy, successful parents who reap joy from parenthood, rather than grief.

Marriage and parenthood were often seen as synonymous in the past. Having children was the major goal of marriage. Many popular assumptions that supported this goal included the following:

- Marriage means children.
- Having children is the essence of a woman's self-realization.
- Reproduction is a woman's biological destiny.
- All families have a duty to produce children to replenish the society.
- Children prove the manliness of the father.
- Children prove the competence and womanliness of the mother.
- Humans should be fruitful and multiply.
- Having children is humanity's way to immortality; children extend one into the future.
- Children are economic assets, providing necessary labor.

The beliefs of the past often remain to encumber the present, long after the original reasons for the belief have disappeared. So it is with reproduction and parenthood. For thousands of years, humans had to reproduce—had to be parents—for the species to survive. And they had little choice in the matter, because the pleasure of sexual relations often meant pregnancy.

The dogma of the past has quietly become the liability of the present; uncontrolled reproduction today may lead to Earth's overpopulation and the ultimate demise of the species, not its survival. In the past, high infant mortality, uncontrolled disease, famines, and war meant that society had to pressure all families to reproduce at a high level, to maintain the population. Women were pregnant during most of their fertile years. Even today in many developing countries, infant mortality runs as high as 200 per 1,000 live births; because of malnutrition and disease, many families see fewer than half of their children reach maturity. By comparison, infant mortality in the United States is 6.8 per 1,000 live births (CDC 2003).

What Effects Do Children Have on a Marriage?

The most often-heard words from parents to couples thinking about becoming parents for the first time are: "Your life will never be the same again." "We know, we know," replies the couple contemplating parenthood. "Oh no, you don't," reply the parents. And the parents are right. Parenthood is constant. It is demanding. One can divorce a problem spouse, but one can't divorce a problem child. Parenthood lasts to the end of your life.

Although generally considered a joyful anticipated event, the birth of a child represents a highly stressful occurrence in the lives of many families. Adjustments related to issues such as family roles, time management, sexuality, financial obligations, and the physical demands of children make the addition of a child an especially challenging transition in a family's career. (Peterson and Hawley 1998, 221)

Is one still a parent to a 35-year-old? Russ received an urgent call from his 35-year-old son living 1,000 miles away. The son asked Russ to come over as soon as possible because he and his wife were breaking up and he needed his dad's help and support. Fortunately, Russ was retired, so he left the next day. He arrived just in time to see the wife and her brothers driving off with all the furniture that Russ had given to his son and daughter-in-law as a wedding gift. He and his son ended up sleeping on the floor in the empty house for a week, until the son got things sorted out.

The point of Russ's story is that parenting does not end when children are 21 years old, when they are married, or when they live at some distance from their parents. Parenthood is a lifetime commitment and obligation.

Most literature talks about how parents affect their children, but the parent-child relationship is best described as a dual process, in which there is an interchange of influence between parents and their children (G. Brody 1994, 359). The stress that children place on their parents and the effect they have on their parents' marriage are complicated and involve many factors (McBride et al. 2002; Nomaguchi and Milkie 2003). Parenthood may actually enhance some couple relationships, undermine others, and have little effect on still others (M. Cox et al. 1999, 612). Certainly, the readiness of the couple for pregnancy and ensuing parenthood is crucial. Studies indicate that both husbands and wives coping with a child from an unplanned pregnancy tend to suffer more marital dissatisfaction than those with planned pregnancies (M. Cox et al. 1999, 621). Becoming a parent is likely to be the biggest and most permanent decision in an individual's life (Twenge et al. 2003). It certainly is not to be taken lightly. We have already discussed the importance of planning parenthood, so that the child is wanted. Yet even parents who want a child will experience ambivalence when they near the point of actually having one. Are we really ready? How much freedom will I lose? Can we afford a child? Will I be a successful parent? Such questions will arise constantly for any expectant parent.

We had expected to feel differently about ourselves once we became parents, but we were startled by what felt like a major upheaval in our sense of ourselves. We had assumed that there would be some changes in the "who does what" of our life. We hadn't known that becoming parents might lead us to feel more distant from each other based on different levels of involvement in work inside and outside the family. We hadn't anticipated that having a baby could revive long-buried feelings of gratitude or disappointment about how loved we had felt as children, or realized that our disagreements about whether the baby needed to be picked up and comforted or left alone to "cry it out" would actually have more to do with our own needs than they did with the baby's. And, having always viewed the important issues of life similarly, we certainly did not anticipate how differently we could see the "same" things after becoming parents. (Cowan and Cowan 1992, 2)

Unplanned pregnancies make such questions infinitely more difficult. Couples often feel trapped into parenthood by an unexpected pregnancy. They are not ready for and have not consented to parenthood. They are often angry and resentful—especially the prospective father, because he has the least control over the pregnancy. An unwed father, for example, has virtually no legal rights in deciding what shall be done about an unwanted pregnancy. Yet, he is legally responsible for the newborn child, as is the mother.

Men may have more trouble working out parenthood-readiness questions than women do because society, unfortunately, often views fathers as less important to

WHAT DO YOU THINK?

1. At what age do you think parenthood will end if you have children?

2. Would you do what Russ did to help his son? Why? Why not?

children than mothers. Yet, a man may worry about losing his wife or baby during childbirth, facing increased responsibilities, perhaps being replaced in his wife's life by the newborn, and just how good a father he will be.

Although women seem better at handling readiness-for-parenthood questions, they too must confront meaningful life choices. In the past, women did not have much choice; the parenting-homemaking role was often the only role open to them. In those days, however, the homemaking role was much larger and more varied than it is today. Women were productive members of farm and craft teams, along with their husbands. Children either shared in the work of the household or were left to amuse themselves. These mothers were usually not lonely or isolated, because the world came into their homes in the form of farmhands, relatives, customers, and so on. Such women had no reason to complain of the boredom and solitude of spending 10-hour days alone with their children. In addition to being mothers, they contributed to their family and society in other important and productive ways.

Today, the scope of the homemaking role has been drastically reduced. The general affluence—especially of the American middle-class family—the small nuclear family structure, the influence of feminism, and the transfer of economic production from the home to a workplace all have combined to decrease the satisfactions derived from the now narrow and exaggerated maternal-homemaking role. The modern woman has many other career choices available to her, which make the decision of whether she is ready for parenthood increasingly difficult (see Chapters 4 and 13).

The idea that a mother must remain at home with her children at all times is losing credence as more mothers join the workforce (Hewlett and West 2002; W. D. Manning and Lamb 2003; Steinberg 1996). The negative effects that absent mothers, babysitters, and child-care centers often have on children are reduced when good child care is available (see Chapter 13). The key word here is good. Unfortunately, good child care for working mothers has not kept pace with the need, and neglected children seem to be more of a problem than ever before (Blankenhorn 1995; Meier 1995; Whitehead and Popenoe 2003). Full-time parents still remain the best child-care givers.

Regardless of the state of preparation and readiness, the actual transition to parenthood involves a number of costs to the parents:

1. *The physical demands associated with caring for the child are usually far greater than parents anticipate.*
2. *Unforeseen strains are placed on the husband-wife relationship.* Many studies suggest that the presence of children in the family lowers the marital happiness or satisfaction of the parents. Many couples report that the happiest times in their marriage occurred before the arrival of the first child and after the departure of the last. Marital satisfaction seems to follow a U-shaped curve: higher at the beginning of marriage, lowest when children are teenagers, rising again when the children leave home. However, more recent research seems to negate the idea of a U-shaped curve, indicating instead that marital dissatisfaction occurs rather rapidly during the first 10 years of marriage and then levels off, regardless of the presence of children (Bradbury et al. 2000, 184). The effect that children will have on the marital relationship will depend on the couple's understanding and tolerance of the natural demands and strains that children create. The effects children have on a marriage also vary greatly with the ages of the children (see "The Growing Child in the Family," pages 317–325).

Some of the negative effects that children appear to have on the marital relationship may actually be caused by the parents' expectations about the changes children will cause in the marital relationship. Parents who anticipate the changes correctly do not experience the degree of marital dissatisfaction that other parents report. (MacDermid et al. 1990)

3. *The personal elements of the marital relationship—friendship, romance, and sex—tend to become less satisfying as the relationship becomes focused more on day-to-day obligations.* Such obligations include managing and running the family, integrating family routine and work schedules, and so on. The time that the childless couple has used to nurture their personal, emotional relationship may be displaced by the time demands of the child.

Sometimes, especially in equal partnership marriages, the baby becomes the primary love object of both parents, who participate equally in child care. This couple may form a parental bond, but at the expense of the couple bond. (Sharpe 2000, 71)

4. *New parents complain about the limits children place on their social lives, particularly their freedom to travel or be spontaneous.*

5. *Parents also complain about the monetary cost of rearing children.* Even though younger children cost less than teenagers, the total cost of raising a child to 18 years of age will certainly be over $100,000; if college or professional school costs are added, the total arises to well over $200,000.

Despite the negativity of some research findings, most parents express overall satisfaction with children and their parenting role. In one large study, more than three-quarters (77 percent) of parents said that their children were "the main satisfaction in my life" (Blankenhorn et al. 1990, 75). Also, parents tend to do better with a second child or children born later (S. D. Whiteman et al. 2003). It seems obvious that parents will be more confident and relaxed with children born later, because they are experienced with the firstborn child.

A number of positive themes appear in new parents' lives. Parents derive emotional benefits from the joy, happiness, and fun that accompany child care. New parents often report feelings of self-enrichment and personal development when undertaking parental responsibilities. Parents also report an increased sense of

HIGHLIGHT

The Highs and Lows of Parenthood

Colleen (Monday afternoon): I never realized that an infant could be so demanding. I never get a moment's peace when she's awake, not even to go to the bathroom. I can't even say "wait a minute" because she is too young to understand (10 months). I feel as though my whole life has been taken over by a tyrant. All she does is demand—me! me! me! I've lost my identity, my individuality, to a word—*mother*.

Colleen (Tuesday morning): I'm glad I have her. It's neat to know she is a part of me and of Bob. She is a real little person who is perfect and loves me and needs me. She makes me feel

successful at something—helping another. It makes me feel good to know that I was able to produce a child and be a parent (LaRossa and LaRossa 1981, 177).

Colleen: I do still have some time to myself, such as when she (the baby) is taking her nap. But it isn't really totally free time for me to use any way I wish. I'm still on call. Bob tries to give me free time by staying home with her and letting me go out. But if he has been at work all day, he also needs some free time, so it is hard for him, too.

family cohesiveness, and strengthened relationships between themselves and their extended family. Grandparents-to-be are often as thrilled with new grandchildren as they were with the birth of their own children (see Chapter 11).

At first glance, parents' expression of satisfaction with their children and family life may seem to contradict the general evidence of reduced marital satisfaction when children enter the family. In fact, both the negative and positive expressions are genuine. In many ways, parenthood is a paradox. On the one hand are the problems of contending with a demanding child; on the other hand are the joys of being close to and caring for another person. Parenthood exhibits this paradoxical character not only because it includes both positive and negative experiences, but also because the lows and highs of parenting tend to be extreme. Children often make you want to cry one moment and laugh the next.

The fact is that children add enormously to the complexity of family relationships. For example, a couple must contend with communication in only two directions. Add one child, and this becomes six-way communication; add two children, and there is now 12-way communication (Figure 10-1).

Prospective parents can become better prepared for parenthood by reading and taking classes. Yet, even with a thorough preparation, bringing the first baby home is an exciting, happy, and frightening experience for parents. No matter how much they have read about children, they can never be sure that what they are doing is correct. If the baby cries, they worry. If the baby doesn't cry, they still worry. Parents also have to adjust to the feeding routine. As the next section explains, most parents become more traditional, with the mother becoming the primary caregiver. This means she cannot venture far from the baby, especially if she is breast-feeding unless she uses a breast pump to store extra milk. The father has to adjust to taking second place, at least for a while, to the newborn.

Traditionalization of the Marital Relationship

Examining a couple's relationship after a child enters the family provides a clear illustration of the overriding change that occurs. In almost all cases, except for two-career families that can afford to place their child almost entirely in the hands of others, the parents' relationship moves in the traditional direction; that

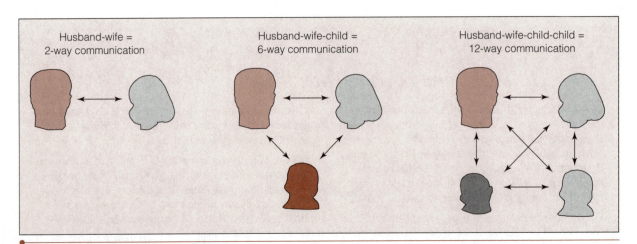

FIGURE 10-1   *How Adding Children Complicates Communication*

is, the division of labor becomes more gender differentiated (Cowan and Cowan 2000; Huston 2000, 304; MacDermid et al. 1990; Walzer 1998). The changes in the marital relationship are more pronounced for wives. Mother assumes most of the parenting and household roles (even if she works), while the father turns toward the work world. This change occurs even in couples who believe in and work for an egalitarian homemaking relationship.

Several theories try to account for this traditionalization phenomenon. Some suggest a physiological explanation, in which hormonal changes in women during pregnancy, birth, and breast-feeding establish "a clear link between sexuality and maternalism" (Azar 2000; Rossi 1977, 17). New mothers receive "erotogenic pleasure" from nursing their infants, which means, according to Rossi, that "there may be biologically based potential for heightened maternal investment in the child, at least through the first months of life, that exceeds the potential for investment by men in fatherhood" (Rossi 1978, 24).

Personality theories of socialization can also be used to account for traditionalization. Early sex-role socialization is usually modeled upon an individual's parents. Because our society provides little or no formal education for parenthood, we tend to act out these parenting models when we have our own children. As young people among our peers and early in marriage, we may be egalitarian in philosophy. But when children arrive, we fall into behavior patterns to which we were socialized early; in most cases, these tend to be traditional.

Sociological explanations of traditionalization examine the social constraints placed on parents, due to the nature of the human infant. The human infant is born in the most dependent state of any animal. A human baby is so dependent that, without continuous care from an adult, it will not survive. The infant's demands are also nonnegotiable: It cannot be told to "wait until tomorrow." Thus, during the early life of the child, continual adult supervision is essential. This demand clashes with the social realities. Economic reality for most couples is that the man can earn more money. The birth of a child increases the economic burden on the family. If the husband can earn more than the wife, who should go out to work, and who should stay home to care for the child? The answer is obvious.

If the mother breast-feeds and the society frowns on breast-feeding in public, social pressure keeps her at home with her infant. If workplaces do not have infant-care facilities (most do not), how can a mother go to work? If she does, much of what she earns is absorbed by the costs of outside child care (see Chapter 13).

The infant's need for continuous adult care also means that when one partner "wins" (is free to pursue his/her own interests outside parenting), the other must "lose" (forgo his/her interests for the sake of the baby). This constant time demand by the young child causes much of the conflict between parents. Loss of free time is the aspect of parenthood that bothers new parents the most.

Parental Effectiveness

Most people take on the job of parenthood, assuming they will be successful: "Of course, we can be good parents." Yet, raising children is an extremely demanding job. The idea that it is easier to build a strong child than to repair a broken adult is really the essence of parenting. But what qualifications do expectant parents have? They all have parents, and all have been children. But, as we know from our own experience as children, all parents fail in their job to some degree.

Calvin and Hobbes

by Bill Watterson

Some failure, especially in the eyes of one's children, seems to be an integral part of parenting. No parent is ever totally adequate for the job of being a parent, and there is no way not to fail at it to some degree, especially in the eyes of one's adolescent children. No parent ever seems to have enough love or wisdom or maturity.

Much advice flows to new parents from grandparents, physicians, clergy, popular magazines, and child experts, all of whom claim to know best how to rear children. Benjamin Spock, known to millions of parents for his advice about child rearing (his *Baby and Child Care* has sold more copies than any other book except the Bible), has observed that:

A great many of our efforts as professionals to help parents have instead complicated the life of parents—especially conscientious and highly sensitive parents. It has made them somewhat timid with their children—hesitant to be firm. They are a little scared of their children because they feel as parents they are being judged by their neighbors, relatives, and the world on how well they succeed. They are scared of doing the wrong thing. (Spock 1980)

And, of course, a great deal of advice comes from the children themselves, who are as yet unblessed with their own children. What 15-year-old doesn't know exactly how bad his/her parents are and how to be the perfect parent?

The problem with all this advice is that there is little agreement about the best way to rear children. In a sense, children come to live with their parents; their parents do not go to live with them. Parents need not think they must relinquish their lives completely in favor of the child's development. Parental growth and development go hand in hand with child growth and development.

Good parents love their children but are also honest enough to know that there will be days when they are angry, unfair, and in other ways inadequate. The ideal of parents treating each child in their family equally is unrealistic, even though equal treatment is a proper, if unobtainable, goal. Age, gender, and personality make each child a unique individual (Tucker et al. 2003). It is difficult but necessary to treat all children in a family as equally as possible and also to recognize each child's individuality. Secure, accepted, and loved children will survive parents' occasional anger, unequal treatment, and other lapses.

Despite the common thought that parenting is usually accomplished by a husband and wife living together, the statistics describing American families tell a different story. If trends continue, demographers predict that more than half of children born in the 1990s will spend at least part of their childhood in a single-parent home. These statistics do not imply that all the other children live in two-parent homes. For example, some children live with foster parents, grandparents, or other relatives—often on a transitory basis (J. Fields 2003).

Although there is discussion of cultural variations in parenting among American ethnic groups, research comparing White, Black, Hispanic, and Asian American parents suggests that parents from all groups are more similar than they are different, when it comes to rearing their children. Differences seem to arise more from a family's economic level than from their ethnicity (Demo and Cox 2000).

Acceptance of gay and lesbian parenthood is increasing. Some of these parents have children from previous heterosexual marriages and relationships. Others have opted for adoption, although ease of adoption for gay parents varies by state. Artificial insemination is also used by would-be lesbian parents and surrogate mothers, by gay men (Elias 2001). The little research that has been done on these families has found the children to be well adjusted; they also haven't seemed more likely to be gay as adults (Demo and Cox 2001; Patterson 1992, 2000; Pruett 2000; Savin-Williams and Esterberg 2000).

There is no single, correct way to rear children. If children are wanted, respected, and appreciated, they will be secure. If they are secure, they usually will also be flexible and resilient. What is important is that the parents are honest and true to themselves. Small children are very empathetic; that is, they have the ability to feel as another is feeling. They respond to their parents' feelings as well as to their actions, even though the children may not be cognitively advanced enough to understand the reasons why their parents are acting as they do. Parents who are naturally authoritarian will fail in coping with their children if they attempt to act in a *laissez-faire* or permissive style simply because they read that permissiveness is a beneficial way of rearing children. The children will feel the tension in their parents and will respond to that rather than to the parents' overt actions. Parents who attempt to be something other than what they are, no matter how theoretically beneficial the results are supposed to be, will generally fail. Thus, parental sincerity is one of the necessary ingredients of successful child rearing. By and large, the pervasive emotional tone used by the parents affects the children's subsequent development more than either the particular techniques of child rearing (such as *laissez-faire* or permissive, *authoritarian* or restrictive, punishment or reward) or the cohesiveness of the marital unit (whether it is stable or broken by divorce or death).

In general, studies have found that the *egalitarian parenting* or *authoritative* (warm, firm, and fair) relates most closely to producing competent children. This style emphasizes the development of autonomy and independence in children, with reasonable parental limits. Parents who (1) focus on cues given by the child as to his/her needs, (2) have extensive knowledge about child development and rearing, (3) consciously consider child-focused goals and develop plans of action to reach these goals, and (4) provide opportunities for the child to be self-directing in some situations seem to do the best job rearing competent children (Cooke 1991, 11; Gray and Steinberg 1999). Remember that the end result of good parenting is a healthy, happy, successful adult.

There are several key dimensions of mothering and fathering: physical contact, attentiveness, verbal stimulation, material stimulation, responsive care, and some

TABLE 10-1	
Children Living in Single-Parent Households, March 2002 (%)	
Whites	
Mother only	16
Father only	4
Blacks	
Mother only	48
Father only	5
Hispanics	
Mother only	25
Father only	5
Asians and Pacific Islanders	
Mother only	13
Father only	2

SOURCE: J. Fields 2003.

HIGHLIGHT

Diversity in Child-Rearing Values and Practices

Although families of all kinds tend to rear their children in similar manners, there are differences. For example, families in poverty tend to value obedience, to issue commands, to be restrictive, and to use physical punishment with their children more than their affluent counterparts.

Asian American parents typically exercise control over their children's friendship choices and extracurricular activities, and retain this control through high school. Chinese American children are taught traditions that emphasize harmonious relations with others, loyalty and respect for elders, and subordination in hierarchical relationships, especially in father-son, husband-wife, and older brother–younger brother relationships. Japa-

nese American children tend to be closely supervised by parents, who simultaneously teach them two different, but overlapping, sets of values: one rooted in Japanese culture, and another that helps them assimilate into mainstream American culture.

Mexican American children, often reared across extensive kinship networks, are taught to value cooperation, family unity, and solidarity over competition and individual achievement.

These generalities about child rearing by various groups serve to point out how general cultural values impinge on child-rearing values and practices. (See Demo and Cox 2000, 2001 for a general review of this research.)

restrictiveness. Above all, acceptance seems to be the most important, regardless of culture, ethnicity, race, gender, or social class (Khaleque and Rohner 2002).

Stimulation of the child is necessary for the development of basic behavioral capacities. Early deprivation generally leads to slower learning later in life. Early stimulation, on the other hand, enhances development and later learning. For example, research suggests that, given the present state of knowledge, the best physical environment in which to rear children is one that gives them experience with a variety of physical objects. Broad experience gives children an opportunity to develop basic motor and mental skills that they must have to be successful in more specific learning situations later.

The emphasis on early childhood education and intervention, beginning with the Head Start Program in the early 1960s, reflects recognition of the importance of early environmental stimulation for children. Although some controversy exists over the effectiveness of Head Start experiences and the contribution they make to lasting changes in the child, the preschool years are undoubtedly important to the cognitive and emotional development of the child (Goleman 1998). Many researchers feel that early stimulation is the most important factor in brain development. Indeed, some are suggesting that government should make preschool mandatory.

Some stress in a child's life is beneficial to development of stress tolerance. This doesn't mean that parents should deliberately introduce stress into their children's lives, but rather, that parents should relax and be less concerned if their children are placed in a stressful situation. Children who are secure in the love and warmth of their parents will be able to survive and, in fact, grow in the face of stress. For example, moving away from a neighborhood, from familiar places and friends, can be upsetting. Yet, studies of long-distance moves find they have little negative effect on young children, especially when the child is prepared ahead and help is given in the new location with finding friends and adapting to the new neighborhood and school. The children seem to make friends easily, the school change is not difficult, and any disturbance in their behavior dissipates quickly. In general, adolescent children seem to have the most trouble (Seppa 1996). Families who are able to move out of poor or crime-ridden neighborhoods generally find that their children, and also the parents, report higher levels of mental and emotional well-being (DeAngelis 2001a).

Overconcern and overprotection can cause problems for children. Children need increasing degrees of freedom if they are to grow into independent adults. They must have freedom to fail as well as to succeed. They need to experience the consequences of their actions, unless the consequences are dangerous to their well-being. Consequences teach children how to judge behavior. Parents who always shield their children from failure are doing them a disservice. The children will not be able to modify their behavior to make it more successful, because they will be ignorant of the results. The best protection for a child visiting a river, lake, or ocean is water knowledge and skill. Parents who scream hysterically at their children when they move toward the water are doing only one thing—teaching them to fear the water. Any lifeguard knows that fear of the water may lead to panic if there is trouble and that panic is the swimmer's worst enemy. Parents need to be aware of children's activities near water, but parents should show their concern by helping children learn about the river, lake, or ocean and by showing them how to have fun in the water and gain confidence in their abilities, rather than by making them fearful.

Neglect and lack of parenting is a far bigger problem than overprotection. Without enough parental contact, there can be little or no proper behavior for a child to model. Children raised in married, two-parent families generally fare better than children raised in single-parent and stepparent families (Acs and Nelson 2001; S. Brown 2001; DeLeire and Kalil 2002; Hao and Xie 2001; W. D. Manning and Lamb 2003). Because of the increase in unmarried and single-parent families, it is equally, if not more, important for society to support such families in their parenting roles.

It is also important to remember that the family and the child are not isolated from the broader society. Parents are the first to be blamed for their children's faults. Yet, the influences on the child from sources outside the family become increasingly powerful as the child grows older. School, peers, friends, and the mass media exert influence on both parents and children. For example, parents will find it difficult to enforce a rule against alcohol use when their teenager's friends are using it and the media run numerous advertisements depicting beautiful young people having great fun drinking beer. But if their child has alcohol problems, the parents are usually blamed, seldom the peers or the media.

Parents have the most difficult task in the world—to rear happy, healthy children to become competent, successful adults—and they must do this in an incredibly complex environment. The vast majority of parents love their children and do the best job they can. Our society puts the tremendous responsibility of child rearing almost entirely on the parents, even though parents are only one among many influences on the development of children. Our economic system, schools, religious organizations, mass media, and many other institutions and pressure groups impinge on the child's world. Often, these influences are positive and help the parent in the task of rearing and socializing children. Unfortunately, these influences can sometimes be negative, countering the direction parents wish their children to take.

Parenting, then, is not something done only by parents. How parenting is accomplished and the results of the parenting are influenced both by the family and the larger society. Parents must pay attention to the society as well as to their children. To expect children to become adults who reflect only the values and be-

---

WHAT DO YOU THINK?

1. Do you want children at some time in the future? Why, or why not?

2. If you want children, how many do you want? Why?

3. In what ways do you think your parents were successful in their parenting? How were they unsuccessful?

4. How would you change the ways in which your parents raised you? Why?

5. How do you plan to raise your children, if you have any?

Show me how, Dad!

haviors of their immediate family is unrealistic. The effective parent must also be a concerned citizen and work to better the society at large.

The Father's Role in Parenting

The movement toward traditionalization in new parents' relations means that the ideal of sharing (if it was an ideal before birth) may be slowly lost. However, the past 3 decades have seen fathers increasing their involvement in parenting, although mothers continue to spend more time caring for children than the fathers do. Compared to data collected 30 years ago, when fathers provided only about 25 percent of the time that mothers provided, current data show that fathers provide about 75 percent of the care of children that mothers provide (Pruett 2000, 9). A new, more active fathering role seems to be emerging, especially on weekends. Latino fathers are becoming even more involved with their children than either White or Black fathers, both of whom seem to be increasing their parenting roles in intact families (Yeung et al. 2001).

In fact, we are now learning much more about the importance of father's presence and the positive ways in which fathering differs from mothering. Fathers tend to choose child-rearing tasks they prefer, tasks that need not be done everyday, and tasks that are fun. They play more with their children (Pruett 2000, 27) and tend to be more involved with sons than with daughters (Starrels 1994). Fathers' involvement with their children yields numerous positive results. It helps the child gain self-control, contributes to the child's emotional intelligence, and gives male children a role model for future family and work roles (Gottman 1997; Parke and Brott 1999; Pruett 2000).

Many fathers and mothers indicate that providing economic support for the family is the most important part of the father's parenting roles. Those fathers who cannot successfully fulfill this obligation often see themselves as a burden to their family, and their level of family involvement may be reduced out of shame and embarrassment.

The successful, happy father must accept the reality of having a child and being a father. Even though he loses personal and economic freedom and some of his wife's attention, he must feel that the gains of fatherhood offset such losses. The new father must also alter his role in his extended family. He moves from being primarily a son to being a father. Good fathering is less about playing a role and more about the hard work of parenting in a way that builds connections across generations —creating a new generation of caring, socially involved adults capable of carrying on and improving the culture in which they will live.

Today, there is a deficit model of fathering. American men, including fathers, are too often portrayed on TV and in

Never too young to learn.

the movies as being bumbling, or deadbeats, or even dangerous to women and children (Parke and Brott 1999). Television sitcoms usually depict fathers in one of these ways. Some researchers even suggest that fathers are unnecessary (and mothers, too) to successful parenting (Harris 1998a, 1998b; Silverstein and Auerbach 1999). Fathers are often described by the media as mostly absent, abusive to both the mothers of their children and their children, and irresponsible in their support of the family. These negative characteristics especially stereotype the minority father.

By concentrating on the negative, we have failed to recognize the great contributions fathers give and can give to their children. We have forgotten to remind fathers that parenting is also men's most important line of work, and that no other form of labor can offer men as much meaning and satisfaction (Snarey 1997, xii). Even though the American father has not shared equally in the parenting role, some research indicates that many minority fathers who do not live with their children still make an effort to be present in their lives via participation in the community surrounding the child (Pruett 2000; Salem et al. 1998, 331).

This trend needs to be encouraged by showing men a more positive picture of fathering. The historical characterization of the harsh, austere, removed, disciplinarian father is being replaced by a more humanistic, affectionate, caring father. Kyle Pruett discusses how involved fathering changes the man. More responsible, more gentle, more emotional, nicer, mellow, and settled, seem to be common characteristics of men who become very involved in fathering their children (Pruett 2000, 14, Chapter 8). David Eggebeen and Chris Knoester (2001) found that fathers who actively participate in rearing their children, especially their own biological children, were more satisfied with life and more active in their communities and service organizations than fathers who were not very active in the rearing of their children.

Liberation of the woman's role also works to liberate the father's traditional role, freeing him to be more loving than he has been in the past. For example, fathers now tend to participate in childbirth. Those who do tend to hold and rock their infants more. The experience seems to lead to earlier bonding between a father and his newborn. As more women move into careers, more men will have to share a larger portion of the parenting role. A few men are experimenting with fulfilling the home and parent role that used to be limited to their wives, while their wives move out into the work world (Shapiro 1993). Although the percentage is small, the number of children raised by their father as primary caretaker is increasing steadily. In 1970, only about 1 percent of families were headed by single fathers, whereas by 2000, father-headed families accounted for 5 percent of all families (Bianchi and Casper 2000, Table 4). Although these examples are limited, they are indicative of a trend to expand the father's role to include more parenting. Most experts agree that this would be rewarding for all concerned because research indicates that fathers more involved in parenting have both happier wives and more stable marriages (Kalmun 1999; Pruett 2000).

We have stressed the importance of commitment to each other in the healthy, well-functioning family. It is assumed that a mother is automatically committed to her children (even though this is not always the case) because she bears the children. Fathers, however, may have to learn that commitment to their children is a high priority that will yield the high rewards. *Fatherwork* means promising to care for, connect with, and provide for a child throughout his/her life. It means that fathers pledge themselves to their partners, to their children, and to their community so that whatever circumstances may arise, they will, as much as possible, provide a secure environment for the next generation (Dollahite et al. 1997, 30; Pruett 2000). When a father emphasizes the joy of successful fathering, it helps to create a sense of connection with his child and a sense of accomplishment. By supporting the successful growth of a tiny infant into a mature and capable man or woman, men can find the satisfaction that good mothers have always found. The negative depiction of American men as husbands and fathers by the more radical women's movement has been detrimental both to fathers and to their children. Earlier welfare rules that drove the father away kept many men from close contact with their families. It is hoped that painting a positive picture of fatherhood in the media, and emphasizing the joys and rewards that can be derived from fatherhood, will serve as a self-fulfilling prophecy for men.

For men to become involved fathers, society must also offer support. Women (mothers) have urged employers to become more family oriented by creating flexible work schedules, better sources of child care, better family-leave policies, and the like (see Chapter 13). All these family supports are equally important for fathers. In the past, the hard-working young father seeking economic security had little, if any, time for parenting because the society did not support time for parenting. Then by midlife, when at last he had time, he found that his children were young adults starting their lives and had no time for him. Finally, he became a father with time—but for his grandchildren, not his children. This scenario explains, in part, why so many fathers have felt somewhat distant from their children, yet have been able to build close relationships with grandchildren.

The rise in divorce, remarriage and stepfamilies, and unwed pregnancies has increased the difficulties men have in becoming successful fathers. (These problems will be discussed later in this chapter and in Chapters 15 and 16.) The emphasis of this book is always the building of strong and healthy relationships, marriages, and families. Many researchers have found that it is the quality of the marriage, whether reported by the husband or wife, that is "the most consistently powerful predictor of paternal involvement and satisfaction with the fathering role" (Gottman 1997; Parke and Brott 1999; Pruett 2000).

Some of the more important research findings about fathers' involvement with their children include the following:

Don't tip the boat, Dad!

© Frank Cox

- Children of fathers who are very involved in parent-teacher associations complete more years of schooling and have higher wages and family incomes as adults than children whose fathers are not involved in school activities.

- Contrary to popular belief, most unmarried fathers in their mid- to late 20s and early 30s maintain a close relationship with at least one of their children.

- Children of mother-only families exhibit more behavioral problems and have lower mathematical and reading ability than children from father-present families.

- African American fathers of 3-year-olds in low-income, urban areas who are satisfied with parenting, contribute financially to the family, and are nurturing to their children have children with better cognitive and language development.

- When fathers are not present in their children's lives, their sons are more likely to become fathers themselves when they are teenagers and to live apart from their children (Byrne 1997; Horn 1997).

- Infants who have been well fathered during the first 18 to 24 months tend to be more curious, less hesitant, and less fearful (Pruett 2000, 41).

- Preschool children with highly involved fathers tend to have higher verbal skills, compared to preschoolers without much father involvement (Pruett 2000, 43).

- Children with close ties to both their fathers and mothers tend to exhibit less gender-role stereotyping (Pruett 2000, 48).

- Young children with highly involved fathers tend to exhibit more self-control and less impulsivity (Pruett 2000, 50).

Television, Video Games, and the Internet as Surrogate Parents

The major outside influence on children is the electronic screen. As Figure 10-2 indicates, watching screens takes up approximately 4½ hours (281 minutes) per day of children's time. Television leads the way. By age 17, most children have spent more time watching television than going to school. Judging from the controversy surrounding unsupervised television viewing, violent video games, and pornography on the Internet, it is clear that many Americans see these as negative influences. In February 1992, the Children's Television Act of 1990, the first federal law to regulate children's programming, went into effect. In 1997, a TV program rating system was instituted to help parents better monitor TV viewing by their children. The "3-Hour Rule" asks commercial broadcasting stations to air at least 3 hours of programming specifically designed to educate and inform children 16 years of age and younger, to gain an expedited license renewal (FCC 1996; Jordan 2000). The V-chip (V stands for violence), which can block out television programs with certain ratings, became mandatory in all television sets manufactured since January 1, 2000.

In most American homes, three parents mind the children: the father, the mother, and the television or video games. Indeed, with the increasing incidence of single-parent (especially female-headed) families, television may have replaced father for many children. Certainly, the generations born in the 1960s and later are the product of the TV age.

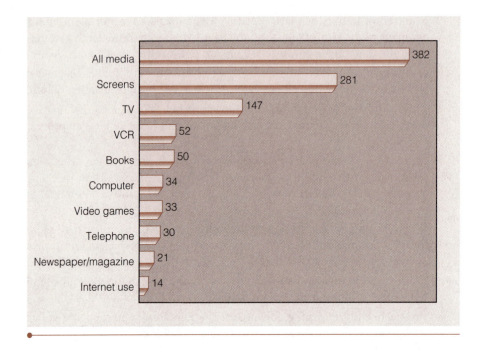

For some children, television has become a surrogate parent. The child can be taught and entertained by television as well as by the parent. Television can free the parent from child-care responsibilities for short periods of time, allowing the parent time for self-improvement and revitalization. Ninety-eight percent of American families have TV sets, with the average family owning 2.75 sets (Woodard and Gridina 2002). Twenty percent of 2- to 7-year-olds, 46 percent of 8- to 12-year-olds, and 56 percent of 13- to 17-year-olds have TVs in their bedrooms (Gentile and Walsh 2002). Fifty-seven percent of single parents report children who have their own TV, compared to 45 percent of households with multiple caretakers present (Woodard and Gridina 2002, 14–15).

Television is a wondrous medium that allows the viewer to walk on the moon with the astronauts, enjoy the wild animals via a safari through the African Serengeti Plain, watch the birth of a baby, and visit the seven wonders of the world. Perhaps we are fortunate that TV programming is often inferior and second rate. If it were always as good as it could be, viewers would never turn off the television set.

Because TV viewing has a strong impact on viewers, especially children, it is important for parents to consider its place in their home. How can it be used to promote social, rather than antisocial, behavior? How can it improve children's cognitive development, rather than retarding it? A few concerned parents who object to the negative influence of television opt to have no television. Yet, as their children grow and begin to visit friends, this strategy can backfire because the children end up watching at friends' houses, thus limiting parental control over TV input. Because 98 percent of American homes contain television, the best course of action for parents is to set up supervised viewing for their children. First, parents will need to consider the role of television in their own lives. For instance, if a mother has the television turned on all day for company or the father watches sports all weekend, they will need to decide whether they want to continue these practices.

Once parents determine their preferences about television's role in their own lives, they will need to set appropriate ground rules for their children. These rules should include the following:

1. How much time per day and per week the children may devote to television.
2. The actual time of the day or evening that television may be watched.
3. The kinds of programs that may be watched.
4. The amount of adult attention and discussion to be given during and after TV viewing (such parental interaction with the child has been found to mitigate the most negative effects of television).
5. Whether television will be used as a reward to influence other kinds of behavior besides TV viewing.

Such rules can also be applied to computer and video game usage.

Simply to dismiss television as harmful to children is to miss out on using one of the most powerful modern inventions for the good of children. The potential for good in television is every bit as large as its potential for harm if it is properly supervised.

What negative effects does all of this TV viewing have on children? Essentially, there are four major problems: the effect of commercials, the content of programs (especially violence and sex), the time spent in passive observation, and the possible developmental effects on very young children.

Small children accept commercial messages noncritically because their thinking processes are not advanced enough to make judgments about truth and fantasy. Because young children believe everything they see in commercials, they are easily manipulated into wanting everything they see. The parents are, therefore, placed in the position of resisting the constant demands of their children, created by TV advertising.

The content of TV programs is even more controversial. As early as 1952, congressional hearings investigated the amount of violence on television. Over the years, a great deal of research has been conducted on the violence question, yet the answers remain somewhat clouded (C. A. Anderson and Bushman 2002; Blum et al. 2000; Bushman and Anderson 2001; Ferguson 2000). Nevertheless, most researchers now conclude that there is at least a moderate causal relationship between TV/video game violence and later aggressive behavior (Thomas 1995). Unfortunately, the level of violence on television has not dropped. Although many parents rely on the various media rating systems to guide their children's movie, TV, and other media watching, the ratings system often entices, rather than discourages, children to watch if their parents are not monitoring their TV behavior (Bushman and Cantor 2003).

A child is not born knowing instinctively the mores, rules, and laws by which his/her society is governed. Each child must be taught by adults the necessary behaviors that will allow the society to function. Researchers suggest that, by watching a great deal of violence on TV in the early years, a child may come to accept violence as a normal part of life. A child who accepts violence as normal and, therefore, does not internalize social rules against violence, obviously stands a good chance of acting in a violent manner.

It appears that a minimum amount of socialization against violence must take place early in a person's life, or a sense of social responsibility about reducing violence cannot be instilled. Violent adults who come into contact with authorities have a conspicuous lack of parenting. They have often been raised in a series of foster homes and have had little relationship with their parents if, in fact, they know who both of their parents are.

In addition to watching TV, playing video games, and visiting the Internet, a number of trends that affect children may be combining to reduce parenting of American children, such as increasing unwed pregnancy and more fatherless homes, the movement of mothers into the workforce, and the high divorce rate. Although actual child abuse makes the headlines, child neglect is the greater problem in modern America.

Parenting is increasingly being done by babysitters, day-care programs, television, and the child's peers. Babysitters and day-care programs can certainly care for children and can also work to socialize children. But the day-to-day caring and intimacy of the family that gives a child love, security, and an identity may be missing. In this case, especially for older children, their peers may become their source of identity, as evidenced in adolescent gangs. Unsocialized peers cannot socialize other children. This can only be done by a responsible, socialized adult.

Compare the top seven public school problems listed by teachers in the 1940s with the top seven problems listed by teachers today. The comparison is frightening because our children are the future of America.

1940s	1990s
1. Talking out of turn	Drug abuse
2. Chewing gum	Alcohol abuse
3. Making noise	Pregnancy
4. Running in the halls	Suicide
5. Cutting in line	Rape
6. Dress code infractions	Robbery
7. Littering	Assault

Television, video games, and Internet sexual content available to children are other areas of concern for most parents. As many as 64 percent of television shows during the 2001–2002 season contained sexual content (in 1997, about 56 percent contained sexual content). In addition, almost half of the sexually in-volved characters had just met, and had no established romantic relationship (Ballie 2001b; Kunkel 2003). Only about 15 percent of these shows addressed safety during sex. This is up from 9 percent in 1997. Among shows with sexual content involving teenagers, about 34 percent include a safer sex reference, which is up from 18 percent in 1997. Because many teens say that TV plays a helpful role at least some of the time in learning about sexual health, it may be more important to improve TV shows' sexual content than simply to reduce the amount of sexual content, as is so often suggested (Kunkel 2003).

Internet pornography (estimated to be 260 million Internet pages; see pages 252–253) is easily available, unless parents take steps to have it blocked. Although pornographic sites are "pay for entry," the site ads themselves are immediately available and depict every sexual activity imaginable. Sex chat rooms, and the possibility of a child or teen being lured into an actual meeting with someone they met at a chat site, make it imperative that parents closely monitor their children's use of the Internet.

Violence and sex aside, the content of TV programming and the Internet allow children access to all of the social and human problems from which they were shielded to some extent in the past. In a way, television is doing away with childhood. What makes one group (adults) different from another group (children) is the knowledge they possess. When knowledge was gained through the written word, a child had to learn to read before newspaper stories of violence or corruption could be understood. Today, via pictures and spoken words, children can understand immediately. The constant reporting of all the negative happenings in the world leads to skepticism about the worth of adult society. Children who view a lot of television are more likely to view the world as a mean and scary place. To what extent does viewing so much negativity in the world undermine a child's belief in adult rationality, in the possibility of an ordered world, in a hopeful future? Without such hopes and beliefs, a person is unlikely to work to better the human condition.

Also, the way in which TV content is presented poses difficulties for the child. Everything is handled quickly. For example, many books and hundreds of research papers have been written about AIDS, but when television does an in-depth study of the AIDS epidemic, it may last 45 seconds on each of four different nights. Thus, a major social problem is condensed into 3 minutes, interspersed among commercials. How important can the problem be? The classroom teacher who must ask pupils to concentrate so they can learn is hard put to compete against the excitement and constant change of TV programming. Many teachers report that holding students' attention has become their major classroom problem. There is some recent evidence that 1- to 3-year-old children who watch TV daily appear more at risk for attention problems by age 7 than children who watch no TV at all (Christakis 2004; D. Jackson 2004, G3).

Hours of passive observation deprive the child of social interaction and of the physical activity necessary to grow in a healthful manner. Children may now enter school without social skills because they have had so little social interaction. Many children also appear in poor physical condition and are too often overweight, apparently because they have little active play and exercise time. Some families who voluntarily gave up TV viewing report that their children played and read more, that siblings fought less, that family activities became more common, and that mealtimes were longer.

Child Rearing, Discipline, and Control

Most new parents tend to deny that their parenting methods are much the same as their own parents used. Yet, research indicates that quality of parenting received as a child; satisfaction with the parent-child relationship; education; and various parenting beliefs predict the parenting of both mothers and fathers (Simons et al. 1993, 91).

Often, parents accomplish control of their children in a haphazard manner. If the parents liked their own parents, they tend to copy their child-rearing methods. If the parents disliked their own parents' methods, they tend to do the opposite. In either case, the parents' own experiences as children influence how they parent.

Parents overwhelmingly use negative techniques such as scolding, spanking, and threatening to control their children. Unfortunately, negative control methods tend to have negative side effects, because they serve as model behaviors (Straus 1999; Straus and Stewart 1999). A child who is screamed at tends to become a child who screams to get his/her way. A child who is punished violently tends to learn that violence is the way to change another's behavior. Hostility, low

WHAT DO YOU THINK?

1. How important do you think parenting is to the success of a society? Explain.

2. Do you know any people who seem unsocialized and appear to lack values and a conscience?

3. If you do, do you think the person was lacking in parenting as a child?

4. Why do you think it is necessary to instill values at an early age in children, if they are to grow up to be responsible adults?

5. If and when you have children, how will you control their TV viewing and video game playing?

self-esteem, and feelings of inferiority and insecurity are frequent reactions of children who are reared by basically negative methods.

Rather than having only a few control techniques and using them automatically, parents must develop a variety of well-understood child-rearing methods. Each child and each situation will be unique, and parents should try to react in accordance with the specific situation.

The parent should first try to understand the child and the specific problem, then identify what changes are necessary, and finally accomplish the changes in the best possible manner. Thus, child rearing becomes a rational, thoughtful, directive process rather than an irrational, reactive process.

In general, children can be controlled by using many different methods that vary in intensity from mild to strong. Parents often use a strong method such as punishment when, in fact, a mild method such as distraction might have worked equally well. By first trying milder methods, the negative side effects of punishment can be eliminated or at least reduced. Table 10-2 outlines a continuum of mild to strong methods. One must consider the age of the child when using this table. Two-year-olds will not understand item 3c, an appeal to their sense of fair play, because they are too young to grasp this ethical concept.

TABLE 10-2

Making Decisons:
Mild to Strong Child-Control Methods

1. Supporting the child's self-control
   a. Signal interference (catch the child's eye, frown, say something)
   b. Proximity control (get physically close to the child)
   c. Planned ignoring (children often misbehave to get attention, and if they don't get it, they will cease the behavior because they are not getting what they want)
   d. Painless removal (remove the child from the problem source)
2. Situational assistance
   a. Giving help
   b. Distraction or restructuring a situation
   c. Support of firm routines
   d. Restraint
   e. Getting set in advance
3. Reality and ethical appraisals
   a. Showing consequences of behaviors
   b. Marginal use of interpretation
   c. Appealing to sense of reason and fair play (not useful until child is intellectually able to understand such concepts)
4. Reward and contracting
   a. Rewards (payoffs) should be immediate
   b. Initial contracts should call for and reward small pieces of behavior (a reward for picking up toys rather than a reward for keeping room clean for a week)
   c. Reward performance after it occurs
   d. Contract must be fair to all parties
   e. Terms of contract must be clear and understood
   f. Contract must be honest
   g. Contract should be positive
   h. Contract must be used consistently
   i. Contract must have a method of change to cope with failures
5. Punishment (see "Making Decisions: Using Discipline and Punishment Effectively")

MAKING DECISIONS

Using Discipline and Punishment Effectively

Whether one agrees with the use of punishment as a means of teaching children, studies of parental control indicate that it is still a major method used by most parents. Unfortunately, punishment does not always work, and it causes anger and hostility.

If punishment is used, the parent needs to understand a few simple principles that help maximize its usefulness and minimize the negative side effects. If mild punishment is to be effective, an alternative behavior should be open to the child. For example, if a child is punished for turning on the television rather than dressing for school, the punishment will work better if the child knows when he/she can watch television. Then the child knows that there is another way to do what he/she wants that will not result in punishment. In addition, punishment works best if the child is not highly motivated. For example, a child eats just before dinner. If the child missed lunch, chances are he/she is very hungry (highly motivated), and punishment will probably not be very effective in keeping the child from eating. If there are alternatives and if the child is not highly motivated, following the guidelines based on psychological learning theory set forth here will help achieve desired goals, while keeping punishment to a minimum.

1. Consider the individual child and the potential negative side effects. *Example:* Randy reacts strongly to punishment, and the reaction lasts for a considerable length of time. He is better controlled by reward or distraction. If punishment is used, it is mild. Michelle is not sensitive to punishment. When she is punished, her behavior changes, and there are no lasting side effects.
2. Punish as soon after the act as possible. *Example:* Children are bright, but they have difficulty associating an act with punishment that comes many hours after the act. The mother who at 10 a.m. tells the child to "Just wait until your father comes home, you will get it," does little more than turn father into an ogre. By the time the father arrives home, little change of behavior will be derived from punishment.
3. If possible, let the punishment flow from the act. *Example:* If a child is constantly warned when he reaches toward a hot stove, behavior change takes time. The child who touches the hot stove is immediately punished by his own action, understands what the word *hot* means, and has

learned in one trial. Obviously, a parent cannot burn a child's hand, but if a nonharmful situation where punishment flows from the child's own behavior occurs, it will be the most efficient in changing behavior.

4. Be sure that children understand what their alternatives are. *Example:* Many young children do not understand exactly what they have done and what new behavior is desired. Mary was punished when her mother found her at the cookie jar before dinner. Later Mary found some cookies on the counter and was again punished. Mary did not understand that she was being punished for eating cookies the first time. She thought her mother did not want her around the cookie jar for fear she might break it. Also, she did not understand that she could have cookies after dinner.
5. Keep the punishment mild and devoid of emotion. *Example:* Jimmy's mother lost her temper and spanked him. By losing her own temper, she increased the emotional atmosphere, causing Jimmy to become even more upset. She also modeled overt anger.
6. Try to punish the act, not the child. *Example:* "You are a bad boy. We don't love you when you are bad." This is a threatening and upsetting statement to a child and really is unnecessary. Generally, it is the particular act that is bad, rather than the child. When the child changes the behavior, there is no longer a need for punishment, and the child's relationship should immediately return to normal. When punishment is directed only at the act, the child will not continue to be punished by thinking he/she as an individual is bad and unworthy of love ("I don't like your behavior, so please change it. I like you, however.").

What Do You Think?

1. What were the major methods of control used by your parents when you were a child?
2. How did you react to these methods as a child?
3. What do you think of them as an adult? Did they work? Why, or why not?
4. Which methods would you use with your children? Why?
5. What will you do if you and your spouse disagree on control methods?

Directing a child's behavior ahead of time is preferable to, and much easier than, being unprepared and surprised by a child's behavior and then trying to react properly. For example, if you give a child a difficult task, stay close (1b, proximity control) so that you can offer help if it is needed (2a).

Firm routines (2c) reduce conflict and the number of overt decisions required. For example, an orderly routine at bedtime accomplishes toothbrushing, elimination, getting into pajamas, story reading, and lights off. In a sense, the child is on automatic pilot, and little conflict arises once the routine is established. Such routines are only helpful in certain areas of life, and probably work best with the preschool and early school children. Distraction (2b) is helpful with small children; they usually have short attention spans and are easily shifted from one focus to another. Saying no to a child immediately creates a confrontation. In contrast, presenting an alternative and saying "do this" does not necessarily create such a confrontation. When a child is touching an expensive art book one can say, "Why don't you color with these crayons?" while handing them to the child. Normally, the child will turn his/her attention to the crayons rather than the art book. Parents in strong families emphasize positive behaviors, attitudes, and moral character and avoid overemphasis on minor details. In general, building on a child's strengths (asset building) is more rewarding than focusing only on problems and the child's weaknesses.

Spanking

The use of spanking to control children is as old as history itself. However, over the past several decades, M. Straus (1994, 1996, 1999) and others have suggested that any kind of corporal punishment is detrimental to the child. They link childhood corporal punishment to about every kind of adult problem imaginable: depression, suicide, drug abuse, spousal violence, child abuse, and so on. Some child guidance books go so far as to discuss spanking under the heading of child abuse. Such extreme rejection of any spanking fails to differentiate between kinds of spanking. Mild spanking with parental emotional support, no anger, and no injury probably causes little harm. It is clear that corporal punishment becomes physical abuse when it results in injury to the child, is done without provocation by the child, or occurs when the parent is out of control—perhaps because of drug or alcohol use. Review of a broader spectrum of research on corporal punishment yields a more balanced perspective (Gershoff 2002).

One study that followed children from preschool to age 14 suggests that mild, occasional spanking can be an effective method to stop a child from misbehaving (Ballie 2001a). Yet, the researchers go on to suggest that when parents are loving and firm and communicate well with their children, the children will be well adjusted whether or not the parents spanked them mildly as preschoolers (Baumrind 1996a, 1996b). A more recent study, comparing the results of spanking young Hispanic, Black and White children, found no evidence that the relation between spanking and behavior problems is related to race or ethnicity. The researchers did find that a high level of parental emotional support moderated the impact of mild spanking, by influencing the child's interpretation of physical discipline. "The child may be less likely to view spanking as harsh, unjust, and indicative of parental rejection when relations with the parent are generally warm and supportive (McLoyd and Smith 2002, 51). However, they pointed out that the moderating effect of parental warmth and support at the time of spanking does not negate some of the reasons cited by other researchers for discouraging parental reliance on physical punishment. If a parent chooses to use some corporal punishment, then following some of the learning principles suggested in this section will help to reduce negative side effects. However, using some of the methods suggested in Table 10-2 would most likely eliminate the need of corporal punishment altogether. The overall incidence of spanking has been reduced over the past several decades, both by the negative judgments about it and by the increasing parental understanding of other child-control methods.

The Growing Child in the Family

The essence of children is growing, changing, maturing, and becoming, rather than sameness. The essence of parenting is caring for and contributing to the life of the next generation. Parents must constantly change in their relationship to the growing child. Just about the time they have adapted and learned to cope with a totally dependent infant, they will need to change and learn to deal with a suddenly mobile yet still irresponsible 2-year-old. Then come the school years, when the increasing influence of peers signals declining parental influence. Puberty and adolescence introduce the launching stage, when once-small, dependent children go into the world on their own, ultimately to establish new families and repeat the cycle. As a child grows and changes, the family also changes. For example, the mother usually remains close to the child during its infancy. During the elementary school years, the parents often become chauffeurs—taking the child to a friend's home, music lessons, after-school sports, and so on. Thus, change and growth in the child mean parental and family change.

One way of viewing these changes is to see them as a series of social and developmental situations involving encounters with the environment. These situations involve normal problems that children must solve if they are to function fully. This section focuses on psychosocial stages, rather than biological developmental stages, though the two are interrelated.

Erik Erikson (1963) identifies eight psychosocial developmental stages, each with important tasks that describe the human life cycle, from infancy through old age (Figure 10-3). Erikson's stages are theoretical, of course, and there is some controversy over the exact nature of developmental stages. Nevertheless, the idea of stages is useful in helping parents understand the changing nature of the growing child and themselves. In each of these stages, children must establish new orientations to themselves and their environment, especially their social environment. Each stage requires a new level of social interaction and can shape the personality in either negative or positive ways. For example, if children cope successfully with the problems and stress in a given stage, they gain additional strengths to become fully functioning. On the other hand, if children cannot cope with the problems of a particular stage, they will, in effect, invest continuing energy in this stage, becoming fixated to some extent or arrested in development. For example, an adult who always handles frustration by throwing a temper tantrum has failed to move out of the early childhood stage, when temper tantrums were the only manner of handling frustration. Such an individual is not coping successfully with the stress and problems of adult life.

Notice in Figure 10-3 that each stage can be carried down the chart into adulthood. For example, trust versus mistrust influences all succeeding stages. A person who successfully gains basic trust is better prepared to cope with the ensuing developmental stages. The reverse is also true. A mistrusting individual will have more trouble coping with ensuing stages than a trusting person.

Infancy: The First Year

The Oral-Sensory Stage: Trust versus Mistrust   In the first years, children are completely dependent on parents or other adults for survival; they cannot contribute to the family because their responses to the environment are quite limited. Thus, their development of trust depends on the quality of care they re-

	Trust vs. Mistrust (mothering, feeding)	Autonomy vs. Shame, Doubt (toilet training, self-control)	Initiative vs. Guilt (increased freedom and sexual identity)	Industry vs. Inferiority (working together, school)	Identity vs. Role Diffusion (adult role)	Intimacy vs. Isolation (love and marriage)	Generativity vs. Self-Absorption (broadening concerns beyond self)	Integrity vs. Despair
1 Infancy Oral-sensory 1st year	Trust vs. Mistrust (mothering, feeding)							
2 Toddler Muscular-anal 2 to 3 years		Autonomy vs. Shame, Doubt (toilet training, self-control)						
3 Early Child Locomotor-genital 4 to 5 years			Initiative vs. Guilt (increased freedom and sexual identity)					
4 School Age Latency 6 to 11 years				Industry vs. Inferiority (working together, school)				
5 Puberty and Adolescence 12 to 18 years					Identity vs. Role Diffusion (adult role)			
6 Young Adult						Intimacy vs. Isolation (love and marriage)		
7 Adulthood, Middle Age							Generativity vs. Self-Absorption (broadening concerns beyond self)	
8 Maturity Old Age								Integrity vs. Despair

FIGURE 10-3 *Erikson's Eight Developmental Stages*

SOURCE: Erikson 1963.

ceive from their parents or the adults who care for them. The prolonged period of dependence makes the child more amenable to socialization (learning the ways of the society). Sometimes, however, a child will be too strongly or wrongly socialized in the early stages and thus will suffer from needless inhibitions as an adult. Such inhibitions are easily recognized by modern youth as *hang-ups*.

The first year, when the infant must have total care to survive, is a difficult adjustment for many new parents because they previously enjoyed more personal freedom and had time to devote to each other. A newcomer usurps that freedom overnight and has first call on their time. Not only does the infant demand personal time and attention, but many other considerations also arise. Going to a movie now entails the additional time and cost involved in finding and paying a babysitter. Taking a Sunday drive means taking along special food, diapers, a car seat, and so forth. But for the couple who really want children, the challenge

New mothers meet to discuss their problems and joys.

© Pamela Cox

and fun of watching a new human grow and learn can offset many of the problems entailed in their new parental roles.

During this first year, the husband must adjust to sharing his wife's love and attention. As we mentioned earlier, it is not uncommon for a new father to feel some resentment and jealousy over this change in his position. It is important for the new parents to arrange times to be by themselves, to pay attention to their own relationship. In the past, this was easier to do because relatives who could watch the infant were often close by. Today, spending time together usually means paying someone to take care of the child. Such money is well spent if it gives the couple an opportunity to improve their own relationship. Their relationship is the primary one; if it is good, the chances are greater that the parents-children relationship will be good also.

Children learn trust through living in a trusting environment. This means that their needs are satisfied on a regular basis and that interactions with the environment are positive, stable, and satisfying. Parents who have a good relationship are better able to supply this kind of environment.

The first year of life is called the oral-sensory stage, because eating and the infant's mouth and senses are his/her major means of knowing the world. New parents are often concerned that everything seems to go into the infant's mouth for the first year or so.

The Toddler: 2 to 3 Years of Age

The Muscular-Anal Stage: Autonomy versus Shame and Doubt
As children develop motor and mental capacities, opportunities to explore and manipulate the environment increase. From successful exploration and manipulation emerges a sense of autonomy and self-control. If the child is unsuccessful or made to feel unsuccessful by too-high parental expectations, feelings of shame and doubt may arise. This is a time of great learning. Children learn to walk, talk, feed, partially dress themselves, say no, and so forth. If parents thought the in-

fant was demanding, they quickly learn what demanding really can be with their 2- and 3-year-old. There never seems to be a minute's peace. Parents seem to spend an inordinate amount of time chasing the newly mobile child. The relationship between the parents can suffer during this time because of the fatigue and frustration engendered by the child's insatiable demands. On the other hand, it is exciting to watch the child's skills rapidly developing. First steps, first words, and curiosity all make this period one of quick change for child and parent alike.

During this stage, parents should try to create a stimulating environment for the child. As mentioned earlier, stimulation appears to enhance learning skills. Alphabet books, creative toys that are strong enough to withstand the rough treatment given by most 2- and 3-year-olds, picture books, and an endless answering of questions all help stimulate the child. Toilet training, which also occurs during this period, needs to be approached positively and with humor if it is to be easily accomplished. The achievement of toilet training is a great relief to parents, of course, because the messy job of diaper changing and cleaning is over.

The name of this stage, muscular-anal, denotes the increasing activity of the child and the toilet-training tasks.

Early Childhood: 4 to 5 Years of Age

The Locomotor-Genital Stage: Initiative versus Guilt
During early childhood, children become increasingly capable of self-initiated activities. This expansion of children's capabilities is a source of pride for parents, and exciting to see. With each passing month, the children seem to be more mature, have more personality, and become more fun to interact with. As their capabilities and interests expand, so must the parents'. Play and pretend foster physical fitness and also social and cognitive development (Azar 2002a, 2002c). The extent of children's energy is a constant source of amazement and bewilderment to often-tired parents. However, school is just around the corner, and with it comes a little free time.

Children who do not increase their capabilities, perhaps because of accident or illness, or because of overprotective parents, may feel guilty and inadequate about their school performance.

Locomotor indicates that the child is now very active in the environment; genital refers to interest in and exploration of sex organs. This latter characteristic is sometimes unnecessarily upsetting to parents who find their 4-year-old boy or girl naked, playing doctor or "I'll show you mine if you show me yours" with a 4-year-old of the opposite sex who lives next door.

School Age: 6 to 11 Years of Age

The Latency Stage: Industry versus Inferiority
At last the children are in school, and for a few hours a day, the house is peaceful. Now the children's peers begin to play a more active role in the family's life. Relationships broaden considerably as parents also become PTA members, den mothers, or Little League coaches. This stage is often a period of relative family tranquility as far as the child is concerned.

The children's increasing independence also affects the parents. They find that what other parents allow their children to do becomes an important influence on their own children. "Mom, everyone else can do this, why can't I?" becomes a constant complaint of children trying to get their own way.

The children have new ideas, a new vocabulary, and broader desires, all of which can conflict with parental values. Reports about the children's behavior may also come from other parents, teachers, and authorities. How the children are doing at school becomes a source of concern.

Children become increasingly expensive as they grow. They eat more, their clothes cost more, and they need more money for school and leisure activities. As the latter expand, many parents find themselves juggling schedules to meet the demands of after-school sports, PTA meetings, Little League, and music lessons.

Yet, for most families, the elementary school years go smoothly. Children become more interesting, more individual, and increasingly independent. Important for the parents is their own increased freedom, because the children are now away from home for part of the day.

During this time, it is essential for parents to work together in child rearing. Children at this stage are aware and insightful and can play their parents against each other to achieve their ends. Children can also cause conflict between their parents, especially if the parents differ widely in their philosophy of child rearing. Often, a well-functioning, supportive husband-wife relationship can buffer or inhibit the negative impact of a difficult child. Couples who do not enjoy a supportive relationship may find that fighting about the children is one of the major points of disruption in their relationship.

Latency refers to the general sexual quietness of this stage, although there is more sexual activity than was first thought.

Puberty-Adolescence: 12 to 18 Years of Age

Identity versus Role Diffusion The tranquility of the elementary school years is often shattered by the arrival of puberty. The internal physiological revolution causes children to requestion many earlier adjustments. **Puberty** refers to the biological changes every child, regardless of culture, must pass through to mature sexually. **Adolescence** encompasses puberty and also the social and cultural conditions that must be met to become an adult. The adolescent period in Western societies tends to be exaggerated and prolonged, with a great deal of ambiguity and marked inconsistencies of role. Adolescents are often confused about proper behavior and what is expected of them. They are not yet adults, but at the same time they are not allowed to remain children. For example, an 18-year-old boy may enter the armed services and participate in battle, yet in many states he may not legally drink beer. The fact that prolonged adolescence is a cultural artifact does not lessen the problems of the period.

Historically, the boy who left home at 15 years of age to make it on his own was a hero of sorts in the society if, indeed, he did make it. Today, the parents might be arrested for child neglect. Longer schooling, postponing work, and living at home for a longer period all combine to increase the need of parental support for a much longer period of time than was necessary historically (A. Booth et al. 1999). Thus, the parental stress of adolescence has been increasingly prolonged in modern America.

The problems of puberty and adolescence fall into the following four main categories:

1. Accepting a new body image and appropriate sexual expression.
2. Establishing independence and a sense of personal identity.
3. Forming good peer-group relations.
4. Developing goals and a philosophy of life.

puberty
Biological changes a child goes through to become an adult capable of reproduction

adolescence
The general social and biological changes the child experiences in becoming an adult

H I G H L I G H T

To Tattoo or Not to Tattoo: Out Damned Spots!

Tattooing by adolescents and young adults became very popular in the 1990s. Unfortunately, for those who indulged this fad and later as adults wish to remove them, there is a rude awakening. Doctors are receiving increasingly more requests from adults who wish to remove a tattoo. A study done by the University of North Carolina suggests that anywhere between 15 and 25 million Americans have tattoos, and about half of these individuals want them removed. Dermatologists say that this is the fastest growing area of their business.

The good news for individuals wanting their tattoo removed: "Yes, it can be removed." The bad news is that removal will take up to 2 years and sometimes more, depending on the colors and if it is a good, professionally drawn one. Removal will cost several thousand dollars. Removal is painful. An average tattoo the size of the hand palm requires a visit to the dermatologist every month or 6 weeks for the entire 2 years. The pain level in the treated area is described as feeling like it has been repeatedly snapped by a rubber band. The treated area will scab or blister much like a burn does.

Tattooing a girlfriend's or boyfriend's name is especially problematic for a young person. What does he/she do when there is a breakup—find another person with the same name?

In general, parents should counsel their children against tattooing and body piercing, because these decisions made when young are often regretted in adulthood. In addition, social rejection and possible health problems also arise (Holleman 2003).

During this stage, peer influence becomes stronger than parental influence. What friends say is more important than what parents say. For example, friends and the culture put pressure on young girls to create a thin body. This can create eating disorders, such as anorexia and bulimia, both of which can be life-threatening health problems (DeAngelis 2002; Haworth-Hoeppner 2000).

The major problem facing parents now is how to give up their control, how to have enough faith in the child to let go. However, letting go is made more difficult by the fact that adolescent children and young adults tend to have more trouble with the law than older persons. Cigarettes, alcohol, drugs, sexual expres-

F I G U R E 10-4 *Portion of the Individual Differences in Health-Risk Behaviors Explained by Demographics*

Source: Blum et al. 2000.

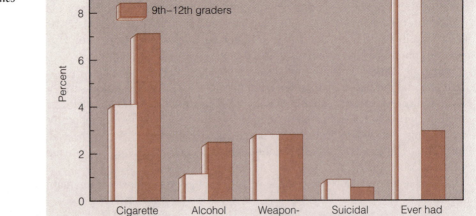

sion leading to possible pregnancy, criminal activity, vehicular accidents, and gang participation on the part of their children often cause parents grief (Figure 10-4; Blum et al. 2000; Daw 2001; U.S. Department of Health and Human Services 2003).

Parental monitoring of preadolescent behavior is helpful in setting a basic behavioral foundation for young people. It is hoped that preadolescents will then make proper behavioral choices outside of parental purview (Longmore et al. 2001).

Parents, who by now are entering middle age, are also requestioning and re-ordering their lives (see Chapter 11). They must begin to think about coping with the "empty-nest" period of their lives as their adolescent children grow into young adulthood and leave home to establish their own families. Most research suggests that this is the stage of greatest family stress. Families lose some cohesion at this time, although enduring family qualities of friendship and emotional support tend to continue across generations (Baer 2002; Cui et al. 2002). In general, parents who report higher competence in rearing adolescents have children who report more parental monitoring and responsiveness and less psychological control. Parents tend to equate adolescent openness to socialization and normal stress as signs of competent parenting (Bogenschneider et al. 1997).

The Young Adult: 19 to 30 Years of Age

Intimacy versus Isolation   Although some children leave the family in their late teens, many remain longer, especially if they choose to go on to higher education. Seeking a vocation and a mate are the major goals of this period. In the past, success in these two tasks ended children's dependence on the family. Today, however, parents often continue to support their children for several more years of schooling, and may support a beginning family if a child marries during school. This can be a financially strained period for the family. In addition, studies indicate that the presence of unmarried sons and daughters over age 18 in the home strains their parents' relationship. Both husbands and wives often report that the period when grown children are living at home is a dissatisfying time in their marriage. However, most families seem to cope well with the return of older children to the parental home (Mitchell and Gee 1996), even though

---

WHAT DO YOU THINK?

1. Have you ever caused trouble between your parents? Explain.

2. What could your parents have done to avoid the trouble you caused?

3. Overall, do you think that having children improved your parents' relationship and marriage? Why, or why not?

4. Assuming that you have children when you marry, what are some of the ways that they might positively influence your relationship with your spouse? Negatively influence your relationship?

HIGHLIGHT

Adolescent Hormones and Brain Functioning

Adolescent behavioral problems, fast-changing moods, poor decision making, unhappiness, and rebellion have been blamed on "raging hormones" that accompany the physical transition to adulthood. Yet, recent research now finds that these adolescent behaviors are the product of two factors: a flood of hormones and also a lack of the cognitive controls needed for mature behavior.

By using high-powered magnetic resonance imaging (MRI) to map brain activity, scientists have discovered that the brain continues to develop (not in size, but in specialized regions) much longer than was previously thought. The best estimate at this time for when the brain is truly mature is 25 years of age. Thus, some youthful behaviors that were thought to be just hormone related (sensation seeking, poor decision making, lack of responsible behavior, and so on) now appear more related to the failure of certain parts of the brain to reach full adult development.

SOURCE: Adapted from Kraft-Corbis 2004, 55–65.

For Better or For Worse® by Lynn Johnston

there will be relational strains. Older adult children who return to live with their parents usually do so out of their own needs, not the needs of their parents. Economic worries and problems brought on by divorce are two major reasons adult children return home to live with their parents.

Ideally, by the time their children are grown, parents like and love them and take pride in a job well done. Now, after 20 or more years, the marital relationship again concerns just two people, husband and wife (assuming all the children have left the nest). Sometimes, parents discover that their relationship has been forfeited because of the urgency of parenthood. In this case, they face a period of discovery and rediscovery, of building a new relationship, or of emptiness. Parents must work to maintain their relationship throughout the period of child rearing. If they fail to do this, their relationship may become simply "for the children." When the children leave, the marital relationship loses its foundation and may fail. The last two of Erikson's stages are discussed in Chapter 11.

HIGHLIGHT

Oh No, John Is Back Home Again

John is 24 and living at home with his parents, William and Jan. John works and contributes a little toward the food budget. However, his parents find that his living at home strains their relationship. John has a new car and insists that because it is new, it should have a place in the garage. His father often transports clients in his car, and though his car is older, he thinks it should be parked inside so it will remain clean. Jan often intercedes in her son's behalf, "Oh, William, it's his first new car, let him use the garage." This angers William, and he sometimes feels as though it's two against one. He also has wanted a den for years and would like to convert John's bedroom into one. When he encourages John to look for his own

place, Jan says that he is "throwing John out." "We should be happy that he wants to be with us," she adds. However, Jan does find it difficult when John has his friends over. They take over the living room and television, leaving her and William with little to do but retreat to their bedroom.

What Do You Think?

1. Who should keep their car in the garage? Why?
2. Do you still live with your parents? What kind of problems does this pose for you? Your parent?
3. How are the problems of living with parents changed when a child returns home after being on his/her own?

WHAT RESEARCH TELLS US

Young Adults Living with Parents

More young adults are living at home with their parents (14.4 percent of family households had their own children, 18 and older, living at home; U.S. Census Bureau 1998c, 2001a). Between 1960 and 2000, the proportion of men aged 18 to 24 who lived in their parents' home increased from 43 to 56 percent. In 2000, 43 percent of women aged 18 to 24 lived at home with their parents. Of these, 4 percent had children of their own also living in the home. Even among "older" young adults, aged 25 to 34, there was an increase from 8 percent of men living with their parents in 1970 to 12 percent in 2001. About 5.4 percent of women in this age group lived with their parents in 2001 (U.S. Census Bureau 2001a).

Broader Parenting

Perhaps the decreased marital happiness experienced by some parents results from the American nuclear family, in which the parents (or one of them, in a single-parent home) are expected to give a child total parenting. But most societies do not expect one father and one mother to supply 100 percent of a child's needs. Grandmothers and grandfathers, aunts and uncles (blood relatives or not), older siblings, and many others also supply parenting to children.

In many American suburban families, children and their parents (especially children and their mothers) are basically alone together. The parents cannot get away from the small child for a needed rest and participation in adult activities, and the child cannot get away from the parents. The nuclear family is often isolated from friends and relatives who might occasionally serve as substitute parents.

What if parents find they can't be good parents for a particular child? The child will be trapped in the setting for years. In the past, this child might have lived with relatives who were better suited to act as his/her parents. Children were sometimes traded between families for short periods, or spent summers on a farm, and were assisted in growing up by numerous adults and older children. This extended family meant that no one person or couple was responsible for total parenting. Some critics of the nuclear family suggest that the nuclear family pattern of parents and children always alone together creates problems in children, rather than preventing problems. In that the reality of their own nuclear family is the only reality small children may know, they cannot correct misconceptions. Because they have no other basis with which to compare the actions of adults, recognizing problems and reorienting themselves is difficult.

Studies find that there is parenting diversity provided by grandparents (Chapter 11), aunts and uncles, and even friends and neighbors more often than might be realized (Hunter 1998). Black children are much more likely than White children to live in, and benefit from, extended family arrangements (Demo and Cox 2001, 106). One in five Black families are extended, compared to one in 10 White families (Glick 1997).

Broader parenting might be supplied by trading care of children with other families, setting up volunteer community nursery schools, establishing business-supplied day-care centers for workers, and expanding the nuclear family to include relatives. The idea of mentoring young people by older adults is important, especially for those children being raised in a one-parent family. Your author had a wonderful mentor when he was a young teenager, who taught your author to ski and sail. Later on, when your author was just married, he helped by teaching our new family about the importance of investments and offering other guidance that proved invaluable.

Parents without Pregnancy: Adoption

The 2000 census reported that adopted children under age 18 years accounted for approximately 2.5 percent of all children under age 18 years. If adoptees over age 18 years are also included, there are about 2,058,915 adopted children living with their adopted parents or parent (U.S. Census Bureau 2003a, Table 1).

There are several types of adoptions:

1. *Public*. Children in the public child welfare system are placed in permanent homes by government-operated agencies, or by private agencies contracted by a public agency. Approximately 15 to 20 percent of adoptions are public adoptions.
2. *Private*. Children are placed in nonrelative homes through the services of a nonprofit agency, which may be licensed by the state in which it operates.
3. *Independent*. Children are placed in nonrelative homes directly by the birth parents, who use a facilitator, medical doctor, clergy member, or attorney. Adoption types two and three account for about 38 percent of all adoptions.
4. *Kinship*. Children are placed in relatives' homes, with or without the services of a public agency.
5. *Stepparent*. Children are adopted by a stepparent. Adoption types four and five account for approximately 42 percent of all adoptions (National Adoption Information Clearinghouse 2004).

Adopting from a foreign country that has a well-organized program usually speeds the adoption timetable; the adoptive parents often receive the child within 1 year. The number of immigrant visas issued to orphans coming to the

Adoption helps many adults to have children and many children to have parents.

© David Young-Wolff/PhotoEdit

United States for adoption increased from about 7,000 in 1990 to nearly 18,000 in 2000. About half of these children are from Asia, about one-third from Latin America, and another 16 percent from Europe. Korea was the largest single-country source of foreign-born, adopted children in 2000 (Selman 2002; U.S. Census Bureau 2003a). An international treaty (Hague Convention on Intercountry Adoption) that creates adoption ground rules has helped to regulate some of the abuses that can be tied to an international adoption (N. Anderson 2000).

The continuing trend toward more openness (knowing the birth parents) in adoption; the increase in international, interracial, and special-needs adoption; and the growing number of single-parent and gay-lesbian adoptive families have generated more interest (and controversy) over the role of adoption in America (March and Miall 2000, 359). (The entire October 2000 issue of *Family Relations* is devoted to adoption.)

People adopt for many reasons. Mainly, couples adopt because they want children but cannot have children of their own. When parents are unable to care for their children, friends or relatives may adopt the children to give them a home. Couples who feel strongly about the problems of overpopulation may decide to adopt, rather than add to the population. A husband or wife may wish to adopt his/her spouse's children by a prior marriage, to become their legal parent as well as their stepparent.

The choice of adoption by the birth mother is difficult, because the mother has carried the child to term and actually given birth. She may feel grateful toward the adopting family, but she also may feel resentment for the new parents' ability to enjoy the child in her stead. However, adolescents placing their children for adoption tend to complete more schooling, are more likely to be employed, and also postpone further reproductive behaviors longer than parenting adolescents (B. Donnelly and Voydanoff 1996). Most often, the child will be better off living in a more mature and complete family.

The choice to adopt a child is just that—a reasoned decision, a choice made by a couple after a lot of deliberation and thought. As such, the decision-making process leading to adoption makes an ideal model that all couples desiring children can follow. Adoption takes time, and the prospective parents must meet certain requirements, such as family financial stability and proper housing. Although many older children and children with special problems are available for adoption, healthy infants are in highest demand.

Adoptive parents have both advantages and disadvantages, compared with natural parents. Adoptive parents may choose their child. To some degree, they can select the genetic, physical, and mental characteristics of the child. They can bypass some of the earlier years of childhood, if they desire. On the other hand, they do not experience pregnancy and birth, which help focus a couple on impending parenthood. Of course, some may consider this another advantage of adoption.

A unique parenting problem faced by adoptive parents is that of deciding whether and when to share the knowledge of adoption with the child. Experts believe the best course is to inform the child from the beginning, but this is sometimes difficult for parents to do. Most adoption agencies, such as the Children's Home Society, counsel prospective adoptive parents on how to tell the child. The parent-child relationship can easily be harmed if the child finds out about the adoption from others. The child's basic trust in the parents may be weakened, or even destroyed, if the parents aren't the ones to tell the child about the adoption.

At some time in their lives, many adoptive children feel the need to know something about their natural parents. Those adoptees who become preoccupied with finding their biological parents may do so for a number of reasons. Generally, girls are more concerned with their adoptions than are boys (Kohler et al. 2002). If the preoccupation is too strong, it may well cause trouble in the child's relationship with the adoptive parents, although this depends greatly on the strengths of the relationships within the adoptive family.

Although their situation is seldom mentioned, parents who give up their child for adoption also often feel the need to know what has become of their child. Parents who give a child up to an adoption agency, rather than using a direct adoption, may never know if the child is actually adopted.

For many years, legal adoption was a long, drawn-out, and costly process in which prospective parents went through a strenuous screening process to establish their parental suitability. Recently, such screening has been minimized. Federal legislation (the Adoption Promotion and Stability Act of 1996) includes a tax credit of up to $5,000 to couples who adopt a child. In addition, the bill makes it easier for couples to adopt children who have different racial and ethnic backgrounds (*Santa Barbara News Press* 1996). In a few cases, single persons and gay or lesbian couples are being allowed to adopt. The adoption of older children and those with disabilities tends to be more difficult. The children have often experienced numerous foster families, have various problems that have kept them from being adopted earlier, and may be less able to fit into the new family, although this is not always the case (Glidden 2000; Groze 1996; Ward 1997).

Cooperative adoption, in which the biological parents and the adoptive parents mutually work out the adoption, is also being tried. The idea is that the child is gaining a family, rather than losing a family, when adoption occurs. Needless to say, cooperative adoptions must be entered carefully so that problems do not arise later between the two sets of parents.

Most experts today agree that adoptees need to know some of the basic facts about their natural parents. Such knowledge is important to shaping the child's identity. Also, the increasing interest in and need for family medical and genetic information may be encouraging adoptive families to find out about the birth parents, to better understand the child's health risks (Lebner 2000).

Although very few adoptions are ever challenged by the biological parents, recent highly publicized court decisions have frightened adoptive or potentially adoptive parents. For example, the Illinois Supreme Court gave custody of Baby Richard to the biological father he had never met, taking the child from the adoptive parents who had raised him all of his 4 years (*National Law Journal* 1996, A10; *Santa Barbara News Press* 1995). There are continuing legal battles over whether children of one race should be placed with adoptive parents of another race (Bartholet 1993, 1994). Such litigation, although uncommon, acts to make adoption more uncertain.

In general, adoptions work out well for all concerned. Adopted children are usually better off, as are the mothers who relinquish their child. As pointed out, adoption means becoming a parent after making a thoughtful and deliberate choice, which is as it should be with all people choosing to have a child (A. Fisher 2003).

The Single-Parent Family

The single-parent family was the fastest-growing family type in the United States during the 1990s (see Figure 10-5). The proportion of children under age 18 who live with one parent has increased from 12 percent in 1970 to 28 percent in

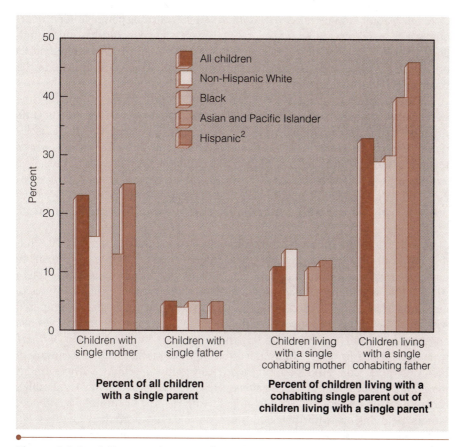

FIGURE 10-5 *Children with Single Parents and Proportion with Cohabiting Single Parent, March 2002*

[1] The parent is the householder or partner in an unmarried partner household. Single means the parent has no spouse in the household.

[2] People of Hispanic origin may be of any race.

SOURCE: U.S. Census Bureau, Annual Demographic Supplement to the March 2002 Current Population Survey.

2002; 23 percent live with the mother only and 5 percent live with the father only. Another 4 percent of children live in households with neither parent present, such as a foster family (U.S. Census Bureau 2001a). The fact that the situation is only temporary for many single parents does not minimize the dramatic change in American family life brought about by this phenomenon.

Although family form (single-parent, nuclear, three-generation, foster, and so on) affects children, the belief that only the traditional two-parent, father-mother family does a good job in rearing children is passing. Still, there is little doubt that single parenting has more inherent problems than two-parent child rearing. In many ways, the problems of the single-parent family are exaggerated versions of the working mother's problems (Chapter 13). In practical terms alone, a single working parent just can't give the time, attention, and guidance to a child that two parents can give.

The majority of single-parent families are women and their children, and they account for over 50 percent of poor families in the United States. The single mother must cope with low pay, child-care problems, and the overburden of working and continuing to shoulder household and child-rearing duties. Child support from the father, as noted elsewhere, is inconsistent at best and often absent altogether, especially for unwed mothers. The single father may face these same problems, except that his income is usually much better.

Single-parent families headed by women with low earning power suffer numerous logistic problems. Finding affordable child care is a major concern. Finding adequate housing in a satisfactory neighborhood is often difficult. Larger amounts of welfare aid go to these families than to any others, which is indicative of the financial difficulties they face.

Social isolation is one of the problems faced by any single parent. Juggling work, home maintenance, and child-care duties usually leaves the parent little time for social interaction and/or self-improvement activities. Emotional isolation is a second major problem. Having no other adult in the home with whom to interact often leads to feelings of loneliness and a sense of powerlessness. The emotional isolation of a separated, divorced, or unmarried parent may be increased by the social stigma often attached to these statuses. The parent who is widowed early, on the other hand, will experience sympathy and support from society.

Many criticize the missing father for his lack of economic support of his children and call for drastic measures (garnishing of wages, jail time, and so on) to ensure that support. Although economic difficulty is the most obvious problem when father is missing, his parenting duties are also missed. When it comes to discipline, it is a combination of mother and father that yields justice tempered with mercy. Mothers typically discipline children moment to moment and have an emotional umbilical cord that lets them read the individual child. Fathers usually discipline by rules. Children learn from their mothers how to be aware of their emotional side. From dad, they learn how to live in society. Fathers are far more important to their children's success than simply providing for the family's economic needs (page 306). Unfortunately, with divorce or unwed parenthood, fathers' relationships with their children tend to suffer more than mothers' do. Custody is almost always awarded to the mother. This usually limits the father's contact with his children.

Many single parents attempt to alleviate their isolation and reduce expenses by sharing living arrangements. Having a roommate can help reduce the parental load on the single parent. Living with family gives a single parent the support

HIGHLIGHT

Father Absence in African American Homes

The Morehouse Conference on African-American Fathers recently released an important statement entitled *Turning the Corner on Father Absence in Black America* (1999). William Raspberry, the keynote speaker, said, "Are black fathers necessary? Damn right we are!" The conference called on the nation as a whole to support the movement to "reconnect fathers and strengthen families." The statement urges Congress to pass legislation that would authorize more funds for community-based fatherhood programs promoting both marriage and marriageability, especially for young, poorly educated, low-income men. The funds would also be used for employment and parental-skills training. The statement exhorts the federal, state, and local governments and the leadership of the African American community to recognize the high priority of restoring the African American family. It encourages civil rights organizations to move this issue to the top of their agendas and bring the same intense dedication to rebuilding African American families as they have to obtaining basic civil rights.

Statement coauthor Enola Aird said, "As an African American mother, all I can say is, it's about time. The heartrending crisis of African American father absence that African American children suffer has cultural, economic, and spiritual roots. Addressing all of these to strengthen marriage and fatherhood in the African American community should be our most urgent priority."

and help with child rearing that are lost when living alone. Approximately 10 percent of children who live with a single mother are grandchildren of the householder (see the section on grandparenting, Chapter 11).

Many families identified as single-parent families are not, in fact, a single mother or father living alone with their child or children, but rather, a cohabiting couple. In 2002, 11 percent of children living with a single mother lived in a household with their mother and an unmarried partner. Thirty-three percent of children were living with a single father who was sharing his household with an unmarried partner (U.S. Census Bureau 2001a). It is assumed that the unmarried partner also participates in parenting of the children. In addition, a small but growing number of gay or lesbian households have the child of one partner living with them.

It will be a long time before all of the evidence is in about the ability of the single-parent family to rear children successfully. Certainly, the job is much harder for the single parent, but whether the increase in single-parent families will result in increasing numbers of problem children and hence, problem adults, is still open to debate. However, more Americans find that a combination of divorce, births to unmarried women, dual-worker families, and a preoccupation with self mean that more American children are neglected and troubled.

There are people who choose to become single parents, such as some high-profile persons in the entertainment business (Drummond 2000). In general, the women making this choice are monetarily successful, often in professional careers, nearing the end of their childbearing years, and uninterested in developing an intimate relationship with the father. Indeed, they may become pregnant through artificial insemination or through a strictly business relationship with the father, who agrees to play no future role with the child. Such women wish to experience motherhood and have the maturity and the economic means to successfully rear their child. Having a child does not trap them into a marginal lifestyle, and remaining free of an intimate relationship with the father protects their independence. It is interesting to note that many of these women are given favorable publicity (especially if they are high-profile entertainment personalities) and find favor with the more radical feminists. These economically successful women who choose to have a child on their own cannot be equated with the large majority of single mothers who struggle with poverty, lack of freedom, and marginal life styles. Minimal data exist on single men who adopt or use a surrogate mother to have a child. Here again, such men are financially well off and cannot be economically equated with most single fathers.

Summary

1. Children have a major effect on their parents' lives. The newborn tends to move the parents toward a more traditional relationship.

2. Although the father's role in parenting is sometimes downplayed, it is clear that his role is an important one. Too often, television, video games and the Internet take his place.

3. Child rearing is a difficult job that requires discipline and control of the child, especially in the early years.

4. As the child grows, the challenges faced by the parents evolve and change.

5. Adoption is an alternate method of having children. Adoptive parents need to understand the special characteristics of adoption if an adoption is to be successful for both the parents and child.

Resources on the Internet

Companion Website for *Human Intimacy: Marriage, the Family, and Its Meaning,* Tenth Edition

http://sociology.wadsworth.com/cox10e/

Gain an even better grasp on this chapter by going to the companion website to take one of the tutorial quizzes, use the flash cards to master key terms, or check out the many other study aids you'll find there. You'll also find special features such as GSS data, Sociology Online, and Census 2000 information that will put data and resources at your fingertips to help you with that special project or to help you as you do some research on your own.

InfoTrac College Edition

http://www.infotrac-college.com/wadsworth/

You can access reliable resources anytime, anywhere, with InfoTrac College Edition, the online library. This fully searchable database offers more than 20 years' worth of full-text articles (not abstracts) from almost 5,000 diverse sources, such as top academic journals, newsletters, and up-to-the-minute periodicals, including *Time, Newsweek, Science, Forbes, The New York Times,* and *USA Today*. You can conduct electronic key word searches using key terms from this chapter to supplement your reading and learning experience. To aid in your search and to gain useful tips, see the Student Guide to InfoTrac College Edition, which you can access through the companion website for this book.

Should Parents Stay Home to Rear Their Children?

YES

As more mothers with small children enter the workforce, there has been an increasing call for child-care centers to help them cope with their children. Certainly, from the adult perspective, day care is the most obvious answer to the child-care quandary. But is it an answer that works for the child?

A great deal of research and numerous theories support the idea that the young child needs meaningful adult attention for proper development. Although such attention need not come from a parent, it is clear that the adult who gives it must be around the child enough for a meaningful relationship to develop. What is enough is not yet clear. It seems obvious, however, that the 2-year-old child who is placed in a day-care center at 7:30 a.m. and picked up at 5:30 p.m. by a tired, overworked parent will have less chance to form a meaningful, intimate relationship with parents than a child not so treated. Even in an excellent facility, the child-care workers will have to divide their time and attention among many children, rather than devoting it to just one or two children, as a nonemployed parent can. At least in a child-care center, there are adults to interact with the children. Older children are often *latchkey* children (children who have no supervision except, perhaps, at school and who must await the return of their working parents alone).

Perhaps most troublesome is the possibility that many American children simply are not getting enough parenting to become socialized, to internalize the ethics, values, and mores that make it possible for a society to operate smoothly.

After two years of doing child day care, it was obvious that there was a problem inherent in day care itself. The children were too young to be spending so much time away from their parents. They were like young birds being forced out of the nest and abandoned before they could fly, their wings undeveloped, unready to carry them into the world. (Dreskin and Dreskin 1993)

The high level of violence among some segments of American youth, the drug and alcohol use, and the sometimes frightening lack of conscience and empathy for their neighbors make one wonder if, perhaps, a segment of our youth is becoming sociopathic from lack of parenting. Has neglect become the worst child abuse of all?

Helburn and colleagues (1995) conducted a long-term study of day-care centers and rated only one in seven as providing good-quality care. They found that many day-care centers, especially for infant and toddler care, provided such poor care that the children's emotional and intellectual development was harmed.

There are obligations involved in having children. Too often, child-care facilities seem to be created to help the parent avoid these obligations, rather than to provide for the good of the child. No one ever claimed that having children was convenient (see Chapter 13, "Stay at Home Moms").

NO

It might be wonderful for all parents to stay home with their children until the children are grown. The fact is that most parents today must work to support their families (see Chapter 13). Criticizing outside child care is unrealistic. Such energies would be better spent on being realistic—understanding that increasingly, both parents must work and that with the high divorce rate in America, the number of single parents will grow. Hence, improving child care, not denying that we need it, is the real issue.

The National Institute of Child Health and Development Study of Early Child Care, the largest, most comprehensive longitudinal study ever done on child care in the United States, examined the effect of day care on the mother-child emotional bond. It compared children reared exclusively at home with groups of children receiving daily, substitute care. The similarities between the groups far outweighed the differences, thus leading to the conclusion that the mother-child bond was not weakened nor was development harmed by day care in the first 3 years (Azar 2000; National Institute of Health 1998). Other studies (such as C. L. Booth et al. 2002) tend to support these conclusions. For some parents, the social learning aspects of day care are an important experience for their child; thus, such parents select a day-care center rather than a single babysitter.

Early intervention projects such as Program Head Start show that day care for preschool children from low-income homes can greatly enrich the children's environment and improve later school performance.

Studies of day care have discovered that the maternal attitude toward work and day care has an effect on the child's adjustment to day care. Children whose mothers were working and wanted to work scored significantly higher on several measures of adjustment and competence than children whose mothers were working but did not want to do so. There seems to be no significant relationship between a child's successful school adjustment and whether or not the mother was employed outside the home.

A program providing affordable child care is amply justified by the fact that it is an indispensable part of care for children of working parents and is a partial cure for child poverty, which afflicted about 17 percent of American children in 2002 (Helburn and Bergman 2002, 221; Population Reference Bureau 2004a).

What Do You Think?

1. If your parent(s) worked when you were young, did you spend time in day-care centers? If you did, what influence do you think this has had on your life?

2. If you were going to place your young child in a day-care center, what kinds of questions would you want to ask of the center personnel to be sure that it was a good day-care center?

3. What do you think of the current governmental idea that all children should be placed in preschool, and why?

4. If you had (or have) young children, ideally, how would you want to raise them? Would you be a stay-at-home dad or mom, work and place the child (children) in a day-care center, use relatives to help care for the children while at work, alternate between father and mother staying home if work allowed, perhaps trade babysitting with another family or . . . ?

© Antonio Mo/Taxi/Getty Images

Family Life Stages: Middle Age to Surviving Spouse

Chapter

11

Questions to reflect upon as you read this chapter:

- Do you think you cope with stress in your life in a positive manner?

- Is the proportion of older people growing or shrinking in relation to younger people in America?

- What is meant by the Sandwich Generation?

- What are some steps a person should take to make the transition to retirement more easily?

- List several advantages you see in the grandparenting role.

- What kind of relationship, if any, do you have with your grandparents?

There is a revolution in the life cycle. Americans are living longer than ever. Fifty is now what 40 used to be. Sixty is what 50 used to be. The growth and change in America's older population (those over 65 years of age) rank among the most important demographic developments of the twentieth century.

The 2000 census counted nearly 35 million people in the United States 65 years of age or older, about one in every eight Americans. By 2030, demographers estimate that one in five Americans will be age 65 or older, which is four times the proportion of elderly 100 years earlier. The effects of this older age profile will reverberate throughout the American economy and society in the next 50 years. (Himes 2001, 3)

As important demographically, if not more so, than the rapidly increasing American population aged 65 and over is the gender disparity. Women outnumber men in this population by 6.2 million (Hetzet and Smith 2001, 3; D. Smith 2003). Because men tend to die earlier than women, this gender difference becomes crucial with widowhood (see pages 350–354).

The flush of love and the excitement of exploring a new relationship effectively keep most newly married couples from thinking about the later stages of their relationship. After all, who can think about children leaving home when no children have yet arrived? What possible relevance can retirement have for a 23-year-old man receiving his first job promotion? And what newly married woman can be thinking about the very real likelihood that she will spend the last 10 to 20 years of her life as a widow, alone and without this man she now loves so much? Yet these are important questions that almost all couples must grapple with at some time in their lives.

The young adult reader may find it difficult to think about such questions, because the answers will not be needed until far into the future. Perhaps one way to make such questions relevant is to consider them in the context of your parents' and grandparents' lives. Their experiences may serve as a preview for some of the changes that will come into your own life.

If a couple has children, for example, eventually those children will leave home. The aging process cannot be avoided. Couples who think about and prepare for these inevitable family life changes before they occur stand a better chance of adapting to them in a creative and healthy manner.

Marriage and family relationships change, just as individuals within the relationship can and do change. Change does not mean the end of a relationship. In-

WHAT RESEARCH TELLS US

Increasing Life Expectancy

In 1900, life expectancy was 48.2 years for White males and 51.1 years for White females. Now the figures are 74.5 years and 80.0 years, respectively. For Black males, life expectancy is 67.6 years, and for Black females, it is 74.8 years (NCHS 2000c, 2). Much of the racial difference in life expectancy between Black and White populations stems from higher mortality in younger Blacks. By age 65, White American life expectancy has dwindled to only about 2 years greater than Black American life expectancy. Japan has the highest overall life expectancy, 81.0 years (Population Reference Bureau 2000b), while the United States' overall life expectancy is the highest it has ever been at 76.9 years (Himes 2001, 11). As noted, the percentage of the population over age 65 is also increasing rapidly.

deed, lack of change over time is usually unhealthy and may ultimately lead to the demise of the relationship.

This chapter deals with later-life families, the "young-old." These are families who are beyond the child-rearing years and have begun to launch their children, families who are facing retirement or are retired, families who are in the grandparenting stage of the life cycle, and couples who will experience the death of one of the partners sometime in the foreseeable future.

Families can change for better or worse. If too much of the change is for the worse, the marriage may end. On the other hand, individuals and families can grow in positive directions, becoming stronger, more intimate, more communicative, more fulfilling of needs, more supportive, and more loving. Because change cannot be avoided, the real question is, can I (we) deal with the changes, crises, and stresses that occur in all lives and families in a positive and healthful manner?

Dealing with Change: Another Characteristic of the Strong, Healthy Family

As we describe changes across the life cycle, it is important to remember that changes take place not only within the family, but also in the society. Many of the adjustments that successful families must make are adjustments to outside societal forces. Technological advances such as e-mail, which allows more people to work out of their home rather than going to the office every day, represent one example of social change to which the family is adapting.

Each family life stage presents challenges to family members, The birth of children adds parenting challenges to the couple (see Chapter 10); children leaving home (the empty nest) means new adjustments; and retirement makes the family face a new situation again. Other changes include grandparenting, the passing of a spouse (leading to living alone or with adult children), and possible remarriage late in life (in an increasing number of cases). Thus, success in dealing with change must be another characteristic of the strong, healthy family.

Where strong, healthy families seem to differ from weak, unsuccessful families is in their ability to deal with the problems and crises that come their way. Strong families deal with difficulties from a position of strength and solidarity. They can unite and pull together as a team to cope with problems. They do not squander energy on intrafamily differences and conflicts, but focus on the problem at hand. Each family member's attitude seems to be "What can I do to help?"

By planning and working together, strong families often head off crises by preparing ahead for changes, such as paying for children's college or their own retirement. Thus, life runs more smoothly and is generally free of emergencies.

Strong families also seem to remain flexible, and this adaptability helps them weather the storms of life and the inevitable changes that occur in every family and every relationship. Strong families bend and adapt; when the storm is over, they're still intact. A crisis is a turning point for such a family, rather than a disaster. Because the family is strong, it not only resolves the crisis, but also learns and emerges stronger than before, as a result of having dealt with the situation successfully.

It is clear that such families have a pool of resources to draw on when times become difficult. In contrast, unhealthy families are worn out and depleted by the daily stress of poor relationships and poor planning. When a crisis comes along, the unhealthy family must add the new problem to the burden they already carry. It is no wonder that a crisis sometimes destroys such a family.

Strong families expect problems, anticipate changes, and consider them to be a normal part of family life. Strong families seem to have the ability to find the positive, even in the most negative situations, and do not expect the family to be perfect.

Because the strong family recognizes that problems and change are a normal part of life, they develop problem-solving and decision-making skills (see Chapter 1). They also recognize when they may need outside help, and are able to seek assistance (see Chapter 17).

Successful families work out ways to cope with the minor irritations before they grow into major confrontations. Naturally, part of the solution is the respect and appreciation that each family member has for the others in the family. Obviously, the family that learns to cope with stress, frustration, and day-to-day irritations is likely to have the greatest chance of dealing successfully with all kinds of changes and will, therefore, remain intact.

The Graying of America

As we saw in Chapter 10, a developmental stage approach has long been used in studying children. Dividing family life into stages, or cycles, will similarly help us gain a better understanding of the changes that people go through as they move from birth to death.

What are the stages in the life of a maturing marriage? If we examine the general kinds of problems faced at various marital stages, we find that the average American couple goes through six important periods in a long-term marriage: (1) newly married, (2) early parenthood, (3) later parenthood, (4) middle age (empty nest), (5) retirement, and (6) widowed singleness. We have discussed the first three stages in earlier chapters.

The remaining three stages—middle age, retirement, and widowed singleness—have all been greatly affected by increased life expectancy. The combination of increased life expectancy and declining infant mortality has a number of effects on the family:

- It is no longer necessary to have two babies to produce one adult, as it is in many underdeveloped countries.
- The number of living grandparents is greatly increased for children, and three-, four-, and even five-generation families are more common.

- Marriages potentially can last much longer, so that more couples experience middle age and prolonged retirement together.

- Greater survival advantages for women, relative to men, have increased the period of widowhood at the end of life.

- The number of elderly persons depending on middle-aged children is increasing.

- The graying of Americans means that more methods of care for infirm, elderly persons must be made available.

© Lawrence Migdale/Stone/Getty Images

Later life brings time to pursue new interests.

Middle Age: The Empty Nest

For most couples, middle age starts when the children become independent and ends when retirement draws near. Although these figures are arbitrary, according to the Census Bureau, most people are in the middle-age stage between the ages of 45 and 65. It is better, however, to define this stage by the kinds of changes and problems that occur. For example, a very young mother might face empty-nest changes by the time she is 35. With increased life expectancy, middle age may now be extended to 70 years or, perhaps, even later.

Historically, middle age is a relatively new stage in marriage. Owing to the much shorter life spans before 1900, most wives buried their husbands before the last child left home.

Men and women may face somewhat different changes during middle age. For most women who have not had a career, adjustment centers on no longer being needed by their children. This is, of course, not true for childless women. Men's problems usually revolve around work and their feelings of achievement and success.

For both partners, however, middle age reactivates many of the questions that each thought had been answered much earlier in their lives. Indeed, many of the questions that arise resemble those struggled with by adolescents: Who am I? Where am I going? How will I get there? What is life all about? How do I handle my changing sexuality? In general, the more of these questions middle-aged people can answer positively, the more they feel a sense of control in their lives and the healthier and happier they are (Azar 1996a). The term **middlessence** has been coined to describe this stage.

The problems of middle-aged transition can become a vehicle for growth. Marriage enrichment for the middle-aged couple is not necessarily about improving the marriage; it's about improving the humans in it. What really counts is not so much what actually happens to us in middle age, as our attitude toward it. We all experience some kind of changes in this stage, as in other stages, but how do we use these changes? If we use them for further growth and expansion, our mar-

middlessence

The second adolescence, experienced in middle age, usually involving reevaluation of one's life riages, our relationships, and we ourselves may all benefit, and the second half of our lives may be even more fulfilling than the first half.

There is infinite variety in the way individuals face the questions that arise when they realize that life is finite. Some people simply look the other way and avoid the questions. Others try to change their external world by relocating, finding a new spouse, or starting a new job. Others seek internal changes, such as a new philosophy of life.

Obviously, a person's life circumstances influence the questions and their answers. A child-centered mother out of the work world for 20 years may feel panic as she realizes that her children will soon not need her. Another mother who has always worked in addition to raising her children may be happy that her children will no longer need her, and that she will be free to pursue long-neglected interests. Because of these individual differences, our discussion of midlife changes must remain general. Some individuals will undergo the experiences we discuss, others may not, and still others may experience things that are not discussed. Despite these limitations, examining the middle years of life is worthwhile because we all pass through them.

For the traditional wife and mother, the midlife transition may be a crisis, although crisis is probably too strong a term. Perhaps a better term might simply be "midlife change" or "midlife correction" (A. Stewart and Ostrove 1998). She has to face her partial failure as a parent (all parents fail to some degree, as we saw in Chapter 10), and her feelings of loss and uselessness as the children leave the home. She may have general feelings of dissatisfaction with her marriage and her life, and may experience periods of increased introspection as she seeks new life goals. Her feelings are somewhat akin to a worker's feelings on retirement. Today, women must face the questions of changing identity and roles brought on by the women's movement. As more women become career oriented (Chapter 13), the launching of the children will probably denote release and freedom, rather than a sense of loss and uselessness (A. Stewart and Ostrove 1998, 1190).

Any woman, whether traditional or not, faces the biological boundary of the end of her childbearing years. She faces the fact: "No more children, even if I wanted them." It is true that advancing technology means there is still a possibility of childbearing after nature has declared an end, but this is still very rare (Chapter 9). The end of her childbearing years, along with the accent on youth in America, will cause her to reevaluate her sexuality. "Am I still attractive to men? Is there more to sex than I have experienced?" Some women may welcome the end of their reproductive capacities because it frees their sexuality from the constant worry about possible pregnancy. Many women report renewed and intensified sexual appetite once they have completed menopause.

A wife, and also her husband, will become aware of death in a personal way, as their parents die one by one. These events cause many people to direct their thoughts for the first time to the inevitability of death and the realization that their lives are finite.

Thus, the midlife changes faced by women are broad and profound. Furthermore, both the woman and her husband also experience the turmoil the other feels during this stage.

The typical husband's midlife changes revolve around his work, rather than his family. This is particularly true for highly successful men. In our competitive economic society, a person must devote a lot of energy to his/her work, to achieve success. The male is often forced to handle two marriages: the first to his work, and the second (perhaps in importance, also) to his family.

HIGHLIGHT

How a Highly Trained Engineer Became a Bicycle-Repair Man

Jeff Johnson retired from his engineering job, saying that he would be happy doing nothing for the first time in his life. Within 6 months, he was so restless that his wife, Janice, encouraged him to seek out new horizons. One afternoon, a 12-year-old neighbor boy was frustrated while working on his bicycle brakes. Jeff gave him a hand and soon repaired the brakes. He began to help other children in the neighborhood repair their bicycles. It was so interesting that he began to watch for old broken bicycles; soon, he had a garage full of bikes to recondition. It wasn't long before he had a thriving bicycle-repair business.

What Do You Think?

1. What interests or hobbies do you have that might become important activities for you during retirement?
2. If either of your parents are retired, how have your parents' lives changed? Your grandparents?
3. Do you think the changes have been for the good? For the bad?

America's emphasis on youth presents problems for the man. Competitive, younger men become a threat to his job security. Subtle comments about retirement may take on a personal significance.

A husband's midlife transition centers around letting go of some of the dreams of his youth. He begins to recognize that he may not achieve all of his dreams. For the few men who have fulfilled their dreams, the midlife question becomes, what do I do now? Most men, however, must cope with some disillusionment, that is, the sense that the dream was counterfeit, the vague feeling of having been cheated—that the dream isn't really what they thought it would be—and the growing awareness that, perhaps, they will never ever achieve all their dreams. Yet, as Gail Sheehy (1995, 6) suggests, "The day you turn 45 is now the infancy of another life." With increased life expectancy and improved health at later ages, people at midlife who remain open to new vistas of learning can dream new dreams and anticipate conquering new goals. For example, some retired couples have become very active in public service, even joining organizations such as the Peace Corps.

Unfortunately, if there is disillusionment, it may carry over into the man's family. He doubts himself and also doubts his family. He may even blame them for his failure to achieve the dreams of his youth.

Along with general self-doubts come doubts about sexuality. Unlike his wife's, though, the husband's doubts usually revolve less around physical attractiveness than around performance. His wife's sexuality is at its peak, and she may be more aware and assertive of her own sexuality. Her assertiveness can heighten his self-doubts. Self-doubt can be a man's greatest enemy in achieving satisfactory sexual relations. The "other woman" may become a problem at this time, as the man seeks to prove his sexuality (Wright 1994).

For many middle-aged wives, the midlife crisis for their husbands has only one meaning: a married man having an affair, usually with a younger woman. No matter how good the marriage is, living with another person is never perfect. When an aging spouse meets a new person of the opposite sex, it is ego building to be found interesting and attractive.

Often, the extramarital affair in midlife breaks up the marriage (see pages 207–210). This is usually less common for the middle-aged wife because the pool of possible new male partners is far more limited for her than is the pool of eligible women available to her husband.

In some relationships, a partial exchange of roles takes place at this time of life. The woman becomes more interested in the world outside her family and more responsive to her own aggressiveness and competitive feelings. She returns to the work world or to school, or becomes interested in public affairs and causes. Her husband becomes more receptive to his long-repressed affiliative and loving urges. He renews his interest in the family and in social issues outside his work. He becomes more caring in the way that his wife has been caring within the family. Unfortunately, the children are now seeking their own independence and are less receptive to his newfound parental caring. A mother's new interests in the world outside her family may be impeded by the fact that the empty nest is increasingly refilled by returning adult children, grandchildren, or the couple's elderly parents.

The Sandwich Generation: Caught in the Middle

During the twentieth century, young adults (defined by the Census Bureau as 18 to 34 years old) left home earlier with each succeeding generation, but the brakes went on sometime in 1970. Now, young adults are more likely to be living in the homes of their parents than they were in 1970. In 2000, 56 percent (7.5 million) of men 18 to 24 years old lived at home with one or both of their parents. Although women typically marry at younger ages, a sizable proportion (43 percent) of this age group lived at home with at least one parent (J. Fields and Casper 2001). About one in seven parents over 65 years of age has a child living at home (L. White 2000). This situation helps put the middle-aged parents into what is called the *sandwich generation*.

A number of factors probably contribute to the higher proportion of young adults living with their parents. Marriage is being postponed, increasing emphasis is being placed on advanced education, and housing costs are high in many areas of the country. The high divorce rate among young couples also prompts increased numbers of adult children to return home, at least temporarily.

Research indicates that a young adult in the home contributes to a high level of parental dissatisfaction (see Chapter 10). With more young adults remaining or returning home, parents find themselves having to share their home and lives with other adults, rather than being alone and able to concentrate on themselves and their relationship. The continuing lack of privacy and their children's failure to contribute monetarily or share in the day-to-day running of the home are parents' major complaints about having adult children living at home. In turn, the young adults complain about lack of freedom and parental interference.

Young adults remaining or returning home represent one side of a sandwich. Longer life expectancy has also increasingly led to a middle-aged couple's empty nest being refilled by the responsibility for aging parents—the other half of the sandwich. Even if elderly parents do not actually live with the couple, most couples concern themselves with caring for and helping their aging parents. In addition, modern medicine is keeping older people with chronic illnesses, like Alzheimer's disease, alive for years longer. Thus, the likelihood that a couple will have to help care for aging parents continues to increase.

Families face several ethical dilemmas (Kahana and Biegel 1994) when a relative becomes more dependent, as follows:

- What is a family's obligation, and to what extent should care be provided?

- Caregiving should most fairly be divided among family members, but how should the family handle caregiving when only one or a few family members participate?

- How do caregivers balance the obligations they have to other family members, and those they have to outside personal commitments?

- How do caregivers decide how much money to invest before the caregiving begins to impoverish their own family?

- When should caregivers step in to make decisions that older relatives can no longer make themselves?

- How can the caregiver take into account the needs and autonomy of the care receiver, while taking care of their immediate family needs?

Caring for the elderly has traditionally been women's work, a natural extension of their role as homemakers and nurturers. Numerous studies find that daughters or daughters-in-law, rather than sons, are the primary caregivers to aging parents (Bromley and Blieszner 1997; Himes 2001, 23). Female caregivers tend to report greater feelings of burden, stress, and depression than do male caregivers (Raschick and Ingersoll-Dayton 2004; Yee and Schulz 2000), probably because the responsibility for elder care falls mainly on their shoulders. In addition, there are cultural differences in caregiving for elders. For example, Asian American families are more likely to provide support for elders. This is usually attributed to the Asian concept of filial obligation, although financial, religious, and social factors play important roles (Ishii-Kunta 1997; Wachwithan et al. 1998). About 41 percent of Caucasian, 52 percent of Hispanic, and 40 percent of African American adult children provide informal care for their elderly parents (70 years or older).

It must also be noted that distance plays a large role in the amount of care provided by adult children to aging parents. Children who live far away from their parents obviously cannot give as much personal care to aging parents as children

For Better or For Worse®   by Lynn Johnston

who live close by. However, it is often true that aging parents move closer to their offspring.

Americans can expect to spend more years caring for their parents than they did caring for their children (see Figure 11-1). About 22 percent of Americans in their 50s have older, frail parents in need of care or supervision (National Institute on Aging 1993).

Although most of the literature discusses the stress put on adult children by care for aging parents, the problems of adult children also cause distress to their parents. Parents whose adult children experience serious problems report increased stress and depression in their own lives (Pillemer and Suitor 1991).

Many times, the care given by adult children to their aging parents is reciprocated by the parents, so caregiving by the adult children is not a one-way street. Monetary aid and/or advice, help with grandchildren, help around the house, and other such things are often given back to the adult children by the cared-for parents. This help from care recipients is an important component of caregiver rewards. Although such help given by the elderly parents to their caregiving children is generally appreciated, it sometimes makes the caregiver feel dependent (monetary aid in particular) or inadequate, depending on how it is given (Ingersoll-Dayton et al. 2001; Raschick and Ingersoll-Dayton 2004).

For the woman joining the workforce or renewing her interest in the world outside the family after the children have left home, shouldering the burden of caring for elderly parents can be particularly frustrating, especially if she must quit her job to care for her aging parents.

Conflict among siblings over caring for aging parents is a major source of dissatisfaction for caregivers. The more-involved siblings often experience frustrations and anger toward those who are less involved in caregiving. In addition, the less-involved siblings may feel guilty about not assuming their fair share of responsibility. The anger and frustration is especially strong when the caregiving sibling asks for help from brothers and sisters, and is rebuffed (Ingersoll-Dayton et al. 2003). In this situation, sibling fighting over the parents' estate is often a result. The caregiving sibling(s) feel(s) entitled to a greater share.

Most children do not want to put their aging parents into institutional homes. Despite the public stereotype of the elderly living in nursing homes, only a very small percentage actually live in institutions. Indeed, making the decision to intervene with frail, elderly parents is extremely difficult for most children. How do you tell a parent or parents that their home is now too much for them to care for, that they can't drive, or that it might be wise for them to move into a retirement home?

Although we often hear that Americans callously abandon their elders, the opposite is true. More children then ever are helping to provide care to their parents when it becomes necessary. They are providing this care for longer periods than in the past, due mainly to increased life expectancy.

Some research suggests that the family with young children still in the home seems to have less interpersonal strain between the adult parents and their parents for whom they are caring. It seems that grandchildren are frequently identified as secondary caregivers or regular helpers in caring for their aging grandparents. About 4 percent of Caucasian, 6 percent of Hispanic, and 10 percent of African American grandchildren give informal aid to their grandparents (Himes 2001, 23). This help may reduce the burden on their parents for the care of the children's grandparents (Stull et al. 1994).

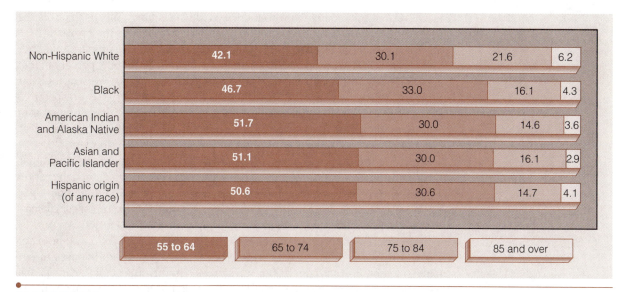

	55 to 64	65 to 74	75 to 84	85 and over
Non-Hispanic White	42.1	30.1	21.6	6.2
Black	46.7	33.0	16.1	4.3
American Indian and Alaska Native	51.7	30.0	14.6	3.6
Asian and Pacific Islander	51.1	30.0	16.1	2.9
Hispanic origin (of any race)	50.6	30.6	14.7	4.1

FIGURE 11-1 *Percent Distribution of People 55 Years and Over by Race, Hispanic Origin, and Age, 2002*
SOURCE: U.S. Census Bureau, Annual Demographic Supplement to the March 2002 Current Population Survey.

There seems to be a positive relationship between aid parents give to their children and help they receive from their children when they, the parents, age. However, parents who hold high expectations for help in their later years from their children often express disappointment, even when the children do aid them (G. Lee et al. 1994).

Thus, middle age may mean an empty nest and a renewed emphasis on the primary couple, or it may not. For many middle-aged couples, balancing work, child care, and helping their elders are the major challenges faced.

For both men and women, however, midlife usually means rethinking and reevaluating life. It is a time of restructuring. The resulting turmoil may threaten the marriage and family, or it may result in a revitalized marriage. By recognizing and squarely confronting the issues of midlife, the chances increase that the results will be positive and growth-enhancing, rather than negative and destructive.

The husband and wife may draw closer during the middle-aged stage. In earlier marital years, when many wives devote their energies to the children and increasingly to work outside the home, and their husbands devote themselves to their work, husband and wife may well grow apart emotionally. When the children are gone, a couple may renew their life together and reinvest in each other. Renewing marital vows as a way of refreshing a couple's marriage is becoming more common among long-married couples (Uyehara 2000). Such renewal of a couple's relationship helps to lay the foundation for the next marital stage, retirement.

Retirement

During this century, the increasing life expectancy, automation, and America's emphasis on youth (older people remaining active in youthful activities) have combined to almost double the length of retirement in the average worker's life. People now spend at least a quarter of their adult lives in retirement. The falling age of retirement, combined with increased life expectancy, means that this segment of peoples' lives can become as long, or even longer, than their working lives.

Early retirement has become so widespread that a smaller proportion of individuals are working beyond age 65 than in the past. In 1950, 68+ percent of men aged 55 years and older and 45+ percent aged 65 and older were in the labor force. By 1994, the percentages had dropped to 39+ and just 16+, respectively (Treas 1995, 21–22). The average age of retirement for men has dropped from 66.9 years of age in 1950–1955 to 61.7 years of age in 2000–2005. The corresponding ages for women are 67.7 and 61.2 years of age at retirement.

For retirement to take place, an economy must produce enough to support those retiring, those retiring must have worked and contributed enough that the general society agrees to their support, and there must be some method—personal savings, pension plans, or government support (for example, social security)—whereby their support is assured (Morgan and Kunkel 1998).

Early retirement is especially affected by the state of the economy. For example, during recessions, many businesses offer early retirement bonuses to employees in an effort to reduce their workforces. On the other hand, many people who are eligible for retirement continue to work, out of fear that they would be unable to survive financially in an inflationary economy if they retired.

The retiring worker faces the problem of adjusting to leisure, after years of basing much of his/her self-worth on working and income production. Much of an individual's identity is determined by his/her work. I am a teacher, mechanic, businessman or businesswoman, and so on. Under the American work ethic, stature is gained by work, not leisure, so retirement equals obsolescence in the view of American society. For the productive worker, retirement presents a choice: adjust or be miserable.

In a sense, retirement for a worker is similar to the empty-nest or launching stage for the traditional, child-centered wife. The retiree must cope with a lack of purpose and feelings of uselessness. Those who cannot develop substitute goals may find retirement an unhappy period. For some, it is literally a short-lived period; death often arrives shortly after retirement for those who cannot adjust. Retirement may mean a loss of contact with significant others, such as work colleagues, if one moves to a new area. It also means making new relationships, friends, and acquaintances to replace those lost because of retirement and/or moving.

For some, especially those who have been financially successful, retirement may mean rebirth. It may signal a new beginning, expanding interests, rediscovery of the marital partner, making new and exciting friends, and seeking new activities and goals.

Retirement is best characterized as a process, rather than an event. People may partially retire, such as when they reduce time spent working on the job. They may retire from one job, but start another. They may draw retirement, such as social security, yet not consider themselves retired. Thus, being retired means different things to different people.

Retirement can be a wonderful time of catching up on all the activities that a person or couple has put off during the working part of their lives. Findings vary about the relationship between retirement and marital satisfaction. Some studies suggest that marital satisfaction increases in later years, whereas other studies find little or no relationship. Some older couples may experience a honeymoon phase after retirement because they are no longer hampered by the demands of work and can spend more time with each other. Other couples may find that retirement increases conflict (Moen et al. 2001). Those who cannot cope well with the transitions required by retirement may experience depression. Depression is relatively more common in the aging, retired population (Sandberg 2002).

Differing retirement patterns for a couple affect a family's adjustment to retirement. For example, G. Lee and Shehan (1989) found that an employed wife with a retired husband experiences lower marital satisfaction than a wife who retires first, or at the same time as her husband (G. Lee 1996).

In many nonindustrialized societies, retirement is a gradual process. The hardest physical labor is performed by young men and women at the peak of their physical condition. As they age and their offspring grow to maturity, the parents assume more administrative and supervisory duties, while the children take on the harder physical labor. This gradual tapering off of duties, with a smooth transition of tasks from one generation to the next, was often the practice on the farms of rural America. By contrast, in American urban areas, retirement is usually abrupt; it occurs when the worker reaches a certain arbitrary age or has worked for a specified number of years. This abruptness of retirement can cause severe adjustment problems for the newly unneeded worker.

Working spouses are suddenly placed in a totally new role. After 40 to 50 years of going to work, they receive a gold watch and are told to go fishing and enjoy their new leisure time. They are thrust into a new lifestyle that, for some retirees, is characterized by less income, declining health, and increasing loneliness. In contrast, the woman in the traditional mothering role who faces her retirement at the earlier, empty-nest stage finds that it comes gradually, as her children gain independence one by one.

The retired couple are now free to travel and catch up on the activities they missed because they were working, but they also miss their colleagues and the status derived from their jobs. Now when a new acquaintance asks, "What do you do?" the former worker answers, "Nothing, I'm retired." As noted, because much of a career person's self-image is defined by work identity, retirement often brings an identity crisis.

Because retirement comes abruptly for many Americans, it is necessary to prepare ahead, both economically and psychologically. Many retirees take on new activities and responsibilities, and their lives become even busier than when they were working full time. These retirees seem to make the transition from the work world to retirement more comfortably than those retirees who have little to do after they leave the work world.

A few retirees seem able to find happiness in "doing nothing." "It's great not to have to get up every morning and go to work." Yet most cannot "do nothing" easily and happily after years of working. Many retirees continue to do odd jobs or work part time if their occupation permits. For example, a retired schoolteacher may occasionally do substitute teaching. Some retirees change activities entirely and take up a new vocation or avocation. They may decide to open their own business or actively pursue some long-suppressed interest, such as writing or painting. Perhaps they busy themselves with volunteer work. There are numerous examples of people retiring and then returning to the work world. Konrad Adenauer assumed the leadership of postwar West Germany at the age of 73, and actively guided Germany to a powerful world position for 14 years, until he was 87.

Those who have always had broad interests or are able to develop new interests generally make the quickest and best adjustment to retirement. People who age optimally are those who stay active and manage to resist shrinkage of their social world. They maintain the activities of middle age as long as possible, and then find substitutes for the activities they are forced to relinquish. They find substitutes for work, when forced to retire, and substitutes for friends and loved ones lost to death.

Wives of husbands recently retired list time available to do what you want and increased companionship as the two most positive aspects of their husbands' retirement. Financial problems and retired husbands not having enough to do are the two most negative aspects of husbands' retirement, according to their wives.

Although most of the literature on retirement concerns men, questions of retirement are also important for women. Past research has indicated that retirement does not hold the same significance for women as for men, even when the woman has worked full time for most of her life. One possible reason for this is that many women experience several other roles (wife, mother, homemaker) that they perceive to be as important as the role of worker. This pattern may well change, however, as more women seek lifelong careers. Just as men must prepare for retirement, so must working women.

Money and health seem to be the two most important factors influencing the success of both retirement and general adjustment to old age. It takes a lot of money to live free of economic worries. Generally, a person's income is more than halved on retirement. Yet, if a couple are active, spending may well increase, contrary to the popular idea that both income and expenditures are reduced with retirement.

The stereotype of the older retiree fighting off poverty has largely disappeared since social security and pension plans have been indexed to the inflation rate. The Census Bureau reports that the overall economic position of those over 65 years of age has improved significantly. In 1959, 35+ percent of those over 65 years of age were below the poverty line. In 2001, only 10.1 percent of this group was below the poverty line (D. Smith 2003). However, there are major differences in income among population subgroups, defined by such factors as living arrangements, marital status, race, ethnicity, educational attainment, and former occupational status. For example, elderly Caucasian married-couple families have fared economically the best over the past decade (Bureau of Labor Statistics 2000). The major asset for most retired persons is their home. Approximately 80 percent of persons 65 years and older own their own home, and the majority of these homes are free and clear monetarily (U.S. Census Bureau 2000a).

For those who still suffer financially in retirement, it is ironic that when they finally retire and have free time for the activities they want to pursue, they have little money available for more than subsistence. Financial status is directly related to the quality of life for the retired couple. Can they travel, indulge their hobbies, and satisfy their interests? They can, but only if they are financially comfortable.

Due to increased life expectancy, retirees must not only prepare economically for retirement, but must also plan for a potentially long period of retirement. As we saw earlier, people may now live up to 25 percent of their lives in retirement. Inflation is the retired person's greatest enemy. Those retired on an essentially fixed income fall further behind with each passing day of inflation. A couple financially well off at retirement may drop into poverty 10 years later, due to continued inflation and the shrinking value of the dollar. Indexing retirement income to the inflation rate works to cancel this trend, but unfortunately, such indexing also fuels inflation.

Health is the other major influence on people's adjustment to retirement and old age. About 40 percent of those over age 65 have chronic conditions. Although government programs such as Medicare help the elderly cope financially with health problems, most retirees find that they need private medical insurance as well as long-term care insurance. Thus, lack of money may compound health problems and further reduce the quality of life for the retired. Given the high costs of medical care, few people can be completely safe from financial disaster brought on by prolonged or severe health problems.

The health problems of seniors are often viewed as hypochondria by the general public. Objective studies of medical use by elderly persons, however, indicate that they do not make disproportionately large numbers of visits to physicians. Considering the physical problems that accompany increasing age, elderly persons do not seem to be any more prone to hypochondria than the general population.

Assuming that monetary and health problems are not overwhelming, many couples report that the period of retirement is one of enjoyment and marital happiness. They are able, often for the first time, to be together without jobs and children making demands on them. They can travel and pursue hobbies and long-neglected interests. They can attend to each other with a concentration not available since they first dated. In some ways, a successful retirement is like courting. After years of facing the demands of work and family and growing apart in some ways, couples find that retirement provides the time needed to renew the relationship, make discoveries about each other, and revive the courtship that first brought the couple together. In addition, most retirees express general satisfaction with family relationships and do not feel neglected or abandoned by their relatives (Brubaker 1990, 960; Treas 1995, 29–30).

Indeed, the increasing longevity of Americans has created the phenomenon of the four-generation family. An estimated half of all persons over age 65 with living children are members of four-generation families. This does not mean that they live with their families, but that they are great-grandparents. This role is relatively new in society. Great-grandchildren can be as great a source of joy and fulfillment as grandchildren. At the same time, the generational gap is so large that great-grandchildren can also be a source of bewilderment, because their lifestyles are so different from those of the oldest generation.

Elderly parents who live with their children may put those children in the position of parenting their parents, as we saw earlier. This role reversal can be difficult for both the elderly parents and their children. The frustration that may occur when elderly parents live with their children creates the possibility of elder abuse (see Chapter 14).

Because children no longer automatically care for their aging parents, government agencies have had to be created to help the elderly. Government-regulated retirement programs and the Social Security Administration have been established to help economically. Medicare and nutrition programs such as Meals on Wheels assist in the area of health care. The National Council on Aging publishes a directory of special housing for the elderly. Licensing and supervision of institutional facilities for the aged are being tightened.

The elderly are also organized as a power bloc (for example, the American Association of Retired Persons [AARP] and the Gray Panthers), to work to improve care and opportunities for people in their later years. Programs such as the Retired Senior Volunteer Program, which pays out-of-pocket expenses to those involved in community activities and projects, and the Senior Corps of Retired Ex-

---

WHAT DO YOU THINK?

1. If you have living grandparents, do they receive personal help from either or both of your parents? Do they get economic help? Explain.

2. Where do they live?

3. Are they independent, or dependent on others for day-to-day living?

4. What kind of relationship do you have with your grandparents?

5. Do your grandparents give monetary or other help to your parents?

ecutives (SCORE), which pays former managers to counsel small businesses, are springing up to keep the elderly active and useful. All these developments bode well for people as they move into the later years of their lives.

The family, as the center of intimate relationships, remains the place to age gracefully and die with dignity and care. In fact, the percent of people 65 years and older living in nursing homes has declined from 5.1 percent in 1990 to 4.5 percent in 2000 (Hetzet and Smith 2001, 7).

Widowhood as the Last Stage of Marriage

Inevitably, one of the marital partners dies, so the final marital stage is usually a return to singleness, widowhood. The average age at widowhood is 70 years for women and 72.3 years for men. About 40 percent of women 65 years and older live alone, compared to only 17 percent of men in this age range (J. Fields and Casper 2001, Table 6).

There are approximately six times more widows than widowers in the United States. Thus, widowhood is a far greater possibility for women than for men. Although the percentage of widowed persons in the general population has dropped during this century, the number of years that widowed singleness may be expected to last has risen dramatically. For example, half the women widowed at age 65 can expect 15 more years of life, while men live approximately eight years as widowers (Treas 1995, 28). Thus, this final marital stage can be especially lengthy for women.

Being widowed is predominantly an older person's problem. Younger widowed persons tend to remarry, but remarriage becomes increasingly remote with advancing age. For the widowed person who does remarry, being widowed may not be the last stage of marriage. Remarriage will reinstate an earlier marital stage.

Regardless of age, the death of a spouse arouses all the emotions that occur whenever one loses a loved one. Grief, feelings of guilt, despair, anger, remorse, depression, both turning away and toward others—all are normal reactions (see Chapter 14). Given time, most people can overcome their emotional distress at losing a spouse and go on with their lives. But, for a few long-married couples, death of one precipitates the death of the other. This often occurs when a woman has derived most of her self-identity from her husband. Because American marriages involve an especially close relationship between husband and wife, the loss

Older widows outnumber older widowers.

© Bill Aron/PhotoEdit

HIGHLIGHT

Doctor's Wife

My grandmother died only a few months after my grandfather, even though she was in good health and had seldom been sick in her life. My grandfather was a strong, independent man who worshiped my grandmother and took especially good care of her. He never allowed her to work or to want for anything and remained deeply in love with her, often publicly displaying his affection.

He was an old-fashioned, family doctor who made house calls and regarded his patients as his family. My grandmother's entire identity revolved around being the "doctor's wife" (that is how she often referred to herself). Her life was his life, and in hindsight, I realize she never developed any interests of her own. In fact, she seemed to have no interests apart from his interests. As the doctor's wife she took care of him, the family, and the house. When the children became independent, she became even more attentive to him and didn't develop any other interests to replace the missing children.

With the arrival of grandchildren, she became the happy mother all over again, caring for the grandchildren as often as she could. When grandfather died, we all tried to visit her often and invited her to visit our families. She told us to give her a little time to adjust and said that for the present she preferred to stay home. About 3 months later, I found her lying in grandfather's bed, having passed away from an apparent heart attack. In retrospect, I think that she had died in spirit when grandfather passed away. The death certificate records "heart attack" as the cause of death.

When I think about it, I had no grandmother. My grandmother was the doctor's wife. When he died, her identity died and soon thereafter, her body. I prefer to substitute "broken heart" for "heart attack" on the death certificate.

---

of the spouse may be more devastating than in societies where a married couple is more embedded in the broader social network, such as an extended family or clan (Lopata 1996, 217–218).

The mortality rate for widows is only a little higher than for married women. Widows appear to have better support systems than widowers (more friends, companions, and interaction with children and grandchildren). This may stem from the fact that wives generally pay more attention to the couple's social relationships, and this social activity simply extends into widowhood.

Widowers generally exhibit more severe problems of disorganization than widows. The men have higher rates of suicide, physical illness, mental illness, alcoholism, and accidents, although, in part, these differences are characteristics of all males. Widowers' overall mortality rate is considerably higher than for married men with the same traits (age, schooling, smokers, nonsmokers, and so on). Remarriage by widowers dramatically lowers their mortality rates. This tends to support research showing that marriage promotes health (see Chapter 2).

Despite the negative findings about the problems of widowers, the social problems caused by aging center on the much larger numbers of widows in the American society. The problems of widowhood are, for the most part, the same problems faced by the elderly in general, but the widow faces them alone, without a spouse.

In rural America a century ago, many retired parents, especially those who were widowed, received some support from their grown children. People grew old on the farm within the family setting. Grandparents gradually turned the farm over to their children, but remained there—giving advice, fulfilling the grandparent role, and being active family members until they died. Thus, much of the widowed person's normal role in the family remained intact.

As America urbanized, however, children often moved away from their parents and established independent households. Often, both partners worked at jobs that took them out of the home. Living space diminished, so that room was available for only the immediate nuclear family. Thus, urbanization slowly moved the care of the elderly out of some children's reach.

As we have seen, many Americans own a home, which often gives them refuge and a major asset in their later years. Only 14.4 percent of retired homeowners have a mortgage, so their home is free and clear. Thus, the majority have no mortgage payment, but rather, a valuable asset against which they may borrow if necessary (Bureau of Labor Statistics 2000). More than two-thirds of the elderly remain in their home until death. Remaining in the home reduces the impact of spousal loss, because the widowed person is not forced to adjust to new living quarters while he/she must cope with the loss. Another 25 percent of persons over age 65 live with a child. The remaining elderly live in a variety of circumstances: with relatives other than children, with roommates, in nursing homes, and in rented quarters. Perhaps half a million are sufficiently well-off monetarily to buy or lease living quarters in exclusive retirement communities.

The Adjustment Process and Remarriage

We pointed out earlier that most people do finally make an adjustment to the death of a loved one. Generally, the surviving spouse goes through three stages in the grieving process. First is the crisis-loss phase, when the survivor is in a state of chaos. Fortunately, the survivor usually has a lot of support from friends and family immediately after the loss of the loved one. The second stage is the transition phase, when the survivor attempts to create a new life. The final stage entails the establishment and continuation of a new lifestyle (see Chapter 14). The reorganization may result in a new life as a single, widowed person (particularly for surviving wives), or a remarriage may occur (particularly for surviving husbands).

In general, about 50 percent of widows report that it takes a year or more to adjust to the loss of their spouse. A few widows report that they seem unable to establish a new life, but only time will tell if their belief is correct. As more married women actively pursue roles outside the home, they may be able to more easily start a new life after the death of a spouse. In the past, many women had never been alone in a home, having gone from the home of their parents directly into the home they established with marriage. Many of them had no occupational skills, never having worked except, perhaps, in a few odd jobs before marriage. Their traditional socialization was to be passive about the world outside their home environment. Thus, their socialization and consequent life experiences did not adequately prepare them to start a new life alone. The emphasis by the women's movement's on individual identity for all women, even when married, should also help widows cope better with the loss of a spouse.

The upsurge of interest in death and bereavement has helped both widows and widowers cope with their new roles. More understanding and empathy have been extended to them. For example, programs in which widows counsel other widows have been helpful in reducing the adjustment period. It behooves widows and widowers to learn to browse the Internet because there are numerous sites that can help direct them about many subjects, such as grandparenting, money control and investment, and support groups with which to interact.

Remarriage is, perhaps, the best solution for the widow or widower (McMahon 1995). Most elderly who remarry face a more complex marriage, because the children and grandchildren of each newly married spouse often play roles in the new marriage. Children may be concerned that their parent's new spouse will

HIGHLIGHT

Remarriage after the Loss of a Spouse

Lloyd and Luceda had not seen each other since they double-dated to their high school senior prom (each with another person) 50 years ago. Neither had ever attended a class reunion after marrying. Lloyd had three children and Luceda had two children. Lloyd lost his wife to cancer about 2 years before the anniversary of his fiftieth high school graduation. Luceda's husband died in an auto accident about 8 years before that anniversary. Although they lived a great distance apart, and far from Kansas where they had attended high school, old high school friends persuaded them to attend the class reunion.

They greatly enjoyed seeing each other after so many years, and planned to meet again when they were both visiting children who happened to live in the same Florida town. One thing led to another, and they married a year after the reunion. However, before marrying, they traveled to various parts of the country to introduce each other to their children and their families. The children were supportive of their plans, and at the wedding all five children and the eight grandchildren hit it off. In fact, the two families got on so well that they all planned to get together for one big family Christmas the following December.

Lloyd and Luceda have now been married for 4 years and say they wouldn't have missed it for anything.

What Do You Think?

1. Have any of your grandparents remarried after the loss of their partner? If "yes," how has it worked out? Are their children happy, or disappointed in the remarriage? Why?
2. How would you feel about your mother or father remarrying at a later age, after the death of the partner? Explain.

take away their inheritance, or worry that the new spouse is a con artist preying on the widowed. The children will also have to interact with a whole new set of relatives. Despite these potential problems and the complexity of remarriage, most elderly who remarry after the loss of their spouse report high levels of satisfaction (Sheehy 1995, 363–366).

For widows, the lack of available, older men is the major obstacle to remarriage. The sex ratio for people aged 65 to 74 is 83 men per 100 women; for people aged 75 to 84, the sex ratio is 67 men for every 100 women (D. Smith 2003, Figure 1). Some people suggest that polygyny might be an appropriate solution. If the few available older men were allowed more than one wife, older widows would have more opportunities to remarry. Another way to alleviate the problem would be for women to marry men 8 to 10 years younger than themselves, but this would require a change in the long-standing American tradition of women marrying men 1 to 2 years older than themselves. Nevertheless, marriages in which the woman is older seem to be on the increase. Approximately 23.5 percent of American brides marry younger men. That figure jumps to 41 percent for women ages 35 to 44 (Borg 1996, B4).

Regardless of one's age, it is important to plan ahead against the loss of a spouse. Yet, few people do—probably because it means contemplating the death of a loved one, and this is always unpleasant and often avoided. Nevertheless, married people need to be able to answer a number of practical questions if their spouse should pass away, as follows:

- Is there a will? If so, where is it, and what are its contents?
- Is there life insurance? If so, how much is it, and what must the survivor do to collect it?
- What are the financial liabilities and assets?
- Does one have access to safety deposit boxes and know where the keys are?
- How much cash can be raised in the first 60 days to keep the family going?

Death of a Young Wife: A Young Father Alone

Jim was only 28 when Jane was killed in an auto accident while returning home from her job. He was left with their son, Mike, who was 4 years old. Jim tells his story.

"At first, everyone rushed over and wanted to keep Mike for me. Some of our friends almost couldn't accept no for an answer. Yet, I felt that Mike needed to be with me, and certainly I needed him.

"It is also amazing to remember who came over the first few days. People whom we had hardly known brought food. At first, when they asked what they could do, I replied, 'Nothing.' Yet, they were so obviously disappointed that I finally tried to think of something for them to do. This seemed to make them feel better, although I'm not sure it helped me. In a way, I needed things to do—things to distract me from my grief at least for a short time.

"Although I certainly mourned at first, I found that I was often too angry to mourn. I wanted revenge on the other driver, and I was particularly mad at the city for allowing such a dangerous intersection to exist without stop signs. At times, I was even angry at my wife for driving so poorly, but this always made me feel guilty. Actually, I think I was angry at the whole world. Why did events conspire to take her away from me?

"I had a lot of remorse for things I had put off doing with and for Jane. We really should have spent the money and gone home to see her parents last Christmas. Why hadn't I told her I loved her more often? In many ways, at first, I felt I had failed her.

"As time passed, however, I realized that perhaps she had failed me just a little also. At first, I could only think of the good things. To think of bad things between us when she wasn't there to defend herself just seemed terrible. Gradually, though, I have been able to see her and our relationship as it really was, with both good and bad. I want to preserve her memory for myself and our son, but I want it to be a realistic memory, not a case of heroine worship. I do fantasize about her, especially when I'm alone, and it really helps me to relive some of the memories, but I know that they can't substitute for the present. I must keep living and going forward for myself and our son.

"Although I'm not dating yet, I will in the future. She'd not want me to remain alone the rest of my life. Right now, though, I prefer to be alone with our son, with my thoughts and memories. I need time to understand what has happened, time to be sad. I need time for grief, time to adjust to my new single-parent role, time to ease the pain. I'll be ready for a new relationship only after I have laid the old one gently and lovingly to rest. When the time comes, I will look forward to marrying again."

Satisfactory answers to these questions can greatly ease the practical transition to widowhood. This, in turn, allows the surviving spouse more time to deal with the emotional and psychological transition.

The Grandparenting Role

Many people speak nostalgically of the extended family of rural nineteenth-century America. Parents, grandparents, and grandchildren happily living and working together on the farm is a wonderful, romantic ideal. In reality, the short life expectancy during the nineteenth century meant that this arrangement was relatively rare.

However, between 1900 and 2000, the probability of three or four grandparents being alive when a child reached age 15 increased greatly. Today, three-quarters of older Americans are grandparents, and of these, 40 percent are great-grandparents.

Greater life expectancy has also drastically changed the image of grandparents. Grandparents are no longer the little, old, white-haired couple sitting contentedly in their rocking chairs. Grandparents are often still actively employed, living full and active lives of their own. In fact, the term *young-old* seems an appropriate way to describe many grandparents. Often, couples who became pregnant as teenagers will become grandparents in their 30s. Individuals may now spend up to half their lives as grandparents (C. Henry and Plunkett 1996, 14).

Granddad reading to little one.

Grandma with daughter's new twins.

For some retired, widowed, and elderly persons, fulfilling the role of grandparent can bring back many of the joys and satisfactions of their own early family life. Grandchildren can mean companionship, renewal of intimate contact, the joy of physical contact, and a sense of being needed and useful. One source of family strength reported by older couples is the support offered to them by their children and grandchildren. This is especially true of divorced or single people with children.

Like the retirement and widowed stages of marriage, grandparenting has been greatly extended by increased life expectancy. Most children have a relationship with one or more grandparents throughout their youth. It is quite possible today for grandparenting to be longer than the period in which one's own children were at home. Of course, younger grandparents still married, employed, and living in their own homes do not have the same need to grandparent as older, retired, and perhaps widowed grandparents.

For some older grandparents, the role might be described as "pleasure without responsibility." They can enjoy the grandchildren, without the obligations and responsibilities they had to shoulder for their own children. The grandparenting role gives people a second chance to be an even better parent: "I can do things for my grandchildren that I could never do for my own kids. I was too busy to enjoy my own, but my grandchildren are different." Association with grandchildren can also yield a great deal of physical contact for the widow or widower. Such contact is one of the things most often reported to be missed after a spouse dies.

Physical contact (other than sexual) is an important source of intimacy and emotional gratification, regardless of age. Much physical warmth can be had with grandchildren. Hugging, kissing, and affection gained from grandchildren can go a long way toward replacing the physical satisfaction found earlier in marriage.

Many grandparents also report feelings of biological and psychological renewal from interacting with their grandchildren. "I feel young again," they report, and "I see a future."

This picture of grandparenting is only one of many. The kinds of relationships that grandparents have with their grandchildren vary considerably. Much of the relationship depends on how close grandparents and grandchildren live to each other (King and Elder 1997). If a grandchild lives far away, there may not be a relationship. If one set of grandchildren lives nearby and all the others live at some distance, the grandparents may have a relationship only with those nearby. Some grandchildren may live with parents who don't get along with their own parents, and as a result there may be little or no relationship. There is often selective investment in grandchildren. Grandparents may only relate to one or two of many grandchildren, remaining detached or passive with the others. In addition, the kinds of relationships established by grandmothers and grandfathers tend to differ.

We have discussed the kinds of satisfactions grandparents can derive from their grandchildren. At the same time, grandparents can give a lot to their grandchildren and also to the overall family. Just being there is an important grandparenting function. Family members can draw support and feelings of well-being just by knowing that their grandparents are alive and available (Gutowski 1994). Since grandparents normally are not responsible for grandchildren and, therefore, do not have to set the rules and administer punishment, they can become friends and sometimes allies of their grandchildren. They can give more unconditional love. The grandparents can provide a buffer against family mortality, act as a deterrent against family disruption, serve as arbitrators in family disputes, and provide a place to go to escape marital difficulties. Children whose parents divorce can seek support and permanence with their grandparents. Grandparents can become surrogate parents for grandchildren who don't get along with their parents or are missing a parent because of death or divorce. The media sometimes refer to grandparents as the "silent saviors," "the second line of defense," and the "safety net" for those grandchildren whose parents divorce or are troubled by drugs or other problems that make it impossible for them to parent successfully (Jendrek 1993, 609). In fact, recognizing the value of grandparents led to Congress designating 1995 as the "Year of the Grandparent" (Public Law 103-368, 103rd Congress; Kornhaber 1996, 9).

For those who must suddenly revert to the parenting role at an older age, the task may be difficult, especially if they must take over full-time care of their grandchildren. Freedom is lost, long-made plans for retirement may have to be altered or set aside, and the grandchild/children may be troubled and unruly (Fingerman 1998). For many retired grandparents, raising a grandchild on a fixed income may cause financial hardship. In fact, about 19 percent of grandparent caregivers have incomes below the poverty level (J. M. Simmons and Dye 2003). Grandparental health problems also may be exacerbated by the rearing of grandchildren.

In any case, having raised their own children already, grandparents may feel a certain amount of unfairness in having to partially or totally rear the grandchildren. In such a case, the parents (for whatever reason) shirk their parental responsibilities and hand them over to their own parents by default. Grandparents who assume responsibility for their grandchildren find that their other family relations are also changed. There may be problems between the parent and/or son/daughter or son-/daughter-in-law over such questions as disciplining the children or providing economic support of them.

There are many positive, and also negative, aspects of grandchild care by grandparents (Beaton et al. 2003; Gattai and Musatti 1999). Grandparents who must assume full care for grandchildren so that the children can avoid foster care in a strange home obviously have far more problems than the grandparents who babysit a grandchild a few hours a week.

Across the United States, grandparents representing all classes, races, and cultural backgrounds reported that they were providing day-to-day care of their grandchildren (Allen et al. 2000, 920; Fuller-Thompson et al. 1997; J. M. Simmons and Dye 2003). They take on the parenting role for a number of reasons such as parental divorce, death, substance abuse, and incarceration. With far more two-parent families now in the workforce, as well as unmarried mothers,

Becoming Parents Again after Your Children Are Raised

Mary and her husband Tom are in their early 60s. Their daughter, Suzanne, recently divorced. Because Suzanne had dropped out of high school (against her parents' wishes) to marry, her lack of education made finding a good-paying job impossible. She asked her parents to help her return to school and to care for her two children, aged 2 and 3. Mary and Tom have been caring for Suzanne's children for over 2 years now, with no end in sight because Suzanne has decided to go on to college when she finishes her high school education.

Mary and Tom have had to postpone their travel plans. Although they planned financially for their retirement, the added economic burden of helping Suzanne and caring for her two children has meant a much tighter budget than they had anticipated. "Of course I love the children, but I've been deprived of my golden years and I resent it," says Mary.

What Do You Think?

1. Do you think that Mary and Tom are obligated to help their adult daughter raise her two children?
2. Have you or any of your friends been basically raised by your/their grandparents? What do you think about it?
3. Would you assume responsibility for your grandchildren, if necessary? Why or why not?

---

grandparents are often taking on part-time and even full-time child care for their grandchildren (Baydar and Brooks-Gunn 1998; J. M. Simmons and Dye 2003). Grandparents also play a large role in state foster care programs, accepting more than 60 percent of grandchildren taken from abusive, drug-addicted, and neglectful parents (Banks 2000). The Census Bureau estimates that the number of grandchildren living with their grandparents has increased from 3.0 percent in 1970 to approximately 7+ percent in 2003. Black children are more likely to live with their grandparents than are White or Hispanic children. Approximately 13 percent of Black children, 6 percent of Hispanic, and 4 percent of White children live in the home of their grandparents (J. M. Simmons and Dye 2003; U.S. Census Bureau 1997b, 1998b).

Some grandparents actually have legal custody of their grandchildren, in cases where the parents have died or have been legally declared incompetent. It can be lifesaving for grandchildren to have grandparents who love and care for them when their parents cannot, but it changes the grandparents' lives drastically and, perhaps, unfairly.

The grandparenting role differs by ethnicity or subculture (Kivett 1993; J. M. Simmons and Dye 2003). Among Mexican Americans, relations between grandparents and grandchildren usually remain close. The families often reside nearby, and there is considerable contact between generations. Asian American families also maintain close relationships between grandparents and grandchildren, and have the highest percentage of relatives (usually grandparents) living with the primary family (Barringer et al. 1993, 149–151; S. Lee 1998, 20–21). Another striking contrast between groups is that a larger percentage of African Americans over the age of 60 report raising children other than their own (J. M. Simmons and Dye 2003), compared to Whites and Hispanics.

Increasing life expectancy and the high divorce rate have combined to add a number of new dimensions to the grandparenting role. The increasing life expectancy means that more grandparents have their own elderly parents still living and, perhaps, will have to shoulder some responsibility for them.

Divorce of grandparents or death of one and remarriage of the other may lead to new stepgrandparents for grandchildren. Also, especially when a grandfather remarries, it could mean new children for the grandparents. Of course, remar-

WHAT DO YOU THINK?

1. Do you have a relationship with either set of grandparents?

2. If so, what role do they play in your life?

3. What kinds of problems do your grandparents face?

4. Do you have living great-grandparents? Do you have any interaction with them?

5. Do you think it is fair for children to ask their parents to take care of the grandchildren? Explain.

riage of the grandparents' own children may also mean the grandparents must relate to a new set of stepgrandchildren.

Divorce also raises the question of visitation rights for grandparents and spouses. Because children live with the mother in 89 percent of divorces, it is the paternal grandparents who usually lose contact with their grandchildren. Detached or passive grandparents may not be distressed by this situation, but it can be devastating for active grandparents. At the behest of grandparents, all 50 states now have statutes granting grandparents legal standing to petition for legally enforceable visitations with their grandchildren—even over parental objections (Cory 1994; Kornhaber 1996, 177–189; Purnell and Bagby 1993).

Older but Coming on Strong

The growth of the older population is stunning. As the years proceed, we can expect to see less of the traditional American focus on youth. From our studies of the elderly, however, it is abundantly clear that the life one leads as a young person affects one's prospects in older age. Leading a healthy life when young tends to translate into a healthy old age. Handling money effectively and planning for the older years while still young usually means a comfortable life after retirement.

Gerontologists have no sure-fire prescription for staying physically healthy longer, but they do make some strong recommendations:

- Cut back on drinking.
- Avoid smoking.
- Avoid foods rich in cholesterol and fat.
- Eat high-fiber foods such as whole-grain cereals.
- Stay out of the sun.
- Exercise at least three times a week, including both aerobics and strength training.

The last point is very important because older people tend to lose muscle; this muscle loss, in turn, weakens them and makes them more sedentary (J. Brody 1994).

One hundred years ago, men and women were considered elderly and inactive at age 60. Today, Americans in their 60s are still considered middle-aged. Improved medicine and healthier lifestyles are combining to make the later years of life more active, more fulfilling, and happier for the great majority of Americans.

Summary

1. *Families, like the people within them, span a long period of time.* As individuals change over time, so do families.

2. *Young families who learn to resolve conflicts and handle the stresses and strains of life will have the best chance of becoming enduring, lasting families.* They are most likely to reach the later stages of the family life cycle.

3. *When the last child leaves home, the middle-aged (empty-nest) stage begins*. This is usually a time of reassessment, when both husband and wife consider how their lives will change now that the children are gone. What will be their relationship with their adult children? Between themselves? How will they fulfill the grandparent role?

4. *Retirement enters one's thoughts during the middle years. When it finally comes, the family will have adjustments to make*. Time demands will change abruptly, monetary circumstances may shift downward, and one or both of the spouses leaving the work world that has been part of their identity for so long may have to cope with feelings of uselessness.

5. *Eventually, death must be faced as one spouse (usually, the husband) dies*. Thus, many people return to singleness in their later years. In fact, the period of widowhood is becoming longer as life expectancy continues to rise.

6. *The grandparenting role is taking on more importance as life expectancy increases*. The role is also becoming more complex as families divorce and remarry. Grandparenting can bring small children back into one's life and yield new satisfactions to those aging within the family.

Resources on the Internet

Companion Website for *Human Intimacy: Marriage, the Family, and Its Meaning*, Tenth Edition

http://sociology.wadsworth.com/cox10e/

Gain an even better grasp on this chapter by going to the companion website to take one of the tutorial quizzes, use the flash cards to master key terms, or check out the many other study aids you'll find there. You will also find special features such as GSS data, Sociology Online, and Census 2000 information that will put data and resources at your fingertips to help you with that special project or to help you as you do some research on your own.

InfoTrac College Edition

http://www.infotrac-college.com/wadsworth/

You can access reliable resources anytime, anywhere, with InfoTrac College Edition, the online library. This fully searchable database offers more than 20 years' worth of full-text articles (not abstracts) from almost 5,000 diverse sources, such as top academic journals, newsletters, and up-to-the-minute periodicals, including *Time*, *Newsweek*, *Science*, *Forbes*, *The New York Times*, and *USA Today*. You can conduct electronic key word searches using key terms from this chapter to supplement your reading and learning experience. To aid in your search and to gain useful tips, see the Student Guide to InfoTrac College Edition, which you can access through the companion website for this book.

Should Physicians Help Terminally Ill Patients Commit Suicide?

Assisted suicide, mercy killing, active euthanasia, passive euthanasia, physician-assisted death, no matter what you call it, it's usually illegal. In 1994, a limited right-to-die measure barely passed in Oregon. It allowed physicians to prescribe—but not to administer—a deadly dose of medication to terminally ill patients. Numerous other states have introduced similar bills in their legislatures, but none have passed to date. In addition, suits have been filed to overturn the states' bans on assisted suicide. The debate about this issue may, in the future, begin to rival the debate over abortion.

YES

Dr. Jack Kervorkian has often been in the news in the past few years for his controversial assistance in the suicide of terminally ill patients. The Hemlock Society believes that an individual should have control of his/her own life and, thus, control of her/his own death. The Dutch allow physicians to assist suicide, again under very specific guidelines. About 5 percent of the total number of Dutch people that die each year choose to get help in doing so (Girsh 1999).

Most public opinion polls find that 60 to 80 percent of those polled favor legalization of assisted suicide. Advanced medical technology can now keep patients alive for months, whereas in the past, they would have died. Many people are leery of such life-extending treatments, feeling that life may not be worth living if the quality of that life is insufficient. For example, acute cancer patients often must go through treatments that are toxic and cause a great deal of discomfort, with little or no chance of more than postponing the inevitable death.

To die with dignity is far better, to many persons, than being artificially kept alive in a miserable state via breathing and heart machines and numbed by drugs. Another consequence of legalizing assisted suicide would be to reduce the number of desperately ill persons who try by other means, such as a gun, to end their lives. These persons sometimes fail and end up in worse shape and/or traumatize their families.

If the historical American emphasis on and respect for individual rights is considered, then allowing a person the decision to die in the face of debilitating disease or accident seems correct.

Note: The lower house of the Dutch Parliament approved a bill legalizing euthanasia in November 2000, and the upper house approved it in 2001. The following guidelines were set forth:

- The physician must be convinced the patient's request is voluntary and well-considered.
- The physician must be convinced that the patient is facing unremitting, and unbearable, suffering. The patient does not have to be terminally ill.
- The patient must have a correct and clear understanding of his situation and prognosis.
- The physician must reach the conclusion, together with the patient, that there is no reasonable alternative acceptable to the patient. The decision to die must be the patient's own.
- The physician must consult at least one other independent doctor who has examined the patient.
- The doctor must carry out the termination of life in a medically appropriate manner (Deutsch 2000).

NO

Where would the line be drawn? Who is to define terminally ill, and with ever-improving medical technology, can one ever conclude with certainty that an illness is terminal? How does one define a standard of life so low that death would be the better option? Once the door is opened to assisted suicide, might the decision be lost to the ill individual and made by others for whom the individual has become too heavy a burden? What about the elderly or infirm who can't make the decision for themselves? How sure can physicians be that the person in a coma will never come out of it?

The potential ethical, religious, and legal questions are too numerous to mention. The line between voluntary and nonvoluntary euthanasia cannot hold in practice. Once suicide and assisting suicide are okay for reasons of "mercy," then delivering others that a society may deem in need of mercy to the afterlife, whether chosen by said individual or not, becomes a possibility. Life is sacred and should be regarded as such. Once the doors are opened to allow death upon request, how long will it be before death upon order is permissible for all the "right" reasons: to put the pained out of their misery, to reduce the monetary drain on the society, to rid society of those who are social burdens, and so forth?

What Do You Think?

1. Under what circumstances, if any, could you personally assist another person's suicide?

2. How do you feel religiously and/or ethically about physicians assisting the suicide of their patients who request their help?

3. If you believe in assisted suicide, do you think physicians alone should have this authority, or should family members also be allowed to assist suicide? Explain.

4. If a person believes in assisted suicide, does this mean that a person also believes in the death penalty for certain classes of criminals? Explain.

5. Can a person rationally believe in assisted suicide and not believe in the death penalty? Explain.

© Bob Daemmrich/PhotoEdit

The Importance of Making Sound Economic Decisions

Chapter

12

Questions to reflect upon as you read this chapter:

- How important do you think money and economics are to the success of the family?

- Do you use a credit card? Do you run a monthly balance? If so, what interest do you pay on this balance?

- Does advertising affect the way you spend? How so?

- If you were to marry, how would you manage your money with your partner?

- What are some financial maneuvers you can do to guard against inflation?

- If you were married, had small children, and owned a house, what kinds of insurance would you need?

- What role does systematic saving have in your life?

- How much thought have you given to investments?

Marital quality, job satisfaction, and money matters are all closely related. Although few people directly ascribe difficulties in their marriages and families to their work situation and/or money problems, the fact is that work, money, and marital quality are closely intertwined in most families. Trouble in one area almost guarantees trouble in the other two areas. For example, the spouse who works two jobs to earn enough money to pay all the bills almost certainly comes home overtired and, perhaps, somewhat disinterested in working on his/her marital relationship.

In the majority of today's families, both husband and wife work full time (Moen 2002; L. White and Rogers 2000); thus, time for togetherness, sharing, play, enjoying the children, making love, and other intimate family activities is more limited.

Unemployment within a family brings worry and stress as bills accumulate and money dwindles. Worry and stress are particularly troublesome enemies of successful, intimate relationships. Economic success certainly does not guarantee an intimate, healthy, happy family, but economic failure almost always leads to family problems and breakdown. Divorce, separation, and desertion are highest among the poor.

As a software research consultant earning more than $50,000 a year, 34-year-old Dave Johnson could be your typical affluent yuppie. Unfortunately, his income does not go as far as one might expect, and he finds himself increasingly frustrated by his inability to provide for his family as he would like.

After paying rent and other expenses, Dave, his wife, and their two young children have little money left over to save toward the purchase of a house.

"I'm making more money than my dad ever did in his life, but I can't afford the home he could." Noting that his car payments of $300 a month were more than his dad's house payments, he said, "It's frustrating. I've got a good and stable job with potential for upward mobility, yet with just one income, getting a nice home will be difficult."

Dave and his family are experiencing the frustrations felt by many Americans as they try to cope with the economic realities of providing a living and finding economic satisfaction. Spillover between marital quality and job satisfaction affects all families (J. J. Rogers and May 2003). For most families, the married single-earner family has been replaced by the dual-earner family, because over half of all wives are now in the workforce, usually to help make ends meet (see Chapter 13).

Most courting couples today pay almost no attention early in the relationship to each other's financial values. Money is often a more taboo subject than sex. These couples may discuss their prior sex lives, but rarely raise the question of economic histories. After all, talking about one's potential income, use of credit cards, or feelings about saving or indebtedness is not very romantic. Yet, money matters are the topics most commonly discussed, and money is the most common source of conflict for America's married couples (Cohen 1995; Kluwer et al. 1996; Madanes and Madanes 1994).

What should we buy? When should we buy? Who should buy? Who should make the spending decisions? Who will pay the bills? Should we pool our money, or maintain separate accounts? Such questions become particularly troublesome if the partners have divergent attitudes about money. For example, consider a person who comes from a background of thrift and practicality and takes pleasure in making a good buy. This individual becomes excited about buying a used car at wholesale, rather than retail prices and will probably brag about the purchase. Any minor problems with the car will not necessarily be upsetting, because it was such a good buy. Suppose this person's partner comes from a luxurious environment, where emphasis is placed on obtaining precisely what one wants and success is often measured in economic terms. He/she believes in buying a new car of the appropriate model and considers a used car, especially a "steal," a mark of poor taste and economic failure. The married life of such a pair may be filled with conflict because of their different monetary attitudes. Again, as with so many aspects of life, it is one's expectations and attitudes, this time about money, that cause conflict (see Chapter 7).

Money and work affect everything we do, both as individuals and as family members. We are involved with work and money on a daily basis. Even the retired who no longer work must attend to finances, as must the wealthy. (How are my investments doing? Will my retirement and/or social security check arrive today?) Successful work and efficient money management are two of the major foundations of family success, regardless of economic level.

As we think about the characteristics of strong, healthy families, it is clear that economic stability is necessary for the development of family strength. Economic stability and security enable families to turn their attention from issues of basic survival to enhancing the quality of life. Family strengths can evolve only if economic survival is ensured. Given the complexity and uncertainty of the economic climate, families need to learn all that they can about the economic system if they are to be successful.

WHAT RESEARCH TELLS US

GNI per Capita in Various Countries

In 2003, per capita gross national product (GNI, gross national income per person) in America was $36,110. Of the major countries of the world, only Norway ($36,690) had a higher per capita GNI. In many nations, it is much lower. For example, Congo and Tanzania have per capital GNIs of less than $800. The average per capita GNP in the 16 countries making up western Africa is only $1,070 (Population Reference Bureau 2004c).

WHAT RESEARCH TELLS US

Median Income of Households and Families

The median income of households was $41,994 as measured in census 2000. Median family income was $50,946 with married-couple families having a higher median income ($57,345). For females with no husband present, the median income was $25,458. With a male householder and no wife present, the median income was $35,141. Real median household income and family income are now at their highest levels since the Census Bureau started compiling these estimates in 1967. Asians continued to have the highest median household income. White households had the second highest, followed by Hispanic households and Black households (Clark et al. 2003). The poverty rate also dropped to its lowest rate in 2 decades (11.8 percent), although it was still above its historical low of 11.3 percent in 1973 (Holmes 2000, A1). It might appear from the data that marriage may increase the odds of affluence. (See Hirschl et al. 2003 for a discussion of this hypothesis.)

Work also produces by-products other than money. Both personal and family status is derived from occupation. A medical doctor and his/her family enjoy high status (at least in America; the status of occupations varies among cultures), which influences the family's lifestyle. Individuals' sense of self-esteem also relates closely to their work. Self-esteem is one of the first things threatened by unemployment. Furthermore, one's sense of identity is based largely on what is done in the workplace. Often, the first question asked of a new acquaintance is "What do you do?" A person's (family's) circle of friends is drawn, in part, from the work world. Thus, the work world and the family world are intimately related, and each influences the other in many ways. Just as the work world will influence a person's marriage and family life, so a person's marriage and family life can affect how he/she does in the work world (Grzywacz et al. 2002; J. J. Rogers and May 2003). Despite Americans' apparent wealth, dissatisfaction with money matters remains high.

Many families, regardless of their actual income, feel that everything would be fine if they could earn $10,000 to $20,000 more per year. In most families, needs and desires are, in fact, increasing even faster than income. Thus, monetary dissatisfaction is often higher for families earning $50,000 a year than it is for families earning $25,000.

Americans may earn more money than most others, but they also spend more than other people and are more deeply in debt. Most money is spent by family units to support family members, so personal income is an important measure of family function. In the last few years, personal income has increased, but so also has personal spending and consumer debt, reducing the savings rate to just 2 percent of after-tax income in 2003 (Powell 2004).

Not all Americans share equally in the wealth. For example, married couples do considerably better monetarily than average. (Table 12-6 indicates some of these differences by various groups.)

The modern American family is the basic economic unit of society because it is the major consumption unit. In the early years of our country, the family was also the major production unit. At that time, 90 percent of the population worked in agriculture, on family-owned farms. Over the years, however, most farm workers became factory workers and then service workers. Thus, today's average family is not directly involved in economic production in the home or on their own land.

As a consuming unit, the family exerts great economic influence. A couple anticipating marriage and children are also anticipating separate housing from their parents. This means a refrigerator, range, furniture, dishes, television, and so on. And how will this new family acquire all of these items? Probably by using credit—not an American invention, but certainly an American way of life. Approximately 85 percent of all American households owe money at some time in each year. Unfortunately, credit and debt are directly associated with bankruptcy.

"Buy now, pay later!" "Why wait? Only $15 per week." These are the economic slogans of modern American society. The extension of credit to the general population has produced a material standard of living the likes of which the world has never seen. On the promise of future payment, we can acquire and do almost anything we desire: material goods, travel, education, and services such as medical and dental care. If the credit system were suddenly abolished, the degree to which it supports the economy would become glaringly clear. Traffic congestion would end as the majority of autos disappeared from the roads. Many buildings, both business and residential, would become empty lots or smaller, shabbier versions of themselves. Thousands of televisions would disappear from living rooms. If debtors' prisons were reestablished at the same time, practically the entire population would be incarcerated!

It is imperative that individuals plan their economic destiny carefully. Those who do not plan are doomed to lose control of their economic lives. Knowing how to budget, spend, save, borrow, and invest are important skills for personal and family stability and happiness. These skills help individuals and families cope with an ever-changing economy. When economic times are good, we tend to forget the raging inflation of 1979–1980 and the recession of 2002–2003, but such changes in the economy have always occurred.

Even in good economic times and with good economic planning, married couples often quarrel about money. As we shall see later, these arguments over money essentially revolve around allocation of resources and control of spending. These arguments stem mainly from the attitudes about money that each person brings to the partnership (see "Making Decisions: Compare Your Attitudes About Money").

Although futurists are predicting that most people will probably have several jobs over their lifetime rather than just one, as was prevalent in the past (Moen 2002), finding the right job and avoiding unemployment are important to a person and his/her family. One way to increase work security is to select a vocation that will grow in the future. The Bureau of Labor Statistics has prepared projections of the U.S. economy through the year 2008 (Table 12-1). By this time, the bureau expects the labor force will have expanded by 14.4 percent and will have become increasingly minority and female. Work will require constant learning and upgrading of one's skills. There will be fewer nine-to-five jobs, and the hiring of temporary employees will increase. Careful review of Table 12-1 will indicate where the bureau expects job expansion. (See also the U.S. Department of Labor's *Occupational Outlook Handbook*, 2004–2005 edition.)

Slowly Drowning in a Sea of Debt

For many young couples, marriage actually means a drastic reduction in standard of living. Accustomed to living at home and sharing their parents' standard of living (usually created by 30 years or more of their parents' hard work), young newlyweds are now on their own economically. Entry-level jobs may be scarce, and pay is low compared with their parents' current earnings. If the newlyweds

TABLE 12-1

Fastest Growing Occupations, 1998–2008 (%)

Occupation	Percent	Occupation	Percent
Computer scientists	118	Surgical technologists	42
Computer engineers	108	Securities, commodities, and financial services sales agents	41
Computer support specialists	102	Occupational therapy assistants and aids	40
System analysts	94	Correctional officers	39
Computer, mathematical, and operations research	93	Speech-language pathologists and audiologists	39
Database administrators	77	Clinical laboratory technologists	39
Paralegals and legal assistants	62	Social and recreation workers	38
Medical assistants	58	Social workers	36
Personal care and home health aides	58	Teachers, special education	34
Social and human services assistants	53	Medical and health services managers	33
Physician assistants	48	Emergency medical technicians and paramedics	32
Social residential counselors	46	Models, demonstrators, and product promoters	32
Electronic semiconductor processors	45	Law enforcement occupations	31
Medical records and health information technicians	44	Computer programmers	30
Physical therapy assistants and aids	44	Camera operators, television, video motion picture	29
Other health services workers	44	Health services occupations	29
Respiratory therapists	43	Registered nurses	22
Dental assistants	42		

SOURCE: Braddock 1999, Tables 2 and 3.

fail to understand this and attempt to maintain a standard of living comparable to their parents, they are likely to become so trapped in the economic system that they may never be able to gain economic freedom.

Many young Americans, especially men, have love affairs with automobiles, and this love often starts them down the road of debt. In the United States in 2000, approximately 221 million cars and trucks were in use for a population of 294 million. Americans have more vehicles for its population than any country in the world (U.S. Census Bureau 2002b, Table 1062). The average price of a new car in 2001 was $21,605, compared to $14,371 in 1990 and $3,500 in 1970 (U.S. Census Bureau 2002b, Table 1010).

Buying a car in the United States by using credit is simple, and this is often the young person's first indebtedness. The costs of car ownership are much higher than many people realize. The Census Bureau reports that the average yearly cost in 2000 for transportation was $7,417 (U.S. Census Bureau 2002b, Table 650). Hertz Corporation reports yearly on per-mile costs for its fleet of rental vehicles. For example, an economy car costs about 38 cents per mile to operate. This sum includes all costs such as insurance, depreciation, gas, and maintenance. A luxury car will cost about 85 cents per mile. Since the automobile is one of the products in the American economy that usually quickly declines in value, it quickly becomes worth less than what is owed on it.

A young person moving away from home is taking a big economic step. He/she is now independent, free of constant parental supervision, and enters an economic world that may not be completely understood. All too often, the basic necessities such as food and rent also include television and DVD sets, a washer and dryer, and so forth. Rather than postponing buying or buying used, some newly independent persons buy on credit and find themselves trapped economically be-

WHAT RESEARCH TELLS US

How Much Do Americans Owe?

- Consumer debt includes automobile loans and credit cards, but excludes real estate. Americans owed approximately $1.98 trillion, or approximately $18,700 per household, for outstanding consumer credit in 2003 (Powell 2004).

- Approximately $447 billion of the above was for the purchase of automobiles.

- Total credit card debt in 2001 was $735 billion, projected to rise to $985 billion in 2005. This works out to nearly $7,000 per household. Average interest rate was 18+ percent or $1,800+ a year in interest cost on a $10,000 balance (Powell 2004; U.S. Census Bureau 2002b, Table 1165).

- Home mortgage debt includes home-equity lines of credit. Americans owed approximately $5.385 trillion (U.S. Census Bureau 2002b, Table 678).

- Personal bankruptcies hit an all-time high in the 12 months ending June 2003 (1,613,097), up 100+ percent over 1990 (*Santa Barbara News Press* 2003).

fore they realize it. Often it is the credit card, with its high interest, that places them thousands of dollars in debt. Gradually, one's income is claimed before it is even received. Economic freedom is lost.

As debt increases, there may come a day when a person's salary doesn't quite cover monthly expenses. Loan companies offer what appears to be a way out, but it is often only a way into deeper debt: "Borrow all the money you need to get completely out of debt." Of course, combining all small debts into one giant debt that reduces overall payments isn't getting out of debt. In actuality, the interest on such loans is usually high, thus placing the borrower deeper in debt. Rather than continuing this sad story, suffice it to say that within a few years, some young people (and many old enough to know better) may be sucked so deeply into debt that bankruptcy court is the only way out. In most bankruptcy cases, the indebted person has not gambled and lost on a big-investment speculation; he/she has simply slowly drowned in a rising sea of debt.

Contrary to popular opinion, most families that go bankrupt are lower- and middle-class families. Moreover, going through bankruptcy does not seem to help people become more economically prudent. Of those who file for bankruptcy, 80 percent use credit and are in debt trouble again within 5 years. Hence, it is critical for everyone to know as much as possible about credit and borrowing so that they can control their use of credit, rather than being controlled by it.

Consumer-counseling agencies help clients who are having financial troubles by teaching them better money management. These counselors say that many who find themselves insolvent didn't even realize they were getting into financial difficulty. How can you tell if you are headed down the road to insolvency? Consider the following:

- You are spending more than you make. Bills are piling up, and creditors are sending second and third notices.

- You find that you are tucking away nothing for a rainy day. You have no regular savings plan.

- You are making no plans for retirement (even if you are just starting to work, such plans are important).

- You have many credit cards, all of which run a considerable balance.

- You have no liquid assets. All your money is tied up in your home or business, and you cannot easily come up with cash.

MAKING DECISIONS

Compare Your Attitudes about Money

Answer the following questions without discussing them with your partner. Then have your partner answer them. If you answer the questions differently, it may indicate points of attitudinal difference and possible conflict. You should discuss the reasoning behind your answers and how your two positions can be reconciled.

1. Are you comfortable living without a steady income?
2. Did your parents have a steady income?
3. Do you consider yourself to have come from an economically poor, average, or wealthy background?
4. Do you think that saving is of value in America's inflationary economy?
5. Do you have a savings account? Do you contribute to it regularly?
6. In the past, have you postponed buying things until you had saved the money for them, or did you buy immediately when you desired something?
7. In the past, have you often bought on installment?
8. Do you have credit cards? How many? Do you use them regularly?
9. Do you have money left over at the end of your regular pay period?

10. Do you think positively about making a really good buy or finding a real bargain?
11. If it were possible to save $100 a month for a year, what would you do with the money?
12. Possible answers to question 11 are listed below. Rank them in order of importance, using 1 for what you would most likely do with the $1,200 and 10 for what you are least likely to do:
   a. Save it for a rainy day.
   b. Save it to buy something for cash rather than on credit.
   c. Invest it.
   d. Use it for recreation.
   e. Use it for payments for a new car.
   f. Use it for travel and adventure.
   g. Use it to buy a home or property.
   h. Use it to improve your present living place.
   i. Divide it in half and let each spouse spend it as he/she chooses
   j. Use it for an attractive wardrobe, eating out, and entertaining.

- You can never say no to your wants and desires.
- You think your financial problems will take care of themselves and somehow magically disappear.

Making Good Credit, Borrowing, and Installment-Buying Decisions

Credit buying has allowed the average American a higher standard of living than most people in the past ever dreamed possible. The ability to buy only on the promise of future payment is a relatively modern invention. Credit has given Americans the means for a healthier, more fulfilling life, but it can also enslave people subtly and at great psychological cost. This entrapment and loss of freedom usually occur because people are ignorant of the system and blindly accept persuasive and seductive advertising urging them to use credit to purchase everything they want. An understanding of economics and the relationship of debt and credit to personal freedom will help you make the system work for you, instead of against you (see "Debate the Issues" at the end of this chapter).

Modern economic entrapment is far more seductive than historical systems of slavery based on power. In real slavery, slaves know who the enemy is and where to direct their hostility. But in the American economic system, one places oneself in the slavery of debt and thus has no one else to blame for the monetary predicament.

People borrow for two basic reasons: to buy consumer goods and services, and to invest in tangible assets. Consumer debt is high-priced money, because it is used for consumable goods such as cars, furniture, and clothing that diminish in value with time. **Discount interest** is usually charged for consumer debt. This kind of interest is charged on the total amount of the loan for the entire time period.

Investment debt, or real-property debt, is lower-priced money, because it is used for tangible assets such as real estate or businesses, where value is more permanent. If for some reason the debt is not paid, the creditor may assume ownership of the asset and sell it to regain the loaned money. Simple interest is charged for investment debt. This kind of interest is charged only on the unpaid balance of the loan. (See Table 12-2 for comparative costs of consumer credit.)

discount interest
Interest charged on the total amount of a loan for the entire time period
investment debt
Money borrowed at a lower interest rate than discount interest because it is used for tangible assets, such as real estate, where value is more permanent

Discount Interest: Consumer Purchases

If an individual borrows $1,000 for 3 years at 10 percent interest per year, he/she must pay $100 each year for the use of the money ($0.10 \times \$1,000 = \$100$). Eight dollars and 33 cents is paid each month in interest ($100/12 months = $8.33). In addition, the principal of $1,000 will be paid back in 36 equal monthly installments, so that it is all paid off at the end of the 3 years. The monthly principal payment will be $27.77 ($1,000/36 = $27.77). Thus, the total monthly payment is $36.10; interest plus the principal payment ($8.33 + $27.77 = $36.10). (Interest rates now change so rapidly, that the rates used in these examples may not accurately reflect the rates in effect when you are reading this text. The general principles hold, however.)

TABLE 12-2

Comparative Costs of Credit

Lender	Type of Loan	Annual Percentage Rate [1]	Remarks
Banks	Personal loans (consumer goods) Real-property loans General loans	7–18 [2]	60% of all car loans, 30% of other consumer-good loans; real-property loans have lower interest rates because property retains value, which may cover defaulted loans.
Credit cards	Personal loans Cash loans	17–21 [2]	Used as convenience instead of cash; credit is approved for the card rather than individual purchase; no interest charged if bills are paid in full each month; cash in varying amounts, depending on individual's credit rating, may also be borrowed against the approved line of credit.
Credit unions	Personal loans Real-property loans	6–15	Voluntary organizations in which members invest their own money and from which they may borrow.
Finance companies	Personal loans Real-property loans	12–40	Direct loans to customers; also buy installment credit from retailers and collect rest of debt so that retailers can get cash when they need it.
Savings and loan companies	Real-property loans	7–15 [2]	Low interest rates because real property has value that may cover defaulted loans.

[1] Interest rates vary because of pressures of inflation and recession.

[2] If the institution charges interest on the *face amount* of the loan over the entire period of the loan, double the listed interest rate to estimate the actual interest rate.

Although $100 interest will be paid each year on the $1,000 loan, the individual does not actually have the use of the full $1,000 for the entire 3 years. Each month, $27.77 of the loan is paid back. At the end of a month (after one payment), the debt is only $972.23 ($1,000 minus the principal payment of $27.77 = $972.23) on the loan. Each month, what is actually owed on the loan is reduced by the principal payment, until at the end of the 3 years (36 payments), the loan is paid off.

All an individual has to remember is that on a discount interest loan, he/she pays interest on the full amount of the loan each year, even though part of the loan has been paid back. Such interest is usually figured for the full term of the loan, and is added to the face amount of the loan at the time a person makes the loan. Thus, in our example, an individual would sign for a $1,300 debt ($1,000 principal + $300 interest = $1,300), but only receive $1,000. Because that person has less of the $1,000 with each payment, yet pays interest on the entire $1,000 throughout the 3 years, he/she is really paying a much higher interest than 10 percent on the actual money borrowed. As a rule of thumb, to figure the actual interest rate on this type of loan, simply double the stated interest rate. When computed accurately, however, the interest rate will slightly exceed the doubled figure.

Be careful using credit. Debt can build up quickly.

Credit Card Use   If a person maintains a balance on a credit card, rather than paying promptly at the end of each month, he/she is charged interest at a very high rate. The interest rates charged by the top-ten credit card lenders range from 15 to 21 percent. No interest is charged if payment is made within 30 days. Thus, if used properly, a credit card gives you 30 days of open credit; that is, funds can be used without having to pay interest. On the other hand, carrying a large balance is extremely costly. About 60 percent of cardholders neglect to pay the full balance each month (Powell 2004).

Many high school students, and even more college students, now have their own credit cards. The credit card seems like real money to many young people, who frequently have trouble controlling their credit card spending. Many students now graduating from college are already greatly in credit card debt. If loan debts for college attendance are added, many students graduate thousands of dollars in debt.

When Antonio first got a credit card, he often went on buying sprees. Using the credit card didn't seem the same as spending real money. Just before he married, he had a total of $4,500 outstanding balances on all his credit cards combined. Try as they might after marrying, Kimberly and Antonio just couldn't seem to reduce this debt. In fact, it gradually increased, because both of them now used the credit cards. Do you have a credit card(s)? Do you run a balance each month? If so, how much on the average? How much interest do you pay on the average per month? Per year?

Simple Interest: Home Loans

Because home loans are invested in a tangible asset, the interest charged is usually at a lower percentage rate and is **simple interest** (that is, charged only on the outstanding balance). An individual decides to buy a $100,000 home (home

simple interest
Interest charged only on the outstanding balance of a loan

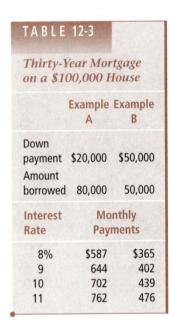

TABLE 12-3

Thirty-Year Mortgage on a $100,000 House

	Example A	Example B
Down payment	$20,000	$50,000
Amount borrowed	80,000	50,000

Interest Rate	Monthly Payments	
8%	$587	$365
9	644	402
10	702	439
11	762	476

prices vary greatly from area to area; see page 389). Twenty thousand dollars is put down, and a loan of $80,000 for 30 years at 10 percent simple interest is taken (interest rates for fixed home loans have varied dramatically in the past years from lows of 5.0+ percent to highs of 12 percent). For example, to pay off the $80,000 loan and interest in 30 years, payments will be $702 per month. Each month, the individual pays less interest and more principal. At the end of 30 years when the $80,000 loan is paid off, about $172,735 in interest will have been paid! At present, one can deduct home mortgage interest charges from taxable income, which somewhat reduces the actual interest cost.

Table 12-3 shows what monthly payments would be at different rates of interest on a $100,000 house when $20,000 is put down ($80,000 owed), or when $50,000 is put down ($50,000 owed). The term of the loan is 30 years.

Financial Problems and Marital Strain

It is clear that financial difficulties affect all areas of a relationship, not just the economic realm. A couple's constant worry about overdue bills, whether their savings are sufficient to cover a medical emergency, disappointment because they cannot take a vacation or buy a new dining set, the wide gap between their marginal economic survival and the plush lifestyles shown on television, and discontent with work or the way income is spent—all combine to lower marital satisfaction.

Economic entrapment can quickly engulf young couples who do not control their spending and credit use. Antonio and Kimberly become so indebted they finally declare bankruptcy. Under a court-supervised payment plan, Antonio and Kimberly are slowly paying off their debts, including those incurred by the birth of their second child. They resent being economically trapped because every cent of Antonio's salary seems to be earmarked for the old debts.

Kimberly keeps asking why Antonio doesn't get a better job. At one time, he investigated going back to school to qualify for a market manager position. But he would need to spend another 2 years in school, and Kimberly cannot earn enough to support the family during those 2 years, much less pay off past debts. Instead, Antonio has taken a second job as a night watchman. He is usually so tired when he comes home that the smallest noise is painful, and he constantly yells at the children to be quiet or at Kimberly to make them be quiet. He is usually too tired to make love at night. Why, he wonders, did he get married in the first place? Kimberly sometimes wonders the same thing about herself. Why didn't they wait before making major purchases? Why didn't they postpone children until they were on their feet economically? In one of the wealthiest nations in the world, how could this couple have been economically strangled?

Financial pressures have obviously put a great strain on this marriage. Taking a second job, especially one that takes Antonio away at night, greatly increases problems between him and Kimberly. Research indicates that among men with children, married less than 5 years, working nights makes separation or divorce six times more likely relative to working days (Presser 2000, 93). In view of this pressure and the fact that they married in their teens, it is quite likely that Antonio and Kimberly's marriage will end in divorce.

The Seductive Society: Credit and Advertising

Advertising and the availability of credit make spending and consumption hard to resist. Modern spending behavior in American society bears little resemblance to the economic truisms one learns in the family and/or at school. Buying and spending have quietly replaced thrift and saving. Traditional values, though still preached, are often no longer practiced. And in a productive, inflationary economy such as ours, they are no longer even virtues. Spending is important in a credit-oriented, inflationary society. Goods must be kept moving. The failure of consumers to buy can immediately have dire consequences for the economy. Hence, the financial section of the newspaper includes such statements as "Strong growth in consumption reflects rising income and bodes well for the economy." Such attitudes encourage individual and family spending for the good of the overall economy and for the family.

A slowdown in consumer spending in any one of the basic industries (such as automobiles) affects the whole economy, not just the particular industry. The impact is on literally hundreds of subsidiary industries, and many other major industries such as steel, rubber, and aluminum. If consumers fail to buy, production must be cut back, which means laying off workers. The problem is then compounded by the loss of these workers' buying power as consumers. Therefore, to keep goods flowing, stimulation and creation of wants and desires in the consumer is necessary.

In his classic work *The Affluent Society* (1958), John Kenneth Galbraith observes that the theory of consumer demand in America is based on the following two broad propositions:

1. The urgency of wants does not diminish appreciably as more of them are satisfied.
2. Wants originate in the personality of the consumer, and are capable of indefinite expansion.

These two propositions go a long way toward explaining why actual income bears little resemblance to a family's feeling of economic satisfaction. Although Americans command a better standard of living than most people, many Americans are dissatisfied with the amount of money they make. Economic contentment appears to be more closely related to one's attitudes and values than to one's actual economic level.

Advertising and need stimulation have become essential parts of the American economy. A family or individual has to be made to want new material goods for more than rational, practical reasons. For example, though a well-maintained automobile can last 10 years, such longevity for the average car would greatly upset automobile production. The auto industry has met this problem by changing models often. In all fairness to the auto industry, it should be noted that many model changes are improvements. However, change has also been made so that older models will appear less desirable.

Today's youth are growing up in a different economic atmosphere from that of their great-grandparents. The society they know is an affluent society. Even a period of relative job scarcity and recession has little effect on spending habits. Buying, spending, credit, and debt are now familiar accompaniments to marital life. The advantages of such a system cannot be denied. Yet, to use the system to fullest advantage, one must also be aware of the dangers. The power of advertising and the ability to satisfy one's needs or desires immediately are formidable and seductive forces for even mature adults. How well can a young couple resist the invitation to use a store's credit again, when they have almost paid off their bill? An official-looking check arrives in the mail, announcing that they can obtain an additional $500 worth of merchandise for nothing down and no increase in the monthly payments that were otherwise about to end. If they thoroughly understand what exercising their desires in this manner means, they can make use of some or all of the offer with no danger. On the other hand, another purchase could be the proverbial backbreaking straw when it is added to the rest of their financial debt.

Personal freedom and indebtedness vary inversely. The more debt one assumes, the less personal freedom one has. Some experts suggest that families should limit their debt total to no more than 20 percent of their spendable income.

In *The Affluent Society*, Galbraith pointed out that a direct link between production and wants is provided by the institutions of modern advertising and salesmanship. These cannot be reconciled with the notion of independently determined desires, for their central function is to create desires—to bring into being wants that previously did not exist (1958, 155).

Vance Packard, in another classic book, *The Hidden Persuaders* (1958), exposed the extent to which advertising influences the public's attitudes, values, and behavior. He questioned the morality of advertising techniques that manipulate the consumer into buying, regardless of the consequences. In concluding his book, he asked the following series of provocative questions that young married couples might well consider:

1. What is the morality of the practice of encouraging [homemakers] to be irrational and impulsive in buying family food?
2. What is the morality of manipulating small children even before they reach the age where they are legally responsible for their actions?
3. What is the morality of playing upon hidden weaknesses and frailties—such as our anxieties, aggressive feelings, dread of nonconformity, and infantile hangovers—to sell products?
4. What is the morality of developing in the public an attitude of wastefulness toward national resources by encouraging the "psychological obsolescence" of products already in use?

Approximately $80 billion was spent on advertising in 2002 by the top-100 advertisers. See Table 12-4 for the top-10 advertisers. To give you an idea of national advertising costs, a full-page ad in *Time* magazine on a single day cost $150,000 to $200,000 in 2003. The majority of this advertising went toward creating wants and desires that will add new frustrations to American families, many of whom are already monetarily unhappy, even though they live at one of the highest material levels in the world.

Effective Money Management

Few dimensions of family life are as important, yet as difficult, as the management of family financial resources. One-third of all married couples consider money their number one area of conflict (Pybrum 1995, vii).

TABLE 12-4

Ten Leading National Advertisers, 2002

Rank	Company	Advertising Dollars (in Billions)[1]
1.	General Motors	3.65
2.	Time-Warner	2.92
3.	Procter and Gamble	2.67
4.	Pfizer	2.57
5.	Ford	2.25
6.	Daimler Chrysler	2.03
7.	Walt Disney	1.80
8.	Johnson and Johnson	1.80
9.	Sears	1.66
10.	Unilever	1.64

[1] Based on measured media expenditures only; does not include local advertising coupons, direct mail, premiums, trade shows, or product sampling.

SOURCE: *Advertising Age* 2003.

The most important step in effective money management is to determine ahead of time how most monetary decisions will be made. A family can handle monetary decisions in at least six ways:

1. The husband can make all the decisions.
2. The wife can make all the decisions.
3. They can make all decisions jointly.
4. One spouse can control the income, but give the other a household allowance.
5. Each spouse can have separate funds and share agreed-on financial obligations.
6. The spouses can have a joint bank account on which each can draw, as necessary.

Once a couple reaches an agreement on the system by which they will handle family finances, most day-to-day monetary decisions can be handled automatically.

To Pool or Not to Pool Family Money?

The question of whether earnings should be pooled seems to have no right answer. Couples who favor pooling their earnings seem neither more nor less satisfied with their money management than couples who insist on keeping money separate. Nevertheless, both types of couples felt that their system was right for them.

Both systems have advantages and disadvantages. The newly married couple often finds it difficult to avoid pooling. If one partner suggests separating the money, the other is apt to interpret the suggestion as a lack of commitment to the relationship.

Pooling is simpler than having separate funds, because there are fewer accounts to balance. Also, each spouse can see what the other spouse is doing monetarily. On the other hand, spouses may lose their feelings of independence. Pooling can also lead to confusion; for example, a joint checking account can be overdrawn if each partner writes a check unknown to the other. Despite how a couple may feel about pooling their funds when they first marry, pooling tends to become the method of choice as the relationship persists.

Pooling is more highly favored by married couples than cohabiting couples, probably because married couples feel more permanent than cohabiting couples.

People who remarry after divorce are more apt to maintain separate funds for two reasons: (1) they may feel less permanence in the new relationship due to the divorce trauma, and (2) they are very apt to bring to the new marriage assets that were accumulated prior to the relationship—assets that belong not to the new relationship, but to themselves as individuals (and perhaps to any children of the first marriage).

Allocation of Funds: Who Makes the Spending Decisions?

If funds are separate, some kind of joint responsibility for family spending must be established. Does one partner assume monetary responsibility for rent or house payments, while the other pays for food? Do they both contribute from separate funds to a single household account that is used to make family payments? If so, who controls the account? Are both partners free to spend their separate funds in any way they wish? If one partner has greater income or assets than the other, does that partner assume more of the responsibility for family expenses?

The answers to such questions are related to power and control within a relationship. Generally, the partner who supplies the primary monetary support also claims the majority of the power in a relationship. "After all, it is my money, and I can spend it any way I please!" "Remember that I earn the money in this family, and you will spend it the way I want you to!" Power through monetary control is greatest in the single-earner family. Obviously, in the past, the power most often accrued to the male partner because it was he who worked outside the home to supply the necessary funds. The rapidly growing contribution of wives to family finances has wrought a revolution in the family power structure (see Chapter 13).

Answers to the questions posed above will vary greatly between single- and dual-earner families. In either case, the answers must be satisfactory to both partners (role equity) if the family is to function smoothly and efficiently. No evidence suggests that a particular manner of monetary allocation is most desirable. What is important is that the partners agree and are comfortable with the manner of allocation chosen.

Budgeting: Enlightened Control of Spending

The next step in reducing monetary conflict is to agree on a budget. A budget is actually a plan of spending to ensure that what is needed and wanted is attained. For example, a family's income must cover such basic necessities as housing, food, clothing, and transportation and leave some money for discretionary expenditures, such as vacations and recreation. How Americans spend their money is shown in Figure 12-1.

To many people, budgeting sounds boring and uninteresting. Yet, we all budget, at least informally. The decision to put off buying a new music system when cash is short is budgeting. To budget formally is to gain control over one's financial life.

Budgeting and long-term monetary planning lead to financial independence. Financial independence means having the economic wherewithal to say to your-

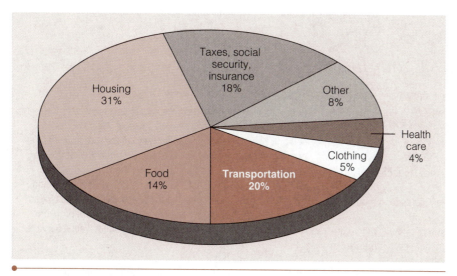

FIGURE 12-1 *Annual Expenditures for a Four-Person, Medium-Income Family*
SOURCE: U.S. Census Bureau 2002b, Table 651.

self: "I may not be rich, but I am financially free to do what I want." In a other words, financial independence means freedom.

The first step is to allot money for necessities. Whatever money is left over can be divided among the family's other wants. For example, it is estimated that the spontaneous food shopper spends approximately 10 to 15 percent more for food than the shopper who has a planned food budget and a shopping list of needed items.

By living within a budget (see "Making Decisions: How to Budget Your Income"), you can minimize or even avoid monetary problems. In addition, by allotting even a small amount to savings, investments can also be made. Saving is really only deferred spending. If immediate spending is deferred, it becomes possible to use the money to earn additional income.

A budget should only be used for a specified time, and then it should be updated to reflect changing family circumstances. For example, a newly married couple who both work may feel well off. They have two incomes and minimal expenses. But danger lies ahead if they become accustomed to using up both incomes. If they decide to have children, the wife may want to give up her income, at least for a while. Expenses also increase when children are added to the family. The combination of lower income and rising expenses can throw a family into an economic crisis unless they have planned economically for both eventualities.

Another eventuality that some families must plan for is college education for the children. College usually means a drastic rise in expenditures. Today, the cost of putting a child through college may be as high $100,000+.

After children become independent, a couple can usually enjoy a comfortable period of relative affluence. They must, however, plan carefully for their coming retirement. Without such planning, they may live out their older years struggling financially, especially if inflationary pressures exist. Thus, good budgeting remains important across the lifespan.

As dull and uninteresting as budget planning may seem, the family that does not put time into planning its finances may face increasing monetary strain, even destruction. This is especially true when the inflation rate is high.

MAKING DECISIONS

How to Budget Your Income

Once you decide to do some positive money management, you must figure out a budget and try to stick to it. The budget is a planning tool to help you reduce undirected spending.

Steps in Budget Making

Create a spending plan by following these four basic steps:

1. Analyze past spending by keeping records for a month or two.
2. Determine *fixed expenses* such as rent and any other contractual payments that must be made—even if they are infrequent, such as insurance and taxes.
3. Determine *flexible expenses,* such as food and clothing.
4. Balance your fixed plus flexible expenditures with your available income. If a surplus exists, you can apply it toward achieving your goals. If there is a deficit, reexamine your flexible expenditures. You can also reexamine fixed expenses with a view to reducing them in the future.

Note that so-called fixed expenses are only fixed in the short run. In the longer run, everything is essentially flexible or variable. Fixed expenses can be adjusted by changing one's standard of living, if necessary.

The Importance of Keeping Records

Budget making, whether you are a college student, a single person living alone, or the head of a family, will be useless if you don't keep records. The only way to make sure that you are carrying out your budget is by keeping records of what you are actually spending. The ultimate way to maintain records is to write everything down, but that becomes time-consuming and therefore costly. Another way to keep records is to write checks for everything. Records are also important in case of problems with faulty products or services or the Internal Revenue Service.

General Budgeting

On the following page is a monthly general-budget form that encompasses both estimated and actual cash available and fixed and variable payments.

You will note that the savings category is located under the "Fixed Payments" heading. This is because the money in your savings account may be used to pay such fixed annual expenses as auto and life insurance, and it is necessary to plan to save in advance for these expenses.

The key to making a budget work for you is to review your figures every month to see how your monthly estimates compare with your spending.

Saving through Wise Spending

When many families think of saving, they think only of putting money into a savings account. Yet, wise spending is another important way of saving. Wise spending means buying when an item is on sale, seeking out bargains, buying used instead of new, being aware of consumer traps in marketing, studying seasonal price fluctuations to buy at the best time, and simply being an astute consumer.

Being a wise shopper is not as easy as one might imagine. Salespeople may be more interested in making a sale than in the exact needs of the individual. Also, because salespeople are not necessarily highly knowledgeable about the products they sell, they may not be capable of helping the buyer choose the most appropriate product. But wise salespeople will learn about their merchandise and will take their customers' needs into consideration. By doing this, they ensure themselves of satisfied customers and repeat business.

Extensive research should be done before purchasing big-ticket items such as household appliances and automobiles. For example, objective test reports on various household items (as well as other things) can be found in the magazine *Consumer Reports.* The various automotive magazines describe car tests each month. The Internet also offers information on various goods. For example, one can find out dealer costs on automobiles and

© Jeff Greenburg/PhotoEdit

Wise spending is also saving.

MAKING DECISIONS

Budget Worksheet

Cash Forecast Month of _____	Estimated	Actual
Cash on hand and in checking account, end of previous period	_____	_____
Savings needed for planned expenses	_____	_____

Receipts

	Estimated	Actual
Net pay	_____	_____
Borrowed	_____	_____
Interest/dividends	_____	_____
Other	_____	_____
Total cash available during period	_____	_____

Fixed payments

	Estimated	Actual
Mortgage or rent	_____	_____
Life insurance	_____	_____
Fire insurance	_____	_____
Auto insurance	_____	_____
Medical insurance	_____	_____
Savings	_____	_____
Local taxes	_____	_____
Loan or other debt	_____	_____
Children's allowances	_____	_____
Other	_____	_____

Total Fixed Payments

	Estimated	Actual
	_____	_____

Flexible Payments

	Estimated	Actual
Water	_____	_____
Fuel	_____	_____
Medical	_____	_____
Household supplies	_____	_____
Car	_____	_____
Food	_____	_____
Clothing	_____	_____
Nonrecurring large payments	_____	_____
Contributions, recreation, etc.	_____	_____
Other—credit cards	_____	_____

Total Flexible Payments

	Estimated	Actual
	_____	_____

Total All Payments

	Estimated	Actual
	_____	_____

Recapitulation

	Estimated	Actual
Total cash available	_____	_____
Total payments	_____	_____
Cash balance, end of period	_____	_____

thus better understand what would be a fair consumer cost. Knowing exactly what is wanted before approaching a salesperson is an important part of wise buying. The prepared buyer will not be sold an item that is inferior, fails to meet needs, or is more than the buyer wants or needs.

Avoiding consumer traps is also important to wise shopping. Although not every trap can be discussed here, the following are some of the more common ones:

1. *Bait and switch*. An ad describes a certain product, such as a washing machine, with a very low price. At the time of purchase, the salesperson claims that the product is inferior or that the store is out of the advertised machine, and guides the consumer to a more expensive machine.

2. *Low ball*. This technique is often used in car sales. The salesperson offers a car at a good price, the buyer thinks a deal has been made, but then a higher authority (sales manager) countermands the salesperson's offer. By that time, the customer may be psychologically committed to buying, and the few hundred extra dollars added to the price may seem acceptable.

3. *High ball*. Again, this technique is mainly found in car sales. Here, the salesperson offers an inflated trade-in allowance on the old car. The manager reduces this offer once the customer is psychologically committed to buying.

4. *Telemarketing*. The telephone is used in sales schemes that range from offering low-cost vacations to notification of winning some wonderful prize. Of course, there is always a catch. Some warning signs of a scam are: "Just dial 1-900. . . ." (Calls to 900 numbers are usually very costly.) "Just send $_____ and you'll receive. . . ." "All we need is your credit card or checking account number." "The offer is only good today."

5. *Contest winner*. In this case, the consumer is told he/she has won a contest that hadn't even been entered. It turns out something must be purchased in order to receive the prize.

6. *Free goods*. By purchasing this carpeting, the installation will be free. Usually one is actually paying for the "free" extras by paying a higher price for the other goods.

7. *Off-brand items*. If the consumer is knowledgeable, it may be perfectly all right to purchase a product that is new to the market or is not a well-known brand. In general, however, well-known brands, although usually higher priced, do offer certain advantages (especially to the ignorant buyer). The quality usually varies less, warranties are often better, one knows that the company will be there to honor the warranty, and the company usually has a wider repair network.

8. *Hard sell*. Beware of the salesperson who says one must buy this sale item today because it is the last one, or that the price will go up tomorrow. There is always time to buy, and an attempt to rush the consumer usually is a signal that something's not right.

9. *Home repairs*. People agree to make repairs on a property and ask for money in advance to buy the necessary materials. Once they have the money, they disappear.

10. *Magazines*. The caller claims to be doing a survey: Which magazines does a person like best? That person is then billed for subscriptions not really wanted.

11. *Credit repair*. An ad guarantees that a bad credit record can be erased for a fee, often several hundred dollars. Bankruptcies and judgments can be removed from one's records, it claims. They can't. If information in one's records is

wrong, one must contact the credit company and have corrections made—and there is no charge.

12. *Travel.* An announcement of having won a free vacation or being entitled to large travel discounts is received. To qualify, one must join a travel club for a fee or purchase a companion ticket at an inflated price.

13. *Advance-fee loans.* A company offers to find a loan for an advance fee of several hundred dollars, but often the loan never materializes.

A person or family victimized by one or another of these consumer traps should fight back by lodging complaints with the proper authorities. For example, if you buy something by mail order and find it is defective or not as advertised, you can lodge a complaint with the postal department if the company does not remedy the situation. The Better Business Bureau is also a strong consumer advocate and has offices in almost every city.

The Economy and Black, Hispanic, Asian American, and Single-Parent Families

Table 12-5 compares median family incomes for various groups of people. If we look only at averages, we find that certain family groups do better economically than other groups.

TABLE 12-5

Comparison of Income in 1998 by Selected Characteristics

Characteristic	Median Income
All Races	
All households	$59,067
Married couples	65,203
Female householder, no husband present	24,393
White	
All households	63,887
Married couples	71,102
Female householder, no husband present	27,542
Black	
All households	40,316
Married couples	55,734
Female householder, no husband present	17,737
Hispanic	
All households	44,468
Married couples	40,541
Female householder, no husband present	18,452
Asian	
All households	69,412

SOURCE: U.S. Census Bureau. All households, Table HINC-06, 2003; *Current Population Survey 2004, Annual Social and Economic Supplement*; Lichter and Crowley 2002. These figures vary slightly according to source.

Median income and unemployment are related to educational achievement. Men and women with the highest level of education earn the most. Those groups with the highest percentages graduating from high school also tend to be the groups with the lowest percentages below the poverty level.

Female heads of household (no husband present) of all ethnic groups earn considerably less than others (Table 12-5) and account for over half of all poor families. Early out-of-wedlock pregnancy, dropping out of school, and employment difficulties appear to be significantly related. School-age women who have a child are much more likely to drop out of school. When schooling is incomplete, high-paying employment is much harder to obtain. Poverty rates decrease dramatically as years of schooling completed increase.

As was discussed in Chapter 2, the real economic story of the last 40 years for Blacks has been the emergence of the middle class, whose gains have been real and substantial. Although the media emphasize those in poverty, over one-third of Black families have an annual income at or above the national median. Black married-couple families in which both the husband and wife work have a yearly median income (2002) of almost $55,734—$15,000+ less than White married dual-earner families (Table 12-5). These families tend to have well-educated members. About 8.5 percent of Black married-couple households fall into the poverty category (23.6 percent of all Blacks fall into the poverty category), as compared with 3.8 percent of White married-couple families (7.7 percent of all Whites fall into the poverty category; U.S. Census Bureau 2000b). Overall median income of Black households is substantially lowered by the high number of female householder, usually single-parent, mother-headed families. Thirty-nine percent of these families fall into the poverty category, as compared with 22.5 percent of similar White families (U.S. Census Bureau 2000b). (See Table 12-6 for 2003 poverty income thresholds.)

Many Hispanic families also fall below the official poverty level. Closer analysis, however, reveals large educational differences between various Hispanic groups, which create large differences in the percentage of each group having economic problems. In addition, persons of Hispanic background who have been in the United States for a long period of time are much better off economically than newly arrived immigrants.

Overall, the poverty rate for all Americans has fluctuated from a high of 22.4 percent in 1959 to a low of 11.1 percent in 1973. The poverty rate for all Americans in 2003 was 12.5 percent, or approximately 35.9 million persons. The poverty rate for children was 17.6 percent in 2003.

TABLE 12-6	
2003 Poverty Thresholds	
One person	$ 9,573
Two people related	12,384
Three people related	14,680
Four people related	18,810

Source: U.S. Census Bureau.

Inflation and Recession

Normally, the economy alternates between relatively long periods of inflation (increasing costs) and shorter periods of economic downturn, or recession. Not many years ago, economists thought that inflation and recession were opposites and could never occur together. Yet, this combination has occurred at times. Therefore, effective money management must take into consideration inflation, recession, and a combination of the two. In an inflationary-recessionary economy, every possible bad thing is happening at once. Production is falling, unemployment is rising, and inflation continues. Fortunately, such conditions seldom last for long.

This book is not the place for a detailed discussion of inflation and recession, although some understanding is necessary if you are to make the economy work for you, rather than against you. Inflation, recession, economic growth, and employment are all related, so a change in one tends to produce changes in the others. The successful money manager must understand these trends.

Inflation

Since World War II, the United States has experienced almost constant inflation. Although the rate has varied, inflation has averaged 4 to 6 percent a year. Inflation tends to hurt those families on fixed incomes, retirees, disabled persons, and female-headed families more than others. We are constantly being surprised, dismayed, and angered at the increased **nominal costs** (absolute price for an item) of almost everything we buy. Bread is often more than $2.00 a loaf, yet it seems only yesterday that it was 50 cents. A new car 20 years ago cost about $5,000, tax and license included. Today, the same model is priced above $20,000. "Buy now before the price increases" is an often-repeated advertising slogan that feeds our fears about inflation. Public opinion polls show that inflation is a constant concern of most people. Inflation rates since 1991 have varied between 4.2 percent and 1.6 percent per year; the 2004 rate varied from 1.93 percent in January to 2.65 percent in August (Financial Forecaster 2004).

nominal cost
Absolute price for something

Prices do not always rise. The consumer price index (CPI) is calculated on a fixed market basket of goods and services, as measured in 91 urban areas across the country. Declines in the index have occurred during recessions or depressions in the economy. Overall, however, the CPI has risen over 700 percent in the past 80 years. It should be noted that the CPI does not include taxes. Federal, state, and local taxes consume 35 to 40 percent of incomes, one of the highest levels ever.

Inflation rates by themselves tell only half the story of the U.S. economy. Inflation simply indicates that nominal prices have risen. However, income has also risen during this time. If income rises at the same rate as prices, buying power remains the same. Thus, a more important measure of the economy than the inflation rate is real per capita income. This is computed by subtracting the inflation rate from the percentage increase in per capita income. If income increases 10 percent in a year in which inflation is only 5 percent, real income (buying power) has increased 5 percent. The real income (buying power) of most people has actually increased since World War II. In the last few years, per capita income has risen, although in some past years it has declined.

The CPI is the most common indicator used by the popular media to measure price fluctuations. Because it is an average, the CPI tends to mask actual price fluctuations for a specific item. Thus, it is important for the consumer to look at the relative price of a product, rather than just its nominal or absolute price. Although the nominal prices of most goods have increased greatly, some relative prices have actually declined. Television sets and computers, for example, have become much cheaper. So, even though overall prices as measured by the CPI are going up, the prices of some goods are rising more slowly than others, making these goods relatively better buys. Sometimes, too, nominal prices have declined as technological breakthroughs have lowered production costs in certain industries, such as electronics.

Another way to see the relationship between price inflation and income is to compare the actual amount of income needed to produce the same purchasing power between 1984 and 1995. A family that earned $30,000 in 1984 needed to earn approximately $50,000 in 1995, and will need approximately $65,000 in

Mexico: The Middle-Class Family

Mexico is our nearest southern neighbor. The recent furor over immigration has focused on poorer Mexicans seeking better wages, who enter the United States illegally in hope of finding work. The focus on this segment of the Mexican population has caused the rapidly expanding Mexican middle class to be overlooked. Although Mexico's per capita income (2004: $8,800) is far below America's (2004: $36,110), the middle class has been rapidly expanding, and consumerism is alive among middle-class families (Population Reference Bureau 2004c).

Seventy-one percent of Mexico's population is now urban (in the United States, the population is 75 percent urban), and urban housing has not yet caught up with the rapid change from rural to urban population. Thus, housing square footage for a middle-class Mexican family is smaller than for an American middle-class family. The Mexican middle class is also reducing the number of children they produce, although the overall Mexican birthrate per 1,000 population is higher (25) than that of the United States (14). By reducing the number of children produced, the Mexican middle-class family increases the amount of money available per family.

Middle-class Mexicans' family

Family cooldown. Anything to beat the hot weather.

Enjoying an outdoor market.

Family togetherness at mealtime.

The church is important to many Mexican families.

belongings include most of the things that Americans have: televisions, stereos, VCRs, washing machines, and so on. Automobiles are taxed heavily, thus restricting their ownership to a far smaller percentage of the Mexican population than is found in the American population.

As the middle class grows in Mexico, family life will become increasingly affluent. However, the extended family is of greater importance to the average Mexican than it is to the average American. Thus, care and support of relatives tends to place a greater financial strain on the Mexican family than it does on the average American family. Once the Mexican middle class is well established, the pressure to support relatives should diminish because it is hoped that many relatives will also be moving into the middle class.

If Mexico can thrive economically, the middle class should continue to grow. Combining economic improvement with a declining birthrate should lessen the problem of illegal immigration from Mexico into the United States. The North American Free Trade Agreement (NAFTA) encourages freer trade with North America, which may increase economic activity in Mexico and, in turn, reduce unemployment and the need for families to leave the country to seek economic survival.

2005 to be able to purchase as much as $30,000 purchased in 1984. Another way of looking at inflation is to compare the value of today's dollar with its past value. A dollar as of 2000 is worth approximately 8 cents, compared with the dollar in 1947.

Severe inflation was controlled in 1982, but some degree of inflation is likely to influence the economy for the foreseeable future, and Americans must take it into consideration if they are to be successful economically. Even a 5 percent inflation rate each year means that a dollar will lose over half its value in 10 years. You can combat mild inflation in a number of ways, as follows:

1. *Minimize your cash holdings.* Cash obviously loses value at the rate of inflation. If I bury $1,000 cash to protect it from theft for a year when the inflation rate is 10 percent, inflation robs me of $100. At the end of the year, I have only $900 in purchasing power.

2. *Select high-yield savings accounts whenever possible.* The longer-term accounts often impose substantial penalties if funds are withdrawn before the end of the term. Savings should be spread over a number of different kinds of accounts. For instance, keep a small balance to cover unexpected expenses in a regular passbook account, where withdrawals can be made at any time without penalty. Place some money in a 6-month to 1-year term account, which earns higher interest, although the higher yields may require larger minimum deposits.

3. *Include a cost-of-living clause in employment contracts.* Many unions have been successful in gaining automatic cost-of-living (COL) raises for their members. If inflation increases the CPI by 10 percent, COL clauses take effect, and the worker's income is automatically increased to match. Unfortunately, such raises also help maintain inflation.

4. *Try not to let inflation cause panic buying.* One is constantly told to buy now before prices increase. Yet, some prices may actually decline relative to the CPI, even though they go up in absolute terms.

5. *Learn about investments.* Money earns money. The wise investor can stay ahead of inflation. For example, many stocks in recent years have stayed ahead of inflation. Unfortunately, inflation is accompanied by a certain amount of irrationality. Thus, not all stocks have kept up with inflation. Small investors should study carefully the particular investments that interest them.

6. *Understand that inflation tends to favor the borrower.* During a period of inflation, money borrowed today is paid back in cheaper dollars in the future. For example, by borrowing $10,000 at 10 percent interest per year for a 5-year period during which inflation is 10 percent per year, you are paying essentially nothing for the use of the money. Interest will have been $1,000 per year, or $5,000 in simple interest at the end of 5 years. However, at the end of 5 years, the $10,000 is worth only $5,000 in purchasing power because of the accumulated 50 percent inflation, which has halved the value of your dollars. (Of course, if the $10,000 was spent foolishly, you are not getting ahead financially, regardless of this principle.)

7. *Try to buy wisely.* Watch for bargains, such as year-end sales and seasonal price reductions.

8. *Have more members of the family work.* This suggestion is discussed in more detail in Chapter 13. Higher inflation rates are partially responsible for the increasing number of married women seeking employment.

9. *Conserve and save to accumulate investment funds.*

Periods of Reduced Inflation and Mild Recession

The recession of 1991–1993 and the milder recession of 2001–2002 caused people a great deal of hardship. Unemployment rose, production declined, and government income fell, while the costs of social programs (unemployment benefits, and others) rose. Because such economic problems are much more obvious than the negative effects of creeping inflation, the government comes under great pressure to support at least a mild form of inflation, despite its long-term negative effects. In the face of a recession, people quickly tend to forget about their problems with inflation.

Most likely, one will experience mild inflation more frequently than recessions; therefore, in this book emphasis has been placed on how to cope with inflation. At times, however, the economy will experience slumps. As protection against such times, the prudent money manager will want to do the following:

1. Maintain enough liquidity to cover emergencies.
2. Beware of investments with a large balloon payment due in the near future.
3. If a slowing of inflation and consequent economic downturn is foreseen, try to maintain a larger percentage of assets in cash, so that resulting good buys can be made.
4. Make sure that your financial position is flexible enough so that you can ride out short-term economic downturns. Judging from the inflation history of the United States during the past century, it is safe to assume that some amount of inflation will remain for the foreseeable future. But there will also be periodic, short-term economic downturns, which must be planned for in advance.

Deciding What Insurance Is Needed

Proper use of insurance can protect a family from catastrophic financial setbacks. Every family must have medical insurance, automobile insurance if they own a car, and fire insurance if they own their own home.

Medical Insurance

Medical coverage is an absolute necessity. Medical costs have become so high that no average family can sustain the expense of a prolonged illness. For a young, healthy couple, coverage can be limited to catastrophic illness with a large deductible, perhaps as high as $500 to $1,000. This is the least expensive type of medical coverage. When children arrive, a policy that covers everyday medical problems and has a lower deductible should be sought. A family of four may have to pay $100 to $600 per month for medical coverage, depending on how comprehensive it is. Besides insurance plans such as Mutual of Omaha and Blue Cross, prepaid foundation plans such as the Kaiser Permanente plan are available; these provide full medical coverage at a certain facility, hospital, or clinic for a specified monthly fee. Health maintenance organizations (HMOs) have grown rapidly in recent years. These are also prepaid plans that offer member physicians and hospitals to enrollees at reduced fees.

Many employers offer group medical plans as part of their fringe benefits; such plans help reduce health coverage costs for their employees. In such cases, the family will not need to supply its own medical coverage.

The government is also playing a larger role in the health field with Medicare plans of various kinds, Social Security disability programs, and workers' compensation insurance. Many analysts suggest that health services will one day be a branch of government, but for the time being, you must plan for health emergencies or face potential financial ruin.

Automobile Insurance

Automobile coverage is also essential. In fact, in most states it is illegal to be uninsured. Property damage and liability are the crucial elements. Covering one's own car for damage is less important unless, of course, it is being purchased through an installment loan. In that case, the lender will require coverage for collision damage.

Home Owner's Insurance

Home owners or owners of other real property must have fire coverage. This is a mandatory condition for obtaining a mortgage. Because of inflation, coverage should be increased periodically to keep up with rising construction costs. Home owner's package policies give much more protection than just fire coverage. Usually, they include coverage for such contingencies as theft, personal liability, and wind and water damage (but not flood damage).

Even if you do not own your own home, it is a good idea to have an insurance policy for personal belongings. Such items as stereo equipment, cameras, furniture, and clothing are surprisingly expensive to replace if they are stolen or destroyed in a fire. This type of insurance, on the other hand, is relatively inexpensive.

Life Insurance

Life insurance is also important, although it is not an absolute necessity. Essentially, individuals must protect their earning power, which is their most valuable asset. This asset should be insured against these two potential hazards:

1. *Premature death*. This risk is about 30 percent before age 65 years. It is especially important for married persons with young children to protect themselves.
2. *Long-term disability*. This means economic death, and can be worse than death itself because the victim usually faces large medical costs, but is unable to produce income.

There are basically two types of life insurance plans: term life and cash value. Remember, the purpose of life insurance is to protect one's estate and provide for the family until the children are independent. The best protection for the least money is term insurance. With **term insurance**, a given amount of insurance is bought for a set period of years, usually 5. Every 12 months, the premium is increased slightly to take into account increasing age (increasing risk). For a typical $100,000 policy, the first-year premium for a nonsmoking male, age 25, is $160 to $200 with a waiver of premiums benefit (a small extra charge that covers the premium payment in case the insured is disabled).

Cash value, or **whole-life insurance**, has two parts: death benefits and the cash value, or savings part, of the policy. The annual cost for a $100,000 policy on a nonsmoking male, age 25, is around $900 per year. Although such a plan is

term insurance
A given amount of insurance bought for a set period of years

**cash value,
or whole-life insurance**
This insurance has two parts: first, a death benefit, and second, a savings part that acts as an investment

considerably more expensive than term life insurance, the policy builds savings, or a cash value. In recent years, the saving aspects of such policies have paid such high returns that for some couples, whole-life insurance may now be a better long-term value (if the couple can afford the higher premiums) than straight term insurance. In general, the best policy for a young family is one that insures for a substantial part of the economic loss that would occur at an early death and also fits into their budget. The appropriate insurance may be all term, all whole-life, or a combination of the two.

Basically, the amount of life insurance a couple needs depends on the number and ages of their children, their standard of living, and their other investments. What life insurance must do is protect the family if the major monetary contributor should die. It should cover death costs, taxes, and outstanding debts and should supply enough money to enable the family to continue functioning. Just how much this will be depends on the individual family.

Deciding to Buy a Home

Home ownership has been a way of life for most Americans. The home-ownership rate in the United States is about 69.2 percent for all households (U.S. Census Bureau 2004b, Table 5). This is among the highest in the world. The home is a major source of savings for many retired Americans. Depending on the state of the economy and the location of the home, however, one can argue that the costs of home ownership make the investment less attractive than is commonly believed. If money spent on home ownership is saved and invested wisely, it will probably make more money than would accrue through appreciation of a home, especially in those years when the stock market is surging upward. However, highly desirable areas have recently seen home prices increase at very rapid rates, such that investing in a home in such an area has been one of the better investments. This is especially true if rents continue to rise more slowly than the general inflation rate, as they have in the past. A person's home, however, yields many personal satisfactions beyond the possibility of economic gain. The American dream of one day owning a home is more than merely an economic dream.

The average U.S. home's market value gained 9.4 percent between mid-2003 and mid-2004. That is more than three times the consumer inflation rate. A record 74 metro housing markets had double-digit gains during this period, and no market experienced net losses in average home values (Harney 2004).

Where a person lives is an important determinate of the kind of values that are passed to children. Most privileged families with children select their home with an eye to the neighborhood and the quality of the schools that their children will attend. They realize that the neighbors, the schools, and the children met on the playgrounds all influence their children and help to set up the conditions for the transmission of those values held dearly by the family (Marks 2000, 621). For those families that cannot afford to choose where they live, the influence of poor schools and troubled neighborhoods make it extremely difficult to raise their children and keep them out of trouble. Although school vouchers that help parents have a broader choice of schools is a controversial idea, many parents living in less desirable neighborhoods see vouchers as a way to help their children move to better schools and escape at least some of the neighborhood problems.

Table 12-7 shows the amount of pretax income necessary to qualify for a 30-year, $100,000 mortgage at various interest rates. Home lenders usually allow

Owning a home is still the realization of a dream for most families.

© Ryan McVay/Photodisc/Getty Images

TABLE 12-7

Pretax Income Needed to Qualify for a Thirty-Year $100,000 Mortgage*

Interest Rate	Monthly Payments to Amortize Loan	Monthly Income	Annual Income
6%	$600	$1,708	$20,496
7	665	1,900	22,800
8	734	2,100	25,200
9	805	2,300	27,600

*There are many plans to help one finance a home. In this example, the bank is limiting the payment to 35 percent of gross income.

TABLE 12-8

Monthly Payment: How Much Can You Afford to Spend?

Annual Income	Gross Monthly Income	Affordable Monthly Payment
$ 20,000	$1,667	$ 467
25,000	2,083	583
30,000	2,500	700
35,000	2,917	817
40,000	3,333	933
45,000	3,750	1,050
50,000	4,167	1,167
55,000	4,583	1,283
60,000	5,000	1,400
65,000	5,417	1,517
70,000	5,833	1,633
75,000	6,250	1,750
80,000	6,667	1,867
100,000	8,333	2,333

you to spend 28 percent of your total monthly income to make mortgage payments of principal, interest, taxes, and insurance. Table 12-8 shows how much 28 percent is at various income levels.

The economy has reacted in a number of ways to try to keep home ownership within reach of the American family. Smaller homes, modular homes, mobile homes, condominiums, and cooperatives have all increased in popularity. Government funding of lower-interest mortgages has also been made available. Shared ownership is another way to move into home ownership. Two families can pool their funds and buy a large home, a duplex, or a triplex. With a triplex, the families can rent out the extra apartment, and the income can help with the monthly home owner expenses.

Once the decision is made to buy a home, there are five major considerations to be decided, as follows:

1. *Affordability*. The first decision is to consider what price home is affordable. How big a down payment and what size monthly payments are acceptable?

2. *Location*. Where are the schools, and how good are they? What about shopping? Do you want to live in the city, suburbs, or country?

3. *Size*. Number of children one has or is planning, and does each child need a private bedroom? What about number of bathrooms? Does a family member plan to work at home? How many vehicles will be on the property?

4. *Livability*. How comfortable is the house? Is it fairly easy to maintain? Is the kitchen efficient? Is there room for family activities and hobbies?

5. *Status*. This may seem a strange consideration, but for some occupations it may be important. Will there be a lot of entertaining? If work is done out of the home, will clients and colleagues be spending time in the house?

The Decision to Invest

According to some, an individual's chances of earning $1 million are much less than they were for his/her grandparents. But the number of millionaires today actually far exceeds the number in our grandparents' time. In fact, for the first time in history, America now has a large group of individuals who are billionaires. Granted, $1 million may be worth considerably less in purchasing power today, but it is still a healthy mark of affluence. A gradually inflationary economy is also an economy in which money is more easily accumulated. For most young people, gathering the first small amount of capital is particularly crucial, because the system has a tendency to work against them. In the early years, they must stay alert to keep the system from entrapping them and thereby, canceling their attempt to accumulate initial investment capital. If they can win this battle and start on the road to financial success, they will be using the system to their advantage, rather than being used by it. When costs of living in the United States are compared with those in other countries, it is clear that the United States offers great economic opportunities to its citizens.

Some people can achieve economic success by working in certain jobs and businesses, but there are other avenues to success. Figure 12-2 shows the broad range of investment opportunities from which to choose. They range from a very conservative bank savings account to highly speculative gambles, for high return on such things as mining and oil exploration. You might ask, "How can the average newly married couple even consider investments? It's all they can do to set up housekeeping." This is a legitimate question. However, the couple who plans on investing, even if only at a later date, has the greatest chance of economic prosperity and freedom. A positive attitude toward investment is actually more important than the investment itself. Such an attitude recognizes that money makes money, that there is value in budgeting and staying free of consumer debt, that controlling one's desires in early years can lead to greater rewards later, and that the American economic system can free one from economic worries if used properly.

Even if a couple can put aside only a few dollars a month toward future investments, they stand a chance of improving their economic position compared with that of their friends who have no interest in or knowledge of investing. For example, putting away $20 each month, starting at age 25, is the same as putting away $200 each month starting at age 45 with retirement at age 65. Table 12-9 shows how money grows at 6 percent and at 12 percent compounded-daily interest. The figures are predicated on putting aside $100 per month initially, until you have $12,000.

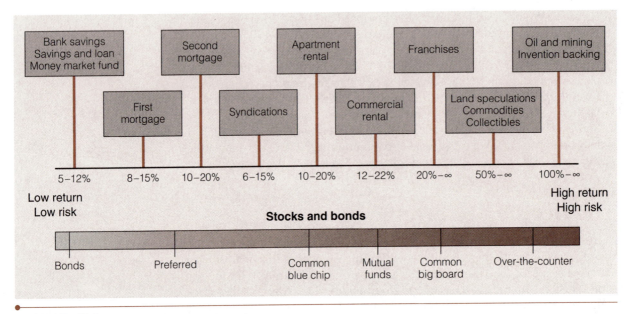

FIGURE 12-2 *The Investment Continuum*

Note that the percentage return in successful investments increases as the risk increases—the chances of striking gold are slim, but if you do, the return is great; percentages also change with economic conditions.

TABLE 12-9				
$100 per Month Invested at 6 Percent and 12 Percent				
Years	**Amount Invested**	**6 Percent**	**12 Percent**	**Difference**
10	$12,000	$ 16,766	$23,586	$6,820
20	24,000	46,791	96,838	50,047
30	36,000	100,562	324,351	223,789
40	48,000	196,857	1,030,970	834,113

In recent years, the government has added greatly to the incentive to save toward retirement by offering several different savings plans in which the money saved, and the interest earned by that money, are tax free until the money is withdrawn at a later age. The best known of these plans is the individual retirement account, better known by its initials IRA. Contributions are now deductible only to a certain level, depending on income. Interest on your contributions remains tax free until withdrawn. The Roth IRA allows you to withdraw your money tax free, although you pay taxes on the money at the time of contribution.

In simple terms, a young person and/or family can make investment a part of their planning by following these steps:

1. Use the credit system wisely to avoid economic entrapment.
   a. Avoid buying consumer goods on credit.
   b. Avoid running balances on credit card accounts.
   c. Pay bills promptly to maintain a good credit rating.

HIGHLIGHT

You Can Still Make a Million Dollars

The stories that follow demonstrate that a person can become wealthy with hard work and creativeness, but the point is not necessarily to explain how to become wealthy. Rather, it is to demonstrate that economic security leading to increased freedom is a possibility for any American.

- To the computer industry, Bill Gates was always seen as a young man in a hurry. The inventor of the operating system now built into every IBM (or clone) home computer, Gates became a business legend before his twenty-first birthday. He founded his company, Microsoft, with money he earned as a teenage programming consultant, and has never had to borrow from a bank. The company's sales exceeded $32 billion in 2003. *Forbes* magazine listed him as the wealthiest single American from 1995 to 2005, with the exception of a year or two.

  Gates has been the model entrepreneur. He has hired good professional managers, and constantly adds innovative products to the company. "We wanted to be a leader with the goal of putting a computer onto every desk in America. It would be a great failure not to see that through."

- Fred DeLuca is not yet 50 years old, but looks like a wide-eyed boy when he describes his favorite sandwich: "Tuna on whole wheat with the works—lots of peppers." Not ex-actly a gourmet's delight, but he is talking about subs. He is cofounder and president of Subway Sandwiches, which now has more outlets than fast-food giants Burger King and Kentucky Fried Chicken. At last count, there were 8,000+ outlets in 14 countries. Sales in 1996 were approximately $2.5 billion.

  DeLuca opened his first Subway in 1965, when he was 17 years old. Backed by a partner and $1,000, he rented a small store and started to sell subs. In 6 months, the store was losing money; instead of closing, however, he opened a second store to create the illusion of success. Within a year, there were three stores, and he hasn't stopped opening stores since.

There are about 2 million millionaires in the United States, and an increasing number of billionaires. Their number has increased greatly in the past few years, with the success of the technology sector of the economy and the growth of the Internet. Such stories as these are possible in this country, and that really is the beauty of the American economic system. America's young millionaires rely less on luck than on talent, tenacity, and creativity. They display enormous energy, patience, and resolve in getting what they want, which often is the personal satisfaction of having done something profitably and well.

---

2. Develop the habit of saving regularly, even if at first you can only save a small amount.
3. Learn about the many types of investments available to individuals in the United States, some of which are shown in Figure 12-2. Such steps can lead to successful investments, which can, in turn, lead to economic freedom.

Many readers of this book have little experience with investments. Most students have neither the money nor the time to invest. However, education is an investment in oneself, and for most people, this investment will pay monetary dividends over the years in higher earnings. If the idea of investment seems foreign, if one thinks investments only come later in life, then learning about investments is an important task to start as soon as possible. The individual who thinks in investment terms early in life stands the best chance of monetary success leading to individual and family freedom. The next Bill Gates or Fred DeLuca may be sitting right in your class. It could be you.

One can learn about investing by reading such magazines as *Forbes*, *Business Week*, and *Money*. In addition, many money management and investment books are designed for the beginner. Marshall Loeb's *Money Guide*; *The Power of Money Dynamics* by Venita Van Caspel; *Personal Finance* by E. T. Garman and R. E. Forgue; *The Complete Guide to Investment Opportunities* by Marshall Blume and Jack Friedman; and *One Up on Wall Street* by Peter Lynch are all worthwhile.

Summary

1. *The family is the major unit of consumption in the United States.* A family must have the economic ability to provide food, shelter, and transportation and meet the needs of its members. The family that is economically successful stands a much better chance of staying together than the family that fails economically.

2. Credit use in the United States has allowed Americans to maintain the world's highest standard of living. Yet, this easy availability of credit can also curtail individual and family freedom when it is abused or misunderstood. *Agreeing to make future payments for present goods or services can lock a person into an inflexible life pattern.*

3. *A thorough understanding of credit, installment buying, interest costs, and budgeting can work to a family's benefit,* allowing them to invest and, perhaps, to achieve not only economic security, but also economic freedom.

4. *The day-to-day handling of money can be a problem in a family if the partners have different values about money.* Conflict can be minimized if the couple plans how monetary decisions will be made. Budgeting will also help them to plan for necessities and see how their income is spent.

5. *Inflation is the primary economic enemy of the newly married couple.* It is important for families to understand inflation so that they can take steps to guard against it. Proper budgeting and good investments are two steps that a family can take to reduce the unwanted effects of inflation.

6. *Insurance should be considered a necessity.* A family with a car and a home needs medical, automobile, and fire insurance. Couples should start with a medical policy that protects them against catastrophic illness and then changes to broader coverage as children arrive. Life insurance is also important, though not a necessity. The couple should buy life insurance, increasing the amount of the coverage as needed to protect family members.

7. *The American dream of home ownership for every family may be fading in the face of increased housing prices,* yet a higher percentage of Americans own their own home than ever before.

8. *Investments are a means of supplementing income and making money work to produce more money.* The family who can save and invest even a small portion of its income is freer of possible economic entrapment and stands a better chance of survival than the family who cannot control wants and desires and spends its entire income.

9. *Investments can be plotted along a continuum from low risk/low return to high risk/high return.* Risk and return increase with such investments as first and second mortgages, syndications, apartment houses, commercial property, and franchises. The rate of return can be very high for speculations such as land, commodities, oil and mining, and invention backing, but the risk is too high for young couples with limited funds. The stock market is another investment outlet. Here again is a continuum from low risk/low return to high risk/high return.

Resources on the Internet

Companion Website for *Human Intimacy: Marriage, the Family, and Its Meaning,* Tenth Edition

http://sociology.wadsworth.com/cox10e/

Gain an even better grasp on this chapter by going to the companion website to take one of the tutorial quizzes, use the flash cards to master key terms, or check out the many other study aids you'll find there. You will also find special features such as GSS data, Sociology Online, and Census 2000 information that will put data and resources at your fingertips to help you with that special project or to help you as you do some research on your own.

InfoTrac College Edition

http://www.infotrac-college.com/wadsworth/

You can access reliable resources anytime, anywhere, with InfoTrac College Edition, the online library. This fully searchable database offers more than 20 years' worth of full-text articles (not abstracts) from almost 5,000 diverse sources, such as top academic journals, newsletters, and up-to-the-minute periodicals, including *Time, Newsweek, Science, Forbes, The New York Times,* and *USA Today.* You can conduct electronic key word searches using key terms from this chapter to supplement your reading and learning experience. To aid in your search and to gain useful tips, see the Student Guide to InfoTrac College Edition, which you can access through the companion website for this book.

Credit: The Way to Economic Well-Being, or a Slavery System?

The Way to Economic Well-Being

For most American individuals and families, credit is a way of life that leads to abundance. In addition, credit may allow a person to travel, school him-/herself, and receive medical services and countless other benefits.

Try to imagine America without credit. What would you have? Where and how would you live? Indeed, credit buying has given the average American a healthier, more fulfilling life, with a higher standard of living than was ever dreamed possible in earlier times. It has given the average American adequate housing in a world that still has difficulty putting even a lean-to roof over the heads of much of its population. Imagine what a person living in Pakistan with a gross national product (GNP) per person of only $1,960 in 2004 would think of a country that sets the poverty level for a family of four at $18,979, as America currently does (2003).

To give you an idea of Americans' affluence, compare the number of labor minutes needed to buy a McDonald's Big Mac hamburger in other countries' cities, with the number of minutes needed in Los Angeles:

Nairobi, Kenya	178 labor minutes
Bombay, India	105
Moscow, Russia	74
Mexico City	66
Rio de Janeiro, Brazil	45
Seoul, Korea	25
Paris, France	19
Sydney, Australia	13
Los Angeles, California	11

Source: Swiss Bank UBS.

Of course, there have always been very wealthy people who lived affluent lives, but historically, the rest of the population has been poor. Very rich and very poor still describe the societies in many third-world countries today. Yet in America, there is a large economic middle class that lives very well, and by and large, it is credit that plays a major role in producing this high standard of living.

A Slavery System

Slavery system appears to be a strong term to describe the economic system of one of the freest countries in the world. Yet, credit and debt are in direct opposition to personal freedom. You may gain all sorts of material possessions for nothing down and $10 per month, but you also become a slave to the promise you make and the legal obligation you assume to pay the $10 per month for the next "X" number of years.

Credit and debt are directly opposed to personal freedom. When an individual contracts to pay for a new automobile over a period of 60 months, for example, he/she gains the use of the automobile but loses a degree of personal freedom. Regardless of the circumstances or what is done with the car, the individual is legally responsible to pay a certain amount each month for the next 5 years. If at the end of 1 year, the individual wishes to take a lower-paying but more satisfying job, or return to school to improve skills, he/she may be unable to do so unless a way to continue payments can be found. Returning the car does not cancel the debt. Even if the finance company sells the car, chances are that something is still owed, because during the first year or two, the car depreciates in value faster than the debt is reduced. Although the remaining debt may be smaller, it still exists, and so does the obligation.

Debt always curtails a certain amount of personal freedom. If you cannot resist going into debt, you may become so deeply obligated as to lose almost all freedom. As you read earlier in this chapter, many people go so far into debt that they must declare bankruptcy. Suddenly, you owe so much that even working two jobs can't make all the monthly payments. Before the paycheck arrives at the first of the month, it is already spent because of the many payments due. Your paycheck is held captive by your overuse of credit and your accumulated debt.

© Buccina Studios /Photodisc Green /Getty Images

The Dual-Worker Family: The Real American Revolution

Chapter

13

Questions to reflect upon as you read this chapter:

- Why do you think the entry of the American wife and mother into the workforce is considered to be a family revolution?

- What are the advantages to the family when the wife/mother goes to work outside the home?

- What are the disadvantages to the family?

- If a family includes young children, how are they cared for when mother enters the labor force?

- What changes can the workplace make that will help families balance work and family life?

- In a dual-career family, what kinds of conflicts arise between the careers of husband and wife?

- How might one resolve these conflicts?

- If you are a woman, do you plan to work after you marry?

- Will you have children?

- If you are a man, will you want your wife to work?

- What about after you have children?

- If you are a woman, what do you think about stay-at-home dads?

The large-scale entrance of American women, especially married women, into the workforce since World War II has wrought significant changes not only in the workforce, but also in the modern American family. It is the most revolutionary change in family life that has occurred since the Industrial Revolution. Most working husbands have working wives, most children have working mothers, and almost half the workforce is now female (Moen 2002). Therefore, the relationship between work and family is now more complex. Work and family interact more intimately. Partners grapple with more of the same problems. Gender roles are less clear and more overlapping. Power within the family is redistributed because both partners contribute monetarily to the family's support. Family activities and time schedules are more restricted. The care of children is more variable, ranging from full-time parental supervision to none at all in the case of the latchkey child, who returns to an empty home after school because both parents are working. Mothers try to become supermoms as they cope with both family responsibilities and an outside career. Fathers may become house husbands if their wives' careers blossom. The American married family has become two jobs, two paychecks, paid child care, and much less time.

In the past, most women in the American labor force were single. The working woman in general, and the working wife-mother in particular, were unusual phenomena. Married women were full-time wives and mothers and were considered negligent in their family duties if they worked outside the home. Today, that has all changed. The number of children at home is no longer strongly related to workforce participation by women. The role of the woman as homemaker-mother only is clearly passing, and the traditional marriage in which the husband works to support the family while the wife remains home caring for the children has be-

WHAT RESEARCH TELLS US

Women in the Workforce

- One in two workers are women.

- Four in five mothers of school-age children work for pay.

- Two in five working women are managers or professionals.

- One in five working women have administrative support jobs.

- One in two people who work more than one job are women.

- One in two working women provide half or more of their household income.

- Seven in 10 married working mothers work more than 40 hours a week.

- More than 58 percent of workers paid by temporary help agencies are women.

- Seventy-two percent of part-time workers are female.

- Three in 10 working women work evenings, weekends, or some combination.

- Three women in five work at or below the minimum wage.

- Women's presence in once male-dominated professions such as medicine, dentistry, and law increased significantly between 1990 and 2000.

- The presence of women in traditionally male, technical and trade occupations remains rare.

- Hispanic mothers are less likely than Black or White mothers to work during a year. Black mothers are much more likely than others to be year-round, full-time workers.

SOURCES: AFL-CIO 2004; *American Woman* 2003–2004; Bureau of Labor Statistics 2003a; Mishel et al. 2003; *Working Woman* 2004.

come a minority pattern. This has created what the Census Bureau now describes as a "husband-primary earner, wife-secondary earner" family—more popularly known as the two-career, two-earner, or dual-worker family.

Women and the Economy

The major restraint to freer role choice in the past has been women's inferior economic position. In general, men earn more than women, regardless of their individual skills. This economic differential in earning power locks each sex into many traditional roles. For example, a father who would prefer to spend more time at home caring for his family usually cannot afford to do so. In most cases, this would mean giving up a portion of his income that his wife cannot completely replace and, perhaps, losing possible job advancement.

More and better job opportunities for women are now available, but the gap between male and female earnings, although decreasing, persists. Basically, the earnings gap is between women with children and men. Economically, single women without children have long held their own with men. Having children usually means career interruptions and the avoidance of jobs that demand long hours, or travel. Even with fathers who shoulder some parenting duties and good child-care facilities, it is mother who still is most responsible for parenting.

The Equal Pay Act of 1963, and the creation of the Equal Opportunity Commission (EOC) under Title VII of the Civil Rights Act of 1964, have helped women and minorities move in the direction of equal pay for equal work.

More married women are working outside the home for a number of reasons, as follows:

1. The inflationary pressures of the American economy and expectations of a rising standard of living have combined to bring many women into the workforce.

The majority of working wives work to help make ends meet. Thus, economic need is the major reason most women go to work.

2. Since World War II, real wages for both men and women have increased dramatically. (**Real wages** are earnings that have been adjusted for inflation.) Because a woman can now earn much more than in the past, the relative cost of staying home with her family has increased, thus drawing more women into the labor force. One might think that increased real family income would encourage women to stay home because they would have less economic pressure to work and greater financial means to enjoy leisure pursuits, but this has not been the case. One reason may be that desires for increasingly higher standards of living have outpaced the increase in real income. Another reason may be that in years of high inflation, real income has not increased rapidly. Indeed, in some years it has actually declined. Thus, despite generally increasing real income, women have not remained home to enjoy it, but have entered the labor market to participate in the higher wages. Increased income also enables families to reduce the amount of unpaid labor in the home, because they can purchase labor-saving devices and domestic help.

3. The number and kinds of jobs available to women have increased tremendously. The importance of physical strength in many industrial jobs has diminished. Service jobs have expanded greatly. The opportunity for part-time work has also increased. Equal opportunity legislation has created demands for women in jobs previously closed to them. For example, women now hold almost half of all managerial positions.

4. Declining birthrates have certainly contributed to the increased numbers of women working outside the home. By postponing children, having fewer children, or having none at all, women have reduced the demands of family work and have thus become freer to enter the labor force.

5. Increasing education has contributed to women's working outside the home. Over half of all college graduates are now women; the same holds true of recipients of master's degrees (see Table 13-1). Of the doctorates awarded to African Americans, 55 percent were given to women in 1993, up from 39 percent in 1977. Better education certainly opens job opportunities. An educated person tends to become more aware of his/her potential. The role of wife-mother becomes only one of many roles for the educated woman as she becomes more aware of her potential.

6. Attitudes about the role of the woman in the family have changed greatly during this century. Today, most women believe that working outside the home is important for personal satisfaction, rather than just for earning additional money, although the desire to work outside the home is by no means universal.

7. In the future, the lower birthrate will reduce the number of future workers available. Thus, as the workforce shrinks, women will become increasingly important in maintaining a sufficient workforce.

Although the surge of women into the workforce is slowing, few experts believe that women who have tasted the satisfaction, and added affluence of a paycheck, will return to being full-time housewives in great numbers. Many employed wives appear to be working not so much because of financial need as because of interest in their jobs. They simply enjoy their employment and derive satisfaction and self-esteem from their work. However, some working women with children are beginning to return to full-time motherhood if they can afford it (Wallis 2004). The ideal of the 1980s supermom, who does it all at home and at work, no longer seems realistic for these women. As one person puts it, "Supermom has come

---

real wages
Earnings that have been adjusted for inflation

TABLE 13-1

Women's Share of Higher Education Degrees, 1985 and 2002

	1985	2002
Undergraduate	53%	58%
Bachelor's degree	51	56
Master's degree	50	59
Doctorate	34	46

Source: Choy et al. 2004.

down with chronic fatigue syndrome." Considerable publicity about the problems of children of divorce and single mothers (Wallerstein et al. 2000), where in most cases the mother must work to make ends meet, have drawn some working mothers out of the workforce and back to full-time mothering.

Job Opportunities for Women

The vast majority of working women are not in the glamorous professions, the upper-management levels of corporate America, or government leadership roles. The work world for women is much the same as it is for most men: 8 a.m. to 5 p.m. workdays, 2-week paid vacations each year, and often mundane duties.

Women have gained access to a greater variety of jobs and to higher-level employment in all areas. More women are physicians, lawyers, corporate executives, and managers than ever before. But these are the exceptions, just as they are among men. Although opening all types and levels of jobs to women is a worthy goal, the reality is that most women and men will not achieve lofty occupational positions. The vast majority of women (and men) in the labor force are more concerned with general nondiscriminatory job availability and good pay than with obtaining top-management positions.

The mass media has called attention to the "glass ceiling" blocking women from upper management. Not a barrier deliberately created, the glass ceiling is the cumulative effect of various practices that place work and family in opposition to each other. The work world generally fails to recognize family needs, and this affects women more than men. For example, many more women than men quit their jobs over such family needs as pregnancy, child rearing, looking after elderly parents, and the like.

Despite the glass ceiling, women are breaking through to high positions in corporate America. In 2000, Hewlett-Packard hired a woman as president and CEO, making it the first Dow 30 company headed by a woman. Just since 1996, the percent of women in senior leadership positions in U.S. firms has risen from 46 to 50 percent, the percent of corporate officers has risen from 10 to 12 percent, and the percent of board of directors has risen from 9.5 to 11 percent (*Santa Barbara News Press* 2000). The following companies are listed by *Working Mother* (2004) as the 10 top companies when it comes to family-friendly programs and policies that mothers need to balance work and family:

Bristol-Myers Squibb Company

Discovery Communication

Eli Lilly and Company

IBM

Johnson and Johnson

J. P. Morgan Chase

PricewaterhouseCoopers, LLP

Prudential Financial, Inc.

S. C. Johnson and Sons, Inc.

Wachovia Corporation

All of these companies have flexible scheduling, generous maternity leave policies, and advancement and leadership training for women and men. They also have other perks; some even have on-site child-care facilities.

The move of women into corporate management has not included many African American, Native American, or Hispanic women. These women make up 33 percent of the U.S. female workforce, but only 15 percent of the managerial female workforce (M. Jackson 1997). However, the number of African American, Hispanic, and Native American women becoming more educated is increasing; with increased education comes increased opportunity for advancement. For example, in science and engineering education, where women still lag behind men, there have been gains for all women—including minorities (D. Smith 2000).

More important than these professional careers, however, are the many categories of jobs now filled by women. From insurance adjusters to real estate brokers to production line assemblers, women make up an increasing percent of the workforce. Although one might be tempted to dismiss such jobs as less important, the opening of these jobs to women has had far more impact than the acceptance of a few women into glamorous, high-level careers.

Most women still work in the jobs and occupations that have historically been open to them. These jobs generally are lower on the pay scale (Bureau of Labor Statistics 2004c). The following 10 occupations employed the most women in the United States in 2000 (U.S. Census Bureau 2003d, Table 3):

- Secretaries
- Elementary and middle school teachers
- Registered nurses
- Cashiers
- Retail sales persons
- Bookkeeping, accounting, and auditing clerks
- Nursing, psychiatric, and home health aides
- Customer service representatives
- Child-care workers
- Waiters and waitresses

Other jobs with a preponderance of women:

- Typists
- Receptionists
- Licensed practical nurses
- Teacher aides
- Textile sewing machine operators
- Bank tellers
- Dietitians
- Information clerks
- Librarians

Despite the concentration of women in certain jobs, the variety and levels of jobs open to women are clearly improving. Because education level and employability go hand in hand, the fact that so many more young women are attaining higher educational levels (see Table 13-1) than in the past means that they will find a greater variety of higher-level jobs available. These better-educated women are finding the labor market friendlier. Such women will also experience less job loss in future recessions because they are more skilled.

One area often overlooked when studying women in the workforce is women-owned businesses, which represent 40 percent of U.S. businesses (*Working Woman* 2000a, 52). Many women who have previously been unable to break through the glass ceiling in corporate America have simply started their own businesses. (See *Working Woman* 2000a for a list of the top 500 women-owned businesses.)

It is interesting to note that over 40 percent of American women working overseas in executive positions find that being female actually helps in male-dominated business dealings, because they are presumed to be the best and brightest (N. Adler 1994, 16).

Pay Differentials between Men and Women

The narrowing of the wage gap between men and women is one of the most important recent economic trends. Between 1980 and 2002, the ratio of women's to men's earnings among full-time, year-round workers increased from 60 to 77 percent, an increase of 27 percent in real earnings (adjusted for inflation) (Bureau of Labor Statistics 2004e; *Working Woman* 2004). The gap is being closed more quickly in some fields, such as technology; in other areas, such as law and chemistry, entry-level pay is the same for men and women. In a few fields such as advertising, library science, and occupational therapy, women routinely earn more than men (*Working Woman* 2000b, 58).

The median income of single-parent families with children dramatically illustrates pay differentials between men and women (see Table 13-2). Single mothers earn considerably less than single fathers. It should be noted that a large proportion of mother-only families are young, never-married mothers with little education and thus few employment skills.

TABLE 13-2

Median Income of Families with Related Children under Age 18, by Race and Hispanic Origin and Family Structure, 1990 and 2001

	(2001 dollars) 1990	2001
Married-couple families	$54,187	$65,203
White	56,952	71,102
Black	46,912	55,734
Hispanic	36,081	40,541
Mother-only families	17,194	21,997
White	19,526	31,879
Black	13,535	19,086
Hispanic	13,319	19,021
Father-only families	33,110	31,932
White	34,366	32,933
Black	27,008	28,645
Hispanic	27,284	27,385

SOURCE: U.S. Census Bureau 2002a, Tables F10A, B, C.

Why does such an earnings gap exist, and why does it persist? The debate over this question is loud and often emotional. Some say the gap is due to discrimination against women, pure and simple. Others suggest that it is the women who have babies and are the primary adults who provide child care. Thus, the working mother must take time off from work and interrupt her career to have her children and to care for them, especially at the young ages. Knowing this, the working mother-to-be often chooses occupations that can better tolerate such interruptions. These occupations tend to be lower paying. Still others suggest that at least in the past, men have been better trained (had more schooling and experience) and have, therefore, been more productive. Still others suggest that men are the primary breadwinners and are, therefore, more committed to their work. Many companies have been reluctant to invest in long-term training for women because the companies fear women employees will get married, become pregnant, and quit work periodically.

Much research has been conducted on pay differentials, but the findings are mixed. Hours worked, amount of past experience and/or length of job tenure, occupation and/or industry (all real productivity measures) account for some of the pay differential.

One might conclude that the remaining difference is, indeed, discrimination against women in the labor force. Again, however, such a conclusion can be disputed. In light of civil rights laws and general societal disapproval, discrimination is not always easy to discover. If women freely make different work choices than men that result in a pay differential, has discrimination occurred?

According to one argument against the existence of job discrimination, family and housekeeping duties frequently lead women to make different choices about work hours, type and amount of work, and occupation than the choices made by men. Because of their home responsibilities, women work fewer hours and thus accumulate less experience than men. Furthermore, because women anticipate lower levels of lifetime employment and lower returns, they devote less effort to acquiring skills and choose jobs that require less training and/or experience.

Women who consistently make plans to work from the time they are in school tend to earn more than those working women who never plan to work. The women who expect not to work (far fewer today than in the past), or who plan to withdraw from the labor force when they have children, have less incentive to invest in work-related skills early in their working lives.

The counterargument suggests that as a result of labor market discrimination, women have more difficulty finding jobs that are full time, offer opportunities for training and advancement, or are in "male" occupations. Sex barriers to high-paying jobs account for significantly more of the pay differential than does differential pay between men and women for doing the same job. Thus, women do not freely choose the less remunerative jobs, but rather, are forced into such jobs because of discrimination.

Regardless of which argument you accept or what the real reasons are for the pay differential, it does exist, and it does influence family life. For example, when a wife-mother goes outside the home to work to help with family expenses, her earnings will probably be lower than her husband's. The ramifications of such a decision for their relationship are great. If society can narrow the pay differential between men and women, families will have more economic choices.

Perhaps more promising than legislation are the signs of improving employment opportunities for women that we have already discussed. Better education, more training, higher percentages of women in historically male-dominated occupations, and more women in managerial positions all should help reduce the pay differential in the long run. Most important may be the sheer numbers of women now in the workforce. As these numbers continue to grow, so will women's power, and with power comes the ability to bring about change.

One very positive change is already occurring. If the pay differential between men and women is broken down according to age, the gap is least among the younger age groups. For example, in the second quarter of 2004, median weekly earnings for full-time wage and salary workers was $397 for men 16 to 24 years of age and $370 for women of the same age. It was $627 for men 25 to 34 years of age, and $553 for women of the same age (Bureau of Labor Statistics 2004d, Table 2). In the first instance, women were earning 92 percent of the men's wages; in the second example, women were earning 88 percent of the men's wages. It appears that the efforts to achieve equal pay for women are paying off in starting wages for women. If these starting wage gains persist, we can expect the overall pay differential between men and women to diminish with time. In fact, young women between the ages of 25 and 30 years who are childless are now earning close to 98 percent of what men earn (*Working Woman* 2004).

Another positive change has occurred in the female-male unemployment differential. Females have historically suffered higher unemployment rates than males. This obviously contributes to lower overall earnings for women and thus to the pay differential. The unemployment gap between men and women started to narrow in 1978 and actually reversed for the first time in 1982. In most years since then, however, the unemployment rate for women has been higher. In 2003, the women's unemployment rate was 5.8, while the men's rate was 4.1 (Bureau of Labor Statistics 2004b, Table 5).

Still another change is the rising number of wives who actually earn more than their husbands. Between the mid-1960s and the mid-1990s, the percentage of wives earning more than their husbands in two-paycheck couples increased from 3 percent to 23 percent (Brennan et al. 2001; Sandroff 1994, 39). College-educated African American women generally earn more than college-educated African American men, and even slightly more than Caucasian women with a similar education (Hayghe 1993, 43; S. Roberts 1994, A9). Earlier theories predicted that women's increased economic autonomy would lead to increased marital dissatisfaction, and this is true in some cases. However, it appears to be untrue when a wife earns more than her husband.

Unfortunately, some of the decrease in the earnings gap between men and women is due to reduced wages for men (Bowler 1999), which mainly results from the loss of high-pay manufacturing jobs in the United States. This trend is also partially responsible for the economic need of some wives to work. The family must have the second paycheck to survive.

Another way of looking at wage differentials is to compare job promotion rates between men and women. There is evidence of a gender gap in promotions for young men and women early in their careers, but this seems to disappear over time. In 1990, men were more likely to report having received a promotion than women. By 1996, the promotion rate of women was slightly higher than for men, even though both had dropped (Cobb-Clark and Dunlop 1999, 34).

Several interesting studies indicate that the increased international trade in manufacturing and deregulation in the banking industry may have helped reduce discrimination against women, at least in these industries (S. Black 1999, 39).

All of the evidence tends to support the continuing improvement in women's earnings and portend the decrease in the earnings differential between men and

WHAT DO YOU THINK?

1. Are women discriminated against in the workplace?

2. How does this discrimination occur?

3. If more women earn high wages, will more men further shirk their family responsibilities?

4. Are there other ways that the pay differential could be reduced?

women. These changes also indicate increasing equity in future family relationships (see Chapter 1).

Making the Decision to Become a Two-Earner Family: The Wife Goes to Work

The woman entering the work world is faced with more complicated and, often, more limited choices than her husband. Basically, she must choose from the following four major work patterns:

Pattern A. She works for a few years until she marries and has children, then settles into the homemaker job for the rest of her life. This was the predominant pattern for Caucasian, middle-class women until World War II. Although many women still follow this pattern, their proportion is declining. Today, such women are apt to be mothers of more than three children, wives of affluent men, or women who have meager opportunities in the job market because they do not have a high school education. Economic necessity may force some stay-at-home mothers to do in-home work, such as providing day care for other children, piecework, and so forth.

Pattern B. Women follow the same career pattern as men; that is, they remain in the paid labor force continuously and full time, through the years between school and retirement. Women most likely to follow this pattern are women without children, African American women, and women in professional and managerial jobs.

Pattern C. A woman works until she has children, then stays home for a certain amount of time (perhaps 5 to 10 years), and returns to the labor force on a basis that will not conflict with her remaining family responsibilities.

Pattern D. The woman remains in the labor force continuously, with short time-outs to have children. She combines family duties equally with work responsibilities.

Most men follow pattern B, and more women are also following this pattern. Patterns A and C are limited by the job opportunities available. Many employers hesitate to place young, unmarried, or newly married women in jobs with long-term advancement potential, or higher-level jobs that require extended training. They fear that such women will soon leave the job by choosing one of the other two patterns. In fact, during the past 10 years, there has been a 15 percent increase in women who choose to leave work and stay home once they have children (see "Stay-at-Home Moms," this chapter; Stahl 2004). Pattern C presents special difficulties for the woman returning to work after a long absence. She often finds that her skills are outdated. A woman who takes a break to have children often earns less after she returns. Higher-level jobs may also demand too much of her attention, causing conflict with her job as mother, homemaker, and wife. Partly because of these problems, women are increasingly choosing pattern D. This pattern, however, involves so much responsibility that it may lead to overwork, stress, and strain. The woman tries to be a superwoman, running the family and holding a job outside the home. Yet, this pattern in which women combine work with family is probably the most prevalent at this time. Business is recognizing that a partnership between work and the family yields the most benefit to both the worker and the business (see pages 416–419).

Increasing work availability for women has also meant increasing independence. This independence means increased freedom within marriage, but it can also mean increased freedom from marriage. There is little doubt that the working woman's ability to support herself has freed her to seek new roles. Part of the reason for the later median age of first marriage for both men and women may well be the greater work opportunities available to women. Greater work availability also allows a woman to escape from an unsatisfying marriage.

In the past, a woman's inability to support herself trapped her in marriage. She had to have a husband to survive. But with wider economic opportunities, this is no longer the case. Now a woman alone can survive financially, even when she has children (although her financial position is often marginal).

Not only did a woman in the past have fewer economic alternatives than her husband, but she also had to derive her status from his success. "Who are you?" she would be asked. "I'm the wife of a doctor," she would proudly reply. Work availability for women, then, frees women to have their own occupational identity.

Work outside the home can improve and enhance a woman's family life. Her earnings can increase the family's standard of living and alleviate the family's monetary restraints. The family can take longer vacations together, afford better housing in a nicer neighborhood, and improve the children's education. Thus, the wife's working can contribute greatly toward the family's well-being and the permanence of the marriage.

Besides the direct economic advantages of having another wage earner in the family, there are numerous other advantages for a wife who works outside the home. The working wife may derive great personal satisfaction from her work, just as many men do. By interacting with other adults outside her family, she may feel more stimulated and fulfilled, especially if she has small children at home. Her self-esteem may increase with the knowledge that she is a more equal partner in the marriage.

The other side of the coin is the possible guilt a working woman may feel toward her children, especially if they are young. Is she being a good mother and a good wife? Does the increased family income make up for the added stress and lack of time for her family and herself?

How the increase in independence will affect a woman who enters the workforce is hard to predict. Each individual will react differently. The point is that this increased independence is now a fact of life. It will enhance and improve family life for many women. It may also postpone or effectively end the marriages of other women.

The Working Wife's Economic Contribution to the Family

Olivia is typical of many wives who, after their children are old enough for day care, go back to work to help the family economically. Olivia has heard that a local company is expanding and needs new employees. She applies for a job, gets it, and suddenly finds herself a full-time working mother. Although her income is relatively small compared with her husband's, she believes her $2,000 per month will not only get the family out of debt, but will also allow them to purchase a few luxuries they have had to forgo. She and her husband, Rich, hope it will also help them save money toward a down payment on a house.

Rich and Olivia learn that her $2,000 per month does not raise the family income by that amount. A number of increased costs are associated with Olivia's return to work. Arrangements must be made for child care. Child care is usually the single biggest expense when a mother with a young child or children returns to work. There are several options. Mrs. Smith, an older mother down the street, also needs some extra money and, for $25 a day, will keep both children at her house during Olivia's working hours. Rich's mother is willing to keep them one day a week for free. A public day-care center near Olivia's workplace will care for the children for $50 to $100 a week. The center's charges are based on the family's income, so the weekly costs vary from family to family. Rich and Olivia decide to leave the children with Mrs. Smith 4 days a week ($100 a week) and with Rich's mother for the remaining day (free). This way, the children will be with people they know and will also be staying in their own neighborhood. If this arrangement doesn't work out, they can put them in the day-care center. Thus, the family's monthly child-care costs are $400. This leaves $1,600 from Olivia's paycheck.

Transportation must also be considered. A bus goes past Olivia's workplace, but the route is circuitous and requires her to leave the house half an hour earlier than if she drove. Suddenly, time for family work and activities has become very short. Riding the bus both ways will take an hour a day and cost $44 a month ($2 a day). If Olivia takes the bus, Rich will have to take the children to the sitter's each morning, which will cause him to be late to work. Olivia cannot find a car pool or ride to work with a colleague, so she and Rich decide that buying an older economy car is probably the best time-saving solution. With the car, Olivia can help deliver and pick up the children and can be more efficient, generally. They borrow $2,000 for 24 months at 12 percent interest per year and buy a used car from a friend. The car is in good condition and requires only a new set of tires. There are additional insurance costs, however. They purchase only liability coverage, thinking that because the car is old, it isn't worth the cost of collision coverage. The total monthly cost for the car is $198. Subtracting $198 more from Olivia's monthly paycheck leaves $1,402.

Taxes and social security are also deducted from Olivia's paycheck. She takes no deductions for the children, and finds that another large bite has been taken from her paycheck. She and Rich will receive a tax refund at the end of the year because of child-care costs and other deductions, but overall, the taxes and social security costs average about $200 per month. Thus, Olivia's monthly check shrinks further, to $1,202. (In the past, income taxes have been biased against the two-worker family, costing working married women more than others. However, the marriage penalty tax was changed in 2002.) There are other miscellaneous costs associated with Olivia's going to work. She doesn't have as much time for food preparation and household work. Food costs increase as she finds herself using more prepared foods. She sends more clothing to the laundry, and she has to buy some new clothes for work. These costs add up to about $200 per month.

The bottom line is that Olivia's $2,000-a-month pay adds only about $1,002 to the family income. This amounts to just over 50 percent of her gross pay. In general, the working mother must spend between 25 and 50 percent of her income in order to work, depending on the age of her children, type of work, and other factors unique to her situation. Because women tend to hold lower-paying jobs, the actual amount of money they contribute to the family tends to be small in relation to the amount of work they must do to earn it.

In the past, the working wife was the exception, and her work was often viewed as a family insurance policy, a buffer against hard times, or a source of "play" money—money used for recreation, luxuries, and extras. Today, a working wife's income has become necessary to family survival. This means that many families no longer have an economic buffer against hard times. They need both incomes to survive; if either is lost, family finances become precarious, if not impossible. As noted, young couples who both work early in their marriage, and who become used to the lifestyle afforded by spending two incomes, may be in financial distress when children arrive and the cost of living escalates.

Community Service and the Working Wife

The "just homemaker" role usually includes more than homemaking activities. Volunteer work and providing informal support to others and the community are other aspects of the homemaker role that are seldom discussed (Hook 2004). It is the "just homemakers" who often do much of the important volunteer work for society. They manage the PTA, organize church events, help a neighbor, and raise funds for the children's school. They serve on community committees and boards, help run the local Red Cross, donate time to political campaigns, and assume political roles. Women's "invisible careers," or volunteer activities for the community, help ensure safety and beautification and help with social problems (J. Anderson 1995; Lopata 1993). Married persons and parents with children under the age of 18 are more likely to volunteer than persons with no children of that age (volunteer rates of 37.5 percent and 24 percent, respectively). About 25 percent of men and 32 percent of women did volunteer work in the year ending in September 2003 (Bureau of Labor Statistics 2003). Women volunteer at a higher rate than men, a relationship that holds across all age groups and education levels. It is interesting to note that teenagers had a relatively high volunteer rate (30 percent), perhaps reflecting an emphasis on volunteer activities in schools.

Among volunteers with children under the age of 18, 47 percent of mothers volunteer for an educational/youth service organization, such as school or after-school sports (Bureau of Labor Statistics 2003). When school budgets become tight, schools increasingly rely upon such volunteers to help in the classroom, on the playing field at recess, and in fund-raising drives.

As more wives enter the formal workforce, our society is experiencing a loss in this informal workforce of volunteers. In Chicago, for example, the number of Girl Scout volunteers dropped from 10,000 in 1980 to 2,100 in 1994 (*Chicago Tribune* 1994). Community service may diminish because the working wife simply won't have time or, if she makes the time as many do, the energy drain may be too great.

Another effect of women entering the workforce in large numbers is that fewer adults are left in residential neighborhoods during working hours. Some mothers who do not work are becoming neighborhood mothers; that is, they serve as temporary mothers for children of working parents until a parent returns home. These neighborhood mothers do not run day-care centers in their homes; they are simply mothers to whom school-age children can come if there is a problem while the parents are working.

America is somewhat unique in the large number of persons who do volunteer work and the large number of persons who donate money and time to charitable organizations. The increased numbers of working women has had a negative effect upon volunteerism because in the past, it was the stay-at-home mom who did much of the volunteer work.

WHAT DO YOU THINK?

1. If your mother does not work, is she involved in any community projects? What are they?

2. If your mother works, is she involved in any community projects? What are they?

3. Do either of your parents play any volunteer or leadership role in your immediate neighborhood? Your school? In the community?

4. Considering all the volunteer work that is done in the United States, is the country gaining or losing as more women join the formal workforce? Why?

Household Activities and Supermothers

It seems strange to hear a mother of two small children answer the question "What do you do?" with "Oh, nothing, I'm just a homemaker." Obviously, a mother with two small children does a great deal of work for her family inside the home. Therefore, when she takes a job outside the home, something has to change inside the home. Mothers with children at home average about 36 hours a week working in the home. Generally, their time is divided into three major household activities: (1) meal preparation and cleanup, about 30 percent; (2) care of family members, 15 to 25 percent; and (3) clothing and regular house care, 15 percent.

What happens to all of this work when a mother takes an outside job? Essentially nothing. It must still be done and the mother still does most of it, although fathers are helping more than in the past. The working mother simply cuts down the amount of time she gives to each task and donates much of her leisure time (weekends and evenings) to household tasks. For the dual-worker family, time becomes a critical resource that must be managed well (Daly 1996; Hochschild 1997). "Balancing the competing and often overwhelming demands of paid work and family commitments is perhaps the most central challenge in women's lives as we enter a new century" (Milkie and Peltola 1999, 476; Spain and Bianchi 1996, x).

With the entrance of the wife and mother into the labor force, one would expect that husbands and other family members would be helping more with child rearing and housework. Husbands are likely to agree that they should do more in the households when their wives work, but in fact, they rarely live up to their professed beliefs or their wives' expectations.

Men report doing a large share of the housework and child care, but when it comes down to actually doing the work, they rarely take as much responsibility for it as their wives do. A husband, for example, may take out the garbage (2 minutes), while the wife does the dishes (15 minutes). To date, although men are doing more to offset the household pressures created by women's increased participation in the labor force, they still do not carry their fair share of household and child-care work when their wives work.

Older children often contribute to housework and younger sibling care when mothers work. However, like fathers, their domestic labor is below their mother's contribution (Manke et al. 1994).

Higher-educated husbands and husbands of higher-earning wives tend to share more home responsibilities (Ross and Mirowsky 1994). This is probably true because education tends (but not always) to move people away from a narrow, traditional view of gender roles. Men who have a more egalitarian gender-role ideology appear to shoulder more household duties than men with more traditional gender ideologies do (Greenstein 1996).

Division of domestic tasks is dynamic, rather than static. For example, when one spouse is under unusual stress, the other spouse often assumes a greater proportion of the housework than usual (Pittman et al. 1996). When children arrive or work schedules change, the division of domestic labor and child care is usually renegotiated in an attempt to achieve a more acceptable balance.

Distribution of family work (housework and child care combined) is a critical issue for most dual-worker families, yet husbands' low level of household labor compared with their wives' is not necessarily negatively perceived. Studies have found that both husbands and wives can feel the division of labor is fair, even when the wife bears most of the household responsibilities. Apparently, the perception by the spouses of the willingness of the other to shoulder responsibility and give emotional support for family work is more important to marital satisfaction than the actual amount of work accomplished (Bianchi and Spain 1996). In other words, the perceived balance (equity, see Chapter 4) between work and family work is perhaps more important to partner satisfaction than the actual balance (Milkie and Peltola 1999, 477; Stevens et al. 2001).

Although husbands don't appreciably increase their share of household work when their wives go to work, overall they are sharing more household work than in the past, whether their wives work or not. This change stems from shifting attitudes about gender roles and the increased emphasis on egalitarian marriage in the United States.

One restraint on husbands sharing family work is their continued earning advantage over women. Because the husband can usually contribute more economically to the family than can his working wife, he or the couple may feel that this makes up, in part, for his share of family work. Married men are more likely than single men to set higher earnings goals and to pursue job-shift patterns that result in wage gains (Gorman 1999). Because these men contribute more economically, the couples may feel he can do less family work proportionately, and thereby maintain an egalitarian relationship at home. Hochschild found, however, that the amount of money men earn above what their wives earn does not necessarily relate to the amount of family work the man does. In her sample, men who earned less than their wives uniformly did little or no family work. She hypothesized that because those men could not demonstrate superiority through higher earnings, having the wife do all of the family work was a way of maintaining power (Hochschild 1989, 221). However, some women may judge their husband's failure to participate in housework and caring for the family as yielding power to them as a family manager. In this case, homework and family care may become a source of satisfaction and pride (Mederer 1993).

For many wives, one result of going to work is overload and strain. They end up doing two jobs—one outside the home, and one inside. Their leisure time is greatly reduced. The quality of their household work declines. Time becomes their most precious commodity, especially spending more time with the family (Blankenhorn et al. 1990, 88; Hochschild 1997).

It is overload of the working wife, especially the working mother, that families most resent. In general, working wives and mothers indicate they do not have enough time for themselves. Fewer working fathers feel this as strongly, although both working parents mention lack of time with family and children and long hours on the job as the greatest strains placed on the family. Constant complaints are made about lack of time for recreation, picnics, vacation trips, children, lovemaking, and plain old doing nothing.

If husbands and fathers do not react positively to their working wives' overload, it will likely cause tension between the partners. The media try to portray the working supermom as the ideal, but such a role is almost impossible for anyone. The work overload leads women to feel resentful and put upon, which they are if their husbands do not give them family help.

Women may try to cope with the unfair burden in a number of ways. They may try to change their husband's behavior directly so that he does more family work. Setting up work schedules and making lists to organize family time is one approach. Wives may also try indirectly to

Dad's participation is important to the success of the dual-worker family.

convince their husband to do more work. Or they may simply give up on their husband and try to be supermoms. When this proves impossible, they may cut back on housework and child care or try to cut back at their job (Bianchi and Spain 1996, 32; Hochschild 1997, 193–199; Parcel and Menaghan 1994).

Although we are discussing the stresses and strains on women, such stresses are also felt by men—regardless of whether or not they share the family work burden. If a husband shares the family work, then he too may feel overburdened. If he does not share, he may feel guilty about the wife's unfair workload. He will certainly feel her resentment. The revolution of the woman entering the paid workforce has not been accompanied by enough of a revolution in how family work is handled, either by her spouse or by society's expectations. Nor has there been enough change in how employers cope with working women with families. And this gap between what happens to women in the workplace and in the home can cause family conflict.

Deciding in Favor of Part-Time Work

Olivia works for 7 months at her new, full-time job. Although her salary enables she and her husband to pay off their debts, she finds that she is increasingly fatigued. She and Rich never seem to have time together. Their sex life seems nonexistent. If she does find some free time, all she wants to do is sleep. The house looks unkempt. She feels guilty about the little time she is able to spend with the children. She is grumpy and unhappy much of the time. Now that they are out of debt, Olivia decides to see if she can find a part-time job. She finally does and, with a sigh of relief, quits her full-time job.

Women who work part time—that is, fewer than 35 hours per week—made up 25 percent of all female wage and salary workers in 2003. In contrast, just 11 percent of men in wage and salary jobs worked part time (Bureau of Labor Statistics 2004c).

Approximately half of all women who work part time give "child care and family obligations" as their reason for preferring part-time work (U.S. Census Bureau 1999b, Table 671). The creation of more and better part-time jobs for mothers with young children would help alleviate the overload experienced by full-time working mothers.

Unfortunately, part-time work reduces the economic contribution to the family, not only because fewer hours are worked, but also because pay standards are lower. On an hourly basis, part-time work generally pays only 75 percent as much as full-time work.

Those who seek part-time employment are usually assumed to be intermittent workers who are not committed to a long-term career. There-

WHEN MURDER SEEMS TOO MILD

Cartoonists & Writers Syndicate

Cartoon by Signe. Reprinted by permission of Cartoonists and Writers Syndicate.

fore, part-time work seldom offers fringe benefits, job protection, or opportunities to advance. This is true for all contingent workers (part-time, temporary-help service employment, contracting out, and home-based work) (Hipple and Stewart 1996). Also, the kinds of jobs available to part-time workers tend to be lower-level, less-challenging jobs (Stahl 2004). Lack of fringe benefits, especially health insurance, combined with low pay in part-time jobs may keep some people on welfare. For example, nonwage compensation (benefits aside from monetary reward) cost employers 28 percent of a worker's total compensation. Health insurance prorated on an hourly basis is more costly for part-time employees than for full-time employees; therefore, it is not usually offered to part-time employees. In all fairness, a fairly large percentage of part-time, married workers turn down health coverage, opting instead to be on their spouse's health plan (Lettau and Buchmueller 1999). Welfare recipients, on the other hand, are eligible for free medical care under the Medicaid program, and in some cases, welfare payments may provide as much income as can be earned on a part-time job.

Ninety percent of part-time employment occurs in the service industries, the so-called "pink-collar" occupations where women predominate. Four out of five waitresses work less than full time. Department store sales work is becoming increasingly part time, and businesses are increasingly turning to temporary help, to avoid much of the government employer–employee red tape and to reduce costs. Beauty salons have always used part-time workers. In the health field, where women make up 75 percent of the workforce (except in upper-level positions), shift work and part-time arrangements are commonplace. Schools are turning increasingly to part-time substitute teachers, to save money. In contrast, part-time and temporary work is seldom found in the industries dominated by men.

As a result of these patterns, most mothers will be downwardly mobile if they work part time, rather than upwardly mobile like their husbands. More and better jobs are open to women before they have families because they can work full time. After a woman has a family, the hours when she is available for work often determine the job she finds. Also, women with families tend to have absence rates from work approximately double the rate of men's absences from work (4.4 percent compared to 2.5 percent) (Bureau of Labor Statistics 2003), due usually to child-care problems. In the face of demands on her time, the young mother is likely to find that the scheduling of her job is the most important single consideration. Her immediate job choice is dictated in large measure by her family time constraints. As more mothers enter the workforce and prove to be good workers, however, part-time jobs may take on more of the advantages that come with full-time work.

In our example, Olivia moved from full-time to part-time employment and experienced a reduction in family conflict. However, women moving into part-time work from nonemployment (full-time homemaker role) may actually experience more conflict than if they had moved into full-time employment. Perhaps this is because they bring such a small amount of extra money into the family. "It's not worth having you work. We lose more than we gain," suggest some husbands. When the wife is contemplating taking a full-time job, the couple are more apt to make a considered decision than when the wife takes on a part-time job. Often, a wife just slides into part-time work because it has become available. Little thought is given to the effect that her part-time work has on the family. It should also be noted that women account for 65 percent of all workers who combine several part-time jobs into a full-time schedule (working 35 or more hours per week) (J. Gardner 1996).

WHAT DO YOU THINK?

1. What do you see as the advantages to Olivia and the family when she works part-time?

2. What are the disadvantages?

3. How easy do you think it would be for her to find part-time work?

Child Care and Parental Leave

As more families become dual-worker families, parental leave for pregnancy, elder care, child sickness, and increasing child-care options become important. Since 1975, the labor force participation of mothers with children under 18 years has grown from 47 to 72 percent. The biggest increase in labor force participation among mothers has occurred among women with children under age 3. Fully 61 percent of this group was in the labor force in 2002, compared to 34 percent in the late 1970s (Bureau of Labor Statistics 2004e). Today, only one in three mothers stays home and provides full-time care for her children, although this number is increasing among women in the middle and upper economic classes (Stahl 2004).

Affordability and quality are the two major problems confronting parents who seek child care help while they work (Helburn and Bergmann 2002). (See "Debate the Issues," Chapter 10, for a more detailed discussion of child care outside the family.) The quality of care provided for children of working mothers is a matter of both public and private concern. When both the mother and father work outside the home, other arrangements must be made. Children must be left with babysitters, with relatives, or at child-care centers. Although professionally run child-care centers might seem to be the best solution for working mothers, many women cannot afford the $20 to $30 per day ($300 to $500 per month) that good private centers charge. Consequently, in recent years, women have argued and worked for family-leave legislation and for government and business funding of day-care centers.

One objection to government funding is that the quality of child care may be poor. Consistency and predictability are crucial to the development of young children. In the past, parents provided this guidance. Profamily groups argue that the chances for good care are still best with the child's own parents: They know the child best and are ego-involved with the child. In day-care centers, the quality and philosophy of child care and the sort of person interacting with the children are often unknown to the parents.

Researchers who observed 400 child-care centers in several different states concluded that only one in seven centers were of good quality, where children enjoyed close relationships with adults and teachers who focused on individual needs of the children (Vobejda

© Comstock Images/Getty Images

The business that offers family support to its employees will increase its productivity.

1995). These same researchers found also that the parents greatly overestimated the care that was being given to their children. Such reports support the importance of working parents checking carefully on the child-care provider they choose.

In 1993, the first comprehensive, federal family-leave bill (Family and Medical Leave Act) was passed. The law grants workers in companies with 50 or more workers up to 12 weeks' unpaid leave for family emergencies. Maternity leaves, elder care, and child-care leaves are included. Also, an employee would be guaranteed his/her job or an equivalent job upon returning from the leave. Since the passage of the act, businesses have increased leave coverage for their employees (Waldfogel 1999). In general, women tend to take more advantage of leave time than men do.

Hofferth and Deich (1994) compared this new family leave legislation with similar legislation in Germany, France, Sweden, and Hungary. They found that the European countries spend a higher proportion of their gross domestic product on child care and family benefits. One major difference, for example, is the level of wage replacement for parental leaves. This level varies from 60 to 100 percent for 15 months in Sweden, 18 months in Germany, 16 weeks in France, and up to 3 years in Hungary, as compared with 12 weeks of unpaid leave allowed under the U.S. legislation.

Traditionally, only a small minority of mothers with children under age 6 have employment that allows them to care for their young children. Yet, effective child care by someone other than the true biological mother does not necessarily lead to severe problems in children (Azar 1996b, 2000). Actual effects of substitute child care are just as difficult to uncover as are the effects of natural parenting. Toby Parcel and Elizabeth Menaghan (1994) point out that many factors affect the child or children of working mothers, so it is difficult to predict just what the effects will be. They list the following factors and their interactions that influence the effect a working mother will have on her children:

- Type of child care
- Family characteristics
- Number of children
- Dual-career family or not
- Parental pay
- Work hours
- Type of job
- Child or children's personal characteristics
- Reasons for mother going to work
- Acceptance of mother's working by family members

The quality of substitute child care can be excellent. It can provide for the child both physically and psychologically. A loving, caring babysitter is often an important and happy influence on a child. It would seem, however, that other family members would be more likely than strangers to give the child adequate love and attention. Thus, substitute child care cannot be wholly praised or condemned, but must be judged on its specific merits. It can be good or bad for a child, just as the child's biological parents can be. Table 13-3 shows the approximate use percentage of different child-care providers. Single mothers tend to use other relatives more often than married mothers do. Also, many parents use a combination of child-care providers, rather than just one.

Individual children will react differently to the partial loss of their mother and the substitution of another type of care. Their reaction may be positive or negative. A hyperactive, disruptive child may experience much negative feedback in a large day-care center, but may thrive with an attentive babysitter.

Stay-at-Home Moms

In the professional and managerial classes, where higher incomes permit more choices, a reluctant revolt is underway. Today's women execs are less willing to play the juggler's game (juggling full time work and family obligations), espe-

TABLE 13-3

Child-Care Arrangement for Preschool Children

Provider	Percentage of Children Cared For
Relatives	41
Child-care centers	30
Family day care	17
Child goes to work with mother	6
In-home caregiver	5
Other arrangements	1

SOURCE: Merrill 1995.

cially in its current high speed mode, and more willing to sacrifice paychecks and prestige for time with their family. (Wallis 2004, 52)

Among the group of women who are well-educated, married to successful husbands, White and over 30 years of age, there has been a significant increase in those leaving full-time work to remain home with their young children. For mothers with a child younger than 1 year old, there has been over a 6 percent drop in workplace participation in the past 5 years. The Census Bureau reports a 15 percent increase in stay-at-home moms in the past 10 years. Harvard's graduate business school reports that only 38 percent of their graduate women of childbearing age are in the workforce full time (Stahl 2004). Jill Kiecolt (2003) found in a broad study of over 24,000 respondents that women had become less likely to find work a haven and that home life had become relatively more satisfying.

It seems that younger women who are successful economically and in their career are more interested in balancing their work and family responsibilities, rather than centering mainly on their careers. Perhaps they learned from their own mothers that combining a full-time, high-pressure job and family to become a supermom comes at too high a cost. More young women and men are rating personal and family goals higher than career goals. Many of these stay-at-home moms also express the desire to return to work in the future, because they miss the challenges and rewards of being successful in their careers.

Perhaps an inflexible work world motivates many stay-at-home moms, rather than the joys of child rearing. Workplace inflexibility, lack of challenging part-time jobs, lack of decent child-care facilities, too few retraining venues, and other workplace problems serve to further motivate new mothers to leave the workplace. This is, of course, a loss of skilled labor for business and probably increasingly detrimental to businesses as the low American birthrate further reduces the skilled labor force.

Employers, Pregnant Employees, and Working Mothers

If the family is to survive both parents' working outside the home, and if business is to have satisfied, productive workers, the obligations of family will have to be integrated with the obligations of work. Organizing the workplace so that a company can both successfully compete in the global marketplace and provide personal growth for its employees and their families is a major challenge facing U.S. employers.

Many employers feel that women with children are not as likely to make the same commitments to their careers as men are. As we noted earlier, more women quit their jobs than men. Also, there is the trend of highly skilled women dropping out, at least temporarily, to become stay-at-home moms. Yet, with women making up almost half of the U.S. workforce, and with the majority of mothers with children under 18 years of age already in the workforce, women are clearly an integral and necessary part of America's labor supply. It is true that mothers bear the major responsibility for child care and thus are more torn between their families and their work than men are. Yet, by taking this into account rather than simply complaining about it, employers can improve workers' performance while supporting family stability.

Better leave policies, more flexible working hours, job sharing, on-site child-care facilities, and, with the advent of computer networking, increased use of the home as a workplace are all possible ways to improve the relationship between family and work. Programs and policies such as these help workers resolve the conflict between personal and professional responsibilities, leading to increased profits. Research shows that such policies can, in fact, improve a company's fortunes, primarily by increasing worker loyalty, thereby reducing costly turnover and stepping up productivity (Lawlor 1996; Moskowitz 1997).

For example, American Telephone and Telegraph (AT&T) has a workforce that is 55 percent women. AT&T offers many perks to its employees. Work at home, job sharing, compressed workweeks, job benefits such as health care for part-time workers who work a minimum of 25 hours per week, and training for managers on alternative schedules and how to implement them all serve to allow workers greater flexibility in meeting the demands of their jobs and families. **Comp time**, or compensatory exchange or trade of overtime pay for time off, is another method of allowing workers increased flexibility (Stolba 2001). Childbirth leave with some pay, phase-back for new mothers, adoption aid, assisting workers in finding appropriate child and elder care, and scholarships for employees' children are some of the family benefits offered. Perhaps most unique is AT&T's Family Care Development fund, launched in 1990. The fund's purpose is to improve and expand child- and elder-care programs in communities where AT&T workers live. Some of the activities funded have been before- and after-school, holiday, and summer programs for children.

comp time
Trade of overtime pay for time off

Flexible schedules probably do the most to help workers with children at home. In 1985, about 12.4 percent of workers had flexible work schedules, but by mid-2001, this had risen to 28.8 percent. Note that flexible arrangements for many workers are informal. About 11.1 percent of the total have formal employer-sponsored flextime work schedules. Flexible schedules are more common among managerial and professional specialty occupations, with almost one-half of executives, administrators, and managers able to vary their work hours (Bureau of Labor Statistics 2002b).

More companies are establishing programs such as those discussed above. However, such programs in and of themselves are not productive unless management actively supports them. For example, flexible work schedules don't help if a supervisor doesn't believe in them and does not help his/her employees to work out the best possible schedule. A list of some of the larger companies that offer many of the programs described above may be found on page 401.

To illustrate how such family support programs work, consider companies that offer mothers shortened shifts within school hours. These shifts allow mothers time to be with their children and maintain their home, and yet contribute economically to the family. These mothers can participate in the world outside the home and maintain a career. Such flexible time schedules help working parents cope with the demands of the home, while maintaining high productivity on the job.

Unfortunately, some evidence shows that employees fail to take advantage of such work opportunities, for fear of being thought less committed to their work and, therefore, being passed over for promotion (Phipps 1996). To the degree that this is true, such alternative work schedules do not serve the purposes for which they were instituted.

With the changes suggested above, parents would have more alternatives and would derive more satisfaction from their work. Businesses would have better workers, because employees who feel worthwhile and productive in all areas of life do a better job and are less prone to quit. Fair pay and a humane work system that supports, rather than hinders, home life are worthwhile social goals.

When women ran the home and reared the children and men supported their families by working outside the home, employers did not need to concern themselves with these types of accommodations to keep their workers satisfied and productive. The days of such separation between family and work are probably gone forever. A century and a half ago, work and family were integrated on the American farm, but they were separated by the Industrial Revolution. Now, as women have become essential to the American workforce, the family and work will have to become integrated and supportive of each other again. As employers team up with the family instead of engaging in conflict with it, American productivity will be stimulated to reach new heights.

cathy®   **by Cathy Guisewite**

Home-Based Work

The inventions of the fax machine, computer, modem, e-mail, pager, and cell phone are making working at home more practical for some jobs. AT&T estimates that approximately 16,000 of their employees work at home. Self-employed individuals, sales workers, managers, and professionals are the most likely to work at home. Blue-collar workers are much less likely to work at home. Workers who reported working at home at least once per week account for 15 percent of total employment. Half of those who usually work at home are wage and salary workers who take work home from the job on an unpaid basis. Another 17 percent have a formal arrangement with their employer to be paid for the work they do at home. The remainder who work at home (about 30 percent) are self-employed (Bureau of Labor Statistics 2002b). Women and men are about equally likely to do some job-related work at home.

It may sound idyllic for a mother or father to work at home, thereby combining work with family care. However, it is not easy to do. There must be time and a work space for working. Small children demand constant attention. Thus, even when working at home, it may be necessary to use child care of some kind. Other family members must realize and respect the work needs of the family member working at home, so that there is not constant disruption. There are costs involved, such as buying any necessary devices and maintaining communication lines with the office. Despite these difficulties, people like Craig McCaw (1994), the founder of McCaw Cellular, see working outside the office, yet remaining in constant contact with the office, as the wave of the future.

As with any change, there are conflicting viewpoints about home-based work. On the positive side, workers gain flexibility and better control of their time. Working at home allows family members to care for children or for elderly or disabled relatives, while participating in the labor market. In fact, mothers and fathers with preschool children tend to be the most positive about working at home. Home-based work also facilitates employment of the disabled.

Critics point out the increased possibility of worker exploitation for those who work at home. The chances of low wages, few benefits, and perhaps, substandard working conditions are increased. For example, reports of an expansion in illegal sewing of women's apparel by immigrant labor has increased the criticism of working at home (L. Edwards and Field-Hendrey 1996). The biggest criticism by those working at home was the blurring of boundaries between work and family (E. Edwards and Edwards 1996; E. Hill et al. 1996): "I cannot leave the office anymore." "Even on vacation, my work follows me." "I don't have a specific hour when I can know that I am finished with work."

Marital Satisfaction in the Two-Earner Family

As with so many areas we have discussed, the question of marital satisfaction in a two-earner family is double-edged. The family may gain satisfaction through the wife's economic contribution, or the family may lose satisfaction because she can no longer supply all the caring and services of a full-time wife and mother. Economic strain may be reduced when she works, but psychological and physical strain may be increased.

The research evidence on marital satisfaction when the wife works is mixed. After reviewing many studies, researchers have concluded that couples in which the wife works from choice rather than economic necessity (and the husband views his wife's employment favorably), couples in which a wife works part-time, and couples with more equitable levels of power and influence tend to score higher on marital-satisfaction measures. It is clear that there is a multifaceted relationship between marital satisfaction and job satisfaction. Early hypotheses about the spillover between the two suggested that the stress and strain of work would have negative consequences for the family. This is sometimes true; however, later studies have suggested that spillover from the marital relationship to job satisfaction is more meaningful. Positive spillover to job satisfaction from increases in marital satisfaction occurs as does the reverse; that is, marital discord significantly relates to job dissatisfaction (J. J. Rogers and May 2003). Other research suggests that the benefits from combining work and family outweigh the burdens and strains (Barnett 1994, 1998; Greenhaus and Parasuraman 1999; Grzywacz et al. 2002).

When wives are not employed, marital discord significantly increases the likelihood that they will enter the labor force. It is not clear whether wives enter the labor force to prepare for divorce, or just to improve their own lives (and perhaps their marriages) by seeking personal challenges (S. Rogers 1999).

How couples cope with the wife working seems to be individually determined by each couple. Once a couple works out the new routines and relationship changes, marital satisfaction seems to return to its normal level.

Although we have spoken mainly of the working wife's economic contribution to the family, it is clear that she may also receive psychological dividends from her participation in the world of work. Work may allow her to use some of her skills that are unused in the homemaker role. She will meet and interact with a wider variety of adults. She will gain more power in relation to her husband by contributing economically to the family. Her feelings of integrity, self-respect, competence, self-determination, and accomplishment may increase if she enjoys her work and has successfully solved the problems of working and caring for her family.

The evidence of husbands' marital satisfaction when their wives work is also mixed, but tends to indicate that they are less satisfied than the wives. It seems that some husbands accept their wives' work grudgingly, and that some men may have more trouble than women in adapting to nonstereotypical roles. In going to work, a woman is frequently expanding into a new role—one that is higher in status than that of homemaker—while a husband who assumes homemaking functions is adopting a role of lower status, which may strain not only his sense of status and identity, but also his feeling of competence. If a husband sees his wife working as a statement about his inability to care for the family, he may feel inadequate or resentful. Furthermore, a busy wife may not be able to provide the same level of physical and emotional support that a full-time homemaker can, so a husband may well come to feel he is losing out on all fronts.

On the other hand, a second income can provide the husband additional freedom. He may cut down on moonlighting or overtime work. He may be able to take a temporary reduction in pay to enter a new career or job he finds more satisfying. He may participate more in child care.

Because of the importance of expectations in human relations (see Chapter 7), what one thinks or expects about something is often as important as what actually happens. Research on marital quality in families where the wife works indicates that happiness with the relationship is more closely related to the congruence between the role expectations of one spouse and the role performance of the other spouse, than to any particular pattern of roles. It is not simply a matter of whether a woman's working has an impact on marital adjustment, but rather, the extent to which that behavior violates her own and her family's role expectations. If a woman expects to be a homemaker, if her husband expects her to stay home, and if significant others (parents, in-laws, children, and others) in her environment have negative attitudes about her working, the chances are great that both her and her family's satisfaction will drop when she goes to work (Lye and Biblarz 1993).

As the working wife has become the norm for our society, negative attitudes have diminished. Children of working mothers tend to be more supportive of the idea of wives working. Dual-career families tend to produce children whose attitudes are more egalitarian and who prefer dual-career families themselves. This being the case, the incidence of families in which the wife works should continue to increase, as should the level of marital satisfaction in such families. You will remember Olivia reducing her work from full time to part time. When she was a full-time homemaker, Olivia found that her two children were always underfoot. She never seemed to have a minute's peace. They called, "Mommy, Mommy, Mommy" so often that she almost never paid attention to them. In fact, sometimes when she heard the children, she would deliberately hide to escape their constant pressure. Now that she works part time, she finds she enjoys spending time with the children. She looks forward to the weekends so she can do projects with them and give them her undivided attention. Olivia suggests that having more time available for the family, while at the same time contributing economically, has improved both her job satisfaction and marital quality.

Work and Family: Sources of Conflict

Ideally, work and family should complement and support each other; in reality, these two arenas of life often conflict. The conflicts tend to differ for husbands and wives, in large part because of the historical division of labor between the sexes. The work world tends to intrude on a husband's family life. If he is asked to work overtime, for example, he must usually do so to keep his job, regardless of what plans he may have made with his family. The opposite is true for the wife; the family is more likely to intrude into her work world. For example, if a child is sick, it has traditionally been her responsibility to tend to the child. A husband's family usually must bend to his work demands; a wife's work must usually bend to her family's demands. Sanctions against a man for poor performance have traditionally been greater on the job, whereas sanctions against a woman for poor performance have traditionally been stronger in the family.

As more wives and mothers enter the labor force, however, it becomes imperative that the conflicts between job and family be reduced. The work world and the family realm must be balanced. The work world cannot ask men and women to forgo having families. With increasing numbers of two-earner families, parenthood becomes one of the costs of doing business. Family members must work to support themselves and keep society functioning. But to enable work to proceed efficiently, employers must recognize and accept family concerns.

Time becomes a highly valuable commodity to both partners in the dual-worker family, especially to the wife, and particularly if there are children. Since the husband in the dual-worker family usually does not shoulder his fair share of household tasks, as we saw earlier, time is clearly more of a constraint on the working woman. This is also true of the single-parent family. Returning home from a long day's work to prepare dinner, houseclean, and perform other family chores is hardly an inviting prospect. If there are children, how does one find any time for them, much less summon up enough energy to make it quality time? The increased pressure on home time, due to long hours at work and inflexible work schedules, means that much less of a couple's time can be devoted to their children; to each other; and to recreation, play, and self-renewal activities.

The dual-worker family with children suffers the most from lack of time, especially parenting time. Parents are working more, and thereby spending less time with their children. Many American parents suggest that spending less time with their families is the most important cause of the fragmentation and stress in contemporary family life.

When both husband and wife work, they will not necessarily have less time together, but the time they have together will be consumed by daily necessities. Too often, the important time for intimate caring and working on the relationship is lost. When a couple works exactly the same hours, there is no lost "togetherness" time. But if one mate starts work an hour earlier than the other (off-scheduling), then the couple loses an hour of togetherness time.

We should also note that time at home is not necessarily family time. A spouse may well bring work home, either concretely or psychologically. The dawn of computers, fax machines, and cellular phones has made it possible to work anywhere, anytime (see page 419). This can be beneficial sometimes, and harmful at other times. A career-oriented spouse is much more apt to bring work home than is a family-oriented partner. A career-oriented partner usually puts his/her career ahead of the family. This can cause family resentments because of the perceived lack of commitment to the family. "You're married to your job" is a criticism often made by other family members.

Strain on the job can cause difficulty at home, or vice versa. "Don't ask your father to borrow the car the minute he gets home," a teenage child might hear. "Give him time to relax." This is an example of work strain causing family difficulty. Obviously, the stresses and strains of the work world take their toll emotionally on the working family member, and will also be felt at home.

Although the strain on the family due to unemployment is often discussed, little is studied about the effects of job insecurity. Jeffery Larson and his colleagues (1994, 138) suggest that the continuing restructuring and downsizing by business, required to compete globally, not only causes unemployment but also increases job insecurity for those still employed. The debate over **outsourcing**, sending jobs overseas where employee costs are less than in the United States, increases job insecurity even more. It should be noted that the debate seldom mentions the increasing numbers of jobs provided by importing jobs from other countries. Almost all of the major, foreign automakers have plants and produce many of their automobiles in the United States.

Insecurity results not only from the threat of potential job loss; there also may be salary/benefit cutbacks, relocation, or loss of potential for promotion. Relocation and residential change affect as many as 10 percent of Americans every year. Because such moves are often initiated to further men's careers, rather than from unemployment, husbands tend to evaluate relocation experiences more favorably than their wives do. Their wives tend to exhibit higher levels of

outsourcing
Sending jobs overseas where employee costs are less than those in the United States

stress—experiencing boredom, loneliness, and feelings of loss (Frame and Shehan 1994, 196).

Less often discussed is the effect of family strain on the job. Many companies considering a person for promotion evaluate both job performance and family life. One might object to this as an invasion of privacy, but the job performance of a worker undergoing severe family strain (illness, divorce, and so on) will be hampered.

An enlightened employer will recognize legitimate family needs and concerns and will try to arrange work accordingly. The parent who is concerned about who is caring for the children, who will meet with a child's teacher, or who will stay home with a sick child may also falter at work. As we saw earlier, an employer who offers help with such family concerns improves the productivity of the workforce.

Specific patterns of role behavior on the job and in the family may be incompatible. For example, the male managerial role tends to emphasize self-reliance, emotional stability, aggressiveness, and objectivity. Family members, on the other hand, may expect that same person to be warm, nurturing, emotional, and vulnerable in interactions with them. A person must be able to adjust her/his behavior to comply with the expectations of different roles. To the degree she/he cannot adjust, conflict is likely to ensue.

Such role disparities are especially difficult to cope with when the home-oriented spouse does not share in the other spouse's work world. In the traditional husband working–wife homemaking family, failure to know and understand each other's domains causes conflict. In fact, in extreme cases, the husband lives in a work world totally unknown to his wife and has little understanding or empathy for her problems with the family. A husband and, perhaps, father is legally in the family, but for all practical purposes is not a functioning member of the family, other than supplying economic support. In general, however, the traditional family has less conflict between work and family, because one spouse assumes most of the family responsibility while the other spouse assumes most of the economic responsibility.

Overload is another problem faced by members of dual-worker families, especially the working mother. Because working women still perform most of the family work, they are, indeed, often overloaded. As discussed earlier, whether this will be a serious problem seems to be determined not by the workload in itself, but rather, by the perception that it is "too much" and by the perceived unwillingness of the spouse to share the domestic role.

Often overlooked is the employment overload. Many workers—especially those who are self-employed, have management positions, or have voluntarily taken on extra shifts or jobs—work far longer than 40 hours a week. Corporate downsizing is also leading to longer work weeks. The idea of Sunday as a day off was laid to rest years ago in America. Obviously, the longer the workday or workweek, the less time one can devote to the family.

Jobs, Occupations, and Careers

Up to this point, we have been discussing women taking jobs in the labor market. However, a short- or even long-term job is not the same as a long-term career. Essentially, we can place work on an attitudinal continuum, according to the degree of commitment (Kahn and Weiner 1973, 153):

Basic Attitude Toward Work	Basic Additional Value
1. Interruption	Short-run income
2. Job	Long-term income; some work-oriented values (working to live)
3. Occupation	Exercise and mastery of gratifying skills; some satisfaction of achievement-oriented values
4. Career	Participating in an important activity; much satisfaction of work-oriented, achievement-oriented, advancement-oriented values
5. Vocation (calling)	Self-identification and self-fulfillment
6. Mission	Near fanatic or single-minded focus on achievement or advancement (living to work)

Most people in the labor force occupy one of the first three levels. A far higher percentage of men than women fall into the latter three categories, however. As attitudes about gender roles have changed, more career opportunities have opened to women. The dual-career family is becoming a more visible reality. A career may be denoted by (1) a long-term commitment, including a period of formal training; (2) continuity (one moves to increasingly higher levels, if successful); and (3) mobility, to follow career demands. More self-employed women fall into the latter three categories, and their number is increasing. As of 1975, women represented about one in four self-employed workers. By 1998, they accounted for one in three. Hispanic and Black women and men have substantially lower self-employment rates (U.S. Census Bureau 2002b, Tables 718–723). Self-employed women are more likely to be married than their wage and salary counterparts. They also tend to have more years of education (Devine 1994).

Dual-Career Families

In the families of most career-oriented men, the man's career dictates much of the couple's life. Where and how the family lives depends on his career demands. These demands are met relatively easily if the wife is a homemaker or works at one of the first three levels. A dual-career family, however, may encounter conflict between the partners over career demands, and other kinds of problems that occur in any family when both partners work. For example, what happens when one spouse is offered an important promotion, but it means moving to another location? Will moving harm the other partner's career? If the new location is not too far away, should one spouse commute? Should they take up two residences? What will this living arrangement do to their relationship? Each time one partner has a major career change, a series of such questions will have to be answered. When the couple strive for career equality, the answers are not easy.

Commuter Marriage and/or the Weekend Family

The dean of a local community college is married to a high-level school administrator in that district. She is offered a new position as president of a college several hundred miles away. After much discussion, the couple decides that they should continue their own careers. They are now a weekend family. Each spends the week at the job, and they take turns visiting each other on the weekends. Modern technology such as the cell phone, pager, and e-mail have helped such couples maintain contact. In fact, big phone bills are an important part of the commuter marriage (Glotzer 2000).

Past literature on the dual-career family reported that the impact of dual-career stress is felt mostly by women. This is true not only in the woman's family life, but also in her work. She takes more career risks, sacrifices more, and makes more compromises in her career ambitions in attempting to make the dual-career pattern work. Stress is also related to the number of children in the family.

Stress is reduced if the career woman has a supportive husband who is willing to leave his job and relocate to advance his wife's career. Strain is also greatly reduced if the couple is free of child-rearing responsibilities and each partner has a flexible work schedule. A flexible work schedule is especially important if the dual-career couple have young children.

If both partners are successful in their careers, it usually means that major career decisions will have to be made periodically throughout the relationship. Each new decision may upset the balance that the couple has worked out. Scaling back and lowering expectations is one strategy dual-worker couples use to cope successfully with the demands of work and family. Limiting the number of children, reducing social commitments and service work, having less leisure time, reducing expectations for housework, and trying to limit work hours are all scaling-back techniques that dual-worker couples may use (Becker and Moen 1999, 999).

About 1 million couples in the United States are believed to have commuter marriages. Of course, such a lifestyle is not suitable for families with small children. Couples that choose this lifestyle tend to be free of child-rearing responsibilities; they are either young and without children, or older and have children who no longer live at home (Becker and Moen 1999). They also tend to have established careers, high educational levels, high-ranking occupations, and high income levels.

WHAT DO YOU THINK?

1. What are the disadvantages of such a relationship?

2. Are there any advantages?

3. Can you think of things such a couple can do to maintain closeness in their relationship?

Summary

1. Many consider the number of women entering the workforce to be the major change affecting the American family since the Industrial Revolution. In the past, the woman's (especially the mother's) place was in the home. *This is no longer true. In most families today, the woman is an active participant in the economic support of her family.*

2. Women are now a permanent and large part of the American labor force, yet they *still earn a disproportionately lower income than their male counterparts. The earnings gap is closing.*

3. *Career opportunities are still narrower for women than for men.* Many steps have been taken to change these inequities. Although change is slow, there are many hopeful signs that the future work world will be as advantageous for women as for men.

4. *The working mother is often overburdened, performing the duties of her job and being the major worker in the home.* This situation occurs because husbands of working wives do not yet shoulder their fair share of family work.

5. *Increasing job availability has made women more independent than they have been in the past.* This independence puts pressure on many husbands, because women now have a realistic alternative to a bad marriage—moving out and supporting their families themselves. But a woman's increasing independence can also reap rewards for her family. Her additional earnings may help the family invest and start the economy working for them, ease the pressure on the husband to be the only breadwinner, and generally lead to a more egalitarian relationship within the marriage.

6. *The lower incomes often earned by women blunt some of the possible advantages of working.* Due to the costs of working—including child care, clothes, transportation, and increased taxes—the woman's real economic contribution is often small and at times even "not worth it." However, as women improve their skills and become an indispensable part of the workforce, gain political power, and become more career oriented, the pay differential between male and female workers decreases.

Resources on the Internet

 Companion Website for *Human Intimacy: Marriage, the Family, and Its Meaning,* Tenth Edition

http://sociology.wadsworth.com/cox10e/

Gain an even better grasp on this chapter by going to the companion website to take one of the tutorial quizzes, use the flash cards to master key terms, or check out the many other study aids you'll find there. You will also find special features such as GSS data, Sociology Online, and Census 2000 information that will put data and resources at your fingertips to help you with that special project or to help you as you do some research on your own.

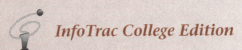 ***InfoTrac College Edition***

http://www.infotrac-college.com/wadsworth/

You can access reliable resources anytime, anywhere, with InfoTrac College Edition, the online library. This fully searchable database offers more than 20 years' worth of full-text articles (not abstracts) from almost 5,000 diverse sources, such as top academic journals, newsletters, and up-to-the-minute periodicals, including *Time, Newsweek, Science, Forbes, The New York Times,* and *USA Today.* You can conduct electronic key word searches using key terms from this chapter to supplement your reading and learning experience. To aid in your search and to gain useful tips, see the Student Guide to InfoTrac College Edition, which you can access through the companion website for this book.

Is Sexual Harassment in the Workplace Pervasive?

Sexual harassment occurs when one employee makes continued, unwelcome sexual advances, requests sexual favors, or engages in other conduct of a sexual nature toward another employee, against his/her wishes. According to a current issues update from the U.S. Equal Employment Opportunity Commission (EEOC), sexual harassment occurs "when submission to or rejection of this conduct explicitly or implicitly affects an individual's employment, unreasonably interferes with an individual's work performance or creates an intimidating, hostile or offensive work environment."

YES

The number of sexual harassment cases brought to the EEOC over the past 10 years has varied from a high of 15,889 in 1997 to a low of 13,566 in 2003. In 2003, $50 million in damages were awarded against employers (EEOC 2004b). Eighty-five percent of these charges were brought by females. These numbers make it clear that sexual harassment in the workplace is alive and well.

For those readers who need convincing that sexual harassment is indeed a major problem, the following are but a small sample of some of the thousands of sexual harassment cases filed with the EEOC (2004a):

Reed and Bull Information Systems v. Steadman (1999). Ms. Steadman resigned her position as secretary to the marketing manager because of his frequent, sexually provocative remarks and suggestive behavior toward her, which continued even after she informed him that his behavior was offensive. Although she did not make a formal claim, the company became aware of her problems through other colleagues, but took no action. The tribunal found that it was not necessary for a woman to make a public fuss to indicate disapproval. Provided that any reasonable person would understand, whether by her words or conduct, that she had rejected the behavior in question, then continuation of this behavior is harassment.

Driskell v. Peninsula Business Services (2000). Ms. Driskell, an advice line consultant, alleged that her department manager, who had a history of making sexually explicit comments and banter, advised her to wear a short skirt and a see-through blouse when arranging to interview her for a more senior post. The company investigated her complaints, but rejected them. She then indicated to the company that she could no longer work for the department manager and was consequently dismissed. The tribunal found that sexual banter does amount to sexual harassment and is especially intimidating when directed toward a woman, rather than toward another man.

Johnson v. Gateway Food Markets (1990). Harassment of a female general manager by her female supervisor was determined to be sexual harassment.

Gates v. Security Express Guards (1993). Same as above, but involving two men.

The term *sexual harassment* has been broadened into the idea of a sexually hostile environment. A local U.S. Forest Department branch in California was found guilty of creating a sexually hostile environment. The women employees had endured years of harassment and had felt they must tolerate it if they wanted to keep their jobs. Such things as suggestive, semi-nude pictures of women adorned many of the department's vehicles, and lewd jokes were often part of the general conversation, contributing to a sexually hostile environment.

Although the vast preponderance of harassment reports are made by women reporting male harassment, there are reports of males being harassed by females, and also reports of same-sex harassment among homosexuals, as in the cases presented. For example, two brothers fired from clerical jobs at an airline filed a lawsuit, claiming they were sexually harassed by two female co-workers. They contended that the women graphically described sex acts, exposed themselves, and talked about the size of male workers' genitals. When they complained to their supervisors, they were allegedly fired from their jobs (*Santa Barbara News Press* 1994).

It appears from these and many other stories that sexual harassment is pervasive, not only in the workplace, but throughout society (B. Murray 1998).

NO

The original intent of sexual harassment laws was to protect employees from feeling forced to provide sexual favors to those with power over them in order to keep their jobs, get promoted, and so on. This is relatively straightforward and easy to prove, compared with proving a generally sexually hostile environment in which people work.

Unfortunately, hostile environment laws are increasingly becoming an excuse for old-fashioned censorship. Indeed, in the extreme, they directly deny people their First Amendment rights—namely, freedom of speech (Paglia 1998, 54).

An example of the extremes to which people now go in claiming sexual harassment is a California case involving a 49-year-old fireman reading *Playboy* magazine in a firehouse, whose crew was all men except for one woman. In 1992, the Los Angeles County Fire Department implemented a sexual harassment doctrine to comply with state and federal law. This doctrine banned sexually explicit magazines such as *Playboy*, posters, and other things that might be deemed offensive to the 11 women among the 2,400 members of the county fire department. The fireman took the fire department to court, contending that the sexual harassment policy violated his First Amendment right to read. After both sides spent about $400,000 to fight the case, a federal court agreed with the fireman and struck down the ban. After winning his case, the fireman noted that the amount spent on the case would buy a new fire truck or pay for five new firefighters (Moreau 1994).

Millions of dollars are being awarded on a vague hostile environment standard, with no requirement to show harm and with heavy emphasis on the feelings and subjective perception of the complainant. An offense is becoming what a complainant says it is. The most hypersensitive worker gets to set the threshold of offensiveness, and unguided juries do what they wish with awards. Is it any wonder that increasingly more people seem to find discomfort in their workplace environment because of sexual harassment? In fact, of the 13,566 cases of sexual harassment brought before the EEOC in 2003, 46.1 percent of them were rejected as being without reasonable cause (EEOC 2004b).

As the noted jurist Alan Dershowitz (1994a) comments,

> They [radical feminists] are using the pretext of sexual harassment in an attempted end-run around the First Amendment. It shouldn't work. But if it does, the entire theory of the First Amendment is in danger. Female employees would be given an "offensiveness veto" over freedom of speech—a veto that would know no bounds. (349)

> The time has come to end this new tyranny of censorship. Sexual harassment should be limited to quid pro (direct threat) efforts to obtain sexual favors from subordinates. Sexual harassment laws must not be allowed to become speech codes designed by radical feminists to circumvent the First Amendment. (353)

What Do You Think?

1. As a woman, do you feel sexually harassed if you hear a sexual joke? If you see a man reading a magazine such as *Playboy*? If you see a poster of a seminude young woman advertising something?

2. As a man, do you feel that any of the above behaviors constitute sexual harassment?

3. As a woman, how would you define sexual harassment in the workplace?

4. As a man, how would you define sexual harassment in the workplace?

Family Crises

Chapter

14

- What are some ways by which you and your family handle stress?
- Do you think that you and your family handle stress successfully?
- What are the usual stages a person goes through in reacting to the death of a loved one?
- In what ways might unemployment affect a family?
- What kind of problems do military families face during times of war?
- In what ways are drug and alcohol abuse harmful to a family?

crisis

Any event that upsets the smooth functioning of a person's life

One of the characteristics of all strong families is the ability to handle crises. A **crisis** is any event that upsets the smooth functioning of a person's life. As defined here, it is an emotionally significant event or a radical change of status in a person's life. It need not be negative. The birth of a child is a positive event for most couples, and yet, it is a turning point—a crucial change in the couple's life. Unemployment, moving to a new location, divorce, remarriage, illness, natural catastrophes, injury, and death of a loved one are all turning points in life. Developing crisis-management skills is a top priority for all individuals and all families, because everyone will face such periodic crises as they go through life. Although a crisis may directly affect only one individual in a family, it will indirectly affect the entire family. In a strong family, family members rally to help each other in times of crisis.

When we speak of crises, most people think only of negative events, but a positive event can also be a crisis if an individual or family is upset by it. Certainly, the marriage of a grown child is a positive and happy event, but perhaps this is the last child living at home. Thus, when he/she marries and leaves home, the parents will have to adjust to being alone again. If the adjustment to this change is smooth and problem free, there is no crisis. If the parents are upset and have difficulty adjusting, the happy event can also become a crisis.

Not all turning points are crises. Whether a change is a crisis depends on the family—its crisis-management skills, its resources, and the way it views the turning point. To one family, short-term unemployment may be highly disruptive and become a severe crisis. Another family may have planned for the unemployment and saved enough funds to last until the family member is reemployed; this family, which views unemployment as an opportunity to do other important things for a short time, may not consider the change a crisis.

Some families seem to experience many crises. When such families are studied, the researchers often discover that the families lack some of the eight qualities found in strong families (see Chapter 1). Crisis-prone families are often troubled families, long before a crisis arises. A poorly functioning family—one having day-to-day unresolved problems—is in an already weakened condition. When a crisis arises, such a family may simply not have the resources and energy available to cope with it.

Whatever crises families may face, keep in mind the eight characteristics of strong families, which are (1) commitment, (2) appreciation, (3) good communication skills, (4) desire to spend time together, (5) a strong value system, (6) the ability to deal with crises and stress in a positive manner, (7) resilience, and (8) self-efficacy. By working to build these characteristics into your own family, you are taking the first and most important step toward successful crisis management. A strong family can survive crises because the family members have the resources, the sense of unity, and the sense of direction that enables them to work together to overcome problems.

Coping with Crises

Depending on a family's viewpoint and resources, many different kinds of events are apt to become crises. We term such crisis-provoking events **stressor events**. Stressors vary in several ways; they may come from within or outside of the family, be predictable or unexpected, and so forth.

stressor event
A crisis-provoking event

Stress in one area of life also tends to spill over into other areas of life. A person who is experiencing problems at work will probably bring those problems home to her/his family. This may cause increased stress, because the family then becomes upset. As noted earlier, the severity of stress situations varies, according to how the family views the situation and the resources the family has to cope with the stress. Nevertheless, most families find certain situations to be more stressful than others. Research indicates various events can be ranked, according to the relative severity of the stress produced.

Stress: Healthy and Unhealthy

There are often popular articles about the damage, both psychological and physical, that stress can cause. These articles also usually emphasize the importance

---

HIGHLIGHT

Types of Stressor Events

- **Internal**   Events that arise from someone inside the family, such as alcoholism, suicide, or running for election.

- **Normative**   Events that are expected over the family life cycle, such as birth, launching a young adult, marriage, aging, or death.

- **Ambiguous**   The facts surrounding the event are uncertain. They are so unclear that you're not even sure that the crisis is happening to you and your family.

- **Volitional**   Events that are wanted and sought out, such as a freely chosen job change, a college entrance, or a wanted pregnancy.

- **Chronic**   A situation that has long duration, such as diabetes, chemical addiction, or racial discrimination.

- **Cumulative**   Events that pile up, one right after the other, so that there is no resolution before the next one occurs. A dangerous situation in most cases.

- **External**   Events that arise from someone or something outside the family, such as earthquakes, terrorism, the

inflation rate, or cultural attitudes toward women and minorities.

- **Nonnormative**   Events that are unexpected, such as winning a lottery, getting a divorce, dying young, war, or being taken hostage. Often, but not always, disastrous.

- **Nonambiguous**   Clear facts are available about the event: what is happening, when, how long, and to whom.

- **Nonvolitional**   Events that are not sought out but just happen, such as being laid off or the sudden loss of someone loved.

- **Acute**   An event that lasts a short time but is severe, such as breaking a limb, losing a job, or flunking a test.

- **Isolated**   An event that occurs alone, at least with no other events apparent at that time. It can be pinpointed easily.

SOURCE: From Pauline Boss, *Family Stress Management*, page 40. Copyright © 1988 Sage Publications, Inc., Newbury Park, CA. Reprinted by permission of Sage Publications, Inc.

of reducing stress in your life. Everyone faces frustration, disappointment, and the resulting stress as life goals are pursued. A healthy person is not necessarily free of stress, but copes with stress when it arises.

Tolerance to stress varies greatly between people. What may be stressful to one may not be stressful to someone else. The amount of stress tolerance one develops is unique to the individual. An individual can learn to understand his/her own personal reactions to stress and to cope with stress in healthful ways. Some stress in life can be healthful. For example, proper stress on muscles causes them to grow stronger. Weight lifting is healthful stressing of the body. Individuals who experience a moderate amount of realistic fear before major surgery have milder emotional reactions after the operation. Their reaction is often, "Oh, it was not as bad as I feared it would be."

Research has also found that moderate stress, especially during childhood, may be related to later achievement. A classical study (Goertzel and Goertzel 1962) of over 400 famous twentieth-century men and women—including such individuals as the author Pearl Buck, the inventor Alexander Graham Bell, and the musician Louis Armstrong—found the following:

1. Three-fourths of the individuals were troubled as children. Among the problems experienced were poverty, divorced parents, rejecting or domineering parents, physical disabilities, and parental dissatisfaction over the children's failure at school or vocational choices.
2. One-fourth of the sample had experienced disabilities such as blindness, deafness, being crippled, having a speech defect, being homely, or being undersized or overweight.

In the ideal environment, stress will occur within a healthful range, but this range varies according to a person's adaptability and stress tolerance. Stress is harmful only when it becomes so strong that it causes an individual to behave in an unhealthful manner. Regardless of the source of stress, its physical and psychological symptoms are similar (Table 14-1). Moreover, individuals exhibit a pattern of response to stress called the *general-adaptation syndrome*. This response occurs in three phases: (1) alarm, (2) resistance, and (3) recovery or exhaustion.

In the first phase, the *alarm reaction*, a person experiences physical and psychological changes. These changes, which are largely automatic, are controlled by the sympathetic or parasympathetic nervous systems (Figure 14-1). These two systems usually operate in opposition to each other; that is, if one system activates a response, the other counteracts the response.

If exposure to stress continues, the alarm reaction is followed by the second phase, *resistance*. During this stage, various physical responses appear to return to normal, because the body has built up resistance to the stress.

The third phase, *exhaustion* or *recovery*, follows if stress continues. During this stage, exhaustion occurs if the original symptoms return. The person often becomes physically ill. On the other hand, if the stress has been reduced to a tolerable level, the symptoms do not come back and recovery ensues.

Crisis Management

How one defines and reacts to an event largely determines whether it is a crisis. Almost everyone would define the death of a husband, wife, or child as a crisis. Being stricken by a severe illness such as AIDS or cancer would certainly qualify as a crisis in most people's minds. Remember that the word crisis has its roots in

TABLE 14-1

Signs of Stress

Physical Signs	Psychological Signs
Pounding of the heart; rapid heart rate	Irritability, tension, or depression
Rapid, shallow breathing	Impulsive behavior and emotional instability; the overpowering urge to cry or to run and hide
Dryness of the throat and mouth	
Raised body temperature	Lowered self-esteem; thoughts related to failure
Decreased sexual appetite or activity	Excessive worry; insecurity; concern about other people's opinions; self-deprecation in conversation
Feelings of weakness, light-headedness, dizziness, or faintness	
Trembling; nervous tics; twitches, shaking hands and fingers	Reduced ability to communicate with others
Tendency to be easily startled (by small sounds and the like)	Increased awkwardness in social situations
High-pitched, nervous laughter	Excessive boredom; unexplained dissatisfaction with job or other normal conditions
Stuttering and other speech difficulties	Increased procrastination
Insomnia—i.e., difficulty in getting to sleep or a tendency to wake up during the night	Feelings of isolation
Grinding of the teeth during sleep	Avoidance of specific situations or activities
Restlessness, an inability to keep still	Irrational fears (phobias) about specific things
Sweating (not necessarily noticeably); clammy hands; cold hands and feet; cold chills	Irrational thoughts; forgetting things more often than usual; mental "blocks"; missing of planned events
Blushing; hot face	Guilt about neglecting family or friends; inner confusion about duties and roles
The need to urinate frequently	
Diarrhea; indigestion; upset stomach, nausea	Excessive work; omission of play
Migraine or other headaches; frequent unexplained earaches or toothaches	Unresponsiveness and preoccupation
Premenstrual tension or missed menstrual periods	Inability to organize oneself; tendency to get distraught over minor matters
More body aches and pains than usual, such as pain in the neck or lower back; or any localized muscle tension	Inability to reach decisions; eratic, unpredictable judgments
	Decreased ability to perform different tasks
Loss of appetite, unintended weight loss; excessive appetite, sudden weight gain	Inability to concentrate
	General ("floating") anxiety; feelings of unreality
Sudden change in appearance	A tendency to become fatigued; loss of energy; loss of spontaneous joy
Increased use of substances (tobacco, legally prescribed drugs such as tranquilizers or amphetamines, alcohol, other drugs)	Nightmares
Accident proneness	Feelings of powerlessness; mistrust of others
Frequent illnesses	Neurotic behavior; psychosis

the Greek verb *krinein*, meaning "to judge" and "to choose." Therefore, a crisis is a moment when you must choose from among various alternatives and opportunities that present themselves. The following are steps in crises management:

1. *Describe the event in realistic terms and determine whether it is a crisis for you, your family, or your friend.* It is important to face the facts squarely. Often, the first reaction to crisis is disbelief and denial. "This isn't happening to me." "They aren't really going to lay me off from work." As long as one is in a state of denial, it is impossible to handle the crisis.

   Sometimes, defining the event can cause trouble within a family. Family members may describe and define an event differently. The mate of the per-

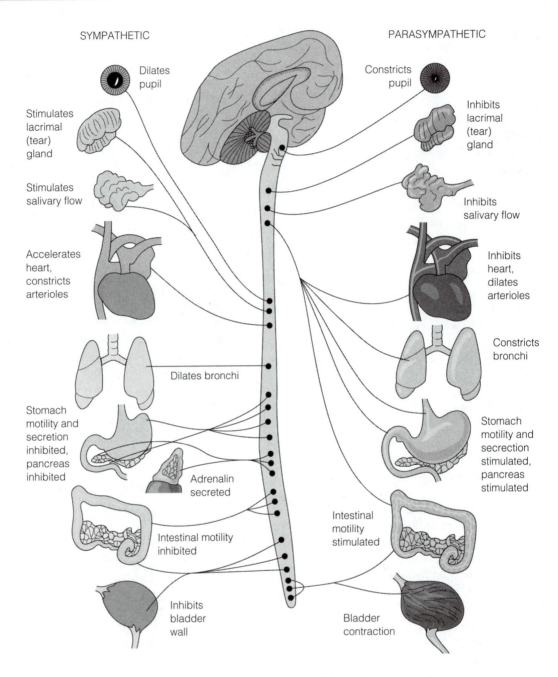

SYMPATHETIC

PARASYMPATHETIC

Dilates
pupil

Constricts
pupil

Stimulates
lacrimal
(tear)
gland

Inhibits
lacrimal
(tear)
gland

Stimulates
salivary flow

Inhibits
salivary flow

Accelerates
heart,
constricts
arterioles

Inhibits
heart,
dilates
arterioles

Constricts
bronchi

Dilates bronchi

Stomach
motility and
secretion
inhibited,
pancreas
inhibited

Stomach
motility and
secretion
stimulated,
pancreas
stimulated

Adrenalin
secreted

Intestinal
motility
stimulated

Intestinal motility
inhibited

Inhibits
bladder
wall

Bladder
contraction

Notice that the sympathetic and parasympathic systems work in relation to one another.
For example, the sympathetic system dilates the eye pupil while the parasympathic would constrict it.

FIGURE 14-1 *Autonomic Nervous System*

SOURCE: F. Cox 2002, 455.

son who is laid off may define the layoff as a crisis, because he/she feels the family cannot survive financially. This definition of the situation will cause conflict if the partner sees his/her own unemployment as an opportunity to return to school and/or change careers, regardless of the financial hardship of unemployment.

2. *Examine emotions and reactions to the crisis situation.* Uncontrolled emotions can disrupt the decision-making process. Emotions tend to disrupt rational thinking unless they are used in a positive manner.

    Crisis situations are stressful, but stress can be helpful, not just disruptive. Stress can heighten awareness and may increase efficiency. Learning to control and use stress constructively will also be of great help in managing crises.

3. *Seek support and help from friends and family in order to see alternatives.* This step is especially important if one is extremely emotionally upset. Often, one's first reaction in a crisis is to feel trapped and think there is no way out, no answer. Friends and family can help one recognize alternatives and find solutions to ease the crisis.

4. *Consider all possibilities and take decisive action to resolve the crisis.* Using outside resources may be necessary to resolve a crisis successfully. **Therapy** is a broad term used to describe actions taken to cure or solve any problem. Taking an aspirin is therapy for a headache. Counseling from a hospice after the death of a loved one is also therapy. Many groups and organizations help individuals cope with crises in their lives. Outside individuals and groups have the advantage of being objective. Although therapists and counselors can empathize with a client's feelings, they are not emotionally involved and thus are able to see the problems more clearly.

therapy
A broad term used to describe actions taken to cure or solve any problem

    In the past, families and friends were primarily responsible for helping in times of crisis. Although this is still true, the mobility of people means that family and friends may not be near enough to help at crucial times. Therefore, the role of community agencies has become more important. Many agencies have resources for finding help. For example, the Catholic Charities and Jewish Family Services offer counseling services to families. Family Service of America, an organization found in many communities, also helps families. Teachers and school counselors can help directly and can also suggest other sources of assistance in the community. Often, the public library can provide this information too.

    Crisis situations tend to bring out the best in people. For example, during times of natural catastrophes—such as floods, earthquakes, and fires—not only family and friends, but also strangers often offer help and support (see "Highlight: Agencies That Can Help with Various Crises").

    Past experience with crises can be of great help when working through a new crisis. Having coped successfully with past crises in life, the crisis-management skills previously learned will help with future crises.

    Remember that crisis management is the ability to make sound decisions and to solve problems constructively. This is one of the characteristics of a strong family. The better that family relations are, and the stronger one's family, the better one will be able to handle crises that arise in life.

Defending against Harmful Stress

We cannot always master every situation in which we find ourselves. Coping devices are ways of handling or dealing with stress, frustration, pain, fear, anxiety, and any other problems arising from stressful situations. Some coping devices are more helpful than others.

    When confronting a stressful situation and emotions are aroused, it helps to focus one's energy. The emotional reactions caused by stress are often undirected. Under stress, one may simply feel generally upset, angry, or fearful without directing emotions toward anything in particular.

HIGHLIGHT

Agencies That Can Help with Various Crises

Resource	Clientele	Services
Alcoholics Anonymous	Victims of alcoholism	Free local, self-help groups that follow rules of strict anonymity in overcoming alcoholism
Al-Anon and Alateen	Families of victims of alcoholism	Local support groups for parents, children, and friends of alcoholic victims
American Red Cross	Provides services to families in the community	Provides food, shelter, and other essentials for recovery to disaster victims and families in emergencies; promotes health and safety awareness
Battered Women's Center	Provides help for victims of family violence	Offers shelter, counseling, public education, and advocacy for victims; centers provide shelter, food, and clothing for victims and their children
Big Brothers and Big Sisters	Provides adult companionship for children from single-parent families	One-to-one relationship between adult volunteers and children, ages 7 to 15, from one-parent families; adult provides guidance, social contact, and academic enrichment through regular weekly contacts
Children's Protective Services (agency name may vary)	Children who are in need of protection due to abuse, neglect, abandonment, or sexual exploitation	This government agency assigns social workers to supervise families against whom there have been complaints of abuse and/or neglect; in severe cases, the agency will remove children from the home and provide substitute care, supervision, and related services for the children
Family Counseling Center(s)	Individuals, families, and groups needing counseling	Family, marital, parent-child, and individual counseling is provided; efforts are made to improve interpersonal relationships and family living
Visiting Nurse	The general public	Provides general help with injury and illness
Hospice	The general public	Offers help in coping with terminal illness
YMCA and YWCA	The general public	Programs include child care, camping, health and fitness groups, informal education, older youth programs, parent-child programs, senior citizen programs, and other related services

anxiety

A generalized fear without a specific object or source

Anxiety is a generalized fear, without a specific object or source. For example, if one fears heights, one knows what is causing the fear. In a state of anxiety, the source of the fear is unclear, yet the same emotional reactions are present. Anxiety often accompanies stress and frustration. If one focuses emotional reactions on the source of stress, anxiety can be limited. For example, honestly feeling "I am angry at the factory supervisor for firing me," rather than "I am angry at the world because I have been fired," will focus your energy where it belongs.

It is important, yet difficult, to relax in the midst of stress. By relaxing, one can reduce emotional arousal and can then make better decisions.. The following steps may be used to achieve relaxation:

1. Assume a comfortable position, ideally in a quiet environment and with eyes closed.
2. Breathe deeply.
3. Deliberately start to relax each muscle in the body, starting with the feet and working up to the neck and face. This is called *progressive relaxation*.

4. Maintain a quiet attitude, letting thoughts come and go, and allow relaxation to proceed at its own pace.

5. Once relaxed, remain in this state for 15 to 20 minutes.

Some people may call this self-hypnosis, meditation, yoga, or any of a number of other names for self-induced relaxation.

Defense Mechanisms

Defense mechanisms are the methods an individual uses to deny, excuse, change, or disguise behaviors that cause stress or anxiety. They are unconscious, unlike the coping strategies we have been discussing (Cramer 1998, 2000, 2001). However, some researchers feel this dichotomy does not always hold true (Erdelyi 2001; Newman 2001). There is also controversy over whether defense mechanisms are helpful in coping with stress. Research indicates that defense mechanisms can be both detrimental and helpful to successful adaptation to stress, depending how they are used and the situation (Cramer 2000).

Lou uses the defense mechanism of rationalization in a detrimental way. He has a severe alcohol problem. Although all of his friends and family recognize that he is an alcoholic, he tells them that he really is only a social drinker and can quit anytime he wishes. Whenever his drinking causes a problem, he finds some reason or excuse for his behavior (rationalization). It is never his fault. He claims it is the fault of others or bad luck. It just happened, and he has no idea why. He is sure that he is not responsible. Assuming that Lou really is an alcoholic, why do you think he doesn't recognize it, even though his friends and family do?

As we see in Lou's case, defense mechanisms can be unproductive ways of fooling ourselves. According to this interpretation, until Lou stops making excuses for his drinking habit, he will have no reason to change. Lou's defense mechanism is denying responsibility for himself, thus locking him into his behavior. "It's not my fault, so why should I change?" he maintains. In other words, defensive behavior keeps Lou from seeing the reality of his situation or behavior. This makes coping with his drinking problem impossible.

Nevertheless, defense mechanisms used in a moderate and recognizable manner can contribute to satisfactory adjustments. The following are potential positive uses of defense mechanisms:

1. *Defense mechanisms can give time to adjust to a problem that might at first be overwhelming.* For example, if one can find an excuse (rationalization) for an inappropriate anger outburst, one may feel more comfortable and, perhaps, better able to work on controlling the anger.

2. *Defense mechanisms may lead to experimentation with new roles.* For example, Tracy has low self-esteem and feels worthless. She joins a prestigious club and brags about membership in it to make herself feel better. However, joining the club does actually help her change for the better. Because the club members accept her, Tracy's sense of self-worth increases.

Although they have positive uses, defense mechanisms also tend to block open and truthful communication. Emotional health is tied closely to the ability to communicate well. Stress and the resulting frustration tend to block communication. Thus, the better one understands defense mechanisms, the better one's ability to communicate will be, and the better able one will be able to handle stress.

The following are a few of the more commonly used defense mechanisms.

1. *Repression* at first glance appears similar to forgetting, but it is actually more complex. Repression is an unconscious blocking of whatever is causing the individual stress and frustration. For example, Jamie's mother sends him to the store to buy a number of items, including liver. Jamie has always disliked liver. When he returns home with the groceries, his mother discovers that there is no liver—Jamie has forgotten to buy it. He did not do this deliberately. Forgetting the liver is an example of repression, through which Jamie avoided the stress and frustration of having to eat something he doesn't like.

2. *Displacement* is a straightforward substitution of a less-threatening behavior for another. For example, Richard's boss criticizes him unfairly. Richard is angry, but afraid to speak up for fear of being fired. That evening, he becomes angry with his girlfriend for no apparent reason. It is less stressful and safer for Richard to take his anger out on his girlfriend. He has displaced, or wrongly shifted, his anger from his boss to his girlfriend.

3. *Rationalization*, a common defense mechanism, involves finding an excuse for a behavior that is causing trouble. For example, Jeffrey is rationalizing when he tells his professor "I missed the test because I ran out of gas." Jeffrey is fooling himself. His professor, who sees that Jeffrey is just making an excuse instead of taking responsibility for his behavior, asks, "Why didn't you look at the gas gauge?" Rationalization often appears to others as "making excuses."

4. *Projection* is a defense mechanism whereby one's own characteristics or impulses are imposed upon others. Projection can take two different forms. First, it can justify one's own desires; a rapist says of his victim, "She really wanted it." Or, projection can serve to rid one of unacceptable characteristics or behaviors. Lou sometimes claimed that his friends were drunks, not him. Projection is the major psychological mechanism used in **scapegoating**, whereby a person or group is blamed for the mistakes or crimes of another. Scapegoating may also be evident when a person or group is blamed for some misfortune that is due to another cause.

5. *Sublimation* involves converting a socially unacceptable impulse into a socially acceptable activity. Jose is very aggressive and often feels like fighting. He joins the football team and is praised for his aggressiveness on the field.

6. *Compensation* allows a person to make up for a shortcoming in one area by becoming successful in another area. Art has always liked football, but is too small to make the team. He makes the debate team and becomes a very successful debater.

There are many kinds of defense mechanisms. Understanding and recognizing these mechanisms will lead to better, more open, and honest communication.

Death in the Family
Natural Causes

Almost everyone defines the death of a loved one as a crisis. In Chapter 15, we will examine the crisis of marital failure—namely, divorce—which can arouse many of the same feelings as a death in the family. In both cases, individuals may experience a sense of loss, grief, and loneliness.

Death is the other experience besides divorce that can end a family's existence. Loss of a husband or wife ends that particular family, although the surviv-

scapegoating
Blaming a person or group for the mistakes or crimes of another

ing spouse may start another family. Loss of a child does not end a family, but it creates a family crisis and brings about an extreme sense of loss.

Death can result from many causes. Dying of natural causes, such as old age, may be gradual and expected. In this case, the family can prepare for the death and coming changes in the family. But death by natural causes can also occur suddenly, such as from a severe heart attack. Death can also result from accidents, such as automobile accidents or fires. The sudden death of a loved one creates an immediate and extremely traumatic shock and crisis for the family.

Ambiguous Loss

There are two basic kinds of ambiguous loss. In the first, people are physically absent but psychologically present, because it is unclear whether they are dead or alive. Missing soldiers and kidnapped persons illustrate this type of loss. The second type is when a person is physically present but psychologically absent, such as people with Alzheimer's disease, addictions, and other chronic mental illnesses, or who are in a coma. Of all the losses experienced in personal relationships, ambiguous loss is the most devastating, because it remains uncertain and indeterminate.

Ambiguous loss is confusing. People are baffled and immobilized. They can't problem-solve because they do not yet know whether the problem (loss) is final or temporary. The uncertainty prevents people from adjusting. People tend to become frozen in place, holding on to the hope that things will return to the way they used to be. The Vietnam prisoners of war were first listed as missing, and it was usually some time (often several years) before their families knew if they were alive or dead. It may turn out to be somewhat similar with those reported missing in the previous Iraq conflict with Kuwait, in 1992. Even upon the prisoners' return after many years of captivity and torture (up to 7 years) they were not the same persons who had left their relationships years before, nor were their families. Thus, the frozen picture that family members had held of their relationships was usually broken when the missing soldiers did return ("The American Experience: Return with Honor" 2000).

Ambiguous loss also denies people the symbolic rituals that ordinarily support a clear loss—such as a funeral after a death. The absurdity of ambiguous loss reminds people that life is not always rational or just. Finally, ambiguous loss is a loss that goes on, so that the people experiencing it become exhausted from the relentless uncertainty. As an example, for children, divorce is often an ambiguous loss in that the absent parent is still alive; this leads to the continuing hope that the parents will be back together at some future time. (This section is adopted from Pauline Boss 1999, 2000, 2001.)

Suicide and Homicide

Death of a loved one because of a suicide or homicide is especially traumatic, because survivors are reacting to an intentional death. Suicides or homicides are generally unexpected. They usually cannot be reasonably explained. Family members, especially with a suicide, may feel partially responsible. "I should have known he/she felt this way. Perhaps I could have done something."

Unfortunately, suicide and homicide are not infrequent forms of death in our society. Suicide is the eighth leading cause of death among males, although it has declined for the last several years from 21.5 deaths per 100,000 population in 1990 to 17.7 deaths per 100,000 population in 2000. Suicide rates for females are much lower, 4.8 deaths per 100,000 population in 1990 and 4.0 deaths per 100,000 population in 2002 (NCHS 2004a, Tables 31, 45).

The overall suicide rate in the United States has remained between 10 and 13 suicides per 100,000 population for the past 50 years. However, the suicide death rate among White males aged 15 to 19 has increased during the past 10 years (see Table 14-3). Suicide among young persons is particularly stressful to a family. "Young people have so much to live for." "His whole life was ahead of him," and "It's such a waste." The use of drugs is to blame for many youthful suicides.

Homicide (with legal intervention) is the thirteenth leading cause of death in males, but has been on a downward trend since 1990, declining from 14.8 deaths per 100,000 population in 1990 to 9.0 deaths per 100,000 population in 2000. Homicide rates for women are again much lower, 4.0 deaths per 100,000 population in 1990 and 2.8 deaths per 100,000 population in 2000 (NCHS 2004a, Tables 31, 46). These rates continue to be three to eight times higher than rates in most other industrialized countries. For example, the homicide rate per 100,000 population for Japan and the United Kingdom is less than 1.0. Canada and Australia have rates between 2 and 2.5.

African American males had the highest homicide rates, followed by Hispanic, Native American, White, and Asian American males (NCHS 2004a, Table 45). Homicide rates for minorities have fallen substantially since 1970, contrary to public opinion, narrowing the racial gap in rates. Non-White homicide rates were 10 times higher than those of Whites in 1970, but are now six to seven times higher (see Table 14-2). Homicide and legal intervention is the second leading cause of death for those aged 15 to 24 years (NCHS 2004a, Table 32). Although violent crime rates have dropped dramatically in the past few years, more violent crimes are committed by minors. This has provoked state legal systems into trying more minors as adults, reasoning that juvenile laws are too lenient in cases of violent crime.

TABLE 14-2

Death Rate for Homicide and Legal Intervention, 1980 and 2002 (rate/100,000 population)

	1980	2002
All persons (male)	16.6	9.0
All persons (female)	4.4	2.8
White male	10.4	5.2
White female	3.2	2.1
Black male	69.4	35.4
Black female	13.2	7.1
Hispanic male	27.4 (1990)	11.8
Hispanic female	4.3 (1990)	2.8
Asian male	9.1	4.3
Asian female	3.1	1.7
American Indian male	23.3	10.7
American Indian female	4.6	3.0

SOURCE: NCHS 2004a, Table 45.

TABLE 14-3

Death Rate by Suicide, 1980 and 2002 (rate/100,000 population)

	1980	2000
All persons (male)	19.9	17.7
All persons (female)	5.7	4.0
White male	20.9	19.1
White female	6.1	4.3
Black male	11.4	10.0
Black female	2.4	1.8
Hispanic male	23.5 (1990)	20.2
Hispanic female	2.3 (1990)	1.7
Asian male	10.7	8.6
Asian female	5.5	2.8
American Indian male	19.3	16.0
American Indian female	4.7	3.8

SOURCE: NCHS 2004a, Table 46.

Society tends to view the family as somehow to blame for the suicide of a young person. Typical comments are "They must have been bad parents," "They should have known their son or daughter was using drugs," "Couldn't they see how depressed he/she was?"

Depression, a perceived sense of isolation, lack of personal attachments, and a dearth of coping skills are often uncovered in the suicidal person. Yet, despite availability of antidepressant medication, suicide counseling, and hot lines, the rate of suicide has remained constant, as noted above. The fact is that suicide is not totally understood and therefore not accurately predictable (DeAngelis 2001b).

Some of the questions that are commonly asked about suicide include the following:

1. *Why do young people want to die?* Experts state that, except for a tiny percentage, those who attempt suicide actually want to live. Because of this, suicide may be the nation's number one preventable cause of death among young people.

2. *If suicidal persons want to live, then why would they try to kill themselves?* Suicidal persons usually believe that they are not loved or accepted. They want desperately to know that someone cares. A suicide attempt is really a cry for help; it says, "Look at me, help me, save me!" In fact, some victims are found dead while holding the telephone; others call the police to say they plan an overdose of drugs. The majority of attempts are made in such a way that someone is sure to save them.

3. *If a person's mind is set on suicide, can anyone or anything change it?* The overwhelming majority of people who survive a suicide attempt are glad they are alive. Only a few repeat the attempt and succeed. The wish to die generally lasts only a few hours or days, not a lifetime. If victims can be helped through the crisis period, chances are good that life will go on normally afterward.

4. *Is it hard to face life after attempting suicide?* Of course, adjustment can be hard after an attempted suicide. However, with counseling and support from family and friends, life can be resumed.

5. *Do people who talk about killing themselves just want attention? Is it best to just ignore their threats?* No. When a person talks about suicide, it can lull friends and family into indifference. Meanwhile, the person is actually displaying suicidal intentions. Friends and family members who ignore the suicide talk may even give victims added reason to follow through, to prove the seriousness of the intent.

6. *Are people who try to kill themselves emotionally ill?* Most people who commit suicide suffer from deep despair, loneliness, and hopelessness, but they usually don't have mental disorders. Some people contemplating suicide may be depressed, but others may be happier than they have been in a long while. They have decided on suicide as a way to solve their problems once and for all.

7. *What causes people to attempt suicide?*
   - They may be trying to get attention.
   - They may have an unrealistic, romantic view of death.
   - They may be under too much pressure to succeed.
   - They may not be able to express their anger or pain.
   - They may feel rootless, without the anchor of a strong parent-child relationship.
   - They may not have firm values on which to base life decisions.
   - They may have suffered a loss that has caused unbearable grief.
   - They may have a parent or friend who committed suicide, making the act seem permissible.
   - They may be under the influence of drugs and not realize exactly what they are doing.

8. *How can you tell if someone is about to commit suicide?* Be on the lookout for warning signs. Take seriously signs of severe depression. Watch for standoffish, withdrawn behavior. Watch for the signs of drug use. Take suicide threats seriously.

9. *What can you do to help if you suspect an oncoming suicide attempt?* Take the person seriously and get involved. Don't wait to see what develops, because tomorrow may be too late. Ask outright if the person is planning suicide. Do not be afraid to mention it. Chances are, the person already has the idea in mind, and talking about it with another who is clearheaded can help. Be careful not to deny the seriousness of the intent. If you imply that the friend doesn't mean it, you may unknowingly be daring the person to do it. At the same time, show concern. Try to convey that you know the crisis is major, but offer reassurance that it is temporary. Try to help the person get professional help and counseling. If the person seems on the verge of making a suicide attempt, do the following:
   - Phone a suicide or crisis intervention hotline immediately. If not available, call 911.
   - Stay with the person until help arrives.

10. *What if you fail to prevent a suicide?* Accepting the suicide of someone close can be one of the hardest things in life to face. Still, you cannot change what has happened. No one is responsible for another person's suicide. Emotional support is available to survivors. Learning about the process of grief will also help.

11. *What if you sometimes feel like ending your own life?* Almost everyone feels that way sometimes. Things can seem very bad, but they can also get better, given time. You can talk to someone during bad times. Ask someone to listen

TABLE 14-4

Death Rates for Firearm-Related Injuries (rate per 100,000 population)

	1980	2002
All persons	14.8	10.4
Male	25.9	18.1
Female	4.7	2.8
White male	22.1	15.9
White female	4.2	2.7
Black male	60.1	34.2
Black female	8.7	3.9
Hispanic male	27.6 (1990)	13.6
Hispanic female	3.3 (1990)	1.8
American Indian male	24.0	13.1
American Indian female	5.8	2.9

SOURCE: NCHS 2004a, Table 47.

to you. Most towns have suicide hot lines available 24 hours a day. You can talk with a counselor about your feelings. Enter group counseling with others who have similar feelings. Share your feelings with those who care about you—family and friends (Whitney and Sizer 1988).

Firearm Mortality

In 1998, a total of 30,708 persons died from firearm injuries in the United States. Note that suicide accounted for a greater number of firearm deaths than did homicide, 56.7 percent and 39.4 percent, respectively. Between 1980 and 2000, the death rate for firearm deaths declined 31 percent. The 2000 rate for males was 5.9 times that for females. The rate for Blacks was 2.4 times that of the White population. Between 1980 and 2000, firearm deaths decreased almost 50 percent for Black males and 27 percent for White males. The rate for Black and White females also declined during this time by 45 percent and 65 percent, respectively. Note that rates for all races, both male and female, declined greatly between 1980 and 2000 (see Table 14-4; NCHS 2000c, 2004a, Table 45).

Grief and Bereavement

The death of a loved one almost always causes a crisis in a family. No matter how well one anticipates the death of a parent or spouse, the loss will be disturbing and adjustment will take time. Loss of a child or young adult is even more disturbing. Everyone understands that death comes to us all. Hopefully, when parents pass away, one can look back on their lives and say, "I miss them, but they lived a good life." The death of a child or young adult or spouse is much more difficult to accept because one feels cheated: "My son's (or daughter's or spouse's) life was all ahead of him/her. He/she didn't have a chance to enjoy life."

If death comes slowly to a loved one, as with cancer or AIDS, the survivor may go through the same basic emotional reactions to death that the dying person usually passes through:

1. *Denial and isolation.* A typical first reaction to impending death is an attempt to deny its reality. One tries to ignore any information that points to impending death. One refuses to believe that the loved one is really going to die. "The lab reports are wrong." "Somebody mixed up the X-rays." One may even refuse to discuss the subject of death.
2. *Anger.* "Why my wife or child?" "I hate the world or God for taking this person from me." One finds that anger spills over onto those still living. "Why is my husband being taken, rather than another person?" is a typical reaction. One is angry and envious of the health of others.
3. *Bargaining.* One may find oneself bargaining for the life of a family member or friend, just as those who are dying will often bargain. "Just let her live a little longer" is the plea. "I'll treat her better than I have in the past."
4. *Depression.* Finally, as death draws near, one begins to recognize that it can't be prevented; there is nothing that can be done about it. Such recognition of the inevitability of death usually causes a profound sadness and temporary depression.
5. *Acceptance.* Finally, there is acceptance of the inevitable—the actual death of the loved one—and the process of bereavement begins.

Many, but not necessarily all, people pass through these stages when death slowly approaches. It is interesting to note that most of these reactions accompany any major loss such as a divorce, loss of a job, and so forth.

After a friend, spouse, child, or relative has died, whether the death is sudden or prolonged, grief follows. This reaction is natural, normal, and necessary for adjustment. At first, there is a period of numbness and shock. The person grieving may seem to be adjusting well to the loss because during this period, little emotion is felt. This phase is followed fairly quickly by tears and the release of the bottled-up emotions.

This initial shock is followed by sharp pangs of grief. These are episodes of painful yearning for the dead person. During this period, agitated distress alternates with silent despair, and suffering is acute.

These sharp feelings of loss gradually give way to a prolonged period of dejection and periodic depression. Life seems to have lost meaning. There is a large gap in life that can't seem to be filled. Although the mourner is usually able to resume work and normal life ("life must go on"), listlessness, lack of energy, and difficulty concentrating continue. Gradually over time, maybe several years, life does return to normal for the mourning person. This does not mean that the loss will never again be mourned. But the mourning no longer dominates one's life.

Historically, as modern medicine became predominant, death became less a family affair and less visible as it took place increasingly in the hospital or institutionalized setting. As the medical emphasis on prolonging life became the central focus, it became more difficult to discuss death with those concerned. Today, with the growing hospice movement, death is again being openly discussed. Relatives of the dying individual are encouraged to participate in providing comfort and care prior to the death. The goal is now to provide a comfortable, dignified death for the individual and to support family members as they deal with anticipatory grief and bereavement. Current family health policy related to end-of-life decision making includes the use of advanced directives, which are legal documents outlining choices for medical and life-prolonging treatments (Hoppough and Ames 2001; Machir 2001).

Accidents, Injuries, and Catastrophic Illness

Accidents, injuries, and catastrophic illness are other kinds of major crises that can happen to any family. Accidents (unintentional injuries) are the number one cause of death and injury for young people aged 1 to 44. The accidents most likely to cause death for this age group involve motor vehicles, drowning, firearms, poison, and fires and burns. Considering all ages, accidents are the fourth leading cause of death for males and the eighth for females (NCHS 2004a, Table 31). For the 15 to 24 years age group, motor vehicle accidents are the highest cause of accidental death (NCHS 2004a, Table 44).

A family with a child or parent who is severely injured because of an accident can be thrust into a crisis situation without warning. Many times, a severe illness such as cancer will develop gradually, so that the family can plan ways to cope with the crisis. An accident, however, happens without warning, without time to prepare.

After the adjustment to death is made, the family can usually resume their everyday activities, even though feelings of loss may persist. This is not necessarily so with an accident.

Often, we consider accidents as just that—accidents, or unfortunate events that occur unexpectedly or by chance. We become philosophical and say something about "being in the wrong place at the wrong time." In truth, accidents are not always chance or random events; they can be caused by ignorance or carelessness. Thus, many accidents are preventable. Statistics indicate that accidents do follow patterns, such as the following:

- Males have more of every kind of accident than do females, for every age after 1 year.
- People under age 40 have more accidents than those over 40.
- Most accident victims live in urban areas.
- Most accidental deaths involve motor vehicles.
- Most accidental injuries occur in the home.
- Accident rates peak on certain days, such as holidays.

By recognizing such patterns, preventive steps can be taken. For example, to help reduce the number of automobile accidents due to alcohol, police may stop cars at various checkpoints on holidays to check for drunk drivers. Motor vehicle fatalities are highest among 16-year-old drivers, so a number of states have made it harder for new drivers to earn their license (G. Johnson 1998).

Several factors increase the chances of becoming an accident victim. Physical conditions, such as fatigue, can lower a person's awareness and good judgment. Emotions such as anger, grief, depression, or joy can preoccupy people and thus increase the chances of an accident. Certain personality traits—including overconfidence in one's abilities, exaggerated self-importance, and impulsiveness—can lead people to make faulty decisions that lead to accidents.

The consequences of an accident or illness that does not lead to death may cause a prolonged family crisis. Everyday activities may require adjustments that continue for the rest of the family's life. Having a spinal cord injury, as Christopher Reeve did, may confine a person to a wheelchair. Catastrophic illness—such as stroke, heart attack, AIDS, and cancer—is any illness that is life threat-

MAKING DECISIONS

The Case of the Weinstein Family

The Weinsteins are a middle-class family with two teenage children, Walt and Kara. They have some savings in the bank. They are all covered by Mr. Weinstein's health coverage at work. Walt has an interest in cross-country mountain cycling. He has worked at a part-time job and bought, with his own money, a good mountain bike.

One day, riding in the hills with a group of his friends, Walt loses control of his bicycle and goes over the side of the trail, dropping some 50 feet into a ravine. The search-and-rescue team retrieve him from the ravine. He has a broken neck and is paralyzed.

Although the family's health coverage pays for the immediate expenses, it does not pay for prolonged physical therapy or for at-home nursing care. Walt recuperates from the immediate injuries, but will be confined to a wheelchair the rest of his life. He needs constant care to function.

What Do You Think?

1. If the family's insurance doesn't cover at-home nursing care, how do you think the family will handle Walt's need for supervision?
2. What do you see as the family's most important adjustment to Walt's accident?
3. Do you know a family in which a child or a parent has a serious, long-term illness or injuries? How do they cope with the ongoing crisis?
4. List the many adjustments that Walt's family will likely have to make as a result of his accident.

ening. It may disable a person for a long period of time, perhaps for the rest of the person's life. In this case, the family will have to find some way to provide long-term care for the person, as the Weinstein family had to do. Placing the person in some type of long-term care facility is often financially impossible for the average family. In that case, home care will be necessary. Home care means that family routines will have to change; this increases family stress.

When accidents and injuries cause a crisis in a family, family members experience many emotions and are worried about the person who has been hurt. Might the person be crippled or die? Will the person recover and, if so, how long will it take?

If a parent is injured and unable to work, the family may suffer economically. Usually, there is some short-term employer protection against loss of income. An employee will have a number of sick days that can be taken (with pay) to recover. Workers injured on the job may draw workers' compensation payments for a set period of time. Some families will have disability insurance coverage that pays them in case of injury or accident.

The family may face huge medical bills. Hospitalization can cost up to $1,000 or more a day, depending on the treatment. This may not even include doctors' bills. Many families have health coverage from their employment or from private insurance. For families who do not, medical expenses can spell financial ruin. Approximately 16 percent of Americans under age 65 have no health insurance (NCHS 2004a, Table 129). This percentage is much higher among Blacks (19.3 percent), Hispanics (34.8 percent), and American Indians (33.4 percent). The common belief is that such persons cannot receive medical help, but this is untrue. Persons without medical coverage do receive medical attention via emergency rooms and various charity organizations.

Accidents and severe illness are unexpected. No one knows when or if an accident is going to happen or a severe illness will be contracted. When these crises occur, decision-making skills become very important, because the family will have to make decisions quickly. These first decisions may be followed by a long period of adjustment or, perhaps, continual adjustment for the entire family.

Family Violence

Violence within a family almost always leads to a crisis. Violence may occur within a family for many reasons. Family members who cannot talk to each other, don't listen to each other, and simply lack sufficient communication skills to make themselves understood are more likely to resort to violence. Children are often physically violent because they haven't yet learned how to communicate. Adults who cannot communicate remain like children and often express themselves physically, rather than verbally.

The family can be an important source of love, caring, and emotional support. Yet, the possibility for violence and abuse within a poorly functioning family also exists. The family is an emotional hothouse. Although the finest emotions can be fully expressed in the strong family, the emotions of hate and anger can lead to violence in a poorly functioning family.

Family violence is difficult to measure and document, because most of it occurs in the privacy of the home, away from public view. However, family violence is coming increasingly into public view as newspapers, magazines, and television devote more attention to battered wives and abused children. Violence is reported between husbands and wives, between courting couples (see Chapter 6), and between parents and children. Definitions of abuse and violence vary. Abuse does not always refer to major physical or sexual abuse. Abuse means different things to different people. Some people consider corporal punishment, such as spanking, a form of parent-child abuse (Straus 1999). Others regard shoving as spousal abuse. A single exposure to an exhibitionist may be considered child sexual abuse. Hence, the statistics on the amount of family abuse must be interpreted cautiously.

Violence between Partners

The most life-threatening situation a police officer can enter is a family dispute. Emotions run high; the family members usually see their problems as a private matter and consider the police officer an unwanted intruder. Homicide rates between husbands and wives are high. The number of wives killing husbands and the number of husbands killing wives are about equal. Although women account for only 10 percent of defendants charged with murder, women account for almost half of defendants in spousal murder (Brownstein 1994; Dershowitz 1994a, 311). Most physical violence between spouses involves the man hurting his wife or girlfriend, although 25 to 35 percent of arrests for domestic assault are women (Goldberg 1999). Abuse of husbands by wives most often is verbal, rather than physical. Alcohol and/or drug abuse often trigger emotional outbursts that culminate in violent behavior. Violence also occurs between dating couples and cohabiting partners, as we saw in Chapter 6. Although little discussed, violence between homosexual partners, especially men, is not uncommon (Bergen 1998).

Alfred DeMaris and his colleagues (2003) found the following factors tend to elevate the risk of spousal violence: a short relationship duration, forming a union at a younger age, substance abuse, frequent disagreements, a more heated disagreement style, cohabiting, living in an economically disadvantaged neighborhood, having more children in the household, and a nontraditional woman matched with a traditional man. Their findings lend support to the idea that myriad forces converge to affect the risk of violence in relationships. In other words, no one factor is a primary predictor of such violence.

Most major cities now have battered women's shelters, where woman can go temporarily to escape an abusive relationship. The shelters usually have programs available to intervene in violent family situations. The criminal courts try to help by issuing restraining orders or jailing the violent partner for short periods. However, some suggest that court actions simply enrage the partner further and, in some cases, act as a license to kill. Unfortunately, some battered women return to the abusive relationship, even after seeking justice from the courts or after staying in a shelter.

Family violence tends to follow a domino pattern; that is, those who experience courtship violence tend to experience spousal violence. Those experiencing spousal violence tend to abuse their children. Abused children tend to abuse each other and become abusive adults. Do not take this to mean that every person who has ever been abused, in turn, abuses others. The fact is that most people who are abused do not necessarily follow the above pattern. (See Jasinski and Williams 1998 for a comprehensive discussion of partner violence.)

Child Abuse

Mistreatment of children by parents in America hardly seems compatible with mom, apple pie, and Sunday family outings. Yet, some parents do emotionally and physically abuse their children. Increasing interest in child abuse has also revealed more sexual abuse and incest than previously had been thought to exist.

The lack of a commonly accepted definition of child abuse, especially sexual abuse, makes it difficult to quantify (Haugaard 2000). Although numbers are hard to verify, more female children are abused than male children (U.S. Census Bureau 1997a). Some suggest that as many a one to three million children are abused or neglected each year in the United States ("Child Abuse and Neglect" 1997) but this is probably high. According to data about child abuse and neglect cases known to child protective services in the United States in 2001, about 903,000 children experienced or were at risk for child abuse and/or neglect. Most of these children suffered neglect (59 percent), while 19 percent were physically abused, 10 percent were sexually abused, and 7 percent were psychologically abused (National Center for Injury Prevention and Control 2004). Alan Dershowitz (1994a, 127) suggests that child abuse is the most underreported crime, while at the same time it is the most falsely reported crime.

The actual amount of child abuse is difficult to pin down, but whatever the exact number is, any abuse of children is too much. We tend to think of child abuse as actual physical or sexual abuse of a child. Some increase in child sexual abuse can be attributed to the growing number of children in cohabiting situations and blended families (families with stepparents). In such relationships and families, the incest taboo between the live-in partner and child or stepparent and stepchild is weaker than it is between natural parent and child (see Chapter 16).

However, an even greater problem than abuse is simply child neglect, as the numbers above indicated. With the increase in single-parent and dual-worker families, attention to and care of children becomes more difficult and problematic. For example, some cases of sudden infant death syndrome (SIDS) may occur because of neglect.

It is perplexing that the United States accounts for approximately three out of four child slayings in the industrialized world (*San Francisco Chronicle* 1997). Although we tend to believe that child slayings are the work of anonymous kidnappers, most often male, it tends to be parents and relatives that are the perpetrators. The actual number of anonymous strangers kidnapping and killing children is probably no more than 50 to 100 annually (Finkelhor 1990, 1995). When we consider the positive and trusting attitudes toward parenthood, and motherhood in particular, in the United States, it is even more mysterious that mothers kill their children about as often as fathers do. Infants are at greatest risk of homicide during the first week of infancy, with risk being highest on the first day of life (Paulozzi 2002).

With the growing recognition of child abuse, new laws have been passed to protect children (DeLay 2000; Golden 2000). In particular, states have made it mandatory for persons in contact with children, such as teachers and doctors, to report suspected cases of abuse. Such provisions have helped greatly in uncovering cases that remained hidden in the past. These laws also account for the greatly increased numbers of reported child abuse incidents. In fact, child abuse may not actually have increased, but is simply detected more frequently because of such laws.

The flood of publicity about child abuse has had positive and negative effects. One undesirable effect is that parents and school workers are becoming fearful of any kind of physical contact with children. Yet, physical contact with adults in the form of affection and hugs is important to the development of young children. If, in the efforts to protect children from abuse, parents and child-care workers become afraid of any and all physical contact with children, more good may be lost than gained.

The flood of reports brought on by the changes in the law and increased publicity has also greatly increased the number of unfounded reports (Dershowitz 1994a; Fincham et al. 1994; Pezdek 1994). In child custody battles, false accusations of child abuse are becoming more common as a way of discrediting one's spouse. To be falsely accused of child abuse is devastating to both the individual accused and to their family and friends. Small children's reports of abuse under questioning from an adult are often unreliable. Their responses are easily influenced, especially if the questioner assumes the charge of abuse is true, as is usually the case.

At the same time, laws and publicity about abuse seem to have combined to reduce both spousal and child abuse. A large nationally representative study found that child abuse and wife abuse rates have decreased significantly since 1980 (Gelles and Conte 1991).

The following four elements must usually be present in a family for child abuse to occur:

1. *The parent must be a person to whom physical punishment is acceptable.* Often such an abusive parent was abused as a child.
2. *The abusive parent often has unrealistic expectations for the child.* The parent expects things of the child that are impossible for the child's level of development.
3. *The parent perceives the child to be difficult and trying.* Perhaps the child cries a lot. Perhaps the child is sick much of the time, or is very energetic and active.
4. *There is usually a crisis of some kind.* The parent may have lost a job, be experiencing marital troubles, or have a lowered tolerance level due to something unrelated to the child, such as alcohol or drug abuse.

Some cities have created telephone hotlines that a parent can use to receive immediate help if he/she feels unable to cope with a child or children.

WHAT DO YOU THINK?

1. What was your parent's major mode of disciplining you as a young child? As an adolescent?

2. Have your parents used corporal punishment with you? If so, what was it?

3. In your opinion, what was the most effective disciplining technique used by your parent? The most ineffective?

4. If your parents used corporal punishment at times with you, how do you now evaluate it?

5. What would be some of your major techniques for disciplining your children?

There is a highly vocal debate over the use of corporal punishment, such as slapping and spanking with children. Murray Straus and his colleagues make a strong case that corporal punishment is detrimental to children, teaching them violent and abusive behavior rather than the opposite. They cite a number of countries such as Sweden, Austria, and Italy that have laws encouraging parents to move away from corporal punishment in dealing with their children (Straus 1999; Straus and Stewart 1999). On the other hand, corporal punishment such as spanking is strongly supported if used properly, as a means of teaching children discipline and controlling undesirable behavior (Larzelere 1994; Larzelere et al. 1996, 1998). Many adults who underwent corporal punishment look back on it as a good thing. Other adults feel it was harmful when it was severe.

Sibling Abuse

Violence between siblings is the most common form of family violence. Physical abuse is especially likely between young children who have not yet learned other ways to express themselves. Young children have less self-control, and thus their frustrations are often expressed by physical means. The rate of violence decreases as children become older and learn other ways to handle their frustrations. In general, boys are more violent than girls. Children in homes where adults are violent tend to copy the violence with each other.

If brother-sister incest occurs, the effects on the children depend on a number of factors. Sex play among very young children, such as playing doctor ("I'll show you mine if you show me yours") does not seem harmful, as long as both children consent. However, sex play involving actual intercourse or continuous sexual episodes, occurring between children of widely different ages, or taking place without mutual consent is likely to have negative future consequences. In the future, the child may feel ashamed and guilty and may have problems relating to his/her sexuality as an adult.

 HIGHLIGHT

Repressed Memories of Incest

The idea that early childhood sexual abuse can be totally unconscious (forgotten), yet cause a vast array of problems in adulthood, is the basis of a major controversy. Some therapists suggest the importance of recovering such memories to help solve adult problems. The popularity of this belief has led to lengthy lists of characteristics and behaviors indicative of a person who suffered early sexual abuse. One popular book on this subject offers a list of 34 items covering almost every imaginable behavior, from multiple personality and the wearing of a lot of clothing in the summer to swallowing and gagging sensitivity and limited tolerance for happiness (Blume 1990). There is also the suggestion, on the part of those believing in recoverable memories, that some of the child sexual abuse was part and parcel of organized satanic cults.

There is not room here to discuss the merits of this idea. The empirical evidence supporting repressed memory of incest is so lacking that one must be careful in accepting it (R. Gardner 1992; C. Gorman 1995). Perhaps it is true and perhaps it isn't. However, in cases where a recovered memory is false, great harm is done to everyone involved—the person accused of incest, family, friends, and also the accuser. Several lawsuits against therapists suggesting falsely to a client that there is a history of forgotten childhood incest have been won (*California Lawyer* 1995; "Judicial Notebook" 1994, 1997; *Trial* 1994). Some go so far as to suggest that the whole practice of uncovering repressed memories is a fad that is damaging the mental health field (Ofshe and Waters 1994).

Peer Abuse

Although little studied, peer abuse and maltreatment during school years (5 to 18 years of age) is far more frequently recalled by adults than is parental abuse (Ambert 1994). Tonja Nansel and her colleagues studied over 15,000 students in grades six through 10 and found that 17 percent of the students reported having been bullied "sometimes" or more frequently during the school term. Nineteen percent reported bullying others "sometimes" or more often. Six percent reported both bullying and having been bullied (2002).

Nansel and colleagues also found the following:

- Bullying occurs more frequently for grades six to eight.
- Males are more likely to be bullies and victims of bullying than females.
- Bullies and victims of bullying have difficulty adjusting to their environments, both socially and psychologically. Victims of bullying have greater difficulty making friends and are lonelier.
- Bullies are more likely to smoke and drink alcohol, and to be poorer students.

Peer teasing about personal imperfections or failures, exclusion from peer groups, the negative consequences of failure to make a sports team or a cheerleading team, the failure to date or be taken to the junior prom, and so forth are often extremely stressful to the school-age child. This stress is often carried into the family by the child (Ladd 1992). The child may not want to go to school; avoids other children whom the parents think are the child's friends; and is depressed, moody, angry, and unlike him-/herself. In most cases, the peer-abused child makes the adjustment to his/her peers, either by finding a way to join with an approving group of peers, or by rejecting the peers in favor of more individual pursuits.

Sexual harassment, especially of girls, is reported by a high percentage of high school students. According to the AAUW report, *Hostile Hallways: Bullying, Teasing, and Sexual Harassment in Schools*, a third of the students fear being sexually harassed in school, and more than 25 percent report they often experience sexual harassment (D. Smith 2001).

Whereas boys are often physically abusive to some of their peers, girls tend to be relationally aggressive. Gossiping, withdrawing affection and friendship to get what they want, and using social exclusion to retaliate against a friend are all examples of this behavior (DeAngelis 2003a). Most parents with girls have seen such behavior starting at very young ages.

Parental Abuse by Children

Although abuse against parents sounds improbable, there are cases where children physically attack, and even kill, their parents. Although physical abuse of parents by their children is generally rare, verbal and psychological abuse is common. Children and adolescents tend to react verbally when they are frustrated. Verbal abuse heaped on parents by children is relatively common during adolescence.

Abuse of elderly relatives is another form of child-parent abuse. Estimates suggest that numerous elderly people are abused, neglected, or exploited. The most likely victims of elder abuse tend to be the very old and infirm. Often, the victims are wheelchair bound and/or suffer from mental problems such as Alzheimer's disease. In some cases, the child abusers are financially dependent on elderly parents, thus finding themselves in a double bind (must care for the parents to receive financial support). Such abuse is often caused by the frustration felt by an adult child who finds it necessary to care for an elderly relative. Increasing life expectancy means that more children will have to cope with elderly parent care.

Factors Associated with Family Violence

1. *The cycle of violence.* One of the consistent conclusions of domestic violence research is that individuals who have experienced violent and abusive childhoods are more likely to grow into violent adults. In other words, violence begets violence.
2. *Socioeconomic status.* There is an inverse relation between parental income and parental violence. Those with incomes below the poverty line have the highest rates of violence. Those struggling the hardest to support a family are more likely to be frustrated and unhappy. However, family violence occurs at all socioeconomic levels.
3. *Stress.* Another consistent finding is that family violence rates are directly related to economic and social stress in families. Unemployment, financial problems, unwanted pregnancy, single-parenthood difficulties, and alcohol abuse all relate to higher incidents of violence in the family.
4. *Social isolation.* Social isolation increases the risk that severe violence will be directed either at children or between spouses. Families who have religious affiliations, a large circle of intimate friends, or who participate in community and social activities report less family violence.
5. *Traditional male-role orientation.* Participants in violent relationships tend to hold a traditional view of the social roles of men and women. An abusive male has been described as one who is very dominant, controlling, and macho in his outlook toward women. He feels superior to his wife and daughters.
6. *Low self-esteem, understanding, patience, and tolerance.* People who abuse, and those abused, tend to lack self-esteem, understanding, patience, and tolerance. They often do not have acceptable ways to express their emotions.
7. *Alcohol and drug use.* Alcohol and drugs are often associated with all types of interpersonal abuse.

Poverty and Unemployment

Throughout its history, the United States has struggled with the paradox of poverty amidst affluence. Why do so many people struggle economically in a nation blessed, by almost any international standard, with abundant opportunities? Are the poor themselves to blame? Or are they victims of unequal educational opportunities, racism, sexism, or an economic system that favors the rich over the poor? As a rich society, how can we help poor families without fostering economic dependency, unwed childbearing, or other unintended consequences that may perpetuate further poverty? (Lichter and Crowley 2002)

Family well-being is intimately tied to economic well-being. A family that has economic problems experiences frustration and stress. The family members must constantly worry that tomorrow may bring an emergency.

In 2000, 11.3 percent of the U.S. population was in poverty. This was the lowest poverty rate since the 1970s (Weinberg 2003). The 2001 to 2003 recession caused the number of persons in poverty to rise slightly, to 12.5 percent in 2002

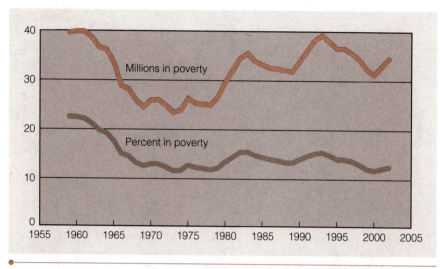

FIGURE 14-2 *Number and Percent of Americans in Poverty, 1959–2000*

SOURCE: U.S. Census Bureau, "Poverty Status of People by Family Relationship, Race, and Hispanic Origin: 1959 to 2000" (www.census.gov/hhes/poverty/histpov/hstpov2.html; accessed March 29, 2002).

TABLE 14-5		
Percent All People and Family Members below 100% of Poverty		
	2002	
	All People	**People in Families**
All races	12.5%	10.8%
Male	11.2	9.8
Female	13.7	11.7
White	10.5	8.7
Male	9.5	8.1
Female	11.5	9.4
Black	24.4	23.1
Male	22.0	20.6
Female	26.5	25.2
Hispanic	22.5	21.5
Male	20.6	19.0
Female	24.4	22.7
Asian	11.8	9.8
Male	11.6	9.8
Female	12.0	9.9

SOURCE: U.S. Census Bureau 2004a.

(see Figure 14-2 and Table 14-5). Note in Table 14-5 that people in families, regardless of race, had lower poverty rates, compared to rates for all people. In 2000, the poverty rate for female-headed families was 32.5 percent compared to a rate of 4.7 percent for married-couple families (J. Fields and Casper 2002). Statistics such as these lead to the idea that encouraging marriage will reduce the overall poverty level. Indeed, the U.S. government, and some state governments,

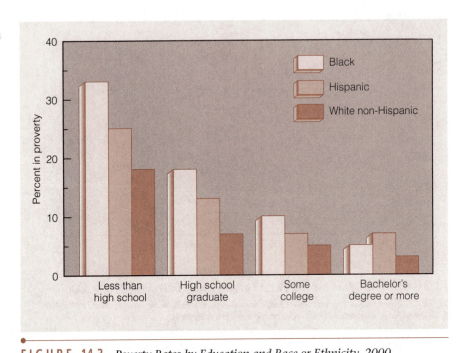

FIGURE 14-3 *Poverty Rates by Education and Race or Ethnicity, 2000*

SOURCE: U.S. Census Bureau, "Years of School Completed by People 25 Years and Over, by Age, Race, Household Relationship, and Poverty Status: 2000" (http://ferret.bls.census.gov/macro/032001/pov/new07_000.htm; accessed March 29, 2002).

have allocated funds to promote marriage programs. Note that increasing education levels among Americans also reduces poverty levels (see Figure 14-3).

Some of this reduction in poverty has resulted from passage of welfare reform (Personal Responsibility and Work Opportunity Reconciliation Act [PRWORA]) in 1996. Since its passage, the number of families receiving welfare has declined 50 percent.

In the United States, poverty is usually defined in a relative way, and what Americans regard as poor hardly qualifies for the term in much of the rest of the world. Yet, even though being poor in the United States does not mean that you will necessarily starve to death or do without television or even an automobile, being poor does cause great hardship to the poor individual or family.

Most poor families move in and out of poverty, depending on family composition and work patterns. The increased unemployment that occurs during economic downturns causes many families to drop into poverty. Once the economy recovers, many of these families move out of poverty again.

During the period of unemployment, however, the stress and strain on a family are great. Joblessness of a major income producer seriously affects the well-being of everyone in that person's family. The constant worry, the loss of self-esteem and self-respect on the part of the unemployed family member, the declining psychological and sometimes physical health of family members are all by-products of unemployment.

Changes in family composition also trigger movement in and out of poverty. For example, divorce throws many women and their children into poverty for a period of time (see Chapter 15). A young man moving from home and setting up his own independent household may find himself beneath the poverty threshold, at least temporarily. A single woman giving birth may drop into poverty as a single parent.

TABLE 14-6

Percent Child Poverty Rates in Metro and Rural Settings

Race/Ethnicity	Metropolitan	Rural
All children	16%	19%
Black	32	42
Hispanic	19	24
Asian	14	14
American Indian	27	36

SOURCE: Population Reference Bureau 2004b.

Many people associate poverty only with unemployment, but in fact, many working people live beneath or close to the poverty threshold. They cannot earn enough to escape poverty status. These people are termed the **working poor**. They are usually members of families with children. More than half of the people who are poor live in families with at least one worker.

working poor
Employed people who live below the poverty threshold

A number of public programs attempt to help people and families in poverty. Unemployment payments help families survive the period of joblessness until work is found. Retraining programs help workers improve their skills so that a wider variety of jobs will be open to them.

Children and Poverty

Table 14-6 indicates that poverty rates are much higher for children, especially those under 5 years, than for the general population. Poverty rates are highest for children in rural areas. Although only about 19% of children live in rural areas, this still accounts for 14 million children (Table 14-6). These children, however, are disproportionately poor, less educated, and their parents tend to be underemployed. They face the same problems as poor children in metropolitan areas, but the problems are exacerbated by the rural isolation and limited access to support services (O'Hare and Johnson 2004).

Because a large proportion of families in poverty are single-parent, female-headed families, various laws have been passed to increase the level of child support by absent fathers. Child-care assistance, public subsidies for housing, food stamps, health-care assistance, and various tax credits that reduce taxes are all societal attempts to help the poor. Although efforts are being made to offer poor people economic help, many Americans feel that much more can be done. Many have felt that monetary help alone may work to entrap some people in poverty, rather than help them regain economic independence; such concern led, in part, to the passage of welfare reform in 1996 (PRWORA). The creation of more and better paying jobs and improving educational opportunities and standards are two important goals that American society must set, if poverty is to be erased in the United States.

The Military Family in the Time of War

Since the military became an all-volunteer force (1973), the number of military family members has outnumbered the services members 60 percent to 40 percent. ("Service members" describes those persons actually in the armed

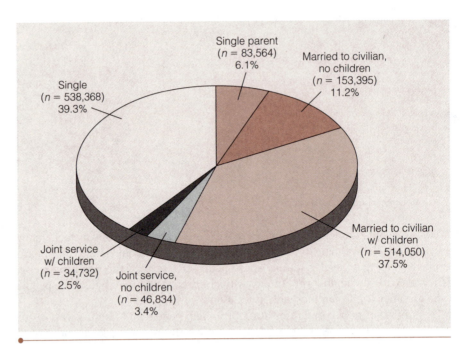

FIGURE 14-4   *Various Family Relationships in the Military*

SOURCE: Military Family Resource Center.

services, not their families.) About 55 percent of service members are married. Many have children, the majority (73 percent) of whom are 11 years of age or younger. Figure 14-4 shows the percentage of service personnel in various family relationships.

Today's military family faces a lifestyle with more frequent deployments (witness the deployments to Afghanistan and Iraq), and thus increased family separations. In addition, there are normal tours of duty for specified amounts of time and then a change to another posting. Moving and resettlement of families are now an ever-present part of military life.

Before discussing the stresses placed on all military families, especially in times of war, it is important to note the advantages of military life. First, the entire military acts as a surrogate family. For example, routines on a base are essentially the same for all families. The goals of the military, regardless of branch, are similar, thus contributing to a high degree of camaraderie. Housing or an allowance is provided, which helps the families when moving from place to place. New assignments bring new friends and a great deal of support from those service families at the new posting.

On the other hand, there are drawbacks for families in the military:

- Frequent moves and loss of friends
- Parental absences with deployment
- Pressure to adjust and fit into the group
- Lack of control over one's future
- Housing, although often provided, looks like everyone else's and close living makes privacy difficult.
- Parenting difficulties when a parent is deployed. In essence, the remaining parent becomes a functional single parent.

- Stress for the trailing party. The job determines the working hours and conditions for the military member. The family has no choice but to adjust (Egan 2002; Segal and Segal 2004).

Deployment and the family. Deployment to areas where war is taking place is extremely stressful to the soldier's family (Pittman et al. 2004). The major worry is obviously the safety of the service member while deployed into a life-threatening environment. Unit (the specific unit to which the soldier is attached) support and informal support from all members of the military and their families lead to a sense of community. The sense of community is especially strong for those families living on base (Bowen et al. 2003). "We are all in this together" provides strong support to the families whose military members have been deployed into combat.

The soldier missing in combat leads to ambiguous loss, which is more difficult to cope with than actual loss through death (see "Ambiguous Loss").

Also, as seen recently in the Iraq conflict, the calling up of reserves is more difficult for the reservists' families than the deployment of regular armed service members' families. Many in the reserves don't really expect to be called to active duty. If they are, it means leaving a job and family, for many reservists. The call-up of reservists for an uncertain length of time is difficult for the person called to active duty, the family, and the employer who now must replace a worker for an unknown period of time. Legally, the employer must return the active-duty reservist to his/her job once the service period is finished.

Although formal family-support programs have been developed by the armed services, they tend to be underutilized, partially because there is a stigma attached to their use by military families, and the programs tend to be underfunded and underevaluated (Drummet et al. 2003).

The family tension and stress induced by deployment tend to increase abusive behavior. Studies show that strict adherence to a code of absolute right and wrong contributes to abuse (Egan 2002). This code also suggests the family knows what is right and should be able to handle its own problems—hence, the stigma attached to seeking help via counseling and other support services.

Although the military has been slow to realize the importance of family support to both the service personnel and their family's morale, the military has put into place a wide variety of family support agencies. Via family support centers, the following services are freely provided to all active-duty military, retirees, and family members (Albano 2002):

- Relocation assistance
- Personal financial management services
- Spouse employment services
- Emergency financial aid
- Information and referral services
- Support during duty separation (pre-deployment, deployment, and reunion support)
- Family life skills education
- Transitions to a civilian career assistance

Drug and Alcohol Abuse

Good nutrition means eating the foods your body needs to be healthy. However, people ingest other substances that have little to do with nutrition, such as drugs and alcohol. Many of these substances can be harmful, both physically and psychologically, even though some may be taken for health reasons.

drug
Any substance taken for medical purposes or for pleasure that affects bodily functions

Defining the term *drug* is difficult. One definition of a **drug** is any substance taken for medical purposes or for pleasure that affects bodily functions. Thus, penicillin is a drug taken to fight disease, and morphine is a drug taken medically to relieve pain. Alcohol is a drug taken for pleasure and to ease tension, and marijuana is an illegal drug taken for the same purposes. Vitamins and minerals are drugs taken to promote and maintain health.

We are concerned here with the abuse of drugs and alcohol, which leads to family crises. Many family problems are related to the misuse of both. For example, much (perhaps most) spousal and child abuses occur when the abusing mate or parent is drinking or using drugs.

Drugs and Drug Abuse

Susan was an honor student at the university. She worked evenings and Saturdays to help pay her tuition. One Saturday when she was especially tired, one of her co-workers offered her a little cocaine, saying it would make her feel better. Although a bit frightened, Susan was curious. Shortly after using it, her fatigue left her and she felt energetic and able to do anything.

Gradually, Susan started using the drug before going to work. She found the effects did not last long, and soon was using it on her work breaks. As she needed increasingly more to maintain the high it gave her, the cost went beyond what she earned. Her co-worker who was supplying the drug offered to give it to her free if she'd go out with him. She soon found that he was not offering free cocaine at all, but was trading it for sex with her.

It wasn't long until she lost her job and stopped going to classes, because seeking and consuming drugs now occupied all her time. Some months later, she was arrested for shoplifting. Shortly after her release from custody, she was picked up again—this time for prostitution. She is now in a rehabilitation program and hopes to return to the university.

WHAT DO YOU THINK?

1. Have you any friends who have become consumed by a drug habit? How did they start using drugs?

2. If they are still using drugs, what do you think are their chances of kicking the habit? Why?

3. Why do you think some people are able to try a drug once or twice and not get hooked, whereas people like Susan seem to become addicted immediately? What kind of person are you?

Unfortunately, this is a common story about the use of addictive drugs. Most Americans express great concern about illegal drugs, such as cocaine, LSD, speed, and marijuana. Yet in many ways, Americans may be more threatened by legal drugs. For example, a person may go to sleep at night with the help of a sleeping pill, drink coffee as a stimulant to wake up the next morning, smoke a cigarette after lunch to ease tension, take a diet pill to reduce appetite for dinner, and drink several glasses of wine to be more relaxed and sociable at an evening party.

As these examples illustrate, defining drug abuse is difficult. This book will use the following definition: *Drug abuse is the persistent and excessive use of any drug that results in psychological or physical dependence, or that the society labels as dangerous or illegal.* Drug abuse can occur with legal drugs. If a person must follow the routine described above or any part of it to function normally, he/she is abusing drugs.

The use of drugs and alcohol among young persons aged 12 to 17 years decreased considerably through the 1980s into the mid-1990s. In 1999, 9.8 percent of this population used an illicit drug; by 2001, this had increased to 10.8 percent. The percentage of males in this age group was only about 1 percent to 1.5 percent higher than females. Marijuana use among this age group was 7.2 percent in 1999, moving to 8.0 percent in 2001. Cocaine use also increased from 1.7 per-

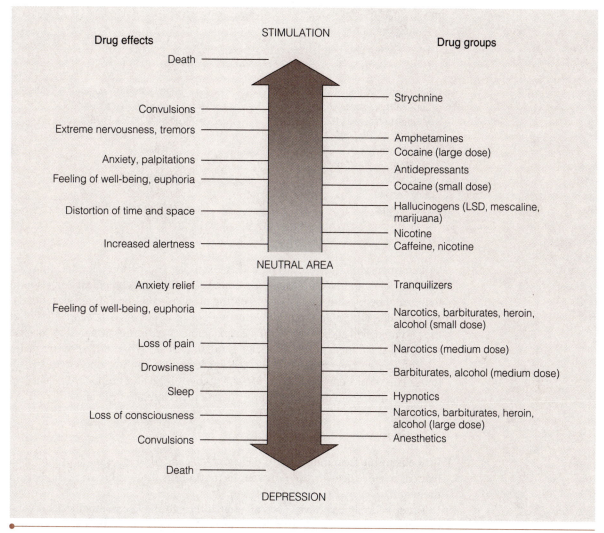

FIGURE 14-5 *Continuum of Drug Action*

cent in 1991 to 4 percent in 1999. Illicit drug use was highest among those aged 18 to 24 years, 16.4 percent in 1999 and 18.8 percent in 2001. Marijuana use was also highest in this age group, 16 percent in 2001 (NCHS 2004a, Table 62).

The major drug groups may be classified along a continuum, from stimulating to depressing (Figure 14-5). As the figure shows, drugs are capable of altering attention, memory, judgment, time sense, self-control, emotion, and perception.

When a person must use a drug to maintain bodily comfort, a physical addiction or dependence exists. The body often adapts to a drug and increases its drug tolerance, which means that an ever-increasing amount of the drug is necessary to maintain comfort. Once a physical addiction exists, any attempt to discontinue the drug usually causes both physical and mental withdrawal symptoms. Withdrawal from some drugs, such as alcohol and heroin, can be extremely difficult. Quitting heroin, for example, causes violent, flu-like symptoms of nausea, vomiting, diarrhea, chills, sweating, and cramps.

Many people believe drugs that cause only psychological addiction are easier to give up than those causing physical addiction. Yet, psychological addiction can

TABLE 14-7	
Well-Known Personalities Who Have Died of Drug Complications or Suicide Related to Drugs	
Rob Pilatus (Milli Vanilli)	Brian Epstein (manager, The Beatles)
Chris Farley (*Saturday Night Live*)	Mary Tyler Moore's son
Michael Hutchence (INXS)	Sid Vicious (Sex Pistols)
Kurt Cobain (Nirvana)	Brian Jones (Rolling Stones)
River Phoenix	John Belushi
Jim Morrison (The Doors)	Marilyn Monroe
Janis Joplin	Paul Newman's son
Judy Garland	Elvis Presley
Jimi Hendrix	Len Bias (basketball)

be extremely difficult to overcome. For example, the eating disorder anorexia nervosa does not involve a physical addiction, yet the psychological addiction to the idea of being slim causes severe eating problems that lead to physical deterioration. Many therapists working with anorexic or bulimic young people find that the psychological addiction to faulty eating habits is extremely difficult to correct.

One of the dangers of doing drugs is mixing drugs. When different drugs are taken together, the effects of the drugs increase and confuse one's bodily reactions. The results can sometimes be fatal. Table 14-7 lists a number of celebrities who have died from drug abuse, or from suicide related to drug abuse.

Alcohol

Bruce is in his freshman year in college. He is elated to have made the football team as a wide receiver. He now spends most of his time with other members of the team.

Bruce's family has never used alcohol, although they occasionally serve wine to friends who come to dinner. Bruce tasted both beer and wine some time ago. He found them too bitter and sour for his taste, and didn't like the aftertaste.

He has attended several parties with other members of the football team, at which a great deal of beer was consumed. Because he did not like the taste of beer, he did not join in the drinking. His friends teased him, calling him a "wimp" and a "nerd." Finally, to shut them up, he joined them in their beer drinking at a party. He found the experience unpleasant and has now decided that he will just avoid those parties where beer is served.

Alcohol use has remained at about 50 percent of high school seniors reporting having used alcohol in the past month (CDC 2000b; NCHS 2003a, Table 62). Alcohol use among persons aged 12 to 17 years increased only slowly from 1999 (16.5 percent) to 2001 (17.3 percent). Alcohol use was highest in the 26 to 34 year age group (59.9 percent in 2001). Male and female rates of alcohol use are approximately the same.

Alcohol is by far America's biggest drug problem. It is estimated that 14 to 18 million Americans have a serious drinking problem. Alcohol abuse and binge drinking (defined as at least five drinks in a row for males and four for females) are problems, especially with college students. It has been found that heavy drinking (binge drinking) by teenagers may actually damage brain development, ham-

WHAT DO YOU THINK?

1. If Bruce really didn't like beer, why do you think he finally joined his friends in drinking it?

2. How would you suggest Bruce handle his friends' pressure to drink at parties?

3. Do you agree with his decision to stop attending parties where beer is served? Why, or why not?

pering both memory and learning abilities (Ballie 2001). Indeed, there are several deaths from binge drinking on college campuses each year.

Contrary to popular belief, alcohol is not a stimulant. Actually, it slows or depresses bodily functions, such as reaction time. Small amounts of alcohol reduce mental control and produce feelings of relaxation. Alcohol also harms sexual performance, especially for males.

Deaths in alcohol-related motor vehicle accidents are one of the most tragic side effects of drinking alcohol. Groups such *as* Students Against Driving Drunk (SADD) work to educate other students about the dangers of drinking and driving. Fortunately, such groups and the increasingly severe laws against driving and drinking have helped. Deaths from alcohol-related motor vehicle accidents have dropped from 57 percent of all traffic fatalities in 1982 to less than 50 percent in 1995.

Peer pressure played a large part in Bruce's drinking beer at the party, as we saw. If you ask people who smoke, drink, or do drugs how and why they started, many will blame it partially on their friends. "My boyfriend was doing drugs and wanted me to try it," one will say. "Everybody in my group smoked, so it seemed like the thing to do," says another. "On Saturday nights, my friends competed to see who could drink the most beer, and I felt out of it if I did not participate," relates a third.

Substance abuse can begin early in life. Part of maturity is making independent decisions for oneself. Young children must have decisions made for them. Adults who allow others to make important decisions about lifestyle and healthful living for them are acting like children. When Bruce drank the beer because of his friends, he was avoiding responsibility for his own behavior. To smoke, drink, or do drugs just because others are doing it is giving someone else power over one's own important life decisions.

Saying no to alcohol, cigarettes, or drugs when friends offer them can be difficult, but it is necessary, so that the individual is making decisions about these behaviors, not his/her friends.

Summary

1. *Stress occurs often in everyone's life.* Mild stress *per se* is neither good nor bad. It is how one handles stress that is important.

2. *People usually go through three phases in reacting to stress.* The first phase is alarm (the body reacts). The second phase is resistance, and the third phase is recovery or exhaustion.

3. *Death of loved ones is a crisis that all people must face at times in their lives.* Natural death from old age is far easier to accept than death from homicide, suicide, or accident—especially if such a death occurs to a youthful loved one.

4. *Mourning and grief are natural reactions to a death.* Usually, the survivors first experience numbness and shock, followed by tears and emotional release after a few days or weeks. The initial shock is followed by pangs of grief and episodes of painful yearning for the dead person. Gradually, the survivors move into a prolonged period of dejection and periodic depression. Finally, life returns to normal, although one may periodically feel the loss.

5. *Injuries and prolonged illness can cause lasting changes in family life.* Unlike death, family members may have to live with the results for years.

6. *Family violence can come in many forms: spousal abuse, child abuse, sibling abuse, and parental abuse by children.* Five general factors have been associated with family abuse, as follows:

   • Violence begets violence. In other words, those who come from a violent environment are more apt to be violent themselves.

   • Family violence is greater among families below the poverty line because they experience greater stress and frustration.

   • High levels of stress contribute to family violence.

   • Social isolation increases the risk of family violence.

   • An orientation to the traditional male role tends to increase the risk of family violence.

7. *Poverty and unemployment put greatly increased strains on the family.*

8. *The military family is unique and experiences high levels of stress* with the deployment of the service member into a war zone.

9. *Drug and alcohol abuse are contributing factors to many family problems.*

Resources on the Internet

Companion Website for *Human Intimacy: Marriage, the Family, and Its Meaning,* Tenth Edition

http://sociology.wadsworth.com/cox10e/

Gain an even better grasp on this chapter by going to the companion website to take one of the tutorial quizzes, use the flash cards to master key terms, or check out the many other study aids you'll find there. You will also find special features such as GSS data, Sociology Online, and Census 2000 information that will put data and resources at your fingertips to help you with that special project or to help you as you do some research on your own.

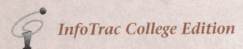

InfoTrac College Edition

http://www.infotrac-college.com/wadsworth/

You can access reliable resources anytime, anywhere, with InfoTrac College Edition, the online library. This fully searchable database offers more than 20 years' worth of full-text articles (not abstracts) from almost 5,000 diverse sources, such as top academic journals, newsletters, and up-to-the-minute periodicals, including *Time, Newsweek, Science, Forbes, The New York Times,* and *USA Today.* You can conduct electronic key word searches using key terms from this chapter to supplement your reading and learning experience. To aid in your search and to gain useful tips, see the Student Guide to InfoTrac College Edition, which you can access through the companion website for this book.

Should Drugs Be Legalized?

YES

The prohibition against various drug use simply isn't working, just as in the past, when prohibition of alcohol failed to work. The government simply cannot legislate certain human behaviors successfully. People who want to get high will find ways around the law. Drug pushers are the same kinds of people that bootleggers were during prohibition.

Legalizing drugs would remove the profitability of illicit sales. Since prices would drop, two kinds of drug-related crime would also fall. First, users who have an expensive habit often resort to petty crime to support their habits. Legalization and the resultant cheaper availability of drugs should reduce the need of addicts to engage in criminal behavior to support their habit. Second, the violence connected to the drug trade would be reduced, because the huge profits would no longer be available. Remember that the gangster era tied to prohibition of alcohol partially disappeared when alcohol again became legally available.

Decriminalization of drugs would free the legal system to pursue real criminals. The government spends huge amounts of money fighting a losing battle against drugs. Freeing the government of the drug battle would free government energy and money to fight other social problems that could be won, if enough resources were devoted to them.

Finally, but perhaps most important, legalizing drugs would make it possible to tie the availability of drugs to some kind of help, such as counseling. Thus, addicts would become known to the society, which, in turn, could help them overcome their addictions. During the presidential election of 2000, California passed a ballot measure requiring counseling help for those arrested on drug charges. Several states, including California, have recently legalized marijuana for medical purposes, although federal law still bans such use.

NO

Although drugs are fairly easy to obtain despite their illegality, legalization would make drug use more widespread, just as alcohol consumption rose following the repeal of prohibition in 1933. This would be particularly devastating to the poor, inner-city families who fight a daily war against drugs.

Those wanting to legalize drugs suggest that much petty crime committed by addicts because of high drug prices would disappear, but there is no empirical evidence for this idea. In fact, studies of the criminal careers of heroin and other drug users document that while drug use tends to intensify criminal behavior, most of their criminal careers were established before the onset of drug use.

Although some kinds of crime might be diminished with drug legalization, the more widespread usage would perpetuate other kinds of violence, such as spousal and child abuse. For example, the availability of crack—commencing about 1986—in New York City has led to a 225 percent increase in child neglect and abuse cases involving drugs, and a dramatic rise in the number of infants abandoned in city hospitals. In addition, there has been a large increase in the number of crack-addicted babies born to these mothers.

Lastly, for society to condone that which it knows is harmful both to the individuals using drugs and the society as a whole is immoral. The recent legalization of marijuana for medical purposes in some states was really only a ploy and a first step to the total legalization of drugs. The law should be repealed.

What Do You Think?

1. What is your intellectual position on the legalization of now-illegal drugs, such as cocaine and marijuana? What is your emotional reaction to the proposal?

2. Do you think legalization would:
   a. Reduce crime?
   b. Increase drug use?
   c. Reduce violence?
   d. Be helpful curing the addict?
   e. Help parents keep their children drug free?

© Novastock /PhotoEdit

The Dissolution of Marriage

Chapter

15

Questions to reflect upon as you read this chapter:

- List several reasons for the high divorce rate in the United States.

- Which reason do you think is most important?

- Why are the emotions of divorce compared with the emotions of a loved one dying?

- Why is a marriage with children not ended completely by divorce?

- List several problems the newly divorced person faces.

- Which problem do you think is the most difficult for the newly divorced person?

- List several types of child custody.

- What problems do children face when their parents divorce?

- List several advantages of divorce mediation.

- What suggestions can you make to reduce divorce rates?

*I*n *sickness and in health, till death do us part." This traditional part of the marriage ceremony might well be changed to the following in modern America: "In happiness and in good health, till divorce do us part."*

Let No One Put Asunder

It seems obvious that the "love match" marriage based on romance, self-gratification, and happiness—in other words, on the fulfillment of one's own needs—must suffer from a high rate of failure. In the past, a good marriage was measured by how well each spouse fulfilled the socially prescribed role of husband or wife. Today, Americans ask a great deal more of marriage—and the higher the stakes, the higher the chances of failure.

In addition, the greatly increased life span means that marriages are now expected to endure much longer than ever before. One hundred years ago, 30 out of every 1,000 marriages were ended each year by the death of one of the spouses. Only 3 in 1,000 were ended by divorce. Today, divorce, separation, and desertion—rather than death—end most marriages. Is it realistic to ask a marriage to last for 50 or 60 or more years?

The divorce rate in the United States rose throughout the twentieth century. In 1900, there was about one divorce for every 12 marriages in a given year. By 1922, there was one divorce for every eight marriages; in the late 1940s, there was one divorce for every three and a half marriages. This peak was probably due to the dislocations arising from World War II. Between 1950 and 1964, the ratio of divorce to marriage leveled off at approximately one divorce for every four marriages. But the divorce rate started to rise again in 1967, and by 1984 approximately one divorce was occurring for every two marriages in a given year. Today, the median duration of marriage until a divorce is approximately 11 years.

crude divorce rate

Ratio of divorces to each thousand persons

The **crude divorce rate**—the ratio of divorces to each 1,000 persons within the population—is really a better, though less dramatic, measure of marital stability than the ratio of divorces to marriages in a given year (see Table 15-1). Note that the crude divorce rate has remained quite stable since 1985, fluctuating between 3.8 and 4.9. If one considers all marriages, only about 2 to 3 percent of marriages end in divorce in a given year.

However, the overall divorce statistics do not yield a complete picture of the incidence of broken marriages. Legal separation claims another 3 percent of all marriages. Of those who separate, 75 percent divorce within 2 years and 90 percent divorce within 5 years (Bramlett and Mosher 2001). Desertion, another manner of breaking a marriage, is especially prevalent among the poor, although statistics on desertion are difficult to obtain because they rarely appear in the records. Usually, it is the husband who leaves, although more wives are now deserting than in the past.

In any case, dissolution of American marriages seems to be relatively commonplace. America's divorce rate is high, but as Table 15-2 indicates, it is not the

TABLE 15-1

The Number of Divorces and Divorce Rates

Year	Number of Divorces	Rate/1,000 Total Population
2003*		3.8
2002		4.0
2001		4.0
1998	1,135,000	4.3
1996	1,158,000	4.4
1990	1,175,000	4.7
1985	1,172,000	4.9
1980	1,182,000	5.2
1970	773,000	3.7
1960	393,000	2.2
1950	385,000	2.6
1946	610,000	4.3
1940	293,000	2.2

*NCHS ceased reporting total number of divorces in 1999 because not all states report this statistic.

SOURCE: Munson and Sutton 2004; NCHS 2000d.

WHAT RESEARCH TELLS US

Divorce Facts

Although many breakups occur during the first few years of marriage, with the peak occurring about 3 years after marriage, nearly 40 percent of all broken marriages have lasted 10 or more years. The median duration of a first marriage at the time of divorce is 11 years for both men and women. Remarriages ending in divorce average 7.4 years. Nationally, all marriages ending in divorce last an average of 9.8 years. The average age of men at time of first divorce is 35 years and for women, 33 years (NCHS 2000d). About 4 percent of divorcing couples have been married less than 1 year (NCHS 2000d).

highest in the world. Some family experts call American marriages "throwaway marriages." A better name is **serial marriages;** that is, Americans tend to marry, divorce, and remarry. In about one-half of all American couples marrying in a given year, one or both partners have previously been married. A high divorce rate does not necessarily mean that Americans are disenchanted with marriage. The divorced remarry in great numbers and relatively quickly, depending on their age at the time of the divorce (see Chapter 16).

To conclude that the institution of marriage and the family is in a state of decay and breakdown, based on divorce statistics alone, is not valid. Many marriages still last a lifetime. Marriage is still high on most Americans' list of values, although the high rate of marital breakup and the increasing granting of marital-like rights to unmarried couples appear to be weakening the value of marriage. What the high divorce rate tells us is that (1) Americans, in general, are more accepting of divorce and (2) when they become disenchanted with their marital partner, they will leave that partner. Whether this is good or bad is another question. On the surface, it may sound good for adults, because it appears acceptable to change partners in an effort to improve one's marital satisfaction. However, as we shall see, marital disruption is not good for some of the children.

Reasons for America's High Divorce Rate

The reasons for America's high divorce rate are many and varied, as follows:

1. Most divorcing couples cite personal reasons, such as unhappiness, a breakdown in communication, sexual failure, and general dissatisfaction.
2. Americans ask a great deal of modern marriage, perhaps too much. High expectations often lead to disappointment and failure. If you ask nothing and receive nothing, "nothing" is not disappointing. If you ask a great deal and receive only a little, unhappiness often follows (see Chapter 7).
3. Tied closely to Americans' high expectations of marriage is the relative freedom of individuals to make marital choices. The second basic principle of this book—that a free and creative society will offer many structural forms by which family functions can be fulfilled—is another reason for America's high divorce rate. Many choices breed a certain amount of dissatisfaction. Is the grass greener on the other side of the fence? Might some alternative be better? Being surrounded by married friends who view marriage similarly and who are committed to it adds strength and durability to one's own marriage. Being surrounded by those who don't support one's concept of marriage—who suggest or live alternative lifestyles, and/or who deride marriage—is disruptive of one's own marital patterns.
4. Changing gender roles are part of the American interest in the general concept of change and its benefits. All those who question traditional gender roles place pressure on the institution of marriage. For example, a woman who decides that the role of mother is not for her, seeks a career, and leaves the care of their children to her husband is bound to face some disapproval from her family and friends. Certainly, the same holds true for the husband who decides at age 40 to quit his secure job, cease supporting his family, and begin writing adventure stories. Although gender-role changes may ultimately be liberating to individuals, the transitory result will be marital disruption for some people.

serial marriages
Marrying, divorcing, and marrying again; a series of legal marriages

TABLE 15-2	
Divorce Rates Worldwide	
Country	Divorce as a Percentage of All Marriages
Russia	65
Sweden	64
Finland	56
Britain	53
United States	49
Canada	45
France	43
Germany	41
Israel	26
Greece	18
Spain	17
Italy	12

SOURCE: Kirn 2000, 76.

5. Greater economic independence of women also encourages separation and divorce. When a wife had no economic alternatives, she was forced to tolerate an unsatisfactory marriage, in return for economic security for herself and her children. Now, there are economic alternatives to an unsatisfactory marriage that allow women to fight back or leave such a marriage (see Chapter 13).

6. Another reason for the high divorce rate is America's heterogeneity. With so many kinds of people and so many beliefs, attitudes, and value systems, family and marriage will naturally mean many and different things to different individuals. Even though people tend to marry people with similar backgrounds, there will still be differences in beliefs and attitudes. For example, consider the case of a female college graduate interested in pursuing both a family life and a career; she marries an engineer from a traditional family background who believes the wife's place is in the home. Conflict seems inevitable. America's heterogeneity also leads to a higher incidence of mixed marriages. People of differing religions, ethnicity, and race are more apt to marry in America, simply because they are here and freedom of marital choice is encouraged. When such persons marry, their differences make it more difficult to build a successful and enduring marriage.

7. The general mobility of Americans may also contribute to increased marital breakup. Families that move often may not create support networks. Married friends, relatives, and membership in institutions such as churches all tend to support a marital relationship. A high degree of mobility tends to weaken such supports.

8. A list of the general reasons for marital failure would be incomplete if it failed to include social upheaval, economic problems, and the general health of the society. Certainly, job insecurity brought on by increased unemployment and recession strain the family institution. Continuing economic worries have also brought failure to many American marriages. Marital failure is highest among the poor and becomes progressively lower as economic status rises (South 2001). As noted earlier, desertion tends to be high among the poor because it does not entail the costs of a divorce.

9. Acceptance of divorce by Americans is another important factor in the rising divorce rates. The stigma of divorce has largely vanished over the past 30 years. In fact, less stigma is probably attached to a 40-year-old divorced individual than to a person of the same age who has never been married. General social acceptance is also noticeable in the trend toward more lenient divorce laws and the increase in economic alternatives to marital dependency, provided through various forms of government assistance. The mere fact of so many divorces in society also adds to the acceptance. In studying the long-term effects of divorce, some researchers suggest, "It is fair to assume that divorce since 1970, when it became comparatively commonplace, unselective, and almost normative, had a fundamentally different character than it did when it was far more rare and stigmatized." Thus, examining the long-term effects of divorce on those divorced prior to 1970 presents different patterns than examining it thereafter (Braver and Cookston 2003, 314).

10. Another reason for increasing marital failure is the personal inadequacies, failures, and problems that contribute to each individual divorce. Regardless of the magnitude of the social problems and pressures that disrupt marriage, the ultimate decision to end a relationship is made by one or both spouses. The specific reasons given by divorcing couples often bear little resemblance

to the real reasons they are divorcing. For example, many couples list sexual incompatibility as a reason for divorce. Yet, sexual incompatibility is more often a symptom of other problems, such as lack of communication.

11. A number of other individual risk factors tend to predict marital instability. Those marrying at a young age are more prone to divorce. Premarital pregnancy, children born out of wedlock (Teachman 2002), parental divorce, and remarriage all relate to a greater probability of divorce.

The reverse of the 11 factors contributing to divorce work to maintain a marriage. A number of other influences also work to support marriage. Religiosity, especially when the spouses share religious values, strongly influences lasting marital relationships. Good communication skills (see Chapter 5), a large circle of relatives and friends supportive of one's marital relationship, and having children all tend to improve the chances for a lasting marriage.

Emotional Divorce and the Emotions of Divorce

Divorce is not such a spontaneous, spur-of-the-moment act, as marriage can be. In most cases, dissolution occurs slowly, and divorce is the culmination of a prolonged period of gradual alienation. In many cases, several years elapse between a couple's first serious thought of divorce and the decree. The divorce process often runs something like the following:

1. First, there are clouds of doubt, early signs that the marriage is getting into trouble. There may be a disturbance in the sex life and affectional response of the couple. Rapport is lost and emotional separation begins.

2. This is followed by a variety of distancing behaviors as the couple become more dissatisfied. The partners increasingly avoid each other, spending a diminishing amount of time together.

3. One of the partners begins to imagine living apart, or substituting a new partner for the old partner.

4. The possibility of divorce is first mentioned. This tends to clarify the relationship somewhat, with the initiator taking the lead and the partner remaining passive through the divorce cycle. Often, one partner may be totally surprised by the other's suggestion of divorce, not having recognized the early problem signs.

5. The appearance of solidarity is broken before the public. The fiction of solidarity is important as a face-saver. Once it is broken, the marriage cannot be the same again.

6. The decision to divorce is made, usually after long discussion and much vacillation. In fact, the couple may move apart, mentally deciding to divorce, yet reconcile and try again to make the marriage work. This may occur several times, even during the next stage of actual physical separation (Binstock and Thornton 2003).

7. The crisis of actual physical separation follows. Severing a meaningful relationship is a traumatic experience at best, even after much thought and discussion has been put into it.

8. Final severance comes with the actual divorce. This may come after a long period of delay and separation. Although the actual legal procedure is usually

Sadness is a common emotion associated with divorce.

thought of as closing the case, stage 9 is necessary before the final adaptation can begin.

9. A period of mental conflict and reconstruction follows the divorce. The former partners enter new social worlds, and full estrangement takes place.

10. If children are present, the divorced partners' relationship may never end, because the problems of coparenting remain.

The actual facts or events that lead up to the divorce are as varied as the individuals who marry, but one thing always happens: one partner begins to concentrate on the other partner's weaknesses, shortcomings, and failures, rather than on the person's strengths. The "20 percent I dislike you" becomes the focus of attention rather than the "80 percent I love you" (see Chapter 7). The inability to accept the partner the way he/she is, or to accept unwanted change in the partner, is at the root of most divorces.

Almost always, there is an initiator who unilaterally begins the uncoupling process, consciously or unconsciously. This partner has the power in the breakup process, because he/she knows earlier than the other partner that the relationship is in trouble and may have to be changed. Thus, the initiator often begins the process of withdrawing from the relationship, building a new and separate life, and preparing a new world for him-/herself long before the partner suspects that the relationship is in real trouble. When the initiator drops hints of dissatisfaction, the unknowing partner simply takes them to mean that there are problems to be worked out. "Every marriage has its ups and downs," that partner thinks. Because the relationship is not in any serious trouble for the partner, he/she may remain blind to the growing discontent and the strength of the initiator's dissatisfaction. Many partners report shock and dismay when the initiator finally makes it clear that she/he wants to end the relationship. "I was totally surprised when my wife indicated that she had seen a lawyer and was suing me for divorce," reported one husband. "I had no idea that she was so unhappy and that our relationship was about to come to an end." In contrast, the initiator may say, "I have worked for years to make this relationship successful, and it is not. I've known for a long time that our marriage was a mistake." Thus, spouses often experience the greatest degree of emotional distress at different points in the divorce process (Amato 2000, 1272).

It is obvious that there is a communication breakdown between the spouses. It seems strange that couples newly in love often indicate that it is the wonderful, open communication that draws them together. "We are so often in harmony together. We seem to always share the same thoughts." How can a couple who have been in love, intimately close, perhaps for years, begin not to communicate well enough to recognize that their relationship has severe problems? The fact is, that the intensity of falling in love and courtship simply cannot be maintained in a long-term relationship. The emotional cost is too high. As relationships age, much of the daily interaction becomes routine. Systems are set up to simplify and make life comfortable, efficient, and convenient, thus conserving energy and time. Once a good relationship has been established, the partners tend to assume

HIGHLIGHT

Not One Divorce but Six

A classical way of viewing divorce is to see divorce as a series of divorces, rather than a single breakup of the married couple.

- *Emotional divorce* is the same as steps 1 and 2 in the divorce process, in that the couple begin to pull apart and distance themselves from each other.

- *Legal divorce* is the court-ordered termination of the marital relationship.

- *Economic divorce* is the property settlement in which the couple's properties are divided between the husband and wife. It usually means a reduced standard of living for both, although often more so for the woman.

- *Parental divorce.* If children are present, the marriage ends, but parenting usually does not. The issues of child custody and support are among the most difficult to resolve and often cause continuing conflict, long after the legal divorce is granted.

- *Community divorce* is the reaction of friends and extended family to the couple's divorce. Who will remain friends with the wife, and who with the husband? What role will grandparents play in the lives of their grandchildren after the divorce?

- *Psychological divorce* is regarded by many as the most important divorce. "When am I finally able to let go of my past marriage and spouse and build a new, independent, stable, and fulfilling life for myself?"

SOURCE: Bohannan 1970.

that all is going well, unless there are very strong indications to the contrary. The initiator does not come quickly and suddenly to the conclusion that the relationship must end, so there are usually no obvious clues to draw the partner's attention. Because every relationship has problems at times, the initiator's complaints are usually dismissed as trivial, and certainly not indicative of serious relational problems. Even when the initiator does get the partner's attention, the partner usually assumes that changes can be made and the relationship saved. Unfortunately for the relationship, by the time the partner is aware of the seriousness of the situation, it is often too late to make the changes that might save the relationship. The initiator has already partially withdrawn and may have substituted other sources of need gratification, so that the relationship is no longer as important to him/her as it is to the unknowing partner. Generally, the initiator has the better chance of coming out of the divorce a winner. However, in Judith Wallerstein's classic, longitudinal study of divorced couples (Wallerstein 1984; Wallerstein and Blakeslee 1989, 40; 2003; Wallerstein et al. 2000), there was only a minority of the divorced couples where both husband and wife were able to reconstruct happier, fuller lives after the divorce. These data have been superseded by more recent studies indicating that many couples are able to reconstruct their lives and move on to happier relationships as time passes, even though painful memories may remain (Hetherington 2003; Kelly and Emory 2003, 359).

Despite those who romanticize divorce as a happy rebirth and return to freedom, many negative emotions will be experienced, including a sense of loss similar to the grief felt at the death of a loved one, and the loss of feelings of psychological well-being. A lucky few—perhaps those divorcing quickly after marriage—do not experience grief and mourning. But most will go through a period of denial ("This really isn't happening"), followed by grief, mourning, and a mixture of the following emotions:

- Self-pity: Why did this happen to me?
- Vengeance: I'm going to get even!

- Despair: I feel like going to sleep and never waking up again.
- Wounded pride: I'm not as great as I thought I was.
- Anguish: I don't know how I can hurt so much.
- Guilt: I'm really to blame for everything.
- Loneliness: Why don't our friends ever call me?
- Fear: No one else will want to marry me. I'll never be able to support myself.
- Distrust: He/she is probably conniving with attorneys to take all the property.
- Withdrawal: I don't feel like seeing anyone.
- Relief: Well, at least it's over, a decision has finally been made.
- Loss of feelings of psychological well-being: I feel awful, depressed, nervous, suicidal.

Unfortunately, our society offers no ritual or prescribed behaviors for the survivor of divorce. The community does not feel it necessary to help, as it does with bereavement. Indeed, the fact that the spouse still lives often prevents the divorced person and the children of divorce from coming to a final acceptance of the breakup—there is always a chance, no matter how small, of recovering the spouse and the lost parent (see "Ambiguous Loss," Chapter 14). Some rejected spouses exaggerate this remote chance, and make it the sustaining theme of their lives for a long time after the legal divorce. Even in cases where marital breakup is hostile and bitter, the loneliness following the breakup may spur a mate to wish the spouse were home, even if only to have someone to fight with and ease the loneliness.

No matter how much a person thinks about and prepares for a separation, there is still a shock when the actual physical separation occurs. Many persons,

 HIGHLIGHT

Divorce: Constant Self-Questioning

"How can I be missing her after all the fighting and yelling we've done the past year or two? It's so nice to be in my own place and have peace and quiet. Damn, it sure is lonely. It's great to be a bachelor again, but after 15 years of married life, who wants to chase women and play all those games? Wonder why the kids don't call. Keeping house is sure a drag. I wonder what she's doing. Do you suppose she is nicer to her dates than she was to me?

"What really went wrong? Two nice kids, a good job, nice home, and I certainly loved her when we married. But she had been so unaffectionate and cold over the years. She never seemed to have time for me. Or was it that I never had time for her? When we dated, she was so flexible—she'd do anything with me. But after our first child, she seemed to become so conservative. She wouldn't do anything daring. I had so little leisure time, with working so hard to give the family a good life. Why couldn't she do what I wanted when I was free? It's really

all her fault. Of course, I could have included her more in my work world. Maybe if I had shared more of my business problems with her, she'd have been more understanding. It's true that I am awfully short-tempered when the pressure is on. Certainly, I didn't listen to her much after a while. It seemed as if she only complained. I get enough of that at the office. You don't suppose there was someone else? Maybe it was all my fault. A failure at marriage, that's me. Never thought it could happen to me. When did it start? Who first thought of divorce? Maybe the idea came because all our friends seemed to be divorcing."

On and on go the thoughts of this newly divorced man. Anger, guilt, frustration, conflict, insecurity, and emotional upheaval are the bedmates of divorce. "Our culture says that marriage is forever, and yet I failed to make a go of it. Why?" The whys keep churning up thoughts, and endless questioning follows marital breakdown. There are so many questions, doubts, and fears.

especially those who have been left, suddenly panic and feel abandoned. Although separation panic comes in many forms, most people experience it as apprehensiveness or anxiety. They feel physically and psychologically shaky and have great difficulty concentrating on any complex task.

Such feelings recede with time for many, but a divorced spouse may later feel a resurgence of hurt, hostility, and rejection on learning of the former spouse's new relationships. To hear that one's former mate is remarrying is often upsetting; it may arouse past hostilities and regrets, even though considerable time has passed. Much to the surprise of those who believe that time always heals, there are some divorced persons and children of divorced parents for whom the hurt, hostility, and anger doesn't disappear.

It is probably important for anger to be a part of divorce, because anger may be the means of finally breaking the emotional bonds that remain between the former spouses. Not until these bonds are broken will each be free. Free really is not the proper word. Few divorced persons are ever completely free of the earlier relationship, as we shall see later in the chapter. Fortunately, many divorcing couples come to a turning point, when their energies can finally be channeled from destruction to construction once again.

Divorce but Not the End of the Relationship

Divorce is the death of a marriage, but strangely enough, it is not the end of the relationship in most cases. Divorce can be the rebirth of an individual. And rebirth carries problems with it, just as death does. Once the mourning, grief, and anger begin to subside, the newly divorced individual faces important choices about life directions. What does one do when newly alone? Seek the immediate security of a new marriage? Prepare to live alone the rest of one's life? Make all new friends, or keep old friends? Maintain ties to the past relationship? Escape into the work world? Seek counseling? Experiment with sexual involvements? More counseling professionals and also divorced persons are saying, "Use the pain and suffering of divorce to learn about yourself. Seek the rebirth of a new, more insightful, more capable person out of the wreckage of failure." This is a wonderful goal. Some even talk about creative divorce and try to paint divorce as a positive experience that will bring growth and improvement to one's life, if properly approached.

Yet, the attitude that divorce is simply a growth experience is a far cry from the way most people feel at the breakup of their marriages. Nevertheless, using divorce as a learning opportunity rather than seeing it only as a failure from which nothing good can be derived makes a great deal of sense. If divorced persons do not use the insight into themselves and their past relationships that they have gained, they are doomed to make the same relational mistakes again; unfortunately, this is often exactly what they do.

However, a romantic myth of divorce hardly prepares the divorcing couple for the traumas and problems they will experience in reaching their goals. It may not even encourage people to reevaluate their broken marriages and themselves, so that they can make more successful future marriages.

To get over a marriage, one has to go back to it. To go forward to a new life, one has to go backward and try to understand what went wrong in the past so that the same mistakes are not repeated in a new relationship.

MAKING DECISIONS

What Changes Must Be Made with Divorce?

There are many decisions to be made when considering divorce, including the following:

- Changing the economic relationship. Who will get the house, car, furniture; what share of the money, investments, retirement benefits?

- Making the practical changes necessary to accomplish the above, such as changing wills, property titles, credit cards, addresses, and the like.

- How will each spouse support him-/herself? Will both work? Will spousal support be paid by one spouse to the other? If so, for how long?

- What about health insurance, designation of children for income tax purposes, definition of beneficiaries on life insurance policies?

- If children under 18 years are present, what type of child custody is best? How will visitation rights be arranged? How will the children be supported economically? Where to set the boundaries that clearly define the former partner as a coparent? (Madden-Derdich et al. 1999, 588.)

- What will be the relationship of each spouse to friends and extended family? How will new friends be made? How will dating be handled? What will be the relationship to the former spouse after the divorce? When one or the other remarries?

- How will sex lives be handled after divorce? Does one avoid sexual relationships? Seek them out? Can or should one have sex with the ex? How is one's new sex life handled with regard to children?

WHAT DO YOU THINK?

1. What do you think is the single most important thing that a newly divorced person should avoid doing if he/she is to find the road to recovery? Why?

2. What are some other things the newly divorced person should avoid doing if she/he wants to speed recovery?

3. If you were counseling a newly divorced person, what would you advise? Why?

Many people who contemplate divorce think it will end their misery by ending forever a relationship that has become intolerable. Yet, this is not always true; it is certainly almost never true if children are involved. Divorced people often remain bound to each other by children, love, hate, revenge, friendship, business matters, dependence, moral obligations, the need to dominate or rescue, or habit. Increasingly, it is being recognized that divorce does not necessarily dissolve a family unit, but simply changes its structure. This is especially true for children. Divorce—rather than reducing conflict for divorcing parents—may actually increase conflict as the parents try to decide on visitation rights, child support, and other day-to-day decisions about how to rear their children.

Every state has provisions for modifying judgments made by the court at the time of divorce. Requests for change of custody, support, or alimony can be made by either spouse at any time. If an individual believes marital problems end with divorce, he/she should attend "father's day" in court. Many larger cities set aside specific days when motions are heard relative to an errant father's failure to pay child support or a mother's refusal to permit her former husband to visit their children. Sometimes, these hearings—years after the divorce—are emotionally packed scenes, replete with name-calling, charges, and countercharges.

On the positive side, some former spouses become good friends after the pain of divorce fades. Some divorced couples remain business partners. Even after one or both remarry, there may be friendly interaction between the couples and their new spouses.

It must be remembered that one's former spouse still has all the characteristics that attracted one to him/her in the first place. Therefore, it seems unrealistic to expect that one will totally dislike the divorced spouse forever, though many divorced persons claim that they will. The best divorce adjustment is attained through an *amicable divorce*, where there is a minimum of conflict and the marital relationship gradually is transformed into a friendship relationship.

Problems of the Newly Divorced

The newly divorced face several major dangers, as follows:

- They may make a prolonged retreat from social contact.
- Some may jump quickly into a new marriage.
- A few may base their life on the hope that the spouse will return.
- Some may organize their life around hostility and getting even with the former mate.

Certainly, the first reaction after any loss may be a momentary retreat, as the individual turns inward to contemplation and avoids all situations reminding him/her of the hurt, disappointment, guilt, and shame of failure. Unfortunately, though, some divorced people make this reaction their lifestyle. In essence, such people die psychologically and end their lives in every way but physically.

A larger group of divorced people seek a new relationship as quickly as possible. Discounting those people who had a satisfying relationship with someone else before their divorce and are now fulfilling it as soon as possible, many people simply can't stand the thought of failure or being alone. They rush into the first available relationship. These people usually have not had time to reassess themselves or their motives. The idea of facing themselves and the challenges of becoming an independent person are simply too frightening. Often, these people have never really been alone. They married early, and in a sense they may never have grown up psychologically. Even though their first marriage may have been unsatisfying, they still prefer marriage to assuming responsibility for themselves. Rushing into a new marriage, they are likely to repeat past mistakes.

We have already looked at some of the problems of attempting to hold onto a past relationship. Living on the basis of false hope creates a prolonged situation, in which hope and disappointment alternate.

A life based on continuing anger and harassment of a former spouse may be the most destructive of all, because both the former spouse and the mate seeking revenge are harmed. In the usual scenario, one divorced spouse takes the other back to court repeatedly, in an effort to punish him/her. Fortunately, few divorced persons continue this behavior indefinitely.

Community divorce refers to the problems and changes in one's lifestyle and in one's community of friends. Divorce affects not just the divorcing couple and their children, but also their friends, their extended families, and their workplace. When an individual marries, single friends are gradually replaced by married-couple friends. When a couple divorce, another change of friends usually occurs. Changes of friendship after divorce tend to be more complicated, both for the newly divorced and for the couple's friends. Some friends may side with one or the other member of a divorcing couple. Remaining friends with a newly divorced person is difficult, for at least two reasons. First, the divorce causes the couple's married friends to reexamine their own marriages. Often, a divorcing couple will inadvertently cause trouble in their friends' marriages. In fact, sometimes a domino effect occurs: When one couple among a group of friends separates, other couples who are unhappy gain psychological support to also separate. Second, the married friends may experience conflict about which partner will remain a friend. In addition, still-married friends may view the newly single person as a threat, or at least as a fifth wheel. Generally, the newly divorced person will find that old friendships tend to fade and be replaced by new ones. Often, those new friends will also be newly divorced.

Most divorced people start to date within the first year after their separation. The new dating partners bring with them a new circle of friends. Thus, most divorced people will experience a gradual change in their larger community of friends and contacts.

Economic Consequences

The bottom line is that divorce brings about severe financial changes for the divorcing couple. For most divorcing couples with children, the man becomes single, but the woman becomes a single parent; as we have seen, poverty often begins with single parenthood (see Chapter 14).

Statistically, divorced men experience a rise in their standard of living in the first year after divorce, while divorced women (and their children) experience a decline. Two factors tend to account for this disparity; men earn more than women, and ex-husbands often don't pay child support. Men have always earned more than women (see Chapter 13). This fact does not affect married women, who traditionally share in the man's resources. Once a woman is divorced, however, she no longer shares the man's higher earning power and must rely on child or spousal support to make up the difference between her earning power and her former husband's.

About 60 percent of custodial parents are awarded child support (custodial mothers 63 percent, and custodial fathers 38 percent). Of those not awarded child support, about 32+ percent do not feel the need to have a legal child support agreement, while another 20 percent don't want their child/children to have contact with the other parent (Grall 2003). Among custodial parents who were due support payments in 2001, 80 percent received at least partial payments, a proportion unchanged since 1993. The average annual amount was $4,300 and did not differ between mothers and fathers. It is interesting to note that about 59 percent of child support agreements have provisions for health care insurance for children. Unwed fathers have the lowest rate of child-support payments, in part, because they also tend to have the lowest incomes.

Child support is awarded based on the child's or children's needs and the supporting parent's ability to pay. Unfortunately, child-support awards vary greatly from state to state, and seldom cover all of the real expenses of growing children. For example, a divorced mother living with two children aged 7 and 13, where the combined income of father and mother was $2,500 per month, received anywhere from $251/month child support in Mississippi to a high of $782/month in Arizona (Pirog et al. 1998, 294–295).

Spousal support, or alimony, requires a showing of financial need and ability to pay, but the duration of the support may be reduced by expectations of the dependent spouse's economic rehabilitation. The move toward no-fault divorce and away from fault-finding has tended to reduce alimony awards because they are no longer used to punish a guilty spouse.

If women have equal rights with men, theoretically they should be able to support themselves; the courts now tend to use this rationale as the basis for awarding spousal support after divorce. This would be fine if women were, indeed, able to earn as much as men; but that is not yet the case, particularly for women who must care for children and who have been long-term homemakers (see Chapter 13). Many states have amended their laws and changed procedures so that permanent, or at least long-term, alimony can again be awarded in some cases of special need. Only 50 to 75 percent of women awarded alimony actually receive payment from their ex-husbands.

These trends have led to an increase in child poverty in the United States, paralleling the increased number of divorces. Combining unmarried birth with the current rate of divorce, it is estimated that about 60 percent of children born today will likely spend some of their childhood in a single-parent family, thus increasing their chances of living in poverty for at least some period of their lives (see Chapter 14).

In light of the poor child- and spousal-support payment records of noncustodial spouses, state, local, and federal governments have tried to enforce such payments through legislation (Haynes and Ross 1994). Those efforts have worked to a degree—the proportion of custodial parents receiving every payment increased from 37 percent to 46 percent between 1993 and 2001 (Grall 2003). Congress unanimously approved legislation to strengthen child-support collection, through mandatory income withholding and the interception of federal and state income tax refund checks to cover past-due support. A federal Parent Locator Service allows custodial parents access to various federal records, such as Internal Revenue Service information, to find an errant spouse. Finding the spouse is, of course, the first step in enforcing support payments. Other legislation allows states to provide information to consumer credit agencies on past-due child support when the arrears are more than $1,000. Late-payment penalties may also be applied. Payments may be made through a state agency, if requested by either parent, and mandatory wage assignments are required in most states as soon as child support is ordered by the court. Maine is the first state to pass a law that takes away business, professional, and/or driver's licenses from parents who don't support their children (Goodman 1994a). In 1998, the California Supreme Court overturned precedent and ruled that a parent can be jailed for refusing to seek available work to pay child support (Egelko 1998). In October 1997, the National Directory of New Hires started listing every person newly hired by every employer in the country, so that federal and state investigators can track down parents who owe money to their children (Pear 1997). There is fear, however, that such a wide database not only places an unfair paperwork burden on employers, but also that it can be misused. All these enforcement tools have helped increase support payments. In those few jurisdictions that impose jail sentences on errant, noncustodial parents, there seems to be quicker compliance with the court-ordered payments.

The single-parent family, created either by divorce or by having a child out of wedlock, experiences not only economic problems, but also severe time problems. To work and rear children successfully without an additional adult requires an almost superhuman effort. The complex logistics of working and providing child care at the same time, especially when money is in short supply, make it difficult for the single-parent family to function.

Many volunteer groups have sprung up to help newly divorced and single-parent families. We Care and Parents Without Partners are two such organizations. The latter is a nationwide organization that has a goal to help alleviate some of the isolation that makes it difficult for single parents to provide a reasonably normal family life for themselves and their children.

Fortunately, the single-parent family tends to be temporary. As we shall see in Chapter 16, the majority of divorced people remarry. Although having children might seem to make it harder to remarry, studies show that this is not necessarily true.

Children and Divorce

America's high divorce rate has been an adult-oriented phenomenon. "I am not happy in this marriage. I have a right to be happy and fulfilled. This right transcends any commitments and/or obligations I might have to others, even my children," says the divorce culture. In the adult-oriented world of divorce, commitment to others, especially children, is lost in the commitment to oneself.

I do my thing, and you do your thing,
I am not in this world to live up to your expectations
And you are not in this world to live up to mine,
You are you and I am I,
And if by chance, we find each other, it's beautiful.
If not, it can't be helped.
(Fritz Perls in the 1960s, as quoted in Whitehead 1997, 49–50)

Perhaps this "prayer" of the 1960s best sums up America's modern belief about the role of divorce in our lives. The problem with this self-oriented view is that the adult's role as parent is lost. Commitment to one's children—to ensure that caring, loving, competent adults are produced to carry on the society successfully—plays second fiddle to the parent's needs. Although adults, aided by psychologists and other professionals, have been led to believe that leaving a "bad" marriage (in the adult marriage-partners' view) is good for the children, increasing amounts of evidence suggest that this is not always true (Ahrons and Tanner 2003; Hetherington 2003; R. Simons et al. 1996; Wallerstein and Blakeslee 2003; Wallerstein et al. 2000; Whitehead 1997). Even in the most successful of divorces (in the adult view), it is the children who often lose. It is also true that children coming from divorced homes are at increased risk for their own future divorces (Amato 1996, 2003; Hetherington 2003, 325; Wallerstein and Blakeslee 2003; Wallerstein et al. 2000).

In the past, it has been marriage that frees a person from their parents and further teaches the young adult to share and compromise and become a competent adult. It has been marriage that links a man and woman to their children. It has been through marriage that a child claims support and affection from a mother and a father. Marriage has established the economic basis of the family and provided the mechanism whereby resources are transferred from the older to the younger generation. Marriage establishes a circle of relatives that support the young, growing family. Marriage attaches the biological father to his children, as well as the mother, and secures a sustained and regular investment in the children (Whitehead 1997, 137). It is these things that are most often lost to the children of divorce. And if we look only at father's participation in child rearing, it becomes clear that single parenthood and divorce most often remove the father from the children's lives.

The idea of staying together for the sake of one's children has lost validity in America, although there is some evidence that the idea is reviving (Ulriksen 2000; Wallerstein and Blakeslee 2003; Wallerstein et al. 2000). In observing the many negative effects of divorce on some children, it might well be worth remaining in a struggling relationship for a few extra years (providing high levels of conflict and/or abuse can be avoided), if it helps the children to better cope with the family breakup. In the 1950s, only 6 out of every 1,000 children experienced parental divorce in a given year. Today, 27.6 percent of all children under age 21 live with only one parent, usually the mother (84.4 percent; Grall 2003). Most of these single-parent households result from divorce, but a growing number result from unmarried reproduction (see Chapter 10).

The first problem for parents contemplating divorce is telling the children. There is no easy way to do this. Judith Wallerstein and her colleagues (Wallerstein and Blakeslee 2003; Wallerstein et al. 2000) report that 80 percent of the children studied were completely unprepared for their parents' separations. In a disturbingly high number of cases, a parent would simply disappear while a child was sleeping or away from home. Sharing an impending separation with one's children is an unpleasant task, but it is important to the child's well-being that it be done.

The effects of divorce on children vary, but tend to be negative when compared to children living in happy, intact families. Certainly, the immediate effects are unsettling, but the long-term effects are probably mixed. Some children may suffer long-term damage. Others may be better off after a divorce if their conflicting parents were abusive. Children in high-conflict marriages that do not break up seem to have even greater increases in behavioral problems than children from broken families (Morrison and Coiro 1999, 626). Such data lead to the idea that an unhappy marriage is an unhappy home for children; if divorce promotes parental happiness, one would think that the children should also benefit. Although this folk wisdom sounds reasonable, long-term study of divorced families generally does not support it. What these studies do show is that the children of divorce are at risk, but many are resilient enough to survive the problems and grow up to be contributing and healthy adults (Hetherington and Kelly 2002; Pasley 2003, 313).

Judith Wallerstein and her colleagues (Wallerstein and Blakeslee 2003; Wallerstein et al. 2000) are doing a long-term study of 60 families who have gone through divorce. They interviewed them close to the time of divorce and at 18 months, 5 years, 10 years, 15 years, and 25 years after the divorce.

Our overall conclusion is that divorce produces not a single pattern in people's lives, but at least three patterns, with many variations. Among both adults and children five years afterward, we found about a quarter to be resilient (those for whom the divorce was successful), half muddling through, coping when and as they could, and a final quarter to be bruised: failing to recover from the divorce or looking back to the pre-divorce family with intense longing. Some in each group had been that way before and continued unchanged; for the rest, we found roughly equal numbers for whom the divorce seemed connected to improvement and to decline. (Wallerstein 1984, 67)

But it is in adulthood that the children of divorce suffer the most. The impact of divorce hits them most cruelly as they go in search of love, sexual intimacy, and commitment. Their lack of inner images of a man and a woman in a stable relationship and their memories of their parents' failure to sustain the marriage badly hobbles their search, leading them to heartbreak and even despair (Wallerstein et al. 2000, 299).

As noted earlier, these conclusions are based on parents who divorced many years ago, and their children. As divorce has become more acceptable, such dire consequences for the children of divorce may have softened (Braver and Cookston 2003, 314). Not surprisingly, children with strong, well-integrated personalities who are well-adjusted before the divorce of their parents tend to make the best post-divorce adjustments. Children also do significantly better when both parents continue to be a part of their lives on a regular basis (F. Walsh 1998, 81).

WHAT RESEARCH TELLS US

The Children of Divorce

In general, past research has examined only the immediate consequences of divorce on children. Paul Amato (1993) examined 180 studies regarding outcomes of children whose parents divorced, and found mixed results. Kim Leon reviewed 24 studies on young children's adjustment to parental divorce, and also found mixed results (2003). The October 2003 volume of the *Journal of Marriage and the Family* (Kay Pasley, editor) was devoted to the legacy of divorce upon the members of divorced families. Because divorcing families differ, and children differ from each other even in the same family, mixed results are to be expected. However, some generalities emerge. For example, studying a national sample of 17,110 children under age 18, Dawson (1991) found that children living with single mothers or with mothers and stepfathers were more likely than those living with both biological parents to have experienced the following:

- Repeated a grade in school.

- Been expelled from school.

- Been treated for emotional or behavioral problems.

- Have elevated scores for behavioral and health problems.

Judith Wallerstein and her colleagues (2000) report the following from their long-term study of divorced families:

- The children of divorce do not experience just one huge loss of their intact family, but often suffer many continuing losses as their parents go in search of new relationships.

- Most youngsters feel rejected by and angry toward at least one parent, usually the father, and this often persists into adulthood.

- Some grow up in settings in which the parents are continually at war with each other, long after the actual divorce. Indeed, high-conflict marriages often lead to high-conflict families after divorce, so that many children really don't escape conflict, even after divorce.

- Two-thirds of the girls, many of whom seemed to sail through the initial crises of family breakup, suddenly become deeply anxious in adolescence, appear unable to make lasting commitments as adults, and are fearful of betrayal in later intimate relationships.

- Both the men and women of divorced families who choose not to marry, or so state, indicate they are afraid of marriage based on what they know of their own history and of the number of broken marriages they have seen.

- Boys appear to experience more negative consequences from divorce than girls (A. Hines 1997, 379; Morrison and Cherlin 1995). Many boys, who were more overtly troubled in the post-divorce years, failed to develop a sense of independence, confidence, or purpose (Wallerstein et al. 2000).

- The severity of the child's reaction at the time of the divorce does not predict how the child will fare 5, 10, 15, and 25 years later.

- Many of the children from families reporting abusive relationships formed abusive relationships themselves as young adults.

- Most states limit court-ordered child support to the child's first 18 years. When Wallerstein compared financial support for college, less than 30 percent of the youngsters from divorced families received full or even partial support for college, compared to almost 90 percent of youngsters from

This is true only when the parents were able to work out a satisfactory and nondestructive post-divorce relationship (King and Heard 1999, 393–394). Aside from these two factors, it is difficult to predict which children will do well and which will not.

It is important to realize that both the age and the sex of the child must be considered in understanding divorce effects. Younger children seem to have more severe immediate reactions to the divorce of their parents than adolescent children. Yet, 10 years after the divorce, these same children seem better adjusted than their older siblings. Research has also found that the younger the age of the child at the time of separation, the lower their subsequent parental attachment, and the more likely they are to perceive both their mother and father as less caring and more overprotective when they are older (Woodward et al. 2000, 162). In general, the younger siblings in divorced families are buffered, as the older children bear the brunt of their parents' distress (Wallerstein and intact families. The fact of nonsupport for college attendance from divorced parents, when a youngster's father and/or mother could actually afford to help, leads to extreme anger and hostility on the part of the child, which often lasts well into adulthood.

Essentially, what the studies are finding is that there is simply a reduction of parenting when a divorce takes place. When marriage breaks down, most men and women experience a diminished capacity to parent. They give less time, provide less discipline, and are less sensitive to their children, being caught up themselves in the personal maelstrom of divorce and its aftermath.

This reduction in parenting suggests that increasing divorce will lead to an increasing number of unattached and fatherless (most children of divorce live with their mothers) children (Whitehead 1997).

- Research indicates that fatherless children are far more vulnerable to poverty, violence, lawbreaking, drugs, and precocious sexual behavior; receive less schooling and are more apt to fail in school; and suffer

other social pathologies (Amato 2003; Blankenhorn 1995; A. Hines 1997, 379; McLanahan and Sandefur 1994; Wallerstein et al. 2000).

- When the father is out of the home after divorce, father visitations seem to have mixed association with child well-being (Ahrons and Tanner 2003; Amato 1993; King 1994; Whitehead 1997). Apparently, such factors as the amount of post-divorce conflict between the parents, frequency of contact by noncustodial parent, and sex of child and noncustodial parent influence child well-being differently.

- Frequency of noncustodial father-child contact relates to a higher percentage of child-support payments (Arditti and Keith 1993).

How much parenting a divorced parent will provide is not easily predictable. Wallerstein and her colleagues (2000) were surprised to find that the quality of parenting before divorce did not predict the kind of parenting that would be offered after divorce. In some cases, a parent who was very loving and responsible toward his/her children ceased being so after the divorce, while others continued to parent as they had before.

One thing is clear: a divorced parent who enlists his/her children to help fight the other parent will probably cause their children harm. One parent may try to turn the children against the former spouse. The visiting parent may use visitation rights to try to lure the child away from the custodial parent. The "Disneyland Dad" syndrome, in which the visiting parent indulges the children, may cause conflict with the custodial parent, who thinks that the "ex" is spoiling the children.

These numerous studies found that, contrary to popular belief, the passage of time does not automatically diminish feelings or memories of hurt, jealousy, and anger that the parents or the children of divorcing parents have. Unless divorced parents can work out some mutually acceptable relationship that holds conflict to a minimum, adjustment will remain difficult for the children. In extreme cases, recurring court battles may be fought over the children, or child snatching by the noncustodial parent may occur—which is frightening, confusing, and possibly dangerous to the child.

Blakeslee 1989, 175–176; 2003; Wallerstein et al. 2000). Preschoolers (2 to 6 years old) react to divorce with fright, confusion, and self-blame. Children 7 to 8 years old seem to blame themselves less, but express feelings of sadness and insecurity. They have a problem expressing anger and a strong desire for parental reconciliation. Nine- to 10-year-old children can express their anger better, but they feel a conflict of loyalty, are lonely, and are ashamed of their parents' behavior. Adolescents express their anger, sadness, and shame most openly and seem better able to dissociate themselves from the parental difficulties.

The role of former grandparents can also be important to children. Divorce removes the children not only from one parent, but also from at least one set of grandparents. Moreover, grandparents who have close and ongoing relationships with their grandchildren and then, because of divorce, are denied access to them are also victims of divorce (although generally unnoticed). Grandparents are one step removed from the conflicts of a divorce and therefore may be able to offer their grandchildren relative security and affection during the trying times. As noted earlier, recognition of the grandparent-grandchild relationship has led most states to protect grandparent visitation rights (see Chapter 11).

Despite the general acceptance of divorce in our society, it remains an unpleasant and traumatic experience for most parents, and a trying and difficult time for most children. Most divorced people remarry, thereby reconstituting the two-parent family. Of course, in this case, children must establish a new successful relationship with the stepparent, which can create problems (see Chapter 16). Children in reconstituted families still, however, do not do as well as children from intact, original families.

Another dimension of divorce for children is the increased necessity to take care of themselves, due to the lack of parental supervision. In addition, a child may have to take over the parenting function of younger siblings if the custodial parent must work full time and/or cannot function well due to the trauma of divorce. G. Jurkovic (1997) calls such a child "parentified." Essentially, such a child loses his/her childhood to the need to parent the siblings. In addition, children may be called upon to parent one or both of their parents who are suffering from the divorce (Mayseless et al. 2004).

Types of Child Custody

Courts have the following four choices when awarding custody of children in a divorce proceeding:

1. The most common is *sole custody*. In this case, the children are assigned to one parent, generally the mother, who has sole responsibility for physically raising the children.
2. In *joint custody*, the children divide their time between the parents, who share the various decisions about their children. It is most prevalent among those of higher income and educational levels, those living in larger cities, and non-Whites (D. Donnelly and Finkelhor 1993). Some states encourage joint custody, with its appealing promise that the children of divorce can keep both parents. This does not mean that the children spend equal amounts of time with each parent. It simply means that the parents share responsibility for major decisions regarding the children's health care, education, and the like. However, enthusiasm toward joint custody has moderated as the courts have realized that custody conditions need to be tailored to each family, rather than trying to use one type of custody in all situations. The perceived advantages of joint custody include the following:

- Both parents continue parenting roles.
- The arrangement avoids sudden termination of a child's relationship with one parent.
- Joint custody lessens the constant child-care burden experienced by most single parents.
- It has reduced the number of litigated custody cases.

Joint custody is a compromise that enables parents to save face by avoiding total loss of custody. However, joint custody also forces the parents to maintain a relationship. If they cannot do this successfully, there will be negative effects on the children. Joint custody can also give the child a sense of homelessness.

A child who lives one week with mother and one week with dad may never be able to establish where his/her home and loyalties lie.

3. A third choice is termed *split custody*. In this case, the children are divided between the parents. In most cases, the father takes the boys and the mother, the girls. This method has the major drawback of separating the children from each other. It does, however, reduce the burden on the parent who might otherwise have sole custody of all the children.

4. The court may award custody to someone other than a parent or parents. This is seldom done, unless both parents are incompetent or offer such a poor environment for the children that the court decides parental custody would be harmful. In most such cases, grandparents or other near relatives gain custody of the children.

Custody and visitation plans often cause conflict between the divorcing couple. Relitigation of a divorce tends to occur more often with parents than with childless couples. Child-support issues followed by visitation rights are the two conflicts most often taken back to court.

In practice, most children are still placed in the physical custody of their mother. Noncustodial fathers tend to gradually drop out of their children's lives (Ahrons and Tanner 2003; Amato and Booth 1996, 364). The following are several factors that play a role in some noncustodial fathers' failure to maintain contact with their children:

Being a single mom or dad is a lot of work.

- Living some distance away from children.
- Conflict with the ex-spouse and/or a stepparent.
- New interests, family, or stepchildren.
- Guilt over nonpayment of support, or failure to fulfill some other obligation to the former spouse and/or the children.
- Teenage children so busy with their own lives that little time is available for the noncustodial parent.

We have earlier discussed the importance of the father role in rearing children (see Chapter 10). Most of the emphasis has been on the father's role in teaching and modeling for boys. Yet, his role is equally important for the girls in the family.

If it is true that a father helps to develop his daughter's confidence in herself and her femininity, that he helps to shape her style and understanding of male-female bonding, and that he introduces her to the outside world and plots navigational courses for her success, then surely it is an indisputable conclusion that the absence of these lessons can produce a severely wounded and disabled woman (Barras 2000 7D; A. Manning 2000).

Because 84.4 percent of divorced mothers have custody or partial custody of the children (Grall 2003), there are few noncustodial mothers; such mothers often have problems that the courts determine make them unfit parents. Although in the past, the courts generally did not demand as much monetary support from a noncustodial mother as they did from a noncustodial father, today the average amount of child support received by the custodial parent does not differ significantly between mothers and fathers (Grall 2003). Noncustodial mothers appear to enjoy nearly as good relations with their adult children as do custodial mothers, in contrast to the deteriorating relationship that noncustodial fathers tend to have with their children.

WHAT DO YOU THINK?

1. Of the various types of child custody, which do you think would have been best for you if your parents had divorced when you were 6 years old?

2. Should parents stay together for the sake of the children? Why, or why not?

3. If a noncustodial parent does not pay court-ordered child support, what do you think should be done?

4. If a parent denies visitation rights to the ex-spouse, what do you think should be done?

5. If visitation rights are denied and the noncustodial parent snatches the child, what do you think should be done?

HIGHLIGHT

Divorce and Dad

Wallerstein and her colleagues (2000), Parke and Brott (1999), Blankenhorn (1995), Whitehead (1997), Madden-Derdich and Leonard (2002) and many other researchers make it clear that the loss of the father (immediate or gradual) in divorcing families has serious, negative ramifications for the children, both male and female. Although some divorced father–child relationships improve with time, interparental conflict, early father remarriage, and low father involvement in the early post-divorce years were associated with worsening relationships, even more than 20 years after parental divorce (Ahrons and Tanner 2003). The following story points out complications of continued parenting as the noncustodial parent, who is usually the father.

Pay but Don't Interfere

I moved out of our home because it seemed easier for Elaine and the kids. After all, I was only one person, and they were three. The children could remain in their schools with their friends and have the security of living in the home they had grown up in. I figured I'd come to visit often. I hoped that our separation and divorce would have minimal impact on the children, and felt that the family remaining in our home was the way to achieve this.

It certainly hasn't worked out as I first imagined. Visiting the children often is no simple matter. Elaine has a new husband; the children have their friends and activities. And I seem to be busier than ever.

At first, I'd just drop by to see the children when time permitted, but this usually upset everyone concerned. Elaine felt I

was hanging around too much, and even accused me of spying on her. Actually, there may have been a little truth to this accusation, especially when she started dating. The children usually didn't have time for me because they had plans of their own. I felt rejected by them.

Next, Elaine and I tried to work out a permanent visitation schedule. I was to take the children one evening a week and one weekend a month. Then she accused me of rejecting the children because I wouldn't commit more time to them. But my work schedule only had a few times when I was sure I would be free and thus able to take the children.

At first, the set visiting times worked out well. I'd have something great planned for the children. Soon, however, Elaine told me that I was spoiling them. She said that they were always upset and out of their routine when I brought them home. Would I mind not doing this and that with them? Gradually, the list of prohibitions lengthened. She had the house, the children, and my money to support the kids, yet I seemed to have fewer and fewer rights and privileges with them.

"Be sure child-support payments arrive promptly, but please leave the children alone"—increasingly, this was Elaine's message.

Since she remarried, my only parental role seems to be financial. I really have no say on what the children do. I feel as though I'm being taken every time I write a child-support check. I feel that I have lost my relationship with my children.

Divorce: The Legalities

Historically, the American attitude toward divorce has been negative, stemming mainly from the majority Christian heritage. The stigma of divorce has a long history. As soon as the Christian Church was established, divorce was formally forbidden and made legally impossible. (There was no civil marriage or civil divorce until recent centuries.) The attitudes of most Christian churches and many Americans, however, have changed; divorce is now a legitimate option. It is considered a right and proper way to end an unsuccessful marriage.

There is strong impetus in the United States to view family relationships as an entirely private matter, something that family members work out largely by themselves. On the other hand, many aspects of family relations are governed by legal restrictions, policies, and procedures (Braver et al. 2002, 325). Remember that marriage is a contract between the marrying couple and the state (see Chapter 7).

Matrimonial law is one of the most dynamic fields of law in this country today. It is ever-changing and differs from most legal fields in that it is based less on historical precedents than on doing the right thing to achieve equity between the parties. The credo of the American Academy of Matrimonial Lawyers is to preserve the best interests of the family and society.

No area of law requires greater emotional strength from the parties, the attorneys, and the judiciary than matrimonial law. There is no area where emotions run higher, where personal feelings obscure the facts more often, and where so many conflicting needs (father, mother, children) must be considered. Divorce is a situation where an all-out effort to win will only lead to losses for all involved.

There is variety in divorce laws because divorce, like marriage, is regulated by individual states. In the past, states laid down certain grounds for divorce—rules that, if broken by one spouse, allowed the other to divorce. Because one spouse had to be proved a wrongdoer, punishments were often established. For example, the American colonies commonly prohibited the guilty spouse from remarrying after a divorce. In the more recent past, payment by one spouse to the other (alimony) was sometimes used to punish the guilty spouse, rather than simply to help a spouse become reestablished.

The necessity of proving one spouse guilty of marital misconduct was the only approach to divorce in America until 1963, when Alaska passed the first **no-fault divorce** laws. By 1985, all states had some form of such laws (Vlosky and Monroe 2002, 323). Thus, the idea that a marriage might be terminated simply because it had broken down, without placing blame on one party or the other, found its way into both the law on the books (how the laws are written) and the law of action (how laws are actually applied). This trend has also been visible in other Western countries (Fine and Fine 1994).

no-fault divorce
Divorce proceeding that does not place blame for the divorce on one spouse or the other

All states have residency requirements for divorce (90 days to 1 year or more). Each state recognizes a divorce granted in any other state, providing the other state's residency requirements have been met. In actuality, even this requirement is not checked in the usual uncontested divorce (only 10 to 15 percent of divorces are ultimately contested). Theoretically, states do not recognize foreign divorces, unless one party actually lives or has lived in the foreign country. But here again, the formal law differs from the informal application, and states take no action about a foreign divorce (even if neither party lives abroad) unless a complaint is received. If a complaint is received and the divorce is contested, however, a state may invalidate a divorce in which residency requirements have not been met. In such cases, spouses who have remarried have sometimes been prosecuted for bigamy when the divorce to the former spouse has been invalidated.

Today, no-fault divorce laws usually ask only if there are irreconcilable differences between the parties, although most states still recognize other grounds for divorce. A few states allow divorce by mutual consent, and a small number grant a divorce if the partners have lived apart for a time period ranging between 1 and 5 years. Since divorce laws are governed by the state and are continually being modified, it is important to check the laws of the state in which one resides if divorce is contemplated.

When a person applies for a no-fault divorce, the court simply asks if his/her marriage has broken down because of irreconcilable differences. The court does not inquire into the nature of the differences, though the court can order a continuance (delay) if, in its opinion, there is a reasonable possibility of reconciliation. No-fault laws often have a unique section that bars evidence of misconduct on the part of either spouse unless child custody is at issue.

Because misconduct of one partner or the other is no longer considered, property settlements, support, and alimony awards are no longer used as punishment.

As long as the parties to the divorce agree and have divided their community property approximately evenly, the courts do not involve themselves in property settlements. The courts do set up support requirements for minor children, deeming such children the responsibility of both parents until they are of age. In practice, this generally means that the children live with the mother and the father pays support.

Although fault is no longer determined, the judge still has the right to award alimony or maintenance, though not as punishment for wrongdoing. In most cases, the nonworking wife will receive rehabilitative alimony until she is able to get back on her feet financially. Often, a wife who has devoted herself exclusively to her family for years has lost what marketable skills she may once have had. If she is older, finding a job is even more difficult. Accordingly, judges are tending to award limited-term or reviewable alimony on the basis of need. **Reviewable alimony** can be extended at the end of the limited term if the need still exists. The amount is generally modified by the court if a material change occurs in either spouse's financial circumstances.

reviewable alimony
An award of alimony that is reviewed periodically and changed if necessary

A number of courts are using the following rough guidelines in awarding alimony. If the marriage has lasted less than 12 years, the period of support will not exceed half the duration of the marriage. In marriages of 20 years or more, awards of support may be permanent, that is, until death or remarriage. Generally, under no-fault proceedings, the amount and duration of alimony have been drastically reduced, on the theory that once the marital partnership has ended and the property is divided, the parties should be able to care for themselves.

The simplified no-fault laws have encouraged a new do-it-yourself trend in divorce. In the past, retaining an attorney to obtain a divorce was mandatory, if for no other reason than to help interpret the complicated divorce laws.

Divorced people often express negative sentiments about attorneys and the adversarial legal system. Sometimes, the lawyer may simply have become a scapegoat for emotional clients with uncertain goals. On the other hand, the legal profession may need clearer guidelines for handling divorce, and also some training in counseling and interpersonal relations. In 1991, the American Academy of Matrimonial Lawyers published *The Bounds of Advocacy*, proposing practice standards for lawyers and clients in matrimonial cases (Friedman 1992). Regardless of whether attorneys are involved, a couple who can work out their differences in an amicable manner before going to court are in a much better position to obtain a fair and equitable divorce. Without such an agreement, a couple puts themselves in the hands of a judge who knows little or nothing of their personal situation. Certainly, a judge's decision will be more arbitrary than a mutual decision by the divorcing couple.

It is important that attorneys who handle divorce cases (1) realize that the termination of marital contracts involves both psychological and legal considerations and (2) attend to these issues as they process cases. Failure to recognize the psychological aspects of this area of law can lead to a multiplicity of problems and further complicate an already difficult situation. This is not meant to suggest that lawyers should become marriage counselors, but to point out that nonlegal issues directly affect a divorce action. The lawyer's job is to shift the focus from emotions to facts, without appearing unsympathetic or becoming too sympathetic. It is a difficult balance to maintain.

Many people are going directly to the courts, following simplified procedures from guidebooks written by sympathetic lawyers or knowledgeable citizens, and obtaining their own divorces for the cost of filing ($100 to $500). However, a do-it-yourself approach to divorce is not recommended unless the partners are relatively friendly; agree on all matters, including child custody and child support; and do not have large incomes, assets, or liabilities. In any case, it is advisable to have a lawyer at least review the settlement, to make sure something important hasn't been omitted. The court cannot be relied upon to do this.

Some Cautions about No-Fault Divorce

The implementation of no-fault divorce procedures has undoubtedly eased the immediate trauma of divorce by focusing on the demise of the marriage, rather than on the guilt of one of the spouses. The fear that easier divorce laws would lead to skyrocketing divorce rates was only partially correct (Glenn 1997, 1999; Nakonezny et al. 1995, 1997, 1999). The divorce rate did increase between 1970 and 1980, by which time almost all states had enacted some similar form of no-fault divorce. However, some researchers suggest that the increased divorce rate was not really the work of no-fault laws, but that it simply represented the ongoing deinstitutionalization of marriage in which the moral, social, and legal barriers to divorce were being removed (Glenn 1997, 1999). As noted earlier however, the divorce rate has remained relatively stable since 1985.

Researchers (Parkman 1992; Weitzman 1985) make a strong case that the no-fault divorce laws have, in practice, caused a severe economic hardship for a large proportion of divorced women and their children. In 1993, about 39 percent of custodial mothers were classified as below the poverty level. Although this has dropped (25 percent in 2001), the percent of custodial fathers below the official poverty level was considerable lower, varying between 10 and 15 percent during this time (Grall 2003, 5). Equal divisions of marital property have caused a dramatic increase in the sale of the family home. Because the home is most couples' major asset, the property cannot be divided equally unless the home is turned into cash via a sale. Sale of the family home adds additional stress and disorganization to the divorcing family, because all parties must now find new housing. Selling the family home and splitting the proceeds usually leaves neither the wife nor the husband with enough money to buy a new home. In addition, if the family must move to a new neighborhood, the children's schooling and friendships will be disrupted at a time they most need continuity and stability. In most states, the divorce judge can award temporary use of the home to the custodial party when children are involved, but judges don't always take this step.

Property settlements reached by divorcing couples under no-fault rules have been found to be incomplete and at times, grossly unfair to one of the pair. An amicable agreement reached by a divorcing couple based on present circumstances may be totally inappropriate at some time in the future.

Another major factor working against custodial parents and their children is the reduction of the age of majority (legal adult status) from 21 to 18. As a consequence, college students who are over 18 are no longer considered children in need of support. Thus, when expenses are apt to be greatest for the custodial parent, the noncustodial parent (the father, in most cases) is under no legal obligation to help financially. Many children of divorce, as a result, have difficulty obtaining a higher education. Wallerstein and her colleagues (Wallerstein and Blakeslee 2003; Wallerstein et al. 2000, 335) found that their sample of children of divorce attended college less than those from intact families and dropped out about twice as often.

Ironically, many of the sex-based assumptions ridiculed by radical feminists a decade ago—assumptions about women's economic dependence, their greater investment in children, and their need for financial support from their ex-husbands—have turned out not to have been so ridiculous after all. Rather, they reflected the reality of married women's lives, and they softened the economic devastations of divorce for women and their children. In the early days of the women's movement, there was a rush to embrace equality in all its forms. Some feminists regarded alimony as a sexist concept that had no place in a society in which men and women were to be treated as equals. Alimony was an insult, a symbolic reflection of the law's assumption that all women were nonproductive dependents. But it soon became clear that alimony was a critical mechanism for achieving the goal of fairness in divorce. To a woman who had devoted 25 years of her life to nurturing a family and at age 50 had no job, career, pension, or health insurance, alimony was not an insult but a lifeline. Today, alimony is viewed in terms of compensatory payment for losses caused by dissolution, such as loss of standard of living, loss of future investment rewards, or disproportional costs of child care (Shehan et al. 2003, 313, 314).

Such criticisms as these are leading some state legislatures and researchers to rethink no-fault divorce and to suggest changes that make divorce more difficult to obtain and fairer to the custodial parent (Gatland 1997; Wagner 1997). Some polls found that 50 percent of those polled thought it should generally be harder to divorce, while 61 percent thought it should be harder to divorce for couples with young children (Kirn 1997).

Divorce under any circumstances—no-fault or not—is difficult. Even if divorce is not a legal trial, it is certainly an emotional one. Remember, too, that the no-fault divorce system does not relate to a particular property-division or support system. Fault can still be found, and damage claims against spouses for such actions as causing injury or the willful infliction of emotional distress are still awarded.

Divorce Counseling and Mediation

Going to court means giving up control of your relationship to a third party—namely, the judge. Generally, a good court settlement is one that leaves both parties feeling shortchanged. In other words, the parties accept a settlement with which neither fully agrees. In contrast, divorce mediation tries to provide a setting in which the divorcing couple can meet, communicate, and negotiate with professional help to reach their own settlement, which can then be presented to the court. In most cases, the court will accept such settlements.

Mediation helps couples share ideas about what would work best for each of them, and their children. The mediator can encourage, instruct, and manage (if necessary) interactions, to stay on track. He/she can diffuse intense or sensitive situations and help each partner to really hear what the other is saying. The mediator can offer different ways to approach issues, and new models that will make it easier to agree on issues, or to accept trade-offs.

Some states have conciliation courts, which attempt to ameliorate the negative effects of marital failure. The first goal of these courts is to save the marriage, but failing that, they try to ensure an equitable divorce. The goal of conciliation courts is to protect the rights of children and to promote the public welfare by preserving, promoting, and protecting family life and the institution of matrimony, and to provide means for the reconciliation of spouses and the amicable settlement of domestic and family controversies.

Some jurisdictions require attendance at a divorce education program, in addition to mediation. Unfortunately, the most common format for such a program is a single, two-hour session comprising a videotape presentation, lecture, and group discussion (Geasler and Blaisure 1998, 167). Although this is not much of an effort, it is at least a start toward having the couple consider carefully their contemplated divorce. The few evaluations of such programs find that the couple's conflict is reduced in many cases (Towes and McKenry 1999).

Divorce mediation involves one or two professional mediators (a lawyer and/or a mental health worker) who meet with both the husband and wife to help them resolve conflicts, reach decisions, and negotiate agreements about the dissolution of their marriage. The mediator acts as a neutral adviser, an expert source of information who suggests options from which the couple are free to choose. The spouses do not abdicate the decision-making responsibility, but are helped to decide for themselves. Both partners must be willing to consent to full disclosure if mediation is to work.

Divorce counseling occurs after the divorce is final, and is aimed at helping the individuals get life started again (see Chapter 16). In a few states, some divorce counseling is available through the courts, and occasionally, courts require such counseling.

Persons using mediation should not expect to find support for their side or simple advice that will resolve their problems. They should also not expect to experience immediate improvement in themselves and their relationship with the former spouse. Rather, they will find clarification and help in assessing their strengths and weaknesses so that they can move toward better understanding and clearer communication. The better the couple's communication and understanding, the less traumatic the divorce will be, both for themselves and for any children they may have.

Reducing Divorce Rates

A number of steps can be taken to help reduce the divorce rate, as follows:

- More effective marriage and family training should be given at home and in the schools. Such courses are gaining popularity, and it is hoped that this interest by young persons will lead to sound partner choices and improved marriages.

- Perhaps more scientific methods of mate selection, than the current haphazard system based largely on emotion, can be developed (Gold-Bikin 1994, C14; Larson 2000; *Time* 1995a, 49).

- Periodic marital checkups through visits to trained marriage counselors should be encouraged. As pointed out earlier, by the time most couples in difficulty seek help, their marriage is beyond help. Facing problems before they are insolvable and building problem-solving techniques into marriage are needed elements in American marriage.

- Changing marriage laws, to encourage couples to take their entry into marriage more seriously, would help reduce the necessity of divorce. Some states empower the court to require premarital counseling for any couple applying for a marriage license, if either party is under 18 years of age and the court deems such counseling necessary.

- Child-care centers would help working couples with young children have more time for each other and more time to work positively on their marriage. Many couples simply do not have time for their marriage in the early child-rearing, work-oriented years. When the children are finally capable of independence and economic stability has been achieved, there is often not enough left of the marriage to be salvaged, much less improved. However, encouraging parents to use child-care centers is somewhat negative for the children, in that parenting of the children is reduced.

- We need to give at least as much attention to the good aspects of marriage as we do to the negative aspects. Publicizing ways in which marriage can be improved (see Chapter 17), highlighting the strengths found in successfully married couples (see Chapter 1), and concentrating on relationship improvement will be of more value than exaggerating the problems inherent in marriage. Using mass media to demonstrate positive relational aspects of marriage and commitment, rather than only the problems of intimate relationships, would be helpful.

- The Family Law Act (1996) in Britain might offer some guidance, as follows:

  a. Three months before actual breakup, couple must go into educational program.

  b. A cooling off period of 9 to 12 months, according to the age of the children, is mandatory before actually filing for divorce.

  c. The children's feelings must be considered in addition to the feelings of the adults (B. Rogers and Pryor 1999a, 1999b).

Divorce May Not Be the Answer

In many cases, divorce as a cure to a problematic marriage is worse than the disease. Although most family therapists see their job as helping people in troubled marriages, they usually remain neutral when the question of divorce is raised. "Whether you stay together or split up is your decision" is the standard answer of a marriage counselor to queries about divorce. This tacit neutrality really signals that divorce is OK and helps move the couple toward it, rather than working on their relationship in an attempt to save their marriage.

However, more family counselors are discovering that the problems surrounding divorce are often far worse than the problems within the marriage. Instead of remaining neutral, some marriage specialists now suggest that counselors should make strong efforts to save troubled marriages headed toward divorce (Maher 2000; Medved 1989; Smolowe 1991; *Time* 1995a; Weiner-Davis 1992).

People seeking a divorce are focused on what they see as unresolvable conflicts within their marriage. Often, they believe the problem resides in their mate. Things are so bad in their relationship that leaving the relationship seems the only way to escape the situation. They believe that divorce means the end of their marital problems. Seldom does the person contemplating divorce anticipate the pain and upheaval divorce leaves in its wake. Seldom do they consider that divorce is usually not the end of the relationship and its problems. This is especially true if children are involved. As we have seen, researchers report that the painful effects of divorce often are felt for years by both the divorced mates and their children.

Michele Weiner-Davis (1992) suggests that:

Battles over parenting issues don't end with divorce, they get played out with children as innocent bystanders or even pawns. Uncomfortable gatherings at future family weddings, bar mitzvahs, graduations, births and funerals provide never-ending reminders that divorce is forever.

I've met children of all ages who, even after both parents remarry, secretly hope their own parents will, someday, reunite. Many well-adjusted adults whose parents separated or divorced when they were children admit an emptiness that never goes away. Most parents recognize the fact that divorce will impact on their children, they just don't anticipate the lasting effects. I've heard too many divorced parents say "I wish I knew then what I know now." Gradually I have come to the conclusion that divorce is not the answer. It doesn't necessarily solve the problems it purports to solve. Most marriages are worth saving.

When you consider that most Americans marry for love, it seems plausible to believe that American marriages start out on a positive basis. The couple is attracted to each other, like each other, and find enough in common to marry. There will be problems, but most problems are solvable. As Weiner-Davis (1992, 14) suggests, "I don't believe in saving marriages. I believe in divorcing the old marriage and beginning a new one—with the same partner." She strongly believes that working to solve marital problems is far less traumatic than the problems surrounding divorce. She asks, "If getting rid of one's problematic spouse is the solution to marital problems, why would 60 percent of second marriages fail?" In fact, as you saw in reading this chapter, divorce entails myriad new problems—financial difficulties, disputes over child custody and visitation rights, loneliness, how to meet potential new mates, how to handle old friends and extended family members, and so forth.

Too often, mates dwell on what is wrong with the relationship, the "20 percent I dislike you," rather than on what is right, the "80 percent I love you" (see pages 203–205). When this occurs, a negative vicious circle commences: I expect the worst, I see the worst, I invite the worst; it becomes a self-fulfilling prophecy. Weiner-Davis suggests that by examining what is positive in the relationship, a troubled couple can slowly, step by step, start a positive self-fulfilling prophecy. Her point is that, since most marriages begin on a happy, positive note, it makes sense to build on the strengths that brought the couple together in the first place, rather than to concentrate on dissecting weaknesses when a couple comes for counseling.

Most couples go through a "honeymoon is over" period (see pages 201–202), a transitional period when the rose-colored glasses come off and expectations are slowly readjusted to reflect reality. This is not an easy process, but the end result—a workable marriage—is worth the effort. The couples who can't or won't make this transition are the couples whose relationship begins to worsen and who often travel the road to divorce.

Because troubled couples tend to become locked into circular action-reaction patterns, some kind of change—some new behavior—is needed to break the cycle. Weiner-Davis finds that even a small change in behavior on the part of one mate precipitates change in the relationship. Too often, both mates spend time diagnosing the problem rather than making behavioral changes. Of course, whether the diagnosis is right or wrong is immaterial to solving the couple's problems, unless concrete behavior changes occur in the relationship. Thus, she suggests starting with changes in behavior, even if the changes are unilateral.

Setting clear, but small and simple goals is extremely important. For example, a couple, after months of arguing, finally agree to be more affectionate. But they don't define what this means in behavioral terms. Several weeks later they come back to the counselor, accusing each other of not trying. Interestingly, each spouse believed that he/she had been trying, but that the other had not tried. How did this happen? Because "being affectionate" had never been clearly defined, they were unaware that they had very different expectations. The wife defined *affectionate* as spending time together as a couple, doing thoughtful things like calling each other to say hello during the day, and giving and getting hugs. The husband defined *affectionate* as sexual foreplay or making love. When he approached her sexually, he felt he was trying, but she recoiled because he hadn't yet shown affection, according to her definition. When she called him at work to let him know she was thinking about him, he seemed oblivious to her affection. They missed each other's attempts to improve their marriage because they could only spot signs of affection based on their own definitions, which they had failed to share with each other (Weiner-Davis 1992, 108). Table 15-3 illustrates how goals can be described in behavioral terms.

Another suggestion is that couples should focus on exceptions to their criticisms of their mates. Many couples coming for counseling make blanket black-and-white statements about each other. "He is never affectionate." "The only thing she ever does is criticize." Yet, in most cases, such statements simply are not true. He is affectionate at times. She does not criticize all the time. By asking couples to think of exceptions to their blanket criticisms, the counselor can start moving the couple toward a more positive focus. Weiner-Davis (1992, 124 – 125) suggests that, by recognizing exceptions,

- The problems shrink because the black-and-white thinking is acknowledged.

- Exceptions demonstrate that people are changeable. If a mate could behave in a desired way once, he/she can behave that way again. Individuals really can change.

TABLE 15-3

Describing Goals in Behavioral Terms

Vague Goals	Action Goals
Be respectful.	You will ask me about my day. I will compliment you about your work.
Be more loving.	You will tell me you love me at least once a week. I will volunteer to watch the kids so you can go out.
Be more sexual.	I will initiate sex once a week. I will suggest we try something different. You will be more verbal when we make love.
Be less selfish.	You will ask what I want to do on weekends. I will check with you before making plans. I will clean up if you make dinner.

Source: Weiner-Davis 1992, 111.

- Exceptions can supply solutions. The exception to the disliked behavior gives the mate an exact description not of what is wrong, but of what can be done to improve the relationship.

- Exceptions empower people. They realize for the first time in months (or years) that, despite the problems, they have been doing some things right. This realization often comes as a complete shock, because most people feel they've tried everything and nothing works. Now they know that something does work.

Those persons suggesting that divorce is not always the answer do not deny that marriages have problems. Of course, even the best relationships suffer at times. Rather, their point is that energy spent on improving a relationship and trying to save a marriage may, in the long run, be more fruitful to the couple than a divorce. As we have seen, a divorce does not end a person's problems. Instead, most divorces introduce additional problems and complicate a person's life. As one researcher puts it, "If Americans treated their marriages as tenderly as their family cars, there might be more commitment to repairing rather than junking marriages" (Whitehead 1997, 191).

Summary

1. Almost every society has some means of dissolving marriage. Divorce is America's mode of ending unsatisfactory marriages. *The divorce rate increased drastically during the past century, probably reflecting Americans' changing expectations for marriage and increasing tolerance of divorce.* Since 1985, however, the divorce rate has stabilized.

2. *There are many reasons for the high divorce rate in the United States.* Americans ask a great deal of marriage, and the more we ask of an institution, the more apt it is to fail. Changing gender roles, emphasis on the individual, heterogeneity of the population, mobility, and poverty all affect marriage and increase the chances for divorce.

3. *The decision to divorce is usually reached slowly and involves emotions similar to the feelings that one goes through after the death of a loved one.* Indeed, long before legal divorce and often long after it, a couple will go through the suffering of an emotional divorce. Divorce is complicated by the presence of children or considerable property.

4. *Divorced people must cope with numerous problems.* They must make changes in their economic situation, their living situation, and their relations with family, friends, children, and co-workers. Most divorced persons eventually remarry. For the remarriage to be successful, divorcing people must learn from their past mistakes. Thus, the divorce should act as a catalyst to further self-insight.

5. *The impact of divorce on children varies with the divorcing family and the individual child. For all children, the impact of divorce is at first negative.* As time passes, the effects of divorce on a given child are difficult to predict.

6. By 1984, *all states had instituted some form of no-fault divorce.* In these proceedings, neither party has to be proved guilty of misconduct, providing grounds (a legal reason) for divorce. The only thing that must be stated is that

the marriage has suffered an irremediable breakdown. No-fault divorce is not perfect, and is currently being questioned. Laws cannot control emotions or eliminate the pain, disruption, and economic hardship that occur when a family breaks up.

7. *Mediation can help couples to better work out their problems, whether the end is divorce or rebuilding of the relationship.* More people are starting to realize that energy spent on saving a marriage may be a far better investment than energy spent trying to rebuild a life after a divorce.

Resources on the Internet

Companion Website for *Human Intimacy: Marriage, the Family, and Its Meaning,* Tenth Edition

http://sociology.wadsworth.com/cox10e/

Gain an even better grasp on this chapter by going to the companion website to take one of the tutorial quizzes, use the flash cards to master key terms, or check out the many other study aids you'll find there. You will also find special features such as GSS data, Sociology Online, and Census 2000 information that will put data and resources at your fingertips to help you with that special project or to help you as you do some research on your own.

InfoTrac College Edition

http://www.infotrac-college.com/wadsworth/

You can access reliable resources anytime, anywhere, with InfoTrac College Edition, the online library. This fully searchable database offers more than 20 years' worth of full-text articles (not abstracts) from almost 5,000 diverse sources, such as top academic journals, newsletters, and up-to-the-minute periodicals, including *Time, Newsweek, Science, Forbes, The New York Times,* and *USA Today.* You can conduct electronic key word searches using key terms from this chapter to supplement your reading and learning experience. To aid in your search and to gain useful tips, see the Student Guide to InfoTrac College Edition, which you can access through the companion website for this book.

Should Couples Stay Together for the Sake of the Children?

All of the evidence discussed in this chapter indicates that children suffer from the divorce of their parents, at least at the time of the divorce. For some children, the negative aftereffects of divorce remain into adulthood and for a few, throughout their lives. Can these negative effects of divorce on children be avoided if parents remain together, postponing divorce until their children reach adulthood?

Parents Desirous of Divorce Should Remain Together until Their Children Reach Adulthood.

Historically, many parents wishing to divorce did remain together until their children were adults and living on their own. Not only does this parental sacrifice for the children prevent their children, when young, from going through the upset of divorce, but also allows the better parenting that two, rather than one, parent can provide. In addition, children can better intellectually understand their parent's differences and thereby better adjust both intellectually and emotionally to the separation of their parents.

When parents remain together for the sake of the children, so does the family and the social network of support for the family. Thus, the children remain within that network. Relatives living nearby, such as grandparents, remain in touch with their grandchildren. The children remain within their neighborhoods and schools, with long-term friends. Family friends remain, serving to further maintain a known and comfortable environment. School friends, adult family friends, and relatives remain available to mediate parental troubles and for the child/children to talk with about family problems.

For these reasons and many more, there is no question that parents wishing to divorce should remain together until their children are grown and on their own. By doing this, many of the negative legacies of divorce suffered by many children can be reduced or avoided entirely.

Parents Wishing to Divorce Should Not Remain Together for the Sake of Their Children Because This Can Cause More Harm Than Good.

Although it sounds wonderful and proper—"We'll stay together until the children are grown, then divorce"—the results for the children may not be ideal. The couple postponing divorce are obviously in trouble, conflicting at some level, and it is probably impossible to hide such troubles. One fact is clear: high conflict among parents is harmful to their children, whether they divorce or not. Children, especially young children, have great empathy for their parents and, although the parents may feel that they are hiding their difficulties, the chances are great that the children will sense their problems and react emotionally. In fact, parents' denial of their problems to the children probably creates greater emotional upset than if parents were honest. When parents are honest, at least the children can understand that their emotional reactions are based in the reality of their parents' conflicts.

Even in the unusual case, when the parents hide their problems so well that the children remain unaware of them, it is very possible that friends and/or relatives give away the parents' secret problems, knowingly or unknowingly. In this case, the parents become liars in the children's eyes or, at least, untrustworthy. It is always better to maintain open and honest communication with children, even if the short-term results may be somewhat negative.

(For some facts on the above debate, see Frank Furstenberg and Kathleen Kiernan, "Delayed Parental Divorce: How Much Do Children Benefit?" *Journal of Marriage and the Family* May 2001, for the results of a study examining the effects of divorce on children 7 to 16 years, 17 to 20 years, and 21 to 33 years of age.)

© Chuck Savage/Corbis

Remarriage: A Growing Way of American Life

Chapter

16

Questions to reflect upon as you read this chapter:

- What are some of the problems a 30-something-year-old divorced person has returning to single life?

- What are some of the problems a couple has in remarriage that they did not face in their first marriages?

- What are some of the problems children in stepfamilies face that they did not face when their biological parents were together?

- What does the stepfamily look like, and how does it operate?

- List a number of potential strengths that might be found in the successful stepfamily.

I
t seems to me that John and Helen were just divorced, and here's an invitation to Helen's wedding."

"Not only that, but I met John and his new girlfriend at lunch yesterday and from the way they were acting, I'll bet we'll soon get an invitation to their wedding."

"It's hard to understand. They were so eager to escape their marriage, and now it seems they can hardly wait to get back into another marriage."

Divorced people as a group are not against marriage. Most divorced people remarry, and they tend to marry other divorced persons (61 percent). Only 35 percent remarry single persons, and another 4 percent marry widowed individuals. Remarriage rates, like first-marriage rates, have dropped overall. However, if the increased rate of cohabitation is considered, finding a partner with whom to build an intimate relationship appears to be as popular among Americans as it has always been.

Stepfamilies are primarily families formed by the remarriage of divorced persons with children. However, one-fourth of all stepfamilies in the United States and fully one-half of all stepfamilies in Canada are formed by cohabitation, rather than remarriage (Cherlin 2004, 849). To demonstrate the complexity of today's family structure, consider the following:

It is not uncommon, especially among the low-income population, for a woman to have a child outside of marriage, end her relationship with that partner, and then begin cohabiting with a different partner. This new union is equivalent in structure to a stepfamily but does not involve marriage. Sometimes the couple later marries, and if neither has been married before, then the union becomes a first marriage with stepchildren. (Cherlin 2004, 849)

In general, men tend to remarry more frequently and more quickly than women do, and this disparity increases with age (Ihinger-Tallman and Pasley 1987; Pasley and Lofquist 1995). The probability of remarriage among divorced Caucasian women is 58 percent within 5 years, 44 percent for Hispanic women, and 32 percent for African American women (NCHS 2002). These percentages have dropped over the past few years, but women entering cohabitation relationships make up for the drop in remarriage. Divorced women who are older, highly educated, and financially independent are less likely to remarry, but again, cohabitation rates have increased for these women.

The median interval before remarriage for previously divorced women is 2.5 years; for previously divorced men, the median is 2.3 years. However, if cohabitation among previously married persons is considered, the interval between divorce and reentering an intimate relationship is shorter. The median age at remarriage after divorce for women is 34.2 years, for men it is 37.4 years (NCHS 2000a).

Because divorced people tend to remarry and the United States has a high divorce rate, it is clear that many (approximately 50 percent) of American marriages will be remarriages for one or both of the partners (Coleman et al. 2000; DeVita 1996, 30; Ganong and Coleman 1994). Although the high rate of remarriage is often presented as a major change in American family structure, remarriage that creates stepfamilies is not new historically. In the past, however, it was the death of one spouse and the ensuing remarriage of the remaining spouse, rather than divorce, that created the stepfamily (Phillips 1997, 7).

Considering the many kinds of stepfamilies created when cohabitation and remarriage are considered, there will be a large diversity of stepfamily types. Keeping the positive perspective of this book, this chapter will concentrate on the characteristics and problems that are similar among stepfamilies, rather than exaggerating the differences.

The prefix step- in stepfather, stepmother, and stepfamily originally meant "orphan." The stepchild of yesteryear had almost always lost a biological parent to death. Most of today's stepchildren have living biological parents, in addition to stepparents.

Age differences between remarried spouses tend to be greater than between first-married couples. For first-married couples, the age difference has consistently been 2 years. In a marriage between a never-married woman and a divorced man, however, the age difference averages 7 years. For remarriages of both spouses, the age difference is 4 years. The only category in which brides are generally older than their grooms (by approximately 1 year, on average) are marriages of previously married women to never-married men. May-December marriages—marriages between older men or women and much younger women or men—are more common among divorced people than among people entering their first marriage. Although we have little empirical data on the success of such marriages, many such couples express high levels of marital satisfaction. The younger partners appreciate the security and stability that the older partners can frequently offer. The older partners express pleasure with the generally high energy level and flexibility of their younger partners.

From the cohabitation and remarriage statistics, it is apparent that divorce serves not so much as an escape hatch from married life, but as a recycling mechanism that gives individuals a chance to improve their intimate relationship (Ganong and Coleman 1994). Considering the rise in divorce rates, the lengthening life span, and the younger ages at which Americans divorce, the incidence of remarriage or cohabitation after divorce appears likely to remain high in the future. What Americans experience is not monogamy, but **serial monogamy**—that is, several spouses or intimate partners over a lifetime, but only one at a time.

serial monogamy

Having several spouses over a lifetime but only one at a time

Although we will concentrate on the remarriage of divorced persons, the widowed also remarry. They stay single longer than the divorced, though, and their remarriage rates are lower because choice is more limited for older people.

Most research on stepparent families has compared such families to intact first-marriage families. Such research has discovered that the problems found in stepfamilies tend to be greater, as common sense would suggest. However, because the stepparent family is becoming increasingly more numerous in the United States, perhaps it is better to try and find ways to improve its functioning, rather than concentrating only on its shortcomings (Gamach 1997).

Returning to Single Life

Many people who have been married for some time, burdened with the responsibilities of a growing family and missing the flush of romance that brought them together with their mate, feel pangs of envy when their friends divorce and reenter the singles' world. Remembering their dating and courting days, they relive nostalgic memories of the excitement of the new date and the boundless energies they expended when they were young, pursuing and being pursued. However, the realities of the singles' world for the divorced are rarely the dreams these married friends imagine.

The return to the singles' world can be frightening for both men and women, especially for those married for some years. "Can I be successful as a single person?" is a question that cannot be answered at first. Most people have experienced a severe blow to their self-esteem with the divorce, and are reluctant to face the potential rejections involved with meeting new people. Those who have been left against their will are more likely to find single life intimidating than those who have voluntarily left their marriage. Those who divorce because they have found someone else while still married may avoid a return to single life altogether.

The point at which divorced persons are ready to return to the single social life varies greatly. Most take about a year after the divorce to get themselves emotionally back together. Those who had an ongoing extramarital relationship are in a different position, and often remarry as soon as legally possible.

Relearning to date and relate to the opposite sex as a single person is especially difficult for anyone who has been married for a long time, because his/her self-image has for so long been that of one member of a couple. Once the divorced (or widowed) persons reenter the social world, they are often surprised at how many people share their newly single status. Except for the very young divorced, most reenter a world of single but formerly married people, rather than one of never-married people. This eases the transition inasmuch as the people they meet have also experienced marital collapse. Divorced persons respond to each other with a certain empathy, which helps the newly divorced person feel more acceptable. In fact, the discovery by the newly divorced that they can meet and interact with people of the opposite sex in their new single role can be exciting and heartening. "Maybe I'm not such a failure after all" is a common response.

A newly divorced person, especially one who was rejected by his/her spouse and did not want a divorce, may initially engage in sexual experimentation. To be desired sexually is a boost to shattered self-esteem. To be close to another person physically—to be held, touched, and sexually pleasured—makes the rejected person feel loved and cared for and also verifies his/her sexual desirability. For these reasons, newly divorced persons are vulnerable to sexual exploitation and may make the error of equating sex with love.

Dating has already been discussed in Chapter 6, but it is important to realize that the dating practices of divorced or widowed persons tend to be different from those of young unmarried individuals. For example, the divorced and widowed have more problems meeting people than young, never-married people. Sometimes friends, relatives, and business associates introduce new acquaintances.

Organizations such as Parents Without Partners and We Care become meeting places. Often, the lofty educational and supportive goals of these groups are secondary to their social functions. The stated goals may make it easier for people to join activities without appearing to be "mate hunting."

Many singles clubs make introductions their major purpose. Such clubs sometimes frighten and intimidate those just recuperating from marital failure. Often, people are uncomfortable attending singles parties that have a "meat market" atmosphere.

Although long used in Europe, the personal newspaper advertisement is relatively new in the United States. Discounting the ads for sexual partners in underground newspapers, legitimate classified advertising is being used, especially by divorced persons, to seek desirable companions and potential mates.

Technology has found a place in the single life of those previously married, in the form of computer and video dating. For a fee, a computer-dating service will search its large bank of personal information on its clients to find an appropriate date. The new client is asked about background, interests, feelings about sexual relations, and the qualities sought in prospective dates. The computer then makes a match based on stated interests, hobbies, habits, education, age, and so on. Computer dating is based on the theoretical notion of homogamy— that those of similar interests, backgrounds, and the like will be attracted to each other.

The client then has the opportunity to look at videotapes of people who share his/her interests. Videotapes have the advantage of giving an overall impression of the person before any meeting. The chance to prescreen dates is the most attractive feature of video dating. Such services can be expensive, and there are few statistics available on success rates of such dating methods.

With the increasing availability of home computers, potential dating partners can contact each other via the Internet. For example, America Online offers chat

cathy®   **by Cathy Guisewite**

© 1991 Cathy Guisewite. Reprinted with permission of Universal Press Syndicate. All rights reserved.

rooms, such as "Romantic Connection" and "Meeting Place." Caution must be used in arranging meetings with people contacted via such channels.

There are many ways to meet new people. The most important step the newly divorced can take is to participate actively in the social world. Unless an individual is out in society, meeting new people is difficult. The newly divorced person in a small town probably stands the poorest chance of meeting someone new.

When divorced and widowed persons date, children are often part of the dating equation. The romantic rendezvous for two frequently includes three or four. With the high divorce rate and increased numbers of single-parent families, many singles now find that dating someone new means getting acquainted with the date's children. "I recently started dating Jean. When I went to pick her up for our first date, I rang the bell and put on my best smile. But it wasn't Jean who opened the door; it was a 10-year-old kid. He seemed to be checking me out the way dates' fathers used to when I was a teenager."

The presence of children after divorce complicates dating and courting. The custodial parent, in particular, must consider the effect dating will have on the children. Dating and serious courtship will drastically affect the children's hopes and dreams of reconciliation between their divorced parents. Children may even directly resist dating on the part of their divorced parents. Thus, a divorced parent has many concerns that do not trouble a single individual: What will the children think of the new date? Would the new date make a good stepparent? How do I handle sexual relations while I am dating? What will the children think if a date stays overnight? What will the children tell my "ex" about my dates? Despite the additional problems children create for the dating divorced mother/father, most research indicates that children do not reduce the chances for remarriage.

Dating and courtship also affect the ex-spouse. Some who hold onto thoughts of reconciliation will regard dating by their former partner as threatening. Others who themselves are dating may be relieved that the ex-spouse is starting to date. Dating by the former spouse can relieve guilt feelings and give the other ex-spouse more autonomy. It can also remove pressure on the former spouse, because the dating spouse now channels some energy toward a new person. In addition, dating can serve as a final break to the old relationship.

From the legal standpoint, dating by a former spouse may give the other an excuse to change the divorce settlement. For example, a noncustodial parent could seek a change of custody or a change in financial support, claiming that the dating custodial parent is shirking parental duties or using money for the dating partner instead of the children.

As these examples illustrate, dating and courtship after divorce differ considerably from dating and courtship before a first marriage. In summary, the dating divorced person will usually differ from the single dating person in several ways:

1. *Age*. The divorced dating person will be older, which offers the advantage of more experience and (it is hoped) wisdom. Yet, this may be offset by a restricted choice of partners, especially for women.
2. *Children*. Although single persons may have a child, the chances are greater that a divorced person has a child or children.
3. *Marital experience*. The divorced person has been through a marriage and has experienced establishing a household and the subsequent breakup of the relationship and household. As we shall see, this experience may make them more knowledgeable about marriage, and thus able to do a better job than

WHAT DO YOU THINK?

1. What do you think the major dating problems would be for the newly divorced man? The newly divorced woman?

2. If you were newly divorced, how would you go about meeting potential dates?

3. Would you use a dating service? Why, or why not?

4. Would you place a personal ad in a newspaper seeking friends? Why, or why not?

during the first marriage. On the other hand, they may more easily divorce if things go wrong, because divorce is no longer an unknown.

Despite the complications of dating after divorce, the high remarriage rates indicate that most people who divorce do meet prospective mates.

Cohabitation as a Courtship Step to Remarriage

As noted earlier in this chapter, cohabitation has played an increasingly important role in the lives of divorced persons. Research on cohabitation among never-married young persons found that its relationship to later marital success was negative, or at best, neutral (see Chapter 6). Contrary to this finding, divorced persons who cohabit seem to increase their chances of success in a remarriage. Lawrence Ganong and Marilyn Coleman (1984, 1994) found that the primary way that people prepare for remarriage is by living together. Fully 59 percent of their sample cohabited before remarriage, and that percentage is probably higher today. Cohabitation had more positive effects on a subsequent remarriage for men than for women. Men reported less conflict and more affection for their wives if they had previously cohabited than did previously noncohabiting men; women reported fewer disagreements although women were less positive overall. The positive effects of cohabitation appear to be limited to the marital relationship. Stepparent-child relationships, parent-child relationships, and extended family effect do not change. Perhaps cohabitation helps those planning to remarry because they have so many additional problems to sort out, especially if children are present. Having failed once at marriage, the person or couple is simply more cautious about entering a new union. By living together before remarriage, they can resolve many of the problems. If the problems are not resolved, the couple can leave the relationship, and thereby avoid another unsuccessful marriage. However, cohabiting and breaking up will add increased stress to any children present.

Remarriage: Will I Make the Same Mistakes Again?

Nothing restores adult self-esteem and happiness after divorce as quickly and as thoroughly as a love affair or a successful second marriage. No matter how badly the men and women in our study were burned by their first marriages, not one turned his/her back on the possibility of a new relationship. (Wallerstein and Blakeslee 1989, 225–226)

"Hope and disillusionment is perhaps the most common story of Stepfamily life" (Bray and Kelly 1999). As noted earlier, high remarriage rates among divorced people indicate that the divorced are still hopeful about marriage and the role of being married. In fact, most activities in the American culture, for better or worse, revolve around the married pair, the couple. High remarriage rates suggest that it is important to have someone with whom to share, to be intimate, to feel closeness, and to experience a part of something larger than oneself.

To love and be loved are important to most Americans. As unhappy as a marriage may have been, most people can recall a time when they experienced love and closeness. Indeed, loss of this intimacy may have been a major reason for leaving the marriage. Finding intimacy is certainly a factor in most remarriages, just as it was in the first marriage.

The route to marriage for young, unmarried Americans is fairly clear. You date, you fall in love, you become engaged, you marry. In remarriages, though, the simplicity of ignorance has been replaced by the knowledge (and for some, the disillusionment) of past experience.

The divorced react to the idea of remarriage in many different ways. Some remarry quickly because they already have another relationship in place at the time of the divorce. Others remarry quickly on the rebound, out of loneliness, out of insecurity, or simply because they know no other way of life but to be married. Such persons often married young, going straight from their parents' homes into marriage, and thus have had no practice at being a single adult. Often, they remarry the *transition person*. This is an individual who, out of friendship, love, and sympathy, helps another person through a difficult period such as divorce. They may temporarily take the place of the missing spouse.

A minority of divorced persons (approximately 20 to 25 percent of divorced men and 25 to 30 percent of divorced women) do not remarry. These percentages appear to be increasing, but because a remarriage can take place at any age, it is too early to tell if this is really a trend toward increasing lifetime singleness, or simply a postponement of remarriage just as first marriage is now being postponed. Or, in both cases, uncounted cohabitation is taking up the slack, and intimate relationships are as sought after as ever.

The reasons for not remarrying are many. Some people may simply enjoy the autonomy and independence of single life. Others may want to remarry, but fail to find an acceptable mate. Still others may have been so hurt in their first marriage that they avoid relationships that might lead to remarriage. A few who choose not to remarry may be psychologically unable to give up their lost spouse. This is especially true of the widowed, who sometimes feel disloyal to the deceased spouse if they form a new relationship. In some cases, the children of a widowed spouse discourage remarriage of their remaining parent, for fear that the stepparent may take what they feel is rightfully theirs.

The most common reaction, however, tends to be a careful, cautious relationship-testing period, leading to remarriage. For persons approaching remarriage slowly, the risk of a second mistake is their main concern. Most divorced persons believe that they were deluded in their first marriage and, therefore, approach a second marriage with extra care, no longer naive about the difficulty of achieving a successful relationship. They realize that they must work out the problems of their first marriage and establish a new, independent, and strong self-image. In this way, they hope that their new relationship will be one of equality and maturity, in contrast to their first immature relationship. They also realize that without such care and work, remarriage will be based on hope, rather than experience, and very apt to fail.

The partners in a remarriage must deal with all the problems any newly married pair faces. In addition, they must deal with attitudes and sensitivities within themselves that were fostered by their first marriage. They may enter remarriage with many prejudices for and against the marital relationship. They need to divest themselves of these attitudes if they are to face the new partner freely and build a new relationship that is appropriate to both. In a remarriage, the mate is new and must be responded to as the individual she/he is, not in the light of the past spouse. An additional task in every remarriage, then, is the effort partners must make to free themselves from inappropriate attitudes and behaviors stemming from the first marriage. In essence, second marriages are built on top of first marriages. Prior spouses remain to haunt remarriages. Indeed, prior spouses were termed *ghosts at the table* in the National Institute of Child Health and

HIGHLIGHT

Bob, Carol, Ted, and Alice

Bob had been married for 12 years to Alice, had two children, and was established economically when the marriage ended in divorce. Two years later, he married Carol, 8 years his junior, who had one child by her previous husband, Ted.

Bob and Carol both approached their marriage carefully, giving much thought to their relationship. Both agree that their new marriage is a big improvement over their past marriages. They find that their biggest problem is making sure they react to each other as individuals, rather than on the basis of their past relationships. This is not always easy.

Bob's past wife, Alice, is emotionally volatile, which both attracted and repulsed him. He liked Alice's displays of happiness and enthusiasm, but hated her fits of temper and general unhappiness.

Carol is placid and even-tempered. In fact, these personality traits were part of what drew Bob to her. When they do things together, however, he keeps asking her if she is having fun, is she enjoying herself? He asks so often that Carol becomes upset at what she regards as nagging. One day she blew up at him over this. Bob reacted strongly to her negative emotional display. Once everything was calm again, they both discovered that the problem grew out of his past marriage. Bob simply expected Carol to show her enjoyment in the same way Alice had. He was not relating to Carol as an individual, but was reacting in light of his past experiences with Alice. When Carol blew up at him, his reaction was much stronger than necessary. Her emotional blast activated Bob's past dislike of Alice's temper tantrums.

Human Development's supported longitudinal, multiyear study comparing 100 remarried families with 100 nuclear, first-married families (Bray and Kelly 1999).

As in the story of Bob and Carol, the ghost-like prior mates often dictate behaviors to the newly remarried pair, if not directly through the courts and divorce settlements, then indirectly via years of previous interaction.

A remarriage between divorced persons is more difficult than a first marriage for a number of reasons:

1. Each mate may have problems of low *self-esteem* stemming from the divorce.
2. The divorced are *less apt to tolerate a poor second marriage*. They have been through divorce and know that they have survived. Life after divorce is not an unknown any longer and is, therefore, less threatening than before. Divorced persons tend to end an unhappy remarriage more quickly than they ended their first marriage.
3. *The past relationship is never really over*. Even if a couple overcomes the kind of dynamics illustrated by Bob, Carol, Ted, and Alice, the past marriage can still directly affect the new marriage. For example, payments to a former spouse may be resented by the new spouse, especially if the current marriage seems shortchanged monetarily. First marriages also indirectly affect remarriage. Remarried couples go to considerable lengths to differentiate marital styles between their first and second relationships. This very effort to change relationship styles suggests that the second relationship is influenced by reactions to their first marriage.
4. A *remarriage that involves children will experience a great many more complications*, as we shall see in the next section. Family law is also inadequate in dealing with the blended family (Ganong and Coleman 1997, 90). For example, there are no provisions for balancing husbands' financial obligations to spouses and children from current and previous marriages. What are the support rights of stepchildren in stepfamilies or in stepfamilies that end in divorce? The courts have traditionally held that first families take priority over second families (Fine 1989, 1997).

5. *The society around the remarrying person tends to expect another failure.* "He/she couldn't make it the first time, so he'll/she'll probably fail this time, too," onlookers predict. "After all, most divorced people don't learn; they usually remarry the same kind of person as their earlier spouse." Or, "Once a failure always a failure." This lack of support can create a climate of distrust in the minds of the remarried couple themselves, and can lead to a self-fulfilling prophecy.

James Bray and John Kelly (1999) found in their long-term study that the major problems facing the stepfamily boiled down to the following four:

1. Because 90 percent of stepfamilies are stepfather families, integrating the stepfather into the child's (children's) life (lives) is crucial.
2. The creation of a satisfying second marriage and separating it from the first marriage.
3. The successful management of change.
4. Dealing with nonresident parents and former spouses.

Remarried families are often strapped financially because so much money must go to support the prior family. In some cases, the former wife uses the new wife's earnings as grounds for requesting an increase in support payments.

Jill and Michael have been married for 6 years and have two children. Michael was married previously and has two children by his former wife. Although Jill works full time, support payments to Michael's first family make it nearly impossible for them to survive financially. As Michael's first children have grown, his former wife has returned to court twice, seeking higher child-support payments because older children cost more. The court did not order an increase the first time because Michael's income had not changed. Jill was expecting their first child and was not working at the time. However, the second time Michael's ex asked for an increase, it was granted. Jill was working full time, and the court considered Michael and Jill's combined incomes—her income becoming a mere extension of her husband's. Even though the higher support payments drastically cut into Michael's ability to provide for his second family, the court considered the needs of only his first family.

Despite the increased problems faced in remarriage, those remarrying seem to take few active steps to increase the chances of success, other than moving slowly. Only about 25 percent of men and 38 percent of women receive counseling. Very few divorced persons attend a support group of any kind. Some do, however, report reading self-help materials and books.

It is important for people considering remarriage to understand how remarriages differ from first marriages. The most troublesome unrealistic expectation of those remarrying is that stepfamilies will function just like the nuclear intact family they left. Rigidly holding to unrealistic expectations of what the stepfamily should be locks the newly remarried couple into an inflexible position that will intensify conflict. This is because they are not open to change and examination of the real differences that exist between first and second marriages (Bray and Kelly 1999; Ganong and Coleman 1997, 99).

What are the statistics on success and failure of second marriages? Unfortunately, the statistics do not present a clear picture. Most studies comparing the divorce rates of first marriages with those of second marriages report that a remarriage is more likely to break up than a first marriage (Bray and Kelly 1999; NCHS 2002). The differences are small; most studies indicate about 60 percent

WHAT DO YOU THINK?

1. Have you ever reacted to a new boy-/girlfriend incorrectly because of a past relationship? Explain.
2. Have you ever been reacted to by a boy-/girlfriend inappropriately because of a past relationship in that person's life?
3. What are some suggestions you can make to help avoid such problems in future relationships?

of remarriages will end in divorce. These studies, however, do not take into account the small group of divorce-prone people who marry and divorce often; these repeated divorces tend to overinfluence the remarriage-divorce figures. It appears that remarriage divorce rates decline with increasing age at time of remarriage. However, remarriages that end with divorce end more quickly than first marriages. Most divorces of remarried couples occur within the first 2 years of the marriage (Bray and Kelly 1999).

Perhaps more important than comparisons of divorce rates are the subjective evaluations made by those remarrying. Marital satisfaction studies also show mixed results, with the differences again being small. People in first marriages indicate slightly greater marital satisfaction than those in remarriages. Men tend to be more satisfied than women with their remarriages. Stepfathers and stepmothers indicate about the same level of remarriage satisfaction (Ganong and Coleman 1994). Most remarried couples report a moderate to high degree of marital happiness. Remarriages of divorced persons that do not end quickly in divorce are, on the whole, almost as successful as intact first marriages (Bray and Kelly 1999).

Remarriages seem to be judged by different criteria than first marriages, perhaps because they are based on different factors. None of the variables that have often served as good predictors of marital satisfaction in first marriages (presence of children, age at marriage, social class, similarity of religion, and so on) seem to be strongly related to remarriage satisfaction. Perhaps the romantic illusion is gone for those remarrying successfully. Overall, at least for some people, divorce and remarriage seem to be effective mechanisms for replacing poor marriages with good ones and for keeping the level of marital happiness fairly high.

Couples in successful remarriages often state that their remarriage is different from their first marriage. Most important, they feel they have married the right person, that is, "someone who allows me to be myself." In other words, they have chosen a better mate for themselves. Because of this better choice, they feel that their remarriage is better than their first marriage. They feel this is true because they have learned to communicate differently and now handle conflicts more maturely. Better communication also leads to better decision making. Because both partners feel they are more equal, the division of labor in remarriage tends to be more equitable. There is a strong need to justify a second marriage as a happy second marriage, because this wipes away some of the guilt and self-blame over the failure of the first marriage. This need to justify may color the partners' positive descriptions of their second marriages and make true assessment difficult.

Those who feel only anger and hostility forever toward their former mate forget that they freely chose their mate and were at one time in love with that person. All the qualities that drew them to the person in the first place probably still exist. Thus, to have only negative feelings toward a former mate is somewhat unrealistic, except in the cases of abuse or violence.

Even though the statistics are mixed on the success of remarriages, a great many are clearly successful, despite the extra problems involved. As divorce has become more prevalent and acceptable, the problems facing those wishing to remarry have diminished. Perhaps social support for remarriage will continue to grow in the future as greater numbers of people marry more than once during their lifetimes.

Family Law and Stepfamilies

Like all law, family law is a balancing of individual liberty and the protection of the common good. Most everyone wants to promote healthy families and the best interest of children. Yet, America's diversity and ever-changing reproduction technologies make keeping family law up to date and fairly balanced between the individual and the common good an extremely difficult task. (See Henderson and Monroe and the special edition of *Family Relations* devoted to family law, 2002.)

One example of laws changing to meet changing family circumstances was the enactment of grandparent visitation statutes. As family breakup became more widespread, grandparents often lost contact with their grandchildren. Yet, grandparents can be a source of both love and support to grandchildren suffering from the demise of their family. Thus, changes in family law attempted to speak to this fact by creating and supporting grandparent visitation rights (T. Hill 2000).

Mark Fine (1989, 1997) reviewed family law relating to remarriage and stepfamilies. His overall finding was that family law does not currently provide clear and comprehensive rules to define the responsibility of parties to the stepparent-stepchild relationship, but legislatures and courts are becoming more sensitive to this relationship. Presently, the only way that stepparents can assure themselves of the same rights as biological parents is through adoption.

Child-Support Obligations

Only a minority of states have statutes that obligate stepparents to support stepchildren, and these are usually limited in scope, applying only when the stepchildren are living with the stepparents (Mahoney 1997). Although the legal support for such statutes varies, essentially they are based on the common law doctrine of *in loco parentis*. Under this doctrine, a person who intentionally assumes parental obligations (actively participates in child rearing, school, social, and recreational functions, and so on) can be treated as the parent. Unlike biological parents, however, stepparents may terminate this relationship and its responsibilities at any time. While support obligations for biological parents remain after divorce, courts and legislatures have generally not extended such support after the end of a remarriage. Nothing in present family law speaks to the problem of supporting stepchildren when the stepparent has the higher obligation of supporting biological children. The law assumes that the biological parents of stepchildren will supply the support. Statistics about child-support payments refute this assumption. Also, nothing in the law protects children conceived in the remarriage from partial loss of economic support that goes to previous biological children.

© Mark Richards/PhotoEdit

His and ours.

Custody and Visitation of Stepchildren

Because stepchildren are not considered children of the marriage, stepparents usually do not have any custody or visitation rights upon divorce. In a few cases where the biological parent in the remarriage has died, custody has remained with the stepparent, in the best interests of the children. In these cases, the noncustodial biological parent has normally had little interaction with the children for an extended period of time.

States are becoming more liberal in granting "fit" nonbiological parents custody. About half the states have passed legislation that allows third parties to file for custody of minor children, whether or not they are already in their care. In specific cases, a broad range of factors has been considered in custody determinations, including character and resources of parents and third parties, the nature of the relationship between adults and the involved children, the advantages and disadvantages that may accrue to the child from various custody options, and the length of time the adult has lived with the child (Fine and Fine 1992, 336).

Although a stepparent has no innate visitation rights, those stepparents who argue that they acted *in loco parentis* are increasingly being granted visitation rights (Hans 2002). All 50 states now grant some third-party visitation, although the rights of the child's grandparents have been recognized to a greater extent than have those of stepparents. Because most family laws are state laws, a person needs to check with his/her particular state to see what legal rights stepparents and children may have.

His, Hers, and Ours: The Stepfamily

As we saw in Chapter 15, having children no longer seems to impede divorce. For couples divorcing from a first marriage, most have children under 18 years of age. Remarried couples that divorce tend to have fewer young children, because they may not have custody of any children and they are older, so their children will also be older and may already have left home.

Many people believe that the divorced person who has custody of the children stands much less chance of remarriage. In actuality, if age is held constant, having children does not seem to have a significant influence on one's chances of remarriage. In fact, remarriages often involve at least three different sets of children. Each spouse may bring their own children into the new marriage, and they may also have children together. Hence "his, hers, and ours" is often a correct description of the children in a **blended family**. Although this is the common term used for stepfamilies, it is misleading. Blended implies smoothness and seamless, but this is not really true of the stepfamily. It is really more like patchwork quilt (Engel 2000), with porous boundaries as various people enter and leave. As mentioned, because mothers receive custody in most divorces, most stepfamily households are stepfather households.

Literature is replete with many myths about the poor treatment accorded stepchildren. The ogre stepparent, especially the stepmother, is a popular myth in fairy tales like "Cinderella." In reality, little evidence supports this stereotype. Also the myth of instant love suggests that stepparents will automatically love their new stepchildren, and vice versa. Both myths cause problems in stepfamilies (Bray and Kelly 1999; Ganong and Coleman 1997, 99).

blended family
Husband and wife, at least one of whom has been married before, and one or more children from previous marriages

Stepparent-stepchild relations start in midstream; that is, the stepchild (unless an infant) already has related to his/her biological parents. A stepparent relationship begins from scratch and must be built. The new stepparent in the child's (children's) life must build the relationship with the stepchildren slowly. It will take time and patience, and a stepparent will never be totally successful in replacing the biological parent (Wallerstein and Blakeslee 2003, 311–322).

Certainly, the transition for children to a new stepparent is not always easy. One might get the impression from the mass media that more children are now being placed in the custody of their father after a divorce. In absolute numbers, this is true. But the number of children living with a divorced mother has increased at the same rate. About 90 percent of children remain with their biological mother. Thus, most custodial stepparents are stepfathers. Because ties are usually closest to the mother, this is probably an advantage for most children. In fact, most studies suggest that stepfathers seem to have an easier time with stepchildren than stepmothers do (Ganong and Coleman 1994, 83; 1997, 114).

Stepparents often appear to be intimate outsiders, at least early in the new marriage. "Stepparent" evokes the image of stepparents as members of the family, but also as intruders who are not privy to the secrets and knowledge shared by family insiders. The status of being both an insider and outsider can be unsettling and uncomfortable (Bray and Kelly 1999, 122; Ganong and Coleman 1997). This is even truer for cohabiting stepfamilies (Coleman et al. 2000).

It should be noted that the single parent caught up in a new romance is excited, happy, and looking forward to the new family to be created by remarriage. The children, on the other hand, are anxious and, perhaps, afraid of where this new relationship will end and what it will mean in their lives. The longer a family is a single-parent family, the more difficult the children's adjustment to the stepparent. First, there was the adjustment to the breakup of their biological parents, and then the children face the breakup of the single-parent family to form a stepfamily. Also, the position of a child moving into a stepfamily may change. For example, instead of being the oldest child, the child may become a younger child if the new stepparent brings older children into the new, blended family.

Compared with adolescents, young children appear better able to become attached to and benefit from the presence of a stepparent. Remarriage involving adolescent children tends to have more sustained problems. Generally, adolescent girls have more problems than adolescent boys (A. Hines 1997, 381). Also, successful stepfamilies that started the relationship with small children will find that, as the children grow into adolescence, problems will reappear and are often more difficult than the normal adolescent problems in an intact family (Bray and Kelly 1999, 242; Wallerstein and Blakeslee 2003). Although children moving from a biological family through a divorce and into a stepfamily certainly must make many more adjustments than children in intact families, not all research indicates lasting problems. A stepparent can make a positive relationship with his/her stepchildren and may be able to reverse some problems caused by the divorce of the children's biological family (A. Hines 1996; 1997, 383).

Research findings about the effects of remarriage on children are mixed. For example, though there is conflicting evidence, some research indicates that children in stepfamilies don't appear to differ significantly in self-image or personality characteristics from children in their original families. On the other hand, Wallerstein and her colleagues (2000) report serious problems in some children of divorce 10, 15, and even 20 years after the divorce.

Counselors and therapists also report that stepchildren and stepparents have a great deal of trouble in their relationships. Stepparents often feel confusion about their roles; children feel loyalty conflicts; and coparenting of children with former spouses may split parental authority (Ganong and Coleman 1994; Wallerstein and Blakeslee 2003; Wallerstein et al. 2000). Perhaps the data on stepchildren are mixed because each situation is unique and one cannot generalize about the effects of divorce or a stepparent on children. It may be that a well-adjusted child with a healthy personality will cope successfully with the family breakup and become a more mature, independent child. If the child is unstable, however, divorce may cause even greater maladjustment.

Stepparents face problems beyond those of natural parents. To begin with, they must follow a preceding parent. If the child and the natural parent had a positive relationship, the child is apt to feel resentful and hostile toward the stepparent. The child may also feel disloyal to the departed parent if a good relationship is established with the stepparent. This sense of disloyalty may provoke negative behavior toward the stepparent, in an effort to counteract growing feelings of affection.

Giving the stepchild permission to like the stepparent is important. "It is all right to like your new stepparent. It doesn't mean that you don't like Mom (or Dad) anymore. A person can like many people at the same time." Children need to be reassured that having a warm relationship with a stepparent will not endanger the relationship with the biological parent.

Often, a child feels rejected and unloved by the parent who leaves the household. Therefore, the child may cling more tightly to the remaining parent as a source of security and continuity. The remaining parent's subsequent remarriage can be threatening to such a child. "This stepparent is going to take my last parent away from me," the child fears. In this case, the stepparent may be met with anger and hostility.

Children can actually break up a remarriage (the divorce rate is higher in remarriages with stepchildren). This is especially true of adolescents, who can assume a great deal of power in a blended family (Bray and Kelly 1999; Chapman

HIGHLIGHT

How to Ruin a Remarriage

Joyce had two teenage children when she remarried Travis. Her 14-year-old daughter was dead set against the marriage. She claimed to hate Travis and showed it in every way she could. She would not eat at the same table with him. She swore at him and called him names, both privately and publicly. Finally, in frustration, Travis made her come to the dinner table one evening and sit with the family. Without telling anyone, she called the police and reported that her stepfather was molesting and beating her. The police appeared at the door shortly after dinner, took Travis down to headquarters, and charged him with child abuse. Although the charges were completely untrue and quickly dropped, the episode caused such trauma to the family that Travis decided he could take no more and left the home. After a long period of counseling, the daughter came to terms with her mother's divorce and remarriage. But the damage done to the couple's relationship was never to be repaired.

What Do You Think?

1. Do you know any stepfamilies in which a child so dislikes the stepparent that they deliberately cause trouble in the marriage? Explain, without identifying them.
2. Can you think of any suggestions for Joyce and Travis that might help them cope with the resentful daughter?

1991; Ganong and Coleman 1994; Wallerstein and Blakeslee 2003). Children can create divisiveness between spouses and siblings by acting in ways that accentuate differences between them. Children have the power to set parent against stepparent, siblings against parents, and stepsiblings against siblings.

When the child's relationship with the departed parent was not good, hostility remaining from this prior relationship can be displaced onto the stepparent. The stepparent simply represents the natural parent in the eyes of the child. On the other hand, a supportive new stepparent may be able to have a good relationship with this child, once the child realizes that he/she is not the same as the departed parent with whom the child was conflicting.

Because children constantly make comparisons between biological parents and stepparents, many stepparents make the mistake of trying too hard, especially at first. It is important for the child to have time to figure out just what the remaining parent's feelings are toward the new mate. Making this adjustment is even more difficult when the stepparent tries to replace the natural parent, especially if the child is still seeing that parent. Probably, the best course for the stepparent is to assume a supplemental role, meeting the needs of the child not met by the noncustodial parent. In this way, the stepparent avoids direct competition with the natural parent (see "The Ten Commandments of Stepparenting," in this chapter).

Approximately half of the women who remarry give birth to another child in their second marriage, most within 24 months of the remarriage (Coleman et al. 2000, 1289; Wineberg 1990, 31). When a remarried family has children of its own, additional problems are apt to arise with stepchildren. The stepchild may feel even more displaced and alienated. The remaining parent may seem to have been taken away, first by the new stepparent and then by the new child. However, having brothers or sisters takes the focus off the stepchild, allowing a more natural adjustment for both parent and child. A new child can be a source of integration in a stepfamily, because everyone finally has someone to whom all are related (Bernstein 1997; Ganong and Coleman 1994, 104–105).

The role of stepparent is difficult. Yet, a successful stepparent can be an additional source of love, support, and friendship and, by making the family a two-parent family again, can solve some of the child-rearing problems of the single parent. If a stepparent enters a child's life when the child is young, the child often comes to look on the stepparent as his/her real parent, which alleviates the child's feelings of loss. The stepparent can bring new ideas into a family that help family members to grow and expand their horizons. Russ was an avid outdoorsman and sports lover. His new wife, Jean, has two adolescent sons from a former marriage who are both very bright and excellent students. Their natural father is an intellectual who has no interest in sports and outdoor activities. He has little relationship with the boys. Through Russ's influence, both boys grew to love the outdoors. Both became active in school sports and earned varsity letters in track and field. After several years, the boys were adopted by Russ.

Although the extent of stepparent adoption of their stepchildren is unknown, it is estimated that 100,000 stepchildren are adopted each year. Generally, the stepfamilies that opt for adoption include those stepparents who develop close relationships with their stepchildren and wish to model themselves after the nuclear family. Adoption might also occur when the nonresident biological parent has little or no relationship with his/her children.

WHAT DO YOU THINK?

1. Do you know any stepfamilies in which the stepparent has made a successful relationship with the stepchild (stepchildren)? Describe?

2. Why do you think this stepfamily is successful?

The Ten Commandments of Stepparenting

The natural family presents hazards enough to peaceful coexistence. Add one or two stepparents, and perhaps a set of ready-made brothers and sisters, and a return to the law of the jungle is virtually ensured. The following are survival guidelines for stepparents:

1. Provide neutral territory. Stepchildren have a strong sense of ownership. The questions "Whose house is it? Whose spirit presides here?" are central issues. Even the very young child recognizes that the prior occupation of a territory confers a certain power. When two sets of children are brought together, one regards itself as the main family and the other as a subfamily; the determining factor is whose house gets to be the family home. Some suggest that when a couple remarries, they should move to a new house. If this is impossible, it is important to provide a special, inviolate place that belongs to each individual child.

2. Don't try to fit a preconceived role. When dealing with children, the best course is to be straight right from the start. Each parent is an individual with all his/her faults, peculiarities, and emotions, and the children are just going to have to get used to this parent. Certainly, a stepparent should make every effort to be kind, intelligent, and a good sport, but that does not mean being saccharine sweet. Children have excellent radar for detecting phoniness, and are quick to lose respect for any adult who lets them walk all over him/her.

3. Set limits and enforce them. Disciplinary measures are one of the most difficult areas for a natural parent and step-

parent living together to agree upon. The natural parent has a tendency to feel that the stepparent is being unreasonable in demanding that the children behave in a certain way. If the parents argue between themselves about discipline, the children will quickly force a wedge between them. It is important that the parents work out the rules in advance and support each other when the rules need to be enforced. The stepparent must move slowly to enforce rules and should probably take the role of supporting the natural parent at first, rather than quickly becoming a disciplinarian. Immediately assuming the disciplinarian role will probably induce rebellion in the stepchildren.

4. Allow an outlet for the children's feelings for natural parents. It is often difficult for the stepparent to accept that his/her stepchildren will feel affection for their natural parent who is no longer living in the household. The stepparent may take this as a personal rejection. Children need to be allowed to express feelings about the parent who is absent. Their feelings should be supported in a neutral way, so that the children do not feel disloyal.

5. Expect ambivalence. Stepparents are often alarmed when children appear to show both strong love and strong hate toward them. Ambivalence is normal in all human relationships, but nowhere is it more accentuated than in the feelings of the stepchild toward the stepparent.

6. Avoid mealtime misery. For many stepfamilies, meals are an excruciating experience. This, after all, is the time when the dreams of blissful family life confront reality. Most individuals believe in the power of food to make people

Weekend Visits of the Noncustodial Child

Thus far, we have concentrated on the problems of the stepfamily with custody of the stepchildren. Periodic visitations of noncustodial children comprise another facet of remarriage. In fact, nonresidential, noncustodial stepmothers are far more common than custodial stepmothers.

Nonresidential stepmothers report more stress, dread, and ambivalent feelings about stepchildren, particularly about their visits, than custodial stepmothers do (Ganong and Coleman 1994, 80–82; Weddle 1994). The biggest problem was having an undefined role with their stepchildren (Bray and Kelly 1999, 153–180).

Time scheduling and lack of time with stepchildren are major problems for noncustodial stepparents. For example, a father has his children two weekends a month, and the month of July in the summer. His children also alternate major holidays between mother and father. What does this mean? It means the following:

1. The stepmother must conform to the visitation plan, which she probably had no part in creating.

happy. Because the mother is most often charged with serving the emotionally laden daily bread, she often leaves the table feeling thoroughly rejected. If the status quo becomes totally unbearable, it is forgivable to decide that peace is more important and to turn a blind eye, at least temporarily, to nutrition. Some suggested strategies include daily vitamins, ridding the house of all junk foods and letting the children fix their own meals, eating out a lot, and letting father do some of the cooking so he can share in the rejection. Stepfathers tend to be less concerned about food refusal, but more concerned about table manners.

7. Don't expect instant love. One of the problems facing a new stepparent is the expectation of feeling love for the child and expecting that love to be returned. Stepparents must acknowledge that it takes time for emotional bonds to be forged, and sometimes this never occurs.

Nonacceptance by the children is often a major problem. Some children make it very clear that "You are not my mother or father!" Of course, they are not. This can be very painful or anger provoking, especially if the stepparent is doing the cooking and laundry and giving allowances. Most children under 3 years old adapt with relative ease. Children over 5 years old have more difficulty.

8. Don't take all the responsibility. The child has some, too. Ultimately, how well the stepparent gets along with the stepchild depends, in part, upon the kind of child he/she is. Like adults, children come in all types and sizes.

Some are simply more lovable than others. If the new stepmother has envisioned herself as the mother of a cuddly little tot and finds herself with a sullen, vindictive 12-year-old who regards her with considerable suspicion, she is likely to experience disappointment. Like it or not, the stepparent has to take what he/she gets. But that doesn't mean taking all the guilt for a less-than-perfect relationship.

9. Be patient. The words to remember here are "Things take time." The first few months, and often years, have many difficult periods. The support and encouragement of other parents who have had similar experiences can be an invaluable aid.

10. Maintain the primacy of the marital relationship. A certain amount of guilt about the breakup of the previous relationship may spill over into the present relationship and create difficulties when there are arguments. The couple needs to remember that their relationship is primary in the family. The children need to be shown that the parents get along, can settle disputes, and most of all, will not be divided by the children. While parenting may be a central element in the couple's relationship, both partners need to commit time and energy to the development of a strong couple relationship; this bond includes, but is greater than, their parental responsibilities.

SOURCE: Turnbull and Turnbull 1983. Later research supports all of these points (Bray and Kelly 1999; Wallerstein and Blakeslee 2003).

---

2. The children aren't around her enough so that an emotional bond may form.

3. The stepmother really doesn't have a well-described mothering role, freely chosen by her.

In addition, the children may resent breaking up their routine to visit their father and this "other person." If the stepmother has her own children living with her and she expresses resentment about the periodic visits of his children, her new husband can feel unjustly accused: "I live with your kids all the time; surely mine should be able to visit and be accepted by you," he might complain.

Weekend visits of noncustodial children can be made more successful by implementing the following suggestions:

1. A child needs a permanent place to keep his/her things and the assurance that things belonging to him/her remain permanently and aren't disturbed by others.

2. Consistent routines, chores, and an assigned place for the child at the eating table work to ensure a sense of belonging.

3. A relaxing environment, without too many special activities, creates a more natural sense of family.
4. Each child in the family, including the part-time visiting child, needs alone time with the parents.
5. Encouraging the child to bring a friend for the weekend makes the visit feel more like home.
6. Use family meetings to get children involved in the workings of the household (Bray and Kelly 1999, 153–180; Wallerstein and Blakeslee 2003). (The book by the latter authors is a very good overview of children in remarriages, and offers many practical hints to the new stepparent.)

Dealing with Sexuality in the Stepfamily

One of the most difficult issues for stepfamily members to deal with is sexuality (Chapman 1991). Although the data are mixed and need to be interpreted cautiously, some studies suggest that incest is more common in stepfamilies (Giles-Sims 1997). Remarried families have what psychologists call a lack of biological incest taboos. In an intact family where father and daughter have known each other since the girl was in messy diapers, the father builds up an immunity to viewing his daughter as sexually attractive. When she enters puberty years later, this immunity protects the adolescent girl and her father. Dad may be aware of the changes in her body, but he is wearing blinders that a newly arrived stepfather rarely has (Wallerstein and Blakeslee 2003, 315).

A stepfamily has not had a long developmental period in which to form intimate parent-child ties and develop strong aversions to incest. The biological family interacts with its children from birth onward. The blended family may suddenly have adolescent children with whom one partner has had no previous experience. Thus, incest aversion has had no time to develop. The new adolescent family member is no more familiar to the stepparent than any other stranger.

The more affectionate sexual atmosphere in the home, during the time when the new couple are more romantically involved, may also contribute to heightened sexual intensities. It is not unusual for stepfamily members to experience sexual fantasies, increased anxiety, distancing behavior, or even anger in responding to and trying to cope with these sexual issues. In more extreme circumstances, a sexual relationship can develop between a stepparent and stepchild, as it did between filmmaker Woody Allen, age 58, and Soon-Yi Previn, age 23, his stepdaughter (J. Brothers 1994), whom he married in 1997. A sexual relationship might also develop between stepsiblings. Research indicates that opposite-sex adolescent stepsiblings tend to be sexually involved with each other much more frequently than opposite-sex biological siblings (Baptiste 1987, 91). Stepsiblings often fight against sexual feeling for each other through open conflict and statements of dislike for each other.

Although all states have laws governing sexual relations between blood relatives, most states do not regulate sexual relations between members of blended families.

In the child-abuse research studies completed to date, the following factors have been found to relate to sexual abuse: low income, high stress, antisocial attitudes and lack of training for protecting children, high conflict, low cohesion, isolation from friends and community, unemployment, drugs, alcohol, and some personality dysfunctions (Giles-Sims 1997, 223). The degree to which these factors generally exist in the blended family, as compared with the intact biological family, is difficult to determine.

How Much Closer Can We Get?

Jane brought her 14-year-old daughter into a remarriage with Rich, who had two children, one a 16-year-old son. All three children lived in the new home with the stepparents. For the first year or so, the two older children fought like cats and dogs, driving the parents to distraction. The parents insisted that the two get along and become friends. The two finally ceased hostilities, much to the relief of the parents, and became inseparable. The parents returned unexpectedly from a weekend trip and discovered the two in bed together. Predictably, the parents were shocked and outraged. Upon being confronted, the daughter commented, "Well, you wanted us to become friends. How much closer can we get?"

What Do You Think?

1. Is sexual interaction between stepsiblings really incest?
2. Do you know anyone who became or is sexually involved with a stepsibling? Explain without identifying.
3. What are some suggestions you can make to help adolescent stepsiblings handle sexual feelings for each other?

The following measures can reduce the likelihood of a sexual relationship developing in a stepfamily:

- Including the stepparent in caregiving encourages bonding.
- Discussing with the new spouse the generally increased sexuality of a new relationship and the possible influence such an environment might have on the children.
- Being prepared to discuss, with all family members, the sexual feelings that might arise in the stepfamily situation.
- Discussing the facts of life with pubescent children.
- Encouraging the children to verbalize their sexual fantasies.
- Resisting the temptation to push opposite-sex stepsiblings to become closer friends than they want or can tolerate at the time.
- Accepting as all right the usual sibling hostilities (Baptiste 1987; Giles-Sims 1997).

The New Extended Family

Although we have spoken only of stepparents, it is important to realize that the blended family will bring another set of kin into the relationships. By and large, the new extended family does not replace the old relationships, but adds to those from the first marriage. For example, there will now be stepgrandparents and the new spouse of the noncustodial parent. A blended family's immediate family tree can be unimaginably complex. As an extreme, imagine the many relationships of the following blended family: Former husband (with two children in the custody of their natural mother) marries new wife with two children in her custody. They have two children. Former wife also remarries man with two children, one in his custody and one in the custody of his former wife, who has also remarried and had a child with her second husband, who also has custody of one child from his previous marriage. The former husband's parents are divorced and both have remarried. Thus, when he remarries, his children have two complete sets of grandparents on his side, plus one set on the mother's side, plus perhaps two sets on the stepfather's side.

The example could extend to any level of complexity; indeed, trying to sort out all the relationships in some blended families is an impossible task. When one considers the complexities of the blended family, it is surprising that any remarriages are successful.

The immediate effect of divorce on relative interaction is that it intensifies contacts between blood relatives and curtails relations with former in-laws. Unless the relatives (mainly grandparents) of the noncustodial parent make real efforts to remain on close terms with the children, contact is slowly lost, just as it is with the noncustodial parent. With remarriage, however, the children's circle of relatives suddenly expands greatly, especially if they have been able to maintain contact with relatives of the noncustodial parent. Whether such expansion occurs or not depends, in part, on the proximity of the relatives. If a remarried family lives a great distance from one set of relatives, that set tends to have less contact with the family than relatives who live close by.

What remarriage does is to add relatives. Individuals have the option, but never the obligation, to define people as relatives when they are not closely related by blood. Kinship is often achieved, rather than ascribed. Remarriage illustrates this principle by creating an enlarged pool of potential kin. It is primarily up to the various parties involved to determine the extent to which new kin will be treated as actual relatives.

Potentially, children in a remarried family can have many sets of grandparents, as we saw in the example. They may have two sets of biological grandparents and four sets of stepgrandparents, if both their biological parents remarry. If all these grandparents live close by and maintain relationships with the blended family, holidays such as Christmas can become logistical feats. At the same time, a wide circle of relatives can also offer much support and love to children in the blended family. In a way, remarriage has brought the idea of extended family (granted, not blood relatives) back into American society.

Building Stepfamily Strengths

A review of popular literature about stepfamilies found discussion of the following potential strengths:

- Stepchildren learn problem solving, negotiation, and coping skills and also become more flexible and adaptable as they adjust to the new stepfamily.
- The presence of a greater number of mature adults adds support and exposes children to a wider variety of people and experiences.
- Additional role models are available to children.
- Stepparents may try harder to be good parents; thus, children gain an additional parent to learn from, to love, and to be loved in return.
- Finances are usually improved.
- Assuming that the remarriage is successful, both adults and children can learn what it is like to be in a happy, enduring relationship.

Although little empirical evidence supports these supposed advantages for stepchildren and their families, it is worthwhile to examine potential strengths that can be built up in blended families. Naturally, all the family strengths enumerated in Chapter 1 will be equally important, or more important to the blended family, because the blended family is more complicated organizationally than a first-marriage family. It also lacks the societal support usually afforded intact nuclear families.

One step that would help build blended-family strength is for society to recognize the blended family as a legitimate alternative to the nuclear family. This would help do away with the wicked stepparent myth that can cause harm to the stepfamily. Such recognition would also facilitate the creation of model roles and rules for the functioning of the stepfamily.

The Prenuptial Agreement

Although we discussed the prenuptial agreement earlier (pages 217–218), it is more important in a remarriage than in a first marriage. The couples entering remarriage often bring with them an extended history, household furnishings, financial investments such as a home, obligations to prior families, and so forth. All of this makes the remarriage much more complicated and thus more prone to conflict. By establishing the rules ahead of time, remarrying couples can head off many potential problems. Naturally, any such agreements cannot nullify state laws concerning marriage. Generally, prenuptial agreements cover what happens to the children and the couple's property if either dies, the division of property in case of divorce, spousal maintenance, and more specific daily living plans.

Because finances and children are the two major sources of conflict in a remarriage, both need to be discussed in detail. Perhaps the most important financial decision is whether to combine money into a single pot, or to keep the partners' money separate. Because people remarrying often bring considerable resources into the new marriage, many couples opt to keep money separate, with both partners donating to a household fund used for daily living. Perhaps the couple is living in a home owned by one party. The home may remain separate property, with the incoming spouse picking up the monthly costs. One spouse may be receiving child-support payments and will keep them separate, to be used only for that purpose. Keeping money separate usually means that a tighter bond will be maintained with the first family and may reduce cohesion in the stepfamily. At the same time, however, separating finances will normally reduce conflicts, especially if the parties entering the remarriage both have considerable resources and prior obligations.

Questions are likely to arise about who will pay for insurance, health benefits, children's illnesses, luxuries, credit cards, prior debts, and so on. If the courts give some freedom in child-custody arrangements, what will be the arrangements? What role will religion play in the family? What changes need to be made in wills? The questions are endless, but the more that are answered ahead of time, the less conflict there will be later and the stronger the remarried family will become.

Mediation to Settle Conflicts and Other Prevention Programs

Remarrying couples who have divorce in their background have probably gone through a great deal of conflict at the time of their divorce. If they enter a new marriage and encounter new conflicts, the relationship will start off negatively. As we have seen, conflict is even more likely in a remarriage than in a first marriage, especially at first. Couples who managed to divorce in an amicable manner will enter a remarriage with an advantage.

Mediation with an objective third party whose goal is to help the new couple solve their problems and reach their own goals can greatly reduce potential conflict. Although few remarrying couples consult mediators or counselors before they marry, such an experience is worthwhile for many couples (Visher et al. 1997). To spend some time and money heading off problems seems infinitely more intelligent than to lose money trying to escape a conflict. Obtaining counseling and mediating problems and conflicts ahead of time usually strengthens the remarriage and stepfamily.

Because remarriages are more complicated and difficult than first marriages, mediation can help reduce potential conflict by using a third, unemotionally involved, professional person to work out agreeable compromises. By helping remarried couples work out problems in an amicable fashion and, perhaps, interjecting a bit of humor, the mediator can contribute to the building of a healthy stepfamily. The mediator can help the stepfamily members assume the role of an audience at times, and thereby share the laughter with the rest of the family at some of the stresses and strains that are taken too seriously, all too often.

Despite the inherent uneasiness, stepfamilies continue to be formed, and stepfamily members continue to struggle with unresolved issues of how best to support family members. Although they often fail, stepfamilies also reap successes as they carve out new patterns of interaction and new ways of being a family. In many ways, stepfamilies can serve as laboratories for how families function. Things that seem natural in long-term, first-married families must be worked out in sometimes awkward and painful fashion in remarried families. It may be hard to imagine, from watching the first painful steps in learning to be a stepparent or stepchild, that there will ever be a moment of triumph. Yet, those moments exist for many. In some ways, the remarried family has the chance of being even stronger than a first-marriage family. Each new spouse brings a wealth of information and experience to a remarriage. By utilizing that experience and knowledge to the fullest, they can create a strong and healthy new family (Ganong and Coleman 1994, 151–153). It is the mediator's job to assist this process.

In addition to mediation help for remarried families, there are numerous programs designed to help the children of divorce. For example, some schools offer such programs as the Children's Support Group and/or the Children of Divorce Intervention Program. These similar 10- to 12-session, school-based programs focus on children's own responses to divorce by providing children with emotional support and helping them to learn coping and control skills (Haine et al. 2003, 398). By helping children to understand and adjust to their parents' divorce, the children are better able to adjust to a new stepfamily status.

In addition, there are numerous other efforts such as custodial mother–focused programs designed to improve parenting skills for divorced mothers. The New Beginnings Program consists of 11 group sessions and two individual sessions studying parenting, father-child contact, negative divorce events, and interparental conflict (Haine et al. 2003, 298; Wolchik et al. 2000).

Programs attempting to help divorced persons and stepfamilies better manage problems inherent to their new situation are too numerous to examine all here. As mentioned, Parents Without Partners is one popular group setting that can guide its members toward specific help programs.

If all goes well with a new stepfamily, the biggest gain for the children might well be returning to the times when extended families were the norm. Now the children have four parents to teach them, love them, and guide them instead of just two parents, as in the nuclear family. In addition, more brothers, sisters, aunts, uncles, and grandparents are there for support.

Summary

1. *High divorce rates do not necessarily mean that Americans are disenchanted with the institution of marriage; high remarriage and cohabitation rates are seen for the divorced.*

2. *The majority of divorced persons remarry or move into a cohabitation relationship, most often within a few years of divorce.* A few remarry as soon as possible, but the rest usually remain single for at least a short period of time. The adjustment to single life is often difficult, especially for those who have been married a long time.

3. Remarriage is sought by most divorced people. Yet, making the actual decision to remarry is often difficult, because the idea of marriage evokes negative attitudes based on their negative experiences with their earlier marriages. *People marrying for a second time carry with them attitudes and expectations from their first marital experience.* In many cases, they still must cope with their first family. Visiting children, child support, and alimony payments may add to the adjustment problems in the second marriage.

4. *Children from prior marriages often add to the responsibilities of second marriage.* Becoming a stepparent to the new spouse's children is not easy. A second family may have children from several sources. Each spouse may have children from his/her previous marriage, and in time they may have children together. Children from previous marriages often mean continued interaction between the formerly married couple when the former mate visits with or takes the children periodically.

5. *About 20 to 30 percent of those divorcing never remarry.* For these persons, single life becomes permanent. As divorce rates rise, however, the likelihood of remarriage also increases, because there are more potential partners. At present, about 50 percent of American marriages involve at least one person who was formerly married. Remarriage, then, has definitely become a way of life for a significant number of Americans.

Resources on the Internet

Companion Website for *Human Intimacy: Marriage, the Family, and Its Meaning,* Tenth Edition

http://sociology.wadsworth.com/cox10e/

Gain an even better grasp on this chapter by going to the companion website to take one of the tutorial quizzes, use the flash cards to master key terms, or check out the many other study aids you'll find there. You will also find special features such as GSS data, Sociology Online, and Census 2000 information that will put data and resources at your fingertips to help you with that special project or to help you as you do some research on your own.

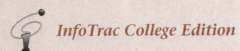

InfoTrac College Edition

http://www.infotrac-college.com/wadsworth/

You can access reliable resources anytime, anywhere, with InfoTrac College Edition, the online library. This fully searchable database offers more than 20 years' worth of full-text articles (not abstracts) from almost 5,000 diverse sources, such as top academic journals, newsletters, and up-to-the-minute periodicals, including *Time*, *Newsweek*, *Science*, *Forbes*, *The New York Times*, and *USA Today*. You can conduct electronic key word searches using key terms from this chapter to supplement your reading and learning experience. To aid in your search and to gain useful tips, see the Student Guide to InfoTrac College Edition, which you can access through the companion website for this book.

Fatherless America: Can a Stepfather Take the Place of a Biological Father?

David Blankenhorn, in his book *Fatherless America* (1995) suggests that the high divorce rate and the increasing number of unwed mothers are combining to produce fatherless children. He further suggests that fathers playing little or no role in the rearing of their children leads to America's most urgent social problems—increased youth violence, domestic violence, child sexual abuse, and child poverty.

However, high remarriage rates of divorced persons and the fact that most unwed mothers eventually cohabit or remarry mean that fathers are, in fact, present. These fathers are substitute fathers—that is, stepfathers. In 1980, about 15 percent of all married-couple households with children contained a stepfather. By 1990, the figure was 21 percent, a 40 percent increase in 10 years (Norton and Miller 1992); by 2005, that figure approximated 40 percent. As mentioned earlier, because the vast majority of mothers maintain physical custody of their children, stepfamily really means a custodial stepfather for most children.

Can these stepfathers do as good a job in rearing children as the biological fathers?

YES

In *Making Peace with Your Stepfamily* (1993), Harold Bloomfield states, "There is no reason why stepparents cannot parent just as effectively as biological parents." Others suggest that children's lives can be enriched by the presence of stepparents and the greater numbers of new relatives that come along with a stepparent. The stepchild can enjoy a variety of lifestyles and select from each that which he/she likes the best.

Stepfamilies offer children a number of benefits that can compensate for the loss of a biological father (Bray and Kelly 1999; Ganong and Coleman 1994; Wallerstein and Blakeslee 1995, 2003), as follows:

- Children gain new role models. A stepfather or stepmother is a new model from which the children can learn. For example, a stepfather interested in music may encourage his stepchildren to appreciate music, play an instrument, or learn to sing.
- Children may be introduced to new ideas and different values. With new living arrangements and new adults, including stepgrandparents, aunts, and uncles now in the family, the children may gain enlarged and/or new perspectives about life.
- Stepparents can become a source of support and security for their stepchildren.
- Children may gain parents who are happily married, rather than unhappily married, as were their biological parents; a happier marriage tends to produce better adjusted, happier children.

When a stepfather enters a child's life very early, such as in infancy, he will probably have no problem becoming the father to the child. The stepfather's role in the child's life will approximate the biological father's role. This is especially true if the biological father is not around, perhaps having deserted the family.

Children of any age may be happier and better adjusted when an abusive biological father is replaced by a kind stepfather. In fact, the stepfather may be viewed as a savior.

NO

Very few children actually believe they have two fathers, or that stepfathers parent just as well as biological parents, or that living in a stepfamily constitutes a new source of stability. These ideas reflect the wishful thinking of adults, rather than the actual circumstances of children.

Despite the list of supposed stepfamily (stepfather) advantages, research is uniformly bleak. In summarizing the research, Frank Furstenberg (1994) concludes that "Most studies show that children in stepfamilies do not do better than children in single-parent families; indeed, many indicate that, on average, children in remarriages do worse." Others agree with his conclusions (Wallerstein and Blakeslee 2003; Wallerstein et al. 2000).

Research uniformly demonstrates that children from single-parent homes are less well adjusted, do more poorly in school, are more apt to engage in antisocial behavior leading to involvement with authorities, and have more difficulty forming close interpersonal relationships. Children living with stepfathers experience outcomes that are not much, if any, better than children in single-parent households (Bray and Kelly 1999; Bray et al. 1992; Furstenberg 1994; Furstenberg and Cherlin 1991; Wallerstein et al. 2000; Zill et al. 1993).

Many stepchildren do not call their stepfathers "Daddy" or "my father" precisely because they do not believe these men are their fathers (Levin 1997, 183; Ritala-Koskinen 1997). These children are right. Stepfathers are not even replacement fathers or second fathers. Remarriage may offer adults a second chance for happiness, but remarriage does not offer children a second chance for fatherhood. In fact, the arrival of a new male in the family means foreclosure of the first-and-only chance for the children to be in a family with both biological parents.

What Do You Think?

1. What do you think the effects are of having no father in a family for a boy? For a girl?

2. What extra problems might a residential stepfather have that a biological father will not have with small children? With adolescents?

3. Will a residential stepfather have more, fewer, or about the same kinds of problems that a residential stepmother will have? Why?

4. If you have or had a residential stepfather, what is your personal evaluation of him as a father to you?

© Frank Cox

Actively Seeking Marital Growth and Fulfillment

Chapter

17

Questions to reflect upon as you read this chapter:

- Do you tend to head off problems before they occur, or do you just react to problems when they occur?

- Why do you think that good management skills would be important in successful families?

- What are some cautions a person needs to consider before joining a program aimed at self-improvement or marital improvement?

- Do you think that the family in America will remain strong and viable?

T*he mutual commitment of wife and husband involves them freely in a growing changing way of life rather than in a static "state of life."* (C. ROGERS 1972)

Marriages grow and change. This is the essence of not only the marital relationship, but of all close relationships. Relationships are not stagnant. They do not remain forever as they were when they were initiated. Because relationships are processes, they must be attended to and worked on, or the relationship will fail. An individual who accuses his/her spouse of changing—"He/she is not the person I married"—has not attended to the relationship. Of course, the person one marries will change, but active participation in the relational changes is what leads to positive growth. Finding ways to vitalize a marriage, or any intimate relationship, should be a central theme for every couple. Successful relationships, including marriages, grow as the couples change and grow. Poor relationships or marriages either stagnate or grow without direction, participation, and guidance from both the partners. Viewing marriage as a process gives a couple cause for hope, even in the midst of relationship problems. It relieves some of the anxiety felt when they are not happy, because they realize that the process of marriage means that their relationship will and can change for the better. Rather than viewing differences and conflicts as signs of incompatibility, couples need to see them as opportunities for growth, for developing skills that they can use to improve their relationship for the future.

This final chapter stresses the idea that every person has the ability to improve his/her intimate relationships and/or marriages. Married couples too often believe that they can't do much about their marriages, so they simply don't take the time to nourish their relationships and make them healthier. Couples also tend to get into trouble because security and comfort may lull them into avoiding risks. Yet, they must be willing to take risks if they expect to grow and maintain positive movement in their lives and relationships. Periodic failures should be viewed as important learning moments, indicating that change is necessary—not that an intimate relationship is necessarily over. But no one should feel pressured to work on their intimate relationships every minute of every day. That would be stifling and take away some of the fun and spontaneity that should exist in every close relationship. However, some time must be devoted to nurturing relationships and/or marriages, if they are to persist.

Most couples in successful relationships treasure lightness and fun. There is playful teasing of each other. Bantering brightens the day. Successful couples often develop a secret language, verbal and bodily, by which they communicate love and share a sense of fun. "Sometimes word play has sexual or erotic overtones, but more often a couple's banter pokes fun at the ups and downs of everyday life, including everyday marriage. Laughter is well worth cultivating" (Wallerstein and Blakeslee 1995, 204). The ability to laugh with others and also at yourself is an important indicator of one's mental health.

We began our journey of marriage and family study by examining the characteristics of strong, successful families. You'll remember that those characteristics are as follows:

- Commitment to each other and the relationship
- Appreciation of all family members
- Good communication
- Spending time together
- Building a value system
- Dealing with problems constructively
- Resilience
- Self-efficacy

© Larry Williams/Corbis

Love and intimacy should mean fun together.

These eight ideal characteristics give us goals—a direction for the changes in our relationships.

Can we commit ourselves to relationships in which we appreciate our mates and develop good communication patterns, so that the time we spend together is fulfilling and growth producing? Can we develop a value system and problem-solving skills that will allow us to deal positively with crises and stress? If we can do these things, we stand a good chance of creating a strong family for ourselves and loved ones. To accomplish this, we must be actively involved in our intimate relationships and in our broader social and cultural networks. Relatives and kin, friends and acquaintances, neighborhoods and schools, and the general societal characteristics all influence us and can serve as support for our more intimate relationships (Bryant and Conger 1999; Julien et al. 1999). We cannot just assume that our intimate relationships will take care of themselves because we love our partners. It is true that the relationship will evolve and change even if we are not active, but will the change be in positive directions that lead to strength, stability, happiness, and fulfillment?

What can couples do to make their intimate relationships more fully functioning? Although everything discussed thus far bears on the question, this concluding chapter will specifically attend to the goal of seeking marital growth and fulfillment.

"And They Lived Happily Ever After"

The American scenario of marriage concludes with "and they lived happily ever after." In other words, once you find the right person, fall in love, and marry, all problems will be over. Of course, this is a fairy tale. All married couples will face problems. Yet this myth persists, even if only at the unconscious level, and it hampers many Americans' efforts to realize the fullest possible potential in their marriages. The reader should examine his/her reaction to the following statement to discover if he/she is influenced by this myth: "All married couples should periodically seek to improve their marriage through direct participation in counseling and marriage-enrichment programs."

What do you think? Some typical reactions include the following:

- "It might be a good idea if the couple is unhappy or having problems."
- "I know couples who need some help, but Jane and I are already getting along pretty well. It wouldn't help us."
- "We already know what our problems are. All we need to do is . . ."
- "I'd be embarrassed to seek outside help for my marriage. It would mean I was a personal failure."
- "We're so busy now, what with work, the children, and social engagements, we wouldn't have time for any of those things."
- "Renaldo is a good husband [Margarita is a good wife]. I really couldn't ask him [her] to participate in anything like that. He [She] would feel I wasn't happy with him [her] or our marriage."
- "I could be happier, but overall, our marriage is fine."

The ability to recognize the need for help and then to seek that help is actually a characteristic of strong families. It means that the family is strong enough to acknowledge its weaknesses and ask for help in time of need. It is true that not all married couples need to seek help from a third party, but it is also true that to be successful, a marriage needs more than just to be maintained.

Although drawing an analogy between marriage and an automobile is superficial and a gross oversimplification, it may clarify this point. A car that has not been maintained quickly malfunctions and wears out; a well-maintained car gives less trouble and lasts longer. Beyond simple maintenance, however, a car can be improved (by buying better tires or changing the carburetion, exhaust, compression, gearing, and so on) and modified to run better (faster, smoother, and more economically). Most Americans spend most of their adult lives married. Yet, they spend little time and energy improving their marriages. If marriage becomes too bad, they leave it to seek a new relationship that they hope will be better. The new relationship may be better for a while, but without maintenance and improvement, it too will soon malfunction.

Some Americans expend a fair amount of energy seeking a new mode of marriage. Some think that living together (cohabiting) and avoiding legal marriage is the answer. Others may decide that communal living is the way to better intimate relationships. Still others think that moving through many intimate relationships will avoid the problems inherent to building a strong, lasting marriage. Yet, often the "improved" alternatives to marriage quickly lead to disenchantment. The new commune member who had problems communicating with his/her spouse finds that communicating intimately with seven other people is even more difficult. The cohabiting couple who thought that limited commitment was the way to avoid the humdrum in their life together may find that a prolonged lack of commitment leads to increased insecurity and discomfort. The individual moving from relationship to relationship may miss the rewards that can come from loving closeness. Perhaps the energy spent seeking some ideal alternative to marriage might be better spent working on marriage itself.

After all, most Americans marry the people they do because they love them, want to be with them, and want to do things for them. They marry by their own decision, in most cases, and start out supposedly with the best of all things—love—going for them. Where does love go? Why isn't love able to solve all problems? Might it be that buying into the fairy-tale ending—"and they lived happily ever after"—keeps them from working to build a better marriage? Do Americans think that love will automatically take care of everything?

In reality, a number of factors combine to keep most Americans from working more actively to improve their marriages. The fairy tale we have been discussing has been called the *myth of naturalism*. This is the idea that marriage is natural and will take care of itself, if we just select the right partner. Many people believe that outside forces may support or hinder their marriage, but that married couples need do little to have a well-functioning marriage, especially when the outside forces are good (full employment, little societal stress, and so on).

Another factor is the general *privatism* that pervades American culture. "It's nobody else's business" is a common attitude about problems in general, and marriages in particular. It's bad taste to reveal our intimate and personal lives publicly. Seeking outside help to improve one's marriage means sharing personal information, which is often viewed as an invasion of privacy.

A third factor is the *cynicism* that treats marriage as a joke and thus heads off attempts to improve it. "You should have known better than to get married. Don't complain to me about your problems." This attitude contradicts the romantic concept of marriage, but operates just as strongly to keep people from deliberately seeking to improve their marriages. "Why would anyone want to improve this dumb institution?" Even though American society is marriage oriented, marriage still evokes a great deal of ridicule and stories about the ball and chain. Facing up to and countering the antifamily themes in American society is an important step toward revitalizing marriage and the family.

Despite these factors, people—especially women (Bringle and Byers 1997)—in growing numbers are actively seeking to improve their marriages. To improve a marriage, it is first necessary to believe that relationships can be improved. In other words, the myth of naturalism must be overcome. A marriage will not just naturally take care of itself. In addition, the privatism and cynicism that surround marriage must be reduced, if effective steps are to be taken to enrich a marriage.

To improve their marriage, a couple must work on three things: (1) themselves as individuals, (2) their relationship, and (3) the environment (especially eco-

 HIGHLIGHT

Economic Success, Marital Failure

Henri and Tara both worked to buy the many things they wanted: a house, fine furnishings, nice clothes, a luxury car, and so on. Henri even held two jobs for a while. Certainly, no one could fault their industriousness and hard work. In time, their marital affluence became the envy of all who knew them. After 7 years of marriage, they divorced. Their friends were surprised. "They had everything, why should they divorce?" Unfortunately, Henri and Tara didn't have much of a relationship, except to say hello and goodbye as they passed on the way to and from work. In addition, they both worked so hard that they had no time for self-improvement. No self-improvement means no growth or change; this eventually leads to boredom with the relationship, and boredom often portends failure.

What Do You Think?

1. What suggestions do you have to help couples keep their relationship alive and growing? When they both work full time?

2. Why do you think that economic success alone was not enough to sustain Henri and Tara's marriage?

HIGHLIGHT

Self-Improvement, Marital Failure

Renaldo and Madison believed that the key to a successful marriage was self-improvement. Both took extension classes in areas of their own interest. They attended sensitivity-training groups and personal expansion workshops. Unfortunately, they could seldom attend these functions together because of their conflicting work schedules. Soon, they were so busy improving themselves that they had little time for each other. The house was a shambles, the weeds in the yard grew, and their relationship disappeared under a maze of "do-your-own-thing" self-improvement. After 7 years, they divorced. Their friends were surprised. "After all, they're so dynamic and interesting, why should they divorce?" Renaldo and Madison had become so self-oriented that their relationship had disappeared and their living environment had become unimportant.

What Do You Think?

1. How can working to improve oneself disrupt an intimate relationship?

2. What could Renaldo and Madison have done to improve themselves, yet save their relationship?

nomic) within which the marriage exists. We have already looked at these elements. For example, in Chapter 7 we discussed the self-actualized person in the fully functioning family; in Chapter 5 we looked at ways to improve communication within a relationship; and in Chapters 12 and 13 we examined marriage as an economic institution and found that the economics of one's marriage will drastically affect the marital relationship.

Although this chapter deals specifically with activities designed to improve marital relationships, it is important that a couple work to improve the other two influences on marriage: themselves as individuals and their economic situation. Neglect of any of these influences or emphasis on only one can lead to marital failure. In fact, a couple can be extremely successful in one of the three areas and still fail miserably at marriage, as the cases demonstrated.

Both of these scenarios happen every day. The second is becoming more prevalent with the growing interest in the human potential movement. The very concepts used in this book—self-actualization, self-fulfillment, and human-growth orientation—can all be taken to such an individual extreme that intimate relationships can be disrupted. Andrew Cherlin (2004, 851) points out that during the last few decades, "expressive individualism" has become a prevalent value in American society. If the importance of individual self-fulfillment ignores the relational aspect of marriage by making self-fulfillment the central goal, it can work to the disadvantage of intimate relationships, each of which requires a great deal of unselfishness, rather than self-centeredness.

More family researchers and also general observers of American society see excessive hedonism as America's greatest enemy. Stressing individuality at the expense of mutuality overlooks the importance of successful human relationships, which many see as a basic human need, even for successful individual functioning. Amitai Etzioni (1983, 1993) suggests that mutuality, the basic need for interpersonal bonds, is not something each person creates on his/her own and then brings to the relationship. Rather, it is constructed by individuals working with each other. This working together is the essence of the healthy family.

To make marriage as rewarding and fulfilling as possible, a couple must be committed first to the idea that effective family relationships do not just happen, but are the result of deliberate efforts by members of the family unit. Then they must be prepared to work on all three facets of marriage: to improve themselves as individuals, to improve their interactional relationship, and to improve their environment (especially the economic). These three areas encompass developing oneself in the physical, social, emotional, intellectual, and spiritual realms.

Such a commitment helps a couple anticipate problems before they arise, rather than simply reacting to them. When both partners are committed to active management and creative guidance of a marriage, the marriage can become richly fulfilling and growth enhancing, both for the family as a social unit and for the individuals within the family. Again, it should be noted that the couple's social network can be an important source of strength. However, a social network that is negative can do just the opposite and harm a couple's personal relationship (Bryant and Conger 1999; Julien et al. 1999). For example, Sandra and Kenny have been married 6 years and have two small children. Kenny's work colleagues are mostly single and like to hit the bars for a few drinks after work. Kenny often joins them, sometimes missing dinner and returning home late. Although he doesn't often drink too much, Sandra is increasingly upset at the amount of time he spends with "the boys." She also begins to hear rumors that his colleagues flirt with women, and she begins to wonder if Kenny joins in the flirting. Things go from bad to worse as the amount of time he spends with the boys increases and his time with the family decreases. In Kenny's case, his social network of single colleagues appears to be harmful to his relationship with Sandra.

WHAT DO YOU THINK?
1. How much free time with "the boys" do you think Kenny should have?
2. Is it OK in your mind for a wife or husband to go out to bars with friends?
3. What about a wife or husband flirting with others?
4. Do you think Kenny's behavior is harmful to his marriage?

Marriage Improvement Programs

Many techniques are now available to help families who seek assistance. Some techniques aim at solving existing problems, and others aim at general family improvement. We will briefly examine some of these techniques in the hope of accomplishing these two goals:

1. To help families seek experiences that benefit them.
2. To alert families to the possible dangers involved in unselective, nondiscriminative participation in some of the popular techniques.

Help for family problems in the past usually came from relatives, friends, ministers, and family doctors. Historically, the idea of enriching family life and improving already adequate marriages simply did not occur to most married people. Marriage traditionally was an institution for child rearing, economic support, and proper fulfillment of marital duties defined in terms of masculine and feminine roles. If there were problems in these areas, help might be sought. If not, the marriage was fine.

Marriage in modern America, however, has been given increasingly more responsibility for individual happiness and emotional fulfillment. The criteria used to judge a marriage have gradually shifted from how well each member fulfills roles and performs marital functions to whether both partners have achieved personal contentment, fulfillment, and happiness (expressive individualism).

Marital complaints now concern sex-role dissatisfaction, loss of individuality, unequal opportunities for growth and personal fulfillment, personal unhappiness and emotional dissatisfaction, and feelings that the marriage is shortchanging the individual partners. In other words, more marriage problems revolve around personal dissatisfactions than around traditional marital functioning. The "me" in today's marriage often seems more important than the "us."

A successful family must find a balance between personal freedom and happiness, on the one hand, and family support and togetherness on the other. The "me" and the "us" must come into an acceptable balance. Balancing connection and autonomy is of primary importance to the successful marriage. You cannot be married and single at the same time. The precise balance will vary from family to family. At one end of the continuum are joint conjugal relationships, in which the balance favors the "us." These couples are close emotionally and share most areas of their lives. Their leisure activities almost always involve each other, and their outside friendships are almost always couple friendships. At the other end of the continuum are separated conjugal relationships, in which the balance favors the individual. Such couples usually have separate leisure activities and friends. Today, with the entry of so many wives into the workforce and the general social support for individual growth and fulfillment, the lifestyles of most American families have moved slowly toward separated conjugal relationships.

Emphasis on emotional fulfillment as the most important aspect of marriage makes an enduring marital union much more difficult to attain. In the past, emotional fulfillment came indirectly, as a result of successfully fulfilling prescribed marital roles. For example, when Australia was first colonized by male convicts, a brisk trade in mail-order wives took place because there were no available women in the country. Most of these marriages seem to have been successful, for the simple reason that the prospective husband and wife expected things of each other that the other could provide. The man needed assistance and companionship of a woman in the arduous task of building a farm, and he wanted sons to help him. He expected certain skills in his wife that all young women reared in rural England were likely to have. Her expectations were similarly pragmatic. She expected him to know farming, to work hard, and to protect her. Neither thought of the other as a happiness machine. If they found happiness together more often than American couples do, it may have been that they were not looking so hard for it. They fulfilled each other because they shared a life; they did not share a life in the hope of being fulfilled (Putney 1972).

The search for emotional fulfillment has led to the development of many new techniques to gain this end. Sensitivity training, encounter groups, family enrichment weekends, sex therapy, sexuality workshops, communication improvement groups, massage and bodily awareness training, psychodrama, women's and men's liberation groups, and many other experiential activities have sprung up in recent years to help Americans enrich their lives.

Although this chapter's overview cannot hope to do justice to the many marriage improvement techniques that are emerging, the following is a brief look at some typical ones. Marriage enrichment will be examined in more detail. (See the special edition of *Family Relations*, October 2004, devoted to marriage education and improvement programs.)

1. *School courses on marriage and the family.* These courses aim to help people better understand the institution of marriage. Many schools offer even more specialized courses, often in the evening, on marital communication, economics of marriage, child rearing, and so on.

2. *Encounter groups.* These consist of group interactions, usually with strangers, where the masks and games used by marital partners to manipulate each other and conceal real feelings are stripped away. The group actively confronts the person, forcing him/her to examine some problems and the faulty methods that may have been used to solve or deny the problems. Participants in such groups release a great deal of emotion. Couples contemplating attending an encounter group should carefully consider the guidelines on pages 532–535.

3. *Family enrichment weekends.* These involve the entire family going to a retreat setting, where they work together to improve their family life. The family may concentrate on learning new activities to share. They may listen to lectures, see films, and participate in other learning experiences together. They may interact with other families and learn through the experiences of others. Family members may participate in exercises designed to improve family communication or general family functioning.

4. *Women's and men's consciousness-raising groups.* These groups center their discussions and exercises on helping people escape from stereotypical gender roles and liberate the parts of their personalities that have been submerged in the gender role. For example, women who believe that the typical feminine role has been too passive may work to become more assertive. Men who feel the typical masculine role has repressed their ability to communicate feelings may work to become more expressive.

5. *Married couples' communication workshops.* These workshops may be ongoing groups or weekend workshops, in which communication is the center of attention. Role playing, learning how to fight fairly, understanding communication processes, and actively practicing in front of the group all help the couple toward better communication. An important aspect of a workshop is the group critique; a couple discusses something that causes a problem for them, and then hears a critique of their communication skills from other group members.

6. *Massage and bodily awareness training.* This training is often included in sexuality workshops. It is aimed at developing the couple's awareness of their own bodies and teaching them the techniques involved in physically pleasuring each other through massage. The art of physical relaxation is part of bodily awareness training.

7. *Psychodrama.* Couples dramatize problems by acting them out with other group members as the players. In the case of marriage enrichment, psychodrama is used to help individuals in the family better understand the roles of other family members. This understanding is accomplished mainly by having the individuals participating in the drama make timely role changes under the direction of the group leader. Shifting roles also helps each player understand how the other persons in the drama see the situation and feel about it.

8. *Sensitivity training.* This training consists of exercises in touching, concentrating, heightening awareness, and empathizing with one's mate. The exercises increase each partner's self-awareness and sensitivity toward the other.

9. *Sex therapy and sexuality workshops.* These workshops focus on a couple's sexual relationship and use sex therapy to overcome sexual problems. The workshops are designed more to help couples improve this aspect of their relationship than to cure severe problems. Such a workshop assumes that the couple has no major sexual problems. The goal is to heighten sexual awareness, so that the couple's sexual relations may be enriched. Films, discussion, mutual exploration, sensitivity, and massage and bodily awareness techniques are all used to reduce inhibitions and expand the couple's sexual awareness.

10. *Marriage counseling and family service organizations.* These services are aimed more at couples with real marital and family difficulties. For example, Parents Without Partners (PWP) is an active volunteer organization to help and support single parents. Planned Parenthood helps families with reproductive problems, such as finding the best contraceptive. The American As-

sociation of Marriage and Family Counselors will help a couple find a reputable marriage counselor. The Stepfamily Association of America is aimed directly at couples who remarry and form stepfamilies.

Because many couples will choose not to seek outside help unless their problems are severe, a couple can still do a number of things together to improve their relationship:

1. Setting aside scheduled times each week to really talk to each other (not necessarily about problems), share, and communicate. Making these intimate times without outside distractions, such as children or television.
2. Perhaps during some of these times, the reading together of self-improvement books, books that discuss relationships, or other books of mutual interest will be helpful.
3. Dating often—doing things that are mutually enjoyable, things that are romantic, things that bring back happy memories.
4. Focusing on "us" as much, if not more, than on "me."

Guidelines for Choosing Marriage Improvement Programs

Unfortunately, the large demand for marriage improvement programs has brought some untrained and unscrupulous people into the fields of marriage counseling and marriage enrichment. For example, it is relatively simple and monetarily rewarding to run a weekend encounter group of some kind. All you need is a place where the people can meet. Some participants have found that not all such experiences are beneficial, or even accomplish what they claim. In a minority of cases, unexpected repercussions—such as divorce, job change, and even hospitalization for mental disturbance—have occurred after some supposedly beneficial group experience. Consider what happened to the following couple because one partner could not tolerate the intensity of the group experience.

David and Melissa have been married for 11 years and have three children. David has always been shy and uncomfortable among people, but nevertheless, has worked out a stable, satisfying relationship with Melissa. More socially oriented than David, Melissa began to attend a series of group encounter sessions out of curiosity. As her interest increased, she decided that David would benefit from a group experience. She asked him to attend a weekend marathon. Unfortunately, the group turned its attention too strongly toward David's shyness, causing him acute discomfort that finally resulted in his fleeing from the group. He remained away from his home and work for 10 days. On returning, he demanded a divorce because he felt that he was an inadequate husband. Fortunately, psychotherapeutic help was available, and David was able to work out the problems raised by the group encounter.

Evaluation studies of the Marriage Encounter program (a popular church-sponsored enrichment program) indicate that few participants experience potentially harmful effects. But because negative effects are possible, it is worthwhile to list what they might be:

1. The perceived benefits may be illusory, or at best, temporary.
2. The emphasis on the relationship may tend to deny individual differences.
3. There may be divisive influences on the couple's relationship with other family members.

WHAT DO YOU THINK?

1. Because David and Melissa have worked out a comfortable relationship, do you think they should take steps to reduce his shyness?

2. Have you ever been to any kind of self-improvement workshop? If so, what kind of experience did you have?

4. The communicative techniques taught may rigidify the couple's communication patterns, and failure to practice the techniques may lead to guilt or resentment.

Because marriage enrichment experiences can be so beneficial to couples, reducing the potential negative aspects to a minimum is important. Couples seeking marital enrichment or help for marital problems are advised to investigate the people offering such services. They should also discuss the kind of experiences they want and make sure that those are the experiences offered. For example, if a couple decide that they would like to improve their sexual relations and wish to do so by seeking some general sensitivity training (learning to feel more comfortable with their bodies, be more aware, and give and accept bodily pleasure), they might be rudely shocked if the group leader conducts a nude encounter group, with the goal of examining each person's emotional hang-ups about sex.

Couples who have a reasonably satisfactory marriage can use the following guidelines in choosing a marriage enrichment activity:

1. Choose the activity together and participate together if possible. If only one partner is willing to participate, it is much more difficult to create positive results (B. Brothers 1997).
2. If only one mate can participate, do so with the consent of the other, and bring the other into the activity as much as possible by sharing your experiences.

HIGHLIGHT

A Family Life Enrichment Weekend

Stephanie and Brian Smith have been married for 15 years. They have two children: Colin, 13, and Beth, 10. Their church recently started a series of family retreat weekends at a nearby mountain camp. After some discussion, the family members agree that it would be fun and rewarding to go on one of the weekends.

At noon, the 20 families gather in the cafeteria/meeting hall for lunch. Everyone is given a name tag, and the leaders are introduced. After lunch, the families introduce themselves. Much to the children's delight, there are many other children present. The leaders assign each family to one of five family subgroups. The family groups meet, and the four families in the Smith's group become better acquainted. The group leader then introduces the first work session, entitled "Becoming More Aware." There are exercises in identification of feelings; attention is given to feelings the participants would like to experience more often, and those they would like to experience less.

After a break, the group leader discusses methods the families might use to reduce unwanted feelings and increase desired feelings. Each family then practices some of these methods, while the others observe. After each family finishes, group members offer a general critique.

The families are free after the work session until dinner. After dinner, short movies on various developmental problems are shown to the children. At the same time, the parents attend a sexuality workshop, where they learn massage techniques designed to relax and give physical pleasure. They are then asked to practice the techniques in their individual cabins. They are assured of privacy because the children will be occupied for at least another hour. Meanwhile, the children form small groups and discuss how the children shown in the films can be helped with the developmental problems portrayed.

The next morning, the first work session is devoted to the theme "Being Free." This involves learning openness in experiencing each other. The children of the four families talk to each other about things they like and don't like, while the parents sit and listen. Then the roles are reversed. Afterward, the families exchange children and are given a hypothetical problem to solve. Each newly constituted family has a half-hour to work on the problem, while the other families observe.

At noon, each family group eats together and then uses the hour recreational period to do something together, such as hiking, boating, or fishing.

The afternoon work session again separates the parents and children. In each case, the assignment is the same. The children are asked to form family groups, in which some children play the parents' roles. They are given problems to work out as a family unit. The parents also form family groups, with some parents taking children's roles.

In the final dinner meeting, all the families come together. Both the families and the leaders try to summarize the experiences of the weekend and their significance. Then the leaders outline several homework assignments, and each family has to choose one and promise to work on it at home.

What Do You Think?

1. What do you see as the major benefits such an experience could provide a family? Why?
2. Would you be willing to participate with your family in such an experience? Why, or why not?
3. Do you have any friends who have participated in any kind of marriage enrichment experience? How did they respond to it?

3. In general, avoid the one-time weekend group; it is often too intense, and no follow-up is available, if needed.
4. Never jump into a group experience on impulse. Give it a lot of thought, understanding that experiences leading to growth may be painful.
5. Do not participate in groups where the people are friends and associates if the group's goal is total openness and emotional expression. What occurs in a group session should be privileged information.
6. Don't remain with a group that seems to have an axe to grind, insists that everybody be a certain type of person, or insists that all must participate in every activity.
7. Participate in groups that have a formal connection with a local professional on whom you can check. The local professional is also a source of follow-up help, if necessary.
8. A group of six to 16 members is optimum size. Too small a group may result in scapegoating; too large a group cannot operate effectively.

Such cautions are not meant to dissuade couples from trying to improve their marriages. They are simply meant to help couples select experiences that are beneficial and supportive, rather than threatening and disruptive. Legitimate marriage counselors throughout the United States are working to upgrade their profession and tighten the rules guiding counseling practices. Many states now have licensing provisions for marriage and family counseling.

An Ounce of Prevention Is Worth a Pound of Cure: Marriage Enrichment

Only recently have those working in the field of marriage and family counseling turned their attention away from marital problems and focused, instead, on marriage enrichment. In the past, marital services tended to be remedial in nature. When a couple had a marital problem, they could seek help from numerous sources. Marriage enrichment places the emphasis on the preventive side, with the object of facilitating positive growth.

Mission Statement: *Promote enrichment opportunities and resources that strengthen couple relationships and enhance personal growth, mutual fulfillment and family wellness. (Association for Couples in Marriage Enrichment 2000)*

In other words, the goal is to help couples with good marriages further improve their relationship and head off potential problems. Marriage enrichment is proactive, rather than reactive.

Marriage enrichment programs are for couples who perceive their marriage as functioning fairly well and wish to make it even more mutually satisfying. Because the emphasis is on education and prevention rather than on therapy, these programs are not for couples with serious problems in their marriage. Enrichment programs are generally concerned with enhancing the couple's communication, emotional life, or sexual relationship; fostering marriage strengths; and developing marriage potential, while maintaining a consistent and primary focus on the couple's relationship. This education is designed to help them develop relationships that meet personal needs and enhance individual development.

Some people in the field make a distinction between marriage enrichment and family life enrichment programs. The latter involve not only the primary couple, but the entire family in the program. Again, they are designed for families without severe problems.

If couples are to direct and improve their marriage, they must increase their awareness. Organizing awareness into these four subcategories is helpful: self, partner, relationship, and topical. Marriage enrichment programs usually spend a lot time helping couples or families to become more aware in each of these categories.

For example, in the category of *self-awareness*, enrichment programs offer sensitivity-training exercises to help focus on internal sensory, cognitive, and emotional processes. Goals include achieving a realistic self-picture, openness to one's feelings, minimal defensiveness, and eliminating some emotional hang-ups.

Partner awareness involves knowing accurately what one's partner is experiencing in terms of his/her own self-awareness. How does this behavior affect one's partner? Is the partner happy, sad, or indifferent? How can one best communicate with his/her partner? What does the partner think or feel about this? Answering such questions accurately is the goal of partner-awareness training.

Relationship awareness shifts the focus from the behavior of one individual to the interactional patterns of the couple, or the entire family. For example, who starts an argument, who continues it, and who ends the interaction? Does each individual contribute self-disclosures, feeling inputs, and negative and positive communications? Do the family members play unproductive games? If so, who initiates the game? What are the rules by which the family interacts?

Every relationship has rules, often outside of direct awareness, that create and maintain meaning and order. People like to conceptualize rules in terms of who can do what, where, when, how, and for what length of time. These rules can be applied to any issue in a relationship. For example, does a family allow personal criticism? Who is allowed to criticize? When? And to what degree? What is a family's mode of handling conflicts? Some families talk directly about issues and make active efforts to solve them. Other families pretend that conflicts don't exist and ignore them, in the hope they will go away. Others deal with the issue in some stereotypical manner that usually fails to solve the conflict, but allows family members to ventilate hostility.

Topical awareness is less important than the three categories just discussed. Topical awareness encompasses references to events, objects, ideas, places, and people—topics that constitute most of everyday conversation. By increasing topical awareness, the couple can focus on their interests and find where they differ and where they coincide. In this way, they can find areas in which they can work and play together. They can also recognize and tolerate areas of their spouse's interests that they don't share.

Another purpose of marriage enrichment programs is to help couples and families develop a game plan for handling disputes and conflicts. What are the rules, how can they be clarified, and what procedure can change them? If you don't have a set of rules for handling conflict, your relationship is likely to degenerate into a series of arguments and squabbles. Thus, most marriage enrichment programs spend a great deal of time on the development of communication skills (see Chapter 5). For example, identifying problem ownership, self-assertion, empathic listening, negotiation, and problem solving are all emphasized.

Esteem building is another area of concern in enrichment programs. We sometimes forget this area when lauding the improvement of communication skills, but better communication can equip a person to be destructive besides constructive. Emphasizing esteem building, communicating with a positive intent or spirit, and valuing both the self and the partner make communication constructive and growth enhancing. Esteem building is particularly difficult for a partner who feels devalued and inferior. Thus, enrichment programs stress the importance of building a relationship that negates such feelings and supports positive feelings of value and high esteem in family members.

The fact that a family is interested in and open to the idea of marriage enrichment is an extremely important strength. Certainly, the kinds of goals sought by the marriage enrichment movement are worthwhile. These goals, however, are not as important as the family's general attitude toward marriage. The family that takes an active role in guiding, improving, and working toward better family relations is the family that stands the greatest chance of leading a long, happy, and meaningful life.

How successful are family and marriage enrichment programs? In general, participation in such programs leads to positive changes in a couple's or family's relationships. Real changes and improvement in interactional behavior are noted. Particular programs vary in their effectiveness, but all the programs investigated

HIGHLIGHT

Marriage Encounter

Marriage Encounter is one of the earliest programs designed to help people improve both their marriage and family life. Started by a Spanish Catholic priest (Fr. Gabriel Calvo) in the 1960s for Catholic families, the movement has broadened to other denominations. Although church related, the program has a long history of success in helping all people, regardless of religious affiliation, build stronger, more successful relationships.

Nine major points are basic to the process of Marriage Encounter:

1. *Discovering oneself.* Before moving toward the "you," the "I" has to be found. This idea was previously examined when we discussed love in Chapter 3. If one is not at peace and accepting of oneself, giving love and establishing an intimate relationship is difficult.
2. *Talking to the other.* This is more than simply conveying information between people. It is a mutuality between persons, in which each person really listens to the other and really tries to understand her/his partner at the deepest level. A husband and wife who truly listen to each other will launch a revolution in their own lives and their relationship. By applying the principles discussed in Chapter 5, couples can improve their communication.
3. *Mutual trust.* Without mutual trust, a close intimate relationship would be impossible to achieve. Such trust is not the work of a single day, but grows out of the little personal confidences between husband and wife that occur all the time in a growing relationship.
4. *Growth in knowledge of each other.* Mutual trust leads to greater knowledge of one another. Couples that fall prey to the illusion that all will be well in their marriage if there is

a strong attraction between them, often fail to increase their knowledge of each other. They are married but are, in fact, strangers to each other.

5. *Understanding each other.* Knowledge of one another leads to a true understanding of each other, and with this understanding can come acceptance.
6. *Acceptance of each other.* So often, a person cannot accept certain characteristics and behaviors in their partners. However, a true acceptance of the other person can give that person the confidence in the relationship that can also enable him/her to change. True acceptance yields confidence.
7. *Helping one another.* Mutual help is a sign of true friendship. One of the advantages of marriage is that it is a partnership—two people are working together to solve the problems of life. Being able to rely on another is a great help and comfort.
8. *Growth in love and union.* All the previous steps lead to a growing relationship and foster a sense of mutual gratitude and profound joy in each other. As the love between the couple strengthens, it broadens to include greater love for the family, kin, and community.
9. *Opening up to others.* The establishment of a loving union gives the couple the security to turn love outward to others and to the broader community.

Regardless of one's religious orientation or lack thereof, these nine qualities are important elements in any successful intimate relationship. It is interesting to note that many studies of successful, enduring families find a strong spiritual element that can be tapped to gain strength in time of adversity (Abbott et al. 1990; Robinson and Blanton 1993).

indicate positive changes (Guerney and Maxson 1990, 1991; Markowski 1991; Mattson 1990; Piercy and Sprenkle 1991; L. Roberts and Morris 1998; Robin 1996; Worthington 1996). Despite this empirical support for the effectiveness of enrichment programs, attendance at such programs is relatively low. Time constraints, fear of the invasion of privacy, cost, and the necessity for both spouses to attend are cited as reasons for nonattendance. Enrichment programs need to be better promoted to overcome such restraints and increase attendance (L. Roberts and Morris 1998).

Although this section deals with marriage enrichment experiences, the idea of prevention is finding its way into our society. Premarital prevention programs are also somewhat successful in improving relationship quality (Carrol and Doherty 2003). For example a number of states, including Texas, Oklahoma, and Florida, have legislation providing a reduction in the marriage license fee to those who complete a premarital preparation course. Other states have established a voluntary "Covenant Marriage," one that includes premarital education, is more difficult to dissolve, and sometimes includes a prolonged waiting period.

Florida requires at least one-half credit in life management skills among the general requirements for high school graduation.

Marriage with Purpose: Effective Management

Popular lore has it that the love marriage simply happens (the myth of naturalism that we discussed earlier). Difficulties will, of course, arise; but they can be worked out successfully by any couple truly in love.

It is almost a sacrilege to suggest that people entering a love relationship should make a conscious effort to guide and build their relationship. Many argue that attempts to guide and control a relationship will, in fact, ruin it. Their advice is to "relax and let it happen." This attitude implies a great tolerance on the part of each individual, because what happens may not be something the other wants. How tolerant can a partner be? Does love mean never judging our mate? Does one accept any behavior from his/her mate in an effort to "just let it happen"?

Most people are tolerant only up to a given point, after which certain behavior becomes unacceptable. Most people are tolerant in some areas of life and intolerant in other areas. Everyone can learn to be more tolerant, but total tolerance of all things is probably impossible. People have many and varying standards. As a result, when a relationship "just happens," it usually isn't long before partners discover some of their intolerances. Then they try to change the relationship with their mate. Conflict usually follows, because the mate may not want to change or have the relationship change in the same direction. Without mutually acceptable ways of handling conflict, unconscious games and strategies may take over, and soon communication will be lost.

Marriage and family require management skills. Work, leisure, economics, emotions, interests, sex, children, eating, and maintaining the household all require effective management in the fully functioning family. Most of these matters have been discussed elsewhere in the book, but tying them all together under the concept of effective management seems a proper way to end.

"Surely you can't be serious?" some will object. "Effective management belongs in business, not in my marriage." Yet, every married couple, especially if children are involved, are running a business. For example, just planning what the family will eat for the next week, buying the food, preparing it, and cleaning up afterward require considerable management and organizational skills, especially if money is in short supply. Furthermore, recurrent personal and family crises are likely to throw off schedules and plans. Just look at a common day in John and Cindy's life:

John and Cindy rise early to pack lunches for the children, after which John prepares breakfast for everyone. Cindy, meanwhile, is helping the children get dressed for school. After breakfast, Cindy showers and dresses for work while John cleans up the kitchen. John then prepares for work while Cindy drives the children to school. They then both go to their respective jobs.

Upon return from work, John picks up the children, dropping their daughter at her piano lesson and their son at Little League. Meanwhile Cindy, on the way home from work, stops at the bank, then at the grocery store to pick up dinner items. While John picks up their daughter and goes to the Little League game, Cindy starts dinner. John returns with the children just as dinner is ready. After dinner, both parents oversee the children's homework. Finally, the children go to bed and John and Cindy have a few minutes alone (adopted from Huston 2000,

301). After this normal family day, can anyone still really believe that family management is unnecessary?

Besides day-to-day management, a family needs to plan long-range goals. For some families, life seems to be a constant struggle from one catastrophe to the next. Other families seem to move smoothly through life, despite the crises that arise periodically. What is the difference? Often, the difference is simply a matter of efficient planning and management, versus lack of planning and management. Compare the attitudes toward money of the following two couples, one believing in planning, the other in luck.

Jim and Chloe went to college with Bill and Sally. They remained close friends after college because Jim and Bill got jobs with the same company. Each couple has two children. Chloe works periodically, and the money she earns is always saved or used for some specific goal, such as a trip or an improvement to the house. Sally works most of the time also, but she and Bill don't care much about things like budgeting. As long as there is enough money to pay the bills, nothing else matters.

"You and Jim are always so lucky. Bill and I have wanted to add a master bedroom for ourselves so we could have a retreat away from the children, but we'll never be able to afford it," complains Sally.

"Luck has nothing to do with it," replies Chloe. "Jim and I have planned to add the bedroom for a number of years. We always budget carefully so our monthly expenses are covered by Jim's salary. That way all of the money I earn, less child-care costs, goes into savings. The new bedroom represents my last 3 years of work," she concludes. "I'd hardly call it luck."

Naturally, Sally isn't interested in hearing Chloe's response. Luck is an easier way to explain her friend's new bedroom. Besides, what Chloe is telling her is, "Manage your money better, and you, too, can add a new bedroom. Luck, after all, is being prepared and ready to take advantage of an unforeseen opportunity." This advice will only make Sally feel guilty and inadequate. "It isn't worth all the trouble to budget money and be tightwads to get a new bedroom," Sally will probably think to herself.

Creative management in all areas helps the family run smoothly and achieve its desired goals. Effective management reduces frustrations and conflicts because it gives family members a feeling of success. Careful planning also helps a family maintain the flexibility necessary to cope with unforeseen emergencies. Such flexibility gives family members a feeling of freedom because they can make choices, rather than having choices forced upon them by events beyond their control.

One reason that many married persons feel trapped is that they do not take the initiative to plan and guide their lives, but simply react to circumstances. Of course, there are times and situations when one can do nothing but react. The poor, in particular, often have so little control over their lives that they give up planning altogether and live by luck and fate.

Family control and rational planning become more difficult as social institutions multiply and infringe on family responsibilities. Some of these external stresses on the family have emerged out of necessity. When a society begins to develop beyond a primitive level, its members soon find that many tasks are better performed by agencies other than the family. The clergy take over the job of interceding with the supernatural; police forces, armies, and fire brigades take over the job of protecting the family from physical harm; and schools undertake to educate children. In the complex modern world, the family often has little say over what kind of work its members will perform, or where, or for how long.

WHAT DO YOU THINK?

1. Do you see a relation between planning and luck? What is it?

2. Why do you think Bill and Sally believe they won't ever be able to add a bedroom to their house?

These matters may be decided by the impersonal forces of the marketplace or by distant corporations, unions, or government bureaus.

In such a situation, it becomes even more important for planning, foresight, and management to be an integral part of family life so that the family can cope successfully with outside pressures. The family that actively takes control of its destiny is most often the family that grows and prospers, thereby helping every member toward self-fulfillment.

In the Future, the Family Will Remain and Diversify

Over the years, many people have tried to predict the future of the American family. These predictions vary from an early death to the emergence of new and improved families that will be havens of fulfillment for their members. Those suggesting an early death point to the continuing high divorce rate, the increased incidence of cohabitation, the high number of births outside of marriage, and the increasing number of children being raised in single-parent homes due to both divorce and the out-of-wedlock births.

As we have seen, despite these symptoms of marital demise, 90 percent of Americans eventually marry. Surveys of high school students conducted annually since 1976 show no decline in the importance they attach to marriage (Cherlin 2004). The percentage of young women who respond that they expect to marry has stayed constant, at roughly 80 percent, and has increased for young men from 71 percent to 78 percent. Approximately the same percent respond that "having a good marriage and family life" is extremely important (Cherlin 2004; Thornton and Young-DeMarco 2001). In addition, the large and fancy wedding, bridal showers, bachelor parties, wedding receptions, and the honeymoon have all increased in popularity. The recent desire of homosexuals to legally marry also indicates an active interest in marriage and family by Americans. Thus, it appears that marriage and family will not soon disappear from the American culture. What seems likely is that there will increasingly be multiple forms of intimate relationships and families, rather than a single, ideal family structure.

Andrew Cherlin, in his insightful article about the deinstitutionalization of marriage (2004), suggests that the wedding is becoming a status symbol and marriage a capstone to a relationship, rather than the family starting point. The wedding is seen as an important status symbol of the partners' personal achievements and successful relationship (Bulcroft et al. 2000; Cherlin 2004). "The couples in our study wanted to make a statement through their weddings, a statement both to themselves and to their friends and family that they had passed a milestone in the development of their self-identities. Through wedding ceremonies, the purchase of a home, and the acquisition of other accoutrements of married life, individuals hoped to display their attainment of a prestigious, comfortable, stable style of life" and expected marriage to provide some enforceable trust (Cherlin 2004, 857).

All families face both external environmental pressures and internal stresses and strains. Older generations will probably always think they are seeing the deterioration of the family because their children choose different lifestyles from their own. Yet, differences in lifestyle and family structure do not necessarily mean deterioration. Perhaps with increased affluence and education, each individual will be able to choose from a wider variety of acceptable lifestyles, thereby increasing the chances that the family will be satisfying and fulfilling to its members.

The family has always been with us and always will be. And it will also always change. Indeed, the flexibility of the family allows it to survive. The family is flexible because humans are flexible and create institutions that meet their purposes at a given time. When the basic flexibility of humans is forgotten, people feel threatened by changes in their institutions, even though the changes may, in fact, be helping people meet their needs. Mary Mason and her colleagues (1998) suggest that the current obsession by many scholars and policy makers with the breakdown of the family is oversimplification and erroneous (Giles-Sims 2000). As your author has suggested throughout this book, it is important to understand and try to support the positive characteristics of successful intimate relationships, including marriage, rather than concentrating on the negatives. The concept of family has always been controversial, but it has also always been a central concept whenever people have congregated into a society. Change is an integral part of the concept of family. As Elise Boulding some years ago suggested:

The family has met fire, flood, famine, earthquake, war, and economic and political collapse over the centuries by changing its form, its size, its behavior, its location, its environment, its reality. It is the most resilient social form available to humans. (1983, 259)

Rather than debating whether or not the American family is in decline, emphasis should be on how best to strengthen the family and improve its functioning, regardless of form.

The following suggestions might help the family to better meet the challenges of this new century.

1. The workplace must be made "family friendly." Chapters 12 and 13 offer numerous suggestions for accomplishing this end. If employers join in partnership with families, both the families and the employers will benefit.
2. Family life education must begin early and young people should be taught the art of healthy communication and family relationships. A pro-family bias needs to be encouraged. For example, divorce need not be stigmatized; yet, it should be presented as a course of last resort, to be used only after all efforts to save a relationship have failed.
3. The image of marriage and family conveyed in the popular culture and media needs to be improved. For example, people need to advocate television programming that portrays realistic role models and suggests that conflicts can be resolved, if family members join to iron out difficulties.

Stable families provide opportunities for personal growth and hold the key to society's future through the socialization of children. For these reasons, Americans should be willing to assert our cultural preferences for traditional norms, such as marriage and the two-parent home, while accepting, accommodating, and reaching out to those who have chosen alternative life settings. There is no need to feel nostalgia for the mythical nuclear family of the 1950s. At the same time, there is no need to redefine the family by regarding all possible living arrangements as equally preferable. Rather, we need to advocate programs that encourage responsible preparation for marriage and parenthood. These programs should communicate that if society acknowledges family cohesion as a desirable personal and communal goal, it is the responsibility of society to support families (Bayme 1990, 258).

Healthy intimate relationships lead to healthy families. Healthy families lead to healthy individuals. Healthy individuals, in turn, build healthy relationships that are crucial to the building of a healthy society. Toward this end, the government, in 2001, directed $1 billion over 5 years into the welfare-to-work budget to promote marriage among low-income people. Despite considerable debate over the government's attempt to promote marriage, numerous programs have been developed and are being evaluated (DeAngelis 2004; Kersting 2004).

A healthy family is more than the happy possession of an individual couple and their children. It is this unit, which represents humans at our civilized best, that shapes children into adults. More than any other human institution, family is the vehicle for transmitting our values to future generations. Ultimately, it is our loving connections that give life meaning. Through intimate relationships we enlarge our vision of life and diminish our preoccupation with self. We are at our most considerate, our most loving, and our most selfless within the orbit of a good family. Only within a satisfying relationship can a man and woman create the emotional intimacy and moral vision that they alone can bequeath to their children.

Perhaps a proper way to end a discussion of intimate relationships, marriages, parenting, and families—a field so often fraught with problems—is to remind ourselves of the characteristics found in survivors. Survivors usually:

- Have insight into themselves and the world in which they live.
- Can take the initiative and take charge of their problems, rather than simply reacting to the challenges of life.
- Have a core set of values and beliefs to sustain them and direct their behavior under adversity.
- Have a kind of humor that can find the comic in the tragic, make them laugh at their own mistakes and shortcomings, and remind them that the joy in life most often resides within, rather than without.

It is usually one's spirit, one's attitude toward life's ups and downs, that makes life successful or unsuccessful, not the outside forces that press on all individuals. Life, in most cases, is what you make it.

Summary

1. The American dream "and they lived happily ever after" remains only an unfulfilled dream *unless the newly married couple commit themselves to the idea of working to not only maintain, but also improve, their relationship.*

2. *Unfortunately, a number of factors work against such commitment to the improvement of a relationship.* One factor is the *myth of naturalism,* which claims love will take care of any problems that arise if you marry the right mate. Second is the general *privatism* that pervades American culture and precludes sharing the very intimate problems that arise in marriage. The third factor is the *cynicism* that often surrounds the idea of marriage.

3. *To improve an intimate relationship, a couple must work on three things: (1) themselves as individuals, (2) their relationship, and (3) the general environment within which their relationship exists (especially the economic environment).* Ignoring any of these areas increases the chances that the relationship will experience problems.

4. *There are now many marriage improvement programs from which couples may choose.* However, couples need to use caution in choosing such programs. Al-

though most programs are beneficial, there is always the potential for damage to the relationship if the couple is unprepared for the experience, or if the particular program is poorly presented.

5. *Marriage and family enrichment programs are aimed at families that do not have serious problems.* Enrichment programs are for families that are getting along well, but want to improve their relationships and the quality of their lives.

6. *As unromantic as it may sound, a fulfilling marriage and family life is largely based on good management.* The family that remains in control of its day-to-day life and its future is most apt to be successful. Good planning and follow-through are essential to maintaining such control.

7. Although the family today faces many problems just as it has in the past, *the flexible nature of the family allows it to survive as the major institution for intimate interaction.* Just what form the family will take in the future remains to be seen, but it will continue to survive.

Resources on the Internet

Companion Website for *Human Intimacy: Marriage, the Family, and Its Meaning,* Tenth Edition

http://sociology.wadsworth.com/cox10e/

Gain an even better grasp on this chapter by going to the companion website to take one of the tutorial quizzes, use the flash cards to master key terms, or check out the many other study aids you'll find there. You will also find special features such as GSS data, Sociology Online, and Census 2000 information that will put data and resources at your fingertips to help you with that special project or to help you as you do some research on your own.

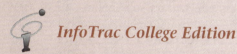

InfoTrac College Edition

http://www.infotrac-college.com/wadsworth/

You can access reliable resources anytime, anywhere, with InfoTrac College Edition, the online library. This fully searchable database offers more than 20 years' worth of full-text articles (not abstracts) from almost 5,000 diverse sources, such as top academic journals, newsletters, and up-to-the-minute periodicals, including *Time, Newsweek, Science, Forbes, The New York Times,* and *USA Today.* You can conduct electronic key word searches using key terms from this chapter to supplement your reading and learning experience. To aid in your search and to gain useful tips, see the Student Guide to InfoTrac College Edition, which you can access through the companion website for this book.

Sexually Transmitted Diseases

Genital Herpes

Of the various kinds of STDs, HIV/AIDS and genital herpes are the most discussed because they are presently incurable. About 22 percent of persons over 12 years of age are currently infected with genital herpes. The number of infected women is higher than infected men. Yet, this sexually transmitted disease was practically unknown to the public just a few years ago. Of the five types of human herpes, types 1 and 2 are at the center of the current epidemic. Type 1 is most familiar as the cause of cold sores; type 2 causes genital lesions. The two types are similar, with type 1 also being capable of causing genital problems.

The first symptom of genital herpes is usually an itching or tingling sensation. Blisters appear within 2 to 15 days after infection. The moist blisters ooze a fluid that is extremely infectious. After 1 to 3 weeks, the blisters gradually dry up and disappear. The infected area is extremely sensitive and sore to the touch. For this reason, sexual activity is precluded when genital herpes is in an active state. In addition, unless the herpes sores are completely healed, the friction of sexual activity can reactivate them.

Because the herpes virus remains in the body once it has been contracted, it can be activated at any time. The exact causes of reactivation are not known, but stress, sunshine, and nutritional and environmental changes are clearly involved. Indeed, herpes is so closely related to a person's moods and emotions that learning to remain calm and under emotional control is one of the most effective preventives against its reactivation.

At this time, there is no cure for genital herpes, only treatment for its symptoms. Acyclovir (trade name, Zovirax), a creamy salve, alleviates symptoms and speeds healing. Unfortunately, it is less effective on subsequent episodes and does nothing to reduce the frequency of outbreak. Some evidence, however, indicates that when taken in oral form, it reduces both the severity and frequency of outbreak.

Genital herpes infection increases the likelihood of HIV/AIDS transmission and acquisition. There is a potential for neonatal infection that is potentially fatal if a pregnant woman is infected.

Because herpes is not yet curable, it is especially important that infected persons share this fact with their partners. In fact, failure to do so is leading people into court. Numerous damage suits have been brought by partners of infected persons who were not apprised of the infected partner's condition.

Chlamydial Infections

Chlamydial infections are caused by the bacteria *Chlamydia trachomatis*. In 2002, the overall rate of chlamydial infection among women (455.9 cases per 100,000 females) was over three times the rate among men (130.1 cases per

100,0000 males). This infection is the most common bacterial STD in the United States. It is the leading cause of pelvic inflammatory disease, one of the most common causes of infertility and ectopic pregnancy in women. Among other problems, such an infection can cause urethritis in men. Fortunately, chlamydial infections are susceptible to inexpensive, readily available antibiotics, so infected persons can be effectively treated. Unfortunately, many infected people, especially females, are asymptomatic and therefore fail to get treatment. An estimated one in 10 adolescent girls and one in 20 women of reproductive age are infected. Young men can be screened for pus in their urine by a relatively simple and inexpensive test.

Pelvic Inflammatory Disease

Pelvic inflammatory disease (PID) affects more than 750,000 women. The patient experiences pain and tenderness involving the lower abdomen, cervix, and uterus, combined with fever, chills, and an elevated white blood cell count. If untreated, PID can progress to the point where the Fallopian tubes are scarred; this, in turn, can lead to female infertility. One potential fatal complication of PID is **ectopic pregnancy**, an abnormal condition that occurs when a fertilized egg implants in a location other than inside a woman's uterus. It is estimated that ectopic pregnancy has increased about fivefold over the past 20 years. Among African American women, ectopic pregnancy is the leading cause of pregnancy-related deaths.

ectopic pregnancy
Implantation of the fertilized egg in one of the Fallopian tubes

Genital Warts

About 3 million cases of genital warts caused by the human papilloma virus (HPV) are diagnosed each year. The hardest hit group seems to be young women in their teens and 20s. Approximately 25 types of HPV can infect the genital area. Infection with high-risk types is a risk factor for cervical cancer, which causes approximately 4,500 deaths among women each year (Office of Women's Health 1999).

Gonorrhea

There were 361,705 new cases of gonorrhea reported in 2001. Teenage girls have the highest infection rate, about 22 times higher than women aged 30 and older. Teenage boys are not far behind. The symptoms in men include an increased need to urinate, some discomfort when urinating, and a purulent urethral discharge. Women may experience abnormal vaginal discharge, abnormal menstruation, and painful urination. Gonorrhea can also infect the rectum during anal intercourse and the throat during fellatio. Here again, women especially—but also some men—may be asymptomatic.

Syphilis

There were 31,575 new cases of syphilis diagnosed in 2001, occurring more frequently in low-income heterosexual populations. Congenital syphilis (the infant becomes infected during birth) occurs in about 1 in 10,000 pregnancies (Hatcher et al. 1998, 206–207; 2004). Rates among females are twice as high as rates among males in the 15 to 19 age group. The symptoms occur in three stages:

Primary. The classical chancre is a painless ulcer, located at the site of exposure.

Secondary. Patients may have a highly variable skin rash, swollen glands, and other signs.

Latent. Patients have no clinical signs of infection.

Hepatitis B

An estimated 150,000 new cases of HBV (hepatitis B virus) are transmitted sexually each year, with another 200,000 people infected by other means, despite the availability of a preventive vaccine. Some cases end in cirrhosis and/or cancer of the liver. Heterosexual intercourse is now the predominant mode of transmission.

Most HBV infections are unapparent. Symptoms, when present, may include moderate liver enlargement with tenderness, vomiting, headache, fever, dark urine, jaundice, skin eruptions, arthritis, and anorexia (Hatcher et al. 1998, 200–201).

HIV/AIDS*

In the newspaper, on the radio, on television, and in your friends' discussions, you frequently hear the terms HIV and AIDS. These terms evoke anxiety, because HIV/AIDS leads to death, in most cases, and at present there is no known cure. AIDS involves the most intimate aspects of your life, namely, sexual behavior. It is also related to drug use, blood transfusions, pregnancy, and birth. Since there are no HIV/AIDS vaccines at this time, your best protection against this deadly disease is information.

How can I live a full life, but avoid contracting and spreading HIV/AIDS? To answer this question, you must know what HIV/AIDS is, how it is transmitted, and what steps you can take to prevent its spread. HIV/AIDS involves intimate aspects of your life, so the answers involve moral and ethical judgments, about yourself and your relationships with others. Because HIV/AIDS is a pandemic around the world, the World Health Organization has designated December 1 of each year World AIDS Day.

What Is AIDS?

We use the letters "AIDS" to denote a deadly group of diseases caused by newly discovered viruses. The letters stand for a long name: **A**cquired **I**mmune **D**eficiency **S**yndrome. *Acquired* means that the conditions are not inherited, but are acquired from environmental factors, such as virus infections. *Immune deficiency* means that AIDS causes deficient immunity, often reflected in poor nutrition and low resistance to infections and cancer. *Syndrome* means that AIDS

---

*This material is adapted from Frank Cox, *The Aids Booklet*, published by McGraw-Hill, 2000. The statistics come from the Centers for Disease Control and Prevention's "HIV/AIDS Surveillance Report," Vol. 14, March 2, 2004.

TABLE A-1

Sexually Transmitted Diseases (STDs)

Disease	Cause	Incubation Period[1]	Characteristics	Treatment[2]
Primary syphilis	Bacteria	7–90 days (usually 3 weeks)	Small, painless sore or chancre, usually on genitals but also on other parts of the body	Penicillin and broad-spectrum antibiotics
Secondary syphilis	Untreated primary		Skin rashes or completely latent, enlarged lymph glands	Same
Tertiary syphilis	Untreated primary		Possible invasion of central nervous system, causing various paralyses; heart trouble; insanity	Same
Gonorrhea	Bacteria	3–5 days	Discharge, burning, pain, swelling of genitals and glands; possible loss of erectile ability in males who delay treatment, with chance of permanent sterility; 80% of infected females asymptomatic	Same
Chancroid	Bacteria	2–6 days	Shallow, painful ulcers, swollen lymph glands in groin	Sulfa drugs, broad-spectrum antibiotics
Lymphogranuloma venereum	Virus	5–30 days	First, small blisters, then swollen lymph glands; may affect kidneys	Broad-spectrum antibiotics
Genital herpes	Virus	Unknown	Blisters in genital area; very persistent	Pain-relieving ointments
Chlamydia	Bacteria	10–20 days	Men: early morning watery discharge, hot or itchy feeling within penis. Women: may be similar to men or symptom-free. Untreated causes pelvic inflammatory disease leading to infertility	Tetracycline
Genital warts	Virus	Unknown	Observable in genital areas	Removed by freezing or ointment
Hepatitis B	Virus	Varies	Most infections are clinically unapparent. When present, symptoms include nausea, vomiting, headache, fever, dark urine, jaundice	No specific therapy exists; only STD for which vaccine exists
AIDS	Virus	5–15 years	Many symptoms	Treat symptoms

NOTE: As soon as you suspect any sexually transmitted disease or notice any symptoms, consult your doctor, local health clinic, or local state health department. Both syphilis and gonorrhea, in particular, can be easily treated if detected early; if not, both can become recurrent, with dire results.

[1] If STD is diagnosed, all sexual partners during the infected person's incubation period and up to discovery of the disease should be examined medically.

[2] Local health clinics or local and state health departments will have more detailed information on current treatment and follow-up.

causes several kinds of diseases, each with characteristic clusters of signs and symptoms.

The virus that causes AIDS has different names, but the preferred term that most scientists are now using is HIV (human immunodeficiency virus). Although we usually speak of one AIDS virus, there are, in fact, several related viruses that can cause AIDS, AIDS-related conditions, and cancers in human beings.

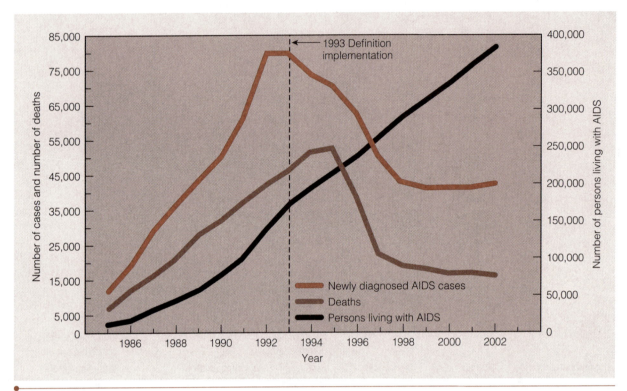

FIGURE A-1 *AIDS Cases, Deaths, and Persons Living with AIDS in the United States, 1985–2002*
SOURCE: CDC 2003a.

People afflicted with HIV/AIDS usually suffer with varying combinations of severe weight loss, many types of infections, and several kinds of cancer. Several symptoms may occur at one time. The usual result is that AIDS-afflicted persons go through long, miserable illnesses that end in death 1 to 2 years after first diagnosis. The person does not actually die from HIV/AIDS, but from one of the opportunistic diseases that strike because of his/her damaged immune system. This general description does not apply to every person with AIDS. Some persons with AIDS alternate between periods of sickness and health, at least in the early period after diagnosis. Some people die within a few months. A few persons have now lived 15 or more years after the initial diagnosis. See Figure A-1.

Varying symptoms are associated with HIV/AIDS because, as the name implies, the AIDS virus (HIV) impairs the body's ability to fight infection. HIV does this by destroying **lymphocytes**, or "lymph cells." These cells are the most common kind of white blood cells and form the basis of the body's **immune system** (its biological defense system). Lymphocytes help prevent cancers by controlling cell growth, and guard against infections by producing antibodies that fight infection. HIV destroys the ability of the lymphocytes to perform these functions. Consequently, persons with AIDS appear undernourished and wasted, often have cancer, and lack protection against infections.

lymphocyte
A type of white blood cell that is active in the body's immune system
immune system
The body's biological defense system that wards off disease and illness

General Symptoms

1. Loss of appetite, with weight loss of 10 or more pounds in 2 months or less
2. Swollen glands (lymph glands) in the neck, armpits, or groin that persist for 3 months or more

3. Severe tiredness (not related to exercise or drug use)
4. Unexplained, persistent/recurrent fevers, often with night sweats
5. Persistent, unexplained cough (not from smoking, cold, or flu), often associated with shortness of breath
6. Unexplained, persistent diarrhea
7. Persistent white coating or spots inside the mouth or throat, which may be accompanied by soreness and difficulty in swallowing
8. Persistent purple or brown lumps, or spots on the skin that look like small bruises on fair-skinned people and appear darker than the surrounding skin on dark-skinned people
9. Nervous system problems, including loss of memory, inability to think clearly, loss of judgment, and/or depression
10. Other problems such as headaches, stiff neck, and muscle numbness or weakness may occur

Any of these symptoms may be caused by diseases other than AIDS, and this makes self-diagnosis difficult. However, if such symptoms persist or several appear at the same time, you must suspect exposure to HIV and should immediately see a physician familiar with the disease.

Incubation Period and Usual Course of Development

Within a few weeks after first exposure, some HIV-infected persons develop an illness that lasts 7 to 14 days. Symptoms include enlargement of the lymph glands, sore throat, fever, muscle aches, headache, and a skin rash that may look like measles. HIV can be detected in circulating blood lymphocytes. The early symptoms usually disappear, or are so mild that they are not even remembered. *However, the infected person is now contagious for the remainder of his/her lifetime and can transmit HIV to other persons.*

Months or years may pass without any signs of AIDS in the infected person. However, usually 2 to 15 years after initial exposure, the infected person can expect to come down with one or more of the characteristic signs of AIDS. Just how long the incubation period can be for any one individual is currently unknown. Some researchers predict that an incubation period as long as 20 years may be possible. Be that as it may, once AIDS appears, death will usually follow within the next several years. At this time, 61 percent of all persons who have developed full-blown AIDS have died (approximately 501,669 as of January 1, 2003; CDC 2004).

Occurrence

Growth in the number of AIDS cases in the United States has leveled since 1998, leading the Centers for Disease Control (CDC) to predict that approximately 20,000 to 30,000 people will be diagnosed with AIDS each year. Researchers estimate that there are currently up to 5 million people who are HIV positive, but show no symptoms of the disease. Such a person is termed **seropositive** and *can spread the disease.* As of January 1, 2003, approximately 384,906 persons are living with AIDS (77% men and 23% women). Those persons aged 25 to 34 years represented 29.8 percent of all new AIDS diagnoses in 2002.

seropositive

Blood acts positively to a test for a disease or illness, indicating that the person is carrying the disease germ or virus

As of 2002, AIDS cases can be divided into the following categories:

Category	Percentage of AIDS Cases
Sexually active homosexual and bisexual men (or men who have had sex with another man since 1977)	39
Homosexual and bisexual men who are also intravenous (IV) drug users	6
Present or past users of illegal IV drugs	29
Heterosexual men and women (including sex partners of persons with AIDS or at risk for AIDS) and people born in countries where heterosexual transmission is thought to be common.	27

HIV/AIDS is found in all races. In the United States, approximately 41 percent of persons diagnosed with AIDS since 1998 are Caucasian, 39 percent are African American, and 18 percent are Hispanic. Children under 5 years of age account for less than 1 percent of all persons with AIDS.

How Do You Get AIDS?

You've read and heard about many different ways people can get HIV/AIDS. Some of the stories are true, but many are false. Unfortunately, you may be afraid of getting AIDS when you need not be. Even more unfortunate, you may not be afraid of getting HIV/AIDS when you should be. Fear is usually not very helpful when you are trying to understand and make logical decisions. Knowing the actual ways HIV/AIDS is transmitted and the ways it is not transmitted is vital to your health.

High-Risk Behaviors

Until now, HIV/AIDS has infected very specific groups of persons in the United States. By studying these groups and their behavior patterns, we have gained a better understanding of how the causative viruses are spread and the kinds of precautions that all people must take to avoid contracting or spreading the disease. Essentially, these groups tell us that certain sexual behaviors and drug-related activities are the major means by which HIV/AIDS is transmitted in the United States. These groups tell us that engaging indiscriminately in sexual or drug activities with multiple partners is particularly dangerous.

HIV is primarily spread through the sharing of virus-infected lymphocytes in semen (the thick, whitish fluid secreted by the male during ejaculation) and in blood. For example, if you have sex or share drug needles with two other persons, each of whom have also shared these activities with two other persons, and so on, you quickly find that you have interacted with literally hundreds of people. *In other words, when you have sex or share needles with someone, you are having sex and sharing needles with everyone he/she has ever done these things with, and also the partners of those persons.* HIV/AIDS is also spread in a few other ways: during pregnancy and birth (to the newborn), through HIV-infected lymphocytes in a mother's milk, and occasionally, through transfusion of blood or blood products.

Sexual Behavior   *HIV is primarily a sexually transmitted disease.* Thus, it is transmitted in much the same way as other STDs, such as syphilis and gonorrhea. Sexual contact with persons who have the disease or carry the virus is currently the most common way by which HIV/AIDS is transmitted. Avoidance of sexual contact (**abstinence**) is the best way to keep from being infected. The safer sex that you have been hearing so much about involves using condoms and other methods to avoid sharing bodily fluids (semen in particular). However, so-called safer sex does not eliminate the danger completely, and avoidance of sexual contact outside of long-term, mutually monogamous relationships (the practice of having only one sexual partner) is the only really safe sex.

Certain sexual practices are very liable to transmit AIDS and therefore should be avoided. *Receptive anal-rectal intercourse* (allowing the penis to enter one's rectum) appears to be the most dangerous sexual practice. Because the rectum is lined by a thin, single-cell layer, HIV-infected lymphocytes in semen can migrate through the lining. This area of the body is richly supplied with blood vessels, and insertion of the penis or other objects into the rectum is very apt to result in tearing and bleeding.

Fellatio (insertion of the penis into the partner's mouth) may also be a means of HIV/AIDS transmission. Persons with tooth or gum infections are particularly at risk. The possibility that teeth may accidentally break the skin of the penis is another risk of fellatio. Because the semen of an infected person has a high concentration of HIV, it is probably not a good idea to swallow it, although there is no current evidence of transmission in this manner.

Vaginal intercourse can transmit HIV/AIDS to either men or women, although researchers are currently unsure what the mechanism for transmission is. The lining of the vagina itself is thick and difficult for HIV-infected lymphocytes to penetrate. In addition, the normally acidic (acid) environment of the vagina is not hospitable to the HIV/AIDS virus. However, the uterus is lined with only a single layer of cells, which is more easily penetrated. Semen is probably most apt to reach the uterus with prolonged or repeated acts of intercourse. It also appears that due to the high concentration of the virus in semen, HIV is more easily transmitted to the woman than from the infected woman to her male partner. The use of condoms with a spermicide such as nonoxynol-9, attention to adequate lubrication, abstinence from sex during menstruation, and mutually monogamous relationships can all help prevent the spread of HIV/AIDS.

Drug Abuse *Intravenous* (IV) drug users shoot drugs directly into their veins, using syringes and needles. IV drug users who share syringes and needles are the second largest group of people with AIDS in the United States. In some parts of the country, such as New York and New Jersey, IV drug users constitute the majority of infected persons. IV drug users are important because they represent the major bridge by which HIV/AIDS spreads to women and children, and thereby, enters the general population.

IV drug users are extremely hazardous because, unlike homosexual men who have organized to fight the spread of HIV/AIDS, drug users are engaged in illegal behavior and thus cannot openly organize to fight HIV/AIDS. Many women IV drug users resort to prostitution to support their drug habits. If infected, they add to the risk of spreading HIV/AIDS via sexual relations. Also, when under the influence of drugs, people are often unaware of exactly what they are doing. Thus, they may not practice lower-risk behaviors and may use unsafe methods when **shooting up** (injecting drugs into the body with a needle and syringe). Sharing infected blood in needles is the major way HIV/AIDS is transmitted between IV drug users.

abstinence
Avoidance of sexual contact

fellatio
Oral copulation of the male

shooting up
Injecting drugs directly into the bloodstream with a needle and syringe

Blood Supply and Transfusions Transfusions and the use of contaminated blood products account for only a small portion of HIV/AIDS cases in the United States. Fortunately, the chances of infection have become smaller because the possibility of transmission through infected blood has been recognized. The CDC suggests the chances of infection via blood transfusion are about 1 in 60,000, although some researchers suggest the chances of infection are higher.

Casual Transmission One of the most important things for you to understand about HIV/AIDS is that the chances of *casual transmission* (the transmission of HIV/AIDS without sexual contact or intravenous drug use) are very low. This means that HIV/AIDS is transmitted in very specific and limited ways, most often involving sexual contact or the sharing of needles among drug users, as was discussed. Outside these specific means of transmission, AIDS does not seem to move from the infected to the noninfected person easily. If HIV/AIDS were casually transmitted, it would be found in much larger numbers throughout the population by this time. Being around a person with HIV/AIDS is not considered dangerous, unless you have sex or share needles and drugs. Although we cannot be absolutely certain that casual transmission never happens, so far nearly all cases have been traced to the sharing of HIV-infected lymphocytes in semen, blood, or blood products. In general, no one is likely to get HIV/AIDS by shaking hands, patting a friend on the shoulder, superficial kissing, drinking from the same cup, working in the same office, eating food served by infected persons, being sneezed or coughed on, using public restrooms, and other casual contacts.

Preventing HIV/AIDS

The following precautions will help prevent you from contracting or spreading HIV/AIDS and other STDs:

1. Practice sexual abstinence, especially when a caring relationship is not involved.
2. Practice sexual fidelity.
3. If you want to have sexual relations but are not in a permanent, monogamous relationship, use mechanical barriers to prevent the exchange of potentially infectious body fluids, especially blood, semen, and uterine secretions.
4. Use the following barriers, proven to prevent pregnancy and STDs:
   a. FDA-approved latex condoms can protect a woman and a man from sharing semen and vaginal secretions during conventional vaginal intercourse.
   b. Physician-prescribed and fitted vaginal diaphragms or cervical caps block semen from reaching the uterus.
   c. Spermicides, such as nonoxynol-9, paralyze sperm and migrant lymphocytes that may have gotten past the barrier. A few people will have allergic reactions to specific spermicides.
5. Use barriers strictly in accordance with recommendations supplied by the FDA (Food and Drug Administration) and inserts supplied by the manufacturers. A lot of careful research, time, and effort are embodied in these recommendations.
6. Avoid anal intercourse with or without a condom because this is the most dangerous way to share semen; condoms are not well designed for this means of sexual expression.

7. Avoid sexual relations with persons at great risk for being HIV positive or having AIDS or other transmissible viruses, such as homosexual or bisexual persons, persons who "shoot" drugs, or persons who sell or buy sex.

8. Do not use alcohol or drugs. They interfere with caring for yourself and for other persons; especially avoid the use of drugs that are injected into the veins.

9. Do not share needles used for injecting drugs into the veins, or handle sharp instruments contaminated with the blood of other persons.

10. Donate blood to your community blood banks and encourage your healthy friends to do so as often as possible. Such donations will help save lives and protect the blood supply from possible contamination.

11. If you know well ahead of time that you might need blood during an operation, pre-donate your blood and let your caring friends of similar blood type know that their donations would be appreciated.

12. Continue to build caring, meaningful, and trusting relationships with people with whom you are intimate. Everyone will benefit.

13. Regarding persons with whom you are not intimate, who are sick with any disease, or who are known to have AIDS, your sincere caring will prove to be mutually helpful.

14. Although AIDS seems destined to become one of the most deadly epidemics humans have faced, do not be afraid. Your chances of becoming infected are near zero if you apply the foregoing precautions.

15. Do not pay attention to rumors and hearsay about AIDS. Whenever you have a question about HIV/AIDS or behavior that increases the chances of exposure, seek out answers from health professionals, medical clinics, teachers, counselors, and other knowledgeable adults.

16. Do not be lulled into a sense of complacency or apathy encouraged by the conflicting reports or lack of reports in the news media. Although significant breakthroughs are often reported, you must do all you can to protect yourself from acquiring HIV because all real and permanent cures still remain in the future.

17. When you seek diagnosis or treatment in a health care facility, be aware that you and all persons working there are at increased risk for infectious diseases. Therefore, you would be wise to protect yourself, along with your health care providers, by insisting on adequate hand washing, proper use of protective gloves, better protection for syringes and needles, and no reuse of multiple-dose vials for giving injectable medications. Although your insistence might seem improper, it will lead to improved care and help reduce infectious disease spread.

18. Finally, your health and the health of your community are our greatest national assets. Don't lose your health by being careless.

Contraceptive Methods

The Condom

The condom has become increasingly popular with both men and women because it can play a significant role in the prevention of sexually transmitted diseases such as HIV/AIDS. Despite the protection that condoms offer and their ability to prevent unwanted pregnancy, however, only a minority of sexually active youth report consistent use of condoms during sexual intercourse.

The **condom** is a sheath of very thin latex or animal gut that fits over the penis and stops sperm from entering the vagina when ejaculation occurs. The condom should be placed on the penis as soon as it is erect, to ensure that none of the preejaculatory fluid (which contains some sperm) gets into the vagina. It is best to leave space at the end to allow room for ejaculate.

Care should be taken not to puncture the condom. After ejaculation, the condom should be held tightly around the base of the penis as the man withdraws, so that no semen will spill from the now-loose condom.

Latex condoms, used properly, are close to 100 percent effective against both pregnancy and STDs. So compelling is the evidence about condom effectiveness that since 1987, the U.S. Food and Drug Administration (FDA) has allowed manufacturers to list diseases (HIV/AIDS, syphilis, gonorrhea, chlamydia, genital herpes) that condoms can help prevent.

Barring breakage, the risk of pregnancy with condom use is essentially zero. Most pregnancies occur because of careless removal of the condom. Risks can be reduced by using a vaginal spermicide along with the condom.

Women's Condom A newer alternative is a condom that women can wear. This new device consists of a soft, loose-fitting polyurethane sheath and two diaphragm-like, flexible rings. It is inserted like a tampon and protects the inside of the vagina, with the inner ring covering the cervix and the outer ring holding it in place by pressing against the woman's body. The female condom offers all the advantages of the male variety and gives the woman control over condom use. Unfortunately, it is expensive, costing up to $3 each, and does require manual insertion, which some might find unpleasant. There is a second, slightly different type that can be cleaned and reused (Hatcher et al. 1998, 2004; Seligmann 1992), but as with the first type, it is not in wide use at this time.

The Diaphragm

The **diaphragm** is a dome-shaped cup of thin latex, stretched over a metal spring. It is available with prescription from a physician. Because the sizes of vaginal openings differ, a diaphragm must be carefully fitted, to ensure that it adequately covers the mouth of the cervix and is comfortable. The fitting should be checked every 2 years and after childbirth, abortion, or a weight gain/loss of more than 10 pounds.

Once a woman has been fitted properly, she can insert the diaphragm with the dome up or down, which ever feels more comfortable. Before insertion, she should spread a spermicidal jelly or cream over the surface of the dome that will lie against the cervix.

A woman can insert a diaphragm just before intercourse, or several hours in advance. If she inserts it more than 2 hours before intercourse, however, she should insert additional spermicide into the vagina before intercourse or take the diaphragm out and reapply spermicide. Spermicide should also be added before any further acts of intercourse. The diaphragm must stay in place for 6 hours after the last act of intercourse, to give the spermicide enough time to kill all sperm.

The Cervical Cap

The **cervical cap** is similar to the diaphragm, in that it blocks entrance to the cervix. Although it has been in use for years, especially in Europe, it is still little known in the United States. The cervical cap is much smaller than the diaphragm and is made of flexible rubber. It looks something like an elongated thimble that might fit over one's thumb. Like the diaphragm, it must be fitted to the individual by a medical doctor. However, unlike the diaphragm, the cap may be left in place for up to 2 days. After intercourse, it should not be removed for 8 to 12 hours. Also like the diaphragm, it should be used with a spermicide.

The Intrauterine Device

The **intrauterine device (IUD)** is a stainless steel or plastic loop, ring, or spiral that is inserted into the uterus. A physician uses a sterile applicator to insert the IUD into the cervical canal, and presses a plunger to push the device into the uterus. The protruding threads are trimmed so that only an inch remains in the upper vagina. Usually, the threads can't be felt during intercourse. The best time for insertion (and removal) is during menstruation because the cervical canal is open widest then, and there is no possibility of an unsuspected pregnancy.

Just how or why the IUD works is still unknown. Theories include production of biochemical changes, interference with the implantation of eggs or the movement of eggs or sperm, and spontaneous abortion. The IUD is quite effective, second only to the pill in overall effectiveness. Once inserted, it can be used indefinitely without additional contraceptive measures; it doesn't interrupt sexual activity, it is fully reversible, and its long-term cost is low.

About 10 percent of women who try IUDs spontaneously expel them, and if the woman does not notice the expulsion, an unwanted pregnancy may result. The threads can also be checked periodically to make sure the device is still in place.

A more serious complication that may occur is pelvic inflammatory disease. About 2 to 3 percent of women using the device may develop the disease, usually in the first 2 weeks after insertion. Most of these inflammations are mild and can be treated with antibiotics. In rare cases, the IUD may puncture the wall of the uterus and migrate into the abdominal cavity, requiring surgery. Pelvic infection, usually caused by bacteria, also seems to be more common among IUD users (G. Stewart 1998b, 511–544).

Most of the risk of pregnancy occurs during the first few months of IUD use, so an additional method of contraception should be used during that period. The failure rate tends to decline rapidly after the first year of use. Pregnancy may occur with the device in place, but the rate of spontaneous abortion for such pregnancies is 40 percent (compared with 15 percent for all pregnancies). There is little additional risk of birth defects for babies born of such pregnancies. The device usually remains in place during the pregnancy and is expelled during the delivery. Unfortunately, due to litigation claiming IUD damage to users, IUD availability is limited in the United States. It is, however, used by many women outside of the United States.

Oral Contraceptive

The pill, or **oral contraceptive**, is a combination of the hormones estrogen and progesterone, and must be prescribed by a physician. Daily ingestion of these hormones stops ovulation. Because no mature eggs are released, pregnancy cannot take place. In addition, the hormones thicken the mucus covering the cervix, thus inhibiting sperm entry.

The combination pill is a monthly supply of 20 or 21 tablets. The woman takes pill 1 on day 5 of her menstrual cycle, counting the first day of the cycle as day 1. She takes another pill each day, until all pills have been taken. Menstruation usually begins 2 to 4 days after the last pill. If menstruation does not occur when expected, a new series of pills should be started a week after the end of the last series. If a period doesn't begin after this series, the woman should consult a physician.

During the first month of the first cycle, the woman should use an additional method of contraception, to ensure complete protection. The pill should be taken at about the same time each day. If the woman forgets a pill, she should take it as soon as possible and should take the next pill at the scheduled time. If she forgets two pills, the woman should use an additional form of contraception for the rest of the cycle. If she forgets three pills, withdrawal bleeding will probably start. In this case, the woman should stop taking the pills and start using another method of birth control. She should then start a new cycle of pills on the fifth day after the bleeding starts, but continue to use an additional method of birth control for the first new cycle.

Taken properly, the pill is the most effective method of contraception available today. It is relatively simple to use, does not affect spontaneity, and is inexpensive and reversible. However, it does not afford protection against sexually transmitted diseases. There is some evidence that it may increase the likelihood of HIV/AIDS infection (F. Cox 2000, 26).

The pill was once the most widely used method of contraception in this country. In 1975, as many as 8 million American women, or about 40 percent of all women using birth control, used the pill. However, use of the pill began to drop rapidly thereafter, due to fear of a link between the pill and breast cancer. This fear has proven to be unsupported, however. Studies have found no clear link between oral contraceptives and breast cancer, regardless of the type of pill or the duration of use, up to 20 years (Hatcher et al. 1998, 2004). Indeed, use of the pill appears to decrease benign breast cysts, reduce the chances of ovarian cancer, and protect against acute pelvic inflammatory disease (Hatcher et al. 1998, 458; 2004). On the other hand, it is interesting to note that it was not until June 1999 that the Japanese Ministry of Health deemed the oral contraceptive safe for Japanese women.

The pill's side effects range from relatively minor disturbances to serious ones. Among the minor disturbances are symptoms of early pregnancy (such as morning sickness, weight gain, and swollen breasts), which may occur during the first few months of pill use. Such symptoms usually disappear by the fourth month.

Other relatively minor problems include depression, nervousness, alteration in sex drive, dizziness, headaches, bleeding between periods, and vaginal discharge. Yeast fungus infections are also more common in women taking the pill. The more serious side effects include blood clots and a possible increase in the risk of uterine cancer. Although the incidence of fatal blood clots is low (about 13 deaths among 1 million pill users in 1 year), women with any history of unusual blood clotting, strokes, heart disease or defects, or any form of cancer should not use the pill. Smoking tends to increase the possibility of these problems, especially for women over 40 years of age.

Another form of the pill, the *minipill*, is taken throughout the month, even during menstruation; this eliminates the necessity of counting pills, and stopping and restarting a series. The minipill also eliminates many of the negative side effects related to the estrogen component of the regular pill. Minipills contain only progestin and do not stop ovulation or interfere with menstruation. Instead, they make the reproductive system resistant to sperm or ovum transport. Should fertilization take place, they also impede implantation. They are slightly less effective than the regular pill, however. Women using the minipills appear more susceptible to **ectopic pregnancy**, implantation of the fertilized egg in one of the Fallopian tubes (Hatcher et al. 1998, 2004). As previously noted (see Chapter 9) pelvic inflammatory disease increases the chance of an ectopic pregnancy by 10 times over the normal frequency of about 1 in every 300 pregnancies (Gosden 1999, 194).

Many suggest that it was "the pill" becoming available in 1960 that triggered the sexual revolution, by detaching conception from sexual intercourse. In fact, "the pill" became almost synonymous with the sexual revolution.

Variations of the pill include the following:

Norplant. Norplant is a progestin-only implant that provides the woman with 5 years of protection. The six-capsule Norplant system is usually implanted under the skin, in the woman's inner upper arm. Once implanted, the woman need do nothing more to achieve contraception. Ovulation can be restored by removing the implants. Norplant is relatively expensive at first, but averaged over the 5-year period of effectiveness, it is a cheap form of contraception.

Depo-Provera. Depo-Provera, a contraceptive injection, provides protection for 3 months. It is the most commonly used injectable progestin approved for use in the United States. Like the pill, it is a hormone that stops ovulation, so no egg is available for fertilization.

Morning-after pill. The morning-after pill contains hormones that prevent implantation of the egg if fertilization has taken place. Because some pills contain a high dose of estrogen, about 70 percent of users report nausea and about 30 percent report vomiting. Users also report menstrual irregularities, headaches, and dizziness. In general, morning-after pills should be a method of last resort, because the full effects that the various hormones have on the body may not be fully understood.

Vaginal Spermicides (Chemical Methods)

Spermicides (sperm-killing agents) come as foams, creams, jellies, foaming tablets, and suppositories. Foams are the most effective because they form the densest, most evenly distributed barrier to the cervical opening. Tablets and suppositories, which melt in the vagina, are the least effective.

Foams are packed under pressure (like shaving cream) and have an applicator attached to the nozzle. Creams and jellies come in tubes with an applicator. A short time before intercourse, the applicator is placed into the vagina (like a sanitary tampon), and the plunger is pushed. Vaginal spermicides remain effective for only about half an hour, so another application is necessary before each act of intercourse.

Vaginal contraceptive film (VCF) is a relative new vaginal spermicidal agent. It is a paper-thin, 2-inch-square vaginal insert containing the spermicide nonoxynol-9, polyvinyl alcohol, and glycerine. When placed high in the vagina 5 minutes to 2 hours before intercourse, it dissolves into a sperm-killing gel. VCF is less messy and more comfortable to insert than other vaginal contraceptives. It cannot be felt by either partner, and nothing needs to be removed because it dissolves completely (Hatcher et al. 1998, 2004).

Vaginal spermicides are generally harmless, relatively easy to use, and readily available in most drugstores without a prescription. However, they are not very effective, though their effectiveness can be increased by using them with a diaphragm and/or condom. Their major disadvantage is the necessity of placing the spermicide into the vagina, which makes some people uncomfortable.

Fertility Awareness: The Rhythm Method, or Natural Family Planning

The *rhythm method* of contraception is based on the fact that usually only one egg is produced each month. Because the egg lives for only 24 to 48 hours if not fertilized, and because sperm released into the uterus live only 48 to 72 hours, conception can theoretically occur only during 4 days of any cycle. Predicting this 4-day period is the difficulty. If each woman had an absolutely regular monthly cycle, the rhythm method would be much more reliable than it is. Unfortunately, not all women have regular cycles. In fact, about 15 percent have such irregular periods that the rhythm method cannot be used at all.

To use the rhythm method, a woman charts her menstrual periods for a full year. Counting the day menstruation begins as day 1, she notes the length of the shortest time before menstruation starts again and also the longest time. If her cycle is always the same length, she subtracts 18 from the number of days in the cycle, which gives the first unsafe day. Subtracting 11 gives the last unsafe day. For example, a woman with a regular 28-day cycle would find that the first unsafe day is day 10 of her cycle and the last is day 17. Thus, she should not engage in intercourse from the tenth to the eighteenth day. If a woman's cycle is slightly irregular, she can still determine unsafe days by using the formula. In this case, she subtracts 18 from her shortest cycle to determine the first unsafe day and 11 from her longest cycle to find the last unsafe day.

Basal body temperature (BBT) can also be charted. The BBT is the lowest body temperature of a healthy person, taken upon awakening. By noting this temperature daily, a woman may determine her time of ovulation. This determination is possible because BBT sometimes drops about 12 to 24 hours before ovulation, and almost always rises for several days after ovulation.

There are other natural signs of ovulation, such as changes in the character of cervical mucus prior to ovulation. However, all of the fertility awareness methods of contraception require a woman to have a high degree of knowledge about her reproductive cycle. This takes a great deal of training and planning to be successful.

Withdrawal (Coitus Interruptus)

Withdrawal is simply what the name implies; just before ejaculation, the male withdraws his penis from the vagina. Withdrawal is probably the oldest known form of contraception. It is free, requires no preparation, and is always available. However, it has a high failure rate. It requires tremendous control by the man, and the fear that withdrawal may not occur in time can destroy sexual pleasure for both partners. The woman may also be denied satisfaction if the man must withdraw before she reaches orgasm. In addition, semen leakage before withdrawal can cause pregnancy.

Sterilization

Sterilization is the most effective and permanent means of birth control. Despite the fact that in many cases it is irreversible, Americans are increasingly choosing sterilization as a means of contraception. Female sterilization is currently the most popular method of birth control among married couples, while male sterilization ranks third in popularity (G. Stewart and Carignan 1998, 545–588).

Vasectomy is the surgical sterilization of the male. Although a vasectomy is safer, simpler, and less expensive than the woman's tubal ligation, only about one-third of the sterilization operations are vasectomies. Done in a physician's office under local anesthetic, it takes about 30 minutes. Small incisions are made in the scrotum, and the vasa deferentia, which carry sperm from the testes, are cut and tied.

The man may feel a dull ache in the surgical area and in the lower abdomen after a vasectomy. Aspirin and an ice bag help relieve these feelings. Usually, the man can return to work in 2 days and can have sex again as soon as he doesn't feel any discomfort, usually in about a week.

An additional method of contraception must be used (at least for a period of time) after the operation, because live sperm still remain in parts of the reproductive system. After about 1 to 2 months and a number of ejaculations, the man must return to have his semen examined for the presence of live sperm. If none are found, other birth control methods can be dropped.

Although the man will continue to produce sperm after the operation, the sperm will now be absorbed into his body. His seminal fluid will be reduced only slightly. Hormone output will be normal, and he will not experience any physical changes in his sex drive. Some males, however, experience negative psychological side effects. For example, some equate the vasectomy with castration and feel less sexual. Such feelings may interfere with sexual ability. Postvasectomy psychological problems occur in perhaps 3 to 15 percent of men. On the other hand, many men report they feel freer and more satisfied with sex after a vasectomy.

In about 1 percent of vasectomies, a severed vas deferens rejoins itself, so that sperm can again travel through the duct and be ejaculated (G. Stewart and Carignan 1998, 547–548). Because of this possibility, a yearly visit to a physician to check for sperm in the semen is a good safety precaution.

One of the major drawbacks of vasectomy (or any other method of sterilization) is that in many cases, it is irreversible. Thus, the husband and wife should be sure to discuss the matter thoroughly before deciding on this method of contraception.

If reversal is desired, the tubes can be reconnected in some cases. In some studies, up to 50 to 90 percent of the men who underwent microsurgical vaso-vasostomy (reconnecting the vas) subsequently ejaculated sperm, and three-quarters of those men were able to impregnate their partners (G. Stewart 1998b, 577). However, some vasectomized men develop antibodies to their sperm, which may persist after the reversal and counteract fertilization.

Tubal ligation is the surgical sterilization of the female. Until recently, it has been a much more difficult operation than vasectomy, because the Fallopian tubes lie more deeply within the body than the vasa deferentia. The operation, performed in a hospital rather than a physician's office, requires a general anesthetic and a hospital stay of about 3 to 4 days. One or two small incisions are made in the abdominal wall, the Fallopian tubes are located and severed, a small section of each is removed, and then the two ends of each tube are tied. Incisions can also be made through the vaginal wall (colpotomy). The procedure will not leave a scar and requires shorter hospitalization (but recent pregnancy or obesity make this approach more difficult or impossible). However, colpotomy requires a much more skilled doctor and has a higher complication rate than tubal ligation or laparoscopy. Both procedures take about 30 minutes.

A more widely used procedure is called *laparoscopy*. This, too, is a hospital procedure that requires a general anesthetic. However, the operation takes only 15 minutes and does not require overnight hospitalization. A tiny incision is made in the abdomen, and a small light-containing tube (laparoscope) is inserted to illuminate the Fallopian tubes. The surgeon then inserts another small tube carrying high-intensity radio waves that burn out sections of the Fallopian tubes. The incision is so small that only a stitch or two are needed; hence, the procedure is often referred to as "Band-Aid" surgery. Most women leave the hospital within 2 to 4 hours after surgery.

Both forms of tubal ligation have a failure rate of about 0.1 to 3 percent. Attempts to reverse the sterilization have been about 70 percent successful (G. Stewart 1998b, 576–577). However, the reversal procedure is a major operation lasting several hours and, as such, is very costly ($10,000 to $15,000).

Women who have been sterilized tend to report an increase in sexual enjoyment because they are now free of the fear of pregnancy. A few women report reduced sexual enjoyment because of the loss of fertility, which they may equate with femininity.

Hysterectomy, which is the surgical removal of the uterus, also ends fertility. This is an extreme procedure, however, and should be used only when a woman has uterine cancer or other problems of the uterus, not just for birth control reasons.

RU 486 is a two-drug regimen that must be prescribed by a medical doctor. When taken within 49 days of a woman's last period, it is 95 percent effective in inducing an abortion of the fetus (see the section on abortion in Chapter 9).

Unfortunately, American product-liability laws have discouraged many companies from continuing research for improved contraceptive methods (Gabelnick 1998, 615). There are a few new contraceptive methods on the horizon, such as a pill for men, but we will have to wait and see if any will reach the marketplace.

Glossary

A

abstinence Avoidance of sexual contact.

adolescence The general social and biological changes the child experiences in becoming an adult.

agape Altruistic, giving, nondemanding side of love.

anaphrodisiac A drug or medicine that reduces sexual desire.

androgyny The blending of traits associated with the sexes by society.

anxiety A generalized fear without a specific object or fear.

aphrodisiac A chemical or other substance used to induce erotic arousal or to relieve impotence or infertility.

B

body language Nonverbal communication expressed via the body.

brainstorming Producing as many ideas as possible within a given time period in an effort to solve a problem.

C

cash value, or whole-life insurance This insurance has two parts: first, a death benefit, and second, a savings part that acts as an investment.

castration Removal of the testes.

clitoris A small organ situated at the upper end of the female genitals that becomes erect with sexual arousal; homologous with the penis.

cohabitation A couple living together without being married.

common law marriage In some states, a couple living together for more than a certain number of years can be treated as legally married.

common sense Practical intelligence or ordinary good sense.

communication The sending and receiving of messages, intentional and unintentional, verbal and nonverbal.

communication behavior Verbal and nonverbal actions that a person takes to accomplish his/her goals.

communication motivation A felt need to communicate.

communication skill The ability to accomplish one's communication goals.

companionate love A strong bond that includes tender attachment, enjoyment of the other's company, and friendship.

comp time Trade of overtime pay for time off.

congenital defect A condition existing at birth or before, as distinguished from a genetic defect.

connotative The personal or emotional meaning of a word.

contraceptive Any agent used to prevent conception.

crisis Any event that upsets the smooth functioning of a person's life.

crude divorce rate Ratio of divorces to each thousand persons.

D

dilation and curettage (D&C) An abortion-inducing procedure that involves dilating the cervix and scraping out the contents with a curette.

discount interest Interest charged on the total amount of a loan for the entire time period.

domestic partnership The law recognizes as valid some unmarried and homosexual couples' relationships.

drug Any substance taken for medical purposes or for pleasure that affects bodily functions.

E

ectopic pregnancy Implantation of the fertilized egg in one of the Fallopian tubes.

embryo The developing organism from the second to the eighth week of pregnancy, characterized by the differentiation of the organs and tissues into their human form.

empathy The ability to understand what the other is thinking, put oneself in the other's place, and intellectually understand the other's condition without vicariously experiencing the other's emotions.

endogamy The tendency of people to marry within their own group.

episiotomy A small incision made between the vaginal and anal openings to facilitate birth.

eros Physical, sexual side of love.

estrogen replacement therapy (ERT) Supplying estrogen to a menopausal woman.

exogamy The tendency of people to marry outside their group.

F

false pregnancy (pseudocyesis) Signs of pregnancy occur without the woman actually being pregnant.

family According to the Census Bureau, a group of two or more persons related by birth, marriage, or adoption and residing together.

family of origin The family into which we were born and grew up.

family science The study of marriage and family combining all disciplines that can shed light on marriage and family functioning.

fecund Having the capacity to reproduce.

fellatio Oral copulation of the male.

fertility rate The number of women who report having a child in a 12-month period per 1,000 women aged 15 to 44 years of age.

fetoscopy Examining the fetus through a small viewing tube inserted into the mother's uterus.

fetus The developing organism from the eighth week after conception until birth.

G

gender identity How one views oneself as either a man or a woman.

genetic defect An abnormality in the development of the fetus that is inherited through the genes, as distinguished from a congenital defect.

H

halo effect The tendency for first impressions to influence succeeding evaluation.

homogamy The principle that people are attracted to others who share similar objective characteristics such as race, religion, ethnic group, education, and social class; the tendency of people to marry persons similar to themselves.

hooking up Sex without commitment.

households According to the Census Bureau, all persons who occupy a housing unit.

I

identity A person's sense of who he/she is; his/her inner sameness.

immune system The body's biological defense system that wards off diseases and illness.

impotence Inability to gain or maintain an erection.

infertile Unable to produce viable sperm if a man or become pregnant if a woman.

inhibited sexual desire A pervasive disinterest in sex.

intersexual A person who has biological characteristics of both sexes.

intimacy Experiencing the essence of one's self in intense intellectual, physical, and/or emotional communion with another human being.

intimate Experiencing intense intellectual, emotional, and, when appropriate, physical communion with another human being.

intuition The immediate understanding of something without conscious reasoning or thinking about it.

investment debt Money borrowed at a lower interest rate than discount interest because it is used for tangible assets, such as real estate, where value is more permanent.

J

jealousy The state of being resentfully suspicious of a loved one's behavior toward a suspected rival.

joined at the hip Couples who spend much of their time together but rarely go out on formal dates.

L

ludus Game-playing love.

lymphocyte A type of white blood cell that is active in the body's immune system.

M

mania Possessive/dependent love.

marriage or prenuptial contract Working out the details of a couple's relationship before they wed.

menopause (climacteric) The cessation of ovulation, menstruation, and fertility in the woman; usually occurs between ages 46 and 51.

middlessence The second adolescence, experienced in middle age, usually involving reevaluation of one's life.

miscegenation Marriage or interbreeding between members of different races.

modeling Learning by observing other people's behavior.

N

natural childbirth Birth wherein the parents have learned about the birth process and participate in exercises such as breathing techniques to minimize pain and, therefore, the use of drugs.

no fault divorce Divorce proceeding that does not place blame for the divorce on one spouse or the other.

nominal cost Absolute price for something.

norms Accepted and expected patterns of behavior and beliefs established either formally or informally by a group.

nuclear family A married couple and their children living by themselves.

O

orgasm The climax of excitement in sexual activity.

osteoporosis Progressive deterioration of bone strength.

outsourcing Sending jobs overseas where employee costs are less than those in the United States.

P

phenomenology The study of how people experience the world.

philos Love found in deep, enduring friendships.

placenta The organ that connects the fetus to the uterus by means of the umbilical cord.

principle of least interest The one who cares the least controls the relationship.

propinquity Dating and marrying someone living quite close geographically.

psychopaths Older term for persons who fail to be socialized to their society.

puberty Biological changes a child goes through to become an adult capable of reproduction.

R

real wages Earnings that have been adjusted for inflation.

reconstituted or blended family A family in which one or both of the partners have been married before.

resilience The capacity to rebound from adversity strengthened and more resourceful.

reviewable alimony An award of alimony that is reviewed periodically and changed if necessary.

Rh factor An element found in the blood of most people that can adversely affect fetal development if the parents differ on the element (Rh negative versus Rh positive).

role equity The roles one fulfills are based on the individual strengths and weaknesses, rather than on a set of preordained stereotypical difference between the sexes.

roles Activities that norms require.

S

saline abortion An abortion-inducing procedure in which a salt solution is injected into the amniotic sac to kill the fetus, which is then expelled via uterine contractions.

scapegoating Blaming a person or group for the mistakes or crimes of another.

self-assertion The process of recognizing and expressing one's feelings, opinions, and attitudes while remaining aware of the feelings and needs of others.

self-concept How do I feel about myself, what are my strengths and weaknesses, what is my worth?

serial marriages Marrying, divorcing, and marrying again; a series of legal marriages.

serial monogamy Having several spouses over a lifetime but only one at a time.

seropositive Blood acts positively to a test for a disease or illness, indicating that the person is carrying the disease germ or virus.

shooting up Injecting drugs directly into the bloodstream with a needle and syringe.

simple interest Interest charged only on the outstanding balance of a loan.

socialization The physical and psychological nurturing of children into adulthood; passing society's values on to new members beginning at birth.

sociopaths Persons who fail to be socialized to their society.

statutory rape Adult having consensual sex with a minor.

storge Friendship love.

stressor event A crisis-provoking event.

structure The parts that comprise a family and their relationships to one another.

surrogate mother A woman who becomes pregnant and gives birth to a child for another woman incapable of giving birth.

T

term insurance A given amount of insurance bought for a set period of years.

therapy A broad term used to describe actions taken to cure or solve any problem.

transsexual A person who feel psychologically that he or she is actually of the opposite gender.

transvestite A person who gains sexual pleasure from dressing like the opposite sex.

U

umbilical cord A flexible cordlike structure connecting the fetus to the placenta and through which the fetus is fed and waste products are discharged.

ultrasound Sound waves are directed at the fetus that yield a visual picture of the fetus; used to detect potential problems in fetal development.

V

vacuum aspiration An abortion-inducing procedure in which the contents of the uterus are removed by suction.

vicious circle A pattern of behavior in which a negative behavior provokes a negative reaction, which, in turn, prompts more negative behavior.

W

working poor Employed people who live below the poverty threshold.

References

Abbott, D., M. Berry, and W. Meredith. 1990. Religious belief and practice: A potential asset in helping families. *Family Relations* (October): 443–448.

Acs, G., and S. Nelson. 2001. Honey, I'm home. Changes in living arrangements in the late 1960s. *New Federalism National Survey of America's Families* B38. Washington, DC: Urban Institute.

Adler, N. 1994. Competitive frontiers: Women managers in a global economy. Reported in *Working Woman* (January): 16.

Adler, N., E. J. Ozer, and J. Tschann. 2003. Abortion among adolescents. *American Psychologist* 28(3) (March): 211–217.

Adler, T. 1989. Sex-based differences declining, study shows. *APA Monitor* (March): 6.

Advertising Age. 2003. 100 leading advertisers (June 23).

AFL-CIO. 2004. Facts about working women. Available online at http://www.aflcio.org/issuespolitics/women/factsaboutworking women.efm.

Ahrons, C., and J. Tanner. 2003. Adult children and their fathers: Relationship changes 20 years after divorce. *Family Relations* (October): 340–351.

Alan Guttmacher Institute. 2003. Facts in brief: Teenage sexual behavior and reproductive behavior in developed countries. Retrieved on August 26 from http://www.agi-usa.org/pubs/fb_teens.html.

Albano, S. 2002. What society can learn from the U.S. military's system of family support. *Family Focus on Military Families* (March): F6–F9. Washington, DC: Population Reference Bureau.

Alford-Cooper, F. 1998. *For keeps: Marriages that last a lifetime*. Armonk, NY: Sharpe.

Allen, K., R. Blieszner, and K. Roberto. 2000. Families in the middle and later years: A review and critique of research in the 1990s. *Journal of Marriage and the Family* (November): 911–926.

Allport, G. 1961. *Pattern and growth in personality* 20. New York: Holt, Rinehart, and Winston.

Alternatives to Marriage Project. 2004. Statistics retrieved on May 8 from http://www.unmarried.org/statistics.html.

Amato, P. 1993. Children's adjustment to divorce: Theories, hypotheses, and empirical support. *Journal of Marriage and the Family* (February): 23–38.

———. 1996. Explaining the intergenerational transmission of divorce. *Journal of Marriage and the Family* (August): 628–640.

———. 2000. The consequences of divorce for adults and children. *Journal of Marriage and the Family* (November): 1269–1287.

———. 2003. Reconciling divergent perspectives: Judith Wallerstein, quantitative family research and children of divorce. *Journal of Marriage and the Family* (October): 332–339.

Amato, P., and A. Booth. 1996. A perspective study of divorce and parent-child relationships. *Journal of Marriage and the Family* (May): 356–365.

Amato, P., and S. Rogers. 1997. A longitudinal study of marital problems and subsequent divorce. *Journal of Marriage and the Family* (August): 612–624.

Amato, P. R., and F. Fowler. 2002. Parenting practices, child adjustment, and family diversity. *Journal of Marriage and the Family* 64 (August): 703–716.

Ambert, A. 1994. A qualitative study of peer abuse and its effects: Theoretical and empirical implications. *Journal of Marriage and the Family* (February): 119–130.

American Association of University Women. 2002. *Hostile hallways: Bullying, teasing and sexual harassment in school*. Washington, DC: American Association of University Women.

The American experience: Return with honor. 2000. Los Angeles: KCET.

American Woman. 2003–2004. 253, 256, 258.

Anderson, C. A., and B. J. Bushman. 2002. Media violence and the American public revisited. *American Psychologist* (June/July): 448–450.

Anderson, J. 1995. Volunteer—it can enrich your life. *Parade Magazine* (March 12): 14–16.

Anderson, N. 2000. U.S. ready to back global rules for foreign adoption. *Santa Barbara News Press* (September 22): A14.

Angier, N. 1994. Male hormone molds women, too, in mind and body. *New York Times* (May 3): C1, C13.

Anson, D. 1989. Marital status and women's health revisited: The importance of the proximate adult. *Journal of Marriage and the Family* (February): 185–194.

Anstett, P. 2002. New doubt on hormone therapy. *Hartford Courant* (July 9): A1.

Antonucci, T., H. Akiyama, and J. Lansford. 1998. Negative effects of close personal relations. *Family Relations* (October): 379–384.

Archer, J., and B. Lloyd. 1985. *Sex and gender*. Cambridge, UK: Cambridge University Press.

Arditti, J., and T. Keith. 1993. Visitation frequency, child support payment and the father-child relationship postdivorce. *Journal of Marriage and the Family* (August): 699–712.

Association for Couples in Marriage Enrichment. 2000. Mission statement. Available online at http://bettermarriages.org.

Avery, C. 1989. How do you build intimacy in an age of divorce? *Psychology Today* (May): 27–31.

———. 1996. Modern mating: Attraction or survival? *APA Monitor* (August): 30–31.

Azar, B. 1996a. Project explores landscape of midlife. *APA Monitor* (November): 26.

———. 1996b. Psychologists caution about day-care results. *APA Monitor* (June): 18.

———. 1997. Nature, nurture: Not mutually exclusive. *APA Monitor* (May): 28.

———. 2000. The postpartum cuddles: Inspired by hormones? *Monitor on Psychology* (October): 54–56.

———. 2002a. It is more than fun and games. *Monitor on Psychology* (March): 50–51.

———. 2002b. The debate over child care isn't over yet. *Monitor on Psychology* (March): 32–34.

———. 2002c. The power of pretending. *Monitor on Psychology* (March): 46–48.

Baer, J. 2002. Is family cohesion a risk or protective factor during adolescent development? *Journal of Marriage and the Family* 64 (August): 668–675.

Baker, L. 1983. In my opinion: The sexual revolution in perspective. *Family Relations* (April): 297–300.

Ballie, R. 2001a. Spanking study gets big play in media. *Monitor on Psychology* (December): 56.

———. 2001b. Study shows a significant increase in sexual content on TV. *Monitor on Psychology* (May): 16.

———. 2001c. Teen drinking more dangerous than previously thought. *Monitor on Psychology* (June): 12.

Bandura, A. 1969. *Principles of behavior modification*. New York: Holt.

———. 1997. *Self-efficacy: The exercise of control*. New York: W. H. Freeman.

———. 2000. Self-efficacy. Talk given at the First Annual Conference of the California Council on Family Relations, San Diego, April 28–29.

———. 2001. Self-efficacy and shapers of children's aspirations and career trajectories. In A. Bandura et al. (Eds.), *Child Development* 72(1): 187–206.

———. 2004. *Albert Bandura*. Available online at http://www.emory.edu/EDUCATION/mfp/Bandura.

Banks, S. 2000. Raising their children's children and wondering what went wrong the first time. *Los Angeles Times* (August 13): E1, E3.

Baptiste, D. 1987. How parents intensify sexual feeling between stepsiblings. In J. Belovitch (Ed.), *Making remarriage work* 91–94. Lexington, MA: Heath.

Barbach, L. 1984. *For each other: Sharing sexual intimacy*. Garden City, NY: Doubleday.

Barnes, J. S., and C. E. Bennett. 2002. The Asian population: 2000. *Census 2000 Brief* C2BR/01-16 (February). Washington, DC: U.S. Census Bureau.

Barnett, R. 1994. Home-to-work spillover revisited: A study of full-time employed women in dual-earner families. *Journal of Marriage and the Family* (August): 647–656.

———. 1998. Toward a review and reconceptualization of the work/family literature. *Genetic, Social and General Psychology Monographs* 124: 25–82.

Barras, J. 2000. *What ever happened to Daddy's little girl: The impact of fatherlessness on Black women*. New York: One World-Ballantine.

Barringer, H., R. Gardiner, and M. Levin. 1993. *Asians and Pacific Islanders in the United States*. New York: Russell Sage.

Bartels, A., and S. Zeki. 2000. The neural basis of romantic love. *NeuroReport* 2(17): 12–15.

Bartholet, E. 1993. *Family bonds: Adoption and the politics of parenting.* New York: Houghton Mifflin.

———. 1994. What's wrong with adoption law? *Trial* (February).

Bateson, M. 2000. Intergenerational learning in a changing society. Talk given at the 61st Annual Conference of the National Council on Family Relations, Irvine, CA, November 12–15.

Baumrind, D. 1996a. The discipline controversy revisited. *Family Relations* 45(4) (May): 405–417.

———. 1996b. Response: A blanket injunction against disciplinary spanking not warranted by the data. *Pediatrics* 98: 828–842.

Baxter, L. A. 1987. Self-disclosure and relationship disengagement. In V. S. Derlega and J. H. Berg (Eds.), *Self-disclosure: Theory, research and therapy* 155–174. New York: Plenum Press.

Baydar, N., and J. Brooks-Gunn. 1998. Profiles of grandmothers who help care for their grandchildren. *Family Relations* (October): 385–393.

Bayme, S. 1990. Conclusion: Family values and policies in the 1990s (part 2). In D. Blankenhorn et al. (Eds.), *Rebuilding the nest.* Milwaukee: Family Services of America.

Beaton, J. M., et al. 2003. Unresolved issues in adult children's marital relationships involving intergenerational problems. *Family Relations* (April): 143–153.

Becker, P., and P. Moen. 1999. Scaling back. Dual-earner couples' work-family strategies. *Journal of Marriage and the Family* (November): 995–1007.

Bellah, R. 1985. *Habits of the heart: Individualism and commitment in American life.* Berkeley, CA: University of California Press.

Bem, S. L. 1998. *An unconventional family.* New Haven, CT: Yale University Press.

Bennett, W. 1994. *The index of leading cultural indicators.* New York: Simon and Schuster.

Bensen, E. 2003. Psychology in Indian country. *Monitor on Psychology* (June).

Bergen, R. (Ed.). 1998. *Issues in intimate violence.* Thousand Oaks, CA: Sage.

Berkman, H. 1995. Attacks on women usually by intimates. *National Law Review* (August 28): A14.

Berkow, R., et al. (Eds.). 2000. *Merck manual of medical information* (home edition). White House Station, NJ: Merck and Co.

Bernstein, A. 1997. Stepfamilies from siblings' perspectives. In I. Levin and M. Sussman (Eds.), *Stepfamilies: History, research, and policy* 153–175. New York: Haworth Press.

Beutler, I., et al. 1989. The family realm: Theoretical contributions for understanding its uniqueness. *Journal of Marriage and the Family* (August): 805–816.

Bianchi, S. 1990. America's children: Mixed prospects. *Population Bulletin* (June). Washington, DC: Population Reference Bureau.

Bianchi, S., and D. Spain. 1996. Women, work, and families in America. *Population Bulletin* (December). Washington, DC: Population Reference Bureau.

Bianchi, S. M., and L. M. Casper. 2000. *American Families* (December). Washington, DC: Population Reference Bureau.

Biddle, B. J. 1976. *Role theory: Expectations, identities, and behaviors.* Chicago: Dryden Press.

Binstock, G., and A. Thornton. 2003. Separations, reconciliations, and living apart in cohabiting and marital unions. *Journal of Marriage and the Family* (May): 432–443.

Black, D., et al. 2000. Demographics of the gay and lesbian population in the United States: Evidence from the available systematic data sources. *Demography* 37: 139–154.

Black, S. 1999. Investigating the link between competition and discrimination. *Monthly Labor Review* (December): 39–43.

Blakeway, M., and D. Kmitta. 1998. Conflict resolution education research and evaluation synopsis and bibliography. A special report of the National Institute for Dispute Resolution. Washington, DC: National Institute for Conflict Resolution.

Blankenhorn, D. 1995. *Fatherless America: Confronting our most urgent social problem.* New York: Basic Books.

———. 1999. Is marriage made in heaven? *Christianity Today* (August 9).

Blankenhorn, D. S., et al. 1990. *Rebuilding the nest: A new commitment to the American family.* Milwaukee: Family Services of America.

Block, J. D. 1999. *Sex over fifty.* Paramus, NJ: Reward Books.

Bloomfield, H. 1993. *Making peace with your stepfamily.* New York: Hyperion.

Blum, R. W., et al. 2000. The effects of race, ethnicity, income, and family structure on adolescent risk behaviors. *American Journal of Public Health* 90(12): 1879–1884.

Blume, E. 1990. *Secret survivors: Uncovering incest and its aftereffects in women.* New York: Ballantine.

Blumstein, P., and P. Schwartz. 1983. *American couples.* New York: Pocket Books.

Bogenschneider, K., et al. 1997. Child, parent, and contextual influences on perceived parenting competence among parents of adolescents. *Journal of Marriage and the Family* (May): 345–362.

Bohannan, P. 1970. The six stations of divorce. In P. Bohannan (Ed.), *Divorce and after.* New York: Doubleday.

Bonhaker, W. 1990. *The hollow doll.* New York: Ballantine.

Boonstra, H. 2003. Legislators craft alternative vision of sex education to counter abstinence-only drive. *Guttmacher Report.* Retrieved on May 25 from http://www.agi-usa.org/journals/toc/gr05o2toc.html.

Booth, A., A. Crouter, and M. Shanahan (Eds.). 1999. *Transitions to adulthood in a changing economy: No work, no family, no future?* Westport, CT: Praeger.

Booth, C. L., et al. 2002. Child-care usage and mother-infant "quality time." *Journal of Marriage and the Family* (February): 16–26.

Borg, L. 1996. More women marrying younger men. *Santa Barbara News Press* (September 21): B4.

Boss, P. 1988. *Family stress management.* Newbury Park, CA: Sage.

———. 1999. *Ambiguous loss.* Cambridge, MA: Harvard University Press.

———. 2000. Ambiguous loss: Learning to live with unresolved grief. Lecture given at the Building Family Strengths International Symposium, Lincoln, NE, May 10–12.

———. 2001. Ambiguous loss: Frozen grief in the wake of the WTC catastrophe. *Family focus on dying* 12–14. Minneapolis: National Council on Family Relations.

Boston Women's Health Book Collective. 1973, 1976, 1984, 1992. *The new our bodies, ourselves.* New York: Simon and Schuster.

Boulding, E. 1983. Familia faber: The family as maker of the future. *Journal of Marriage and the Family* (May): 257–266.

Bowen, G., et al. 2003. Promoting the adaptation of military families: An empirical test of a community practice model. *Family Relations* (January): 33–44.

Bowler, M. 1999. Women's earnings: An overview. *Monthly Labor Review* (December): 13–21.

Bradbury, T. N., et al. 2000. Research on the nature and determinants of marital satisfaction: A decade in review. In R. M. Milardo (Ed.), *Understanding families into the new millennium: A decade in review* 183–199. Washington, DC: National Council on Family Relations.

Braddock, D. 1999. Occupational employment projections to 2003. *Monthly Labor Review* (November): 51–77.

Bramlett, M., and W. Mosher. 2001. First marriage dissolution, divorce, and remarriage: U.S. advanced data. *National Vital Statistics Reports* 323. Hyattsville, MD: National Center for Health Statistics.

———. 2002. Cohabitation, remarriage, divorce in the United States. *National Vital Statistics Reports* 23(22). Hyattsville, MD: National Center for Health Statistics.

———. 2003. Cohabitation, marriage, divorce and remarriage in the United States. *National Vital Statistics Reports.* Hyattsville, MD: National Center for Health Statistics.

Braver, S., and J. Cookston. 2003. Controversies, clarifications and consequences of divorce's legacy: Introduction to special collection. *Journal of Marriage and the Family* (October): 314–317.

Braver, S., et al. 2002. Experiences of family law attorneys with current issues in divorce practice. *Family Relations* (October): 325–334.

Bray, J., and J. Kelly. 1999. *Stepfamilies.* New York: Broadway Books.

Bray, J., et al. 1992. Longitudinal changes in stepfamilies: Impact on children's adjustment. *APA Monitor* (August 15).

Bremern, E., et al. 2002. Trends in sexual risk behaviors among high school students: United States 1991–2001. *Morbidity and Mortality Weekly Report* 51(88): 856–859.

Brennan, R. T., R. C. Barnett, and K. C. Gareis. 2001. When she earns more than he does: A longitudinal study of dual-earner couples. *Journal of Marriage and the Family* (February): 168–182.

Bretschneider, J., and N. McCoy. 1988. Sexual interest and behavior in healthy 80- to 101-year-olds. *Archives of Sexual Behavior* 17(2): 109–129.

Brett, K. M., and G. S. Cooper. 2003. Associations with menopause and menopausal transition in a nationally representative U.S. sample. *Maturitus* 45(2): 89–97.

Brimelow, P. 1999. Who has abortions? *Forbes* (October 18): 110.

Bringle, R., and D. Byers. 1997. Intentions to seek marriage counseling. *Family Relations* (July): 299–304.

Brody, G. 1994. Family processes and child and adolescent development. *Family Relations* (October).

Brody, J. 1994. Strength training helps prevent aging's ill effects. *Santa Barbara News Press* (August 16): D3.

Bromley, M., and R. Blieszner. 1997. Planning for long-term care: Filial behavior and relationship quality of adult children with independent parents. *Family Relations* (April): 155–162.

Brothers, B. 1997. *When one partner is willing and the other is not.* New York: Haworth Press.

Brothers, J. 1994. Terrible family secrets. *Parade Magazine* (August 19): 16–17.

Brown, P. 1995. *The death of intimacy.* New York: Haworth Press.

Brown, S. 2001. Child well-being in cohabiting unions. Paper presented at the Annual Meeting of the Population Association of America, Washington, DC, March.

Brown, S. L. 2000. Union transitions among cohabiters: The significance of relationship assessments and expectations. *Journal of Marriage and the Family* 61: 833–846.

———. 2004. Family structure and child well-being: The significance of parental cohabitation. *Journal of Marriage and the Family* (May): 357–367.

Brownstein, H. 1994. Changing patterns of lethal violence by women: A research note. *Women and Criminal Justice* 5(2): 99–118.

Brubaker, T. 1990. Family in later life: A burgeoning research area. *Journal of Marriage and the Family* (November): 959–981.

Bryant, C., and R. Conger. 1999. Marital success and domains of social support in long-term relationships: Does the influence of network members ever end? *Journal of Marriage and the Family* (May): 437–450.

Bulcroft, K., et al. 2000. The management and production of risk in romantic relationships: A postmodern paradox. *Journal of Family History* 25: 63–92.

Bumpass, L., and H. Lu. 2000. Trends in cohabitation and implications for children's family contexts in the United States. *Population Studies* 54: 29–41.

Bureau of Labor Statistics. 2000. Consumer spending during retirement. *Issues in Labor Statistics* (May 30). Washington, DC: U.S. Department of Labor.

———. 2002a. Work at home in 2001. *News* (March). Washington, DC: U.S. Department of Labor.

———. 2002b. Workers on flexible and shift schedules in 2001. *Summary News* (April 18). Washington, DC: U.S. Department of Labor.

———. 2003a. Highlights of women's earnings in 2002. (September). Washington, DC: U.S. Department of Labor.

———. 2003b. Volunteering in the United States, 2003. *News* (December 17). Washington, DC: U.S. Department of Labor.

———. 2004a. Absences from work of employed, full-time wage and salary workers by age and sex. Most requested statistics: Annual averages: Household data. Washington, DC: U.S. Department of Labor.

———. 2004b. Employment status of the population by sex, marital status, and presence of children. *News. Current Population Survey* (October 9): Table 5. Washington, DC: U.S. Department of Labor.

———. 2004c. Highlights of women's earnings, 2003. Washington, DC: U.S. Department of Labor.

———. 2004d. Median usual weekly earnings of full-time wage and salary workers by age, race, Hispanic or Latino ethnicity, and sex, 2nd quarter 2004 averages. *News. Current Population Survey* Table 2. Washington, DC: U.S. Department of Labor.

———. 2004e. Women in the labor force: A databook. BLS home page. Reports and summaries. Washington, DC: U.S. Department of Labor. http://www.bls.gov.

Burleson, B., and W. Denton. 1997. The relationship between communication skill and marital satisfaction: Some moderating effects. *Journal of Marriage and the Family* (November): 884–902.

Burleson, B., et al. 2000. Communication in close relationships. In C. Hendrick and S. Hendrick (Eds.), *Close relationships: A sourcebook* 245–258. Thousand Oaks, CA: Sage.

Burr, W. 1990. Beyond I-statements in family communication. *Family Relations* (July): 266–275.

Bushman, B. J., and C. A. Anderson. 2001. Media violence and the American public: Scientific facts versus media misinformation. *American Psychologist* 56: 477–489.

Bushman, B. J., and J. Cantor. 2003. Media ratings for violence and sex: Implications for policymakers and parents. *American Psychologist* (February): 130–142.

Buss, D. 1994. *The evolution of desire: Strategies of human mating.* New York: Basic Books.

———. 1995. Psychological sex differences. *American Psychologist* (March): 164–168.

———. 1996. The evolutionary psychology of human social strategies. In E. Higgins and A. Druglanski (Eds.), *Social psychology: Handbook of basic principles* 3–38. New York: Guilford.

———. 1999. *Evolutionary psychology: The new science of the mind.* Boston: Allyn and Bacon.

Buss, D., et al. 2001. A half century of mate preferences: The cultural evolution of values. *Journal of Marriage and the Family* 63(2): 491–563.

Buunk, B. 1991. Jealousy in close relationships: An exchange theoretical perspective. In P. Salovey (Ed.), *Psychology of jealousy and envy* 148–177. New York: Guilford.

———. 1997. Personality, birth order and attachment styles as related to various types of jealousy. *Personality and Individual Differences* 23: 997–1006.

Buunk, B., and P. Dijkstra. 2000. Extradyadic relationships and jealousy. In C. Hendrick and S. Hendrick (Eds.), *Close relationships: A sourcebook* 317–329. Thousand Oaks, CA: Sage.

Buunk, B., et al. 1996. Sex differences in jealousy in evolutionary and cultural perspective: Tests from the Netherlands, Germany and the United States. *Psychological Science* 7: 359–363.

Byer, C., and L. Shainberg. 1994. *Dimensions of human sexuality* 256. Madison, WI: Brown and Benchmark.

Bygdeman, M., and K. G. Danielson. 2002. Options for early therapeutic abortion. *Drugs* 62(17): 2459–2470.

Byrne, G. 1997. Father may not know best, but what does he know? *Population Today* 25(10) (October). Washington, DC: Population Reference Bureau.

Calderone, M. S. 1982. Love, sex, intimacy, and aging as a life style. In M. Calderone (Ed.), *Sex, love and intimacy—whose life styles?* New York: SIECUS.

California Lawyer. 1995. Fade away: The rise and fall of repressed memory theory in the courtroom. (March): 36–41, 66.

———. 1997. A tangled family tree. (January): 21.

Canary, D., and K. Dindia (Eds.). 1998. *Sex differences and similarities in communication: Critical essays and empirical investigations of sex and gender interaction.* Mahwah, NJ: Erlbaum.

Cancian, F. M. 1990. The feminization of love. In C. Carlson (Ed.), *History, class, and feminism* 171–185. Belmont, CA: Wadsworth.

Cano, A., and D. O'Leary. 1997. Romantic jealousy and affairs: Research and implications for couple therapy. *Journal of Sex and Marital Therapy* 23(4): 249–275.

Caplan, P., et al. 1985. Do sex-related differences in spatial abilities exist? *American Psychologist* (July): 879–888.

Carbone, L. 1999. Champions of sisterhood or big brother. *Family Policy* (August). Washington, DC: Family Research Council.

Carlson, J., and L. Sperry. 1999. Introduction: A context for thinking about intimacy. In J. Carlson and L. Sperry (Eds.), *The intimate couple.* Philadelphia: Brunner/Mazel.

Carpenter, S. 2001. Does estrogen protect memory? *Monitor on Psychology* (January): 52–53.

Carrol, J., and W. Doherty. 2003. Evaluating the effectiveness of premarital prevention programs: A meta-analytic review of outcome research. *Family Relations* (April): 105–118.

Carroll, J. L. 2005. *Sexuality now.* Belmont, CA: Wadsworth.

Carter, C. S., et al. 2001. Neuroendocrine and emotional changes in the postpartum period. In J. A. Russell, A. J. Douglas, R. J. Windle, and C. D. Ingram (Eds.), *The maternal brain.* Cambridge, MA: Elsevier.

Casler, L. 1969. This thing called love. *Psychology Today* (December).

Casper, L., et al. 1999. How does POSSLQ measure up? Historical estimates of cohabitation. Working Paper 36. Washington, DC: U.S. Census Bureau.

Cate, R., and S. Lloyd. 1992. *Courtship.* Thousand Oaks, CA: Sage.

Cates, W., and C. Ellertson. 1998. Abortion. In R. Hatcher et al. (Eds.), *Contraceptive technology* 679–800. New York: Ardent Media.

Centers for Disease Control. 1997. Teen sex down, new study shows. Washington, DC: Centers for Disease Control.

———. 2000a. Low teen birthrate no cause for celebration when pregnancy and abortion rates are still high says FRC. Retrieved on August 9 from http://www.frac.org/press/080900.htm.

———. 2000b. Teen drug and tobacco use rose in the 90's. *San Francisco Chronicle* (June 9).

———. 2002. *Sexually transmitted disease surveillance report.* Washington, DC: U.S. Department of Health and Human Services.

———. 2002, 2004. *HIV/AIDS surveillance report.* Washington, DC: U.S. Department of Health and Human Services.

———. 2003. New CDC report shows record low infant mortality rate. *National Center for Health Statistics* NVS4 52. Hyattsville, MD: National Center for Health Statistics.

———. 2004. STDs in adolescents and young adults. Retrieved on February 4 from http://www.cdc.gov/std/stats/adal.htm.

Chalkiey, K. 1997. Female genital mutilation. *Population Today* (October). Washington, DC: Population Reference Bureau.

Chapman, S. 1991. Attachment and adolescent adjustment to parental remarriage. *Family Relations* (April): 232–237.

Cherlin, A. 2004. The deinstitutionalization of American marriage. *Journal of Marriage and the Family* (November): 848–861.

Chia, M., and D. Abrams. 1997. *The multiorgasmic man: Sexual secrets the man should know.* San Francisco: Harper Collins.

Chicago Tribune. 1994. Americans losing time to volunteer. As reported in the *Santa Barbara News Press* (November 15): A9.

Choy, W. J., et al. 2004. *The condition of education.* National Center for Education Statistics, U.S. Department of Education. Washington, DC: U.S. Government Printing Office.

Christakis, D. 2004. TV and young children. *Pediatrics* (July).

Christensen, B. J. 1988. The costly retreat from marriage. *Public Interest* 91 (Spring): 59–66.

———. 1994. The costly retreat from marriage. In G. Bird and M. Sporakowski (Eds.), *Taking sides.* Guilford, CT: Dushkin.

Christopher, F., and S. Lloyd. 2000. Physical and sexual aggression in relationships. In C. Hendrick and S. Hendrick (Eds.), *Close relationships: A sourcebook* 331–344. Thousand Oaks, CA: Sage.

Christopher, F., and S. Sprecher. 2000. Sexuality in marriage, dating, and other relationships: A decade review. *Journal of Marriage and the Family* (November): 999–1017.

Ciabattari, T. 2004. Cohabitation and housework: The effects of marital intentions. *Journal of Marriage and the Family* (February): 118–125.

Clark, J. R. 1961. *The importance of being perfect.* New York: McKay.

Clark, L. C., et al. 2003. Comparing employment, income, and poverty: Census 2000 and current population survey. (September). See http://www.census.gov/prod/cen2000/doc/sf3.pdf.

Clements, M. 1994. Sex in America today. *Parade Magazine* (August 7): 4–5.

Cobb-Clark, D., and Y. Dunlop. 1999. The role of gender in job promotions. *Monthly Labor Review* (December): 32–38.

Cohan, C. L., and S. Kleinbaum. 2002. Toward a greater understanding of the cohabitation effect: Premarital cohabitation and marital communication. *Journal of Marriage and the Family* (February): 180–192.

Cohen, S. 1995. Don't let money wreck your marriage. *Parade Magazine* (March 12): 24–25.

Colditz, M., et al. 1995. The use of estrogens and progestins and the risk of breast cancer in postmenopausal women. *New England Journal of Medicine* (June 15): 1589–1593.

Coleman, M., et al. 2000. Reinvestigating remarriage: Another decade of progress. *Journal of Marriage and the Family* (November): 1288–1307.

Collins, F. S. 1999. Shattuck Lecture: Medical and social consequences of the human genome project. *New England Journal of Medicine* 341: 28–37.

Collins, J., and O. Thornberry. 1989. Health characteristics of workers by occupation and sex: United States, 1983–1985. *National Vital Statistics Reports* 168 (April): 25. Hyattsville, MD: National Center for Health Statistics. (April).

Cooke, B. 1991. Thinking and knowledge underlying expertise in parenting: Comparisons between expert and novice mothers. *Family Relations* (January): 3–13.

Coombs, R. 1991. A healthy marriage. *American Demographics* (November): 40–43.

Coontz, S. 2000. Historical perspectives on family studies. *Journal of Marriage and the Family* (May): 283–297.

Corter, C. A., and E. Fleming. 2002. Psychobiology of maternal behavior in human beings. In M. Barnstein (Ed.), *Handbook of parenting.* Hillsdale, NJ: Erlbaum.

Cory, S. 1994. Court allows visitation rights for grandmother. *Santa Barbara News Press* (October 15).

Cowan, C., and P. Cowan. 1992. *When partners become parents: The big life change for couples.* New York: Basic Books.

Cowan, C. P., and P. A. Cowan (Eds.). 2000. *When partners become parents: The big life change for couples.* Mahwah, NJ: Erlbaum.

Cox, D. I., et al. 2003. *The anger advantage.* New York: Broadway/Doubleday.

Cox, F. 1992. *Premarital sex and religion.* Unpublished study, Santa Barbara City College, Santa Barbara, CA.

———. 2000. *The AIDS booklet.* New York: McGraw-Hill.

———. 2002. *Human intimacy: Marriage, the Family, and Its Meaning* 9th ed. Belmont, CA: Wadsworth.

———. 2004. Personal interviews conducted while in India in 1985 and 2004.

Cox, M., et al. 1999. Marital perceptions and interactions across the transition to parenthood. *Journal of Marriage and the Family* (August): 611–625.

Cramer, P. 1998. Coping and defense mechanisms: What's the difference? *Journal of Personality* 66: 895–918.

———. 2000. Defense mechanisms in psychology today. *American Psychologist* (June): 637–646.

———. 2001. The unconscious status of defense mechanisms. *American Psychologist* (September): 762–763.

Crittenden, D. 1999. *What our mothers didn't tell us: Why happiness eludes the modern woman.* New York: Simon and Schuster.

Crosby, J. F. 1980. A critique of divorce statistics and their interpretation. *Family Relations* (January): 51–58.

———. 1985. *Reply to myth: Perspectives on intimacy.* New York: Wiley.

Cui, M., et al. 2002. Parental behavior and quality of adolescent friendships: A social-contextual perspective. *Journal of Marriage and the Family* 64 (August): 676–689.

Curran, D. 1983. *Traits of a healthy family.* New York: Ballantine.

Daly, K. 1996. *Families and time.* Thousand Oaks, CA: Sage.

Daw, J. 2001. Eating disorders on the rise. *Monitor on Psychology* (October): 21.

Dawson, D. 1991. Family structure and children's health and well-being: Data from the 1988 National Health Interview Survey of Child Health. *Journal of Marriage and the Family* (August): 573–584.

DeAngelis, T. 2001a. Moving up. *Monitor on Psychology* (July/August): 70–73.

———. 2001b. Unraveling the mystery of suicide. *Monitor on Psychology* (November): 74.

———. 2002. Promising treatments for anorexia and bulimia. *Monitor on Psychology* (March): 38–41.

———. 2003a. Girls use a different kind of weapon. *Monitor on Psychology* (July/August): 51.

———. 2003b. When anger's a plus. *APA Monitor* (March): 44–45.

———. 2004a. Marriage promotion: A simplistic "fix." *Monitor on Psychology* (September): 42–43.

———. 2004b. The question of marriage and community well-being. *Monitor on Psychology* (September): 38–40.

DeLay, T. 2000. Fighting for children. *American Psychologist* (September): 1054–1055.

DeLeire, T., and A. Kalil. 2002. Good things come in 3s: Single-parent multigenerational family structure and adolescent adjustment. *Demography* 39.

Del Pinal, J., and A. Singer. 1997. Generations of diversity: Latinos in the United States. *Population Bulletin* (October). Washington, DC: Population Reference Bureau.

DeMaris, A. 2001. The influence of intimate violence on transitions out of cohabitation. *Journal of Marriage and the Family* (February): 235–246.

DeMaris, A., and W. MacDonald. 1993. Premarital cohabitation and marital instability: A test of unconventionality. *Journal of Marriage and the Family* (May): 399–407.

DeMaris, A., et al. 2003. Distal and proximal factors in domestic violence: A test of an integrated model. *Journal of Marriage and the Family* (August): 652–667.

Demo, D., et al. 2000. *Handbook on family diversity*. New York: Oxford University Press.

Demo, D. H., and M. J. Cox. 2000. Families with young children: A review of research in the 1990s. *Journal of Marriage and the Family* 62: 876–895. Also reprinted in R. Milardo (Ed.). 2001. *Understanding families into the new millennium: A decade in review* 95–114. Washington, DC: National Council on Family Relations.

Dershowitz, A. M. 1994a. *The abuse excuse*. New York: Little, Brown.

———. 1994b. *Contrary to public opinion*. Berkeley, CA: Berkeley Publications.

Deutsch, A. 2000. Dutch euthanasia bill advances. *Santa Barbara News Press* (November 29): A6.

Devine, T. 1994. Characteristics of self-employed women in the United States. *Monthly Labor Review* (March): 20–34.

DeVita, C. 1996. The United States at mid-decade. *Population Bulletin* (March). Washington, DC: Population Reference Bureau.

Dick-Read, G. 1972. *Childbirth without fear* 4th ed. New York: Harper and Row.

DiGiuseppe, R., and R. C. Tafrate. 2003. Anger treatment for adults: A meta-analysis view. *Clinical Psychology: Science and Practice* 10(1).

Dimitroff, M. 1999. Spiritual intimacy: A Christian perspective. In J. Carlson and L. Sperry (Eds.), *The intimate couple*. Philadelphia: Brunner/Mazel.

Dittmann, M. 2003. Anger across the gender divide. *APA Monitor* (March).

Dollahite, D. A., et al. 1997. Fatherwork: A conceptual ethic of fathering as generative work. In A. Hawkins and D. Dollahite (Eds.), *Generative fathering: Beyond deficit perspective*. Thousand Oaks, CA: Sage.

Donnelly, B., and P. Voydanoff. 1996. Parenting versus placing for adoption: Consequences for adolescent mothers. *Family Relations* (October): 421–434.

Donnelly, D., and D. Finkelhor. 1993. Who has joint custody? Class differences in the determination of custody arrangements. *Family Relations* (January): 57–60.

Dreskin, W., and W. Dreskin. 1993. *The day care decision: What's best for you and your child*. New York: Evans.

Drummet, A., et al. 2003. Military families under stress: Implications for family life education. *Family Relations* (July): 279–287.

Drummond, T. 2000. Mom on her own. *Time* (August 28): 54–55.

Duncan, B., and J. Rock. 1993. Saving relationships: The power of the unpredictable. *Psychology Today* (January/February): 46–51, 80, 95.

Duncan, W. 2001. Domestic partnership laws in the United States: A review and critique. Provo, UT: Brigham Young University Law School.

Dunifon, R., and L. Kowaleski-Jones. 2002. Who's in the house? Race differences in cohabitation, single parenthood and child development. *Child Development* 23: 1249–1264.

Durden-Smith, J., and D. Desimone. 1982. Is there a superior sex? *Playboy* (May).

Dush, D., M. Cohan, and P. R. Amato. 2003. The relationship between cohabitation and marital quality and stability: Change across cohorts. *Journal of Marriage and the Family* (August): 539–549.

Eagly, A. 1995. The science and politics of comparing women and men. *American Psychologist* (March): 145–158.

Eagly, A., and W. Wood. 1991. Explaining sex differences in social behavior: A meta-analytic perspective. *Personality and Social Psychology Bulletin* (June): 306–315.

Earle, R. 1990. Sexual addiction: Understanding and treating the phenomenon. *Contemporary Family Therapy* 12(2): 89–104.

East, P. 1999. The first teenage pregnancy in the family: Does it affect mothers' parenting, attitudes, or mother-adolescent communication? *Journal of Marriage and the Family* (May): 306–319.

Edwards, E., and S. Edwards. 1996. Top 10 problems of working from home. As reported in K. Kaplan, Home sweet home office. *Los Angeles Times* (March 19).

Edwards, L., and E. Field-Hendry. 1996. Home-based workers: Data from the 1990 census of population. *Monthly Labor Review* (November): 26–34.

Edwards, N. K. 2000. Works in progress: Women's transitional journeys in the realm of sexuality. *Dissertation Abstracts* 04194217. University of Minnesota.

Edwards, R. 1995. New tools help gauge marital success. *APA Monitor* (February): 6.

Egan, J. 2002. The military: A special type of family. *Family focus on military families* (March): 4–5, 9. Minneapolis: National Council on Family Relations.

Egelko, B. 1998. Court overturns child support law. *Santa Barbara News Press* (February 3): B3.

Eggebeen, D. J., and C. Knoester. 2001. Does fatherhood matter for men? *Journal of Marriage and the Family* 63 (May): 381–393.

Elias, M. 2001. Growing up with gay parents. *USA Today* (August): D1.

Embree, M., and O. Griego. 2000. Domestic-partner benefits regardless of marital status? *Workforce* (February).

Emery, N. 1999. Angry White women: The backlash against feminism. *Family Policy* (August). Washington, DC: Family Research Council.

Engel, M. 2000. Stepfamilies: Their personal strengths are undermined by public perceptions and policy. Lecture given at the Building Family Strengths International Symposium, Lincoln, NE, May 10–12.

England, P., and R. Horowitz. 1998. *Birthing from within*. Albuquerque: Partera.

Equal Employment Opportunity Commission. 2004a. Sexual harassment cases. Available online at http://www.eeoc.gov.

———. 2004b. Sexual harassment charges: 2003. Available online at http://www.eeoc.gov.

Erdelyi, M. 2001. Defense process can be conscious or unconscious. *American Psychologist* (September): 761–762.

Erikson, E. 1963. *Childhood and society* 2nd ed. New York: Norton.

Etzioni, A. 1983. *An immodest agenda: Rebuilding America before the 21st century*. New York: McGraw-Hill.

———. 1993. *The spirit of community*. New York: Simon and Schuster.

Everett, C., and S. Everett. 1994. *Healthy divorce*. San Francisco: Jossey-Bass.

Faludi, S. 1991. *Backlash: The undeclared war against women*. New York: Crown.

Family Relations. 2004. Reviews of marriage education programs and initiatives. *Family Relations* (October): 421–571.

Farrell, W. 1993. *The myth of male power*. New York: Simon and Schuster.

———. 1994. As reported in Peter Brimelow, Gender politics. *Forbes* (March 14): 46–47.

Fausto-Sterling, A. 1994. As reported in Peter Brimelow, Gender politics. *Forbes* (March 14): 46–47.

———. 2000. The five sexes revisited. *The Sciences* (July/August): 19–23.

Federal Communications Commission. 1996. Policies and rules concerning children's television programming: Revision of programming policies for television broadcast stations. MMDocket No. 93-48.

Feeney, J. P., and C. Ward. 1997. Marital satisfaction and spousal interaction. In R. Sternberg and H. Mahzad (Eds.), *Satisfaction in close relationships* 160–188. New York: Guilford.

Feingold, A. 1994. Gender differences in personality: A meta-analysis. *Psychological Bulletin* 116(3): 429–456.

Ferguson, C. J. 2000. Media violence: A miscast causality. *American Psychologist* (June/July): 446–447.

Fields, J. 2003. Children's living arrangements and characteristics: March 2002. *Current Population Reports* P20-547. Washington, DC: U.S. Census Bureau.

Fields, J., and L. Casper. 2001. America's families and living arrangements. *Population Characteristics* (June). Washington, DC: U.S. Census Bureau.

———. 2002. America's families and living arrangements. *Current Population Reports* P20-537. Washington, DC: U.S. Census Bureau.

Fields, S. 1997. Women don't need an us-against-them mentality. *Santa Barbara News Press* (September 22).

———. 1999. Living together—a cart before the horse. *Insight on the News* (March 8).

Financial Forecaster. 2004. Inflation Data.Com. Capital Professional Services available online at http://inflationdata.com/inflation_rate/currentinflation.asp.

Fincham, F., et al. 1994. The professional response to child sexual abuse: Whose interests are served? *Family Relations* (July): 244–254.

Fincham, F. S., and S. Kemp-Fincham. 1997. Marital quality: A new theoretical perspective. In R. Sternberg and H. Mahzad (Eds.), *Satisfaction in close relationships* 275–305. New York: Guilford.

Fine, M. 1989. A social science perspective on stepfamily law: Suggestions for legal reform. *Family Relations* (January): 53–58.

———. 1997. Stepfamilies from a policy perspective: Guidance from empirical literature. In I. Levin and M. Sussman (Eds.), *Stepfamilies: History, research, and policy* 249–264. New York: Haworth Press.

Fine, M., and D. Fine. 1992. Recent changes in laws affecting stepfamilies: Suggestions for legal reform. *Family Relations* (July): 334–340.

———. 1994. An examination and evaluation of recent changes in divorce laws in five Western countries: The critical role of values. *Journal of Marriage and the Family* (May): 265–277.

Fingerman, K. 1998. The good, the bad, and the worrisome: Emotional complexities in grandparents' experiences with individual grandchildren. *Family Relations* (October): 403–414.

Finkelhor, D. 1990. Sexual abuse in a national survey of adult men and women. *Child Abuse and Neglect* 14(1): 19–28.

———. 1995. Sexual abuse of children. In R. Gilles (Ed.), *Vision 2010: Families and violence, abuse and neglect*. Minneapolis: National Council on Family Relations.

Finkenauer, C., et al. 2004. Disclosure and relationship satisfaction in families. *Journal of Marriage and the Family* (February): 195–209.

Fisher, A. 2003. A critique of the portrayal of adoption in college textbooks and readers on families, 1998–2001. *Family Relations* 32 (April): 154–160.

Fisher, H. E. 1992. *Anatomy of love*. New York: Norton.

———. 2004. *Why we love*. New York: Henry Holt.

Forste, R., and K. Tanfer. 1996. Sexual exclusivity among dating, cohabiting, and married women. *Journal of Marriage and the Family* (February): 33–48.

Foshee, V., K. Bauman, and G. Linder. 1999. Family violence and the perpetration of adolescent dating violence: Examining learning and social control processes. *Journal of Marriage and the Family* (May): 331–342.

Frame, M., and C. Shehan. 1994. Work and well-being in the two-person career: Relocation stress and coping among clergy husbands and wives. *Family Relations* (April): 196–205.

Friedan, B. 1963. *The feminine mystique*. New York: Dell.

———. 1981. *The second stage*. New York: Summit Books.

———. 1992. *The second stage*. New York: Summit Books.

Friedman, J. 1992. Personal correspondence.

Friedrich, D. D. 2001. *Successful aging*. Springfield, IL: Charles C. Thomas.

Frieze, I., et al. 1978. *Women and sex roles*. New York: Norton.

Fromm, E. [1956] 1970. *The art of loving*. New York: Bantam.

Fuller-Thompson, E., M. Minkler, and D. Driver. 1997. A profile of grandparents raising grandchildren in the United States. *The Gerontologist* 37: 406–411.

Furstenberg, F. 1994. History and current status of divorce in the United States. *Future of Children* 4(1).

Furstenberg, F., and A. Cherlin. 1991. *Divided families: What happens to children when the parents part?* Cambridge, MA: Harvard University Press.

Furstenberg, F., and K. Kiernan. 2001. Delayed parental divorce: How much do children benefit? *Journal of Marriage and the Family* (May).

Gabelnick, H. 1998. In R. Hatcher et al. (Eds.), *Contraceptive technology* 615–622. New York: Ardent Media.

Galbraith, J. K. 1958. *The affluent society*. New York: Houghton Mifflin.

Gallagher, W. 1988. Sex and hormones. *Atlantic Monthly* (March): 77–82.

Gallegher, M. 1996. *The abolition of marriage: How we destroy lasting love*. Washington, DC: Regency.

———. 2000. Where have all the grownups gone? *Family Policy* (July/August). Washington, DC: Family Research Council.

Galvin, K., and B. Brommel. 1982, 1986. *Family communication: Cohesion and change*. Glenview, IL: Scott, Foresman.

Gamach, S. 1997. Confronting nuclear family bias in stepfamily research. In I. Levin and M. Sussman (Eds.), *Stepfamilies: History, research, and policy* 412–470. New York: Haworth Press.

Ganong, L., and M. Coleman. 1984. The effects of remarriage on children: A review of the empirical research. *Family Relations* (July).

———. 1994. *Remarried family relationships*. Thousand Oaks, CA: Sage.

———. 1997. How society views stepfamilies. In I. Levin and M. Sussman (Eds.), *Stepfamilies: History, research, and policy* 85–106. New York: Haworth Press.

Gardner, J. 1996. Hidden part-timers: Full-time work schedules, but part-time jobs. *Monthly Labor Review* (September): 43–44.

Gardner, R. 1992. *True and false accusations of child abuse*. Cresskill, NJ: Creative Therapies.

Gatland, L. 1997. Putting the blame on no-fault. *American Bar Association Journal* (April): 50–54.

Gato, S. 2000. Japanese scientists produce cloned clone. *Santa Barbara News Press* (January 25).

Gattai, F., and T. Musatti. 1999. Grandmothers' involvement in grandchildren's care: Attitudes, feelings, and emotions. *Family Relations* (January): 35–42.

Gazzaniga, M. (Ed.). 1995. *The cognitive neurosciences*. Cambridge, MA: MIT Press.

Geasler, M., and K. Blaisure. 1998. A review of divorce education program materials. *Family Relations* (April): 167–175.

Gelles, R. 1995. *Contemporary families*. Thousand Oaks, CA: Sage.

Gelles, R., and J. Conte. 1991. Domestic violence and sexual abuse of children: Review of the research in the eighties. In A. Booth (Ed.), *Contemporary families: Looking forward, looking back*. Minneapolis: National Council on Family Relations.

Gentile, D. A., and D. A. Walsh. 2002. A normative study of family media habits. *Applied Developmental Psychology* 23 (January 28): 157–178.

Gershoff, E. 2002. Corporal punishment by parents and associated child behavior and experiences: A meta-analytical and theoretical review. *Psychological Review* 102: 458–489.

Giles-Sims, J. 1997. Current knowledge about child abuse. In I. Levin and M. Sussman (Eds.), *Stepfamilies: History, research, and policy* 215–230. New York: Haworth Press.

———. 2000. Book review of *All our families* by M. Mason et al. *Journal of Marriage and the Family* (February): 275–276.

Gillmore, M. R., et al. 2002. Teen sexual behavior: Applicability of the theory of reasoned action. *Journal of Marriage and the Family* (November): 885–897.

Gilmore, D., D. Mack, and S. Nock. 1999. Marriage in the making: A report on the courtship customs of young Americans. Working Paper 71. New York: Institute for American Values.

Girsh, F. 1999. Don't make doctors criminals for helping people escape painful terminal illnesses. *Insight on the News* (March).

Glass, L. 1992. *He says, she says: Closing the communication gap between the sexes*. New York: Putnam.

———. 1995. As reported in R. Edwards, New tools help gauge marital success. *APA Monitor* (February): 6.

Glenn, N. 1997. A reconsideration of the effect of no-fault divorce on divorce rates. *Journal of Marriage and the Family* 59: 1023–1024.

———. 1998. The course of marital success and failure in five American 10-year cohorts. *Journal of Marriage and the Family* (August): 569–576.

———. 1999. Further discussion of the effects of no-fault divorce on divorce rates. *Journal of Marriage and the Family* (August): 800–802.

Glenn, N., and E. Marquardt. 2001. *Hooking up, hanging out, and hoping for Mr. Right: College women on dating and mating today*. New York: Institute for American Values.

Glick, P. 1997. Demographic picture of African-American families. In H. P. McAdoo (Ed.), *Black families* 3rd ed., 118–138. Thousand Oaks, CA: Sage.

Glidden, L. 2000. Adopting children with developmental disabilities: A long-term perspective. *Family Relations* (October): 397–405.

Glotzer, R. 2000. Miles that bind: Family strengths and commuter marriage. Paper presented at the Building Family Strengths International Symposium, Lincoln, NE, May 10–12.

Goertzel, V., and M. Goertzel. 1962. *Cradles of imminence*. Boston: Little, Brown.

Goldberg, C. 1999. More women arrested for assaults. *Santa Barbara News Press* (November 14).

Gold-Bikin, L. 1994. Family law section focuses on saving marriages. *National Law Review* (August 8): C14.

Golden, O. 2000. The federal response to child abuse and neglect. *American Psychologist* (September): 1050–1053.

Goleman, D. 1978. Special abilities of the sexes: Do they begin in the brain? *Psychology Today* (November).

———. 1998. Keynote address at the Annual Convention of the American Psychological Association, San Francisco. Reported in *APA Monitor* (October 15).

Goodman, E. 1994a. A deadbeat dad's worst; repo man takes license. *Santa Barbara News Press* (August 2): A11.

———. 1994b. A hi-tech pseudo-orphan's deadbeat dad. *Santa Barbara News Press* (September 16): A9.

———. 1995. The divorce revaluation: The chief culprit in the family's breakup. *Santa Barbara News Press* (April): A7.

———. 1996. Thin models crossing the line. *Santa Barbara News Press* (June 11).

Gordon, T. 1970. *Parental effectiveness training*. New York: Peter Wyden.

Gorman, C. 1992. Sizing up the sexes. *Time* (January 20): 42–51.

———. 1995. Memory on trial. *Time* (April 17): 54–55.

———. 1998. The downside of Viagra. *Time* (May 4): 53.

———. 1999. Bringing home the bacon: Marital allocation of income-earning responsibilities, job shifts and men's wages. *Journal of Marriage and the Family* (May): 476–490.

Gosden, R. 1999. *Designing babies: The brave new world of reproductive technology*. New York: W. H. Freeman.

Gottman, J. 1994. *Why marriages succeed or fail*. New York: Simon and Schuster.

———. 1997. *The heart of parenting*. New York: Simon and Schuster.

Gottman, J., and C. Notarius. 2000. Decade in review: Observing marital interaction. *Journal of Marriage and the Family* 62(4): 927–947.

Gottman, J., J. Coan, S. Carrere, and C. Swanson. 1998. Predicting marital happiness and stability from newlywed interactions. *Journal of Marriage and the Family* (February): 5–22.

Graglia, F. 1998. *Domestic tranquility: A brief against feminism*. New York: Spence Publishing.

Graham, J. 2004. Fertility in deep freeze. *Santa Barbara News Press* (July 5): A10.

Graham, S. 1992. What does a man want? *American Psychologist* (July): 837–841.

Grall, T. 2003. Custodial mothers and fathers and their child support: 2001. *Current Population Reports* P60-225. Washington, DC: U.S. Census Bureau.

Gray, M. R., and L. Steinberg. 1999. Unpacking authoritarian parenting: Reassessing a multidimensional construct. *Journal of Marriage and the Family* 61 (August): 574–584.

Greenhaus, J. H., and P. Parasuraman. 1999. Research on work, family, and gender: Current status and future directions. In G. N. Powell (Ed.), *Handbook of gender and work* 391–419. Thousand Oaks, CA: Sage.

Greenstein, T. 1996. Husband's participation in domestic labor: Interactive effects of wives' and husbands' gender ideologies. *Journal of Marriage and the Family* (August): 585–595.

Gregg, G., J. Bulgar, and B. Marini. 1990. Absence makes the heart grow fonder? College students' perceptions of changes in family rituals. Paper given at the 61st Annual Conference of the National Council on Family Relations, Irvine, CA, November 12–15.

Groat, H., et al. 1997. Attitudes toward child rearing among young adults. *Journal of Marriage and the Family* (August): 568–581.

Groze, V. 1996. *Successful adoptive families: A longitudinal study*. Westport, CT: Praeger.

Grzywacz, J. G., et al. 2002. Work-family spillover and daily reports of work and family stress in the adult labor force. *Family Relations* (January): 28–36.

Guerney, B., and P. Maxson. 1990. Marital and family research: A decade review and a look ahead. *Journal of Marriage and the Family* (November): 1127–1135.

———. 1991. Marital and family enrichment research: A decade review and look ahead. In A. Booth (Ed.), *Contemporary families: Looking forward, looking back* 446–456. Minneapolis: National Council on Family Relations.

Guierrero, L. K., and P. Anderson. 2000. Emotion in close relationships. In C. Hendrick and S. Hendrick (Eds.), *Close relationships: A sourcebook* 171–184. Thousand Oaks, CA: Sage.

Gupka, S. 1999. The effects of transitions in marital status on men's performance of housework. *Journal of Marriage and the Family* 61: 700–711.

Gutowski, C. 1994. *Grandparents are forever: The power of the extended family*. n.p.: Paulist Press.

Guttmacher, A. 1984. *Pregnancy, birth and family planning*. New York: New American Library.

Haine, R., et al. 2003. Changing the legacy of divorce: Evidence from prevention programs and future directions. *Journal of Marriage and the Family* (October): 297–305.

Hales, D. 1999. *Just like a woman*. New York: Bantam.

Hamilton, B. E., P. D. Sutton, and S. S. Ventura. 2003. Revised birth and fertility rates for the 1990s and new rates for Hispanic populations, 2000 and 2001, United States. *National Vital Statistics Reports* 51(12) (August 4). Hyattsville, MD: National Center for Health Statistics.

Handy, B. 1998. The Viagra craze. *Time* (May 4): 50–55.

Hans, J. 2002. Stepparenting after divorce: Stepparents' legal position regarding custody, access, and support. *Family Relations* (October): 301–307.

Hao, L., and G. Xie. 2001. The complexity and endogenity of family structure in explaining children's misbehavior. *Social Science Research* 31: 1–8.

Harney, K. 2004. Home prices amaze, but buyers keep up. *Boston Herald* (September 24).

Harris, J. 1998a. Talk given at the Annual Convention of the American Psychological Association, San Francisco. Reported in *APA Monitor* (October): 9.

———. 1998b. *The nurture assumption*. New York: Free Press.

Harvey, J., and A. Hansen. 2000. Loss and bereavement in close romantic relationships. In C. Hendrick and S. Hendrick (Eds.), *Close relationships: A sourcebook* 359–370. Thousand Oaks, CA: Sage.

Hatcher, R., et al. 1998. *Contraceptive technology* 17th ed. New York: Ardent Media.

———. 2004. *Contraceptive technology* 18th ed. New York: Ardent Media.

Hatfield, E. 1993. Historical and cross-cultural perspectives in passionate love and sexual desire. *Annual Review of Sex Research* 4: 67–97.

Haugaard, J. 2000. The challenge of defining child sexual abuse. *American Psychologist* (September): 1036–1039.

Haworth-Hoeppner, S. 2000. The critical shapes of body image: The role of culture and family in the production of eating disorders. *Journal of Marriage and the Family* (February): 212–227.

Hayghe, H. 1993. Working wives' contribution to family incomes. *Monthly Labor Review* (August): 39–43.

Haynes, M., and C. Ross. 1994. Negotiating the child support maze. *Trial* (February): 47–51.

Heaton, T., and S. Albrecht. 1991. Stable unhappy marriages. *Journal of Marriage and the Family* (August): 747–758.

Helburn, S., et al. 1995. *Cost, quality and child outcomes in child care centers*. Washington, DC: National Association for the Education of Young Children.

Helburn, S. W., and B. R. Bergman. 2002. *America's child care problem: The way out*. New York: Palgrave Macmillan.

Henderson, T., and P. Monroe. 2002. Introduction to special collection on the intersection of families and the law. *Family Relations* (October): 289–292.

Hendrick, S., and C. Hendrick. 1992. *Liking, loving, and relating* 2nd ed. Pacific Grove, CA: Brooks/Cole.

———. 1997. Love and satisfaction. In R. Sternberg and H. Mahzad (Eds.), *Satisfaction in close relationships* 56–77. New York: Guilford.

———. 2000. Romantic love. In C. Hendrick and S. Hendrick (Eds.), *Close relationships: A sourcebook* 204–214. Thousand Oaks, CA: Sage.

Henry, C., and S. Plunkett. 1996. Grandparenting. In S. Price and T. Brubaker (Eds.), *Vision 2010: Families and aging* 14–15. Minneapolis: National Council on Family Relations.

Henry, S. 1996. Coming to terms: A time to die. *California Lawyer* (January): 38–41, 84.

Herdt, G. 2004. The power of love. *Time* (January 19).

Hetherington, E. 2003. Intimate pathways: Changing pattern in close personal relationships across time. *Family Relations* (October): 318–331.

Hetherington, E., and J. Kelly. 2002. *For better or worse: Divorce reconsidered.* New York: Norton.

Hetzet, L., and A. Smith. 2001. The sixty-five years and over population, 2000. *Census 2000 Brief* C2KBR/01-10. Washington, DC: U.S. Census Bureau.

Hewlett, S. A. 1987. *Lesser life: The myth of women's liberation in America.* New York: Warner Books.

———. 1990. Good news: The private sector and win-win scenarios. In D. Blankenhorn et al. (Eds.), *Rebuilding the nest.* Milwaukee: Family Services of America.

———. 1991. *When the bough breaks: The cost of neglecting our own children.* New York: Basic Books.

Hewlett, S. A., and C. West. 2002. Preface: The challenge. In S. A. Hewlett, N. Rankin, and C. West (Eds.), *Taking parenting public: The case for a new social movement.* Lanham, MD: Rowman and Littlefield.

Hill, E., A. Hawkins, and B. Miller. 1996. Work and family in the virtual office. *Family Relations* (July): 293–301.

Hill, T. 2000. Legally extending the family: An event history analysis of grandparent visitation laws. *Journal of Family Issues* 21: 246–261.

Himes, C. 2001. Elderly Americans. *Population Bulletin* (December). Washington, DC: Population Reference Bureau.

Hines, A. 1996. The effect of parent alcohol use and high conflict on adolescent adjustment in divorcing families. Unpublished manuscript.

———. 1997. Divorce-related transitions, adolescent development, and the role of parent-child relationship: A review of the literature. *Journal of Marriage and the Family* (May): 375–388.

Hines, M. 1989. As reported in T. Adler, Early sex hormone exposure studies. *APA Monitor* (March): 6.

Hipple, S., and J. Stewart. 1996. Earnings and benefits of contingent and noncontingent workers. *Monthly Labor Review* (October): 22–30.

Hirschl, T. A., J. Altobelli, and M. R. Rank. 2003. Does marriage increase odds of affluence? Exploring the life course possibilities. *Journal of Marriage and the Family* (November): 927–938.

Hochschild, A., with A. Machung. 1989. *The second shift.* New York: Viking Press.

———. 1997. *The time bind.* New York: Holt.

Hodge, S., and D. Cantor. 1998. Victims and perpetrators of male sexual assault. *Journal of Interpersonal Violence* 13: 222–239.

Hodgkinson, L. 1994. Marriage today. In G. Bird and M. Sporakowski (Eds.), *Taking sides* 10–22. Guilford, CT: Dushkin.

Hofferth, S., and S. Deich. 1994. Recent U.S. child care and family legislation in comparative perspective. *Journal of Family Issues* (September): 424–448.

Hogan, R. 1969. Development of an empathy scale. *Journal of Consulting and Clinical Psychology* 33: 307–316.

Holleman, J. 2003. Out damned spots. *Santa Barbara News Press* (January 14).

Hook, J. L. 2004. Reconsidering the division of labor: Incorporating volunteer work and informal support. *Journal of Marriage and the Family* (February): 101–117.

Hoppough, S., and B. Ames. 2001. Death as normative in family life. *Family focus on death and dying* 1–2. Minneapolis: National Council on Family Relations.

Horn, W. F. 1997. *Father facts 2, revised edition. A compendium of research into fathering.* Gaithersburg, MD: National Fatherhood Initiative.

Horwitz, A. H., and S. Howell-White. 1996. Becoming married and mental health: A longitudinal study of a cohort of young adults. *Journal of Marriage and the Family* (November): 895–907.

Hughes, R., Jr., et al. 1999. Family life in the information age. *Family Relations* (January): 5–6.

Hunt, M. 1959. *The natural history of love.* New York: Knopf.

Hunter, A. 1998. Parenting alone to multiple caregivers: Child care and parenting arrangements in Black and White urban families. *Family Relations* (October): 343–353.

Hupka, R. B. 1977. Social and individual roles in the expression of jealousy. Symposium presented at the Annual Meeting of the American Psychological Association, San Francisco, August.

———. 1981. Cultural determinants of jealousy. *Alternative Life Styles* 4: 310–356.

Hupka, R. B., et al. 1985. Romantic jealousy and romantic envy: A seven-nation study. *Journal of Cross-Cultural Psychology* 16: 423–446.

Huston, T. 2000. The social ecology of marriage and other intimate unions. *Journal of Marriage and the Family* (May): 298–320.

———. 2001. The social ecology of marriage and other intimate unions. In R. M. Milardo (Ed.), *Understanding families into the new millennium* 15–38. Minneapolis: National Council of Family Relations.

Hyde, J., and A. Plant. 1995. Magnitude of psychological gender differences. *American Psychologist* (March): 159–161.

Ihinger-Tallman, M., and K. Pasley. 1987. *Remarriage.* Thousand Oaks, CA: Sage.

Ingersoll-Dayton, B., et al. 2001. Aging parents helping adult children: The experience of the sandwiched generation. *Family Relations* (July): 362–371.

———. 2003. Redressing inequity in parent care among siblings. *Journal of Marriage and the Family* (February): 201–212.

Institute of American Values. 2004. Why marriage matters: Twenty-one conclusions from social scientists. New York: Institute of American Values. http://www.americanvalues.org.

Ishii-Kunta, M. 1997. Intergenerational relationships among Chinese, Japanese, and Korean Americans. *Family Relations* (January): 23–32.

Jackson, D. 2004. Too much TV frying children's circuits. *Santa Barbara News Press* (April 11): G3.

Jackson, M. 1997. Few minority women in top management. *Santa Barbara News Press* (October 23): A1–A2.

Jamieson, L. 1998. *Intimacy: Personal relationships in modern societies.* Cambridge, UK: Pality Press and Blackwell Publishers.

———. 1999. Closing in on intimacy and emotions. Book review in *Family Relations* (April): 217–221.

Janowiak, W. R., and E. F. Fischer. 1992. A cross-cultural perspective on romantic love. *Ethnologic* 31(2): 149–155.

Jasinski, J., and L. Williams (Eds.). 1998. *Partner violence.* Thousand Oaks, CA: Sage.

Jeffords, J., and T. Daschle. 2001. Policy issues: Political issues in the genome era. *Science* 291(12) (February 16): 1249–1251.

Jendrek, M. 1993. Grandparents who parent their grandchildren: Effects on lifestyle. *Journal of Marriage and the Family* (August): 609–621.

Johansen, A., et al. 1996. The importance of child care characteristics to choice of care. *Journal of Marriage and the Family* (August): 759–772.

Johnson, D., T. Amoloza, and A. Booth. 1992. Stability and developmental change in marital quality: A three-wave study. *Journal of Marriage and the Family* (August): 582–594.

Johnson, G. 1998. Accident fatalities are highest among 16-year-old drivers. *Santa Barbara News Press* (April 15).

Johnson, M. P. 1991. Commitment to personal relationships. In W. H. Jones and D. W. Perlman (Eds.), *Advances in personal relationships.* London: Jessica Kingsley.

———. 1999. Personal, moral, and structural commitment in relationships: Experiences of choice and constraint. In W. H. James and J. M. Adams (Eds.), *Handbook of interpersonal commitment and relationship stability* 73–87. New York: Kluwer Academic-Plenum Press.

Johnson, M. P., J. P. Caughlin, and T. L. Huston. 1999. The tripartite nature of marital commitment: Personal, moral and structural reasons to stay married. *Journal of Marriage and the Family* (February): 160–177.

Jordan, A. 2000. *Is the 3-hour rule living up to its potential?* Philadelphia: University of Pennsylvania, Annenberg Public Policy Center.

Joung, I. 1997. The contribution of intermediary factors to marital status and mental health: A longitudinal study of a cohort of young adults. *Journal of Marriage and the Family* (November): 895–940.

Joung, I., et al. 1997. The contribution of intermediary factors to marital status differences in self-reported health. *Journal of Marriage and the Family* (May): 476–490.

Jourard, S. M. 1963. *Personal adjustment*. New York: Macmillan.

Journal of the American Medical Association. 1999. What teens know and don't (but should) about sexually transmitted diseases. *Newsline*. JAMA Information Center.

Judicial Notebook. 1994. Child's disclosure of sexual abuse held tainted by repeated exposure to suggestive book. *APA Monitor* (January): 14.

Julien, D., E. Chartland, and J. Begin. 1999. Social networks, structural independence, and conjugal adjustment in heterosexual, gay and lesbian couples. *Journal of Marriage and the Family* (May): 516–530.

Jurkovic, G. 1997. *Lost childhoods: The plight of the parentified child*. New York: Brunner/Mazel.

Kahana, E., and D. Biegel. 1994. *Family caregiving across the lifespan*. Thousand Oaks, CA: Sage.

Kahn, H., and A. Weiner. 1973. The future meanings of work: Some "surprise-free" observations. In F. Best (Ed.), *The future of work*. Englewood Cliffs, NJ: Prentice Hall.

Kain, E. 1990. *The myth of family decline*. New York: Lexington Books.

Kalmun, M. 1999. Father involvement in childrearing and perceived stability of marriage. *Journal of Marriage and the Family* (May): 409–421.

Kaminer, W. 1993. Feminism's identity crisis. *Atlantic Monthly* (October): 51–68.

Karen, R. 1998. *Becoming attached: First relationships and how they shape our capacity to love*. New York: Oxford University Press.

Kasl, C. 1989. *Women, sex and addiction*. New York: Technor and Fields.

Kass, L. 1998. Human cloning should be banned. In P. Winter (Ed.), *Cloning* 26–48. San Diego: Greenhaven Press.

Kassinove, H., and R. C. Tafrate. 2002. *Anger management: The complete treatment guidebook for practitioners*. Atascadero, CA: Impact.

Kealing, J. 1990. Sexual fantasies of heterosexual and homosexual men. *Archives of Sexual Behavior* 19(5): 461–475.

Kelley, P. 2000. Family matters: Grandparents raising 2nd, even 3rd generations get little help. *Santa Barbara News Press* (October 30): D4.

Kelly, J., and R. Emory. 2003. Children's adjustment following divorce: Risk and resilience perspectives. *Family Relations* (October): 352–362.

Kersting, K. 2003. Teaching children about resilience. *Monitor on Psychology* (October): 44.

———. 2004. High hopes and happy homes. *Monitor on Psychology* (September): 41.

Khaleque, A., and R. P. Rohner. 2002. Perceived parental acceptance—rejection and psychological adjustment: A meta-analysis of cross-cultural and intercultural studies. *Journal of Marriage and the Family* 64 (February): 54–64.

Kiecolt, K. J. 2003. Satisfaction with work and family life: No evidence of a cultural reversal. *Journal of Marriage and the Family* (February): 23–35.

Kiecolt-Glasser, J., et al. 1997. As reported in *APA Monitor* (February): 7.

Kieffer, C. 1977. New depths in intimacy. In R. Libby and R. Whitehurst (Eds.), *Marriages and alternatives: Exploring intimate relationships* 267–293. Glenview, IL: Scott, Foresman.

Kilner, J. 1998. Human cloning would violate Christian ethics. In P. Winters (Ed.), *Cloning* 13–16. San Diego: Greenhaven Press.

Kimura, D. 1985. Male brain, female brain: The hidden difference. *Psychology Today* (November): 50–58.

———. 1992. Sex differences in the brain. *Scientific American* 267(3): 118–125.

King, V. 1994. Nonresident father involvement and child well being: Can dads make the difference? *Journal of Family Issues* (March): 78–96.

King, V., and G. Elder. 1997. The legacy of grandparenting: Childhood experiences with grandparents and current involvement with grandchildren. *Journal of Marriage and the Family* (November): 848–859.

King, V., and H. Heard. 1999. Nonresident father visitation, parental conflicts and mother's satisfaction: What's best for child well-being? *Journal of Marriage and the Family* (May): 385–396.

Kinsey, A., et al. 1948. *Sexual behavior in the human male*. Philadelphia: Saunders.

———. 1953. *Sexual behavior in the human female*. Philadelphia: Saunders.

Kipnis, A. R. 1991. *Knights without armor*. New York: Tarcher/Perigee Books.

Kirn, W. 2000. Should you stay together? *Time* (September 25): 75–82.

Kivett, V. 1993. Racial comparisons of the grandmother role: Implications for strengthening the family support system of older Black women. *Family Relations* (April): 165–172.

Klassen, A., and S. Wilsnack. 1986. Sexual experience and drinking among women in a U.S. national survey. *Archives of Sexual Behavior* 15(5): 363–389.

Klinetob, N., and D. Smith. 1996. Demand-withdraw communication in marital interaction: Tests of interpersonal contingency and gender role hypotheses. *Journal of Marriage and the Family* (November): 945–957.

Kluger, J. 2004. The power of love. *Time* (January 19).

Kluwer, E., et al. 1996. Marital conflict about division of household labor and paid work. *Journal of Marriage and the Family* (November): 958–969.

Knox, D. 1994. *Human sexuality*. St. Paul, MN: West.

Knox, D., and C. Schacht. 1995. As reported in D. Knox and C. Schacht. 2000. *Choices in relationships*. Belmont, CA: Wadsworth.

———. 2000. *Love attitudes inventory*, abridged version. Greenville, NC: East Carolina University.

Knox, D., et al. 1997. College students homogamous preference for a date and mate. *College Student Journal* 31: 445–448.

Kohler, J. K., H. D. Grotevant, and R. G. McRay. 2002. Adopted adolescents' preoccupation with adoption: The impact of adoptive family relationships. *Journal of Marriage and the Family* 64 (February): 93–104.

Kolodny, R. 1985. The clinical management of sexual problems in substance abusers. In T. E. Bratter and G. Forest (Eds.), *Current management of alcoholism and substance abuse*. New York: Free Press.

Kornhaber, A. 1996. *Contemporary grandparenting*. Thousand Oaks, CA: Sage.

Kraft, B. 2004. What makes teens tick? *Time* (May 10): 55–65.

Krantz, J. 1984. Living together is a rotten idea. In F. Cox (Ed.), *Human intimacy* 3rd ed., 111–113. St. Paul, MN: West.

Kraut, R., et al. 1999. Internet paradox: A social technology that reduces social involvement and psychological well-being. *Family Relations* (January): 6.

Kreider, R. M., and T. Simons. 2003. Marital status: 2000. *Census 2000 Brief*. Washington, DC: U.S. Census Bureau.

Kunkel, D. 2003. A biannual report: Sex on TV. Menlo Park, CA: The Henry J. Kaiser Family Foundation.

Kurdek, L. 1994. Areas of conflict for gay, lesbian, and heterosexual couples: What couples argue about influences relationship satisfaction. *Journal of Marriage and the Family* (November): 923–934.

L'Abate, L., and D. Bagarozzi. 1993. *Sourcebook of marriage and family interaction*. New York: Brunner/Mazel.

Labi, N. 1999. A bad start: Living together, a report claims, may be the road to divorce. *Time* (February 15).

LaCayo, R. 2000. Are you man enough? *Time* (April 24): 59–64.

Ladbrook, D. 2000. Relationship quality and the future of death. Talk given at the Building Family Strengths International Symposium, Lincoln, NE, May 10–12.

Ladd, G. 1992. Themes and theories: Perspectives on process in family-peer relationships. In R. Parke and G. Ladd (Eds.), *Family peer relationships: Modes of linkages* 1–34. Hillsdale, NJ: Erlbaum.

Laner, M., and N. Ventrone. 2000. Dating scripts revisited. *Journal of Family Issues* 21(4): 488–500.

Lang, A. 1985. The social psychology of drinking and human sexuality. *Journal of Drug Issues* 15(2): 273–289.

Larimer, M., et al. 1999. Male and female recipients of unwanted sexual contact in a college sample: Prevalence rates of alcohol use and depression symptoms. *Sex Roles* (February).

LaRossa, R., and M. LaRossa. 1981. *Transition to parenthood.* Beverly Hills, CA: Sage.

Larson, J. 2000. *Should we stay together?* San Francisco: Jossey-Bass.

Larson, J., and R. Hickman. 2004. Are college marriage textbooks teaching students premarital predictors of marital quality? *Journal of Marriage and the Family* (July): 385–392.

Larson, J., and T. Holman. 1994. Premarital predictors of marital quality and stability. *Family Relations* (April): 228–237.

Larson, J., S. Wilson, and R. Beley. 1994. The impact of job insecurity on marital and family relationships. *Family Relations* (April): 138–143.

Larzelere, R. 1994. Should the use of corporal punishment by parents be considered child abuse? Response. In M. Mason and E. Gambrill (Eds.), *Debating children's lives: Current controversies on children and adolescents* 217–218. Thousand Oaks, CA: Sage.

Larzelere, R., D. Baumind, and K. Polite. 1998. The pediatric forum: Two emerging perspectives of parental spanking from two 1996 conferences. *Archives of Pediatrics and Adolescent Medicine* 152: 303.

Larzelere, R., W. Schneider, D. Larson, and P. Pike. 1996. The effects of discipline responses in delaying toddler misbehavior recurrences. *Child and Family Therapy* 18: 35–37.

Lasswell, M., and N. Lobsenz. 1980. *Styles of loving.* Garden City, NY: Doubleday.

Lauer, J., and R. Lauer. 1985. Marriages made to last. *Psychology Today* (June): 22–26.

Lauman, E., et al. 1994. *The social organization of sexuality.* Chicago: University of Chicago Press.

Lawler, J. 1996. Bottom line on work-family programs. *Working Woman* (July/August): 54–58, 74, 76.

Lebner, A. 2000. Genetic "mysteries" and international adoption: The cultural impact of biomedical technologies on the adoptive family experience. *Family Relations* (October): 371–377.

Lederer, W. J., and D. D. Jackson. 1968. *The mirage of marriage.* New York: Norton.

Lee, G. 1996. Retirement and marriage. In S. Price and T. Brubaker (Eds.), *Vision 2010: Families and aging* 28–29. Minneapolis: National Council on Family Relations.

Lee, G., and C. Shehand. 1989. Retirement and marital satisfaction. *Journal of Gerontology* 44: S226–S230.

Lee, G., J. Netzer, and R. Coward. 1994. Filial responsibilities, expectations, and patterns of intergenerational assistance. *Journal of Marriage and the Family* (August): 559–565.

Lee, S. 1998. Asian-Americans: Diverse and growing. *Population Bulletin* (June). Washington, DC: Population Reference Bureau.

Legato, M. 1998. Research on the biology of women will improve health care for men, too. *Chronicle of Higher Education* (May 15): B4–B5.

Lehrer, E. L. 2000. Religion as a determinant of entry into cohabitation and marriage. In L. Waite (Ed.), *The ties that bind.* New York: Aldine de Gruyter.

Lemanna, M. A., and A. Reidman. 2003. *Marriage and families: Making choices in a diverse society.* Belmont, CA: Wadsworth.

Lemonich, M. D. 2004. The chemistry of desire. *Time* (January 19).

Leon, K. 2003. Risk and protective factors in young children's adjustment to parental divorce: A review of the research. *Journal of Marriage and the Family* (July): 258–270.

Lerner, J., et al. 2003. *Psychological Science* 14(2) (March).

Lettau, M., and T. Buchmueller. 1999. Comparing benefits costs for full- and part-time workers. *Monthly Labor Review* (March): 30–35.

Levin, I. 1997. The stepparent role from a gender perspective. In I. Levin and M. Sussman (Eds.), *Stepfamilies: History, research, and policy* 177–190. New York: Haworth Press.

Levine, J. A., H. Pollack, and M. E. Comfort. 2001. Academic and behavioral outcomes among the children of young mothers. *Journal of Marriage and the Family* 63 (May): 355–369.

Levine, M., and R. Troiden. 1989. The myth of sexual addiction. In R. Franscolur (Ed.), *Taking sides.* Guilford, CT: Dushkin.

Lewin, T. 1989. Small tots, big biz. *New York Times Magazine* (January 29): 30–31, 89–92.

Lichter, D. T., and M. L. Crowley. 2002. Poverty in America: Beyond welfare reform. *Population Bulletin* (June). Washington, DC: Population Reference Bureau.

Linton, R. 1936. *The study of man.* New York: Appleton.

Lobo, S. 2002. Census-taking and the invisibility of urban American Indians. *Population Today* (May/June): 3–4. Washington, DC: Population Reference Bureau.

London, R. A. 1998. Trends in single mother's living arrangements from 1970 to 1995: Correcting the current population survey. *Demography* 35: 125–131.

Long, E., et al. 1999. Understanding the one you love: A longitudinal assessment of an empathy training program for couples in romantic relationships. *Family Relations* (July): 235–242.

Longmore, J., et al. 2001. Preadolescent parenting strategies and teens' dating and sexual initiation: A longitudinal approach. *Journal of Marriage and the Family* (May): 322–335.

Loomis, L., and N. Landale. 1994. Nonmarital cohabitation and children rearing among Black and White American women. *Journal of Marriage and the Family* (November): 949–962.

Lopata, H. 1993. The interweave of public and private: Women's challenge to American society. *Journal of Marriage and the Family* (February):176–190.

———. 1996. *Current widowhood.* Thousand Oaks, CA: Sage.

Lorenz, F. O., et al. 2002. Effects of spousal support and hostility on trajectories of Czech couples' marital satisfaction and instability. *Journal of Marriage and the Family* (November): 1068–1082.

Lu, H. H., and H. Koball. 2003. The changing demographics of low-income families and their children. *Living at the edge.* Research Brief 2 (August). New York: Columbia University, National Center for Children.

Lye, D., and A. Biblarz. 1993. The effects of attitudes towards family life and gender roles on marital satisfaction. *Journal of Family Issues* 14: 157–158.

Maccoby, E., and C. Jacklin. 1974. *The psychology of sex differences.* Palo Alto, CA: Stanford University Press.

MacDermid, S. T., et al. 1990. Changes in marriage associated with the transition to parenthood. Individual differences as a function of sex role attitudes and changes in the division of household labor. *Journal of Marriage and the Family* (May): 475–486.

MacDonald, T., and M. Ross. 1997. Psychologists explore why relationships last. *APA Monitor* (October): 9.

Mace, D. (Ed.). 1983. *Prevention in family services: Approaches to family wellness.* Beverly Hills, CA: Sage.

———. 1985. Family wellness: The wave of the future. In R. Williams et al. (Eds.), *Family strengths 6.* Lincoln, NE: University of Nebraska Press.

Mace, D., and V. Mace. 1959. *Marriage east and west.* Garden City, NY: Dolphin Books.

Machir, J. 2001. Medicare hospice policy in nursing home settings hinders end-of-life care. *Family focus on death and dying* 9–10. Minneapolis: National Council on Family Relations.

Mackey, R., and B. O'Brien. 1995. Lasting marriages: Men and women growing together. Westport, CT: Praeger.

Madanes, C., and C. Madanes. 1994. *The secret meaning of money.* San Francisco: Jossey-Bass.

Madden-Derdich, D., and S. Leonard. 2002. Shared experiences, unique realities: Formerly married mother's and father's perceptions of parenting and custody after divorce. *Family Relations* (January): 37–45.

Madden-Derdich, D., et al. 1999. Boundary ambiguity and co-parental conflict after divorce: An empirical test of a family system model of the divorce process. *Journal of Marriage and the Family* (August): 588–598.

Maestripieri, D. 2001. Biological basis of maternal attachment. *Current Directions in Psychological Sciences* 10(3).

Magdol, L., et al. 1998. Hitting without a license: Testing explanations for differences in partner abuse between young adult daters and cohabitors. *Journal of Marriage and the Family* (February): 41–55.

Maher, B. 2000. *Divorce reform: Forming ties that bind*. Washington, DC: Family Research Council.

Mahoney, M. 1997. Stepfamilies from a legal perspective. In I. Levin and M. Sussman (Eds.), *Stepfamilies: History, research, and policy*. New York: Haworth Press.

Malone, N. 2003. The foreign-born population: 2003. *Census 2000 Brief* C2KBR-34. Washington, DC: U.S. Census Bureau.

Manke, B., et al. 1994. The three corners of domestic labor: Mother's, father's and children's weekday and weekend housework. *Journal of Marriage and the Family* (August): 657–668.

Manning, A. 2000. Absent dads scar millions of daughters for life. *USA Today* (June 7): 7D.

Manning, W. 1994. Marriage and cohabitation following premarital conception. *Journal of Marriage and the Family* (November): 949–962.

———. 1995. Cohabitation, marriage, and entry into motherhood. *Journal of Marriage and the Family* (February): 191–200.

Manning, W., and D. Lichten. 1996. Parental cohabitation and children's economic well-being. *Journal of Marriage and the Family* (November): 949–962.

Manning, W. D., and K. A. Lamb. 2003. Adolescent well-being in cohabiting, married, and single parent families. *Journal of Marriage and the Family* 65 (November): 876–893.

March, K., and C. Miall. 2000. Adoption as a family form. *Family Relations* (October): 359–362.

Markman, H., S. Stanley, and S. Blumberg. 1994. *Fighting for your marriage*. San Francisco: Jossey-Bass.

Markowski, E. 1991. Temperament differences in enrichment and therapy couples. Paper presented at a meeting of the American Association of Marriage and Family Therapy, Dallas.

Marks, S. 2000. Teasing out the lessons of the 1960s. Family diversity and family privilege. *Journal of Marriage and the Family* (August): 609–622.

Marks, S., and S. MacDermid. 1996. Multiple roles and the self: A theory of role balance. *Journal of Marriage and the Family* (May): 417–432.

Martin, J. A., et al. 2003. Births: Final data for 2002. *National Vital Statistics Reports* 52(10) (December 17). Hyattsville, MD: National Center for Health Statistics.

Martine, P., and E. Midgley. 2003. Immigration: Shaping and reshaping America. *Population Bulletin* (June). Washington, DC: Population Reference Bureau.

Marty, M. 1998. Cloning would violate human dignity. In P. Winter (Ed.), *Cloning* 17–20. San Diego: Greenhaven Press.

Maslow, A. H. 1968. *Toward a psychology of being*. Princeton, NJ: Van Nostrand.

———. 1971. *The farther reaches of human nature*. New York: Viking Press.

Mason, M., H. Skolnik, and S. Sugarman (Eds.). 1998. *All our families: New policies for a new century*. New York: Oxford University Press.

Masten, A. S., and E. M. Berkmaier. 2001. The power of the ordinary: Resilience in development. *Family focus on stress and resilience* FF-10.

Masters, W., and V. Johnson. 1966. *Human sexual response*. Boston: Little, Brown.

———. 1981. Sex and the aging process. *Journal of American Geriatrics Society* (September): 385–390.

Masters, W., V. Johnson, and R. C. Kolodny. 1988. *Human sexuality*. Glenview, IL: Scott, Foresman.

Mathews, T. J., F. Menaker, and M. F. MacDorman. 2003. Infant mortality statistics from 2001 period linked birth/infant death data set. *National Vital Statistics Reports* 52(2) (September 15).

Matthews, L., et al. 1996. Predicting marital instability from spouse and observer reports of marital interaction. *Journal of Marriage and the Family* (August): 641–655.

Mattson, D. 1990. The effectiveness of a specific marital enrichment program: Time. *Individual Psychology* (January): 89–92.

May, R. 1970. *Love and will*. New York: Norton.

Mayseless, O., et al. 2004. "I was more her mom than she was mine": Role reversal in a community sample. *Family Relations* (January): 78–86.

McBride, B. A., S. S. Schoppe, and T. R. Rane. 2002. Child characteristics: Parenting stress and parental involvement: Fathers vs. mothers. *Journal of Marriage and the Family* 64 (November): 998–1011.

McCaw, C. 1994. Talk given at the University of California at Santa Barbara, November 4.

McCubbin, H., and M. McCubbin. 1988. Typologies of resilient families: Emerging roles of social class and ethnicity. *Family Relations* (July): 247–259.

McCubbin, H., et al. 1988. *Family types and strengths: A life cycle and ecological approach* 2nd ed. Edina, MN: Bellwether.

McGuiness, D., and K. Pribram. 1979. The origins of sensory bias in the development of gender differences in perception and cognition. In M. Bortner (Ed.), *Cognitive growth and development*. New York: Brunner/Mazel.

McKinnon, J. 2003. The Black population in the United States: March 2002. *Current Population Reports* P20-541. Washington, DC: U.S. Census Bureau.

McLanahan, S., and S. Sandefur. 1994. *Growing up with a single parent: What hurts. What helps*. Cambridge, MA: Harvard University Press.

McLoyd, V. C., and S. Smith. 2002. Physical discipline and behavior problems in African American, European American, and Hispanic children: Emotional support and moderator. *Journal of Marriage and the Family* 64 (February): 40–53.

McMahon, M. 1995. It's never too late to walk down the aisle of love. *Santa Barbara News Press* (February 12): D1–D4.

Mead, M. 1968. Jealousy: Primitive and civilized. In S. D. Schmalhausen and V. E. Calverton (Eds.), *Women's coming of age*. New York: Morrow.

Mederer, H. 1993. Division of labor in two-earner homes: Task accomplishment versus household management as critical variables in perceptions about family work. *Journal of Marriage and the Family* (February): 133–145.

Medved, D. 1989. *The case against divorce*. Santa Monica, CA: Donald J. Fine.

Meeks, B. S., et al. 1998. Communication, love and relationship satisfaction. *Journal of Social and Personal Relationships* 15: 755–773.

Meese Report. 1986. *Final report of the U.S. Attorney General's Commission on Pornography*. Nashville: Rutledge Hill.

Meier, M. 1995. A revolution is overdue in the way we think about and deliver child care. *Santa Barbara News Press* (March 12): G5.

Merrill, D. 1995. Who's watching the children? *USA Today* (October 13): B1.

Meschke, L., L. Bartholomae, and S. Zentell. 2000. Adolescent sexuality and parent-adolescent process: Promoting healthy teen choices. *Family Relations* 42(2): 143–155.

Michael, R., et al. 1994. *Sex in America: A definitive study*. New York: Little, Brown.

Military Family Resource Center. 4040 North Fairfax Dr., Room 420, Arlington, VA 22203-1635.

Milkie, M., and P. Peltola. 1999. Playing all the roles: Gender and the work-family balancing act. *Journal of Marriage and the Family* (May): 476–490.

Miller, K. 1999. Adolescent sexual behavior in two ethnic minority samples: The role of family variables. *Journal of Marriage and the Family* (February): 85–98.

Mishel, L., J. Bernstein, and H. Boushey. 2003. *State of working women 2002–2003*.

Mitchell, B., and E. Gee. 1996. Boomerang kids and midlife parental satisfaction. *Family Relations* (October): 442–448.

Moen, P. 2002. *PRB reports on America: The career quandary*. Washington, DC: Population Reference Bureau.

Moen, P., et al. 2001. Couples' work/retirement transitions, gender, and marital quality. *Social Psychology Quarterly* 64: 55–71.

Money, J. 1986. *Lovemaps: Clinical concepts of sexual/erotic health and pathology, paraphilia, and gender transposition in childhood, adolescence, and maturity*. New York: Irvington.

Montagu, M. 1972. *Touching: The human significance of skin*. New York: Harper and Row.

Moore, K., and N. Snyder. 1996. Facts at a glance. *Child Trends* (January).

Moore, K., et al. 1989. Nonvoluntary sexual activity among adolescents. *Family Planning Perspectives* 21: 110–114.

———. 2000. Beginning too soon: Adolescent sexual behavior, pregnancy and parenthood: A review of research and interventions. Office of the Assistant Secretary for Planning and Evaluation (Contract DHHS-100-92-0015). Hyattsville, MD: U.S. Department of Health and Human Services. (From Internet copy posted August 23.)

Moreau, H. 1994. Firehouse heat: Are *Playboy* pinups a form of sexual harassment? *California Lawyer* (March): 41–42.

Morehouse Conference on African-American Fathers. 1999. Turning the corner on father absence in Black America (June 16). See "Fatherhood," "read the report" available online at http://www.AmericanValues.org.

Morgan, L., and S. Kunkel. 1998. *Aging: The social context*. Thousand Oaks, CA: Fine Forge.

Morris, D. 1971. *Intimate behavior*. New York: Random House.

Morrison, D., and A. Cherlin. 1995. The divorce process and young children's well-being: A prospective analysis. *Journal of Marriage and the Family* (August): 800–812.

Morrison, D., and M. Coiro. 1999. Parental conflict and marital disruption: Do children benefit when high-conflict marriages are dissolved? *Journal of Marriage and the Family* (August): 626–637.

Moses, S. 1991. Gender gap on tests examined at meeting. *APA Monitor* (December): 38.

Moskowitz, M. 1997. 100 best companies for working mothers. *Working Mother* (October): 18–100.

Mulligan, T., and C. Moss. 1991. Sexuality and aging in male veterans: A cross sectional study of interest, ability and activity. *Archives of Sexual Behavior* 20: 17–25.

Munson, M. L., and P. D. Sutton. 2004. Births, marriages, divorces, and deaths: Provisional data for October 2003. *National Vital Statistics Reports* 52(18) (April 7). Hyattsville, MD: National Center for Health Statistics.

Murdock, G. 1950. Sexual behavior: What is acceptable? *Journal of Social Hygiene* 36: 1–31.

Murphy, M. K., K. Glaser, and E. Grundy. 1997. Marital status and long-term illness in Great Britain. *Journal of Marriage and the Family* (February): 156–164.

Murray, B. 1998. Psychology's voice in sexual harassment law and long-term illness. *APA Monitor* (August): 50.

Murray, T. 1998. Animal cloning may be acceptable even if human cloning is unethical. In P. Winter (Ed.), *Cloning* 54–57. San Diego: Greenhaven Press.

Murstein, B. 1974. *Love, sex, and marriage through the ages*. New York: Springer.

———. 1986. *Paths to marriage*. Beverly Hills, CA: Sage.

Musick, K. 2002. Planned and unplanned childbearing among unmarried women. *Journal of Marriage and the Family* (November): 915–929.

Nakane, C. 1989. *Japanese society*. Tokyo: Charles E. Tuttle.

Nakonezny, P., et al. 1995. The effect of no-fault divorce law on divorce rates across fifty states and its relation to income, education, and religiosity. *Journal of Marriage and the Family* (May): 477–488.

———. 1997. The effect of no-fault legislation on divorce rates: A response to a reconsideration. *Journal of Marriage and the Family* (November): 1026–1030.

———. 1999. Did no-fault divorce legislation matter? Definitely yes and sometimes no. *Journal of Marriage and the Family* (August): 803–809.

Nanninc, D. K., and L. S. Meyers. 2000. Jealousy in sexual and emotional infidelity. *Journal of Sex Research* 37(2): 117–123.

Nansel, T., et al. 2002. Bullying behavior among U.S. youth: Prevalence and association with psychosocial adjustment. *Journal of the American Medical Association* (October): 2094–2100.

Nardi, P. M., and D. Sherrod. 1994. Friendship in the lives of gay men and lesbians. *Journal of Social and Personal Relationships* 11: 185–200.

National Adoption Information Clearinghouse. 2004. Adoption: Numbers and trends. Washington, DC: U.S. Department of Health and Human Services. Retrieved on April 12 from http://www.naic.acf.hhs.gov/pubs/s-number.cfm.

National Center for Health Statistics. 1999a. Births and deaths: Preliminary data for 1998. *National Vital Statistics Reports* 27(25) (October 5). Hyattsville, MD: National Center for Health Statistics.

———. 1999b. Decline in teenage birth rates, 1991–1998: Update of national and state trends. *National Vital Statistics Reports* 47(26) (October 25). Hyattsville, MD: National Center for Health Statistics.

———. 1999c. Trends in twins and triplet births: 1980–1997. *National Vital Statistics Reports* 47(24) (September 14). Hyattsville, MD: National Center for Health Statistics.

———. 2000a. http://www.cdc.gov/nchs/fastats/marriage.htm.

———. 2000b. Births: Preliminary data for 1999. *National Vital Statistics Reports* 48 (August 8): 14. Hyattsville, MD: National Center for Health Statistics.

———. 2000c. Deaths: Final data for 1998. *National Vital Statistics Reports* 48(11) (July 24). Hyattsville, MD: National Center for Health Statistics.

———. 2000d. Divorce. Available online at http://www.cdc.gov/nchswww/nchshome.htm.

———. 2000e. Infant mortality statistics from 1998: Period linked to birth/death data set. *National Vital Statistics Reports* 48(12) (July 20). Hyattsville, MD: National Center for Health Statistics.

———. 2000f. Variations in teenage birthrates, 1991–1998: National and state trends. *National Vital Statistics Reports* 48(6) (July 20). Hyattsville, MD: National Center for Health Statistics.

———. 2001. Trends in cesarean birth and vaginal birth after previous cesarean, 1991–1999. *National Vital Statistics Reports* 49(13) (December 27). Hyattsville, MD: National Center for Health Statistics.

———. 2002. New report sheds light on trends and patterns in marriage, divorce, and cohabitation. *News Reports* 23(22) (July 24). Hyattsville, MD: National Center for Health Statistics.

———. 2003a. *Health, United States: 2003*. Hyattsville, MD: National Center for Health Statistics.

———. 2003b. Teen birth rate continues to decline; African-American teens show the sharpest drop. *News Releases* retrieved April 3, 2004 from http://www.cdc.gov/nchs/releases/03facts/teenbirth.htm.

———. 2004a. *Health, United States, 2003*. Hyattsville, MD: National Center for Health Statistics.

———. 2004b. Marriage and divorce statistics. Available online at http://www.cdc.gov/NCHS/fastats/marriage.htm.

———. 2004c. *Health, United States, 2004* (September): Tables 12, 61, 62. Hyattsville, MD: National Center for Health Statistics.

———. 2004d. Infant mortality: Statistics from the 2002 period linked to birth/infant death data set. *National Vital Statistics Reports* 53(10) (November 24). Hyattsville, MD: National Center for Health Statistics.

National Center for Injury Prevention and Control. 2004. Child maltreatment: Fact sheet. (August 5).

National Council on Family Relations. 2001. *Family focus on stress and resilience*. Washington, DC: National Council on Family Relations.

———. 2003. 10th anniversary of the International Year of the Family. *NCFR Report* (March): 5. Washington, DC: National Council on Family Relations.

National Institute on Aging. 1993. *1992 data for Americans aged 51–61*. University of Michigan.

National Law Journal. 1996. Adoption case sparks press and judiciary feud. (March 4): A10.

Nation's Health. 1997. Child abuse and neglect is still a widespread problem in America. (May/June): 9.

Newman, L. 2001. Coping and defense: No clear distinction. *American Psychologist* (September): 760–761.

Newport, F. 1999. Americans agree that being attractive is a plus in American society. *The Gallup Organization Poll Releases* (September 15).

Newsweek. 1995. Blood and tears. (September 18): 66–68.

New York Times. 2001. Gay couples found to head more homes. (August 22).

Nobile, P., and E. Nadler. 1986. *United States of America vs. sex: How the Meese Commission lied about pornography*. New York: Minotaur Press.

Nock, S. 1995. A comparison of marriages and cohabiting relationships. *Journal of Family Issues* (January): 53–76.

Noller, A., and M. Fitzpatrick. 1991. Marital communication in the eighties. In A. Booth (Ed.), *Contemporary families* 42–53. Minneapolis: National Council on Family Relations.

Nomaguichi, K. M., and M. A. Milkie. 2003. Costs and rewards of children: The effects of becoming a parent on adults' lives. *Journal of Marriage and the Family* 65 (May): 356–374.

North American Menopause Society. 2003. *Menopause guidebook: Helping women make informed healthcare decisions through perimenopause and beyond*. Cleveland: North American Menopause Society.

Norton, A., and I. Miller. 1992. Marriage, divorce and remarriages in the 1990's. *Current Population Reports* (October): 20–180. Washington, DC: U.S. Census Bureau.

Notarius, C., and H. Markmann. 1993. *We can work it out*. New York: Putnam.

O'Connell, S. 1987. Children of working mothers: What the research tells us. In M. Walsh (Ed.), *The psychology of women: Ongoing debates* 367–377. New Haven, CT: Yale University Press.

O'Conner, J. 1998. Human cloning would be unethical. In P. Winter (Ed.), *Cloning*. San Diego: Greenhaven Press.

Office of Women's Health. 1999. Sexually transmitted diseases web page. Atlanta: Centers for Disease Control.

Ofshe, R., and E. Waters. 1994. *Making monsters: False memories, psychotherapy and sexual hysteria*. New York: Scribners.

Ogunwole, S. D. 2002. The American Indian and Alaska native population: 2000. *Census 2000 Brief* C2KBN/01-15. Washington, DC: U.S. Census Bureau.

O'Hanlon, S., and B. O'Hanlon. 1999. Love is a noun (except when it is a verb): A solution-oriented approach to intimacy. In J. Carlson and L. Sperry (Eds.), *The intimate couple* 247–262. Philadelphia: Brunner/Mazel.

O'Hare, W., and K. Johnson. 2004. *Child Poverty in Rural America* (March). Washington, DC: Population Reference Bureau.

Olds, S. M., M. London, and P. Ladeweg. 1998. *Maternal-newborn nursing* 2nd ed. Reading, MA: Addison-Wesley.

Oliver, M., and J. Hyde. 1993. Gender differences in sexuality: A meta-analysis. *Psychological Bulletin* 114(1): 29–51.

Olsen, D., et al. 1986. *Prepare*. St. Paul, MN: University of Minnesota Press.

Owens, D. 1994. How to hold a successful family talk. *Orlando Sentinel* (August 30).

Packard, V. 1958. *The hidden persuaders*. New York: Pocket Books.

Paglia, C. 1998. A call for lustiness: Just say no to the sex police. *Time* (March 23): 54.

Parcel, T., and E. Menaghan. 1994. *Parents' jobs and children's lives*. New York: Aldine de Gruyter.

Parke, R., and A. Brott. 1999. *Throwaway dads*. New York: Houghton Mifflin.

Parkman, A. 1992. *No-fault divorce: What went wrong?* Boulder, CO: Westview.

Parrott, W., and R. Smith. 1987. Differentiating the experiences of envy and jealousy. Paper presented at the Annual Meeting of the American Psychological Association, New York, August.

Pasley, K. 2003. Editorial. *Journal of Marriage and the Family* (October): 313.

Pasley, K., and A. Lofquist. 1995. Remarriage. In D. Levenson (Ed.), *Encyclopedia of marriage and the family* 581–584. New York: Macmillan.

Patenaude, A. F., A. E. Guttmacher, and F. S. Collins. 2000. Genetic testing and psychology. *American Psychologist* 57(4) (April): 271–282.

Patterson, C. 1992. Children of lesbian and gay parents. *Child Development* 63: 1025–1042.

———. 2000. Family relationships of lesbian and gay men. *Journal of Marriage and the Family* 62 (November): 1052–1069.

Paul, P. 2004. The porn factor. *Time* (January 19).

Paulozzi, V. 2002. Variation in homicide risk during infancy— United States, 1989–1998. *Morbidity and Mortality Weekly Report* 51(9): 187–189.

Pear, R. 1997. Database to track deadbeat parents. *Santa Barbara News Press* (September 22).

Pearson, J. C. 1989. *Communication in the family*. New York: Harper and Row.

People. 1994. Turning back the clock. (January 24): 36–42.

Peplau, L., and S. Gordon. 1985. Women and men in love: Gender differences in close heterosexual relationships. In V. O'Leary et al. (Eds.), *Women gender and social psychology*. Hillsdale, NJ: Erlbaum.

Peplau, L. A., and L. R. Spalding. 2000. The close relationship of lesbians, gay men, and bisexuals. In C. Hendrick and S. Hendrick (Eds.), *Close relationships: A sourcebook* 111–124. Thousand Oaks, CA: Sage.

Peterson, J., and D. Hawley. 1998. Effects of stressors on parenting attitudes and family functioning in a primary prevention program. *Family Relations* (July): 221–227.

Pezdek, K. 1994. Avoiding false claims of child sexual abuse: Empty promises. *Family Relations* (July): 261–263.

Phillips, R. 1997. Stepfamilies from a historical perspective. In I. Levin and M. Sussman (Eds.), *Stepfamilies: History, research, and policy*. New York: Haworth Press.

Phipps, P. 1996. Workplace performance. *Monthly Labor Review* (September): 45–46.

Piercy, F., and D. Sprenkle. 1991. Marriage and family therapy: A decade in review. In A. Booth (Ed.), *Contemporary families: Looking forward, looking back* 446–456. Minneapolis: National Council on Family Relations.

Pillemer, V., and J. Suitor. 1991. Will I ever escape my children's problems? Effects of adult children's problems on elderly parents. *Journal of Marriage and the Family* (August): 585–594.

Pimentel, E. 2000. Just how do I love thee? Marital relations in urban China. *Journal of Marriage and the Family* (February): 32–47.

Pines, K. 1992. *Romantic jealousy: Understanding and conquering the shadow of love*. New York: St. Martins Press.

Pirog, M., et al. 1998. Interstate comparisons of child support orders using state guidelines. *Family Relations* (July): 289–295.

Pittman, J., C. Solheim, and D. Blanchard. 1996. Stress as a driver of the allocation of housework. *Journal of Marriage and the Family* (May): 456–468.

Pittman, J., et al. 2004. Internal and external adaptation in army families: Lessons from Operations Desert Shield and Desert Storm. *Family Relations* (April): 249–260.

Pollard, K. M., and W. P. O'Hare. 1999. America's racial and ethnic minorities. *Population Bulletin* (September). Washington, DC: Population Reference Bureau.

Popenoe, D. 1991. Breakup of the family: Can we reverse the trend? *USA Today Magazine* 50–53. © Society for the Advancement of Education.

———. 1993a. As quoted in B. Whitehead, Dan Quayle was right. *Atlantic Monthly* (April): 47–84.

———. 1993b. American family decline, 1969–1990: A review and appraisal. *Journal of Marriage and the Family* (August): 527–555.

———. 1996. Modern marriage: Revising the cultural script. In D. Popenoe et al. (Eds.), *Promises to keep* 247–270. Lanham, MD: Rowman and Littlefield.

Popenoe, D., and B. Whitehead. 1999. *The state of our unions: The social health of marriage in America*. New Brunswick, NJ: Rutgers, The State University of New Jersey, National Marriage Project.

———. 2000. *The state of our unions: The social health of marriage in America*. New Brunswick, NJ: Rutgers, The State University of New Jersey, National Marriage Project.

———. 2002. *Should we live together? What young adults need to know about cohabitation before marriage—a comprehensive review of the research* 2nd ed. New Brunswick, NJ: Rutgers, The State University of New Jersey, National Marriage Project.

———. 2003. *The state of our unions: The social health of marriage in America*. New Brunswick, NJ: Rutgers, The State University of New Jersey, National Marriage Project.

Population Reference Bureau. 2000a. *2000 kids count*. Washington, DC: Population Reference Bureau.

———. 2000b. *2000 world population data sheet*. Washington, DC: Population Reference Bureau.

———. 2003. *2003 world population data sheet*. Washington, DC: Population Reference Bureau.

———. 2004a. *2004 kids count*. Washington, DC: Population Reference Bureau.

———. 2004b. *Reports on America: Child poverty in rural America* (March). Washington, DC: Population Reference Bureau.

———. 2004c. *2004 world population sheet*. Washington, DC: Population Reference Bureau.

Population Today. 1999. Intergroup married couples: 1998. (February): 6. Washington, DC: Population Reference Bureau.

Powell, E. A. 2004. Phenomenal consumers and deplorable savers. *Santa Barbara News Press* (January 6): B4, B8.

Prager, K. 1999. The intimacy dilemma: A guide for couple therapists. In J. Carlson and L. Sperry (Eds.), *The intimate couple* 109–157. Philadelphia: Brunner/Mazel.

———. 2000. Intimacy in personal relationships. In C. Hendrick and S. Hendrick (Eds.), *Close relationships: A sourcebook* 229–244. Thousand Oaks, CA: Sage.

Prager, K. J., and L. J. Roberts. 2004. Deep intimate connection: Self and intimacy in couple relationships. In D. Mashek and A. Aron (Eds.), *Handbook of closeness and intimacy* 43–60. Mahwah, NJ: Erlbaum.

Presser, H. 2000. Nonstandard work schedules and marital instability. *Journal of Marriage and the Family* (February): 93–110.

Pruett, K. D. 2000. *Father need*. New York: Free Press-Broadway Books.

Purnell, M., and H. Bagby. 1993. Grandparents' rights: Implications for family specialists. *Family Relations* (April): 173–178.

Putney, S. 1972. *The conquest of society*. Belmont, CA: Wadsworth.

Pybrum, S. 1995. *Money and marriage: Making it work together*. Santa Barbara, CA: Abundance.

Qu, Z. 1999. Premarital cohabitation and the timing of marriage. *Canadian Review of Sociology and Anthropology* (February).

Quindlen, A. 2002. And now for a hot flash. *Newsweek* (July 29): 64.

Raley, R. K., and W. E. Wildsmith. 2004. Cohabitation and children's family instability. *Journal of Marriage and the Family* (February): 210–219.

Ramirez, R. R., and P. De La Cruz. 2003. The Hispanic population in the United States: March 2002. *Current Population Reports* P20-545. Washington, DC: U.S. Census Bureau.

Raschick, M., and B. Ingersoll-Dayton. 2004. The costs and rewards of caregiving among aging spouses and adult children. *Family Relations* (April): 317–325.

Reeves, T., and C. Bennett. 2003. The Asian and Pacific Islander population in the United States: March 2002. *Current Population Reports* P20-540. Washington, DC: U.S. Census Bureau.

Reiss, I. 1960. Toward a sociology of the heterosexual love relationship. *Marriage and Family Living* (May).

Reiss, I. L. 1988. *Family systems in America* 5th ed. New York: Holt.

Reuters. 1994. Italian court lets father disown his donor-sperm son. *San Diego Union-Tribune* (February 20): A35.

Ridley, C., D. Peterman, and A. Avery. 1978. Cohabitation: Does it make for better marriage? *Family Coordinator* (April): 135–136.

Rieff, P. 1994. *Sexuality and the psychology of love*. New York: Collier.

Riley, D., and J. Steinberg. 2004. Four popular stereotypes about children in self-care: Implications for family life educators. *Family Relations* (January): 55–101.

Ritala-Koskinen, A. 1997. The stepparent role from a gender perspective. In I. Levin and M. Sussman (Eds.), *Stepfamilies: History, research, and policy*. New York: Haworth Press.

Roan, S. 2000. The abortion pill: Finally at hand. *Los Angeles Times* (August 14): S1, S6.

Roberts, L., and M. Morris. 1998. An evaluation of marketing factors in marriage enrichment program promotion. *Family Relations* (January): 37–44.

Roberts, S. 1994. Educated Black women make financial strides over men. *Santa Barbara News Press* (October 3): A9.

Robin, C. 1996. *Equal partners—good friends: Empowering couples through therapy*. London: Routledge.

Robinson, L., and P. Blanton. 1993. Marital strengths in enduring marriages. *Family Relations* (January): 38–45.

Rogers, B., and J. Pryor. 1999a. *Divorce and separation: The outcomes for children*. Layerthorpe, England: York Publishing, Joseph Roundtree Foundation.

———. 1999b. A review of outcomes for children from divorced families (Britain): Academic research or political battleground? Session 122 at the 61st Annual Conference of the National Council on Family Relations, Irvine, CA, November 12–15.

Rogers, C. 1951. Communication: Its blocking and facilitation. Paper read at the Centennial Conference on Communications, Chicago, October 11.

———. 1972. *Becoming partners: Marriage and its alternatives*. New York: Delacorte Press.

Rogers, J. J., and D. C. May. 2003. Spillover between marital quality and job satisfaction: Long-term patterns and gender differences. *Journal of Marriage and the Family* (May): 482–495.

Rogers, S. 1999. Wives' income and marital quality: Are there reciprocal effects? *Journal of Marriage and the Family* (February): 123–132.

Roiphe, K. 1993. *The morning after: Sex, fear, and feminism on campus*. New York: Little, Brown.

Rokachi, A. 1990. Content analysis of sexual fantasies of males and females. *Journal of Psychology* (January): 427–436.

Roleder, G. 1986. *Starting your marriage*. Minneapolis: Augsberg Fortress.

Roscoe, W. 1994. How to become a berdache: Toward a unified analysis of gender diversity. In G. Herdt (Ed.), *Third sex, third gender: Beyond sexual dimorphism in culture and history*. Cambridge, MA: Zone Books.

———. 2002. How to become a berdache: Toward a unified analysis of gender diversity. As reprinted in E. Paul (Ed.), *Taking sides: Sex and gender* 2nd ed. Guilford, CT: Dushkin.

Rosenthal, R., and L. Jacobson. 1968. *Pygmalion in the classroom*. New York: Holt.

Ross, C., and J. Mirowsky. 1994. Women, work and family changing gender roles and psychological well-being. In G. Handel and G. Whitechurch (Eds.), *The psychological interior of the family* 4th ed., 325–340. New York: Aldine de Gruyter.

Rossi, A. 1977. A biosocial perspective on parenting. *Daedalus* (Spring): 1–31.

———. 1978. The biosocial side of parenthood. *Human Nature* (June): 77–79.

Rubinstein, C. 1982. Wellness is all. *Psychology Today* (October): 27–37.

Russell, B. 1957. Our sexual ethics. *Why I am not a Christian* 171–172. New York: Simon and Schuster.

Rutter, V., and P. Schwartz. 2000. Gender, marriage, and diverse possibilities for cross-sex and same-sex pairs. In D. H. Demo et al. (Eds.), *Handbook of family diversity* 59–81. New York: Oxford University Press.

Salem, D., M. Zimmerman, and P. Nataro. 1998. Effects of family structure, family process, family involvement on psychosocial outcomes among African-American adolescents. *Family Relations* (October): 331–341.

Salovey, B. 1985. The heart of jealousy. *Psychology Today* 19: 22–25, 28–29.

Salovey, B., and J. Rodin. 1989. Envy and jealousy in close relationships. In C. Hendrick (Ed.), *Close relationships* 221–246. Thousand Oaks, CA: Sage.

Saltus, R. 1998. Multiple benefits predicted from first cloned cattle. *Santa Barbara News Press* (January 21): A3.

Sandberg, J. G. 2002. A qualitative study of marital process and depression in older couples. *Family Relations* (July): 256–264.

Sandroff, R. 1994. When women make more money than men. *Working Woman* (January).

San Francisco Chronicle. 1997. Violence kills more U.S. kids (February 7): A1.

Santa Barbara News Press. 1991. Woman gives birth to own grandchildren. (October 13).

———. 1995. Court gives Baby Richard back to father. (January 26).

———. 1996. Adoption bill labor of love. (May 20): A10.

———. 2000. Women in management. (September 9).

———. 2003. Personal bankruptcies reach all-time high. (August 19).

Sargent, A. 1977. *Beyond sex roles*. St. Paul, MN: West.

Saunders, W. L. 2000. Disposable human beings. *Family Policy* 13(6) (November/December). Washington, DC: Family Research Council.

Savin-Williams, R. C., and K. G. Esterberg. 2000. Lesbian, gay and bisexual families. In D. H. Demo and K. R. Allen (Eds.), *The handbook of family diversity* 197–215. New York: Oxford University Press.

Schiavi, R., P. Schreian-Engle, and J. Mandell. 1990. Health, aging and male sexual function. *American Journal of Psychiatry* 147: 766.

Schiavi, R. C. 1999. *Aging and male sexuality*. New York: Cambridge University Press.

Schoenborn, C. 2004. Marital status and health: United States, 1999–2002. Advance data from Vital and Health Statistics. Hyattsville, MD: National Center for Health Statistics.

Schuman, W., et al. 1986. Self-disclosure and marital satisfaction revisited. *Family Relations* 34: 241–247.

———. 1998. 19 nations ban cloning of humans. *Santa Barbara News Press* (January 13).

Schwartz, I. M. 1999. Sexual activity prior to coitus initiation. A comparison between males and females. *Archives of Sexual Behavior* 28(1): 63–69.

Science. 2004. Scientists clone human embryos. (February).

Segal, D., and M. Segal. 2002. America's military population. *Population Bulletin* (December). Washington, DC: Population Reference Bureau.

Seligman, M. 1998. Behavioral researchers call for more study on human strengths. *APA Monitor* (April): 11.

Seligmann, J. 1992. A condom for women. *Newsweek* (February 10).

Selman, P. 2002. Intercountry adoption in the new millennium: The quiet migration revisited. *Population Research and Policy Review* 21: 205–225.

Seltzer, J. A. 2000. Families formed outside of marriage. *Journal of Marriage and the Family* 62: 1247–1268.

Seppa, N. 1996. Moving appears hardest on adolescents. *APA Monitor* (June): 4.

Shalit, W. 1999. *A return to modesty: Discovering lost virtue*. New York: Free Press.

Shapiro, D. 1993. The measure of man: *Becoming the father you wish your father had been*. New York: Delacorte Press.

Sharpe, S. 2000. *The ways of love*. New York: Guilford.

Shaver, P., et al. 1988. Love as attachment: The integration of three behavioral systems. In R. Sternberg and M. Barnes (Eds.), *The psychology of love*. New Haven, CT: Yale University Press.

Shaywitz, S. 1995. As reported in G. Kolata, Study finds sexes use brain differently. *Santa Barbara News Press* (February 16).

Sheehy, G. 1995. *New passages: Mapping your life across time*. New York: Random House.

Shehan, C., et al. 2003. Alimony: An anomaly in family social science. *Family Relations* (October): 308–316.

Shotland, R. 1989. A model of the causes of date rape in developing and class relationships. In C. Hendrick (Ed.), *Close relationships*. Thousand Oaks, CA: Sage.

Sillers, A., et al. 1987. Content themes in marital conversations. *Human Communications Research* 13: 495–528.

Silverstein, L., and C. Auerbach. 1999. Deconstructing the essential father. *American Psychologist* (June): 397–407.

Simmons, J. M., and J. Dye. 2003. Grandparents living with grandchildren: 2000. *Census 2000 Brief* C2KBR-31. Washington, DC: U.S. Census Bureau.

Simmons, T., and G. O'Neill. 2001. Households and families: 2000. *Census 2000 Brief*. Washington, DC: U.S. Census Bureau.

Simons, R., et al. 1993. Childhood experience, conceptions of parenting, and attitudes of spouse as determinants of parental behavior. *Journal of Marriage and the Family* (February): 91–106.

———. 1996. *Understanding differences between divorced and intact families*. Thousand Oaks, CA: Sage.

Skolnick, A. 1991. *Embattled paradise: The American family in an age of uncertainty*. New York: Basic Books.

———. 2000. *Family in transition*. Boston: Little, Brown.

Skolnick, A., and J. Skolnick. 1986. *The family in transition*. Boston: Little, Brown.

Smith, C. 1999. An essay for practitioners: Family life pathfinders on the new electronic frontier. *Family Relations* (January): 31–34.

Smith, D. 2000. Women and minorities make gains in science and engineering education. *Monitor on Psychology* (November): 32.

———. 2001. Harassment in the hallways. *Monitor on Psychology* (September): 38–40.

———. 2002. Canceled trial is yielding useful data. *Monitor on Psychology* (October): 52.

———. 2003. The older population in the United States: March 2002. *Current Population Reports* P20-546. Washington, DC: U.S. Census Bureau.

Smith, T. 1994. Attitudes toward sexual permissiveness: Trends, correlates and behavioral connections. In A. Rossi (Ed.), *Sexuality across the life course* 63–97. Chicago: University of Chicago Press.

Smith Bailey, D. 2003. Compulsive cybersex can jeopardize marriage, rest of life. *Monitor on Psychology* (October): 20.

Smolowe, J. 1991. Can't we talk this over? *Time* (January 7): 77.

Snarey, J. 1997. The next generation of work on fathering. In A. Hawkins and D. Dollahite (Eds.), *Generative fathering: Beyond deficit perspectives*. Thousand Oaks, CA: Sage.

Sokalski, H. 1994. Family: Smallest democracy at the heart of society. *National Council on Family Relations Report* (September): 8–9.

Sokolski, D. 1994. A study of marital satisfaction in graduate student marriages. Unpublished doctoral dissertation, Texas Technical University.

Sommers, C. 1994. *Who stole feminism?* New York: Simon and Schuster.

———. 2000. *The war against boys*. New York: Simon and Schuster.

South, S. 2001. The geographic context of divorce: Do neighborhoods matter? *Journal of Marriage and the Family* (August): 755–766.

South, S., and K. Lloyd. 1995. A longitudinal study of marital problems and subsequent divorce. *American Sociological Review* 60: 21–35.

Spain, D., and S. Bianchi. 1996. *Balancing act: Motherhood, marriage and employment*. New York: Russell Sage.

Spanier, G. 1986. Cohabitation in the 1980s. Recent changes in the United States. In K. Davis and A. Grossland-Shiechtman (Eds.), *Contemporary marriage: Comparative perspectives on a changing situation*. New York: Russell Sage.

Sperry, L. 1999. Levels and styles of intimacy. In J. Carlson and L. Sperry (Eds.), *The intimate couple*. Philadelphia: Brunner/Mazel.

Spock, B. 1980. *Baby and child care*. New York: Pocket Books.

Sponaugle, G. C. 1989. Attitudes towards extramarital relations. In K. McKinney and S. Sprecher (Eds.), *Human sexuality: The social and interpersonal context*. Norwood, NJ: Ablex.

Sprey, J. 2000. Theorizing in family studies: Discovering process. *Journal of Marriage and the Family* (February): 18–31.

Stacey, J. 2000. The handbook's tail: Toward revels or a requiem for family diversity. In D. Demo et al. (Eds.), *Handbook of family diversity* 424–439. New York: Oxford University Press.

Stahl, L. 2004. Staying at home. *60 Minutes* (October 10).

Stanton, T., and S. Chang. 2000 Bridging the digital divide. *Access: America's Guide to the Internet* (June 18): 10–12.

Starrels, M. 1994. Gender difference in parent-child relations. *Journal of Marriage and the Family* (March): 148–165.

Steinberg, L. 1996. *Wall Street Journal* (July 11): A14.

Steinberg, L., et al. 1996. *Beyond the classroom: Why school reform has failed and what parents need to do*. New York: Simon and Schuster.

Sternberg, R. 1986. A triangular theory of love. *Psychological Review* 93: 119–135.

———. 1988. Triangulating love. In R. J. Sternberg and M. L. Barnes (Eds.), *The psychology of love* 119–138. New Haven, CT: Yale University Press.

———. 1998. *Cupid's arrow: The course of love through time*. Cambridge, UK: Cambridge University Press.

Stets, J. 1991. Cohabiting and marital aggression: The role of social isolation. *Journal of Marriage and the Family* (August): 660–680.

Stevens, D., G. Kiger, and P. Riley, 2001. Working hard and hardly working: Domestic labor and marital satisfaction among dual-earner families. *Journal of Marriage and the Family* (May): 514–526.

Stewart, A., and J. Ostrove. 1998. Women's personality in middle age. *American Psychologist* (November): 1185–1194.

Stewart, F. 1998. Menopause. In R. Hatcher et al. (Eds.), *Contraceptive technology* 511–544. New York: Ardent Media.

Stewart, G. 1998a. Impaired fertility. In R. Hatcher et al. (Eds.), *Contraceptive technology* 653–678. New York: Ardent Media.

Stewart, G. 1998b. Interuterine devices. In R. Hatcher et al. (Eds.), *Contraceptive technology* 511–544. New York: Ardent Media.

Stewart, G., and Carignan, C. 1998. Female and male sterilization. In R. Hatcher et al. (Eds.), *Contraceptive technology* 545–588. New York: Ardent Media.

Stinnett, N. 1986. Strengthening families: An international priority. Talk given at the International Conference of Family Strengths, Los Angeles, April 20–22.

———. 1997. *Good families*. New York: Doubleday.

Stinnett, N., and J. DeFrain. 1985. *Secrets of strong families*. Boston: Little, Brown.

———. 2000. Personal conversations at the Building Family Strengths International Symposium, Lincoln, NE, May 10–12.

Stinnett, N., and D. James. 2000. Discussion on strengthening families, Building Family Strengths International Symposium, Lincoln, NE, May 10–12.

Stokols, D. 1992. Establishing and maintaining healthy environments. *American Psychologist* (January): 6–22.

Stolba, C. 2001. *Women, work, and family*. Arlington, VA: Independent Women's Forum.

Straus, M. 1994. *Beating the devil out of them: Corporal punishment in American families*. New York: Lexington.

———. 1996. Presentation: Spanking and making a violent society. *Pediatrics* 98: 837–849.

———. 1999. *The benefits of avoiding corporal punishment: New and more definitive evidence*. Durham, NH: Family Research Laboratory.

Straus, M., and J. Stewart. 1999. Corporal punishment by American parents: National data on prevalence, chronicity, severity, and duration in relation to child and family characteristics. *Clinical Child and Family Psychology Review* 2(2): 55–70.

Strong, J. 1983. *Creating closeness*. Ames, IA: Human Communication Institute.

Stull, D., K. Bowman, and V. Semerglia. 1994. Women in the middle: A myth in the making? *Family Relations* (July): 319–324.

Subramanian, S. 1997. Economic considerations in mate selection criteria. Paper presented at the Annual Convention of the American Psychological Society, Chicago.

Sugimoto, F. 1935. *A daughter of the samurai*. Garden City, NY: Doubleday.

Sullivan, A. 2004. If at first you don't succeed. . . . *Time* (July 26): 78.

Sutton, P. D., et al. 2004. Trends in characteristics of birth by states: United States, 1990, 1995, and 2000–2002. *National Vital Statistics Reports* 52(19) (May 10). Hyattsville, MD: National Center for Health Statistics.

Symons, P., et al. 1994. Prevalence and predictors of adolescent dating violence. *Journal of Child and Adolescent Psychiatric Nursing* 7(3): 14–23.

Szuchman, L., and F. Muscarella. 2000. *Psychological perspectives on human sexuality*. New York: Wiley.

Tafrate, R. C., et al. 2002. Anger episodes in high- and low-trait anger community adults. *Journal of Clinical Psychology* 58(12): 310–315.

Tanfer, K. 1987. Patterns of premarital cohabitation among never married women in the U.S. *Journal of Marriage and the Family* (August): 483–498.

Tannen, D. 1990. *You just don't understand: Women and men in conversation*. New York: Ballantine.

Tavris, C. 1989. *Anger: The misunderstood emotion*. New York: Simon and Schuster.

———. 1992. *The mismeasure of woman*. New York: Simon and Schuster.

Tavris, C., and J. White. 2000. *Sexuality, society, and feminism*. Washington, DC: American Psychological Association.

Teachman, J. 2002. Children living arrangements and the intergenerational transmission of divorce. *Journal of Marriage and the Family* (August): 717–729.

———. 2003. Premarital sex, premarital cohabitation, and risk of subsequent marital dissolution among women. *Journal of Marriage and the Family* (May): 444–455.

Tettamanzi, D. 1998. Anthropological and ethical thoughts on whether domestic partnerships should have the same legal status as the family. *L'Osservatore Romano* (September 30): 9.

Thomas, C. 1995. Television: The children speak. *Santa Barbara News Press* (March 11): A15.

Thompson, A. 1983. Extramarital sex: A review of the literature. *Journal of Sex Research* 19: 1–22.

Thornton, A., and L. Young-DeMarco. 2001. Four decades of trends in attitudes toward family issues in the United States: The 1960s through the 1990s. *Journal of Marriage and the Family* (November): 1009–1037.

Thorp, J. M., et al. 2003. Long-term physical and psychological health consequences of individual abortion: Review of the evidence. *Obstetrical and Gynecological Survey* 58(1): 67–79.

Tillich, P. 1957. *Dynamics of faith*. New York: Harper and Row.

Time. 1995a. Should this marriage be saved? (February 27): 48–56.

———. 1995b. The estrogen dilemma. (June 26): 46–53.

———. 2000a. Bridal vows revisited. (July 24): G1–G3.

———. 2000b. The hottest jobs of the future. (May 22).

———. 2000c. Who needs a husband? (August 28): 46–55.

———. 2000d. The future of technology. (June 19): 62–111.

———. 2001. Cloning gets closer. (February 19): 55.

———. 2004. Still sexy after 60. (January 19).

Towes, M., and P. McKenry. 1999. Court-related predictors of parental cooperation and conflict after divorce. Paper presented at the 61st Annual Conference of the National Council on Family Relations, Irvine, CA, November 12–15.

Townsend, J. W. 2003. Reproductive behavior in the context of global population. *American Psychologist* 28(3) (March): 197–204.

Treas, J. 1995. Older Americans in the 1990s and beyond. *Population Bulletin*. Washington, DC: Population Reference Bureau.

Treas, J., and D. Giesen. 2000. Sexual infidelity among married and cohabiting Americans. *Journal of Marriage and the Family* (February): 48–60.

Trial. 1994. Father wins suit against daughter's therapists for "implanting" memories. (July): 95.

———. 1997. Is a fetus a person? Court decisions debate over fetal rights. (June): 13–16.

Tucker, C. J., S. M. McHale, and A. C. Crouter. 2003. Dimensions of mothers' and fathers' differential treatment of siblings: Links with adolescents' sex-typed personality qualities. *Family Relations* 52 (January): 82–89.

Turnbull, S., and J. Turnbull. 1983. To dream the impossible dream: An agenda for discussion with stepparents. *Family Relations* (April): 227–230.

Twenge, J. M., et al. 2003. Parenthood and marital satisfaction: A meta-analytic review. *Journal of Marriage and the Family* (August): 574–583.

Udrey, J. R. 1994. The nature of gender. *Demography* 31(4) (November): 568–573.

Ulriksen, M. 2000. Should you stay together for the kids? *Time* (September 25): 75–82.

Upchurch, D., et al. 1998. Gender and ethnic differences in the timing of first sexual intercourse. *Family Planning Perspectives* (May/June).

U.S. Census Bureau. 1997a. Child abuse and neglect cases substantiated and indicated victim characteristics. Retrieved on December 13 from http://www.census.gov/statb/freg/960347.txt.

———. 1997b. Marital status and living arrangements: March 1996. *Current Population Reports*. Washington, DC: U.S. Census Bureau.

———. 1998a. Household and family characteristics (update). *Current Population Reports* P20-515. Washington, DC: U.S. Census Bureau.

———. 1998b. Marital status and living arrangements: March 1998 (update). *Current Population Reports* P20-514. Washington, DC: U.S. Census Bureau.

———. 1998c. *Statistical Abstract of the United States: 1998*. Washington, DC: U.S. Census Bureau.

———. 1999a. Table MS-2 estimated age of first marriage by sex: 1890 to the present. http://www.census.gov/population/socdemo/ms-la/tatrns-2.text.

———. 1999b. *Statistical Abstract of the United States:1999.* Washington, DC: U.S. Census Bureau.

———. 2000a. *Housing vacancies survey, 3rd quarter, 2000.* Washington, DC: U.S. Census Bureau.

———. 2000b. *Poverty.* http://www.census.gov/hhes/poverty/povdef.html.

———. 2001a. America's family and living arrangement. *Current Population Reports* P20-537. Washington, DC: U.S. Census Bureau.

———. 2001b. The Black population: 2000. *Census 2000 Brief* C2KBR/01-5. Washington, DC: U.S. Census Bureau.

———. 2001c. The Hispanic population: 2000. *Census 2000 Brief* C2KBR/01-3. Washington, DC: U.S. Census Bureau.

———. 2001d. The two or more races population: 2000. *Census 2000 Brief* C2KBR/01-6. Washington, DC: U.S. Census Bureau.

———. 2001e. U.S. adults postponing marriage. *Census Bureau Reports.* Washington, DC: U.S. Census Bureau.

———. 2002a. Historical income tables. *Current Population Reports* Tables F10A, B, C. Washington, DC: U.S. Census Bureau.

———. 2002b. Poverty status of people by family relationship, race, and Hispanic origin: 1959 to 2000. Retrieved on March 29 from http://www.census.gov/hhes/poverty/histpov/hstpov2.html.

———. 2002c. Years of school completed by people 25 years and over, by age, race, household relationship, and poverty status: 2000. Retrieved on March 29 from http://ferret.bls.census.gov/macro/032001/pov/new07_000.htm.

———. 2002d. *Statistical Abstract of the United States: 2002* Tables 718–723. Washington, DC: U.S. Census Bureau.

———. 2003a. Adopted children and stepchildren: 2000 census. *Special Reports* CCN5R-6RV. Washington, DC: U.S. Census Bureau.

———. 2003b. Current Population Survey 2004. Annual social and economic supplement. Table HINC-06. Available online at http://ferret-BLS.census.gov/macro/032004/hhine/new06_000.htm.

———. 2003c. Grandparents living with grandchildren: 2000. *Census 2000 Brief* C2KBR-31. Washington, DC: U.S. Census Bureau.

———. 2003d. *Occupations: 2000* Table 3.Washington, DC: U.S. Census Bureau.

———. 2003e. *Statistical Abstract of the United States: 2002* Tables 650, 651, 678, 1010, 1062, 1165. Washington, DC: U.S. Census Bureau.

———. 2003f. *Marital status 2000.* Washington, DC: U.S. Census Bureau.

———. 2004a. Age and sex of all people, family members, and unrelated individuals iterated below 100% of poverty. *Annual Demographic Survey* (March supplement): Table POV1. Hyattsville, MD: U.S. Census Bureau.

———. 2004b. *Housing vacancy, second quarter, 2004* Table 5. Washington, DC: U.S. Census Bureau.

———. 2004c. Poverty thresholds: 2003. *Current Population Survey.* Washington, DC: U.S. Census Bureau.

U.S. Department of Commerce. 2003. Women edge men in high school diplomas, breaking 13 year deadlock. *U.S. Department of Commerce News* (March 21). Washington, DC: U.S. Department of Commerce.

U.S. Department of Health and Human Services. 2003. Trends in the well-being of America's children and youth. (July 25). Washington, DC: U.S. Department of Health and Human Services.

U.S. Department of Labor. 2004. Labor force statistics from current big population survey. (April 30). Washington, DC: Bureau of Labor Statistics Data.

———. 2000. Annual data: Consumer price index. *Monthly Labor Review* (June): Table 30.

Uyehara, B. 2000. I still do: Couples are renewing their vows and strengthening their commitments. *Los Angeles Times* (September 6).

Vandewater, E., and T. Antonucci. 1998. The family as a context for health and well-being. *Family Relations* (October): 313–314.

Van Matre, L. 1992. Honesty can be the worst policy in an affair. *The Daily Reflector* (August 28): D8.

Ventura, S. J., et al. 2004. Estimated pregnancy rates for the United States, 1990–2000: An update. *National Vital Statistics Reports* 52(23) (June 15). Hyattsville, MD: National Center for Health Statistics.

Verbrugge, L. 1983. Multiple roles and physical health of women and men. *Journal of Health and Social Behavior* 24: 16–30.

———. 1986. National health interview. *American Demographics* (March).

Verbrugge, L., and J. Madans. 1985. Social roles and health trends of American women. *Milbank Memorial Fund Quarterly/Health and Society* 63 (Fall).

Visher, E., J. Visher, and K. Pasley. 1997. Stepfamily therapy from the client's perspective. In I. Levin and M. Sussman (Eds.), *Stepfamilies: History, research, and policy.* New York: Haworth Press.

Vlosky, D., and P. Monroe. 2002. The effective dates of no-fault divorce laws in the 50 states. *Family Relations* (October): 317–324.

Vobejda, J. 1995. Can you really call it care? *Washington Post* (February).

Vohs, H. D., and R. F. Baumeister. 2004. Sexual passion, intimacy, and gender. In D. Mashek and A. Aron (Eds.), *Handbook of closeness and intimacy* 189–200. Mahwah, NJ: Erlbaum.

Volz, J. 2000. In search of the good life. *Monitor on Psychology* (February).

Von Sydow, K. 2000. Sexuality of older women: The effect of menopause, other physical and social partner-related factors. *Arztl Fortbild Qualitatssich* 94(3): 223–229.

Wachwithan, P., M. Hogan, and D. Detzner. 1998. Caregiving for elderly parents in Thai families. Paper presented at the 60th Annual Conference of the National Council on Family Relations, Milwaukee, November 15.

Wagner, D. 1997. Divorce reform: An idea whose time is coming. *Family Policy* (September). Washington, DC: Family Research Council.

Waite, L., and K. Joyner. 2000. Emotional satisfaction and physical pleasure in sexual unions: Time, horizon, sexual behavior and sexual exclusivity. *Journal of Marriage and the Family* (February): 247–264.

Waite, L. J. 2000. Trends in men's and women's well-being in marriage. In L. J. Waite (Ed.), *The ties that bind: Perspectives on marriage and cohabitation.* New York: Aldine de Gruyter.

Waite, L. J., and M. Gallagher. 2000. *The case for marriage: Why married people are happier, healthier, and better off financially.* New York: Doubleday.

Waldfogel, J. 1999. Family leave coverage in the 1990s. *Monthly Labor Review* (October): 13–21.

Walker, B. W., and P. H. Ephross. 1999. Knowledge and attitudes toward sexuality of a group of elderly. *Journal of Gerontological Social Work* 31(1–2): 85–87.

Wallerstein, J. 1984. Children of divorce—preliminary report of a ten year follow-up of young children. *American Journal of Orthopsychiatry* 54: 3.

Wallerstein, J., and S. Blakeslee. 1989. *Second chances.* New York: Ticknor and Fields.

———. 1995. *The good marriage: How and why love lasts.* New York: Houghton Mifflin.

———. 2003. *What about the kids?* New York: Hyperion.

Wallerstein, J., J. Lewis, and S. Blakeslee. 2000. *The unexpected legacy of divorce.* New York: Hyperion.

Wallis, C. 2004. The case for staying home. *Time* (March 22): 51–59.

Walsh, A. 1991. *The science of love: Understanding love and its effects on mind and body.* Buffalo, NY: Prometheus.

Walsh, F. 1998. *Strengthening family resilience.* New York: Guilford.

Walzer, S. 1998. *Thinking about the baby: gender and transitions to parenthood.* Philadelphia: Temple University Press.

Wamboldt, F., and D. Reiss. 1989. Defining family heritage and a new relationship identity: Two central tasks of marriage. *Family Process* 23(3): 315–335.

Ward, M. 1997. Family paradigms and older-child adoption: A proposal for matching parents' strengths to children's needs. *Family Relations* (July): 257–262.

Warren, N. 1995. Work it out. *Focus on the Family* (October): 2–4.

Weddle, S. 1994. The effect of nonresidential children's perceived physical and psychological presence on remarriage adjustment and affect. Unpublished doctoral dissertation (1993).

Weinberg, D. 2003. Income and poverty, 2002. Press briefing (September 26). Hyattsville, MD: U.S. Census Bureau. http://www.census.gov/hhes/incomeo2/prs03asc.html.

Weiner, R. C., and T. Richman. 1998. Sexually explicit speech is protected on the Internet. *APA Monitor* (January): 45.

Weiner-Davis, M. 1992. *Divorce busting*. New York: Summit Books.

Weinstein, S. P., and E. Gottheil. 1986. Cocaine abuse: What the family needs to know. *Medical aspects of human sexuality* 2: 87–89.

Weis, D., and J. Felton. 1987. Marital exclusivity and the potential for future marital conflict. *Social Work* 32: 45–49.

Weitzman, L. 1985. *The divorce revolution: The unexpected social and economic consequences for women and children in America*. New York: Free Press.

West. 1989. *West's annotated California codes* Vol. 21A, sect. 4100. St. Paul, MN: West.

Wetzstein, C. 1998. Who's happier after? *Insight on the News* (September 28).

Wheeler, D. 1998. Researchers explore gender-related differences in response to pain. *Chronicle of Higher Education* (April 24): A18.

White, B. 1984. *The first three years of life: A guide to physical, emotional, and intellectual growth of your baby* 2nd ed. New York: Avon.

White, G. 1980. Inducing jealousy: A power perspective. *Personality and Social Psychology Bulletin* 6: 222–227.

White, L. 2000. Stepfamilies over the life course. Lecture given at the Building Family Strengths International Symposium, Lincoln, NE, May 10–12.

White, L., and S. Rogers. 1997. Strong support but uneasy relationships. Coresidence and adult children's relationships with their parents. *Journal of Marriage and the Family* (February): 62–76.

———. 2000. Economic circumstances and family outcomes: A review of the 1990s. *Journal of Marriage and the Family* 62: 1035–1051.

Whitehead, B. 1997. *The divorce culture*. New York: Knopf.

Whitehead, B., and D. Popenoe. 2003. *The state of our unions*. New Brunswick, NJ: Rutgers, The State University of New Jersey, National Marriage Project.

Whiteman, M. K., et al. 2003. Risk factors for hot flashes in midlife women. *Journal of Women's Health* 12(5): 459–472.

Whiteman, S. D., S. M. McHale, and A. C. Crouter. 2003. What parents learn from experience: The first child as a first draft. *Journal of Marriage and the Family* 65 (August): 608–621.

Whitney, E., and F. Sizer. 1988. Life choices: Health concepts and strategies. St. Paul, MN: West.

Whyte, M. 1990. *Dating, mating, and marriage*. New York: Aldine de Gruyter.

Wickrama, K., et al. 1997. Marital quality and physical illness: A latent growth curve analysis. *Journal of Marriage and the Family* (February): 143–155.

Willetts, M. 2003. An exploratory investigation of heterosexual licensed domestic partners. *Journal of Marriage and the Family* (November): 939–952.

Wineberg, H. 1990. Childrearing after remarriage. *Journal of Marriage and the Family* (February): 31–38.

Wisensale, S. K., and K. E. Hechart. 1993. Domestic partnerships: A concept paper and policy discussion. *Family Relations* (April): 199–204.

Witelsen, S., et al. 1995. Women have greater numerical density of neurons in posterior temporal cortex. *Neuroscience* 15: 3418–3428.

Wolchik, S., et al. 2000. Maternal acceptance and consistency of discipline as buffers of divorce stressors on children's psychological adjustment problems. *Journal of Abnormal Child Psychology* 28: 87–102.

Wolf, N. 1993. *Fire with fire*. New York: Random House.

Wood, J. T. 1988. *But I thought I meant . . . : Misunderstanding in human communication*. Mountain View, CA: Mayfield.

———. 2000a. Gender and personal relationships. In C. Hendrick and S. Hendrick (Eds.), *Close relationships: A sourcebook*. 301–313. Thousand Oaks, CA: Sage.

———. 2000b. He says/she says: Misunderstandings in communication between men and women. In D. O. Braithwaite and J. T. Wood (Eds.), *Case studies in interpersonal communication* 93–100. Belmont, CA: Wadsworth.

———. 2001. *Gendered lives: Communication, gender, and culture*. Belmont, CA: Wadsworth.

Woodard, E., and N. Gridina. 2002. *Media in the home 2000*. Washington, DC: Annenberg Public Policy Center.

Woodward, L., D. Fergusson, and J. Belsky. 2000. Timing of parental separation and attachment to parents in adolescence: Results of a prospective study from birth to age 16. *Journal of Marriage and the Family* (February): 162–174.

Working Mother. 2004. 100 best companies for working mothers. http://www.workingmother.com.

Working Woman. 2000a. The top 500 women-owned businesses. (June): 52.

———. 2000b. 21st annual salary survey. (August).

———. 2004. Facts about working women. Available online at http://www.aflcio.org/issuespolitics/women/factsaboutworkingwomen.efm.

Worthington, E. (Ed.). 1996. *Christian marital counseling: Eight approaches to helping couples*. Grand Rapids, MI: Baker Press.

Wright, R. 1994. The moral animal: Evolutionary psychology and everyday life. New York: Pantheon Books.

Wu, Z. 1995. The stability of cohabitation relationships: The role of children. *Journal of Marriage and the Family* (February): 211–236.

———. 1996. Childbearing in cohabitational relationships. *Journal of Marriage and the Family* (February): 281–292.

Wu, Z., and T. Balakrishman. 1994. Cohabitation after marital disruption in Canada. *Journal of Marriage and the Family* (August): 723–734.

Yates, A., and W. Wolman. 1991. Aphrodisiacs: Myths and reality. *Medical Aspects of Human Sexuality* (December): 58–64.

Yee, J. L., and R. Schulz. 2000. Gender differences in psychiatric morbidity among family caregivers: A review and analysis. *Gerontologist* 40: 147–164.

Yeung, U. J., et al. 2001. Children's time with fathers in intact families. *Journal of Marriage and the Family* 6 (February): 136–154.

Zill, N., et al. 1993. Long-term effects of parental divorce on parent-child relationships, adjustment, and achievement in young adulthood. *Journal of Family Psychology* 7.

Zitner, A., and S. Savage. 2000. Abortion pill could be in the hands of voters. *Los Angeles Times* (September 12).

Zuo, J. 1992. The reciprocal relationship between marital interaction and marital happiness: A three wave study. *Journal of Marriage and the Family* (November): 870–878.

Zusman, M., and D. Knox. 2000. Relationship problems of casual and involved university students. *College Student Journal*. As reported in D. Knox and C. Schacht. 1998, 2000. *Choices in relationships*. Belmont, CA: Wadsworth.

Credits

Author Index

Subject Index

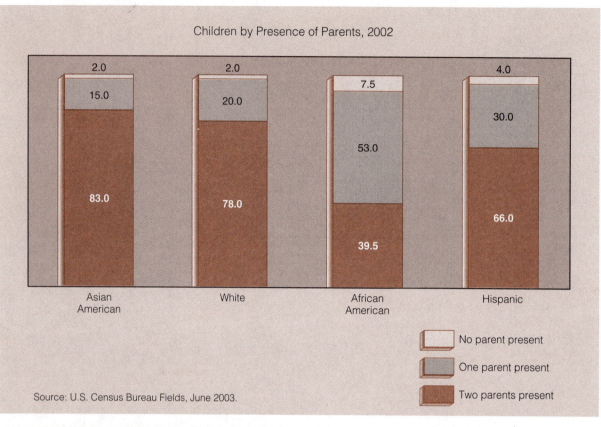

Children by Presence of Parents, 2002

	Asian American	White	African American	Hispanic
No parent present	2.0	2.0	7.5	4.0
One parent present	15.0	20.0	53.0	30.0
Two parents present	83.0	78.0	39.5	66.0

No parent present

One parent present

Two parents present

Source: U.S. Census Bureau Fields, June 2003.

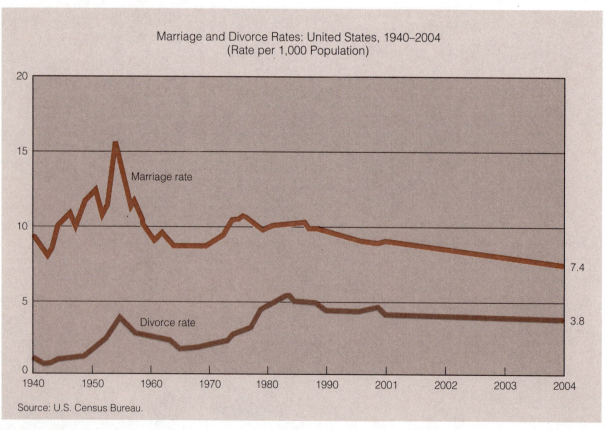

Marriage and Divorce Rates: United States, 1940–2004
(Rate per 1,000 Population)

Marriage rate

Divorce rate

7.4

3.8

Source: U.S. Census Bureau.